SCHAUM'S outlines

Advanced Calculus

Third Edition

Robert Wrede, Ph.D.
Professor Emeritus of Mathematics
San Jose State University

Murray R. Spiegel, Ph.D.
Former Professor and Chairman of Mathematics
Rensselaer Polytechnic Institute
Hartford Graduate Center

Schaum's Outline Series

McGraw Hill

New York Chicago San Francisco Lisbon
London Madrid Mexico City Milan New Delhi
San Juan Seoul Singapore Sydney Toronto

The McGraw·Hill Companies

ROBERT WREDE received his B.S. and M.A. degrees from Miami University, Oxford, Ohio. After teaching there for a year, he attended Indiana University and was awarded a Ph.D. in mathematics. He taught at San Jose State University from 1955 to 1994. He also consulted at IBM, the Naval Radiation Laboratory at Hunter's Point, and with several textbook companies. His text *Introduction to Vector and Tensor Analysis*, first published in 1963, remains available, along with recently published texts *Insights into Geometry*, *Insights into Algebra*, and *Insights into Calculus*. His primary interests have been in tensor analysis and relativity theory. Presently, Dr. Wrede and his wife, Jeanne, live in Aptos, California.

The late **MURRAY R. SPIEGEL** received a M.S. degree in physics and the Ph.D. in mathematics from Cornell University. He had positions at Harvard University, Columbia University, Oak Ridge, and Rensselaer Polytechnic Institute, and served as a mathematical consultant at several large companies. His last position was professor and chairman of mathematics at the Rensselaer Polytechnic Institute, Hartford Graduate Center. He was interested in most branches of mathematics, especially those which involve applications to physics and engineering problems. He was the author of numerous journal articles and 14 books on various topics in mathematics.

Schaum's Outline of
ADVANCED CALCULUS

1 2 3 4 5 6 7 8 9 CUS CUS 0 1 4 3 2 1 0

ISBN 978-0-07-162366-7
MHID 0-07-162366-3

Library of Congress Cataloging-in-Publication Data

Wrede, Robert C.
Schaum's outline of advanced calculus / Robert Wrede, Murray R. Spiegel.—3rd ed.
p. cm.—(Schaum's outline series)
Rev. ed. of: Schaum's outline of theory and problems of advanced calculus. 2nd ed. c2002.
ISBN-13: 978-0-07-162366-7
ISBN-10: 0-07-162366-3
1. Calculus—Outlines, syllabi, etc. 2. Calculus—Problems, exercises, etc. I. Spiegel, Murray R. II. Wrede, Robert C. Schaum's outline of theory and problems of advanced calculus. III. Title. IV. Title: Advanced calculus.
QA303.2.W74 2010
515—dc22 2009049167

Preface to the Third Edition

The many problems and solutions provided by the late Professor Spiegel remain invaluable to students as they seek to master the intricacies of the calculus and related fields of mathematics. These remain an integral part of this manuscript. In this third edition, clarifications have been provided. In addition, the continuation of the interrelationships and the significance of concepts, begun in the second edition, have been extended.

Preface to the Second Edition

A key ingredient in learning mathematics is problem solving. This is the strength, and no doubt the reason for the longevity of Professor Spiegel's advanced calculus. His collection of solved and unsolved problems remains a part of this second edition.

Advanced calculus is not a single theory. However, the various sub-theories, including vector analysis, infinite series, and special functions, have in common a dependency on the fundamental notions of the calculus. An important objective of this second edition has been to modernize terminology and concepts, so that the interrelationships become clearer. For example, in keeping with present usage functions of a real variable are automatically single valued; differentials are defined as linear functions, and the universal character of vector notation and theory are given greater emphasis. Further explanations have been included and, on occasion, the appropriate terminology to support them.

The order of chapters is modestly rearranged to provide what may be a more logical structure.

A brief introduction is provided for most chapters. Occasionally, a historical note is included; however, for the most part the purpose of the introductions is to orient the reader to the content of the chapters.

I thank the staff of McGraw-Hill. Former editor, Glenn Mott, suggested that I take on the project. Peter McCurdy guided me in the process. Barbara Gilson, Jennifer Chong, and Elizabeth Shannon made valuable contributions to the finished product. Joanne Slike and Maureen Walker accomplished the very difficult task of combining the old with the new and, in the process, corrected my errors. The reviewer, Glenn Ledder, was especially helpful in the choice of material and with comments on various topics.

ROBERT C. WREDE

Contents

CHAPTER 1

Numbers

Mathematics has its own language, with numbers as the alphabet. The language is given structure with the aid of connective symbols, rules of operation, and a rigorous mode of thought (logic). These concepts, which previously were explored in elementary mathematics courses such as geometry, algebra, and calculus, are reviewed in the following paragraphs.

Sets

Fundamental in mathematics is the concept of a *set*, *class*, or *collection* of objects having specified characteristics. For example, we speak of the set of all university professors, the set of all letters $A, B, C, D, \ldots, Z$ of the English alphabet, and so on. The individual objects of the set are called *members* or *elements*. Any part of a set is called a *subset* of the given set, e.g., A, B, C is a subset of $A, B, C, D, \ldots, Z$. The set consisting of no elements is called the *empty set* or *null set*.

Real Numbers

The number system is foundational to the modern scientific and technological world. It is based on the symbols 1, 2, 3, 4, 5, 6, 7, 8, 9, 0. Thus, it is called a *base ten* system. (There is the implication that there are other systems. One of these, which is of major importance, is the base two system.) The symbols were introduced by the Hindus, who had developed decimal representation and the arithmetic of positive numbers by 600 A.D. In the eighth century, the House of Wisdom (library) had been established in Baghdad, and it was there that the Hindu arithmetic and much of the mathematics of the Greeks were translated into Arabic. From there, this arithmetic gradually spread to the later-developing western civilization.

The flexibility of the Hindu-Arabic number system lies in the multiple uses of the numbers. They may be used to signify: (a) **order**—*the runner finished fifth;* (b) **quantity**—*there are six apples in the barrel;* (c) **construction**—*2 and 3 may be used to form any of 23, 32, .23 or .32;* (d) **place**—*0 is used to establish place, as is illustrated by 607, 0603, and .007.*

Finally, note that the significance of the base ten terminology is enhanced by the following examples:

$$357 = 7(10^0) + 5(10^1) + 3(10^2)$$

$$.972 = \frac{9}{10} + \frac{7}{10^2} + \frac{2}{10^3}$$

The collection of numbers created from the basic set is called the **real number system.** Significant subsets of them are listed as follows. For the purposes of this text, it is assumed that the reader is familiar with these numbers and the fundamental arithmetic operations.

1. **Natural numbers** 1, 2, 3, 4, . . . , also called *positive integers*, are used in counting members of a set. The symbols varied with the times; e.g., the Romans used I, II, III, IV, ... The *sum* $a + b$ and *product* $a \cdot b$ or ab of any two natural numbers a and b is also a natural number. This is often expressed by

saying that the set of natural numbers is *closed* under the operations of *addition* and *multiplication*, or satisfies the *closure property* with respect to these operations.

2. **Negative integers and zero,** denoted by $-1, -2, -3, \ldots$, and 0, respectively, arose to permit solutions of equations such as $x + b = a$, where a and b are any natural numbers. This leads to the operation of *subtraction*, or *inverse of addition*, and we write $x = a - b$.

   The set of positive and negative integers and zero is called the set of *integers*.

3. **Rational numbers** or *fractions* such as $\frac{2}{3}, -\frac{5}{4}, \ldots$ arose to permit solutions of equations such as $bx = a$ for all integers a and b, where $b \neq 0$. This leads to the operation of *division*, or *inverse of multiplication*, and we write $x = a/b$ or $a \div b$, where a is the *numerator* and b the *denominator*.

   The set of integers is a subset of the rational numbers, since integers correspond to rational numbers where $b = 1$.

4. **Irrational numbers** such as $\sqrt{2}$ and π are numbers which are not rational; i.e., they cannot be expressed as a/b (called the *quotient* of a and b), where a and b are integers and $b \neq 0$.

   The set of rational and irrational numbers is called the set of *real numbers*.

Decimal Representation of Real Numbers

Any real number can be expressed in *decimal form*, e.g., $17/10 = 1.7$, $9/100 = 0.09$, $1/6 = 0.16666\ldots$. In the case of a rational number, the decimal expansion either terminates or if it does not terminate, one or a group of digits in the expansion will ultimately repeat, as, for example, in $\frac{1}{7} = 0.142857\ 142857\ 142\ldots$. In the case of an irrational number such as $\sqrt{2} = 1.41423\ldots$ or $\pi = 3.14159\ldots$ no such repetition can occur. We can always consider a decimal expansion as unending; e.g., 1.375 is the same as 1.37500000 . . . or 1.3749999 . . . To indicate recurring decimals we sometimes place dots over the repeating cycle of digits, e.g., $\frac{1}{7} = 0.\dot{1}\dot{4}\dot{2}\dot{8}\dot{5}\dot{7}$, and $\frac{19}{6} = 3.1\dot{6}$.

It is possible to design number systems with fewer or more digits; e.g., the *binary system* uses only two digits, 0 and 1 (see Problems 1.32 and 1.33).

Geometric Representation of Real Numbers

The geometric representation of real numbers as points on a line, called the *real axis*, as in Figure 1.1, is also well known to the student. For each real number there corresponds one and only one point on the line, and, conversely, there is a *one-to-one* (see Figure 1.1) *correspondence* between the set of real numbers and the set of points on the line. Because of this we often use *point* and *number* interchangeably.

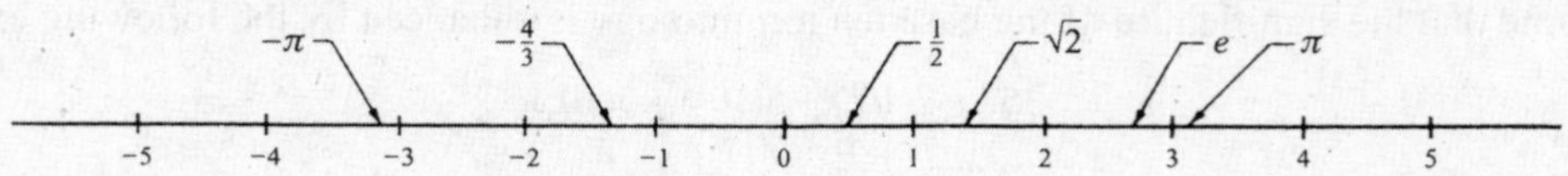

Figure 1.1

While this correlation of points and numbers is automatically assumed in the elementary study of mathematics, it is actually an axiom of the subject (the Cantor Dedekind axiom) and, in that sense, has deep meaning.

The set of real numbers to the right of 0 is called the set of *positive numbers,* the set to the left of 0 is the set of *negative numbers*, while 0 itself is neither positive nor negative.

(Both the horizontal position of the line and the placement of positive and negative numbers to the right and left, respectively, are conventions.)

Between any two rational numbers (or irrational numbers) on the line there are infinitely many rational (and irrational) numbers. This leads us to call the set of rational (or irrational) numbers an *everywhere dense* set.

Operations with Real Numbers

If a, b, c belong to the set R of real numbers, then:

1. $a + b$ and ab belong to R — Closure law
2. $a + b = b + a$ — Commutative law of addition
3. $a + (b + c) = (a + b) + c$ — Associative law of addition
4. $ab = ba$ — Commutative law of multiplication
5. $a(bc) = (ab)c$ — Associative law of multiplication
6. $a(b + c) = ab + ac$ — Distributive law
7. $a + 0 = 0 + a = a$, $1 \cdot a = a \cdot 1 = a$

   0 is called the *identity with respect to addition*; 1 is called the *identity with respect to multiplication*.

8. For any a there is a number x in R such that $x + a = 0$.
   x is called the *inverse of a with respect to addition* and is denoted by $-a$.
9. For any $a \neq 0$ there is a number x in R such that $ax = 1$.
   x is called the *inverse of a with respect to multiplication* and is denoted by a^{-1} or $1/a$.

Convention: For convenience, operations called subtraction and division are defined by $a - b = a + (-b)$ and $\frac{a}{b} = ab^{-1}$, respectively.

These enable us to operate according to the usual rules of algebra. In general, any set, such as R, whose members satisfy the preceding is called a *field*.

Inequalities

If $a - b$ is a nonnegative number, we say that a is *greater than or equal to b* or b is *less than or equal to a*, and write, respectively, $a \geqq b$ or $b \leqq a$. If there is no possibility that $a = b$, we write $a > b$ or $b < a$. Geometrically, $a > b$ if the point on the real axis corresponding to a lies to the right of the point corresponding to b.

Properties of Inequalities

If a, b, and c are any given real numbers, then:

1. Either $a > b$, $a = b$ or $a < b$ — Law of trichotomy
2. If $a > b$ and $b > c$, then $a > c$ — Law of transitivity
3. If $a > b$, then $a + c > b + c$
4. If $a > b$ and $c > 0$, then $ac > bc$

5. If $a > b$ and $c < 0$, then $ac < bc$

EXAMPLES. $3 < 5$ or $5 > 3$; $-2 < -1$ or $-1 > -2$; $x \leq 3$ means that x is a real number which may be 3 or less than 3.

Absolute Value of Real Numbers

The absolute value of a real number a, denoted by $|a|$, is defined as a if $a > 0$, $-a$ if $a < 0$, and 0 if $a = 0$.

Properties of Absolute Value

1. $|ab| = |a|\ |b|$ or $|abc \ldots m| = |a|\ |b|\ |c| \ldots |m|$
2. $|a+b| \leq |a| + |b|$ or $|a+b+c+\cdots+m| \leq |a| + |b| + |c| + \cdots |m|$
3. $|a-b| \geq |a| - |b|$

EXAMPLES. $|-5| = 5$, $|+2| = 2$, $\left|-\dfrac{3}{4}\right| = \dfrac{3}{4}$, $|-\sqrt{2}| = \sqrt{2}$, $|0| = 0$.

The distance between any two points (real numbers) a and b on the real axis is $|a-b| = |b-a|$.

Exponents and Roots

The product $a \cdot a \ldots a$ of a real number a by itself p times is denoted by a^p, where p is called the *exponent* and a is called the *base*. The following rules hold:

1. $a^p \cdot a^q = a^{p+q}$
2. $\dfrac{a^p}{a^q} = a^{p-q}$
3. $(a^p)^r = a^{pr}$
4. $\left(\dfrac{a}{b}\right)^p = \dfrac{a^p}{b^p}$

These and extensions to any real numbers are possible so long as division by zero is excluded. In particular, by using 2, with $p = q$ and $p = 0$, respectively, we are led to the definitions $a^0 = 1$, $a^{-q} = 1/a^q$.

If $a^p = N$, where p is a positive integer, we call a a pth *root* of N, written $\sqrt[p]{N}$. There may be more than one real pth root of N. For example, since $2^2 = 4$ and $(-2)^2 = 4$, there are two real square roots of 4—namely, 2 and –2. For square roots it is customary to define $\sqrt{N}$ as positive; thus, $\sqrt{4} = 2$ and then $-\sqrt{4} = -2$.

If p and q are positive integers, we define $a^{p/q} = \sqrt[q]{a^p}$.

Logarithms

If $a^p = N$, p is called the *logarithm* of N to the base a, written $p = \log_a N$. If a and N are positive and $a \neq 1$, there is only one real value for p. The following rules hold:

1. $\log_a MN = \log_a M + \log_a N$
2. $\log_a \dfrac{M}{N} = \log_a M - \log_a N$
3. $\log_a M^r = r \log_a M$

In practice, two bases are used: base $a = 10$, and the *natural base* $a = e = 2.71828\ldots$ The logarithmic systems associated with these bases are called *common* and *natural*, respectively. The common logarithm system is signified by $\log N$; i.e., the subscript 10 is not used. For natural logarithms, the usual notation is $\ln N$.

Common logarithms (base 10) traditionally have been used for computation. Their application replaces multiplication with addition and powers with multiplication. In the age of calculators and computers, this process is outmoded; however, common logarithms remain useful in theory and application. For example, the Richter scale used to measure the intensity of earthquakes is a logarithmic scale. Natural logarithms were introduced to simplify formulas in calculus, and they remain effective for this purpose.

Axiomatic Foundations of the Real Number System

The number system can be built up logically, starting from a basic set of *axioms* or "self-evident" truths, usually taken from experience, such as statements 1 through 9 on Page 3.

If we assume as given the natural numbers and the operations of addition and multiplication (although it is possible to start even further back, with the concept of sets), we find that statements 1 through 6, with R as the set of natural numbers, hold, while 7 through 9 do not hold.

Taking 7 and 8 as additional requirements, we introduce the numbers $-1, -2, -3, \ldots$, and 0. Then, by taking 9, we introduce the rational numbers.

Operations with these newly obtained numbers can be defined by adopting axioms 1 through 6, where R is now the set of integers. These lead to *proofs* of statements such as $(-2)(-3) = 6$, $-(-4) = 4$, $(0)(5) = 0$, and so on, which are usually taken for granted in elementary mathematics.

We can also introduce the concept of order or inequality for integers, and, from these inequalities, for rational numbers. For example, if a, b, c, d are positive integers, we define $a/b > c/d$ if and only if $ad > bc$, with similar extensions to negative integers.

Once we have the set of rational numbers and the rules of inequality concerning them, we can order them geometrically as points on the real axis, as already indicated. We can then show that there are points on the line which do not represent rational numbers (such as $\sqrt{2}$, π, etc.). These irrational numbers can be defined in various ways, one of which uses the idea of *Dedekind cuts* (see Problem 1.34). From this we can show that the usual rules of algebra apply to irrational numbers and that no further real numbers are possible.

Point Sets, Intervals

A set of points (real numbers) located on the real axis is called a *one-dimensional point set*.

The set of points x such that $a \leqq x \leqq b$ is called a *closed interval* and is denoted by $[a, b]$. The set $a < x < b$ is called an *open interval*, denoted by (a, b). The sets $a < x \leqq b$ and $a \leqq x < b$, denoted by $(a, b]$ and $[a, b)$, respectively, are called *half-open* or *half-closed* intervals.

The symbol x, which can represent any number or point of a set, is called a *variable*. The given numbers a or b are called *constants*.

Letters were introduced to construct algebraic formulas around 1600. Not long thereafter, the philosopher-mathematician Rene Descartes suggested that the letters at the end of the alphabet be used to represent variables and those at the beginning to represent constants. This was such a good idea that it remains the custom.

EXAMPLE. The set of all x such that $|x| < 4$, i.e., $-4 < x < 4$, is represented by $(-4, 4)$, an open interval.

The set $x > a$ can also be represented by $a < x < \infty$. Such a set is called an *infinite* or *unbounded interval*. Similarly, $-\infty < x < \infty$ represents all real numbers x.

Countability

A set is called *countable* or *denumerable* if its elements can be placed in 1-1 correspondence with the natural numbers.

EXAMPLE. The even natural numbers 2, 4, 6, 8, . . . is a countable set because of the 1-1 correspondence shown.

$$\begin{array}{lccccc} \text{Given set} & 2 & 4 & 6 & 8 & \ldots \\ & \updownarrow & \updownarrow & \updownarrow & \updownarrow & \\ \text{Natural numbers} & 1 & 2 & 3 & 4 & \ldots \end{array}$$

A set is *infinite* if it can be placed in 1-1 correspondence with a subset of itself. An infinite set which is countable is called *countable infinite*.

The set of rational numbers is countable infinite, while the set of irrational numbers or all real numbers is noncountably infinite (see Problems 1.17 through 1.20).

The number of elements in a set is called its *cardinal number*. A set which is countably infinite is assigned the cardinal number $\aleph_0$ (the Hebrew letter *aleph-null*). The set of real numbers (or any sets which can be placed into 1-1 correspondence with this set) is given the cardinal number C, called the *cardinality of the contimuum*.

Neighborhoods

The set of all points x such that $|x-a| < \delta$, where $\delta > 0$, is called a δ *neighborhood* of the point a. The set of all points x such that $0 < |x-a| < \delta$, in which $x = a$ is excluded, is called a *deleted* δ *neighborhood* of a or an open ball of radius δ about a.

Limit Points

A *limit point, point of accumulation*, or *cluster point* of a set of numbers is a number l such that every deleted δ neighborhood of l contains members of the set; that is, no matter how small the radius of a ball about l, there are points of the set within it. In other words, for any $\delta > 0$, however small, we can always find a member x of the set which is not equal to l but which is such that $|x-l| < \delta$. By considering smaller and smaller values of δ, we see that there must be infinitely many such values of x.

A finite set cannot have a limit point. An infinite set may or may not have a limit point. Thus, the natural numbers have no limit point, while the set of rational numbers has infinitely many limit points.

A set containing all its limit points is called a *closed set*. The set of rational numbers is not a closed set, since, for example, the limit point $\sqrt{2}$ is not a member of the set (Problem 1.5). However, the set of all real numbers x such that $0 \leqq x \leqq 1$ is a closed set.

Bounds

If for all numbers x of a set there is a number M such that $x \leqq M$, the set is *bounded above* and M is called an *upper bound*. Similarly if $x \geqq m$, the set is *bounded below* and m is called a *lower bound*. If for all x we have $m \leqq x \leqq M$, the set is called *bounded*.

If $\underline{M}$ is a number such that no member of the set is greater than $\underline{M}$ but there is at least one member which exceeds $\underline{M} - \epsilon$ for every $\epsilon > 0$, then $\underline{M}$ is called the *least upper bound* (l.u.b.) of the set. Similarly, if no member of the set is smaller than $\bar{m} + \epsilon$ for every $\epsilon > 0$, then $\bar{m}$ is called the *greatest lower bound* (g.l.b.) of the set.

Bolzano-Weierstrass Theorem

The Bolzano-Weierstrass theorem states that every bounded infinite set has at least one limit point. A proof of this is given in Problem 2.23.

Algebraic and Transcendental Numbers

A number x which is a solution to the *polynomial equation*

$$a_0x^n + a_1x^{n-1} + a_2x^{n-2} + \cdots + a_{n-1}x + a_n = 0 \tag{1}$$

where $a_0 \neq 0, a_1, a_2, \ldots, a_n$ are integers and n is a positive integer, called the *degree* of the equation, is called an *algebraic number*. A number which cannot be expressed as a solution of any polynomial equation with integer coefficients is called a *transcendental number*.

EXAMPLES. $\frac{2}{3}$ and $\sqrt{2}$, which are solutions of $3x - 2 = 0$ and $x^2 - 2 = 0$, respectively, are algebraic numbers.

The numbers π and e can be shown to be transcendental numbers. Mathematicians have yet to determine whether some numbers such as $e\pi$ or $e + \pi$ are algebraic or not.

The set of algebraic numbers is a countably infinite set (see Problem 1.23), but the set of transcendental numbers is noncountably infinite.

The Complex Number System

Equations such as $x^2 + 1 = 0$ have no solution within the real number system. Because these equations were found to have a meaningful place in the mathematical structures being built, various mathematicians of the late nineteenth and early twentieth centuries developed an extended system of numbers in which there were solutions. The new system became known as the *complex number system*. It includes the real number system as a subset.

We can consider a complex number as having the form $a + bi$, where a and b are real numbers called the *real* and *imaginary parts*, and $i = \sqrt{-1}$ is called the *imaginary unit*. Two complex numbers $a + bi$ and $c + di$ are *equal* if and only if $a = c$ and $b = d$. We can consider real numbers as a subset of the set of complex numbers with $b = 0$. The complex number $0 + 0i$ corresponds to the real number 0.

The *absolute value* or *modulus* of $a + bi$ is defined as $|a + bi| = \sqrt{a^2 + b^2}$. The *complex conjugate* of $a + bi$ is defined as $a - bi$. The complex conjugate of the complex number z is often indicated by $\bar{z}$ or z^*.

The set of complex numbers obeys rules 1 through 9 on Pages 3, and thus constitutes a field. In performing operations with complex numbers, we can operate as in the algebra of real numbers, replacing i^2 by -1 when it occurs. Inequalities for complex numbers are not defined.

From the point of view of an axiomatic foundation of complex numbers, it is desirable to treat a complex number as an ordered pair (a, b) of real numbers a and b subject to certain operational rules which turn out to be equivalent to the aforementioned rules. For example, we define $(a, b) + (c, d) = (a + c, b + d)$, $(a, b)(c, d) = (ac - bd, ad + bc)$, $m(a, b) = (ma, mb)$, and so on. We then find that $(a, b) = a(1, 0) + b(0, 1)$ and we associate this with $a + bi$, where i is the symbol for $(0, 1)$.

Polar Form of Complex Numbers

If real scales are chosen on two mutually perpendicular axes $X'\,OX$ and $Y'\,OY$ (the x and y axes), as in Figure 1.2, we can locate any point in the plane determined by these lines by the ordered pair of numbers (x, y) called *rectangular coordinates* of the point. Examples of the location of such points are indicated by P, Q, R, S, and T in Figure 1.2.

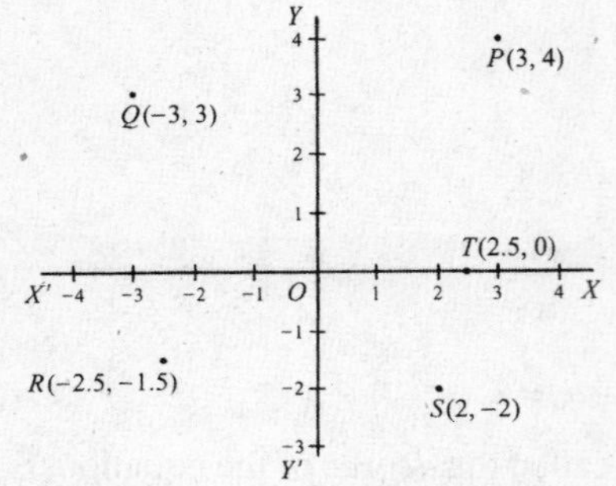

Figure 1.2

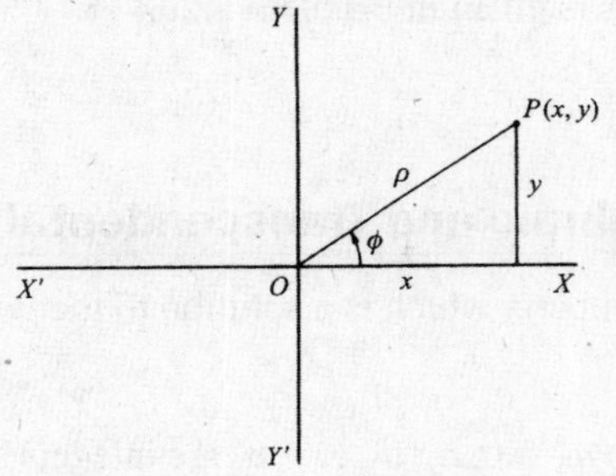

Figure 1.3

Since a complex number $x + iy$ can be considered as an ordered pair (x, y), we can represent such numbers by points in an xy plane called the *complex plane* or *Argand diagram*. Referring to Figure 1.3, we see that $x = \rho \cos \phi$, $y = \rho \sin \phi$, where $\rho = \sqrt{x^2 + y^2} = |x + iy|$ and ϕ, called the *amplitude* or *argument*, is the angle which line OP makes with the positive x axis OX. It follows that

$$z = x + iy = \rho(\cos \phi + i \sin \phi) \tag{2}$$

called the *polar form* of the complex number, where ρ and ϕ are called *polar coordinates*. It is sometimes convenient to write cis ϕ instead of $\cos \phi + i \sin \phi$.

If $z_1 = x_1 + iy_i = \rho_1 (\cos \phi_1 + i \sin \phi_1)$ and $z_2 = x_2 + iy_2 = \rho_2(\cos\phi_2 + i \sin \phi_2)$ and by using the addition formulas for sine and cosine, we can show that

$$z_1 z_2 = \rho_1 \rho_2 \{\cos(\phi_1 + \phi_2) + i \sin(\phi_1 + \phi_2)\} \tag{3}$$

$$\frac{z_1}{z_2} = \frac{p_1}{p_2}\{\cos(\phi_1 - \phi_2) + i \sin(\phi_1 - \phi_2)\} \tag{4}$$

$$z^n = \{\rho(\cos \phi + i \sin \phi)\}^n = \rho^n(\cos n\phi + i \sin n\phi) \tag{5}$$

where n is any real number. Equation (5) is sometimes called *De Moivre's theorem*. We can use this to determine roots of complex numbers. For example, if n is a positive integer,

$$\begin{aligned} z^{1/n} &= \{p(\cos\phi + i \sin\phi)\}^{1/n} \\ &= p^{1/n}\left\{\cos\left(\frac{\phi + 2k\pi}{n}\right) + i \sin\left(\frac{\phi + 2k\pi}{n}\right)\right\} \qquad k = 0, 1, 2, 3, \ldots, n-1 \end{aligned} \tag{6}$$

from which it follows that there are in general n different values of $z^{1/n}$. In Chapter 11 we will show that $e^{i\phi} = \cos \phi + i \sin \phi$ where $e = 2.71828\ldots$. This is called *Euler's formula*.

Mathematical Induction

The principle of *mathematical induction* is an important property of the positive integers. It is especially useful in proving statements involving all positive integers when it is known, for example, that the statements are valid for $n = 1, 2, 3$ but it is *suspected* or *conjectured* that they hold for all positive integers. The method of proof consists of the following steps:

1. Prove the statement for $n = 1$ (or some other positive integer).
2. Assume the statement is true for $n = k$, where k is any positive integer.
3. From the assumption in 2, prove that the statement must be true for $n = k + 1$. This is part of the proof establishing the induction and may be difficult or impossible.

4. Since the statement is true for $n = 1$ (from Step 1) it must (from Step 3) be true for $n = 1 + 1 = 2$ and from this for $n = 2 + 1 = 3$, and so on, and so must be true for all positive integers. (This assumption, which provides the link for the truth of a statement for a finite number of cases to the truth of that statement for the infinite set, is called the *axiom of mathematical induction.*)

SOLVED PROBLEMS

Operations with numbers

1.1. If $x = 4$, $y = 15$, $z = -3$, $p = \frac{2}{3}$, $q = -\frac{1}{6}$, and $r = \frac{3}{4}$, evaluate (a) $x + (y + z)$, (b) $(x + y) + z$, (c) $p(qr)$, (d) $(pq)r$, (e) $x(p + q)$.

(a) $x + (y + z) = 4 + [15 + (-3)] = 4 + 12 = 16$

(b) $(x + y) + z = (4 + 15) + (-3) = 19 - 3 = 16$

The fact that (*a*) and (*b*) are equal illustrates the *associative law of addition.*

(c) $p(qr) = \frac{2}{3}\{(-\frac{1}{6})(\frac{3}{4})\} = (\frac{2}{3})(-\frac{3}{24}) = (\frac{2}{3})(-\frac{1}{8}) = -\frac{2}{24} = -\frac{1}{12}$

(d) $(pq)r = \{(\frac{2}{3})(-\frac{1}{6})\}(\frac{3}{4}) = (-\frac{2}{18})(\frac{3}{4}) = (-\frac{1}{9})(\frac{3}{4}) = -\frac{3}{36} = -\frac{1}{12}$

The fact that (*c*) and (*d*) are equal illustrates the *associative law of multiplication.*

(e) $x(p + q) = 4(\frac{2}{3} - \frac{1}{6}) = 4(\frac{4}{6} - \frac{1}{6}) = 4(\frac{3}{6}) = \frac{12}{6} = 2$

Another method: $x(p + q) = xp + xq = (4)(\frac{2}{3}) + (4)(-\frac{1}{6}) = \frac{8}{6} - \frac{4}{6} = \frac{8}{3} - \frac{2}{3} = \frac{6}{3} = 2$ using the *distributive law.*

1.2. Explain why we do not consider (a) $\frac{0}{0}$ and (b) $\frac{1}{0}$ as numbers.

(a) If we define a/b as that number (if it exists) such that $bx = a$, then $0/0$ is that number x such that $0x = 0$. However, this is true for all numbers. Since there is no unique number which $0/0$ can represent, we consider it undefined.

(b) As in (*a*), if we define $1/0$ as that number x (if it exists) such that $0x = 1$, we conclude that there is no such number.

Because of these facts we must look upon division by zero as meaningless.

1.3. Simplify $\frac{x^2 - 5x + 6}{x^2 - 2x - 3}$.

$\frac{x^2 - 5x + 6}{x^2 - 2x - 3} = \frac{(x-3)(x-2)}{(x-3)(x+1)} = \frac{x-2}{x+1}$ provided that the cancelled factor $(x - 3)$ is not zero; i.e., $x \neq 3$. For $x = 3$, the given fraction is undefined.

Rational and irrational numbers

1.4. Prove that the square of any odd integer is odd.

Any odd integer has the form $2m + 1$. Since $(2m + 1)^2 = 4m^2 + 4m + 1$ is 1 more than the even integer $4m^2 + 4m = 2(2m^2 + 2m)$, the result follows.

1.5. Prove that there is no rational number whose square is 2.

Let p / q be a rational number whose square is 2, where we assume that p / q is in lowest terms; i.e., p and q have no common integer factors except ± 1 (we sometimes call such integers *relatively prime*).

Then $(p / q)^2 = 2$, $p^2 = 2q^2$ and p^2 is even. From Problem 1.4, p is even, since if p were odd, p^2 would be odd. Thus, $p = 2m$.

Substituting $p = 2m$ in $p^2 = 2q^2$ yields $q^2 = 2m^2$, so that q^2 is even and q is even.

Thus, p and q have the common factor 2, contradicting the original assumption that they had no common factors other than ± 1. By virtue of this contradiction there can be no rational number whose square is 2.

1.6. Show how to find rational numbers whose squares can be arbitrarily close to 2.

We restrict ourselves to positive rational numbers. Since $(1)^2 = 1$ and $(2)^2 = 4$, we are led to choose rational numbers between 1 and 2, e.g., 1.1, 1.2, 1.3, . . . , 1.9.

Since $(1.4)^2 = 1.96$ and $(1.5)^2 = 2.25$, we consider rational numbers between 1.4 and 1.5, e.g., 1.41, 1.42,. . , 1.49.

Continuing in this manner we can obtain closer and closer rational approximations; e.g., $(1.414213562)^2$ is less than 2, while $(1.414213563)^2$ is greater than 2.

1.7. Given the equation $a_0x^n + a_1x^{n-1} + \cdots + a_n = 0$, where $a_0, a_1, \ldots a_n$ are integers and a_0 and $a_n \neq 0$, show that if the equation is to have a rational root p / q, then p must divide a_n and q must divide a_0 exactly.

Since p / q is a root we have, on substituting in the given equation and multiplying by q^n, the result is

$$a_0p^n + a_1p^{n-1}q + a_2p^{n-2}q^2 + \cdots + a_{n-1}pq^{n-1} + a_nq^n = 0 \tag{1}$$

or dividing by p,

$$a_0p^{n-1} + a_1p^{n-2}q + \cdots + a_{n-1}q^{n-1} = -\frac{a_nq^n}{p} \tag{2}$$

Since the left side of Equation (2) is an integer, the right side must also be an integer. Then, since p and q are relatively prime, p does not divide q^n exactly and so must divide a_n.

In a similar manner, by transposing the first term of Equation (1) and dividing by q, we can show that q must divide a_0.

1.8. Prove that $\sqrt{2} + \sqrt{3}$ cannot be a rational number.

If $x = \sqrt{2} + \sqrt{3}$, then $x^2 = 5 + 2\sqrt{6}$, $x^2 - 5 = 2\sqrt{6}$, and, squaring, $x^4 - 10x^2 + 1 = 0$. The only possible rational roots of this equation are ± 1 by Problem 1.7, and these do not satisfy the equation. It follows that $\sqrt{2} + \sqrt{3}$, which satisfies the equation, cannot be a rational number.

1.9. Prove that between any two rational numbers there is another rational number.

The set of rational numbers is closed under the operations of addition and division (nonzero denominator). Therefore, $\frac{a+b}{2}$ is rational. The next step is to guarantee that this value is between a and b. To this purpose, assume $a < b$. (The proof would proceed similarly under the assumption $b < a$.) Then $2a < a + b$; thus, $a < \frac{a+b}{2}$ and $a + b < 2b$; therefore, $\frac{a+b}{2} < b$.

Inequalities

1.10. For what values of x is $x + 3(2 - x) \geq 4 - x$?

$x + 3(2 - x) \geq 4 - x$ when $x + 6 - 3x \geq 4 - x$, $6 - 2x \geq 4 - x$, $6 - 4 \geq 2x - x$, and $2 \geq x$; i.e. $x \leq 2$.

1.11. For what values of x is $x^2 - 3x - 2 < 10 - 2x$?

The required inequality holds when $x^2 - 3x - 2 - 10 + 2x < 0$, $x^2 - x - 12 < 0$ or $(x - 4)(x + 3) < 0$. This last inequality holds only in the following cases.

Case 1: $x - 4 > 0$ *and* $x + 3 < 0$; i.e., $x > 4$ and $x < -3$. This is *impossible*, since x cannot be both greater than 4 and less than -3.

Case 2: $x - 4 < 0$ *and* $x + 3 > 0$; i.e., $x < 4$ and $x > -3$. This is possible when $-3 < x < 4$. Thus, the inequality holds for the set of all x such that $-3 < x < 4$.

1.12. If $a \geqq 0$ and $b \geqq 0$, prove that $\frac{1}{2}(a + b) \geqq \sqrt{ab}$.

The statement is self-evident in the following cases: (1) $a = b$, and (2) either or both of a and b zero. For both a and b positive and $a \neq b$. the proof is by contradiction.

Assume to the contrary of the supposition that $\frac{1}{2}(a + b) < \sqrt{ab}$, then $\frac{1}{4}(a^2 + 2ab + b^2) < ab$.

That is, $a^2 - 2ab + b^2 = (a - b)^2 < 0$. Since the left member of this equation is a square, it cannot be less than zero, as is indicated. Having reached this contradiction, we may conclude that our assumption is incorrect and that the original assertion is true.

1.13. If $a_1, a_2, \ldots, a_n$ and $b_1, b_2 \ldots b_n$ are any real numbers, prove *Schwarz's inequality:*

$$(a_1 b_1 + a_2 b_2 + \cdots + a_n b_n)^2 \leqq (a^2_1 + a^2_2 + \cdots + a^2_n)(b^2_1 + b^2_2 + \cdots + b^2_{2\,n})$$

For all real numbers λ, we have

$$(a_1\lambda + b_1)^2 + (a_2\lambda + b_2)^2 + \cdots + (a_n\lambda + b_n)^2 \geqq 0$$

Expanding and collecting terms yields

$$A^2\lambda^2 + 2C\lambda + B^2 \geqq 0 \qquad (1)$$

where

$$A^2 = a^2_1 + a^2_2 + \cdots + a^2_n. \qquad B^2 = b^2_1 + b^2_2 + \cdots + b^2_n, \qquad C = a_1b_1 + a_2b_2 + \cdots + a_nb_n \qquad (2)$$

The left member of Equation (1) is a quadratic form in λ. Since it never is negative, its discriminant, $4C^2 - 4A^2B^2$, cannot be positive. Thus,

$$C^2 - A^2B^2 \leq 0 \quad \text{or} \quad C^2 \leq A^2 B^2$$

This is the inequality that was to be proved.

1.14. Prove that $\frac{1}{2} + \frac{1}{4} + \frac{1}{8} + \cdots + \frac{1}{2^{n-1}} < 1$ for all positive integers $n > 1$.

Let

$$S_n = \frac{1}{2} + \frac{1}{4} + \frac{1}{8} + \cdots + \frac{1}{2^{n-1}}$$

Then

$$\frac{1}{2}S_n = \frac{1}{4} + \frac{1}{8} + \cdots + \frac{1}{2^{n-1}} + \frac{1}{2^n}$$

Subtracting,

$$\frac{1}{2}S_n = \frac{1}{2} - \frac{1}{2^n}$$

Thus,

$$S_n = 1 - \frac{1}{2^{n-1}} < 1 \text{ for all } n.$$

Exponents, roots, and logarithms

1.15. Evaluate each of the following:

(a) $\dfrac{3^4 \cdot 3^8}{3^{14}} = \dfrac{3^{4+8}}{3^{14}} = 3^{4+8-14} = 3^{-2} = \dfrac{1}{3^2} = \dfrac{1}{9}$

(b) $\sqrt{\dfrac{(5\cdot 10^{-6})(4\cdot 10^2)}{8\cdot 10^5}} = \sqrt{\dfrac{5\cdot 4}{8} : \dfrac{10^{-6}\cdot 10^2}{10^5}} = \sqrt{2.5\cdot 10^{-9}} = \sqrt{25\cdot 10^{-10}} = 5\cdot 10^{-5}$ or 0.00005

(c) $\log_{2/3}\left(\frac{27}{8}\right) = x$. Then $\left(\frac{2}{3}\right)^x = \frac{27}{8} = \left(\frac{3}{2}\right)^3 = \left(\frac{2}{3}\right)^{-3}$ or $x = -3$

(d) $(\log_a b)(\log_b a) = u$. Then $\log_a b = x$, $\log_b a = y$, assuming $a, b > 0$ and $a, b \neq 1$.

Then $a^x = b$, $b^y = a$, and $u = xy$. Since $(a^x)^y = a^{xy} = b^y = a$, we have $a^{xy} = a^1$ or $xy = 1$, the required value.

1.16. If $M > 0$, $N > 0$, and $a > 0$ but $a \neq 1$, prove that $\log_a \dfrac{M}{N} = \log_a M - \log_a N$.

Let $\log_a M = x$, $\log_a N = y$. Then $a^x = M$, $a^y = N$ and so

$$\frac{M}{N} = \frac{a^x}{a^y} = a^{x-y} \quad \text{or} \quad \log_a \frac{M}{N} = x - y = \log_a M - \log_a N$$

Countability

1.17. Prove that the set of all rational numbers between 0 and 1 inclusive is countable.

Write all fractions with denominator 2, then 3, . . . , considering equivalent fractions such as $\frac{1}{2}, \frac{2}{4}, \frac{3}{6}, \ldots$ no more than once. Then the 1-1 correspondence with the natural numbers can be accomplished as follows:

$$\begin{array}{lccccccccccc}
\text{Rational numbers} & 0 & 1 & \frac{1}{2} & \frac{1}{3} & \frac{2}{3} & \frac{1}{4} & \frac{3}{4} & \frac{1}{5} & \frac{2}{5} & \ldots \\
 & \updownarrow & \updownarrow & \updownarrow & \updownarrow & \updownarrow & \updownarrow & \updownarrow & \updownarrow & \updownarrow & \\
\text{Natural numbers} & 1 & 2 & 3 & 4 & 5 & 6 & 7 & 8 & 9 & \ldots
\end{array}$$

Thus, the set of all rational numbers between 0 and 1 inclusive is countable and has cardinal number $\aleph_0$ (see Page 6).

1.18. If A and B are two countable sets, prove that the set consisting of all elements from A or B (or both) is also countable.

Since A is countable, there is a 1-1 correspondence between elements of A and the natural numbers so that we can denote these elements by $a_1, a_2, a_3, \ldots$

Similarly, we can denote the elements of B by $b_1, b_2, b_3, \ldots$

Case 1: Suppose elements of A are all distinct from elements of B. Then the set consisting of elements from A or B is countable, since we can establish the following 1-1 correspondence:

$$\begin{array}{lcccccccc}
A \text{ or } B & a_1 & b_1 & a_2 & b_2 & a_3 & b_3 & \ldots \\
 & \updownarrow & \updownarrow & \updownarrow & \updownarrow & \updownarrow & \updownarrow & \\
\text{Natural numbers} & 1 & 2 & 3 & 4 & 5 & 6 & \ldots
\end{array}$$

Case 2: If some elements of A and B are the same, we count them only once, as in Problem 1.17. Then the set of elements belonging to A or B (or both) is countable.

The set consisting of all elements which belong to A or B (or both) is often called the *union* of A and B, denoted by $A \cup B$ or $A + B$.

The set consisting of all elements which are contained in both A *and* B is called the *intersection* of A and B, denoted by $A \cap B$ or AB. If A and B are countable, so is $A \cap B$.

The set consisting of all elements in A but *not* in B is written $A - B$. If we let [B be the set of elements which are not in B, we can also write $A - B = A\overline{B}$. If A and B are countable, so is $A - B$.

1.19. Prove that the set of all positive rational numbers is countable.

Consider all rational numbers $x > 1$. With each such rational number we can associate one and only one rational number $1/x$ in $(0, 1)$; i.e., there is a *one-to-one correspondence* between all rational numbers > 1 and all rational numbers in $(0, 1)$. Since these last are countable by Problem 1.17, it follows that the set of all rational numbers > 1 is also countable.

From Problem 1.18 it then follows that the set consisting of all positive rational numbers is countable, since this is composed of the two countable sets of rationals between 0 and 1 and those greater than or equal to 1.

From this we can show that the set of all rational numbers is countable (see Problem 1.59).

1.20. Prove that the set of all real numbers in $[0, 1]$ is noncountable.

Every real number in $[0, 1]$ has a decimal expansion $.a_1a_2a_3 \ldots$ where $a_1, a_2, \ldots$ are any of the digits 0, 1, 2, . . . ,9.

We assume that numbers whose decimal expansions terminate such as 0.7324 are written 0.73240000 . . . and that this is the same as 0.73239999 . . .

If all real numbers in $[0, 1]$ are countable we can place them in 1-1 correspondence with the natural numbers as in the following list:

$$\begin{array}{ccl} 1 & \leftrightarrow & 0.a_{11}a_{12}a_{13}a_{14}\cdots \\ 2 & \leftrightarrow & 0.a_{21}a_{22}a_{23}a_{24}\cdots \\ 3 & \leftrightarrow & 0.a_{31}a_{32}a_{33}a_{34}\cdots \\ \vdots & & \quad\vdots \end{array}$$

We now form a number

$$0.b_1b_2b_3b_4 \ldots$$

where $b_1 \neq a_{11}$, $b_2 \neq a_{22}$, $b \neq a_{33}$, $b_4 \neq a_{44}$, . . . and where all b's beyond some position are not all 9's.

This number, which is in [0. 1], is different from all numbers in the preceding list and is thus not in the list, contradicting the assumption that all numbers in $[0, 1]$ were included.

Because of this contradiction, it follows that the real numbers in $[0, 1]$ cannot be placed in 1-1 correspondence with the natural numbers; i.e., the set of real numbers in $[0, 1]$ is noncountable.

Limit points, bounds, Bolzano-Weierstrass theorem

1.21. (a) Prove that the infinite set of numbers $1, \frac{1}{2}, \frac{1}{3}, \frac{1}{4}, \ldots$ is bounded. (*b*) Determine the least upper bound (l.u.b.) and greatest lower bound (g.l.b.) of the set. (*c*) Prove that 0 is a limit point of the set. (*d*) Is the set a closed set? (*e*) How does this set illustrate the Bolzano-Weierstrass theorem?

(a) Since all members of the set are less than 2 and greater than –1 (for example), the set is bounded; 2 is an upper bound; –1 is a lower bound.
We can find smaller upper bounds (e.g., 3/2) and larger lower bounds (e.g., $-\frac{1}{2}$).

(b) Since no member of the set is greater than 1 and since there is at least one member of the set (namely, 1) which exceeds $1 - \varepsilon$ for every positive number ε, we see that 1 is the l.u.b. of the set.

Since no member of the set is less than 0 and since there is at least one member of the set which is less than $0 + \varepsilon$ for every positive ε (we can always choose for this purpose the number $1/n$, where n is a positive integer greater than $1/\varepsilon$), we see that 0 is the g.l.b. of the set.

(c) Let x be any member of the set. Since we can always find a number x such that $0 < |x| < \delta$ for any positive number δ (e.g., we can always pick x to be the number $1/n$, where n is a positive integer greater than $1/\delta$), we see that 0 is a limit point of the set. To put this another way, we see that any deleted δ neighborhood of 0 always includes members of the set, no matter how small we take $\delta > 0$.

(d) The set is not a closed set, since the limit point 0 does not belong to the given set.

(e) Since the set is bounded and infinite, it must, by the Bolzano-Weierstrass theorem, have at least one limit point. We have found this to be the case, so that the theorem is illustrated.

Algebraic numbers

1.22. Prove that $\sqrt[3]{2} + \sqrt{3}$ is an algebraic number.

Let $x = \sqrt[3]{2} + \sqrt{3}$. Then $x - \sqrt{3} = \sqrt[3]{2}$. Cubing both sides and simplifying, we find $x^3 + 9x - 2 = 3\sqrt{3}$ $(x^2 + 1)$. Then, squaring both sides and simplifying, we find $x^6 - 9x^4 - 4x^3 + 27x^2 + 36x - 23 = 0$.

Since this is a polynomial equation with integral coefficients, it follows that $\sqrt[3]{2} + \sqrt{3}$, which is a solution, is an algebraic number.

1.23. Prove that the set of all algebraic numbers is a countable set.

Algebraic numbers are solutions to polynomial equations of the form $a_0x^n + a_1x^{n-1} + \cdots + a_n = 0$ where $a_0, a_1, \ldots, a_n$ are integers.

Let $P = |a_0| + |a_1| + \cdots + |a_n| + n$. For any given value of P there are only a finite number of possible polynomial equations and thus only a finite number of possible algebraic numbers.

Write all algebraic numbers corresponding to $P = 1, 2, 3, 4, \ldots$, avoiding repetitions. Thus, all algebraic numbers can be placed into 1-1 correspondence with the natural numbers and so are countable.

Complex numbers

1.24. Perform the indicated operations:

(a) $(4 - 2i) + (-6 + 5i) = 4 - 2i - 6 + 5i = 4 - 6 + (-2 + 5)i = -2 + 3i$

(b) $(-7 + 3i) - (2 - 4i) = -7 + 3i - 2 + 4i = -9 + 7i$

(c) $(3 - 2i)(1 + 3i) = 3(1 + 3i) - 2i(1 + 3i) = 3 + 9i - 2i - 6i^2 = 3 + 9i - 2i + 6 = 9 + 7i$

(d)
$$\frac{-5+5i}{4-3i} = \frac{-5+5i}{4-3i}\cdot\frac{4+3i}{4+3i} = \frac{(-5+5i)(4+3i)}{16-9i^2} = \frac{-20-15i+20i+15i^2}{16+9}$$
$$= \frac{-35+5i}{25} = \frac{5(-7+i)}{25} = \frac{-7}{5} + \frac{1}{5}i$$

(e)
$$\frac{i+i^2+i^3+i^4+i^5}{1+i} = \frac{i-1+(i^2)(i)+(i^2)^2+(i^2)^2 i}{1+i} = \frac{i-1-i+1+i}{1+i}$$
$$= \frac{i}{1+i}\cdot\frac{1-i}{1-i} = \frac{i-i^2}{1-i^2} = \frac{i+1}{2} = \frac{1}{2} + \frac{1}{2}i$$

(f) $|3-4i||4+3i| = \sqrt{(3)^2+(-4)^2}\,\sqrt{(4)^2+(3)^2} = (5)(5) = 25$

(g) $\left|\dfrac{1}{1+3i}-\dfrac{1}{1-3i}\right|=\left|\dfrac{1-3i}{1-9i^2}-\dfrac{1+3i}{1-9i^2}\right|=\left|\dfrac{-6i}{10}\right|=\sqrt{(0)^2+\left(-\dfrac{6}{10}\right)^2}=\dfrac{3}{5}$

1.25. If z_1 and z_2 are two complex numbers, prove that $|z_1z_2| = |z_1||z_2|$.

Let $z_1 = x_1 + iy_1$, $z_2 = x_2 + iy_2$. Then

$$|z_1z_2| = |(x_1+iy_1)(x_2+iy_2)| = |x_1x_2 - y_1y_2 + i(x_1y_2 + x_2y_1)|$$
$$= \sqrt{(x_1x_2-y_1y_2)^2+(x_1y_2+x_2y_1)^2} = \sqrt{x_1^2x_2^2+y_1^2y_2^2+x_1^2y_2^2+x_2^2y_1^2}$$
$$= \sqrt{(x_1^2+y_1^2)(x_2^2+y_2^2)} = \sqrt{x_1^2+y^2}\sqrt{x_2^2+y_2^2} = |x_1+iy_1||x_2+iy_2| = |z_1||z_2|.$$

1.26. Solve $x^3 - 2x - 4 = 0$.

The possible rational roots using Problem 1.7 are ±1, ±2, and ±4. By trial, we find $x = 2$ is a root. Then the given equation can be written $(x - 2)(x^2 + 2x + 2) = 0$. The solutions to the *quadratic equation* $ax^2 + bx + c = 0$ are $x = \dfrac{-b \pm \sqrt{b^2-4ac}}{2a}$ For $a = 1$, $b = 2$, and $c = 2$, this gives $x = \dfrac{-2\pm\sqrt{4-8}}{2} = \dfrac{-2\pm\sqrt{-4}}{2} = \dfrac{-2\pm 2i}{2} = -1 \pm i.$

The set of solutions is $2, -1 + i, -1 - i$.

Polar form of complex numbers

1.27. Express in polar form (*a*) $3 + 3i$, (*b*) $-1 + \sqrt{3}i$, (*c*) -1, and (*d*) $-2 - 2\sqrt{3}i$. See Figure 1.4.

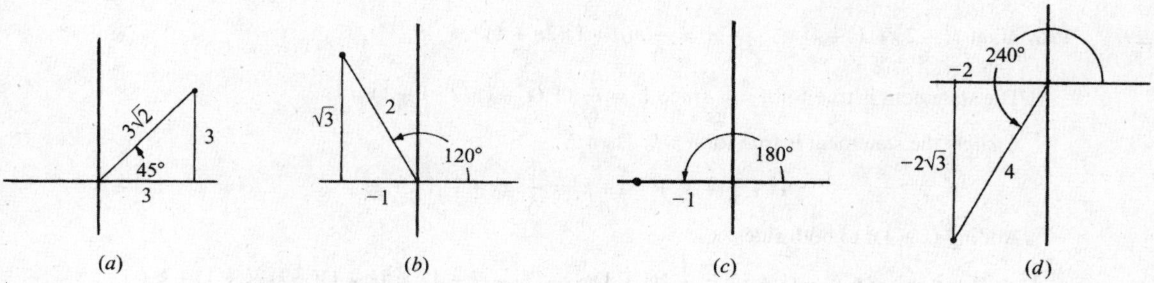

Figure 1.4

(a) Amplitude $\phi = 45° = \pi/4$ radians. Modulus $\rho = \sqrt{3^2+3^2} = 3\sqrt{2}$. Then $3 + 3i = \rho(\cos\phi + i\sin\phi) = 3\sqrt{2}(\cos\pi/4 + i\sin\pi/4) = 3\sqrt{2}\ \text{cis}\ \pi/4 = 3\sqrt{2}e^{\pi i/4}$.

(b) Amplitude $\phi = 120° = 2\pi/3$ radians. Modulus $\rho = \sqrt{(-1)^2+(\sqrt{3})^2} = \sqrt{4} = 2$. Then $-1 + 3\sqrt{3}\,i = 2(\cos 2\pi/3 + i\sin 2\pi/3) = 2\ \text{cis}\ 2\pi/3 = 2e^{2\pi i/3}$.

(c) Amplitude $\phi = 180° = \pi$ radians. Modulus $\rho = \sqrt{(-1)^2+(0)^2} = 1$. Then $-1 = 1(\cos\pi + i\sin\pi) = \text{cis}\ \pi = e^{\pi i}$.

(d) Amplitude $\phi = 240° = 4\pi/3$ radians. Modulus $\rho = \sqrt{(-2)^2+(-2\sqrt{3})^2} = 4$. Then $-2 - 2\sqrt{3} = 4(\cos 4\pi/3 + i\sin 4\pi/3) = 4\ \text{cis}\ 4\pi/3 = 4e^{4\pi i/3}$.

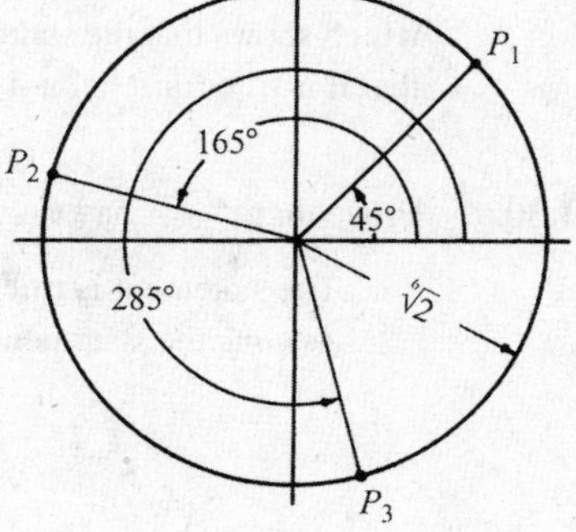

Figure 1.5

1.28. Evaluate (*a*) $(-1+\sqrt{3}i)^{10}$ and (*b*) $(-1+i)^{1/3}$.

(a) By Problem 1.27(*b*) and De Moivre's theorem.

$$(-1+\sqrt{3}i)^{10} = [2(\cos 2\pi/3 + i\sin 2\pi/3)]^{10} = 2^{10}(\cos 20\pi/3 + i\sin 20\pi/3)$$

$$= 1024[\cos(2\pi/3 + 6\pi) + i\sin(2\pi/3 + 6\pi)] = 1024(\cos 2\pi/3 + i\sin 2\pi/3)$$

$$= 1024\left(-\frac{1}{2}+\frac{1}{2}\sqrt{3}i\right) = -512 + 512\sqrt{3}i$$

(b) $-1 + i = \sqrt{2}(\cos 135° + i\sin 135°) = \sqrt{2}[\cos(135° + k\cdot 360°) + i\sin(135° + k\cdot 360°)]$. Then

$$(-1+i)^{1/3} = (\sqrt{2})^{1/3}\left[\cos\left(\frac{135°+k\cdot 360°}{3}\right) + i\sin\left(\frac{135°+k\cdot 360°}{3}\right)\right]$$

The results for $k = 0, 1, 2$ are

$$\sqrt[6]{2}(\cos 45° + i\sin 45°).$$

$$\sqrt[6]{2}(\cos 165° + i\sin 165°),$$

$$\sqrt[6]{2}(\cos 285° + i\sin 285°)$$

The results for $k = 3, 4, 5, 6, 7, \ldots$ give repetitions of these. These complex roots are represented geometrically in the complex plane by points P_1, P_2, P_3 on the circle of Figure 1.5.

Mathematical induction

1.29. Prove that $1^2 + 2^2 + 3^3 + 4^2 + \cdots + n^2 = \frac{1}{6}n(n+1)(2n+1)$.

The statement is true for $n = 1$, since $1^2 = \frac{1}{6}(1)(1+1)(2\cdot 1 + 1) = 1$.

Assume the statement is true for $n = k$. Then

$$1^2 + 2^2 + 3^2 + \cdots + k^2 = \frac{1}{6}k(k+1)(2k+1)$$

Adding $(k+1)^2$ to both sides.

$$1^2 + 2^2 + 3^2 + \cdots + k^2 + (k+1)^2 = \frac{1}{6}k(k+1)(2k+1) + (k+1)^2 = (k+1)[\frac{1}{6}k(2k+1) + k + 1]$$

$$= \frac{1}{6}(k+1)(2k^2 + 7k + 6) = \frac{1}{6}(k+1)(k+2)(2k+3)$$

which shows that the statement is true for $n = k + 1$ *if* it is true for $n = k$. But since it is true for $n = 1$, it follows that it is true for $n = 1 + 1 = 2$ and for $n = 2 + 1 = 3, \ldots$; i.e., it is true for all positive integers n.

1.30. Prove that $x^n - y^n$ has $x - y$ as a factor for all positive integers n.

The statement is true for $n = 1$, since $x^1 - y^1 = x - y$.

Assume the statement is true for $n = k$; i.e., assume that $x^k - y^k$ has $x - y$ as a factor. Consider

$$x^{k+1} - y^{k+1} = x^{k+1} - x^k y + x^k y - y^{k+1}$$

$$= x^k(x - y) + y(x^k - y^k)$$

The first term on the right has $x - y$ as a factor, and the second term on the right also has $x - y$ as a factor because of the previous assumption.

Thus, $x^{k+1} - y^{k+1}$ has $x - y$ as a factor *if* $x^k - y^k$ does.

Then, since $x^1 - y^1$ has $x - y$ as factor, it follows that $x^2 - y^2$ has $x - y$ as a factor, $x^3 - y^3$ has $x - y$ as a factor, etc.

1.31. Prove *Bernoulli's inequality* $(1 + x)^n > 1 + nx$ for $n = 2, 3, \ldots$ if $x > -1, x \neq 0$.

The statement is true for $n = 2$, since $(1 + x)^2 = 1 + 2x + x^2 > 1 + 2x$.
Assume the statement is true for $n = k$; i.e., $(1 + x)^k > 1 + kx$.
Multiply both sides by $1 + x$ (which is positive, since $x > -1$). Then we have

$$(1 + x)^{k+1} > (1 + x)(1 + kx) = 1 + (k + 1)x + kx^2 > 1 + (k + 1)x$$

Thus, the statement is true for $n = k + 1$ if it is true for $n = k$.

But since the statement is true for $n = 2$, it must be true for $n = 2 + 1 = 3 \ldots$ and is thus true for all integers greater than or equal to 2.

Note that the result is not true for $n = 1$. However, the modified result $(1 + x)^n \geqq 1 + nx$ is true for $n = 1, 2, 3, \ldots$

Miscellaneous problems

1.32. Prove that every positive integer P can be expressed uniquely in the form $P = a_0 2^n + a_1 2^{n-1} + a_2 2^{n-2} + \cdots + a_n$ where the a's are 0's or 1's.

Dividing P by 2, we have $P/2 = a_0 2^{n-1} + a_1 2^{n-2} + \cdots + a_{n-1} + a_n/2$.
Then a_n is the remainder, 0 or 1, obtained when P is divided by 2 and is unique.
Let P_1 be the integer part of $P/2$. Then $P_1 = a_0 2^{n-1} + a_1 2^{n-2} + \cdots + a_{n-1}$.
Dividing P_1 by 2, we see that a_{n-1} is the remainder, 0 or 1, obtained when P_1 is divided by 2 and is unique.

By continuing in this manner, all the a's can be determined as 0's or 1's and are unique.

1.33. Express the number 23 in the form of Problem 1.32.

The determination of the coefficient can be arranged as follows:

2) <u>23</u>	
2) <u>11</u>	Remainder 1
2) <u>5</u>	Remainder 1
2) <u>2</u>	Remainder 1
2) <u>1</u>	Remainder 0
0	Remainder 1

The coefficients are 1 0 1 1 1. **Check**: $23 = 1 \cdot 2^4 + 0 \cdot 2^3 + 1 \cdot 2^2 + 1 \cdot 2 + 1$.
The number 10111 is said to represent 23 in the *scale of two* or *binary scale*.

1.34. Dedekind defined a *cut, section,* or *partition* in the rational number system as a separation of *all* rational numbers into two classes or sets called L (the left-hand class) and R (the right-hand class) having the following properties:

I. The classes are non-empty (i.e. at least one number belongs to each class).

II. Every rational number is in one class or the other.

III. Every number in L is less than every number in R.

Prove each of the following statements:

(a) There cannot be a largest number in L and a smallest number in R.

(b) It is possible for L to have a largest number and for R to have no smallest number. What type of number does the cut define in this case?

(c) It is possible for L to have no largest number and for R to have a smallest number. What type of number does the cut define in this case?

(d) It is possible for L to have no largest number and for R to have no smallest number. What type of number does the cut define in this case?

(a) Let a be the largest rational number in L and b the smallest rational number in R. Then either $a = b$ or $a < b$.
We cannot have $a = b$, since, by definition of the cut, every number in L is *less* than every number in R.
We cannot have $a < b$, since, by Problem 1.9, $\frac{1}{2}(a+b)$ is a rational number which would be greater than a (and so would have to be in R) but less than b (and so would have to be in L), and, by definition, a rational number cannot belong to *both* L and R.

(b) As an indication of the possibility, let L contain the number $\frac{2}{3}$ and all rational numbers less than $\frac{2}{3}$, while R, contains all rational numbers greater than $\frac{2}{3}$. In this case the cut defines the rational number $\frac{2}{3}$. A similar argument replacing $\frac{2}{3}$ by any other rational number shows that in such case the cut defines a rational number.

(c) As an indication of the possibility, let L contain all rational numbers less than $\frac{2}{3}$, while R contains all rational numbers greater than $\frac{2}{3}$. This cut also defines the rational number $\frac{2}{3}$. A similar argument shows that this cut always defines a rational number.

(d) As an indication of the possibility, let L consist of all negative rational numbers and all positive rational numbers whose squares are less than 2, while R consists of all positive numbers whose squares are greater than 2. We can show that if a is any number of the L class, there is always a larger number of the L class, while if b is any number of the R class, there is always a smaller number of the R class (see Problem 1.106). A cut of this type defines an irrational number.
From (*b*), (*c*), and (*d*), it follows that every cut in the rational number system, called a *Dedekind cut*, defines either a rational or an irrational number. By use of Dedekind cuts we can define operations (addition, multiplication, etc.) with irrational numbers.

SUPPLEMENTARY PROBLEMS

Operations with numbers

1.35. Given $x = -3$, $y = 2$, $z = 5$, $a = \frac{3}{2}$, and $b = -\frac{1}{4}$, evaluate:

(a) $(2x - y)(3y + z)(5x - 2z)$ (b) $\dfrac{xy - 2z^2}{2ab - 1}$ (c) $\dfrac{3a^2b + ab^2}{2a^2 2b^2 + 1}$ (d) $\dfrac{(ax+by)^2 + (ay-bx)^2}{(ay+bx)^2 + (ax-by)^2}$

Ans. (a) 2200 (b) 32 (c) –51/41 (d) 1

1.36. Find the set of values of x for which the following equations are true. Justify all steps in each case.

(a) $4\{(x-2) + 3(2x-1)\} + 2(2x+1) = 12(x+2) - 2$ (c) $\sqrt{x^2 + 8x + 7} - \sqrt{2x+2} = x + 1$

(b) $\dfrac{1}{8-x} - \dfrac{1}{x-2} = \dfrac{1}{4}$ (d) $\dfrac{1-x}{\sqrt{x^2 - 2x + 5}} = \dfrac{3}{5}$

Ans. (a) 2 (b) 6, – 4 (c) – 1, 1 (d) $-\frac{1}{2}$

1.37. Prove that $\dfrac{x}{(z-x)(x-y)} + \dfrac{y}{(x-y)(y-z)} + \dfrac{z}{(y-z)(z-x)} = 0$, giving restrictions if any.

Rational and irrational numbers

1.38. Find decimal expansions for (a) $\frac{3}{7}$ and (b) $\sqrt{5}$.

Ans. (a) $0.\dot{4}2857\dot{1}$ (b) 2.2360679 . . .

1.39. Show that a fraction with denominator 17 and with numerator 1, 2, 3, . . . , 16 has 16 digits in the repeating portion of its decimal expansion. Is there any relation between the orders of the digits in these expansions?

1.40. Prove that (a) $\sqrt{3}$ and (b) $\sqrt[3]{2}$ are irrational numbers.

1.41. Prove that (a) $\sqrt[3]{5} - \sqrt[4]{3}$ and (b) $\sqrt{2} + \sqrt{3} + \sqrt{5}$ are irrational numbers.

1.42. Determine a positive rational number whose square differs from 7 by less than .000001.

1.43. Prove that every rational number can be expressed as a repeating decimal.

1.44. Find the values of x for which (a) $2x^3 - 5x^2 - 9x + 18 = 0$, (b) $3x^3 + 4x^2 - 35x + 8 = 0$, and (c) $x^4 - 21x^2 + 4 = 0$.

Ans. (a) 3, –2, 3/2 (b) 8/3, $-2 \pm \sqrt{5}$ (c) $\frac{1}{2}(5 \pm \sqrt{17})$, $\frac{1}{2}(-5 \pm \sqrt{17})$

1.45. If a, b, c, and d are rational and m is not a perfect square, prove that $a + b\sqrt{m} = c + d\sqrt{m}$ if and only if $a = c$ and $b = d$.

1.46. Prove that $\dfrac{1+\sqrt{3}+\sqrt{5}}{1-\sqrt{3}+\sqrt{5}} = \dfrac{12\sqrt{5}-2\sqrt{15}+14\sqrt{3}-7}{11}$.

Inequalities

1.47. Find the set of values of x for which each of the following inequalities holds:

(a) $\frac{1}{x} + \frac{3}{2x} \geqq 5$, (b) $x(x + 2) \leqq 24$, (c) $|x + 2| < |x - 5|$, (d) $\frac{x}{x+2} > \frac{x+3}{3x+1}$

Ans. (a) $0 < x \leqq \frac{1}{2}$ (b) $-6 \leqq x \leqq 4$ (c) $x < 3/2$ (d) $x > 3$, $-1 < x < -\frac{1}{3}$, or $x < -2$

1.48. Prove (a) $|x + y| \leqq |x| + |y|$, (b) $|x + y + z| \leqq |x| + |y| + |z|$, and (c) $|x| - y| \geqq |x| - |y|$.

1.49. Prove that for all real x, y, z, $x^2 + y^2 + z^2 \geqq xy + yz + zx$.

1.50. If $a^2 + b^2 = 1$ and $c^2 + d^2 = 1$, prove that $ac + bd \leqq 1$.

1.51. If $x > 0$, prove that $x^{n+1} + \frac{1}{x^{n+1}} > x^n + \frac{1}{x^n}$ where n is any positive integer.

1.52. Prove that for all real $a \neq 0$, $|a + 1/a| \geqq 2$.

1.53. Show that in Schwarz's inequality (Problem 1.13) the equality holds if and only if $a_p = kb_p$, $p = 1, 2, 3, \ldots, n$, where k is any constant.

1.54. If a_1, a_2, a_3 are positive, prove that $\frac{1}{3}(a_1 + a_2 + a_3) \geqq \sqrt[3]{a_1 a_2 a_3}$.

Exponents, roots, and logarithms

1.55. Evaluate: (a) $4^{\log_2 8}$ (b) $\frac{3}{4}\log_{1/8}\left(\frac{1}{128}\right)$ (c) $\sqrt{\frac{(0.00004)(25{,}000)}{(0.02)^5(0.125)}}$ (d) $3^{-2\log_3 5}$ (e) $\left(-\frac{1}{8}\right)^{4/3} - (-27)^{-2/3}$

Ans. (a) 64 (b) 7/4 (c) 50,000 (d) 1/25 (e) –7/144

1.56. Prove (a) $\log_a MN = \log_a M + \log_a N$ and (b) $\log_a M^r = r \log_a M$ indicating restrictions, if any.

1.57. Prove $b^{\log_b a} = a$ giving restrictions, if any.

Countability

1.58. (a) Prove that there is a one-to-one correspondence between the points of the interval $0 \leqq x \leqq 1$ and $-5 \leqq x \leqq -3$. (b) What is the cardinal number of the sets in (*a*)?

Ans. (b) C, the cardinal number of the continuum.

1.59. (a) Prove that the set of all rational numbers is countable. (b) What is the cardinal number of the set in (*a*)?

Ans. (b) $\aleph_0$

1.60. Prove that the set of (a) all real numbers and (b) all irrational numbers is noncountable.

1.61. The *intersection* of two sets A and B, denoted by $A \cap B$ or AB, is the set consisting of all elements belonging to both A and B. Prove that if A and B are countable, so is their intersection.

1.62. Prove that a countable sets of countable sets is countable.

1.63. Prove that the cardinal number of the set of points inside a square is equal to the cardinal number of the sets of points on (a) one side and (b) all four sides. (c) What is the cardinal number in this case? (d) Does a corresponding result hold for a cube?

Ans. (c) C

Limit points, bounds, Bolzano-Weierstrass theorem

1.64. Given the set of numbers 1, 1.1,.9, 1.01, .99, 1.001,.999, . . . , (a) is the set bounded? (b) Does the set have an l.u.b. and a g.l.b.? If so, determine them. (c) Does the set have any limit points? If so, determine them. (d) Is the set a closed set?

Ans. (a) Yes (b) l.u.b. = 1.1.g.l.b. = .9 (c) 1 (d) Yes

1.65. Given the set –.9,.9, –.99, .99, –.999, .999, answer the questions in Problem 1.64.

Ans. (a) Yes (b) l.u.b. = 1, g.l.b. = –1 (c) 1, –1 (d) No

1.66. Give an example of a set which has (a) three limit points and (b) no limit points.

1.67. (a) Prove that every point of the interval $0 < x < 1$ is a limit point. (b) Are there limit points which do not belong to the set in (a)? Justify your answer.

1.68. Let S be the set of all rational numbers in (0, 1) having denominator 2^n, $n = 1, 2, 3, \ldots$ (*a*) Does S have any limit points? (b) Is S closed?

1.69. (a) Give an example of a set which has limit points but which is not bounded. (b) Does this contradict the Bolzano-Weierstrass theorem? Explain.

Algebraic and transcendental numbers

1.70. Prove that (a) $\dfrac{\sqrt{3}-\sqrt{2}}{\sqrt{3}+\sqrt{2}}$, (b) $\sqrt{2}+\sqrt{3}+\sqrt{5}$ are algebraic numbers.

1.71. Prove that the set of transcendental numbers in (0, 1) is not countable.

1.72. Prove that every rational number is algebraic but every irrational number is not necessarily algebraic.

Complex numbers, polar form

1.73. Perform each of the indicated operations:

(a) $2(5 - 3i) - 3(-2 + i) + 5(i - 3)$

(b) $(3 - 2i)^3$

(c) $\dfrac{5}{3-4i} + \dfrac{10}{4+3i}$

(d) $\left(\dfrac{1-i}{1+i}\right)^{10}$

(e) $\left|\dfrac{2-4i}{5+7i}\right|^2$

(f) $\dfrac{(1+i)(2+3i)(4-2i)}{(1+2i)^2(1-i)}$

Ans. (a) 1 – 4i (b) – 9 – 46i (c) $\dfrac{11}{5} - \dfrac{2}{5}i$ (d) –1 (e) $\dfrac{10}{37}$ (f) $\dfrac{16}{5} - \dfrac{2}{5}i$

1.74. If z_1 and z_2 are complex numbers, prove (a) $\left|\dfrac{z_1}{z_2}\right| = \dfrac{|z_1|}{|z_2|}$ and (b) $\left|z_1^2\right| = |z_1|^2$, giving any restrictions.

1.75. Prove (a) $|z_1 + z_2| \leq |z_1| + |z_2|$, (b) $|z_1 + z_2 + z_3| \leq |z_1| + |z_2 + z_3|$ and (c) $|z_1 - z_2| \geq |z_1| - |z_2|$.

1.76. Find all solutions of $2x^4 - 3x^3 - 7x^2 - 8x + 6 = 0$.

Ans. $3, \frac{1}{2}, -1 \pm i$

1.77. Let z_1 and z_2 be represented by points P_1 and P_2 in the Argand diagram. Construct lines OP_1 and OP_2, where O is the origin. Show that $z_1 + z_2$ can be represented by the point P_3, where OP_3 is the diagonal of a parallelogram having sides OP_1 and OP_2. This is called the *parallelogram law* of addition of complex numbers. Because of this and other properties, complex numbers can be considered as *vectors* in two dimensions.

1.78. Interpret geometrically the inequalities of Problem 1.75.

1.79. Express in polar form (a) $3\sqrt{3} + 3i$, (b) $-2 - 2i$, (c) $1 - \sqrt{3}\,i$, (d) 5, and (e) $-5i$.

Ans. (a) $6 \text{ cis } \pi/6$ (b) $2\sqrt{2} \text{ cis } 5\pi/4$ (c) $2 \text{ cis } 5\pi/3$ (d) $5 \text{ cis } 0$ (e) $5 \text{ cis } 3\pi/2$

1.80. Evaluate (a) $[2(\cos 25° + i \sin 25°)][5(\cos 110° + i \sin 110°)]$ and (b) $\dfrac{12 \text{ cis } 16°}{(3 \text{ cis } 44°)(2 \text{ cis } 62°)}$.

Ans. (a) $-5\sqrt{2} + 5\sqrt{2}\,i$ (b) $-2i$

1.81. Determine all the indicated roots and represent them graphically: (a) $(4\sqrt{2} + 4\sqrt{2}i)^{1/3}$, (b) $(-1)^{1/5}$, (c) $(\sqrt{3} - i)^{1/3}$, and (d) $i^{1/4}$.

Ans. (*a*) $2 \text{ cis } 15°, 2 \text{ cis } 135°, 2 \text{ cis } 255°$

(b) $\text{cis } 36°, \text{cis } 108°, \text{cis } 180° = -1, \text{cis } 252°, \text{cis } 324°$

(c) $\sqrt[3]{2} \text{ cis } 110°, \sqrt[3]{2} \text{ cis } 230°, \sqrt[3]{2} \text{ cis } 350°$

(d) $\text{cis } 22.5°, \text{cis } 112.5°, \text{cis } 202.5°, \text{cis } 292.5°$

1.82. Prove that $-1 + \sqrt{3}\,i$ is an algebraic number.

1.83. If $z_1 = \rho_1 \text{ cis } \phi_1$ and $z_2 = \rho_2 \text{ cis } \phi_2$, prove (a) $z_1 z_2 = \rho_1 \rho_2 \text{ cis}(\phi_1 + \phi_2)$ and (b) $z_1/z_2 = (\rho_1/\rho_2)\text{cis } (\phi_1 - \phi_2)$. Interpret geometrically.

Mathematical induction

Prove each of the following.

1.84. $1 + 3 + 5 + \ldots + (2n - 1) = n^2$

1.85. $\dfrac{1}{1\cdot 3} + \dfrac{1}{3\cdot 5} + \dfrac{1}{5\cdot 7} + \cdots + \dfrac{1}{(2n-1)(2n+1)} = \dfrac{n}{2n+1}$

1.86. $a + (a + d) + (a + 2d) + \cdots + [a + (n - 1)d] = \dfrac{1}{2}n[2a + (n - 1)d]$

1.87. $\dfrac{1}{1\cdot 2\cdot 3} + \dfrac{1}{2\cdot 3\cdot 4} + \dfrac{1}{3\cdot 4\cdot 5} + \cdots + \dfrac{1}{n(n+1)(n+2)} = \dfrac{n(n+3)}{4(n+1)(n+2)}$

1.88. $a + ar + ar^2 + \cdots + ar^{n-1} = \dfrac{a(r^n - 1)}{r - 1}, r \neq 1$

1.89. $1^3 + 2^3 + 3^3 + \cdots + n^3 = \dfrac{1}{4} n^2 (n + 1)^2$

1.90. $1(5) + 2(5)^2 + 3(5)^3 + \cdots + n(5)^{n-1} = \dfrac{5 + (4n - 1)5^{n+1}}{16}$

1.91. $x^{2n-1} + y^{2n-1}$ is divisible by $x + y$ for $n = 1, 2, 3, \ldots$

1.92. $(\cos\phi + i\sin\phi)^n = \cos n\phi + i\sin n\phi$. Can this be proved if n is a rational number?

1.93. $\frac{1}{2} + \cos x + \cos 2x + \cdots + \cos nx = \dfrac{\sin(n + \frac{1}{2})x}{2\sin\frac{1}{2}x}, x \neq 0, \pm 2\pi, \pm 4\pi, \ldots$

1.94. $\sin x + \sin 2x + \cdots + \sin nx = \dfrac{\cos\frac{1}{2}x - \cos(n + \frac{1}{2})x}{2\sin\frac{1}{2}x}, x \neq 0, \pm 2\pi, \pm 4\pi \ldots$

1.95. $(a + b)^n = a^n + {}_nC_1 a^{n-1} b + {}_nC_2 a^{n-2} b^2 + \cdots + {}_nC_{n-1} ab^{n-1} + b_n$

where ${}_nC_r = \dfrac{n(n-1)(n-2)\ldots(n-r+1)}{r!} = \dfrac{n!}{r!(n-r)!} = {}_nC_{n-r}$. Here $p! = p(p - 1) \ldots 1$ and $0!$ is defined as 1. This is called the *binomial theorem*. The coefficients ${}_nC_0 = 1$, ${}_nC_1 = n$, ${}_nC_2 = \dfrac{n(n-1)}{2!}, \ldots, {}_nC_n = 1$ are called the *binomial coefficients*. ${}_nC_r$ is also written $\dbinom{n}{r}$.

Miscellaneous problems

1.96. Express each of the following integers (scale of 10) in the scale of notation indicated: (a) 87 (two), (b) 64 (three) (c) 1736 (nine). Check each answer.

Ans. (a) 1010111 (b) 2101 (c) 2338

1.97. If a number is 144 in the scale of 5. what is the number in the scale of (a) 2 and (b) 8?

1.98. Prove that every rational number p/q between 0 and 1 can be expressed in the form

$$\frac{p}{q} = \frac{a_1}{2} + \frac{a_2}{2^2} + \cdots + \frac{a_n}{2^n} + \cdots$$

where the a's can be determined uniquely as 0's or 1's and where the process may or may not terminate. The representation $0.a_1a_2 \ldots a_n \ldots$ is then called the *binary form* of the rational number. (Hint: Multiply both sides successively by 2 and consider remainders.)

1.99. Express $\dfrac{2}{3}$ in the scale of (a) 2, (b) 3, (c) 8, and (d) 10.

Ans. (a) 0.1010101 . . . (b) 0.2 or 0.2000. . . . (c) 0.5252. . . . (d) 0.6666 . . .

1.100. A number in the scale of 2 is 11.01001. What is the number in the scale of 10.

Ans. 3.28125

1.101. In what scale of notation is $3 + 4 = 12$?

Ans. 5

1.102. In the scale of 12, two additional symbols, t and e, must be used to designate the "digits" 10 and 11, respectively. Using these symbols, represent the integer 5110 (scale of 10) in the scale of 12.

Ans. $2e5t$

1.103. Find a rational number whose decimal expansion is 1.636363 . . .

Ans. 18/11

1.104. A number in the scale of 10 consists of six digits. If the last digit is removed and placed before the first digit, the new number is one-third as large. Find the original number.

Ans. 428571

1.105. Show that the rational numbers form a field (see Page 3).

1.106. Using as axioms the relations 1 through 9 on Page 3, prove that (a) $(-3)(0) = 0$, (b) $(-2)(+3) = -6$, and (c) $(-2)(-3) = 6$.

1.107. (a) If x is a rational number whose square is less than 2, show that $x + (2 - x^2)/10$ is a larger such number. (b) If x is a rational number whose square is greater than 2, find in terms of x a smaller rational number whose square is greater than 2.

1.108. Illustrate how you would use Dedekind cuts to define (a) $\sqrt{5} + \sqrt{3}$, (b) $\sqrt{3} - \sqrt{2}$, (c) $(\sqrt{3})(\sqrt{2})$, and (d) $\sqrt{2}/\sqrt{3}$.

CHAPTER 2

Sequences

Definition of a Sequence

A sequence is a set of numbers $u_1, u_2, u_3, \ldots$ in a definite order of arrangement (i.e., a *correspondence* with the natural numbers or a subset thereof) and formed according to a definite rule. Each number in the sequence is called a *term*; u_n is called the *n*th *term*. The sequence is called *finite* or *infinite* according as there are or are not a finite number of terms. The sequence $u_1, u_2, u_3, \ldots$ is is also designated briefly by $\{u_n\}$.

EXAMPLES. 1. The set of numbers 2, 7, 12, 17, . . ., 32 is a finite sequence; the *n*th term is given by $u_n = 2 + 5(n-1) = 5n - 3, n = 1, 2, \ldots, 7$.

2. The set of numbers 1, 1/3, 1/5, 1/7, . . . is an infinite sequence with *n*th term $u_n = 1/(2n-1)$, $n = 1, 2, 3, \ldots$.

Unless otherwise specified, we shall consider infinite sequences only.

Limit of a Sequence

A number l is called the *limit* of an infinite sequence $u_1, u_2, u_3, \ldots$ if for any positive number ϵ we can find a positive number N depending on ϵ such that $|u_n - l| < \epsilon$ for all integers $n > N$. In such case we write $\lim_{n\to\infty} u_n = l$.

EXAMPLE. If $u_n = 3 + 1/n = (3n+1)/n$, the sequence is 4, 7/2, 10/3, . . . and we can show that $\lim_{n\to\infty} u_n = 3$.

If the limit of a sequence exists, the sequence is called *convergent*; otherwise, it is called *divergent*. A sequence can converge to only one limit; i.e., if a limit exists, it is unique. See Problem 2.8.

A more intuitive but unrigorous way of expressing this concept of limit is to say that a sequence $u_1, u_2, u_3, \ldots$ has a limit l if the successive terms get "closer and closer" to l. This is often used to provide a "guess" as to the value of the limit, after which the definition is applied to see if the guess is really correct.

Theorems on Limits of Sequences

If $\lim_{n\to\infty} a_n = A$ and $\lim_{n\to\infty} b_n = B$, then

1. $\lim_{n\to\infty} (a_n + b_n) = \lim_{n\to\infty} a_n + \lim_{n\to\infty} b_n = A + B$

2. $\lim_{n\to\infty} (a_n - b_n) = \lim_{n\to\infty} a_n - \lim_{n\to\infty} b_n = A - B$

3. $\lim_{n\to\infty} (a_n \cdot b_n) = (\lim_{n\to\infty} a_n)(\lim_{n\to\infty} b_n) = AB$

4. $\lim\limits_{n\to\infty} \dfrac{a_n}{b_n} = \dfrac{\lim\limits_{n\to\infty} a_n}{\lim\limits_{n\to\infty} b_n} = \dfrac{A}{B}$ if $\lim\limits_{n\to\infty} b_n = B \neq 0$

   If $B = 0$ and $A \neq 0$, $\lim\limits_{n\to\infty} \dfrac{a_n}{b_n}$ does not exist.

   If $B = 0$ and $A = 0$, $\lim\limits_{n\to\infty} \dfrac{a_n}{b_n}$ may or may not exist.

5. $\lim\limits_{n\to\infty} a_n^p = (\lim\limits_{n\to\infty} a_n)^p = A^p$, for p = any real number if A^p exists.

6. $\lim\limits_{n\to\infty} p^{a_n} = p^{\lim_{n\to\infty} a_n} = p^A$, for p = any real number if p^A exists.

Infinity

We write $\lim\limits_{n\to\infty} a_n = \infty$ if for each positive number M we can find a positive number N (depending on M) such that $a_n > M$ for all $n > N$. Similarly, we write $\lim\limits_{n\to\infty} a_n = -\infty$ if for each positive number M we can find a positive number N such that $a_n < -M$ for all $n > N$. It should be emphasized that ∞ and $-\infty$ are not numbers and the sequences are not convergent. The terminology employed merely indicates that the sequences diverge in a certain manner. That is, no matter how large a number in absolute value that one chooses, there is an n such that the absolute value of a_n is greater than that quantity.

Bounded, Monotonic Sequences

If $u_n \leqq M$ for $n = 1, 2, 3, \ldots$, where M is a constant (independent of n), we say that the sequence $\{u_n\}$ is *bounded above* and M is called an *upper bound*. If $u_n \geqq m$, the sequence is *bounded below* and m is called a *lower bound*.

If $m \leqq u_n \leqq M$ the sequence is called *bounded*. Often this is indicated by $|u_n| \leqq P$. Every convergent sequence is bounded, but the converse is not necessarily true.

If $u_{n+1} \geqq u_n$ the sequence is called *monotonic increasing*; if $u_{n+1} > u_n$ it is called *strictly increasing*. Similarly, if $u_{n+1} \leqq u_n$ the sequence is called *monotonic decreasing*, while if $u_{n+1} < u_n$ it is *strictly decreasing*.

EXAMPLES. 1. The sequence 1, 1.1, 1.11, 1.111, . . . is bounded and monotonic increasing. It is also strictly increasing.
2. The sequence 1, −1, 1, −1, 1, . . . is bounded but not monotonic increasing or decreasing.
3. The sequence −1, −1.5, −2, −2.5, −3, . . . is monotonic decreasing and not bounded. However, it is bounded above.

The following theorem is fundamental and is related to the Bolzano-Weierstrass theorem (Chapter 1, Page 7) which is proved in Problem 2.23.

Theorem Every bounded monotonic (increasing or decreasing) sequence has a limit.

Least Upper Bound and Greatest Lower Bound of a Sequence

A number $\underline{M}$ is called the *least upper bound* (l.u.b.) of the sequence $\{u_n\}$ if $u_n \leqq \underline{M}$, $n = 1, 2, 3, \ldots$ while at least one term is greater than $\underline{M} - \epsilon$ for any $\epsilon > 0$.

A number $\bar{m}$ is called the *greatest lower bound* (g.l.b.) of the sequence $\{u_n\}$ if $u_n \geqq \bar{m}$, $n = 1, 2, 3, \ldots$ while at least one term is less than $\bar{m} + \epsilon$ for any $\epsilon > 0$.

Compare with the definition of l.u.b. and g.l.b. for sets of numbers in general (see Page 6).

Limit Superior, Limit Inferior

A number $\bar{l}$ is called the *limit superior, greatest limit,* or *upper limit* (lim sup or $\overline{\lim}$) of the sequence $\{u_n\}$ if infinitely many terms of the sequence are greater than $\bar{l} - \epsilon$ while only a finite number of terms are greater than $\bar{l} + \epsilon$, where ϵ is any positive number.

A number $\underline{l}$ is called the *limit inferior, least limit,* or *lower limit* (lim inf or $\underline{\lim}$) of the sequence $\{u_n\}$ if infinitely many terms of the sequence are less than $\underline{l} + \epsilon$ while only a finite number of terms are less than $\underline{l} - \epsilon$, where ϵ is any positive number.

These correspond to least and greatest limiting points of general sets of numbers.

If infinitely many terms of $\{u_n\}$ exceed any positive number M, we define lim sup $\{u_n\} = \infty$. If infinitely many terms are less than $-M$, where M is any positive number, we define lim inf $\{u_n\} = -\infty$.

If $\lim_{n\to\infty} u_n = \infty$, we define lim sup $\{u_n\}$ = lim inf $\{u_n\} = \infty$.

If $\lim_{n\to\infty} u_n = -\infty$, we define lim sup $\{u_n\}$ = lim inf $\{u_n\} = -\infty$.

Although every bounded sequence is not necessarily convergent, it always has a finite lim sup and lim inf.

A sequence $\{u_n\}$ converges if and only if lim sup u_n = lim inf u_n is finite.

Nested Intervals

Consider a set of intervals $[a_n, b_n]$, $n = 1, 2, 3, \ldots$, where each interval is contained in the preceding one and $\lim_{n\to\infty} (a_n - b_n) = 0$. Such intervals are called *nested intervals*.

We can prove that to every set of nested intervals there corresponds one and only one real number. This can be used to establish the Bolzano-Weierstrass theorem of Chapter 1. (See Problems 2.22 and 2.23.)

Cauchy's Convergence Criterion

Cauchy's convergence criterion states that a sequence $\{u_n\}$ converges if and only if for each $\epsilon > 0$ we can find a number N such that $|u_p - u_q| < \epsilon$ for all $p, q > N$. This criterion has the advantage that one need not know the limit l in order to demonstrate convergence.

Infinite Series

Let $u_1, u_2, u_3, \ldots$ be a given sequence. Form a new sequence $S_1, S_2, S_3, \ldots$ where

$$S_1 = u_1,\ S_2 = u_1 + u_2,\ S_3 = u_1 + u_2 + u_3, \ldots, + S_n = u_1 + u_2 + u_3 + \cdots + u_n, \ldots$$

where S_n, called the nth *partial sum*, is the sum of the first n terms of the sequence $\{u_n\}$.

The sequence $S_1, S_2, S_3, \ldots$ is symbolized by

$$u_1 + u_2 + u_3 + \cdots = \sum_{n=1}^{\infty} u_n$$

which is called an *infinite series*. If $\lim_{n\to\infty} S_n = S$ exists, the series is called *convergent* and S is its *sum*; otherwise, the series is called *divergent*.

Further discussion of infinite series and other topics related to sequences is given in Chapter 11.

SOLVED PROBLEMS

Sequences

2.1. Write the first five terms of each of the following sequences.

(a) $\left\{\dfrac{2n-1}{3n+2}\right\}$

(b) $\left\{\dfrac{1-(-1)^n}{n^3}\right\}$

(c) $\left\{\dfrac{(-1)^{n-1}}{2\cdot 4\cdot 6\cdots 2n}\right\}$

(d) $\left\{\dfrac{1}{2}+\dfrac{1}{4}+\dfrac{1}{8}+\cdots+\dfrac{1}{2^n}\right\}$

(e) $\left\{\dfrac{(-1)^{n-1}x^{2n-1}}{(2n-1)!}\right\}$

(a) $\dfrac{1}{5}, \dfrac{3}{8}, \dfrac{5}{11}, \dfrac{7}{14}, \dfrac{9}{17}$

(b) $\dfrac{2}{1^3}, 0, \dfrac{2}{3^3}, 0, \dfrac{2}{5^3}$

(c) $1\ \dfrac{1}{2}, \dfrac{-1}{2\cdot 4}, \dfrac{1}{2\cdot 4\cdot 6}, \dfrac{-1}{2\cdot 4\cdot 6\cdot 8}, \dfrac{1}{2\cdot 4\cdot 6\cdot 8\cdot 10}$

(d) $\dfrac{1}{2}, \dfrac{1}{2}+\dfrac{1}{4}, \dfrac{1}{2}+\dfrac{1}{4}+\dfrac{1}{8}, \dfrac{1}{2}+\dfrac{1}{4}+\dfrac{1}{8}+\dfrac{1}{16}, \dfrac{1}{2}+\dfrac{1}{4}+\dfrac{1}{8}+\dfrac{1}{16}+\dfrac{1}{32}$

(e) $\dfrac{x}{1!}, \dfrac{-x^3}{3!}, \dfrac{x^5}{5!}, \dfrac{-x^7}{7!}, \dfrac{x^9}{9!}$

Note that $n! = 1\cdot 2\cdot 3\cdot 4 \ldots n$. Thus, $1! = 1$, $3! = 1\cdot 2\cdot 3 = 6$, $5! = 1\cdot 2\cdot 3\cdot 4\cdot 5 = 120$, etc. We define $0! = 1$.

2.2. Two students were asked to write an nth term for the sequence 1, 16, 81, 256, . . . and to write the 5th term of the sequence. One student gave the nth term as $u_n = n^4$. The other student, who did not recognize this simple law of formation, wrote $u_n = 10n^3 - 35n^2 + 50n - 24$. Which student gave the correct 5th term?

If $u_n = n^4$, then $u_1 = 1^4 = 1$, $u_2 = 2^4 = 16$, $u_3 = 3^4 = 81$, and $u_4 = 4^4 = 256$, which agrees with the first four terms of the sequence. Hence, the first student gave the 5th term as $u_5 = 5^4 = 625$.

If $u_n = 10n^3 - 35n^2 + 50n - 24$, then $u_1 = 1$, $u_2 = 16$, $u_3 = 81$, and $u_4 = 256$, which also agrees with the first four terms given. Hence, the second student gave the 5th term as $u_5 = 601$.

Both students were correct. Merely giving a finite number of terms of a sequence does not define a unique nth term. In fact, an infinite number of nth terms is possible.

Limit of a sequence

2.3. A sequence has its nth term given by $u_n = \dfrac{3n-1}{4n+5}$. (a) Write the 1st, 5th, 10th, 100th, 1000th, 10,000th and 100,000th, terms of the sequence in decimal form. Make a *guess* as to the limit of this sequence as $n \to \infty$. (b) Using the definition of limit, verify that the guess in (*a*) is actually correct.

(a)

$n = 1$	$n = 5$	$n = 10$	$n = 100$	$n = 1000$	$n = 10{,}000$	$n = 100{,}000$
.22222	56000 . . .	.64444 . . .	.73827 . . .	.74881 . . .	.74988 . . .	.74998 . . .

A good guess is that the limit is $.75000\ldots = \dfrac{3}{4}$. Note that it is only for *large enough* values of n that a possible limit may become apparent.

(b) We must show that for any given $\epsilon > 0$ (no matter how small) there is a number N (depending on ϵ) such that $\left|u_n - \dfrac{3}{4}\right| < \epsilon$ for all $n > N$.

$$\text{Now } \left[\frac{3n-1}{4n+5} - \frac{3}{4}\right] = \left[\frac{-19}{4(4n+5)}\right] < \varepsilon \text{ when } \frac{19}{4(4n+5)} < \varepsilon \text{ or}$$

$$\frac{4(4n+5)}{19} > \frac{1}{\varepsilon}, \quad 4n+5 > \frac{19}{4\varepsilon}, \quad n > \frac{1}{4}\left(\frac{19}{4\varepsilon} - 5\right)$$

Choosing $N = \dfrac{1}{4}\,(19/4\epsilon - 5)$, we see that $\left|u_n - \dfrac{3}{4}\right| < \epsilon$ for all $n > N$, so that $\lim\limits_{n\to\infty} = \dfrac{3}{4}$ and the proof is complete.

Note that if $\epsilon = .001$ (for example), $N = \dfrac{1}{4}\,(19000/4 - 5) = 1186\,\dfrac{1}{4}$. This means that all terms of the sequence beyond the 1186th term differ from 3/4 in absolute value by less than .001.

2.4. Prove that $\lim\limits_{n\to\infty} \dfrac{c}{n^p} = 0$ where $c \neq 0$ and $p > 0$ are constants (independent of n).

We must show that for any $\epsilon > 0$ there is a number N such that $|c/n^p - 0| < \epsilon$ for all $n > N$.

Now $\left|\dfrac{c}{n^p}\right| < \epsilon$ when $\dfrac{|c|}{n^p} < \epsilon$; i.e., $n^p > \dfrac{|c|}{\varepsilon}$ or $n > \left(\dfrac{|c|}{\varepsilon}\right)^{1/p}$. Choosing $N = \left(\dfrac{|c|}{\varepsilon}\right)^{1/p}$ (depending on ϵ), we see that $|c/n^p| < \epsilon$ for all $n > N$, proving that $\lim\limits_{n\to\infty} (c/n^p) = 0$.

2.5. Prove that $\lim\limits_{n\to\infty} \dfrac{1+2\cdot 10^n}{5+3\cdot 10^n} = \dfrac{2}{3}$.

We must show that for any $\epsilon > 0$ there is a number N such that $\left|\dfrac{1+2\cdot 10^n}{5+3\cdot 10^n} - \dfrac{2}{3}\right| < \varepsilon$ for all $n > N$.

$$\text{Now } \left|\frac{1+2\cdot 10^n}{5+3\cdot 10^n} - \frac{2}{3}\right| = \left|\frac{-7}{3(5+3\cdot 10^n)}\right| < \varepsilon \text{ when } \frac{7}{3(5+3\cdot 10^n)} < \varepsilon\text{; i.e., when } \frac{3}{7}(5+3\cdot 10^n) > 1/\varepsilon,$$

$3\cdot 10^n > 7/3\epsilon - 5$, $10^n > \dfrac{1}{8}\,(7/3\epsilon - 5)$ or $n > \log_{10}\{\dfrac{1}{3}\,(7/3\epsilon - 5)\} = N$, proving the existence of N and thus establishing the required result.

Note that the value of N is real only if $7/3\epsilon - 5 > 0$; i.e., $0 < \epsilon < 7/15$. If $\epsilon \geqq 7/15$, we see that

$$\left|\frac{1+2\cdot 10^n}{5+3\cdot 10^n} - \frac{2}{3}\right| < \varepsilon \text{ for } all\ n > 0.$$

2.6. Explain exactly what is meant by the statements (a) $\lim\limits_{n\to\infty} 3^{2n-1} = \infty$ and (b) $\lim\limits_{n\to\infty} (1-2n) = -\infty$.

(a) If for each positive number M we can find a positive number N (depending on M) such that $a_n > M$ for all $n > N$, then we write $\lim\limits_{n\to\infty} a_n = \infty$.

In this case, $3^{2n-1} > M$ when $(2n-1)\log 3 > \log M$; i.e., $n > \dfrac{1}{2}\left(\dfrac{\log M}{\log 3} + 1\right) = N$.

(b) If for each positive number M we can find a positive number N (depending on M) such that $a_n < -M$ for all $n > N$, then we write $\lim_{n\to\infty} = -\infty$.

In this case, $1 - 2n < -M$ when $2n - 1 > M$ or $n > \frac{1}{2}(M+1) = N$.

It should be emphasized that the use of the notations ∞ and $-\infty$ for limits does not in any way imply convergence of the given sequences, since ∞ and $-\infty$ are *not* numbers. Instead, these are notations used to describe that the sequences diverge in specific ways.

2.7. Prove that $\lim_{n\to\infty} x^n = 0$ if $|x| < 1$.

Method 1: We can restrict ourselves to $x \neq 0$, since if $x = 0$, the result is clearly true. Given $\epsilon > 0$, we must show that there exists N such that $|x^n| < \epsilon$ for $n > N$. Now $|x^n| = |x|^n < \epsilon$ when $n \log_{10}|x| < \log_{10}\epsilon$. Dividing by $\log_{10}|x|$, which is negative, yields $n > \frac{\log_{10}\epsilon}{\log_{10}|x|} = N$, proving the required result.

Method 2: Let $|x| = 1/(1+p)$, where $p > 0$. By Bernoulli's inequality (Problem 1.31), we have $|x^n| = |x|^n = 1/(1+p)^n < 1/(1+np) < \epsilon$ for all $n > N$. Thus, $\lim_{n\to\infty} x^n = 0$.

Theorems on limits of sequences

2.8. Prove that if $\lim_{n\to\infty} u_n$ exists, it must be unique.

We must show that if $\lim_{n\to\infty} u_n = l_1$ and $\lim_{n\to\infty} u_n = l_2$, then $l_1 = l_2$.

By hypothesis, given any $\epsilon > 0$ we can find N such that

$$|u_n - l_1| < \tfrac{1}{2}\epsilon \text{ when } n > N, \quad |u_n - l_2| < \tfrac{1}{2}\epsilon \text{ when } n > N$$

Then

$$|l_1 - l_2| = |l_1 - u_n + u_n - l_2| \leqq |l_1 - u_n| + |u_n - l_2| < \tfrac{1}{2}\epsilon + \tfrac{1}{2}\epsilon = \epsilon$$

i.e., $|l_1 - l_2|$ is less than any positive ϵ (however small) and so must be zero. Thus, $l_1 = l_2$.

2.9. If $\lim_{n\to\infty} a_n = A$ and $\lim_{n\to\infty} b_n = B$, prove that $\lim_{n\to\infty}(a_n + b_n) = A + B$.

We must show that for any $\epsilon > 0$, we can find $N > 0$ such that $|(a_n + b_n) - (A+B)| < \epsilon$ for all $n > N$. From absolute value property 2, Page 4, we have

$$|(a_n + b_n) - (A+B)| = |(a_n - A) + (b_n - B)| \leqq |a_n - A| + |b_n - B| \tag{1}$$

By hypothesis, given $\epsilon > 0$ we can find N_1 and N_2 such that

$$|a_n - A| < \frac{1}{2}\epsilon \quad \text{for all } n > N_1 \tag{2}$$

$$|b_n - B| < \frac{1}{2}\epsilon \quad \text{for all } n > N_2 \tag{3}$$

Then from Equations (1), (2), and (3),

$$|(a_n + b_n) - (A+B)| < \frac{1}{2}\epsilon + \frac{1}{2}\epsilon = \epsilon \quad \text{for all } n > N$$

where N is chosen as the larger of N_1 and N_2. Thus, the required result follows.

2.10. Prove that a convergent sequence is bounded.

Given $\lim_{n\to\infty} a_n = A$, we must show that there exists a positive number P such that $|a_n| < P$ for all n. Now

$$|a_n| = |a_n - A + A| \leqq |a_n - A| + |A|$$

But by hypothesis we can find N such that $|a_n - A| < \epsilon$ for all $n > N$, i.e.,

$$|a_n| < \epsilon + |A| \qquad \text{for all } n > N$$

It follows that $|a_n| < P$ for all n if we choose P as the largest one of the numbers $a_1, a_2, \ldots, a_N$, $\epsilon + |A|$.

2.11. If $\lim_{n\to\infty} b_n = B \neq 0$, prove there exists a number N such that $|b_n| > \frac{1}{2}|B|$ for all $n > N$.

Since $B = B - b_n + b_n$, we have:

$$|B| \leqq |B - b_n| + |b_n| \tag{1}$$

Now we can choose N so that $|B - b_n| = |b_n - B| < \frac{1}{2}|B|$ for all $n > N$, since $\lim_{n\to\infty} b_n = B$ by hypothesis.

Hence, from Equation (1), $|B| < \frac{1}{2}|B| + |b_n|$ or $|b_n| > \frac{1}{2}|B|$ for all $n > N$.

2.12. If $\lim_{n\to\infty} a_n = A$ and $\lim_{n\to\infty} b_n = B$, prove that $\lim_{n\to\infty} a_n b_n = AB$.

Using Problem 2.10, we have

$$\begin{aligned} |a_n b_n - AB| &= |a_n(b_n - B) + B(a_n - A)| \leqq |a_n||b_n - B| + |B||a_n - A| \\ &\leqq P|b_n - B| + (|B| + 1)|a_n - A| \end{aligned} \tag{1}$$

But since $\lim_{n\to\infty} a_n = A$ and $\lim_{n\to\infty} b_n = B$, given any $\epsilon > 0$ we can find N_1 and N_2 such that

$$|b_n - B| < \frac{\varepsilon}{2P} \text{ for all } n > N_1 \qquad |a_n - A| < \frac{\varepsilon}{2(|B| + 1)} \text{ for all } n > N_2$$

Hence, from Equation (1), $|a_n b_n - AB| < \frac{1}{2}\epsilon + \frac{1}{2}\epsilon = \epsilon$ for all $n > N$, where N is the larger of N_1 and N_2. Thus, the result is proved.

2.13. If $\lim_{n\to\infty} a_n = A$ and $\lim_{n\to\infty} b_n = B \neq 0$, prove (a) $\lim_{n\to\infty} \frac{1}{b_n} = \frac{1}{B}$, (b) $\lim_{n\to\infty} \frac{a_n}{b_n} = \frac{A}{B}$.

(a) We must show that for any given $\epsilon > 0$, we can find N such that

$$\left|\frac{1}{b_n} - \frac{1}{B}\right| = \frac{|B - b_n|}{|B||b_n|} < \varepsilon \quad \text{for all } n > N \tag{1}$$

By hypothesis, given any $\epsilon > 0$, we can find N_1, such that $|b_n - B| < \frac{1}{2}B^2\epsilon$ for all $n > N_1$.

Also, since $\lim_{n\to\infty} b_n = B \neq 0$, we can find N_2 such that $|b_n| > \frac{1}{2}|B|$ for all $n > N_2$ (see Problem 2.11).

Then if N is the larger of N_1 and N_2, we can write Equation (1) as

$$\left|\frac{1}{b_n} - \frac{1}{B}\right| = \frac{|b_n - B|}{|B||b_n|} < \frac{\frac{1}{2}B^2\varepsilon}{|B| \cdot \frac{1}{2}|B|} = \varepsilon \quad \text{for all } n > N$$

and the proof is complete.

(b) From (a) and Problem 2.12, we have

$$\lim_{n\to\infty} \frac{a_n}{b_n} = \lim_{n\to\infty}\left(a_n \cdot \frac{1}{b_n}\right) = \lim_{n\to\infty} a_n \cdot \lim_{n\to\infty} \frac{1}{b_n} = A \cdot \frac{1}{B} = \frac{A}{B}$$

This can also be proved directly (see Problem 2.41).

2.14. Evaluate each of the following, using theorems on limits.

(a) $\displaystyle\lim_{n\to\infty}\frac{3n^2-5n}{5n^2+2n-6}=\lim_{n\to\infty}\frac{3-5/n}{5+2/n-6/n^2}=\frac{3+0}{5+0+0}=\frac{3}{5}$

(b) $\displaystyle\lim_{n\to\infty}\left\{\frac{n(n+2)}{n+1}-\frac{n^3}{n^2+1}\right\}=\lim_{n\to\infty}\left\{\frac{n^3+n^2+2n}{(n+1)(n^2+1)}\right\}=\lim_{n\to\infty}\left\{\frac{1+1/n+2/n^2}{(1+1/n)(1+1/n^2)}\right\}$

$$=\frac{1+0+0}{(1+0)\cdot(1+0)}=1$$

(c) $\displaystyle\lim_{n\to\infty}\left(\sqrt{n+1}-\sqrt{n}\right)=\lim_{n\to\infty}\left(\sqrt{n+1}-\sqrt{n}\right)\frac{\sqrt{n+1}+\sqrt{n}}{\sqrt{n+1}+\sqrt{n}}=\lim_{n\to\infty}\frac{1}{\sqrt{n+1}+\sqrt{n}}=0$

(d) $\displaystyle\lim_{n\to\infty}\frac{3n^2+4n}{2n-1}=\lim_{n\to\infty}\frac{3=4/n}{2/n-1/n^2}$

Since the limits of the numerator and the denominator are 3 and 0, respectively, the limit does not exist.

Since $\dfrac{3n^2+4n}{2n-1}>\dfrac{3n^2}{2n}=\dfrac{3n}{2}$ can be made larger than any positive number M by choosing $n>N$, we can write, if desired, $\displaystyle\lim_{n\to\infty}\frac{3n^2+4n}{2n-1}=\infty$.

(e) $\displaystyle\lim_{n\to\infty}\left(\frac{2n-3}{2n+7}\right)^4=\left(\lim_{n\to\infty}\frac{2-3/n}{3+7/n}\right)^4=\left(\frac{2}{3}\right)^4=\frac{16}{18}$

(f) $\displaystyle\lim_{n\to\infty}\frac{2n^5-4n^2}{3n^7+n^3-10}=\lim_{n\to\infty}\frac{2/n^2-4/n^5}{3+1/n^4-10/n^7}=\frac{0}{3}=0$

(g) $\displaystyle\lim_{n\to\infty}\frac{1+2\cdot10^n}{5+3\cdot10^n}=\lim_{n\to\infty}\frac{10^{-n}+2}{5\cdot10^{-n}+3}=\frac{2}{3}$ (Compare with Problem 2.5.)

Bounded monotonic sequences

2.15. Prove that the sequence with nth $u_n=\dfrac{2n-7}{3n+2}$ (a) is monotonic increasing, (b) is bounded above, (c) is bounded below, (d) is bounded, (e) has a limit.

(a) $\{u_n\}$ is monotonic increasing if $u_{n+1}\geqq u_n$, $n=1,2,3,\ldots$ Now

$$\frac{2(n+1)-7}{3(n+1)+2}\geqq\frac{2n-7}{3n+2}\quad\text{if and only if}\quad\frac{2n-5}{2n+5}\geqq\frac{2n-7}{3n+2}$$

or $(2n-5)(3n+2)\geqq(2n-7)(3n+5)$, $6n^2-11n-10\geqq 6n^2-11n-35$, i.e., $-10\geqq-35$, which is true. Thus, by reversal of steps in the inequalities, we see that $\{u_n\}$ is monotonic increasing. Actually, since $-10>-35$, the sequence is strictly increasing.

(b) By writing some terms of the sequence, we may *guess* that an upper bound is 2 (for example). To *prove* this we must show that $u_n\leqq 2$. If $(2n-7)/(3n+2)\leqq 2$, then $2n-7\leqq 6n+4$ or $-4n<11$, which *is* true. Reversal of steps proves that 2 is an upper bound.

(c) Since this particular sequence is monotonic increasing, the first term -1 is a lower bound; i.e., $u_n\geqq-1$, $n=1,2,3,\ldots$ Any number less than -1 is also a lower bound.

(d) Since the sequence has an upper and a lower bound, it is bounded. Thus, for example, we can write $|u_n|\leqq 2$ for all n.

(e) Since every bounded monotonic (increasing or decreasing) sequence has a limit, the given sequence has a limit. In fact, $\displaystyle\lim_{n\to\infty}\frac{2n-7}{3n+2}=\lim_{n\to\infty}\frac{2-7/n}{3+2/n}=\frac{2}{3}$.

2.16. A sequence $\{u_n\}$ is defined by the recursion formula $u_{n+1}\ \sqrt{3u_n}$, $u_1 = 1$. (a) Prove that $\lim_{n\to\infty} u_n$ exists. (b) Find the limit in (a).

(a) The terms of the sequence are $u_1 = 1$, $u_2 = \sqrt{3u_1} = 3^{1/2}$, $u_3 = \sqrt{3u_2} = 3^{1/2+1/4}, \ldots$.

The nth term is given by $u_n = 3^{1/2+1/4+\cdots+1/2n-1}$, as can be proved by mathematical induction (Chapter 1). Clearly, $u_{n+1} \geqq u_n$. Then the sequence is monotonic increasing.

By Problem 1.14, $u_n \leqq 3^1 = 3$, i.e., u_n is bounded above. Hence, u_n is bounded (since a lower bound is zero).

Thus, a limit exists, since the sequence is bounded and monotonic increasing.

(b) Let x = required limit. Since $\lim_{n\to\infty} u_{n+1} = \lim_{n\to\infty} \sqrt{3u_n}$, we have $x = \sqrt{3x}$ and $x = 3$. (The other possibility, $x = 0$, is excluded, since $u_n \geqq 1$.)

Another method: $\lim_{n\to\infty} 3^{1/2+1/4+\cdots+1/2^{n-1}} = \lim_{n\to\infty} 3^{1-1/2^n} = 3\lim_{n\to\infty}{}^{(1-1/2^n)} = 3^1 = 3$

2.17. Verify the validity of the entries in the following table.

SEQUENCE	BOUNDED	MONOTONIC INCREASING	MONOTONIC DECREASING	LIMIT EXISTS
$2, 1.9, 1.8, 1.7, \ldots, 2-(n-1)/10 \ldots$	No	No	Yes	No
$1, -1, 1, -1, \ldots, (-1)^{n-1}, \ldots$	Yes	No	No	No
$\frac{1}{2}, -\frac{1}{3}, \frac{1}{4}, -\frac{1}{5}, \ldots, (-1)^{n-1}/(n+1), \ldots$	Yes	No	No	Yes (0)
$.6, .66, .666, \ldots, \frac{2}{3}(1-1/10^n), \ldots$	Yes	Yes	No	Yes ($\frac{2}{3}$)
$-1, +2, -3, +4, -5, \ldots, (-1)^n n, \ldots$	No	No	No	No

2.18. Prove that the sequence with the nth term $u_n = \left(1+\frac{1}{n}\right)^n$ is monotonic, increasing, and bounded, and thus a limit exists. The limit is denoted by the symbol e.

Note: $\lim_{n\to\infty}\left(1+\frac{1}{n}\right)^n = e$, where $e \cong 2.71828\cdots$ was introduced in the eighteenth century by Leonhart Euler as the base for a system of logarithms in order to simplify certain differentiation and integration formulas.

By the binomial theorem, if n is a positive integer (see Problem 1.95),

$$(1+x)^n = 1 + nx + \frac{n(n-1)}{2!}x^2 + \frac{n(n-1)(n-2)}{3!}x^2 + \cdots + \frac{n(n-1)\cdots(n-n+1)}{n!}x^n$$

Letting $x = 1/n$,

$$u^n = \left(1+\frac{1}{n}\right)^n = 1 + n\frac{1}{n} + \frac{n(n-1)}{2!}\frac{1}{n^2} + \cdots + \frac{n(n-1)\cdots(n-n+1)}{n!}\frac{1}{n^n}$$

$$= 1 + 1 + \frac{1}{2!}\left(1-\frac{1}{n}\right) + \frac{1}{3!}\left(1-\frac{1}{n}\right)\left(1-\frac{2}{n}\right)$$

$$+ \cdots + \frac{1}{n!}\left(1-\frac{1}{n}\right)\left(1-\frac{2}{n}\right)\cdots\left(1-\frac{n-1}{n}\right)$$

Since each term beyond the first two terms in the last expression is an increasing function of n, it follows that the sequence u_n is a monotonic increasing sequence.

It is also clear that

$$\left(1+\frac{1}{n}\right)^n < 1+1+\frac{1}{2!}+\frac{1}{3!}+\ldots+\frac{1}{n!} < 1+1+\frac{1}{2}+\frac{1}{2^2}+\cdots+\frac{1}{2^{n-1}} < 3$$

by Problem 1.14.

Thus, u_n is bounded and monotonic increasing, and so has a limit which we denote by e. The value of $e =$ 2.71828 . . .

2.19. Prove that $\lim_{x\to\infty}\left(1+\frac{1}{x}\right)^x = e$, where $x \to \infty$ in any manner whatsoever (i.e., not necessarily along the positive integers, as in Problem 2.18).

If $n =$ largest integer $\leqq x$, then $n \leqq x \leqq n+1$ and $\left(1+\frac{1}{n+1}\right)^n \leqq \left(1+\frac{1}{x}\right)^x \leqq \left(1+\frac{1}{n}\right)^{n+1}$. Since

$$\lim_{n\to\infty}\left(1+\frac{1}{n+1}\right)^n = \lim_{n\to\infty}\left(1+\frac{1}{n+1}\right)^{n+1} \Big/ \left(1+\frac{1}{n+1}\right) = e \text{ and } \lim_{n\to\infty}\left(1+\frac{1}{n}\right)^{n+1} = \lim_{n\to\infty}\left(1+\frac{1}{n}\right)^n\left(1+\frac{1}{n}\right) = e,$$

it follows that $\lim_{x\to\infty}\left(1+\frac{1}{x}\right)^x = e$.

Least upper bound, greatest lower bound, limit superior, limit inferior

2.20. Find the (a) l.u.b., (b) g.l.b., (c) lim sup ($\overline{\lim}$), and (d) lim inf ($\underline{\lim}$) for the sequence 2, –2, 1, –1, 1, –1, 1, –1, ...

(a) l.u.b. = 2, since all terms are less than equal to 2, while at least one term (the 1st) is greater than $2-\epsilon$ for any $\epsilon > 0$.

(b) g.l.b. = –2, since all terms are greater than or equal to –2, while at least one term (the 2nd) is less than $-2+\epsilon$ for any $\epsilon > 0$.

(c) lim sup or $\overline{\lim} = 1$, since infinitely many terms of the sequence are greater than $1-\epsilon$ for any $\epsilon > 0$ (namely, all 1's in the sequence), while only a finite number of terms are greater than $1+\epsilon$ for any $\epsilon > 0$ (namely, the 1st term).

(d) lim inf or $\underline{\lim} = -1$, since infinitely many terms of the sequence are less than $-1+\epsilon$ for any $\epsilon > 0$ (namely, all –1's in the sequence), while only a finite number of terms are less than $-1-\epsilon$ for any $\epsilon > 0$ (namely, the 2nd term).

2.21. Find the (a) l.u.b., (b) g.l.b., (c) lim sup ($\overline{\lim}$), and (d) lim inf ($\underline{\lim}$) for the sequences in Problem 2.17.

The results are shown in the following table.

SEQUENCE	l.u.b.	g.l.b.	lim sup or $\overline{\lim}$	lim inf or $\underline{\lim}$
2, 1.9, 1.8, 1.7, . . ., $2-(n-1)/10$ . . .	2	none	$-\infty$	$-\infty$
1, –1, 1, –1, . . . , $(-1)^{n-1}$, . . .	1	–1	1	–1
$\frac{1}{2}, -\frac{1}{3}, \frac{1}{4}, -\frac{1}{5}, \ldots, (-1)^{n-1}/(n+1), \ldots$	$\frac{1}{2}$	$-\frac{1}{3}$	0	0
.6, .66, .666, . . ., $\frac{2}{3}(1-1/10^n)$, . . .	$\frac{2}{3}$	6	$\frac{2}{3}$	$\frac{2}{3}$
–1, + 2, –3, +4, –5, . . ., $(-1)^n n$, . . .	none	none	$+\infty$	$-\infty$

Nested intervals

2.22. Prove that to every set of nested intervals $[a_n, b_n]$, $n = 1, 2, 3, \ldots$ there corresponds one and only one real number.

By definition of nested intervals, $a_{n+1} \geqq a_n$, b_{n+1}, $\leqq b_n$ $n = 1, 2, 3, \ldots$ and $\lim_{n\to\infty} (a_n - b_n) = 0$.

Then $a_1 \leqq a_n \leqq b_n \leqq b_1$, and the sequences $\{a_n\}$ and $\{b_n\}$ are bounded and, respectively, monotonic increasing and decreasing sequences and so converge to a and b.

To show that $a = b$ and thus prove the required result, we note that

$$b - a = (b - b_n) + (b_n - a_n) + (a_n - a) \tag{1}$$

$$|b - a| \leqq |b - b_n| + |b_n - a_n| + |a_n - a| \tag{2}$$

Now, given any $\epsilon > 0$, we can find N such that for all $n > N$

$$|b - b_n| < \epsilon/3,\ |b_n - a| < \epsilon/3, \tag{3}$$

so that from Equation (2), $|b - a| < \epsilon$. Since ϵ is any positive number, we must have $b - a = 0$ or $a = b$.

2.23. Prove the Bolzano-Weierstrass theorem (see Page 7).

Suppose the given bounded infinite set is contained in the finite interval $[a, b]$. Divide this interval into two equal intervals. Then at least one of these, denoted by $[a_1, b_1]$, contains infinitely many points. Dividing $[a_1, b_1]$ into two equal intervals, we obtain another interval—say, $[a_2, b_2]$—containing infinitely many points. Continuing this process, we obtain a set of intervals $[a_n, b_n]$, $n = 1, 2, 3, \ldots$, each interval contained in the preceding one and such that

$$b_1 - a_1 = (b - a)/2,\ b_2 - a_2 = (b_1 - a_1)/2 = (b - a)/2^2, \ldots, b_n - a_n = (b - a)/2^n$$

from which we see that $\lim_{n\to\infty} (b_n - a_n) = 0$.

This set of nested intervals, by Problem 2.22, corresponds to a real number which represents a limit point and so proves the theorem.

Cauchy's convergence criterion

2.24. Prove Cauchy's convergence criterion as stated on Page 27.

Necessity. Suppose the sequence $\{u_n\}$ converges to l. Then, given any $\epsilon > 0$, we can find N such that

$$|u_p - l| < \epsilon/2 \text{ for all } p > N \text{ and } |u_q - l| < \epsilon/2 \text{ for all } q > N$$

Then, for both $p > N$ and $q > N$, we have

$$|u_p - u_q| = |(u_p - l) + (l - u_q)| \leqq |u_p - l| + |l - u_q| < \epsilon/2 + \epsilon/2 = \epsilon$$

Sufficiency. Suppose $|u_p - u_q| < \epsilon$ for all $p, q > N$ and any $\epsilon > 0$. Then all the numbers $u_N, u_{N+1}, \ldots$ lie in a finite interval; i.e., the set is bounded and infinite. Hence, by the Bolzano-Weierstrass theorem there is at least one limit point—say, a.

If a is the only limit point, we have the desired proof and $\lim_{n\to\infty} u_n = a$.

Suppose there are two distinct limit points—say, a and b—and suppose $b > a$ (see Figure 2.1). By definition of limit points, we have

$$|u_p - a| < (b - a)/3 \text{ for infinitely many values of } p \tag{1}$$

$$|u_q - b| < (b - a)/3 \text{ for infinitely many values of } q \tag{2}$$

Figure 2.1

Then, since $b - a = (b - u_q) + (u_q - u_p) + (u_p - a)$, we have

$$|b - a| = b - a \leq |b - u_q| + |u_p - u_q| + |u_p - a| \qquad (3)$$

Using Equations (1) and (2) in (3), we see that $|u_p - u_q| > (b - a)/3$ for infinitely many values of p and q, thus contradicting the hypothesis that $|u_p - u_q| < \epsilon$ for $p, q > N$ and any $\epsilon > 0$. Hence, there is only one limit point and the theorem is proved.

Infinite series

2.25. Prove that the infinite series (sometimes called the *geometric series*)

$$a + ar + ar^2 + \cdots = \sum_{n=1}^{\infty} ar^{n-1}$$

(a) converges to $a/(1 - r)$ if $|r| < 1$, and (b) diverges if $|r| \geq 1$.

Let $$S_n = a + ar + ar^2 + \cdots + ar^{n-1}$$

Then $$rS_n = ar + ar^2 + \cdots + ar^{n-1} + ar^n$$

Subtract $$(1 - r)S_n = a \qquad\qquad - ar^n$$

or $$S_n = \frac{a(1 - r^n)}{1 - r}$$

(a) If $|r| < 1$, $\lim\limits_{n\to\infty} S_n = \lim\limits_{n\to\infty} \dfrac{a(1 - r^n)}{1 - r} = \dfrac{a}{1 - r}$ by Problem 2.7.

(b) If $|r| > 1$, $\lim\limits_{n\to\infty} S_n$ does not exist (see Problem 2.44).

2.26. Prove that if a series converges, its nth term must necessarily approach zero.

Since $S_n = u_1 + u_2 + \cdots + u_n$ and $S_{n-1} = u_1 + u_2 + \cdots + u_{n-1}$, we have $u_n = S_n - S_{n-1}$.

If the series converges to S, then

$$\lim_{n\to\infty} u_n = \lim_{n\to\infty} (S_n - S_{n-1}) = \lim_{n\to\infty} S_n - \lim_{n\to\infty} S_{n-1} = S - S = 0$$

2.27. Prove that the series $1 - 1 + 1 - 1 + 1 - 1 + \cdots = \sum_{n=1}^{\infty} (-1)^{n-1}$ diverges.

Method 1: $\lim\limits_{n\to\infty} (-1)^n \neq 0$; in fact, it doesn't exist. Then by Problem 2.26, the series cannot converge; i.e., it diverges.

Method 2: The sequence of partial sums is 1, 1 − 1, 1 − 1 + 1, 1 − 1 + 1 − 1, . . .; i.e., 1, 0, 1, 0, 1, 0, 1, . . . Since this sequence has no limit, the series diverges.

Miscellaneous problems

2.28. If $\lim\limits_{n\to\infty} u_n = l$, prove that $\lim\limits_{n\to\infty} \dfrac{u_1 + u_2 + \cdots + u_n}{n} = l$.

Let $u_n = \upsilon_n + l$. We must show that $\lim\limits_{n\to\infty} \dfrac{\upsilon_1 + \upsilon_2 + \cdots + \upsilon_n}{n} = 0$ if $\lim\limits_{n\to\infty} \upsilon_n = 0$. Now

$$\frac{\upsilon_1 + \upsilon_2 + \cdots + \upsilon_n}{n} = \frac{\upsilon_1 + \upsilon_2 + \cdots + \upsilon_p}{n} + \frac{\upsilon_{p+1} + \upsilon_{p+2} + \cdots \upsilon_n}{n}$$

so that

$$\left|\frac{\upsilon_1 + \upsilon_2 + \cdots + \upsilon_n}{n}\right| \leq \frac{|\upsilon_1 + \upsilon_2 + \cdots + \upsilon_p|}{n} + \frac{|\upsilon_{P+1}| + |\upsilon_{P+2}| + \cdots + |\upsilon_n|}{n} \qquad (1)$$

Since $\lim_{n\to\infty} v_n = 0$, we can choose P so that $|v_n| < \epsilon/2$ for $n > P$. Then

$$\frac{|\upsilon_{P+1}|+|\upsilon_{P+2}|+\cdots+|\upsilon_n|}{n} < \frac{\varepsilon/2+\varepsilon/2+\cdots+\varepsilon/2}{n} = \frac{(n-P)\varepsilon/2}{n} < \frac{\varepsilon}{2} \tag{2}$$

After choosing P, we can choose N so that for $n > N > P$,

$$\frac{|\upsilon_1+\upsilon_2+\cdots+\upsilon_P|}{n} < \frac{\varepsilon}{2} \tag{3}$$

Then, using Equations (2) and (3), (*1*) becomes

$$\left|\frac{\upsilon_1+\upsilon_2+\cdots+\upsilon_n}{n}\right| < \frac{\varepsilon}{2}+\frac{\varepsilon}{2}=\varepsilon \qquad \text{for } n > N$$

thus proving the required result.

2.29. Prove that $\lim_{n\to\infty}(1+n+n^2)^{1/n}=1$.

Let $(1+n+n^2)^{1/n} = 1+u_n$, where $u_n \geqq 0$. Now, by the binomial theorem,

$$1+n+n^2=(1+u_n)^n=1+nu_n+\frac{n(n-1)}{2!}u_n^2+\frac{n(n-1)(n-2)}{3!}u_n^3+\cdots+u_n^n$$

Then $1+n+n^2>1+\dfrac{n(n-1)(n-2)}{3!}u_n^3$ or $0<u_n^3<\dfrac{6(n^2+n)}{n(n-1)(n-2)}$. Hence, $\lim_{n\to\infty}u_n^3=0$ and $\lim_{n\to\infty}u_n=0$. Thus, $\lim_{n\to\infty}(1+n+n^2)^{1/n}=\lim_{n\to\infty}(1+u_n)=1$.

2.30. Prove that $\lim_{n\to\infty}\dfrac{a^n}{n!}=0$ for all constants a.

The result follows if we can prove that $\lim_{n\to\infty}\dfrac{|a|^n}{n!}=0$ (see Problem 2.38). We can assume $a \neq 0$.

Let $u_n=\dfrac{|a|^n}{n!}$. Then $\dfrac{u_n}{u_{n-1}}=\dfrac{|a|}{n}$. If n is large enough—say, $n>2|a|$—and if we call $N=[2|a|+1]$, i.e., the greatest integer $\leqq 2|a|+1$, then

$$\frac{u_{N+1}}{u_N}<\frac{1}{2},\ \frac{u_{N+2}}{u_{N+1}}<\frac{1}{2},\ \ldots,\ \frac{u_n}{u_{n-1}}<\frac{1}{2}$$

Multiplying these inequalities yields $\dfrac{u_n}{u_N}<\left(\dfrac{1}{2}\right)^{n-N}$ or $u_n<\left(\dfrac{1}{2}\right)^{n-N}u_N$. Since $\lim_{n\to\infty}\left(\dfrac{1}{2}\right)^{n-N}=0$ (using Problem 2.7), it follows that $\lim_{n\to\infty}u_n=0$.

SUPPLEMENTARY PROBLEMS

Sequences

2.31. Write the first four terms of each of the following sequences:

(a) $\left\{\dfrac{\sqrt{n}}{n+1}\right\}$

(b) $\left\{\dfrac{(-1)^{n+1}}{n!}\right\}$

(c) $\left\{\dfrac{(2x)^{n-1}}{(2n-1)^5}\right\}$

(d) $\left\{\dfrac{(-1)^n x^{2n-1}}{1\cdot3\cdot5\cdots(2n-1)}\right\}$

(e) $\left\{\dfrac{\cos nx}{x^2+n^2}\right\}$

Ans. (a) $\frac{\sqrt{1}}{2}, \frac{\sqrt{2}}{3}, \frac{\sqrt{3}}{4}, \frac{\sqrt{4}}{5}$ (d) $\frac{-x}{1}, \frac{x^3}{1\cdot 3}, \frac{-x^5}{1\cdot 3\cdot 5}, \frac{x^7}{1\cdot 3\cdot 5\cdot 7}$

(b) $\frac{1}{1!}, -\frac{1}{2!}, \frac{1}{3!}, -\frac{1}{4!}$ (e) $\frac{\cos x}{x^2+1^2}, \frac{\cos 2x}{x^2+2^2}, \frac{\cos 3x}{x^2+3^2}, \frac{\cos 4x}{x^2+4^2}$

(c) $\frac{1}{1^5}, \frac{2x}{3^5}, \frac{4x^2}{5^5}, \frac{8x^3}{7^5}$

2.32. Find a possible *n*th term for the sequences whose first 5 terms are indicated, and find the 6th term:

(a) $\frac{-1}{5}, \frac{3}{8}, \frac{-5}{11}, \frac{7}{14}, \frac{-9}{17}, \cdots$ (b) $1, 0, 1, 0, 1, \cdots$ (c) $\frac{2}{3}, 0, \frac{3}{4}, 0, \frac{4}{5}, \cdots$

Ans. (a) $\frac{(-1)^n(2n-1)}{(3n+2)}$ (b) $\frac{1-(-1)^n}{2}$ (c) $\frac{(n+3)}{(n+5)}\cdot\frac{1-(-1)^n}{2}$

2.33. The *Fibonacci sequence* is the sequence $\{u_n\}$ where $u_{n+2} = u_{n+1} + u_n$ and u_n and $u_1 = 1$, $u_2 = 1$. (*a*) Find the first 6 terms of the sequence. (*b*) Show that the *n*th term is given by $u_n = (a^n - b^n)/\sqrt{5}$, where $a = \frac{1}{2}(1 + \sqrt{5})$ and $b = \frac{1}{2}(1 - \sqrt{5})$.

Ans. (a) 1, 1, 2, 3, 5, 8

Limits of sequences

2.34. Using the definition of limit, prove that

(a) $\lim_{n\to\infty} \frac{4-2n}{3n+2} = \frac{-2}{3}$ (b) $\lim_{n\to\infty} 2^{-1/\sqrt{n}} = 1$ (c) $\lim_{n\to\infty} \frac{n^4+1}{n^2} = \infty$ (d) $\lim_{n\to\infty} \frac{\sin N}{n} = 0$

2.35. Find the least positive integer *N* such that $|(3n+2)/(n-1) - 3| < \epsilon$ for all $n > N$ if (a) $\epsilon = .01$, (b) $\epsilon = .001$ and (c) $\epsilon = .0001$.

Ans. (a) 502 (b) 5002 (c) 50,002

2.36. Using the definition of limit, prove that $\lim_{n\to\infty} (2n-1)/(3n+4)$ cannot be $\frac{1}{2}$.

2.37. Prove that $\lim_{n\to\infty} (-1)^n n$ does not exist.

2.38. Prove that if $\lim_{n\to\infty} |u_n| = 0$, then $\lim_{n\to\infty} u_n = 0$. Is the converse true?

2.39. If $\lim_{n\to\infty} u_n = l$, prove that (a) $\lim_{n\to\infty} = cu_n = cl$ where *c* is any constant, (b) $\lim_{n\to\infty} u^2_n = l^2$, (c) $\lim_{n\to\infty} u^p_n = l^p$ where *p* is a positive integer, and (d) $\lim_{n\to\infty} \sqrt{u_n} = \sqrt{l}, l \geq 0$.

2.40. Give a direct proof that $\lim_{n\to\infty} a_n/b_n = A/B$ if $\lim_{n\to\infty} a_n = A$ and $\lim_{n\to\infty} b_n = B \neq 0$.

2.41. Prove that (a) $\lim_{n\to\infty} 3^{1/n} = 1$, (b) $\lim_{n\to\infty}\left(\frac{2}{3}\right)^{1/n} = 1$ and (c) $\lim_{n\to\infty}\left(\frac{3}{4}\right)^n = 0$.

2.42. If $r > 1$, prove that $\lim_{n\to\infty} r^n = \infty$, carefully explaining the significance of this statement.

2.43. If $|r| > 1$, prove that $\lim_{n\to\infty} r^n$ does not exist.

2.44. Evaluate each of the following, using theorems on limits:

(a) $\lim_{n\to\infty} \frac{4-2n-3n^2}{2n^2+n}$ (d) $\lim_{n\to\infty} \frac{4\cdot 10^n - 3\cdot 10^{2n}}{3\cdot 10^{n-1} + 2\cdot 10^{2n-1}}$

(b) $\lim_{n\to\infty} \sqrt[3]{\frac{(3-\sqrt{n})(\sqrt{n}+2)}{8n-4}}$ (e) $\lim_{n\to\infty}(\sqrt{n^2+n}-n)$

(c) $\lim_{n\to\infty} \frac{\sqrt{3n^2-5n+4}}{2n-7}$ (f) $\lim_{n\to\infty}(2^n+3^n)^{1/n}$

Ans. (a) –3/2 (b) –1/2 (c) $\sqrt{3}/2$ (d) –15 (e) 1/2 (f) 3

Bounded monotonic sequences

2.45. Prove that the sequence with nth term $u_n = u_n = \sqrt{n}/(n+1)$. (a) is monotonic decreasing, (b) is bounded below, (c) is bounded above, and (d) has a limit.

2.46. If $u_n = \frac{1}{1+n} + \frac{1}{2+n} + \frac{1}{3+n} + \cdots + \frac{1}{n+n}$, prove that $\lim_{n\to\infty} u_n$ exists and lies between 0 and 1.

2.47. If $u_n = \sqrt{u_n + 1}$, $u_1 = 1$, prove that $\lim_{n\to\infty} u = \frac{1}{2}(1+\sqrt{5})$.

2.48. If $u_{n+1} = \frac{1}{2}(u_n + p/u_n)$ where $p > 0$ and $u_1 > 0$, prove that $\lim_{n\to\infty} u_n = \sqrt{p}$. Show how this can be used to determine $\sqrt{2}$.

2.49. If u_n is monotonic increasing (or monotonic decreasing), prove that S_n/n, where $S_n = u_1 + u_2 + \cdots + u_n$ is also monotonic increasing (or monotonic decreasing).

Least upper bound, greatest lower bound, limit superior, limit inferior

2.50. Find the l.u.b., g.l.b., lim sup ($\overline{\lim}$), and lim inf ($\underline{\lim}$) for each sequence:

(a) $-1, \frac{1}{3}, -\frac{1}{5}, \frac{1}{7}, \ldots, (-1)^n/(2n-1), \ldots$ (c) $1, -3, 5, -7, \ldots, (-1)^{n-1}(2n-1), \ldots$

(b) $\frac{2}{3}, -\frac{3}{4}, \frac{4}{5}, -\frac{5}{6}, \ldots, (-1)^{n+1}(n+1)/(n+2), \ldots$ (d) $1, 4, 1, 16, 1, 36, \ldots, n^{1+(-1)n}, \ldots$

Ans. (a) $\frac{1}{3}$, –1, 0,0 (b) 1, –1, 1, –1 (c) none, none, +∞, – ∞ (d) none, 1, +∞, 1

2.51. Prove that a bounded sequence $\{u_n\}$ is convergent if and only if $\overline{\lim}\ u_n = \underline{\lim}\, u_n$.

Infinite series

2.52. Find the sum of the series $\sum_{n=1}^{\infty}\left(\frac{2}{3}\right)^n$.

Ans. 2

2.53. Evaluate $\sum_{n=1}^{\infty}(-1)^{n-1}/5^n$.

Ans. $\frac{1}{6}$

2.54. Prove that $\frac{1}{1\cdot 2}+\frac{1}{2\cdot 3}+\frac{1}{3\cdot 4}+\frac{1}{4\cdot 5}+\cdots=\sum_{n=1}^{\infty}\frac{1}{n(n+1)}=1.$ $\left(\text{Hint: } \frac{1}{n(n+1)}=\frac{1}{n}-\frac{1}{n+1}.\right)$

2.55. Prove that multiplication of each term of an infinite series by a constant (not zero) does not affect the convergence or divergence.

2.56. Prove that the series $1+\frac{1}{2}+\frac{1}{3}+\cdots+\frac{1}{n}+\cdots$ diverges. $\Big($Hint: Let $S_n = 1+\frac{1}{2}+\frac{1}{3}+\cdots+\frac{1}{n}$. Then prove that $|S_{2n}-S_n|>\frac{1}{2}$, giving a contradiction with Cauchy's convergence criterion.$\Big)$

Miscellaneous problems

2.57. If $a_n \leq u_n \leq b_n$ for all $n > N$, and $\lim_{n\to\infty} a_n = \lim_{n\to\infty} b_n = l$, prove that $\lim_{n\to\infty} u_n = l$.

2.58. If $\lim_{n\to\infty} a_n = \lim_{n\to\infty} b_n = 0$, and θ is independent of n, prove that $\lim_{n\to\infty} (a_n \cos n\theta) + b_n \sin n\theta) = 0$. Is the result true when θ depends on n?

2.59. Let $u_n = \frac{1}{2}\{1+(-1)^n\}$, $n = 1,2,3,\ldots$ If $S_n = u_1 + u_2 + \cdots + u_n$, prove that $\lim_{n\to\infty} S_n/n = \frac{1}{2}$.

2.60. Prove that (a) $\lim_{n\to\infty} n^{1/n}$ and (b) $\lim_{n\to\infty}(a+n)^{p/n} = 1$ where a and p are constants.

2.61. If $\lim_{n\to\infty} |u_{n+1}/u_n| = |a| < 1$, prove that $\lim_{n\to\infty} u_n = 0$.

2.62. If $|a| < 1$, prove that $\lim_{n\to\infty} n^p a^n = 0$ where the constant $p > 0$.

2.63. Prove that $\lim \frac{2^n n!}{n^n} = 0$.

2.64. Prove that $\lim_{n\to\infty} n \sin 1/n = 1$. (Hint: Let the central angel θ of a circle be measured in radians.) Geometrically illustrate that $\sin\theta \leq \theta \leq \tan\theta$, $0 \leq \theta \leq \pi$.

Let $\theta = 1/n$. Observe that since n is restricted to positive integers, the angle is restricted to the first quadrant.

2.65. If $\{u_n\}$ is the Fibonacci sequence (Problem 2.33), prove that $\lim_{n\to\infty} u_{n+1}/u_n = \frac{1}{2}(1+\sqrt{5})$.

2.66. Prove that the sequence $u_n = (1 + 1/n)^{n+1}$, $n = 1, 2, 3, \ldots$ is a monotonic decreasing sequence whose limit is e. (Hint: Show that $u_n/u_{n-1} \leq 1$.)

2.67. If $a_n \geqq b_n$ for all $n > N$ and $\lim_{n\to\infty} a_n = A$, $\lim_{n\to\infty} b_n = B$, prove that $A \geqq B$.

2.68. If $|u_n| \leqq |\upsilon_n|$ and $\lim_{n\to\infty} \upsilon_n = 0$, prove that $\lim_{n\to\infty} u_n = 0$.

2.69. Prove that $\lim_{n\to\infty} \frac{1}{n}\left(1 + \frac{1}{2} + \frac{1}{3} + \cdots + \frac{1}{n}\right) = 0$.

2.70. Prove that $[a_n, b_n]$, where $a_n = (1 + 1/n)^n$ and $b_n = (1 + 1/n)^{n+1}$ is a set of nested intervals defining the number e.

2.71. Prove that every bounded monotonic (increasing or decreasing) sequence has a limit.

2.72. Let $\{u_n\}$ be a sequence such that $u_{n+2} = au_{n+1} + bu_n$ where a and b are constants. This is called a second-order difference equation for u_n. (*a*) Assuming a solution of the form $u_n = r^n$ where r is a constant, prove that r must satisfy the equation $r^2 - ar - b = 0$. (*b*) Use (*a*) to show that a solution of the difference equation (called a general solution) is $u_n = Ar^n{}_1 + Br^n{}_2$, where A and B are arbitrary constants and r_1 and r_2 are the two solutions of $r^2 - ar - b = 0$ assumed different. (*c*) In case $r_1 = r_2$ in (*b*), show that a (general) solution is $u_n = (A + Bn)r^n{}_1$.

2.73. Solve the following difference equations subject to the given conditions: (a) $u_{n+2} = u_{n+1} + u_n$, $u_1 = 1$, $u_2 = 1$ (compare Problem 2.34); (b) $u_{n+2} = 2u_{n+1} + 3u_n, u_2 = 5$; (c) $u_{n+2} = 4u_{n+1}, 4u_n$, $u_1 = 2$, $u_2 = 8$.

Ans. (a) Same as in Problem 2.34, (b) $u_n = 2(3)^{n-1} + (-1)^{n-1}$ (c) $u_n = n \cdot 2^n$

CHAPTER 3

Functions, Limits, and Continuity

The notions described in this chapter historically followed the introduction of differentiation and integration. These concepts were established, developed, and applied in the 1700s on a strong mechanical basis but a weak theoretical foundation. In the 1800s, the theoretical inadequacies were resolved with the mathematical invention of limits. Precise definitions of derivatives and integrals were formulated. Many mathematicians, including Bolzano, introduced rigorous proofs free of geometry. Elegant notation, such as the $\varepsilon - \delta$ form of Weierstrass, became available. As a bonus, clear definitions of irrational numbers were made. Also, unexpected properties of infinite sets of real numbers were found by Cantor and other mathematicians.

This chapter sets forth the notion of the limit of a function, concepts that followed, and how these ideas made possible the rigorization of analysis.

Functions

A function is composed of a domain set, a range set, and a rule of correspondence that assigns exactly one element of the range to each element of the domain.

This definition of a function places no restrictions on the nature of the elements of the two sets. However, in our early exploration of the calculus, these elements are real numbers. The rule of correspondence can take various forms, but in advanced calculus it most often is an equation or a set of equations.

If the elements of the domain and range are represented by x and y, respectively, and f symbolizes the function, then the rule of correspondence takes the form $y = f(x)$.

The distinction between f and $f(x)$ should be kept in mind. f denotes the function as defined in the first paragraph. y and $f(x)$ are different symbols for the range (or image) values corresponding to domain values x. However, a common practice that provides an expediency in presentation is to read $f(x)$ as "the image of x with respect to the function f" and then use it when referring to the function. (For example, it is simpler to write $\sin x$ than "the sine function, the image value of which is $\sin x$.") This deviation from precise notation appears in the text because of its value in exhibiting the ideas.

The domain variable x is called the *independent* variable. The variable y representing the corresponding set of values in the range, is the *dependent* variable.

Note: There is nothing exclusive about the use of x, y, and f to represent domain, range, and function. Many other letters are employed.

There are many ways to relate the elements of two sets. (Not all of them correspond a unique range value to a given domain value.) For example, given the equation $y^2 = x$, there are two choices of y for each positive value of x. As another example, the pairs (a, b), (a, c), (a, d), and (a, e) can be formed, and again the correspondence to a domain value is not unique. Because of such possibilities, some texts, especially older ones, distinguish between multiple-valued and single-valued functions. This viewpoint is not consistent with our definition or modern presentations. In order that there be no ambiguity, the calculus and its applications require a single image associated with each domain value. A multiple-valued rule of correspondence gives rise to a collection of functions (i.e., single-valued). Thus, the rule $y^2 = x$ is replaced by the pair of rules $y = -x^{1/2}$

and the functions they generate through the establishment of domains. (See the following section on graphs for pictorial illustrations.)

EXAMPLES. 1. If to each number in $-1 \leqq x \leqq 1$ we associate a number y given by x^2, then the interval $-1 \leqq x \leqq 1$ is the domain. The rule $y = x^2$ generates the range $-1 \leqq y \leqq 1$. The totality is a function f.

The functional image of x is given by $y = f(x) = x^2$. For example, $f\left(-\frac{1}{3}\right) = \left(-\frac{1}{3}\right)^2 = \frac{1}{9}$ is the image of $-\frac{1}{3}$ with respect to the function f.

2. The sequences of Chapter 2 may be interpreted as functions. For infinite sequences, consider the domain as the set of positive integers. The rule is the definition of u_n, and the range is generated by this rule. To illustrate, let $u_n = \frac{1}{n}$ with $n = 1, 2, \ldots$. Then the range contains the elements $1, \frac{1}{2}, \frac{1}{3}, \frac{1}{4}, \ldots$. If the function is denoted by f, then we may write $f(n) = \frac{1}{n}$.

As you read this chapter, reviewing Chapter 2 will be very useful.

3. With each time t after the year 1800 we can associate a value P for the population of the United States. The correspondence between P and t defines a function—say, F—and we can write $P = F(t)$.
4. For the present, both the domain and the range of a function have been restricted to sets of real numbers. Eventually this limitation will be removed. To get the flavor for greater generality, think of a map of the world on a globe with circles of latitude and longitude as coordinate curves. Assume there is a rule that corresponds this domain to a range that is a region of a plane endowed with a rectangular Cartesian coordinate system. (Thus, a flat map usable for navigation and other purposes is created.) The points of the domain are expressed as pairs of numbers (θ, ϕ), and those of the range by pairs (x, y). These sets and a rule of correspondence constitute a function whose independent and dependent variables are not single real numbers; rather, they are pairs of real numbers.

Graph of a Function

A function f establishes a set of ordered pairs (x, y) of real numbers. The plot of these pairs $[x, f(x)]$ in a coordinate system is the graph of f. The result can be thought of as a pictorial representation of the function.

For example, the graphs of the functions described by $y = x^2$, $-1 \leqq x \leqq 1$, and $y^2 = x$, $0 \leqq x \leqq 1$, $y \geqq 0$ appear in Figure 3.1.

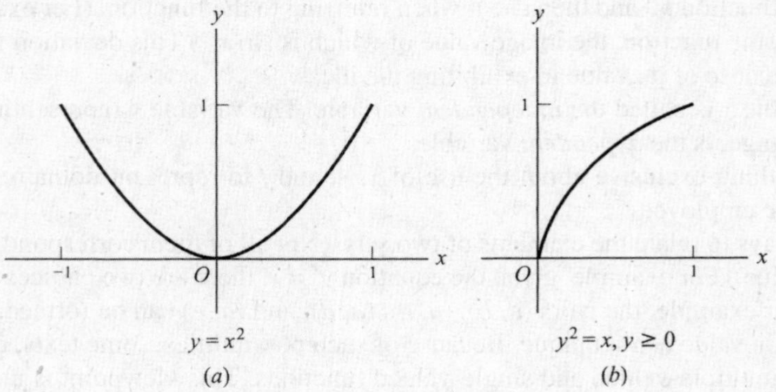

Figure 3.1

Bounded Functions

If there is a constant M such that $f(x) \leqq M$ for all x in an interval (or other set of numbers), we say that f is *bounded above* in the interval (or the set) and call M an *upper bound* of the function.

If a constant m exists such that $f(x) \geqq m$ for all x in an interval, we say that $f(x)$ is *bounded below* in the interval and call m a *lower bound*.

If $m \leqq f(x) \leqq M$ in an interval, we call $f(x)$ *bounded*. Frequently, when we wish to indicate that a function is bounded, we write $|f(x)| < P$.

EXAMPLES. 1. $f(x) = 3 + x$ is bounded in $-1 \leqq x \leqq 1$. An upper bound is 4 (or any number greater than 4). A lower bound is 2 (or any number less than 2).

2. $f(x) = 1/x$ is not bounded in $0 < x < 4$, since, by choosing x sufficiently close to zero, $f(x)$ can be made as large as we wish, so that there is no upper bound. However, a lower bound is given by $\frac{1}{4}$ (or any number less than $\frac{1}{4}$).

If $f(x)$ has an upper bound, it has a *least upper bound* (l.u.b.); if it has a lower bound, it has a *greatest lower bound* (g.l.b.). (See Chapter 1 for these definitions.)

Monotonic Functions

A function is called *monotonic increasing* in an interval if for any two points x_1 and x_2 in the interval $x_1 < x_2$, $f(x_1) \leqq f(x_2)$. If $(f(x_1) < f(x_2)$, the function is called *strictly increasing*.

Similarly, if $f(x_1) \geqq f(x_2)$ whenever $x_1 < x_2$, then $f(x)$ is *monotonic decreasing*, while if $f(x_1) > f(x_2)$, it is *strictly decreasing*.

Inverse Functions, Principal Values

Suppose y is the range variable of a function f with domain variable x. Furthermore, let the correspondence between the domain and range values be one-to-one. Then a new function f^{-1}, called the *inverse function* of f, can be created by interchanging the domain and range of f. This information is contained in the form $x = f^{-1}(y)$.

As you work with the inverse function, it often is convenient to rename the domain variable as x and use y to symbolize the images; then the notation is $y = f^{-1}(x)$. In particular, this allows graphical expression of the inverse function with its domain on the horizontal axis.

Note: f^{-1} does *not* mean f to the negative one power. When used with functions, the notation f^{-1} always designates the inverse function to f.

If the domain and range elements of f are not in one-to-one correspondence (this would mean that distinct domain elements have the same image), then a collection of one-to-one functions may be created. Each of them is called a *branch*. It is often convenient to choose one of these branches, called the *principal branch*, and denote it as the inverse function f^{-1}. The range values of f that compose the principal branch, and hence the domain of f^{-1}, are called the *principal values*. (As will be seen in the section on elementary functions, it is common practice to specify these principal values for that class of functions.)

EXAMPLE. Suppose f is generated by $y = \sin x$ and the domain is $-\infty \leqq x \leqq \infty$. Then there are an infinite number of domain values that have the same image. (A finite portion of the graph is illustrated in Figure 3.2(a). In Figure 3.2(b) the graph is rotated about a line at 45° so that the x axis rotates into the y axis. Then the variables are interchanged so that the x axis is once again the horizontal one. We see that the image of an x value is not unique. Therefore, a set of principal values must be chosen to establish an inverse function. A choice of a branch is accomplished by restricting the domain of the starting function, $\sin x$. For example, choose $-\frac{\pi}{2} \leqq x \leqq \frac{\pi}{2}$. Then there is a one-to-one correspondence between the elements of this domain and the images in

$-1 \leq x \leq 1$. Thus, f^{-1} may be defined with this interval as its domain. This idea is illustrated in Figure 3.2(*c*) and (*d*). With the domain of f^{-1} represented on the horizontal axis and by the variable *x*, we write $y = \sin^{-1} x$, $-1 \leq x \leq 1$.

If $x = -\dfrac{1}{2}$, then the corresponding range value is $y = -\dfrac{\pi}{6}$.

Note: In algebra, b^{-1} means $\dfrac{1}{b}$ and the fact that bb^{-1} produces the identity element 1 is simply a rule of algebra generalized from arithmetic. Use of a similar exponential notation for inverse functions is justified in that corresponding algebraic characteristics are displayed by $f^{-1}[f(x)] = x$ and $f[f^{-1}(x)] = x$.

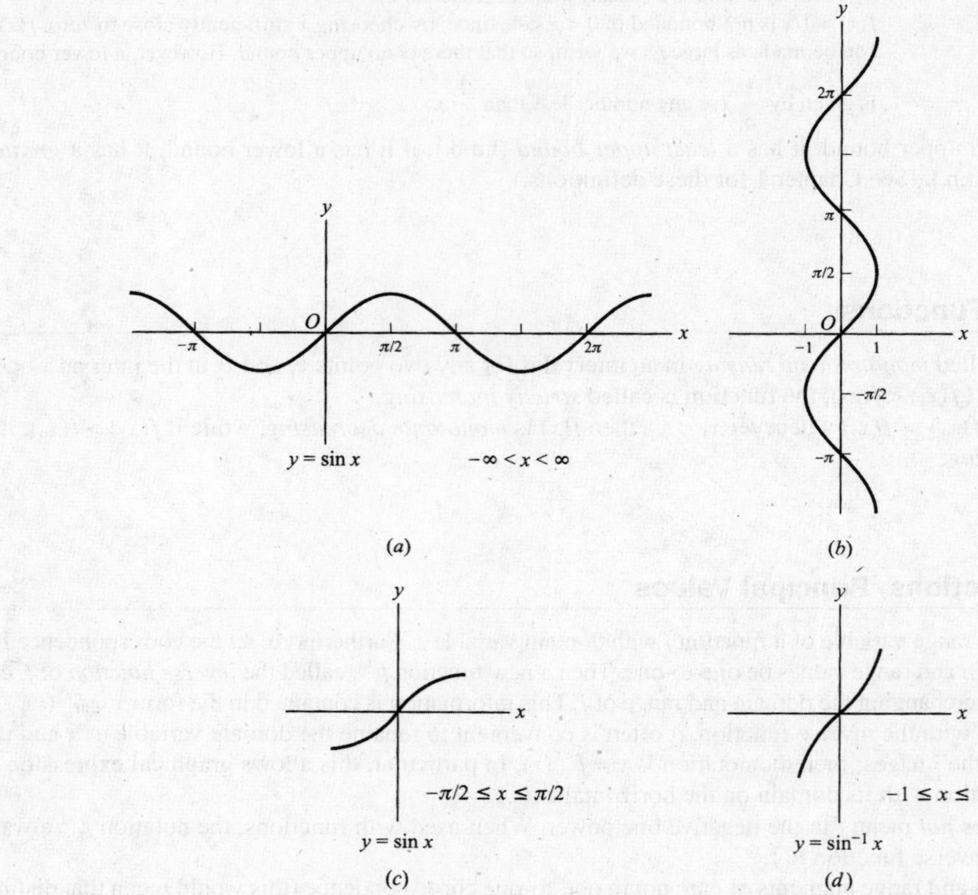

Figure 3.2

Maxima and Minima

The seventeenth-century development of the calculus was strongly motivated by questions concerning extreme values of functions. Of most importance to the calculus and its applications were the notions of *local extrema*, called the *relative maximum* and *relative minimum*.

If the graph of a function were compared to a path over hills and through valleys, the local extrema would be the high and low points along the way. This intuitive view is given mathematical precision by the following definition.

Definition If there exists an open interval (a, b) containing c such that $f(x) < f(c)$ for all x other than c in the interval, then $f(c)$ is a *relative maximum* of f. If $f(x) > f(c)$ for all x in (a, b) other than c, then $f(c)$ is a *relative minimum* of f. (See Figure 3.3.)

Functions may have any number of relative extrema. On the other hand, they may have none, as in the case of the strictly increasing and decreasing functions previously defined.

Definition If c is in the domain of f and for all x in the domain of the function $f(x) \leq f(c)$; then $f(c)$ is an *absolute maximum* of the function f. If for all x in the domain $f(x) \geq f(c)$, then $f(c)$ is an *absolute minimum* of f. (See Figure 3.3.)

Note: If defined on closed intervals, the strictly increasing and decreasing functions possess *absolute extrema*.

Absolute extrema are not necessarily unique. For example, if the graph of a function is a horizontal line, then every point is an *absolute maximum* and an *absolute minimum*.

Note: A *point of inflection* is also represented in Figure 3.3. There is an overlap with relative extrema in representation of such points through derivatives that will be addressed in the problem set of Chapter 4.

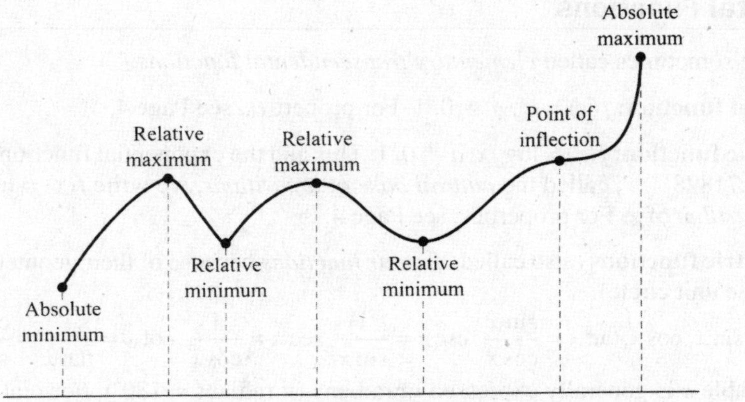

Figure 3.3

Types of Functions

It is worth realizing that there is a fundamental pool of functions at the foundation of calculus and advanced calculus. These are called *elementary functions*. Either they are generated from a real variable x by the fundamental operations of algebra, including powers and roots, or they have relatively simple geometric interpretations. As the title "elementary functions" suggests, there is a more general category of functions (which, in fact, are dependent on the elementary ones). Some of these will be explored later in this book. The *elementary functions* are described as follows.

1. **Polynomial functions** have the form

$$f(x) = a_0x^n + a_1x^{n-1} + \cdots + a_{n-1}x + a_n \tag{1}$$

where $a_0, \ldots, a_n$ are constants and n is a positive integer called the *degree* of the polynomial if $a_0 \neq 0$.

The *fundamental theorem of algebra* states that in the field of complex numbers every polynomial equation has at least one root. As a consequence of this theorem, it can be proved that every nth-degree polynomial has n roots in the complex field. When complex numbers are admitted, the polynomial theoretically may be expressed as the product of n linear factors; with our restriction to real numbers, it is possible that $2k$ of the roots may be complex. In this case, the k factors generating them will be quadratic. (The corresponding roots are in complex conjugate pairs.) The polynomial $x^3 - 5x^2 + 11x - 15 = (x - 3)(x^2 - 2x + 5)$ illustrates this thought.

2. **Algebraic functions** are functions $y = f(x)$ satisfying an equation of the form

$$p_0(x)\, y^n + p_1(x)\, y^{n-1} + \cdots + p_{n-1}(x)\, y + p_n(x) = 0 \tag{2}$$

where $p_0(x), \ldots, p_n(x)$ are polynomials in x.

If the function can be expressed as the quotient of two polynomials, i.e., $P(x)/Q(x)$ where $P(x)$ and $Q(x)$ are polynomials, it is called a *rational algebraic function*; otherwise, it is an *irrational algebraic function*.

3. **Transcendental functions** are functions which are not algebraic; i.e., they do not satisfy equations of the form of Equation (2).

Note the analogy with real numbers, polynomials corresponding to integers, rational functions to rational numbers, and so on.

Transcendental Functions

The following are sometimes called *elementary transcendental functions*.

1. **Exponential function:** $f(x) = a^x$, $a \neq 0, 1$. For properties, see Page 4.
2. **Logarithmic function:** $f(x) = \log_a x$, $a \neq 0, 1$. This and the exponential function are inverse functions. If $a = e = 2.71828\ldots$, called the *natural base of logarithms*, we write $f(x) = \log_e x = \text{In } x$, called the *natural logarithm* of x. For properties, see Page 4.
3. **Trigonometric functions** (also called *circular functions* because of their geometric interpretation with respect to the unit circle):

$$\sin x, \cos x, \tan x = \frac{\sin x}{\cos x}, \csc x = \frac{1}{\sin x}, \sec x = \frac{1}{\cos x}, \cot x = \frac{1}{\tan x} = \frac{\cos x}{\sin x}$$

The variable x is generally expressed in radians (π radians = 180°). For real values of x, $\sin x$ and $\cos x$ lie between –1 and 1 inclusive.

The following are some properties of these functions:

$$\sin^2 x + \cos^2 x = 1 \quad 1 + \tan^2 x = \text{sce}^2 x \quad 1 + \cot^2 x = \csc^2 x$$

$$\sin(x \pm y) = \sin x \cos y \pm \cos x \sin y \qquad \sin(-x) = -\sin x$$

$$\cos(x \pm y) = \cos x \cos y \mp \sin x \sin y \qquad \cos(-x) = \cos x$$

$$\tan(x \pm y) = \frac{\tan x \pm \tan y}{1 \mp \tan x \tan y} \qquad \tan(-x) = -\tan x$$

4. **Inverse trigonometric functions.** The following is a list of the inverse trigonometric functions and their principal values:

(*a*) $y = \sin^{-1} x$, $(-\pi/2 \leqq y \leqq \pi/2)$ (*d*) $y = \csc^{-1} x = \sin^{-1} 1/x$, $(-\pi/2 \leqq y \leqq \pi/2)$

(*b*) $y = \cos^{-1} x$, $(0 \leqq y \leqq \pi)$ (*e*) $y = \sec^{-1} x = \cos^{-1} 1/x$, $(0 \leqq y \leqq \pi)$

(*c*) $y = \tan^{-1} x$, $(-\pi/2 < y < \pi/2)$ (*f*) $y = \cot^{-1} x = \pi/2 - \tan^{-1} x$, $(0 < y < \pi)$

5. **Hyperbolic functions** are defined in terms of exponential functions as follows. These functions may be interpreted geometrically, much as the trigonometric functions but with respect to the unit hyperbola.

(*a*) $\sinh x = \dfrac{e^x - e^{-x}}{2}$ (*d*) $\text{csch } x = \dfrac{1}{\sinh x} = \dfrac{2}{e^x - e^{-x}}$

(*b*) $\cosh x = \dfrac{e^x + e^{-x}}{2}$ (*e*) $\text{sec h } x = \dfrac{1}{\cosh x} = \dfrac{2}{e^x + e^{-x}}$

(*c*) $\tanh x = \dfrac{\sinh x}{\cosh x} = \dfrac{e^x - e^{-x}}{e^x + e^{-x}}$ (*f*) $\coth x = \dfrac{\cosh x}{\sinh x} = \dfrac{e^x + e^{-x}}{e^x - e^{-x}}$

The following are some properties of these functions:

$\cosh^2 x - \sinh^2 x = 1$ $\quad 1 - \tanh^2 x = \operatorname{sech}^2 x$ $\quad \coth^2 x - 1 = \operatorname{csch}^2 x$

$\sinh(x \pm y) = \sinh x \cosh y \pm \cosh x \sinh y$ $\quad \sinh(-x) = -\sinh x$

$\cosh(x \pm y) = \cosh x \cosh y \pm \sinh x \sinh y$ $\quad \cosh(-x) = \cosh x$

$\tanh(x \pm y) = \dfrac{\tanh x \pm \tanh y}{1 \pm \tanh x \tanh y}$ $\quad \tanh(-x) = -\tanh x$

6. **Inverse hyperbolic functions.** If $x = \sinh y$, then $y = \sinh^{-1} x$ is the *inverse hyperbolic sine* of x. The following list gives the principal values of the inverse hyperbolic functions in terms of natural logarithms and the domains for which they are real.

(*a*) $\sinh^{-1} x = \ln(x + \sqrt{x^2+1})$, all x $\quad$ (*d*) $\operatorname{csch}^{-1} x = \ln\left(\dfrac{1}{x} + \dfrac{\sqrt{x^2+1}}{|x|}\right), x \neq 0$

(*b*) $\cosh^{-1} x = \ln(x + \sqrt{x^2-1}), x \geqq 1$ $\quad$ (*e*) $\operatorname{sech}^{-1} x = \ln\left(\dfrac{1+\sqrt{1-x^2}}{x}\right), 0 < x \leqq 1$

(*c*) $\tanh^{-1} x = \dfrac{1}{2}\ln\left(\dfrac{1+x}{1-x}\right), |x| < 1$ $\quad$ (*f*) $\coth^{-1} x = \dfrac{1}{2}\ln\left(\dfrac{x+1}{x-1}\right), |x| > 1$

Limits of Functions

Let $f(x)$ be defined and single-valued for all values of x near $x = x_0$ with the possible exception of $x = x_0$ itself (i.e., in a deleted δ neighborhood of x_0). We say that the number l is the *limit of $f(x)$ as x approaches x_0* and write $\lim_{x \to x_0} f(x) = l$ if for any positive number ϵ (however small) we can find some positive number δ (usually depending on ϵ) such that $|f(x) - l| < \epsilon$ whenever $0 < |x - x_0| < \delta$. In such a case we also say that $f(x)$ approaches l as x approaches x_0 and write $f(x) \to l$ as $x \to x_0$.

In words, this means that we can make $f(x)$ arbitrarily close to l by choosing x sufficiently close to x_0.

EXAMPLE. Let $f(x) = \begin{cases} x^2 & \text{if } x \neq 2 \\ 0 & \text{if } x = 2 \end{cases}$. Then as x gets closer to 2 (i.e., x approaches 2), $f(x)$ gets closer to 4. We thus *suspect* that $\lim_{x \to 2} f(x) = 4$. To *prove* this we must see whether the preceding definition of limit (with $l = 4$) is satisfied. For this proof, see Problem 3.10.

Note that $\lim_{x \to 2} f(x) = f(2)$; i.e., the limit of $f(x)$ as $x \to 2$ is not the same as the value of $f(x)$ *at* $x = 2$, since $f(2) = 0$ by definition. The limit would, in fact, be 4 even if $f(x)$ were not defined at $x = 2$.

When the limit of a function exists, it is unique; i.e., it is the only one (see Problem 3.17).

Right- and Left-Hand Limits

In the definition of limit, no restriction was made as to how x should approach x_0. It is sometimes found convenient to restrict this approach. Considering x and x_0 as points on the real axis where x_0 is fixed and x is moving, then x can approach x_0 from the right or from the left. We indicate these respective approaches by writing $x \to x_0+$ and $x \to x_0-$.

If $\lim_{x \to x_0} +f(x) = l_1$ and $\lim_{x \to x_{0^-}} f(x) = l_2$, we call l_1 and l_{l2}, respectively, the *right- and left-hand limits* of f at x_0 and denote them by $f(x_0+)$ or $f(x_0 + 0)$ and $f(x_0-)$ or $f(x_0 - 0)$. The ϵ, δ definitions of limit of $f(x)$ as

$x \to x_0+$ or $x \to x_0-$ are the same as those for $x \to x_0$ except for the fact that values of x are restricted to $x > x_0$ or $x < x_0$, respectively.

We have $\lim_{x\to x_0} f(x) = l$ if and only if $\lim_{x\to x_0+} f(x) = \lim_{x\to x_{0-}} fs(x) = l$.

Theorems on Limits

If $\lim_{x\to x_0} f(x) = A$ and $\lim_{x\to x_0} g(x) = B$, then

1. $\lim_{x\to x_0} (f(x) + g(x)) = \lim_{x\to x_0} f(x) + \lim_{x\to x_0} g(x) = A + B$

2. $\lim_{x\to x_0} (f(x) - g(x)) = \lim_{x\to x_0} f(x) - \lim_{x\to x_0} g(x) = A - B$

3. $\lim_{x\to x_0} (f(x)g(x)) = \left(\lim_{x\to x_0} f(x)\right)\left(\lim_{x\to x_0} g(x)\right) = AB$

4. $\lim_{x\to x_0} \dfrac{f(x)}{g(x)} = \dfrac{\lim_{x\to x_0} f(x)}{\lim_{x\to x_0} g(x)} = \dfrac{A}{B} \quad \text{if } B \neq 0$

Similar results hold for right- and left-hand limits.

Infinity

It sometimes happens that as $x \to x_0$, $f(x)$ increases or decreases without bound. In such case it is customary to write $\lim_{x\to x_0} f(x) = +\infty$ or $\lim_{x\to x_0} f(x) = -\infty$, respectively. The symbols $+\infty$ (also written ∞) and $-\infty$ are read "plus infinity" (or "infinity") and "minus infinity," respectively, but it must be emphasized that they are not numbers.

In precise language, we say that $\lim_{x\to x_0} f(x) = \infty$ if for each positive number M we can find a positive number δ (depending on M in general) such that $f(x) > M$ whenever $0 < |x - x_0| < \delta$. Similarly, we say that $\lim_{x\to x_0} f(x) = -\infty$ if for each positive number M we can find a positive number δ such that $f(x) < -M$ whenever $0 < |x - x_0| < \delta$. Analogous remarks apply in case $x \to x_0+$ or $x \to x_0-$.

Frequently we wish to examine the behavior of a function as x increases or decreases without bound. In such cases it is customary to write $x \to +\infty$ (or ∞) or $x \to -\infty$, respectively.

We say that $\lim_{x\to +\infty} f(x) = l$, or $f(x) \to l$ as $x \to +\infty$, if for any positive number ϵ we can find a positive number N (depending on ϵ in general) such that $|f(x) - l| < \epsilon$ whenever $x > N$. A similar definition can be formulated for $\lim_{x\to -\infty} f(x)$.

Special Limits

1. $\lim_{x\to 0} \dfrac{\sin x}{x} = 1 \qquad \lim_{x\to 0} \dfrac{1 - \cos x}{x} = 0$

2. $\lim_{x\to \infty} \left(1 + \dfrac{1}{x}\right)^x = e \qquad \lim_{x\to 0+} (1 + x)^{1/x} = e$

3. $\lim_{x\to 0} \frac{e^x - 1}{x} = 1 \qquad \lim_{x\to 1} \frac{x-1}{\ln x} = 1$

Continuity

Let f be defined for all values of x near $x = x_0$ as well as at $x = x_0$ (i.e., in a δ neighborhood of x_0). The function f is called *continuous* at $x = x_0$ if $\lim_{x\to x_0} f(x) = f(x_0)$. Note that this implies three conditions which must be met in order that $f(x)$ be continuous at $x = x_0$:

1. $\lim_{x\to x_0} f(x) = l$ must exist.
2. $f(x_0)$ must exist; i.e., $f(x)$ is defined *at* x_0.
3. $l = f(x_0)$.

In summary, $\lim_{x\to x_0} f(x)$ is the value suggested for f at $x = x_0$ by the behavior of f in arbitrarily small neighborhoods of x_0. If, in fact, this limit is the actual value, $f(x_0)$, of the function at x_0, then f is continuous there.

Equivalently, if f is continuous at x_0, we can write this in the suggestive form $\lim_{x\to x_0} f(x) = f(\lim_{x\to x_0} x)$.

EXAMPLES. 1. If $f(x) = \begin{cases} x^2, & x \neq 2 \\ 0, & x = 2 \end{cases}$ then from the example on Page 45 $\lim_{x\to 2} f(x) = 4$. But $f(2) = 0$.

Hence, $\lim_{x\to 2} f(x) \neq f(2)$ and the function is not continuous at $x = 2$.

2. If $f(x) = x^2$ for all x, then $\lim_{x\to 2} f(x) = f(2) = 4$ and $f(x)$ is continuous at $x = 2$.

Points where f fails to be continuous are called *discontinuities* of f and f is said to be *discontinuous* at these points.

In constructing a graph of a continuous function, the pencil need never leave the paper, while for a discontinuous function this is not true, since there is generally a jump taking place. This is, of course, merely a characteristic property and not a definition of continuity or discontinuity.

Alternative to the preceding definition of continuity, we can define f as continuous at $x = x_0$ if for any $\epsilon > 0$ we can find $\delta > 0$ such that $|f(x_0) - f(x_0)| < \epsilon$ whenever $|x - x_0| < \delta$. Note that this is simply the definition of limit with $l = f(x_0)$ and removal of the restriction that $x \neq x_0$.

Right- and Left-Hand Continuity

If f is defined only for $x \geqq x_0$, the preceding definition does not apply. In such case we call f *continuous* (*on the right*) at $x = x_0$ if $\lim_{x\to x_{0+}} f(x) = f(x_0)$, i.e., if $f(x_0 +) = f(x_0)$. Similarly, f is *continuous* (*on the left*) at $x = x_0$ if $\lim_{x\to x_{0-}} f(x) = f(x)_0$, i.e., $f(x_0 -) = f(x_0)$. Definitions in terms of ϵ and δ can be given.

Continuity in an Interval

A function f is said to be *continuous in an interval* if it is continuous at all points of the interval. In particular, if f is defined in the closed interval $a \leqq x \leqq b$ or $[a, b]$, then f is continuous in the interval if and only if $\lim_{x\to x_0} f(x) = f(x_0)$ for $a < x_0 < b$, $\lim_{x\to x_{a+}} f(x) = f(a)$, and $\lim_{x\to b-} f(x) = f(b)$.

Theorems on Continuity

Theorem 1 If f and g are continuous at $x = x_0$, so also are the functions whose image values satisfy the relations $f(x)+g(x), f(x)-g(x), f(x)g(x)$, and $\dfrac{f(x)}{g(x)}$, the last only if $g(x_0) \neq 0$. Similar results hold for continuity in an interval.

Theorem 2 Functions described as follows are continuous in every finite interval: (*a*) all polynomials; (*b*) $\sin x$ and $\cos x$; and (*c*) a^x, $a > 0$.

Theorem 3 Let the function f be continuous at the domain value $x = x_0$. Also suppose that a function g, represented by $z = g(y)$, is continuous at y_0, where $y = f(x)$ (i.e., the range value of f corresponding to x_0 is a domain value of g). Then a new function, called a *composite function*, $f(g)$, represented by $z = g[f(x)]$, may be created which is continuous at its domain point $x = x_0$. (One says that *a continuous function of a continuous function is continuous.*)

Theorem 4 If $f(x)$ is continuous in a closed interval, it is bounded in the interval.

Theorem 5 If $f(x)$ is continuous at $x = x_0$ and $f(x_0) > 0$ [or $f(x_0) < 0$], there exists an interval about $x = x_0$ in which $f(x) > 0$ [or $f(x) < 0$].

Theorem 6 If a function $f(x)$ is continuous in an interval and either strictly increasing or strictly decreasing, the inverse function $f^{-1}(x)$ is single-valued, continuous, and either strictly increasing or strictly decreasing.

Theorem 7 If $f(x)$ is continuous in $[a, b]$ and if $f(a) = A$ and $f(b) = B$, then corresponding to any number C between A and B there exists at least one number c in $[a, b]$ such that $f(c) = C$. This is sometimes called the *intermediate value theorem.*

Theorem 8 If $f(x)$ is continuous in $[a, b]$ and if $f(a)$ and $f(b)$ have opposite signs, there is at least one number c for which $f(c) = 0$ where $a < c < b$. This is related to Theorem 7.

Theorem 9 If $f(x)$ is continuous in a closed interval, then $f(x)$ has a maximum value M for at least one value of x in the interval and a minimum value m for at least one value of x in the interval. Furthermore, $f(x)$ assumes all values between m and M for one or more values of x in the interval.

Theorem 10 If $f(x)$ is continuous in a closed interval and if M and m are, respectively, the least upper bound (l.u.b.) and greatest lower bound (g.l.b.) of $f(x)$, there exists at least one value of x in the interval for which $f(x) = M$ or $f(x) = m$. This is related to Theorem 9.

Piecewise Continuity

A function is called *piecewise continuous* in an interval $a \leqq x \leqq b$ if the interval can be subdivided into a finite number of intervals in each of which the function is continuous and has finite right- and left-hand limits. Such a function has only a finite number of discontinuities. An example of a function which is piecewise continuous in $a \leqq x \leqq b$ is shown graphically in Figure 3.4. This function has discontinuities at x_1, x_2, x_3, and x_4.

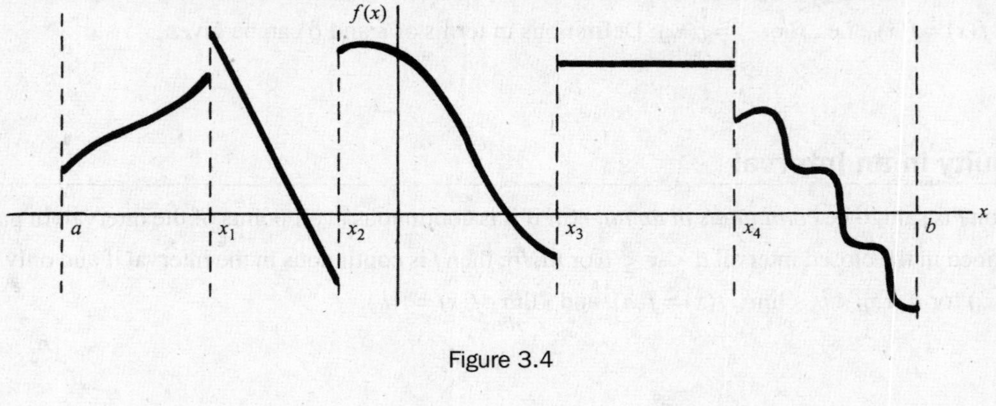

Figure 3.4

Uniform Continuity

Let f be continuous in an interval. Then, by definition, at each point x_0 of the interval and for any $\epsilon > 0$, we can find $\delta > 0$ (which will in general depend on both ϵ and the particular point x_0) such that $|f(x) - f(x_0)| < \epsilon$ whenever $|x - x_0| < \delta$. If we can find δ for each ϵ which holds for all points of the interval (i.e., if δ depends *only* on ϵ and *not* on x_0), we say that f is *uniformly continuous* in the interval.

Alternatively, f is uniformly continuous in an interval if for any $\epsilon > 0$ we can find $\delta > 0$ such that $|f(x_1) - f(x_2)| < \epsilon$ whenever $|x_1 - x_2| < \delta$ where x_1 and x_2 are any two points in the interval.

Theorem If f is continuous in a *closed* interval, it is uniformly continuous in the interval.

SOLVED PROBLEMS

Functions

3.1. Let $f(x) = (x - 2)(8 - x)$ for $2 \leqq x \leqq 8$. (a) Find $f(6)$ and $f(-1)$. (b) What is the domain of definition of $f(x)$? (c) Find $f(1 - 2t)$ and give the domain of definition. (d) Find $f[f(3)]$, $f[f(5)]$. (e) Graph $f(x)$.

(a) $f(6) = (6 - 2)(8 - 6) = 4 \cdot 2 = 8$
$f(-1)$ is not defined since $f(x)$ is defined only for $2 \leqq x \leqq 8$.

(b) The set of all x such that $2 \leqq x \leqq 8$.

(c) $f(1 - 2t) = \{(1 - 2t) - 2\}\{8 - (1 - 2t)\} = -(1 + 2t)(7 + 2t)$ where t is such that $2 \leqq 1 - 2t \leqq 8$; i.e., $-7/2 \leqq t \leqq -1/2$.

(d) $f(3) = (3 - 2)(8 - 3) = 5$, $f[f(3)] = f(5) = (5 - 2)(8 - 5) = 9$.
$f(5) = 9$ so that $f[f(5)] = f(9)$ is not defined.

(e) The following table shows $f(x)$ for various values of x.

x	2	3	4	5	6	7	8	2.5	7.5
$f(x)$	0	5	8	9	8	5	0	2.75	2.75

Plot points (2, 0), (3, 5), (4, 8), (5, 9), (6, 8), (7, 5), (8, 0), (2.5, 2.75), (7.5, 2.75). These points are only a few of the infinitely many points on the required graph shown in the adjoining Figure 3.5. This set of points defines a curve which is part of a *parabola*.

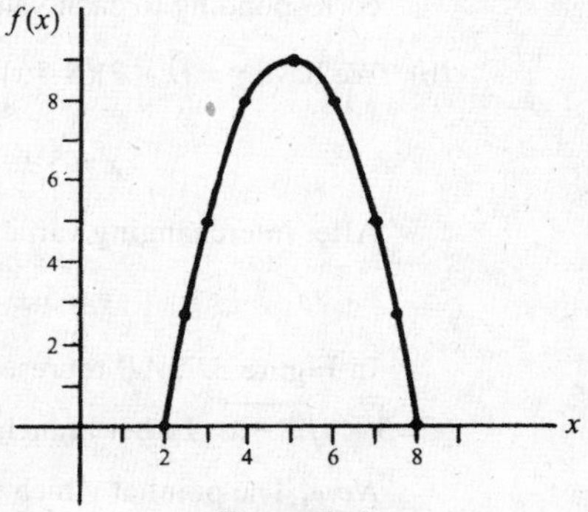

Figure 3.5

3.2. Let $g(x) = (x - 2)(8 - x)$ for $2 < x < 8$. (a) Discuss the difference between the graph of $g(x)$ and that of $f(x)$ in Problem 3.1. (b) What are the l.u.b. and g.l.b. of $g(x)$? (c) Does $g(x)$ attain its l.u.b. and g.l.b. for any value of x in the domain of definition? (d) Answer parts (b) and (c) for the function $f(x)$ of Problem 3.1.

(a) The graph of $g(x)$ is the same as that in Problem 3.1 except that the two points (2, 0) and 8, 0) are missing, since $g(x)$ is not defined at $x = 2$ and $x = 8$.

(b) The l.u.b. of $g(x)$ is 9. The g.l.b. of $g(x)$ is 0.

(c) The l.u.b. of $g(x)$ is attained for the value of $x = 5$. The g.l.b. of $g(x)$ is not attained, since there is no value of x in the domain of definition such that $g(x) = 0$.

(d) As in *(b)*, the l.u.b. of $f(x)$ is 9 and the g.l.b. of $f(x)$ is 0. The l.u.b. of $f(x)$ is attained for the value $x = 5$ and the g.l.b. of $f(x)$ is attained at $x = 2$ and $x = 8$.

Note that a function, such as $f(x)$, which is *continuous* in a closed interval attains its l.u.b. and g.l.b. at some point of the interval. However, a function, such as $g(x)$, which is not continuous in a closed interval need not attain its l.u.b. and g.l.b. See Problem 3.34.

3.3. Let

$$f(x)=\begin{cases}1, & \text{if } x \text{ is a rational number}\\ 0, & \text{if } x \text{ is an irrational number}\end{cases}$$

(a) Find $f\left(\frac{2}{3}\right)$, $f(-5)$, $f(1.41423)$, $f(\sqrt{2})$. (b) Construct a graph of $f(x)$ and explain why it is misleading by itself.

(a) $f\left(\frac{2}{3}\right) = 1$ since $\frac{2}{3}$ is a rational number

$f(-5) = 1$ since -5 is a rational number

$f(1.41423) = 1$ since 1.41423 is a rational number

$f(\sqrt{2}) = 0$ since $\sqrt{2}$ is an irrational number

(b) The graph is shown in Figure 3.6. Because the sets of both rational numbers and irrational numbers are dense, the visual impression is that there are two images corresponding to each domain value. In actuality, each domain value has only one corresponding range value.

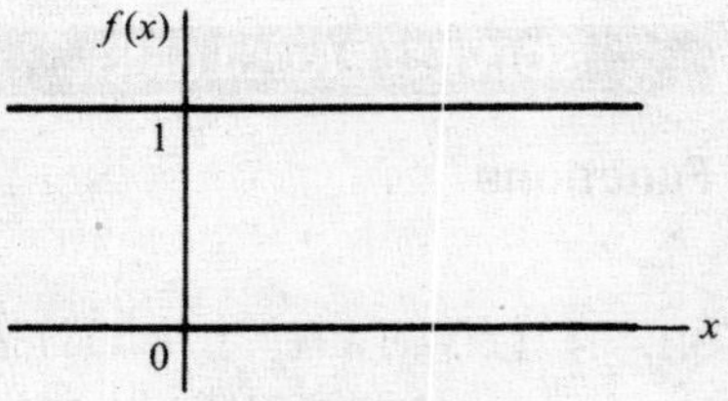

Figure 3.6

3.4. Referring to Problem 3.1: (a) Draw the graph with axes interchanged, thus illustrating the two possible choices available for definition of f^{-1}. (b) Solve for x in terms of y to determine the equations describing the two branches, and then interchange the variables.

(a) The graph of $y = f(x)$ is shown in Figure 3.5 of Problem 3.1(*a*). By interchanging the axes (and the variables), we obtain the graphical form of Figure 3.7. This figure illustrates that there are two values of y corresponding to each value of x, and, hence, two branches. Either may be employed to define f^{-1}.

(b) We have $y = (x-2)(8-x)$ or $x^2 - 10x + 16 + y = 0$. The solution of this quadratic equation is

$$x = 5 \pm \sqrt{9-y}$$

After interchanging variables

$$y = 5 \pm \sqrt{9-x}$$

In Figure 3.7, *AP* represents $y = 5 + \sqrt{9-x}$, and *BP* designates $y = 5 - \sqrt{9-x}$. Either branch may represent f^{-1}.

Note: The point at which the two branches meet is called a *branch point*.

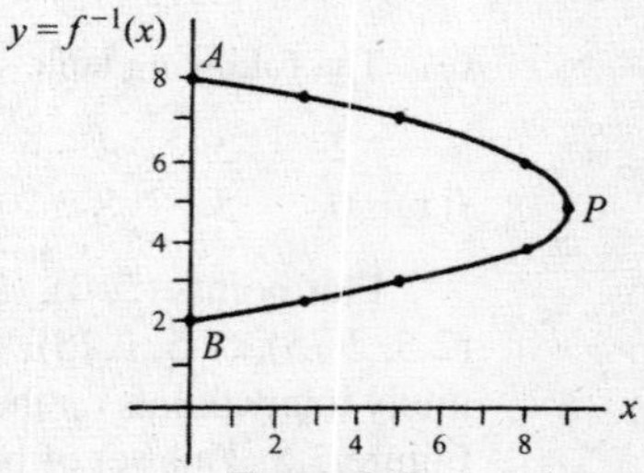

Figure 3.7

3.5. (a) Prove that $g(x) = 5 + \sqrt{9-x}$ is strictly decreasing in $0 \leq x \leq 9$. (b) Is it monotonic decreasing in this interval? (c) Does $g(x)$ have a single-valued inverse?

(a) $g(x)$ is strictly decreasing if $g(x_1) > g(x_2)$ whenever $x_1 < x_2$. If $x_1 < x_2$, the $9 - x_1 > 9 - x_2$, $\sqrt{9-x_1} > \sqrt{9-x_2}$, and $5 + \sqrt{9-x_1} > 5 + \sqrt{9-x_2}$, showing that $g(x)$ is strictly decreasing.

(b) Yes, any strictly decreasing function is also monotonic decreasing, since if $g(x_1) > g(x_2)$ it is also true that $g(x_1) \geq g(x_2)$. However, if $g(x)$ is monotonic decreasing, it is not necessarily strictly decreasing.

(c) If $y = 5 + \sqrt{9-x}$, then $y - 5 = \sqrt{9-x}$ or, squaring, $x = -16 + 10y - y^2 = (y-2)(8-y)$ and x is a single-valued function of y; i.e., the inverse function is single-valued.

In general, any strictly decreasing (or increasing) function has a single-valued inverse (see Theorem 6, Page 52).

The results of this problem can be interpreted graphically using Figure 3.7.

3.6. Construct graphs for the following functions:

$$\text{(a)}\ f(x) = \begin{cases} x \sin 1/x, & x > 0 \\ 0, & x = 0 \end{cases}$$

$$\text{(b)}\ f(x) = [x] = \text{greatest integer} \leqq x$$

(a) The required graph is shown in Figure 3.8. Since $|x \sin 1/x| \leqq |x|$, the graph is included between $y = x$ and $y = -x$. Note that $f(x) = 0$ when $\sin 1/x = 0$ or $1/x =$, $m\pi$, $m = 1, 2, 3, 4, \ldots$, i.e., where $x = 1/\pi, 1/2\pi, 1/3\pi, \ldots$ The curve oscillates infinitely often between $x = 1/\pi$ and $x = 0$.

(b) The required graph is shown in Figure 3.9. If $1 \leqq x < 2$, then $[x] = 1$. Thus, $[1.8] = 1$, $[\sqrt{2}] = 1$, $[1.99999] = 1$. However, $[2] = 2$. Similarly, for $2 \leqq x < 3$, $[x] = 2$, etc. Thus, there are *jumps* at the integers. The function is sometimes called the *staircase function* or *step function*.

3.7. (a) Construct the graph of $f(x) = \tan x$. (b) Construct the graph of some of the infinite number of branches available for a definition of $\tan^{-1} x$. (c) Show graphically why the relationship of x to y is multivalued. (d) Indicate possible principal values for $\tan^{-1} x$. (e) Using your choice, evaluate $\tan^{-1}(-1)$.

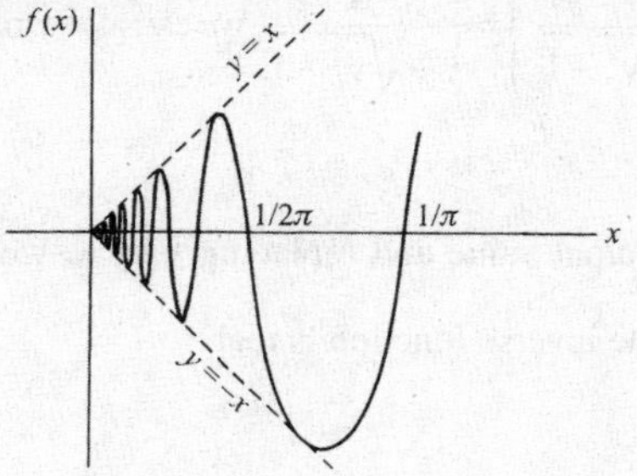

Figure 3.8

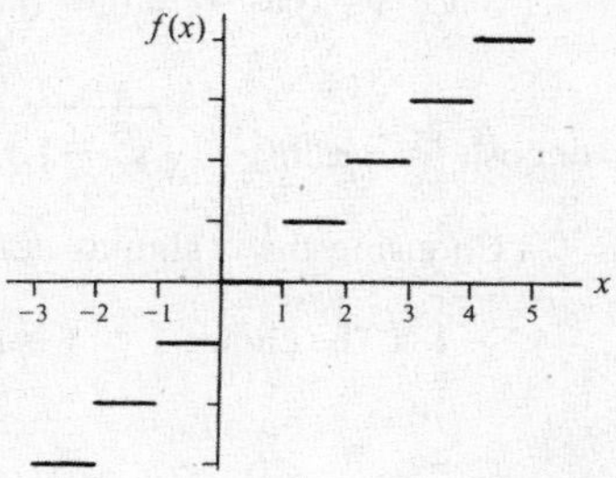

Figure 3.9

(a) The graph of $f(x) = \tan x$ appears in Figure 3.10.

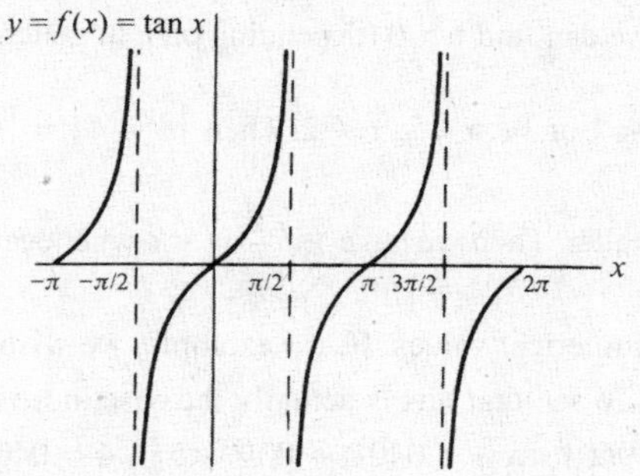

Figure 3.10

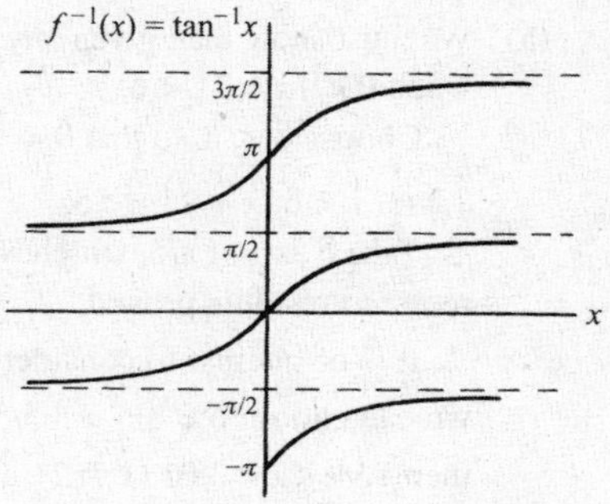

Figure 3.11

(b) The required graph is obtained by interchanging the x and y axes in the graph of (*a*). The result, with axes oriented as usual, appears in Figure 3.11.

(c) In Figure 3.11, any vertical line meets the graph in infinitely many points. Thus, the relation of y to x is multivalued and infinitely many branches are available for the purpose of defining $\tan^{-1} x$.

(d) To define $\tan^{-1} x$ as a single-valued function, it is clear from the graph that we can do so only by restricting its value to any of the following: $-\pi/2 < \tan^{-1} x < \pi/2$, $\pi/2 < \tan^{-1} x < 3\pi/2$, etc. We agree to take the first as defining the principal value.

Note that no matter which branch is used to define $\tan^{-1} x$, the resulting function is strictly increasing.

(e) $\tan^{-1}(-1) = -\pi/4$ is the only value lying between $-\pi/2$ and $\pi/2$; i.e., it is the principal value according to our choice in (*d*).

3.8. Show that $f(x)=\dfrac{\sqrt{x}+1}{x+1}$, $x\neq -1$, describes an irrational algebraic function.

If $y=\dfrac{\sqrt{x}+1}{x+1}$, then $(x+1)y-1=\sqrt{x}$ or, squaring, $(x+1)^2y^2-2(x+1)y+1-x=0$, a polynomial equation in y whose coefficients are polynomials in x. Thus, $f(x)$ is an algebraic function. However, it is not the quotient of two polynomials, so that it is an irrational algebraic function.

3.9. If $f(x)=\cosh x=\frac{1}{2}(e^x+e^{-x})$, prove that we can choose, as the principal value of the inverse function, $\cosh^{-1}x=\ln(x+\sqrt{x^2-1})$, $x\geqq 1$.

If $y=\frac{1}{2}(e^x+e^{-x})$, $e^{2x}-2ye^x+1=0$. Then, using the quadratic formula, $e^x=\dfrac{2y\pm\sqrt{4y^2-4}}{2}=y\pm\sqrt{y^2-1}$.

Thus, $x=\ln\left(y\pm\sqrt{y^2-1}\right)$.

Since $y-\sqrt{y^2-1}=(y-\sqrt{y^2-1})\left(\dfrac{y+\sqrt{y^2-1}}{y+\sqrt{y^2-1}}\right)=\dfrac{1}{y+\sqrt{y^2-1}}$ we can also write $x=\pm\ln(y+\sqrt{y^2-1})$ or $\cosh^{-1}y=\pm\ln(y+\sqrt{y^2-1})$.

Choosing the + sign as defining the principal value and replacing y by x, we have $\cosh^{-1}x=\ln(x+\sqrt{y^2-1})$. The choice $x\geqq 1$ is made so that the inverse function is real.

Limits

3.10. If (a) $f(x)=x^2$ and (b) $f(x)=\begin{cases}x^2, & x\neq 2\\ 0, & x=2\end{cases}$, prove that $\lim_{x\to 2} f(x)=4$.

(a) We must show that, given any $\epsilon>0$, we can find $\delta>0$ (depending on ϵ in general) such that $|x^2-4|<\epsilon$ when $0<|x-2|<\delta$.

Choose $\delta\leq 1$ so that $0<|x-2|<1$ or $1<x<3$, $x\neq 2$. Then $|x^2-4|=|(x-2)(x+2)|=|x-2|\,|x+2|<\delta|x+2|<5\delta$.

Take δ as 1 or $\epsilon/5$, whichever is smaller. Then we have $|x^2-4|<\epsilon$ whenever $0<|x-2|<\delta$, and the required result is proved.

It is of interest to consider some numerical values. If, for example, we wish to make $|x^2-4|<.05$, we can choose $\delta=\epsilon/5=.05/5=.01$. To see that this is actually the case, note that if $0<|x-2|<.01$, then $1.99<x<2.01$ ($x\neq 2$), and so $3.9601<x^2<4.0401$, $-.0399<x^2-4<.0401$, and certainly $|x^2-4|<.05$ ($x^2\neq 4$). The fact that these inequalities also happen to hold at $x=2$ is merely coincidental.

If we wish to make $|x^2-4|<6$, we can choose $\delta=1$, and this will be satisfied.

(b) There is no difference between the proof for this case and the proof in (*a*), since in both cases we exclude $x=2$.

3.11. Prove that $\lim_{x\to 1}\dfrac{2x^4-6x^3+x^2+3}{x-1}=-8$.

We must show that for any $\epsilon>0$ we can find $\delta>0$ such that $\left|\dfrac{2x^4-6x^3+x^2+3}{x-1}-(-8)\right|<\varepsilon$ when $0<|x-1|<\delta$. Since $x\neq 1$, we can write $\dfrac{2x^4-6x^3+x^2+3}{x-1}=\dfrac{(2x^3-4x^2-3x-3)(x-1)}{x-1}=2x^3-4x^2-3x-3$ on cancelling the common factor $x-1\neq 0$.

Then we must show that for any $\epsilon > 0$, we can find $\delta > 0$ such that $|2x^3 - 4x^2 - 3x + 5| < \epsilon$ when $0 < |x-1| < \delta$. Choosing $\delta \leq 1$, we have $0 < x < 2$, $x \neq 1$.

Now $|2x^3 - 4x^2 - 3x + 5| = |x-1|\,|2x^2 - 2x - 5| < \delta|2x^2 - 2x - 5| < \delta(|2x^2| + |2x| + 5) < (8 + 4 + 5)\delta = 17\delta$. Taking δ as the smaller of 1 and $\epsilon/17$, the required result follows.

3.12. Let

$$f(x) = \begin{cases} \dfrac{|x-3|}{x-3}, & x \neq x \\ 0, & x = 3 \end{cases}.$$

(a) Graph the function. (b) Find $\lim_{xf(x)}$. (c) Find $\lim_{x\to 3+} f(x)$. (d) Find $\lim_{x\to 3} f(x)$.

(a) For $x > 3$, $\dfrac{|x-3|}{x-3} = \dfrac{x-3}{x-3} = 1$.

For $x > 3$, $\dfrac{|x-3|}{x-3} = \dfrac{-(x-3)}{x-3} = 1$.

Then the graph, shown in Figure 3.12, consists of the lines $y = 1, x > 3$; $y = -1, x < 3$; and the point (3, 0).

(b) As $x \to 3$ from the right, $f(x) \to 1$; i.e., $\lim_{x\to 3+} f(x) = 1$, as seems clear from the graph. To prove this we must show that given any $\epsilon > 0$, we can find $\delta > 0$ such that $|f(x) - 1| < \epsilon$ whenever $0 < x - 1 < \delta$.

Now, since $x > 1$, $f(x) = 1$ and so the proof consists in the triviality that $|1 - 1| < \epsilon$ whenever $0 < x - 1 < \delta$.

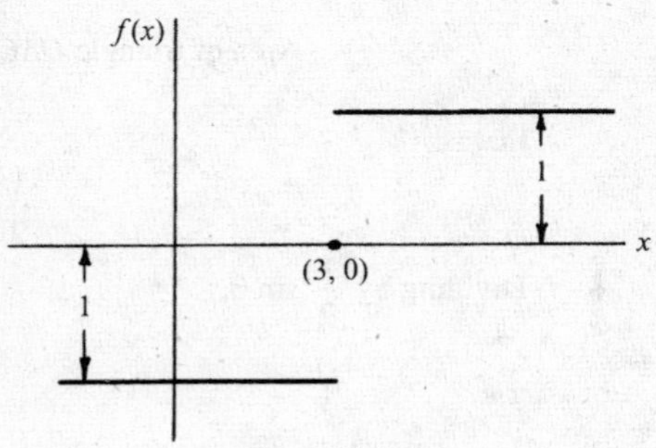

Figure 3.12

(c) As $x \to 3$ from the left, $f(x) \to -1$; i.e., $\lim_{x\to 3-} f(x) = -1$. A proof can be formulated as in (*b*).

(d) Since $\lim_{x\to 3+} f(x) \neq \lim_{x\to 3-} f(x)$, $\neq \lim_{x\to 3-} f(x)$ does not exist.

3.13. Prove that $\lim_{x\to 0} x \sin 1/x = 0$.

We must show that given any $\epsilon > 0$, we can find $\delta > 0$ such that $|x \sin 1/x - 0| < \epsilon$ when $0 < |x - 0| < \delta$.

If $0 < |x| < \delta$, then $|x \sin 1/x| = |x|\,|\sin 1/x| \leq |x| < \delta$, since $|\sin 1/x| \leq 1$ for all $x \neq 0$.

Making the choice $\delta = \epsilon$, we see that $|x \sin 1/x| < \epsilon$ when $0 < |x| < \delta$, completing the proof.

3.14. Evaluate $\lim_{x\to 0+} \dfrac{2}{1+e^{-1/x}}$.

As $x \to 0+$ we *suspect* that $1/x$ increases indefinitely, $e^{1/x}$ increases indefinitely, $e^{-1/x}$ approaches 0, and $1 + e^{-1/x}$ approaches 1; thus, the required limit is 2.

To *prove* this conjecture we must show that, given $\epsilon > 0$, we can find $\delta > 0$ such that

$$\left|\frac{2}{1+e^{-1/x}} - 2\right| < \varepsilon \quad \text{when} \quad 0 < x < \delta$$

Now

$$\left|\frac{2}{1+e^{-1/x}} - 2\right| = \left|\frac{2 - 2 - 2e^{-1/x}}{1+e^{-1/x}}\right| = \frac{2}{e^{1/x}+1}$$

Since the function on the right is smaller than 1 for all $x > 0$, any $\delta > 0$ will work when $e \geq 1$. If $0 < \varepsilon < 1$, then $\dfrac{2}{e^{1/x}+1} < \varepsilon$ when $\dfrac{e^{1/x}+1}{2} > \dfrac{1}{\varepsilon}$, $e^{1/x} > \dfrac{2}{\varepsilon} - 1$, $\dfrac{1}{x} > In\left(\dfrac{2}{\varepsilon} - 1\right)$; or $0 < x < \dfrac{1}{\ln(2/\varepsilon - 1)} = \delta$.

3.15. Explain exactly what is meant by the statement $\lim_{x\to 1} \dfrac{1}{(x-1)^4} = \infty$ and prove the validity of this statement.

The statement means that for each positive number M, we can find a positive number δ (depending on M in general) such that

$$\frac{1}{(x-1)^4} > 4 \qquad \text{when} \qquad 0 < |x-1| < \delta$$

To prove this, note that $\dfrac{1}{(x-1)^4} > M$ when $0 < (x-1)^4 < \dfrac{1}{M}$ or $0 < |x-1| < \dfrac{1}{\sqrt[4]{M}}$.

Choosing $\delta = 1/\sqrt[4]{M}$, the required results follows.

3.16. Present a geometric proof that $\lim_{\theta\to 0} \dfrac{\sin\theta}{\theta} = 1$.

Construct a circle with center at O and radius $OA = OD = 1$, as in Figure 3.13. Choose point B on OA extended and point C on OD so that lines BD and AC are perpendicular to OD.

It is geometrically evident that

$$\text{Area of triangle } OAC < \text{Area of sector } OAD < \text{Area of triangle } OBD$$

that is,

$$\frac{1}{2}\sin\theta\cos\theta < \frac{1}{2}\theta < \frac{1}{2}\tan\theta$$

Dividing by $\dfrac{1}{2}\sin\theta$,

$$\cos\theta < \frac{\theta}{\sin\theta} < \frac{1}{\cos\theta}$$

or

$$\cos\theta < \frac{\sin\theta}{\theta} < \frac{1}{\cos\theta}$$

As $\theta \to 0$, $\cos\theta \to 1$, and it follows that $\lim_{\theta\to 0} \dfrac{\sin\theta}{\theta} = 1$.

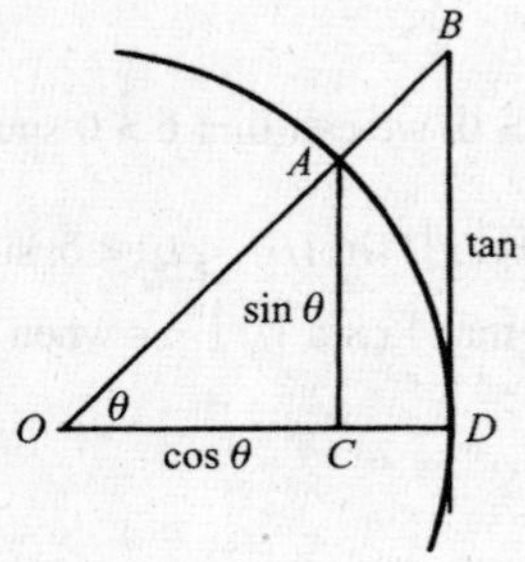

Figure 3.13

Theorems on limits

3.17. If $\lim_{x\to x_0} f(x)$ exists, prove that it must be unique.

We must show that if $\lim_{x\to x_0} f(x) = l_1$ and $\lim_{x\to x_0} f(x) = l_2$, then $l_1 = l_2$.

By hypothesis, given any $\epsilon > 0$ we can find $\delta > 0$ such that

$$|f(x) - l_1| < \epsilon/2 \qquad \text{when} \qquad 0 < |x - x_0| < \delta$$

$$|f(x) - l_2| < \epsilon/2 \qquad \text{when} \qquad 0 < |x - x_0| < \delta$$

Then by the absolute value property 2 on Page 4,

$$|l_1 - l_2| = |l_1 - f(x) + f(x) - l_2| \leqq |l_1 - f(x)| + |f(x) - l_2| < \epsilon/2 + \epsilon/2 = \epsilon$$

i.e., $|l_1 - l_2|$ is less than any positive number ϵ (however small) and so must be zero. Thus, $l_1 = l_2$.

3.18. If $\lim_{x \to x_0} g(x) = B \neq 0$, prove that there exists $\delta > 0$ such that

$$|g(x)| > \frac{1}{2}|B| \qquad \text{for} \qquad 0 < |x - x_0| < \delta$$

Since $\lim_{x \to x_0} g(x) = B$, we can find $\delta > 0$ such that $|g(x) - B| < \frac{1}{2}|B|$ for $0 < |x - x_0| < \delta$.

Writing $B = B - g(x) + g(x)$, we have

$$|B| \leqq |B - g(x)| + |g(x)| < \frac{1}{2}|B| + |g(x)|$$

i.e., $|B| < \frac{1}{2}|B| + |g(x)|$, from which $|g(x)| > \frac{1}{2}|B|$.

3.19. Given $\lim_{x \to x_0} f(x) = A$ and $\lim_{x \to x_0} g(x) = B$, prove (a) $\lim_{x \to x_0} [f(x) + g(x)] = A + B$, (b) $\lim_{x \to x_0} f(x)g(x) = AB$, (c) $\lim_{x \to x_0} \frac{1}{g(x)} = \frac{1}{B}$ if $B \neq 0$, and (d) $\lim_{x \to x_0} \frac{f(x)}{g(x)} = \frac{A}{B}$ if $B \neq 0$.

(a) We must show that for any $\epsilon > 0$ we can find $\delta > 0$ such that

$$|[f(x) + g(x)] - (A + B)| < \varepsilon \qquad \text{when} \qquad 0 < |x - x_0| < \delta$$

Using absolute value property 2, Page 4, we have

$$|[f(x) + g(x)] - (A + B)| = |[f(x) - A] + [g(x) - B]| \leqq |f(x) - A| + |g(x) - B| \tag{1}$$

By hypothesis, given $\epsilon > 0$ we can find $\delta_1 > 0$ and $\delta_2 > 0$ such that

$$|f(x) - A| < \epsilon/2 \qquad \text{when} \qquad 0 < |x - x_0| < \delta_1 \tag{2}$$

$$|g(x) - B| < \epsilon/2 \qquad \text{when} \qquad 0 < |x - x_0| < \delta_2 \tag{3}$$

Then, from Equations (1), (2), and (3),

$$|[f(x) + g(x)] - (A + B)| < \epsilon/2 + \epsilon/2 = \epsilon \qquad \text{when} \qquad 0 < |x - x_0| < \delta$$

where δ is chosen as the smaller of δ_1 and δ_2.

(b) We have

$$\begin{aligned} |f(x)g(x) - AB| &= |f(x)[g(x) - B] + B[f(x) - A]| \\ &\leqq |f(x)||g(x) - B| + |B||f(x) - A| \\ &\leqq |f(x)||g(x) - B| + (|B| + 1)|f(x) - A| \end{aligned} \tag{4}$$

Since $\lim_{x \to x_0} f(x) = A$, we can find δ_1 such that $|f(x) - A| < 1$ for $0 < |x - x_0| < \delta_1$, i.e., $A - 1 < f(x) < A + 1$, so that $f(x)$ is bounded, i.e., $|f(x)| < P$ where P is a positive constant.

Since $\lim_{x \to x_0} g(x) = B$, given $\epsilon > 0$, we can find $\delta_2 > 0$ such that $|g(x) - B| < \epsilon/2P$ for $0 < |x - x_0| < \delta_2$.

Since $\lim_{x \to x_0} f(x) = A$, given $\epsilon > 0$, we can find $\delta_3 > 0$ such that $|f(x) - A| < \frac{\varepsilon}{2(|B|+1)}$ for $0 < |x - x_0| < \delta_2$.

Using these in Equation (4), we have

$$|f(x)g(x) - AB| < P \cdot \frac{\varepsilon}{2P} + (|B|+1) \cdot \frac{\varepsilon}{2(|B|+1)} = \varepsilon$$

for $0 < |x - x_0| < \delta$, where δ is the smaller of $\delta_1, \delta_1, \delta_2, \delta_3$, and the proof is complete.

(c) We must show that for any $\epsilon > 0$ we can find $\delta > 0$ such that

$$\left|\frac{1}{g(x)} - \frac{1}{B}\right| = \frac{|g(x) - B|}{|B||g(x)|} < \varepsilon \quad \text{when} \quad 0 < |x - x_0| < \delta \tag{5}$$

By hypothesis, given $\epsilon > 0$, we can find $\delta_1 > 0$ such that

$$|g(x) - B| < \frac{1}{2}B^2\epsilon \qquad \text{when} \qquad 0 < |x - x_0| < \delta_1$$

By Problem 3.18, since $\lim\limits_{x \to 0} g(x) = B \neq 0$, we can find $\delta_2 > 0$ such that

$$|g(x)| > \frac{1}{2}|B| \qquad \text{when} \qquad 0 < |x - x_0| < \delta_2$$

Then, if δ is the smaller of δ_1 and δ_2, we can write

$$\left|\frac{1}{g(x)} - \frac{1}{B}\right| = \frac{|g(x) - B|}{|B||g(x)|} < \frac{\frac{1}{2}B^2\varepsilon}{|B| \cdot \frac{1}{2}|B|} = \varepsilon \quad \text{whenever} \quad 0 < |x - x_0| < \delta$$

and the required result is proved.

(d) From parts (*b*) and (*c*),

$$\lim_{x \to x_0} \frac{f(x)}{g(x)} = \lim_{x \to x_0} f(x) \cdot \frac{1}{g(x)} = \lim_{x \to x_0} f(x) \cdot \lim_{x \to x_0} \frac{1}{g(x)} = A \cdot \frac{1}{B} = \frac{A}{B}$$

This can also be proved directly (see Problem 3.69).

These results can also be proved in the cases $x \to x_0+$, $x \to x_0-$, $x \to \infty$, $x \to -\infty$.

Note: In the proof of (*a*) we have used the results $|f(x) - A| < \epsilon/2$ and $|g(x) - B| < \epsilon/2$, so that the final result would come out to be $|f(x) + g(x) - (A + B)| < \epsilon$. Of course, the proof would be *just as valid* if we had used 2ϵ (or any other positive multiple of ϵ) in place of ϵ. A similar remark holds for the proofs of (*b*), (*c*), and (*d*).

3.20. Evaluate each of the following, using theorems on limits.

(a)
$$\begin{aligned}\lim_{x \to 2}(x^2 - 6x + 4) &= \lim_{x \to 2} x^2 + \lim_{x \to 2}(-6x) + \lim_{x \to 2} 4 \\ &= (\lim_{x \to 2} x)(\lim_{x \to 2} x) + (\lim_{x \to 2} -6)(\lim_{x \to 2} x) + \lim_{x \to 2} 4 \\ &= (2)(2) + (-6)(2) + 4 = -4\end{aligned}$$

In practice, the intermediate steps are omitted.

(b)
$$\lim_{x \to -1} \frac{(x+3)(2x-1)}{x^2 + 3x - 2} = \frac{\lim\limits_{x \to -1}(x+3)\lim\limits_{x \to -1}(2x-1)}{\lim\limits_{x \to -1}(x^2 + 3x - 2)} = \frac{2 \cdot (-3)}{-4} = \frac{3}{2}$$

(c)
$$\lim_{x \to \infty} \frac{2x^4 - 3x^2 + 1}{6x^4 + x^3 - 3x} = \lim_{x \to \infty} \frac{2 - \frac{3}{x^2} + \frac{1}{x^4}}{6 + \frac{1}{x} - \frac{3}{x^3}}$$

$$= \frac{\lim\limits_{x \to \infty} 2 + \lim\limits_{x \to \infty} \frac{-3}{2} + \lim\limits_{x \to \infty} \frac{1}{x^4}}{\lim\limits_{x \to \infty} 6 + \lim\limits_{x \to \infty} \frac{1}{x} + \lim\limits_{x \to \infty} \frac{-3}{x^3}} = \frac{2}{6} = \frac{1}{3}$$

by Problem 3.19.

(d) $$\lim_{h\to 0}\frac{\sqrt{4+h}-2}{h} = \lim_{h\to 0}\frac{\sqrt{4+h}-2}{h}\cdot\frac{\sqrt{4+h}+2}{\sqrt{4+h}+2}$$

$$= \lim_{h\to 0}\frac{4+h-4}{h(\sqrt{4+h}+2} = \lim_{h\to 0}\frac{1}{\sqrt{4+h}+2} = \frac{1}{2+2} = \frac{1}{4}$$

(e) $$\lim_{x\to 0+}\frac{\sin x}{\sqrt{x}} = \lim_{x\to 0+}\frac{\sin x}{x}\cdot\sqrt{x} = \lim_{x\to 0-}\frac{\sin x}{x}\cdot\lim_{x\to 0+}\sqrt{x} = 1\cdot 0 = 0.$$

Note that in (c), (d), and (e) if we use the theorems on limits indiscriminately we obtain the so-called *indeterminate forms* ∞/∞ and 0/0. To avoid such predicaments, note that in each case the form of the limit is suitably modified. For other methods of evaluating limits, see Chapter 4.

Continuity

(Assume that values at which continuity is to be demonstrated are interior domain values unless otherwise stated.)

3.21. Prove that $f(x) = x^2$ is continuous at $x = 2$.

Method 1: By Problem 3.10, $\lim_{x\to 2} f(x) = f(2) = 4$ and so $f(x)$ is continuous at $x = 2$.

Method 2: We must show that, given any $\epsilon > 0$, we can find $\delta > 0$ (depending on ϵ) such that $|f(x) - f(2)| = |x^2 - 4| < \epsilon$ when $|x - 2| < \delta$. The proof patterns are given in Problem 3.10.

3.22. (a) Prove that $f(x) = \begin{cases} x\sin 1/x, & x \neq 0 \\ 5, & x = 0 \end{cases}$ is not continuous at $x = 0$. (b) Can we redefine $f(0)$ so that $f(x)$ is continuous at $x = 0$?

(a) From Problem 3.13, $\lim_{x\to 0} f(x) = 0$. But this limit is not equal to $f(0) = 5$, so $f(x)$ is discontinuous at $x = 0$.

(b) By redefining $f(x)$ so that $f(0) = 0$, the function becomes continuous. Because the function can be made continuous at a point simply by redefining the function at the point, we call the point a *removable discontinuity*.

3.23. Is the function $f(x) = \dfrac{2x^4 - 6x^3 + x^2 + 3}{x - 1}$ continuous at $x = 1$?

$f(1)$ does not exist, so $f(x)$ is not continuous at $x = 1$. By redefining $f(x)$ so that $f(1) = \lim_{x\to 1} f(x) = -8$ (see Problem 3.11), it becomes continuous at $x = 1$; i.e., $x = 1$ is a removable discontinuity.

3.24. Prove that if $f(x)$ and $g(x)$ are continuous at $x = x_0$, so also are (a) $f(x)$, + $g(x)$, (b) $f(x)g(x)$, and (c) $\dfrac{f(x)}{g(x)}$ if $f(x_0) \neq 0$.

These results follow at once from the proofs given in Problem 3.19 by taking $A = f(x_0)$ and $B = g(x_0)$ and rewriting $0 < |x - x_0| < \delta$ as $|x - x_0| < \delta$, i.e., *including* $x = x_0$.

3.25. Prove that $f(x) = x$ is continuous at any point $x = x_0$.

We must show that, given any $\epsilon > 0$, we can find $\delta > 0$ such that $|f(x) - f(x_0)| = |x - x_0| < \epsilon$ when $|x - x_0| < \delta$. By choosing $\delta = \epsilon$, the result follows at once.

3.26. Prove that $f(x) = 2x^3 + x$ is continuous at any point $x = x_0$.

Since x is continuous at any point $x = x_0$ (Problem 3.25), so also is $x \cdot x = x^2$, $x^2 \cdot x = x^3$, $2x^3$, and, finally, $2x^3 + x$, using the theorem (Problem 3.24) that sums and products of continuous functions are continuous.

3.27. Prove that if $f(x) = \sqrt{x-5}$ for $5 \leqq x \leqq 9$, then $f(x)$ is continuous in this interval.

If x_0 is any point such that $5 < x_0 < 9$, then $\lim_{x \to x_0} f(x) = \lim_{x \to x_0} \sqrt{x-5} = \sqrt{x_0 - 5} = f(x_0)$. Also, $\lim_{x \to 5+} \sqrt{x-5} = 0 = f(5)$ and $\lim_{x \to 9-} \sqrt{x-5} = 2\ f(9)$. Thus the result follows.

Here we have used the result that $\lim_{x \to x_0} \sqrt{f(x)} = \sqrt{\lim_{x \to x_0} f(x)} = \sqrt{f(x_0)}$ if $f(x)$ is continuous at x_0. An ϵ, δ proof, directly from the definition, can also be employed.

3.28. For what values of x in the domain of definition is each of the following functions continuous?

(a) $f(x) = \dfrac{1}{x^2 - 1}$

(b) $f(x) = \dfrac{1 + \cos x}{3 + \sin x}$

(c) $f(x) = \dfrac{1}{\sqrt[4]{10 + 4}}$

(d) $f(x) = 10^{-1/(x-3)2}$

(e) $f(x) = \begin{cases} 10^{-1(x-3)^2}, & x \neq 3 \\ 0, & x = 3 \end{cases}$

(f) $f(x) = \dfrac{x - |x|}{x}$

(g) $f(x) = \begin{cases} \dfrac{x - |x|}{x}, & x < 0 \\ 2, & x = 0 \end{cases}$

(h) $f(x) = x \csc x = \dfrac{x}{\sin x}$.

(i) $f(x) = x \csc x$, $f(0) = 1$.

(a) All x except $x = \pm 1$ (where the denominator is zero)

(b) All x

(c) All $x > -10$

(d) All $x \neq 3$ (see Problem 3.55)

(e) All x, since $\lim_{x \to 3} f(x) = f(3)$

(f) If $x > 0$, $f(x) = \dfrac{x - x}{x} = 0$. If $x < 0$, $f(x) = \dfrac{x + x}{x} = 2$. At $x = 0$, $f(x)$ is undefined. Then $f(x)$ is continuous for all x except $x = 0$.

(g) As in (f), $f(x)$ is continuous for $x < 0$. Then, since

$$\lim_{x \to 0-} \frac{x - |x|}{x} = \lim_{x \to 0-} \frac{x + x}{x} = \lim_{x \to 0-} 2 = 2 = f(0)$$

it follows that $f(x)$ is continuous (from the left) at $x = 0$.

Thus, $f(x)$ is continuous for all $x \leqq 0$, i.e., everywhere in its domain of definition.

(h) All x except $0, \pm\pi, \pm 2\pi, \pm 3\pi, \ldots$

(i) Since $\lim_{x\to 0} x \csc x = \lim_{x\to 0} \dfrac{x}{\sin x} = 1 = f(0)$, we see that $f(x)$ is continuous for all x except $\pm\pi$, $\pm 2\pi$, $\pm 3\pi, \ldots$ [compare (*h*)].

Uniform continuity

3.29. Prove that $f(x) = x^2$ is uniformly continuous in $0 < x < 1$.

Method 1: Using definition.

We must show that, given any $\epsilon > 0$, we can find $\delta > 0$ such that $|x^2 - x_0^2| < \epsilon$ when $|x - x_0| < \delta$, where δ depends *only* on ϵ and *not* on x_0 where $0 < x_0 < 1$.

If x and x_0 are any points in $0 < x < 1$, then

$$|x^2 - x_0^2| = |x + x_0|\,|x - x_0| < 2|x - x_0|$$

Thus, if $|x - x_0| < \delta$, it follows that $|x^2 - x^2{}_0| < 2\delta$. Choosing $\delta = \epsilon/2$, we see that $|x^2 - x_0^2| < \epsilon$ when $|x - x_0| < \delta$, where δ depends only on ϵ and not on x_0 Hence, $f(x) = x^2$ is uniformly continuous in $0 < x < 1$.

This can be used to prove that $f(x) = x^2$ is uniformly continuous in $0 \leqq x \leqq 1$.

Method 2: The function $f(x) = x^2$ is continuous in the closed interval $0 \leqq x \leqq 1$. Hence, by the theorem on Page 48, it is uniformly continuous in $0 \leqq x \leqq 1$ and thus in $0 < x < 1$.

3.30. Prove that $f(x) = 1/x$ is not uniformly continuous in $0 < x < 1$.

Method 1: Suppose $f(x)$ is uniformly continuous in the given interval. Then, for any $\epsilon > 0$ we should be able to find δ, say, between 0 and 1, such that $|f(x) - f(x_0)| < \epsilon$ when $|x - x_0| < \delta$ for all x and x_0 in the interval.

Let $x = \delta$ and $x_0 = \dfrac{\delta}{1+\varepsilon}$. Then $|x - x_0| = \left|\delta - \dfrac{\delta}{1+\varepsilon}\right| = \dfrac{\varepsilon}{1+\varepsilon}\delta < \delta$.

However, $\left|\dfrac{1}{x} - \dfrac{1}{x_0}\right| = \left|\dfrac{1}{\delta} - \dfrac{1+\varepsilon}{\delta}\right| = \dfrac{\varepsilon}{\delta} > \varepsilon$ (since $0 < \delta < 1$).

Thus, we have a contradiction, and it follows that $f(x) = 1/x$ cannot be uniformly continuous in $0 < x < 1$.

Method 2: Let x_0 and $x_0 + \delta$ be any two points in (0, 1). Then,

$$|f(x_0) - f(x_0 + \delta)| = \left|\frac{1}{x_0} - \frac{1}{x_0 + \delta}\right| = \frac{\delta}{x_0(x_0 + \delta)}$$

can be made larger than any positive number by choosing x_0 sufficiently close to 0. Hence, the function cannot be uniformly continuous.

Miscellaneous problems

3.31. If $y = f(x)$ is continuous at $x = x_0$, and $z = g(y)$ is continuous at $y = y_0$ where $y_0 = f(x_0)$, prove that $z = g\{f(x)\}$ is continuous at $x = x_0$.

Let $h(x) = g\{f(x)\}$. Since, by hypothesis, $f(x)$ and $g(y)$ are continuous at x_0 and y_0, respectively, we have

$$\lim_{x\to x_0} f(x) = f(\lim_{x\to x_0} x) = f(x_0)$$

$$\lim_{y\to y_0} g(y) = g(\lim_{y\to y_0} y) = g(y_0) = g\{f(x_0)\}$$

Then

$$\lim_{x\to x_0} h(x) = \lim_{x\to x_0} g\{f(x)\} = g\{\lim_{x\to x_0} f(x)\} = g\{f(x_0)\} = h(x_0)$$

which proves that $h(x) = g\{f(x)\}$ is continuous at $x = x_0$.

3.32. Prove Theorem 8, Page 52.

Suppose that $f(a) < 0$ and $f(b) > 0$. Since $f(x)$ is continuous, there must be an interval $(a, a + h)$, $h > 0$, for which $f(x) < 0$. The set of points $(a, a + h)$ has an upper bound and so has a least upper bound, which we call c. Then $f(c) \leqq 0$. Now we cannot have $f(c) < 0$, because if $f(c)$ were negative we would be able to find an interval about c (including values greater than c) for which $f(x) < 0$; but since c is the least upper bound, this is impossible, and so we must have $f(c) = 0$ as required.

If $f(a) > 0$ and $f(b) < 0$, a similar argument can be used.

3.33. (a) Given $f(x) = 2x^3 - 3x^2 + 7x - 10$, evaluate $f(1)$ and $f(2)$. (b) Prove that $f(x) = 0$ for some real number x such that $1 < x < 2$. (c) Show how to calculate the value of x in (b).

(a) $f(1) = 2(1)^3 - 3(1)^2 + 7(1) - 10 = -4$, $f(2) = 2(2)^3 - 3(2)^2 + 7(2) - 10 = 8$.

(b) If $f(x)$ is continuous in $a \leqq x \leqq b$ and if $f(a)$ and $f(b)$ have opposite signs, then there is a value of x between a and b such that $f(x) = 0$ (Problem 3.32).

To apply this theorem, we need only realize that the given polynomial is continuous in $1 \leqq x \leqq 2$, since we have already shown in *(a)* that $f(1) < 0$ and $f(2) > 0$. Thus, there *exists* a number c between 1 and 2 such that $f(c) = 0$.

(c) $f(1.5) = 2(1.5)^3 - 3(1.5)^2 + 7(1.5) - 10 = 0.5$. Then, applying the theorem of *(b)* again, we see that the required root lies between 1 and 1.5 and is "most likely" closer to 1.5 than to 1, since $f(1.5) = 0.5$ has a value closer to 0 than $f(1) = -4$ (this is not always a valid conclusion but is worth pursuing in practice).

Thus, we consider $x = 1.4$. Since $f(1.4) = 2(1.4)^3 - 3(1.4)^2 + 7(1.4) - 10 = -0.592$, we conclude that there is a root between 1.4 and 1.5 which is most likely closer to 1.5 than to 1.4.

Continuing in this manner, we find that the root is 1.46 to 2 decimal places.

3.34. Prove Theorem 10, Page 52.

Given any $\epsilon > 0$, we can find x such that $M - f(x) < \epsilon$ by definition of the l.u.b. M.

Then $\dfrac{1}{M - f(x)} > \dfrac{1}{\varepsilon}$, so that $\dfrac{1}{M - f(x)}$ is not bounded and, hence, cannot be continuous in view of Theorem 4, Page 52. However, if we suppose that $f(x) \neq M$, then, since $M - f(x)$ is continuous, by hypothesis we must have $\dfrac{1}{M - f(x)}$ also continuous. In view of this contradiction, we must have $f(x) = M$ for at least one value of x in the interval.

Similarly, we can show that there exists an x in the interval such that $f(x) = m$ (Problem 3.93).

SUPPLEMENTARY PROBLEMS

Functions

3.35. Give the largest domain of definition for which each of the following rules of correspondence supports the the construction of a function.

(a) $\sqrt{(3 - x)(2x + 4)}$ (b) $(x - 2)/(x^2 - 4)$ (c) $\sqrt{\sin 3x}$ (d) $\log_{10}(x^3 - 3x^2 - 4x + 12)$

Ans. (a) $-2 \leqq x \leqq 3$ (b) all $x \neq \pm 2$ (c) $2m\pi/3 \leqq x \leqq (2m + 1)\pi/3$, $m = 0, \pm 1, \pm 2, \ldots$ (d) $x > 3$, $-2 < x < 2$

3.36. If $f(x) = \dfrac{3x + 1}{x - 2}$, $x \neq 2$, find:

(a) $\dfrac{5f(-1)-2f(0)+3f(5)}{6}$ (b) $\left\{f\left(-\dfrac{1}{2}\right)\right\}^2$ (c) $f(2x-3)$ (d) $f(x)+f(4/x),\ x \neq 0$

(e) $\dfrac{f(h)-f(0)}{h}\ h \neq 0$ (f) $f(\{f(x)\}$

Ans. (a) $\dfrac{61}{81}$ (b) $\dfrac{1}{25}$ (c) $\dfrac{6x-8}{2x-5},\ x \neq 0,\ \dfrac{5}{2},\ 2$ (d) $\dfrac{5}{2},\ x \neq 0, 2$ (e) $\dfrac{7}{2h-4},\ h \neq 0, 2$ (f) $\dfrac{10x+1}{x+5},\ x \neq -5, 2$

3.37. If $f(x) = 2x^2,\ 0 < x \leqq 2$, find (*a*) the l.u.b. and (*b*) the g.l.b. of $f(x)$. Determine whether $f(x)$ attains its l.u.b. and g.l.b.

Ans. (a) 8 (b) 0

3.38. Construct a graph for each of the following functions.

(a) $f(x) = |x|,\ -3 \leqq x \leqq 3$

(b) $f(x) = 2 - \dfrac{|x|}{x},\ -2 \leqq x \leqq 2$

(c) $f(x) = \begin{cases} 0, & x < 0 \\ \dfrac{1}{2}, & x = 0 \\ 1, & x > 0 \end{cases}$

(d) $f(x) = \begin{cases} -x, & -2 \leqq x \leqq 0 \\ x, & 0 \leqq x \leqq 2 \end{cases}$

(e) $f(x) = x^2 \sin 1/x,\ x \neq 0$

(*f*) $\dfrac{x-[x]}{x}$ where $[x]$ = greatest integer $\leqq x$

(*g*) $f(x) = \cosh x$

(*h*) $f(x) = \dfrac{\sin x}{x}$

(*i*) $f(x) = \dfrac{x}{(x-1)(x-2)(x-3)}$

(*j*) $f(x) = \dfrac{\sin^2 x}{x^2}$

3.39. Construct graphs for (*a*) $x^2/a^2 + y^2/b^2 = 1$, (*b*) $x^2/a^2 - y^2/b^2 = 1$, (*c*) $y^2 = 2px$, and (*d*) $y = 2ax - x^2$, where *a*, *b*, and *p* are given constants. In which cases, when solved for *y*, is there exactly one value of *y* assigned to each value of *x*, thus making possible definitions of functions *f* and enabling us to write $y = f(x)$? In which cases must branches be defined?

3.40. (*a*) From the graph of $y = \cos x$, construct the graph obtained by interchanging the variables and from which $\cos^{-1} x$ will result by choosing an appropriate branch. Indicate possible choices of a principal value of $\cos^{-1} x$. Using this choice, find $\cos^{-1}(1/2) - \cos^{-1}(-1/2)$. Does the value of this depend on the choice? Explain.

3.41. Work parts (a) and (b) of Problem 3.40 for (a) $y = \sec^{-1} x$ and (b) $y = \cot^{-1} x$.

3.42. Given the graph for $y = f(x)$, show how to obtain the graph for $y = f(ax + b)$, where *a* and *b* are given constants. Illustrate the procedure by obtaining the graphs of (a) $y = \cos 3x$, (b) $y = \sin(5x + \pi/3)$, and (c) $y = \tan(\pi/6 - 2x)$.

3.43. Construct graphs for (a) $y = e^{-|x|}$, (b) $y = \ln |x|$, and (c) $y = e^{-|x|} \sin x$.

3.44. Using the conventional principal values on Pages 45 and 46, evaluate:

(a) $\sin^{-1}(-\sqrt{3}/2)$

(b) $\tan^{-1}(1) - \tan^{-1}(-1)$

(c) $\cot^{-1}(1/\sqrt{3}) - \cot^{-1}(-1/\sqrt{3})$

(d) $\cosh^{-1}\sqrt{2}$

(e) $e^{-\coth^{-1}(25/7)}$

(*f*) $\sin^{-1}x + \cos^{-1}x$, $-1 \leq x \leq 1$

(*g*) $\sin^{-1}(\cos 2x)$, $0 \leq x \leq \pi/2$

(*h*) $\sin^{-1}(\cos 2x)$, $\pi/2 \leq x \leq 3\pi/2$

(*i*) $\tanh(\operatorname{csch}^{-1} 3x)$, $x \neq 0$

(*j*) $\cos(2\tan^{-1}x^2)$

Ans. (a) $-\pi/3$ (b) $\pi/2$ (c) $-\pi/3$ (d) $\ln(1+\sqrt{2})$ (e) $\frac{3}{4}$ (f) $\pi/2$ (g) $\pi/2 - 2x$ (h) $2x - 3\pi/2$ (i) $\frac{|x|}{x\sqrt{9x^2+1}}$ (j) $\frac{1-x^4}{1+x^4}$

3.45. Evaluate (a) $\cos\{\pi \sinh(\ln 2)\}$ and (b) $\cosh^{-1}\{\coth(\ln 3)\}$.

Ans. (a) $-\sqrt{2}/2$ (b) $\ln 2$

3.46. (a) Prove that $\tan^{-1}x + \cot^{-1}x = \pi/2$ if the conventional principal values on Pages 45 and 46 are taken. (b) Is $\tan^{-1}x + \tan^{-1}(1/x) = \pi/2$ also? Explain.

3.47. If $f(x) = \tan^{-1}x$, prove that $f(x) + f(y) = f\left(\frac{x+y}{1-xy}\right)$, discussing the case $xy = 1$.

3.48. Prove that $\tan^{-1}a - \tan^{-1}b = \cot^{-1}b - \cot^{-1}a$.

3.49. Prove the identities: (a) $1 - \tanh^2 x = \operatorname{sech}^2 x$, (b) $\sin 3x = 3\sin x - 4\sin^3 x$, (c) $\cos 3x = 4\cos^3 x - 3\cos x$, (d) $\tanh\frac{1}{2}x = (\sinh x)/(1+\cosh x)$, and (e) $\ln|\csc x - \cot x| = \ln\left|\tan\frac{1}{2}x\right|$.

3.50. Find the relative and absolute maxima and minima of (a) $f(x) = (\sin x)/x$, $f(0) = 1$ and (b) $f(x) = (\sin^2 x)/x^2$, $f(0) = 1$. Discuss the cases when $f(0)$ is undefined or $f(0)$ is defined but $\neq 1$.

Limits

3.51. Evaluate the following limits, first by using the definition and then by using theorems on limits.

(a) $\lim_{x\to 3}(x^2 - 3x + 2)$

(b) $\lim_{x\to -1}\frac{1}{2x-5}$

(c) $\lim_{x\to 2}\frac{x^2-4}{x-2}$

(d) $\lim_{x\to 4}\frac{\sqrt{x}-2}{4-x}$

(e) $\lim_{h\to 0}\frac{(2+h)^4-16}{h}$

(f) $\lim_{x\to 1}\frac{\sqrt{x}}{x+1}$

Ans. (a) 2 (b) $-\frac{1}{7}$ (c) 4 (d) $-\frac{1}{4}$ (e) 32 (f) $\frac{1}{2}$

3.52. Let $f(x)=\begin{cases}3x-1, & x<0\\ 0, & x=0.\\ 2x+5, & x>0\end{cases}$

(a) Construct a graph of $f(x)$. Evaluate (b) $\lim\limits_{x\to 2} f(x)$, (c) $\lim\limits_{x\to -3} f(x)$, (d) $\lim\limits_{x\to 0+} f(x)$, (e) $\lim\limits_{x\to 0-} f(x)$, and (f) $\lim\limits_{x\to 0} f(x)$, justifying your answer in each case.

Ans. (b) 9 (c) –10 (d) 5 (e) –1 (f) does not exist

3.53. Evaluate (a) $\lim\limits_{h\to 0+}\dfrac{f(h)-f(0+)}{h}$ and (b) $\lim\limits_{h\to 0-}\dfrac{f(h)-f(0-)}{h}$, where $f(x)$ is the function of Problem 3.52.

Ans. (a) 2 (b) 3

3.54. (a) If $f(x)=x^2\cos 1/x$, evaluate $\lim\limits_{x\to 0} f(x)$, justifying your answer. (b) Does your answer to (a) still remain the same if we consider $f(x)=x^2\cos 1/x$, $x\neq 0$, $f(0)=2$? Explain.

3.55. Prove that $\lim\limits_{x\to 3} 10^{-1/(x-3)^2}=0$, using the definition.

3.56. Let $f(x)=\dfrac{1+10^{-1/x}}{2-10^{-1/x}}$, $x\neq 0$, $f(0)=\dfrac{1}{2}$. Evaluate (a) $\lim\limits_{x\to 0+} f(x)$, (b) $\lim\limits_{x\to 0-} f(x)$, and (c) $\lim\limits_{x\to 0} f(x)$, justifying answers in all cases.

Ans. (a) $\frac{1}{2}$ (b) – 1 (c) does not exist.

3.57. Find (a) $\lim\limits_{x\to 0+}\dfrac{|x|}{x}$ and (b) $\lim\limits_{x\to 0-}\dfrac{|x|}{x}$. Illustrate your answers graphically.

Ans. (a) 1 (b) –1

3.58. If $f(x)$ is the function defined in Problem 3.56, does $\lim\limits_{x\to 0} f(|x|)$ exist? Explain.

3.59. Explain *exactly* what is meant when you write:

(a) $\lim\limits_{x\to 3}\dfrac{2-x}{(x-3)^2}=-\infty$ (b) $\lim\limits_{x\to 0+}(1-e^{1/x})=-\infty$ (c) $\lim\limits_{x\to\infty}\dfrac{2x+5}{3x-2}=\dfrac{2}{3}$

3.60. Prove that (a) $\lim\limits_{x\to\infty} 10^{-x}=0$ and (b) $\lim\limits_{x\to -\infty}\dfrac{\cos x}{x+\pi}=0$.

3.61. Explain why (a) $\lim\limits_{x\to\infty}\sin x$ does not exist and (b) $\lim\limits_{x\to\infty} e^{-x}\sin x$ does not exist.

3.62. If $f(x)=\dfrac{3x+|x|}{7x-5|x|}$, evaluate (a) $\lim\limits_{x\to\infty} f(x)$ (b) $\lim\limits_{x\to -\infty} f(x)$, (c) $\lim\limits_{x\to 0+} f(x)$, (d) $\lim\limits_{x\to 0-} f(x)$, and (e) $\lim\limits_{x\to 0} f(x)$.

Ans. (a) 2 (b) $\frac{1}{6}$ (c) 2 (d) $\frac{1}{6}$ (e) does not exist

3.63. If $[x]$ = largest integer $\leqq x$, evaluate (a) $\lim_{x\to 2+} \{x - [x]\}$ and (b) $\lim_{x\to 2-} \{x - [x]\}$.

Ans. (a) 0 and (b) 1

3.64. If $\lim_{x\to x_0} f(x) = A$, prove that (a) $\lim_{x\to x_0} \{f(x)\}^2 = A^2$ and (b) $\lim_{x\to x_0} \sqrt[3]{f(x)} = \sqrt[3]{A}$. What generalizations of these do you suspect are true? Can you prove them?

3.65. If $\lim_{x\to x_0} f(x), = A$ and $\lim g(x), = B$, prove that (a) $\lim_{x\to x_0} \{f(x) - g(x)\} = A - B$ and (b) $\lim_{x\to x_0} \{af(x) + bg(x)\} = aA + bB$, where a, b = any constants.

3.66. If the limits of $f(x)$, $g(x)$, and $h(x)$ are A, B, and C respectively, prove that (a) $\lim_{x\to x_0} \{f(x) + g(x) + h(x)\} = A + B + C$ and (b) $\lim_{x\to x_0} f(x)g(x)h(x) = ABC$. Generalize these results.

3.67. Evaluate each of the following using the theorems on limits.

(a) $\lim_{x\to 1/2} \left\{\dfrac{2x^2 - 1}{(3x+2)(5x-3)} - \dfrac{2-3x}{x^2 - 5x + 3}\right\}$

(b) $\lim_{x\to\infty} \dfrac{(3x-1)(2x+3)}{(5x-3)(4x+5)}$

(c) $\lim_{x\to -\infty} \left(\dfrac{3x}{x-1} - \dfrac{2x}{x+1}\right)$

(d) $\lim_{x\to 1} \dfrac{1}{x-1}\left(\dfrac{1}{x+3} - \dfrac{2x}{3x+5}\right)$

Ans. (a) −8/21 (b) 3/10 (c) 1 (d) 1/32

3.68. Evaluate $\lim_{h\to 0} \dfrac{\sqrt[3]{8+h} - 2}{h}$. (Hint: Let $8 + h = x^3$.)

Ans. 1/12

3.69. If $\lim_{x\to x_0} f(x) = A$ and $\lim_{x\to x_0} g(x) = B \neq 0$, prove directly that $\lim_{x\to x_0} \dfrac{f(x)}{g(x)} = \dfrac{A}{B}$.

3.70. Given $\lim_{x\to 0} \dfrac{\sin 3x}{x} = 1$, evaluate:

(a) $\lim_{x\to 0} \dfrac{\sin 3x}{x}$

(b) $\lim_{x\to 0} \dfrac{1 - \cos x}{x}$

(c) $\lim_{x\to 0} \dfrac{1 - \cos x}{x^2}$

(d) $\lim_{x\to 3} (x-3)\csc \pi x$

(e) $\lim_{x\to 0} \dfrac{6x - \sin 2x}{2x + 3\sin 4x}$

(f) $\lim_{x\to 0} \dfrac{\cos ax - \cos bx}{x^2}$

(g) $\lim_{x\to 0} \dfrac{1 - 2\cos x + \cos 2x}{x^2}$

(h) $\lim_{x\to 1} \dfrac{3\sin \pi x - \sin 3\pi x}{x^3}$

Ans. (a) 3 (b) 0 (c) $\dfrac{1}{2}$ (d) $-1/\pi$ (e) $\dfrac{2}{7}$ (f) $\dfrac{1}{2}(b^2 - a^2)$ (g) −1 (h) $4\pi^3$

3.71. If $\lim_{x\to 0}\frac{e^x - 1}{x} = 1$, prove that (a) $\lim_{x\to 0}\frac{e^{-ax} - e^{-bx}}{x} = b - a$, (b) $\lim_{x\to 0}\frac{a^x - b^x}{x} = \ln\frac{a}{b}$, $a, b > 0$, and (c) $\lim_{x\to 0}\frac{\tanh ax}{x} = a$.

3.72. Prove that $\lim_{x\to x_0} f(x) = l$ if and only if $\lim_{x\to x_0^+} f(x) = l$.

Continuity

In the following problems, assume the largest possible domain unless otherwise stated.

3.73. Prove that $f(x) = x^2 - 3x + 2$ is continuous at $x = 4$.

3.74. Prove that $f(x) = 1/x$ is continuous (a) at $x = 2$ and (b) in $1 \leqq x \leqq 3$.

3.75. Investigate the continuity of each of the following functions at the indicated points:

(a) $f(x) = \frac{\sin x}{x}$; $x \neq 0$, $f(0) = 0$; $x = 0$ (c) $f(x) = \frac{x^3 - 8}{x^2 - 4}$; $x \neq 2$, $f(2) = 3$; $x = 2$

(b) $f(x) = x - |x|$; $x = 0$ (d) $f(x) = \begin{cases} \sin \pi x, & 0 < x < 1 \\ \ln & 1 < x < 2 \end{cases}$; $x = 1$

Ans. (a) discontinuous, (b) continuous, (c) continuous, (d) discontinuous

3.76. If $[x]$ = greatest integer $\leqq x$, investigate the continuity of $f(x) = x - [x]$ in the interval (a) $1 < x < 2$ and (b) $1 \leqq x \leqq 2$.

3.77. Prove that $f(x) = x^3$ is continuous in every finite interval.

3.78. If $f(x)/g(x)$ and $g(x)$ are continuous at $x = x_0$, prove that $f(x)$ must be continuous at $x = x_0$.

3.79. Prove that $f(x) = (\tan^{-1} x)/x$, $f(0) = 1$ is continuous at $x = 0$.

3.80. Prove that a polynomial is continuous in every finite interval.

3.81. If $f(x)$ and $g(x)$ are polynomials, prove that $f(x)/g(x)$ is continuous at each point $x = x_0$ for which $g(x_0) \neq 0$.

3.82. Give the points of discontinuity of each of the following functions.

(a) $f(x) = \frac{x}{(x-2)(x-4)}$ (c) $f(x) = \sqrt{(x-3)(6-x)}$, $3 \leqq x \leqq 6$

(b) $f(x) = x^2 \sin 1/x$, $x \neq 0$, $f(0) = 0$ (d) $f(x) = \frac{1}{1 + 2\sin x}$

Ans. (a) $x = 2, 4$ (b) none (c) none (d) $x = 7\pi/6 \pm 2m\pi$, $11\pi/6 \pm 2m\pi$, $m = 0, 1, 2, \ldots$

Uniform continuity

3.83. Prove that $f(x) = x^3$ is uniformly continuous in (a) $0 < x < 2$ (b) $0 \leqq x \leqq 2$ and (c) any finite interval.

3.84. Prove that $f(x) = x^2$ is not uniformly continuous in $0 < x < \infty$.

3.85. If a is a constant, prove that $f(x) = 1/x^2$ is (a) continuous in $a < x < \infty$ if $a \geqq 0$, (b) uniformly continuous in $a < x < \infty$ if $a > 0$, and (c) not uniformly continuous in $0 < x < 1$.

3.86. If $f(x)$ and $g(x)$ are uniformly continuous in the same interval, prove that (*a*) $f(x) \pm g(x)$ and (*b*) $f(x)\, g(x)$ are uniformly continuous in the interval. State and prove an analogous theorem for $f(x)/g(x)$.

Miscellaneous problems

3.87. Give an "ϵ, δ" proof of the theorem of Problem 3.31.

3.88. (a) Prove that the equation $\tan x = x$ has a real positive root in each of the intervals $\pi/2 < x < 3\pi/2$, $3\pi/2 < x < 5\pi/2$, $5\pi/2 < x < 7\pi/2, \ldots$

(b) Illustrate the result in (a) graphically by constructing the graphs of $y = \tan x$ and $y = x$ and locating their points of intersection.

(c) Determine the value of the smallest positive root of $\tan x = x$.

Ans. (c) 4.49 approximately

3.89. Prove that the only real solution of $\sin x = x$ is $x = 0$.

3.90. (a) Prove that $\cos x \cosh x + 1 = 0$ has infinitely many real roots. (b) Prove that for large values of x, the roots approximate those of $\cos x = 0$.

3.91. Prove that $\lim\limits_{x \to 0} \dfrac{x^2 \sin(1/x)}{\sin x} = 0$.

3.92. Suppose $f(x)$ is continuous at $x = x_0$ and assume $f(x_0) > 0$. Prove that there exists an interval $(x_0 - h, x_0 + h)$, where $h > 0$, in which $f(x) > 0$. (See Theorem 5, page 52.) (Hint: Show that we can make $|f(x) - f(x_0)| < \frac{1}{2} f(x_0)$. Then show that $f(x) \geqq f(x_0) - |f(x) - f(x_0)| > \frac{1}{2} f(x_0) > 0$.)

3.93. (*a*) Prove Theorem 10, Page 52, for the greatest lower bound m (see Problem 3.34). (*b*) Prove Theorem 9, Page 52, and explain its relationship to Theorem 10.

CHAPTER 4

Derivatives

The Concept and Definition of a Derivative

Concepts that shape the course of mathematics are few and far between. The derivative, the fundamental element of the differential calculus, is such a concept. That branch of mathematics called analysis, of which advanced calculus is a part, is the end result. There were two problems that led to the discovery of the derivative. The older one of defining and representing the tangent line to a curve at one of its points had concerned early Greek philosophers. The other problem of representing the instantaneous velocity of an object whose motion was not constant was much more a problem of the seventeenth century. At the end of that century, these problems and their relationship were resolved. As is usually the case, many mathematicians contributed, but it was Isaac Newton and Gottfried Wilhelm Leibniz who independently put together organized bodies of thought upon which others could build.

The tangent problem provides a visual interpretation of the derivative and can be brought to mind no matter what the complexity of a particular application. It leads to the definition of the derivative as the limit of a difference quotient in the following way. (See Figure 4.1.)

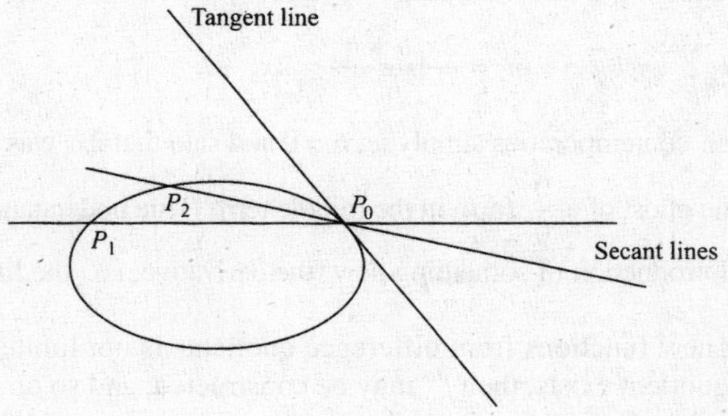

Figure 4.1

Let P_0 (x_0) be a point on the graph of $y = f(x)$. Let $P(x)$ be a nearby point on this same graph of the function f. Then the line through these two points is called a *secant line*. Its slope, m_s, is the difference quotient

$$m_s = \frac{f(x) - f(x_0)}{x - x_0} = \frac{\Delta y}{\Delta x} \tag{1}$$

where Δx and Δy are called the increments in x and y, respectively. Also this slope may be written

$$m_s = \frac{f(x_0 + h) - f(x_0)}{h} \tag{2}$$

where $h = x - x_0 = \Delta x$. See Figure 4.2.

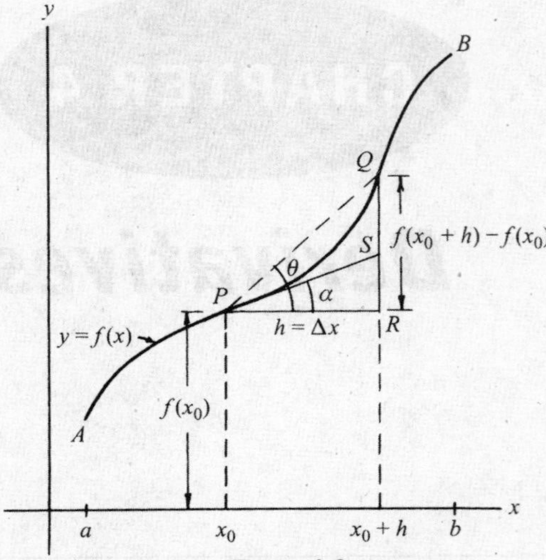

Figure 4.2

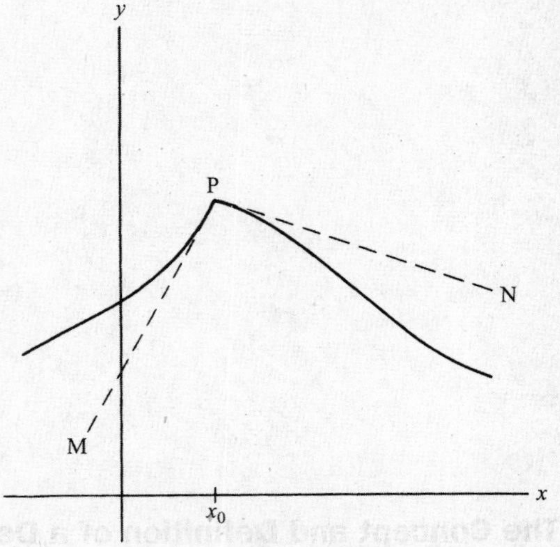

Figure 4.3

We can imagine a sequence of lines formed as $h \to 0$. It is the limiting line of this sequence that is the natural one to be the tangent line to the graph at P_0.

To make this mode of reasoning precise, the limit (when it exists), is formed as follows:

$$f'(x) = \lim_{h \to 0} \frac{f(x_0 + h) - f(x_0)}{h} \tag{3a}$$

As indicated, this limit is given the name $f'(x_0)$. It is called the *derivative* of the function f at its domain value x_0. If this limit can be formed at each point of a subdomain of the domain of f, then f is said to be *differentiable* on that subdomain and a new function f' has been constructed.

This limit concept was not understood until the middle of the nineteenth century. A simple example illustrates the conceptual problem that faced mathematicians from 1700 until that time. Let the graph of f be the parabola $y = x^2$; then a little algebraic manipulation yields

$$m_s = \frac{2x_0 h + h^2}{h} = 2x_0 + h \tag{3b}$$

Newton, Leibniz, and their contemporaries simply let $h = 0$ and said that $2x_0$ was the slope of the tangent line at P_0. However, this raises the ghost of a $\frac{0}{0}$ form in the middle term. True understanding of the calculus is in the comprehension of how the introduction of something new (the derivative, i.e., the limit of a difference quotient) resolves this dilemma.

Note 1: The creation of new functions from difference quotients is not limited to f'. If, starting with f', the limit of the difference quotient exists, then f'' may be constructed, and so on.

Note 2: Since the continuity of a function is such a strong property, one might think that differentiability followed. This is not necessarily true, as is illustrated in Figure 4.3.

The following theorem puts the matter in proper perspective.

Theorem: If f is differentiable at a domain value, then it is continuous at that value.

As indicated, the converse of this theorem is not true.

Right- and Left-Hand Derivatives

The status of the derivative at endpoints of the domain of f, and in other special circumstances, is clarified by the following definitions.

The *right-hand derivative* of $f(x)$ at $x = x_0$ is defined as

$$f'_+(x_0) = \lim_{h\to 0+} \frac{f(x_0 + h) - f(x_0)}{h} \tag{5}$$

if this limit exists. Note that in this case $h(= \Delta x)$ is restricted only to positive values as it approaches zero.

Similarly, the *left-hand derivative* of $f(x)$ at $x = x_0$ is defined as

$$f'_-(x_0) = \lim_{h\to 0-} \frac{f(x_0 + h) - f(x_0)}{h} \tag{6}$$

if this limit exists. In this case h is restricted to negative values as it approaches zero.

A function f has a derivative at $x = x_0$ if and only if $f'_+(x_0) = f'_-(x_0)$.

Differentiability in an Interval

If a function has a derivative at all points of an interval, it is said to be *differentiable in the interval*. In particular, if f is defined in the closed interval $a \leq x \leq b$—i.e. $[a, b]$—then f is differentiable in the interval if and only if $f'(x_0)$ exists for each x_0 such that $a < x_0 < b$ and if both $f'_+(a)$ and $f'_-(b)$ exist.

If a function has a continuous derivative, it is sometimes called *continuously differentiable*.

Piecewise Differentiability

A function is called *piecewise differentiable* or *piecewise smooth* in an interval $a \leq x \leq b$ if $f'(x)$ is piecewise continuous. An example of a piecewise continuous function is shown graphically on Page 47.

An equation for the tangent line to the curve $y = f(x)$ at the point where $x = x_0$ is given by

$$y - f(x_0) = f'(x_0)(x - x_0) \tag{7}$$

The fact that a function can be continuous at a point and yet not be differentiable there is shown graphically in Figure 4.3. In this case there are two tangent lines at *P*, represented by *PM* and *PN*. *PN*. The slopes of these tangent lines are $f'_-(x_0)$ and $f'_+(x_0)$, respectively.

Differentials

Let $\Delta x = dx$ be an increment given to x. Then

$$\Delta y = f(x + \Delta x) - f(x) \tag{8}$$

is called the *increment* in $y = f(x)$. If $f(x)$ is continuous and has a continuous first derivative in an interval, then

$$\Delta y = f'(x)\,\Delta x + \epsilon\Delta x = f'(x)dx + dx \tag{9}$$

where $\epsilon \to 0$ as $\Delta x \to 0$. The expression

$$\text{dy} = f'(x)dx \tag{10}$$

is called the *differential of y or f(x)* or the *principal part of* Δy. Note that $\Delta y \neq dy$, in general. However, if $\Delta x = dx$ is small, then dy is a close approximation of Δy (see Problem 4.11). The quantities dx (called the *differential of x*) and dy need not be small.

Because of the definitions given by Equations (8) and (10), we often write

$$\frac{dy}{dx} = f'(x) = \lim_{\Delta x\to 0} \frac{f(x + \Delta x) - f(x)}{\Delta x} = \lim_{\Delta x\to 0} \frac{\Delta y}{\Delta x} \tag{11}$$

It is emphasized that dx and dy are *not* the limits of Δx and Δy as $\Delta x \to 0$, since these limits are zero, whereas dx and dy are not necessarily zero. Instead, given dx, we determine dy from Equation (10); i.e., dy is a dependent variable determined from the independent variable dx for a given x.

Geometrically, dy is represented in Figure 4.1, for the particular value $x = x_0$ by the line segment SR, whereas Δy is represented by QR.

The geometric interpretation of the derivative as the slope of the tangent line to a curve at one of its points is fundamental to its application. Also of importance is its use as representative of instantaneous velocity in the construction of physical models. In particular, this physical viewpoint may be used to introduce the notion of differentials.

Newton's second and first laws of motion imply that the path of an object is determined by the forces acting on it and that, if those forces suddenly disappear, the object takes on the tangential direction of the path at the point of release. Thus, the nature of the path in a small neighborhood of the point of release becomes of interest. With this thought in mind, consider the following idea.

Suppose the graph of a function f is represented by $y = f(x)$. Let $x = x_0$ be a domain value at which f' exists (i.e., the function is differentiable at that value). Construct a new linear function

$$dy = f'(x_0)\,dx$$

with dx as the (independent) domain variable and dy the range variable generated by this rule. This linear function has the graphical interpretation illustrated in Figure 4.4.

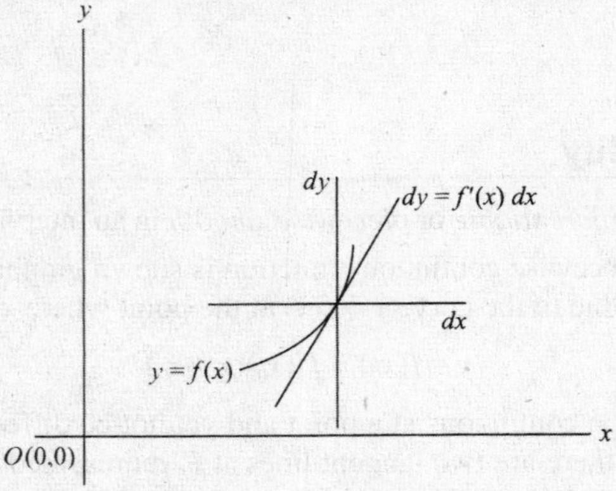

Figure 4.4

That is, a coordinate system may be constructed with its origin at P_0 and the dx- and dy-axes parallel to the x- and y-axes, respectively. In this system our linear equation is the equation of the tangent line to the graph at P_0. It is representative of the path in a small neighborhood of the point, and if the path is that of an object, the linear equation represents its new path when all forces are released.

dx and dy are called differentials of x and y, respectively. Because the preceding linear equation is valid at every point in the domain of f at which the function has a derivative, the subscript may be dropped and we can write

$$dy = f'(x)\,dx$$

The following important observations should be made. $\dfrac{dy}{dx} = f'(x) = \lim_{\Delta x \to 0} \dfrac{f(x+\Delta x) - f(x)}{\Delta x} = \lim_{\Delta x \to 0} \dfrac{\Delta y}{\Delta x}$, thus $\dfrac{dy}{dx}$ is not the same thing as $\dfrac{\Delta y}{\Delta x}$.

On the other hand, dy and Δy are related. In particular, $\lim_{\Delta x \to 0} \dfrac{\Delta y}{\Delta x} = f'(x)$ means that for any $\epsilon > 0$ there exists $\delta > 0$ such that $-\varepsilon < \dfrac{\Delta y}{\Delta x} - \dfrac{dy}{dx} < \varepsilon$ whenever $|\Delta x| < \delta$. Now dx is an independent variable and the axes of x and dx are parallel; therefore, dx may be chosen equal to Δx. With this choice,

$$-\varepsilon \Delta x < \Delta y - dy < \varepsilon \Delta x$$

or

$$dy - \varepsilon\Delta x < \Delta y < dy + \varepsilon\Delta x$$

From this relation we see that dy is an approximation to Δy in small neighborhoods of x, dy is called the *principal part* of Δy.

The representation of f' by $\frac{dy}{dx}$ has an algebraic suggestiveness that is very appealing and appears in much of what follows. In fact, this notation was introduced by Leibniz (without the justification provided by knowledge of the limit idea) and was the primary reason his approach to the calculus, rather than Newton's, was followed.

The Differentiation of Composite Functions

Many functions are a composition of simpler ones. For example, if f and g have the rules of correspondence $u = x^3$ and $y = \sin u$, respectively, then $y = \sin x^3$ is the rule for a composite function $F = g(f)$. The domain of F is that subset of the domain of F whose corresponding range values are in the domain of g. The rule of composite function differentiation is called the *chain rule* and is represented by $\frac{dy}{dx} = \frac{dy}{du}\frac{du}{dx}[F'(x) = g'(u)f'(x)]$.

In the example,

$$\frac{dy}{dx} \equiv \frac{d(\sin x^3)}{dx} = \cos x^3(3x^2 dx)$$

The importance of the chain rule cannot be too greatly stressed. Its proper application is essential in the differentiation of functions, and it plays a fundamental role in changing the variable of integration, as well as in changing variables in mathematical models involving differential equations.

Implicit Differentiation

The rule of correspondence for a function may not be explicit. For example, the rule $y = f(x)$ is *implicit* to the equation $x^2 + 4xy^5 + 7xy + 8 = 0$. Furthermore, there is no reason to believe that this equation can be solved for y in terms of x. However, assuming a common domain (described by the independent variable x), the left-hand member of the equation can be construed as a composition of functions and differentiated accordingly. (The rules for differentiation are listed here for your review.)

In this example, differentiation with respect to x yields

$$2x + 4\left(y^5 + 5xy^4\frac{dy}{dx}\right) + 7\left(y + x\frac{dy}{dx}\right) = 0$$

Observe that this equation can be solved for $\frac{dy}{dx}$ as a function of x and y (but not of x alone).

Rules for Differentiation

If f, g, and h are differentiable functions, the following differentiation rules are valid.

1. $\frac{d}{dx}\{f(x) + g(x)\} = \frac{d}{dx}f(x) + \frac{d}{dx}g(x) = f'(x) + g'(x)$ (Addition rule)

2. $\frac{d}{dx}\{f(x) - g(x)\} = \frac{d}{dx}f(x) - \frac{d}{dx}g(x) = f'(x) - g'(x)$

3. $\frac{d}{dx}\{Cf(x)\} = C\frac{d}{dx}f(x) = Cf'(x)$ where C is any constant

4. $\frac{d}{dx}\{f(x)g(x)\} = f(x)\frac{d}{dx}g(x) + g(x)\frac{d}{dx}f(x) = f(x)g'(x) + g(x)f'(x)$ (Product rule)

5. $\frac{d}{dx}\left\{\frac{f(x)}{g(x)}\right\} = \frac{g(x)\frac{d}{dx}f(x) - f(x)\frac{d}{dx}g(x)}{[g(x)]^2} = \frac{g(x)f'(x) - f(x)g'(x)}{[g(x)]^2}$ if $g(x) \neq 0$ (Quotient rule)

6. If $y = f(u)$ where $u = g(x)$, then

$$\frac{dy}{dx} = \frac{dy}{du}\cdot\frac{du}{dx} = f'(u)\frac{du}{dx} = f'\{g(x)\}g'(x) \tag{12}$$

Similarly, if $y = f(u)$ where $u = g(v)$ and $v = h(x)$, then

$$\frac{dy}{dx} = \frac{dy}{du}\cdot\frac{du}{dv}\cdot\frac{dv}{dx} \tag{13}$$

The results (*12*) and (*13*) are often called *chain rules* for differentiation of composite functions. These rules probably are the most misused (or perhaps unused) rules in the application of the calculus.

7. If $y = f(x)$ and $x = f^{-1}(y)$, then dy/dx and dx/dy are related by

$$\frac{dy}{dx} = \frac{1}{dx/dy} \tag{14}$$

8. If $x = f(t)$ and $y = g(t)$, then

$$\frac{dy}{dx} = \frac{dy/dt}{dx/dt} = \frac{g'(t)}{f'(t)} \tag{15}$$

Similar rules can be formulated for differentials. For example,

$$d\{f(x) + g(x)\} = df(x) + dg(x) = f'(x)dx + g'(x)dx = \{f'(x) + g'(x)\}dx$$

$$d\{f(x)g(x)\} = f(x)dg(x) + df(x) = \{f(x)g'(x) + g(x)f'(x)\}dx$$

Derivatives of Elementary Functions

In the following we assume that u is a differentiable function of x; if $u = x$, $du/dx = 1$. The inverse functions are defined according to the principal values given in Chapter 3.

1. $\frac{d}{dx}(C) = 0$
2. $\frac{d}{dx}u^n = nu^{n-1}\frac{du}{dx}$
3. $\frac{d}{dx}\sin u = \cos u\frac{du}{dx}$
4. $\frac{d}{dx}\cos u = -\sin u\frac{du}{dx}$
5. $\frac{d}{dx}\tan u = \sec^2 u\frac{du}{dx}$
6. $\frac{d}{dx}\cot u = -\csc^2 u\frac{du}{dx}$
7. $\frac{d}{dx}\sec u = \sec u\tan u\frac{du}{dx}$

16. $\frac{d}{dx}\cot^{-1} u = -\frac{1}{1+u^2}\frac{du}{dx}$
17. $\frac{d}{dx}\sec^{-1} u = \pm\frac{1}{u\sqrt{u^2-1}}\frac{du}{dx}\begin{cases} + \text{ if } u > 1 \\ - \text{ if } u < -1\end{cases}$
18. $\frac{d}{dx}\csc^{-1} u = \mp\frac{1}{u\sqrt{u^2-1}}\frac{du}{dx}\begin{cases} - \text{ if } u > 1 \\ + \text{ if } u < -1\end{cases}$
19. $\frac{d}{dx}\sinh u = \cosh u\frac{du}{dx}$
20. $\frac{d}{dx}\cosh u = \sinh u\frac{du}{dx}$
21. $\frac{d}{dx}\tanh u = \sec h^2 u\frac{du}{dx}$
22. $\frac{d}{dx}\coth u = -\text{csch}^2 u\frac{du}{dx}$

8. $\frac{d}{dx}\csc u = -\csc u \cot u \frac{du}{dx}$

9. $\frac{d}{dx}\log_a u = \frac{\log_a e}{u}\frac{du}{dx} \quad a > 0, a \neq 1$

10. $\frac{d}{dx}\log_e u = \frac{d}{dx}\ln u = \frac{1}{u}\frac{du}{dx}$

11. $\frac{d}{dx}a^u = a^u \ln a \frac{du}{dx}$

12. $\frac{d}{dx}e^u = e^u \frac{du}{dx}$

13. $\frac{d}{dx}\sin^{-1} u = \frac{1}{\sqrt{1-u^2}}\frac{du}{dx}$

14. $\frac{d}{dx}\cos^{-1} u = -\frac{1}{\sqrt{1-u^2}}\frac{du}{dx}$

15. $\frac{d}{dx}\tan^{-1} u = \frac{1}{\sqrt{1+u^2}}\frac{du}{dx}$

23. $\frac{d}{dx}\operatorname{sech} u = -\operatorname{sech} u \tanh u \frac{du}{dx}$

24. $\frac{d}{dx}\operatorname{csch} u = -\operatorname{csch} u \coth u \frac{du}{dx}$

25. $\frac{d}{dx}\sinh^{-1} u = \frac{1}{\sqrt{1+u^2}}\frac{du}{dx}$

26. $\frac{d}{dx}\cosh^{-1} u = \frac{1}{\sqrt{u^2-1}}\frac{du}{dx}$

27. $\frac{d}{dx}\tanh^{-1} u = \frac{1}{1-u^2}\frac{du}{dx}, \quad |u| < 1$

28. $\frac{d}{dx}\coth^{-1} u = \frac{1}{1-u^2}\frac{du}{dx}, \quad |u| > 1$

29. $\frac{d}{dx}\operatorname{sech}^{-1} u = \frac{1}{u\sqrt{1-u^2}}\frac{du}{dx}$

30. $\frac{d}{dx}\operatorname{csch}^{-1} u = -\frac{1}{u\sqrt{u^2+1}}\frac{du}{dx}$

Higher-Order Derivatives

If $f(x)$ is differentiable in an interval, its derivative is given by $f'(x)$, y' or dy/dx, where $y = f(x)$. If $f'(x)$ is also differentiable in the interval, its derivative is denoted by $f''(x)$, y'' or $\frac{d}{dx}\left(\frac{dy}{dx}\right) = \frac{d^2y}{dx^2}$. Similarly, the nth derivative of $f(x)$, if it exists, is denoted by $f^{(n)}(x)$, $y^{(n)}$ or $\frac{d^n y}{dx^n}$, where n is called the order of the derivative. Thus, derivatives of the first, second, third, . . . orders are given by $f'(x), f''(x), f'''(x), \ldots$.

Computation of higher-order derivatives follows by repeated application of the differentiation rules given here.

Mean Value Theorems

These theorems are fundamental to the rigorous establishment of numerous theorems and formulas. (See Figure 4.5.)

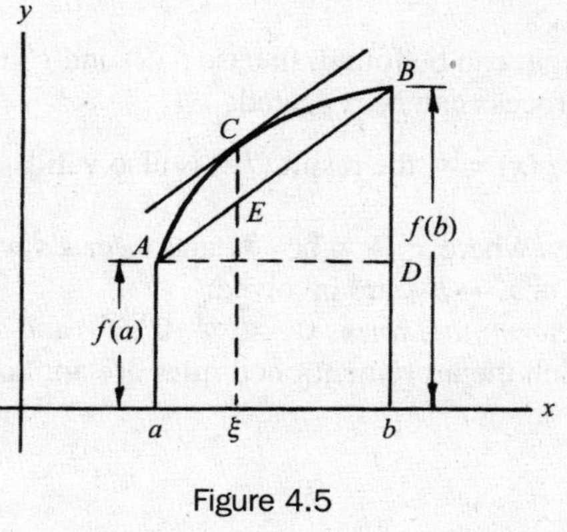

Figure 4.5

1. **Rolle's theorem.** If $f(x)$ is continuous in $[a, b]$ and differentiable in (a, b) and if $f(a) = f(b) = 0$, then there exists a point ξ in (a, b) such that $f'(\xi) = 0$.

Rolle's theorem is employed in the proof of the mean value theorem. It then becomes a special case of that theorem.

2. **The mean value theorem.** If $f(x)$ is continuous in $[a, b]$ and differentiable in (a, b), then there exists a point ξ in (a, b) such that

$$\frac{f(b)-f(a)}{b-a} = f'(\xi) \quad a<\xi<b \tag{16}$$

Rolle's theorem is the special case of this where $f(a) = f(b) = 0$.

The result (16) can be written in various alternative forms; for example, if x and x_0 are in (a, b), then

$$f(x) = f(x_0) + f'(\xi)(x - x_0) \quad \xi \text{ between } x_0 \text{ and } x \tag{17}$$

We can also write result (16) with $b = a + h$, in which case $\xi = a + \theta h$, where $0 < \theta < 1$.

The mean value theorem is also called the *law of the mean.*

3. **Cauchy's generalized mean value theorem.** If $f(x)$ and $g(x)$ are continuous in $[a, b]$ and differentiable in (a, b), then there exists a point ξ in (a, b) such that

$$\frac{f(b)-f(a)}{g(b)-g(a)} = \frac{f'(\xi)}{g'(\xi)} \quad a<\xi<b \tag{18}$$

where we assume $g(a) \neq g(b)$ and $f'(x)$, $g'(x)$ are not simultaneously zero. Note that the special case $g(x) = x$ yields (16).

L'Hospital's Rules

If $\lim_{x\to x_0} f(x) = A$ and $\lim_{x\to x_0} g(x) = B$, where A and B are either both zero or both infinite, $\lim_{x\to x_0} \frac{f(x)}{g(x)}$ is often called an *indeterminate* of the form 0/0 or ∞/∞, respectively, although such terminology is somewhat misleading since there is usually nothing indeterminate involved. The following theorems, called *L'Hospital's rules*, facilitate evaluation of such limits.

1. If $f(x)$ and $g(x)$ are differentiable in the interval (a, b) except possibly at a point x_0 in this interval, and if $g'(x) \neq 0$ for $x \neq x_0$, then

$$\lim_{x\to x_0} \frac{f(x)}{g(x)} = \lim_{x\to x_0} \frac{f'(x)}{g'(x)} \tag{19}$$

whenever the limit on the right can be found. In case $f'(x)$ and $g'(x)$ satisfy the same conditions as $f(x)$ and $g(x)$ given above, the process can be repeated.

2. If $\lim_{x\to x_0} f(x) = \infty$ and $\lim_{x\to x_0} g(x) = \infty$, the result (*19*) is also valid.

These can be extended to cases where $x \to \infty$ or $-\infty$, and to cases where $x_0 = a$ or $x_0 = b$ in which only one-sided limits, such as $x \to a+$ or $x \to b-$, are involved.

Limits represented by the *indeterminate forms* $0 \cdot \infty$, ∞^0, 0^0, 1^∞, and $\infty - \infty$ can be evaluated on replacing them by equivalent limits for which the aforementioned rules are applicable (see Problem 4.29).

Applications

Relative Extrema and Points of Inflection

See Chapter 3, where relative extrema and points of inflection are described and a diagram is presented. In this chapter such points are characterized by the variation of the tangent line and then by the derivative, which represents the slope of that line.

Assume that f has a derivative at each point of an open interval and that P_1 is a point of the graph of f associated with this interval. Let a varying tangent line to the graph move from left to right through P_1. If the point is a relative minimum, then the tangent line rotates counterclockwise. The slope is negative to the left of P_1 and positive to the right. At P_1 the slope is zero. At a relative maximum a similar analysis can be made except that the rotation is clockwise and the slope varies from positive to negative. Because f'' designates the change of f', we can state the following theorem. (See Figure 4.6.)

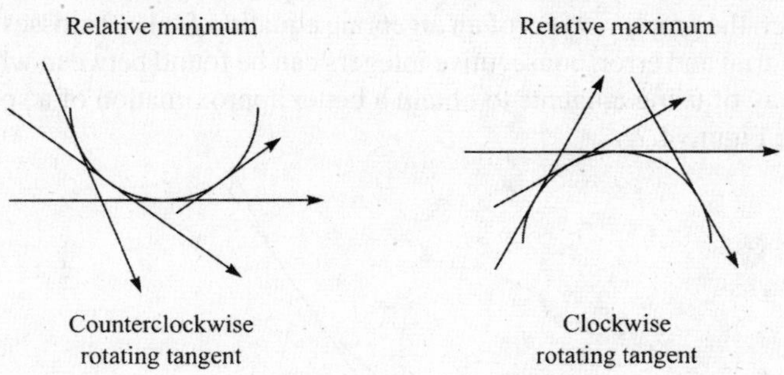

Figure 4.6

Theorem Assume that x_1 is a number in an open set of the domain of f at which f' is continuous and f'' is defined. If $f'(x_1) = 0$ and $f''(x_1) \neq 0$, then $f(x_1)$ is a relative extreme of f. Specifically:

(a) If $f''(x_1) > 0$, then $f(x_1)$ is a relative minimum.

(b) If $f''(x_1) < 0$, then $f(x_1)$ is a relative maximum.

(The domain value x_1 is called a *critical value*.)

This theorem may be generalized in the following way. Assume existence and continuity of derivatives as needed and suppose that $f'(x_1) = f''(x_1) = \ldots f^{(2p-1)}(x_1) = 0$ and $f^{(2p)}(x_1) \neq 0$ (p a positive integer). Then:

(a) f has a relative minimum at x_1 if $f^{(2p)}(x_1) > 0$.

(b) f has a relative maximum at x_1 if $f^{(2p)}(x_1) < 0$.

(Notice that the order of differentiation in each succeeding case is two greater. The nature of the intermediate possibilities is suggested in the next paragraph.)

It is possible that the slope of the tangent line to the graph of f is positive to the left of P_1, zero at the point, and again positive to the right. Then P_1 is called a *point of inflection*. In the simplest case this point of inflection is characterized by $f'(x_1) = 0$, $f''(x_1) = 0$, and $f'''(x_1) \neq 0$.

Particle Motion

The fundamental theories of modern physics are relativity, electromagnetism, and quantum mechanics. Yet Newtonian physics must be studied because it is basic to many of the concepts in these other theories, and because it is most easily applied to many of the circumstances found in everyday life. The simplest aspect of Newtonian mechanics is called *kinematics*, or the *geometry of motion*. In this model of reality, objects are idealized as points and their paths are represented by curves. In the simplest (one-dimensional) case, the curve is a straight line, and it is the speeding up and slowing down of the object that is of importance. The calculus applies to the study in the following way.

If x represents the distance of a particle from the origin and t signifies time, then $x = f(t)$ designates the position of a particle at time t. Instantaneous velocity (or speed in the one-dimensional case) is represented by $\dfrac{dx}{dt} = \lim_{\Delta t \to 0} \dfrac{f(t+\Delta t)}{\Delta t}$ (the limiting case of the formula change in $\dfrac{\text{change in distance}}{\text{change in time}}$ in time for speed when the motion is constant). Furthermore, the instantaneous change in velocity is called *acceleration* and represented by $\dfrac{d^2x}{dt^2}$.

Path, velocity, and acceleration of a particle will be represented in three dimensions in Chapter 7, on vectors.

Newton's Method

It is difficult or impossible to solve algebraic equations of higher degree than two. In fact, it has been proved that there are no general formulas representing the roots of algebraic equations of degree five and higher in terms of radicals. However, the graph $y = f(x)$ of an algebraic equation $f(x) = 0$ crosses the x axis at each single-valued real root. Thus, by trial and error, consecutive integers can be found between which a root lies. Newton's method is a systematic way of using tangents to obtain a better approximation of a specific real root. The procedure is as follows. (See Figure 4.7.)

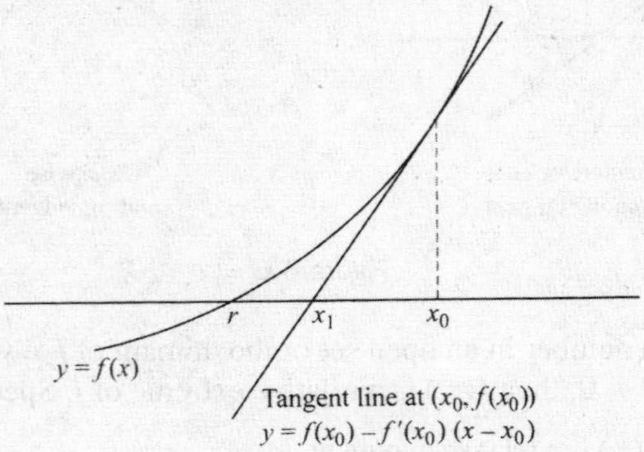

Figure 4.7

Suppose that f has as many derivatives as required. Let r be a real root of $f(x) = 0$; i.e., $f(r) = 0$. Let x_0 be a value of x near r—for example, the integer preceding or following r. Let $f'(x_0)$ be the slope of the graph of $y = f(x)$ at $P_0[x_0, f(x_0)]$. Let $Q_1(x_1, 0)$ be the x-axis intercept of the tangent line at P_0; then

$$\frac{0 - f(x_0)}{x - x_0} = f'(x_0)$$

where the two representations of the slope of the tangent line have been equated. The solution of this relation for x_1 is

$$x_1 = x_0 - \frac{f(x_0)}{f'(x_0)}$$

Starting with the tangent line to the graph at $P_1[x_1, f(x_1)]$ and repeating the process, we get

$$x_2 = x_1 - \frac{f(x_1)}{f'(x_1)} = x_0 - \frac{f(x_0)}{f'(x_0)} - \frac{f(x_1)}{f'(x_1)}$$

and, in general,

$$x_n = x_0 - \sum_{k=0}^{n} \frac{f(x_k)}{f'(x_k)}$$

Under appropriate circumstances, the approximation x_n to the root r can be made as good as desired.

Note: Success with Newton's method depends on the shape of the function's graph in the neighborhood of the root. There are various cases which have not been explored here.

SOLVED PROBLEMS

Derivatives

4.1. (a) Let $f(x) = \dfrac{3+x}{3-x}$, $x \neq 3$. Evaluate $f'(2)$ from the definition.

$$f'(2) = \lim_{h\to 0} \frac{f(2+h) - f(2)}{h} = \lim_{h\to 0} \frac{1}{h}\left(\frac{5+h}{1-h} - 5\right) = \lim_{h\to 0} \frac{1}{h} \cdot \frac{6h}{1-h} = \lim_{h\to 0} \frac{6}{1-h} = 6$$

Note: By using rules of differentiation we find

$$f'(x) = \frac{(3-x)\dfrac{d}{dx}(3+x) - (3+x)\dfrac{d}{dx}(3-x)}{(3-x)^2} = \frac{(3-x)(1) - (3+x)(-1)}{(3-x)^2} = \frac{6}{(3-x)^2}$$

at all points x where the derivative exists. Putting $x = 2$, we find $f'(2) = 6$. Although such rules are often useful, one must be careful not to apply them indiscriminately (see Problem 4.5).

(b) Let $f(x) = \sqrt{2x-1}$. Evaluate $f'(5)$ from the definition.

$$f'(5) = \lim_{h\to 0} \frac{f(5+h) - f(5)}{h} = \lim_{h\to 0} \frac{\sqrt{9+2h} - 3}{h}$$

$$= \lim_{h\to 0} \frac{\sqrt{9+2h} - 3}{h} \cdot \frac{\sqrt{9+2h} + 3}{\sqrt{9+2h} + 3} = \lim_{h\to 0} \frac{9+2h-9}{h(\sqrt{9+2h} + 3)} = \lim_{h\to 0} \frac{2}{\sqrt{9+2h} + 3} = \frac{1}{3}$$

By using rules of differentiation we find $f'(x) = \dfrac{d}{dx}(2x-1)^{1/2} = \dfrac{1}{2}(2x-1)^{-1/2}\dfrac{d}{dx}(2x-1) = (2x-1)^{-1/2}$. Then $f'(5) = 9^{-1/2} = \dfrac{1}{3}$.

4.2. (a) Show directly from definition that the derivative of $f(x) = x^3$ is $3x^2$.

(b) Show from definition that $\dfrac{d}{dx}\sqrt{x}) = \dfrac{1}{2\sqrt{x}}$.

(a) $$\frac{f(x+h) - f(x)}{h} = \frac{1}{h}[(x+h)^3 - x^3]$$

$$= \frac{1}{h}[x^3 + 3x^2h + 3xh^2 + h^2] - x^3] = 3x^2 + 3xh + h^2$$

Then

$$f'(x) = \lim_{h\to 0} \frac{f(x+h) - f(x)}{h} = 3x^2$$

(b) $$\lim_{h\to 0} \frac{f(x+h) - f(x)}{h} = \lim_{h\to 0} \frac{\sqrt{x+h} - \sqrt{x}}{h}$$

The result follows by multiplying numerator and denominator by $\sqrt{x+h} - \sqrt{x}$ and then letting $h \to 0$.

4.3. If $f(x)$ has a derivative at $x = x_0$, prove that $f(x)$ must be continuous at $x = x_0$.

$$f(x_0 + h) - f(x_0) = \frac{f(x_0 + h) - f(x_0)}{h} \cdot h, \quad h \neq 0$$

Then

$$\lim_{h\to 0} f(x_0 + h) - f(x_0) = \lim_{h\to 0} \frac{f(x_0 + h) - f(x_0)}{h} \cdot \lim_{h\to 0} h = f'(x_0) \cdot 0 = 0$$

since $f'(x_0)$ exists by hypothesis. Thus,

$$\lim_{h\to 0} f(x_0+h)-f(x_0)=0 \qquad \text{or} \qquad \lim_{h\to 0} f(x_0+h)=f(x_0)$$

showing that $f(x)$ is continuous at $x = x_0$.

4.4. Let $f(x)=\begin{cases} x\sin 1/x, & x\neq 0 \\ 0, & x=0 \end{cases}$.

(a) Is $f(x)$ continuous at $x = 0$? (b) Does $f(x)$ have a derivative at $x = 0$?

(a) By Problem 3.22(b), $f(x)$ is continuous at $x = 0$.

(b) $$f'(0)=\lim_{h\to 0}\frac{f(0+h)-f(0)}{h}=\lim_{h\to 0}\frac{f(h)-f(0)}{h}=\lim_{h\to 0}\frac{h\sin 1/h-0}{h}=\lim_{h\to 0}\sin\frac{1}{h}$$

which does not exist.

This example shows that even though a function is continuous at a point, it need not have a derivative at the point; i.e., the converse of the theorem in Problem 4.3 is not necessarily true.

It is possible to construct a function which is continuous at every point of an interval but has a derivative nowhere.

4.5. Let $f(x)=\begin{cases} x^2\sin 1/x, & x\neq 0 \\ 0, & x=0 \end{cases}$.

(a) Is $f(x)$ differentiable at $x = 0$? (b) Is $f'(x)$ continuous at $x = 0$?

(a) $$f'(0)=\lim_{h\to 0}\frac{f(h)-f(0)}{h}=\lim_{h\to 0}\frac{h^2\sin 1/h-0}{h}=\lim_{h\to 0}h\sin\frac{1}{h}=0$$

by Problem 3.13. Then $f(x)$ has derivative (is differentiable) at $x = 0$ and its value is 0.

(b) From elementary calculus differentiation rules, if $x \neq 0$,

$$f'(x)=\frac{d}{dx}\left(x^2\sin\frac{1}{x}\right)=x^2\frac{d}{dx}\left(\sin\frac{1}{x}\right)+\left(\sin\frac{1}{x}\right)\frac{d}{dx}(x^2)$$

$$=x^2\left(\cos\frac{1}{x}\right)\left(-\frac{1}{x^2}\right)+\left(\sin\frac{1}{x}\right)(2x)=-\cos\frac{1}{x}+2x\sin\frac{1}{x}$$

Since $\lim_{x\to 0} f'(x)=\lim_{x\to 0}\left(-\cos\frac{1}{x}+2x\sin\frac{1}{x}\right)$ does not exist (because $\lim_{x\to 0}\cos 1/x$ does not exist). $f'(x)$ cannot be continuous at $x = 0$ in spite of the fact that $f'(0)$ exists.

This shows that we cannot calculate $f'(0)$ in this case by simply calculating $f'(x)$ and and putting $x = 0$, as is frequently supposed in elementary calculus. It is only when the derivative of a function is *continuous* at a point that this procedure gives the right answer. This happens to be true for most functions arising in elementary calculus.

4.6. Present an "ϵ, δ" definition of the derivative of $f(x)$ at $x = x_0$.

$f(x)$ has a derivative $f'(x_0)$ at $x = x_0$ if, given any $\epsilon > 0$, we can find $\delta > 0$ such that

$$\left|\frac{f(x_0+h)-f(x_0)}{h}-f'(x_0)\right|<\varepsilon \quad \text{when} \quad 0<|h|<\delta$$

Right- and left-hand derivatives

4.7. Let $f(x) = |x|$. (a) Calculate the right-hand derivatives of $f(x)$ at $x = 0$. (b) Calculate the left-hand derivative of $f(x)$ at $x = 0$. (c) Does $f(x)$ have a derivative at $x = 0$? (d) Illustrate the conclusions in (a), (b), and (c) from a graph.

(a) $$f'_+(0) = \lim_{h\to 0+} \frac{f(h) - f(0)}{h} = \lim_{h\to 0+} \frac{|h| - 0}{h} = \lim_{h\to 0+} \frac{h}{h} = 1$$

since $|h| = -h$ for $h > 0$.

(b) $$f'_-(0) = \lim_{h\to 0-} \frac{f(h) - f(0)}{h} = \lim_{h\to 0-} \frac{|h| - 0}{h} = \lim_{h\to 0-} \frac{-h}{h} = -1$$

since $|h| = -h$ for $h < 0$.

(c) No. The derivative at 0 does not exist if the right- and left-hand derivatives are unequal.

(d) The required graph is shown in Figure 4.8. Note that the slopes of the lines $y = x$ and $y = -x$ are 1 and -1, respectively, representing the right- and left-hand derivatives at $x = 0$. However, the derivative at $x = 0$ does not exist.

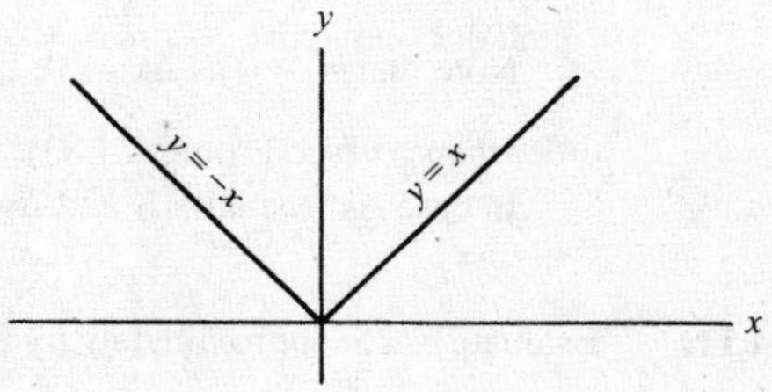

Figure 4.8

4.8. Prove that $f(x) = x^2$ is differentiable in $0 \leqq x \leqq 1$.

Let x_0 be any value such that $0 < x_0 < 1$. Then

$$f'(x_0) = \lim_{h\to 0} \frac{f(x_0 + h) - f(x_0)}{h} = \lim_{h\to 0} \frac{(x_0 + h)^2 - x_0^2}{h} = \lim_{h\to 0} (2x_0 + h) = 2x_0$$

At the endpoint $x = 0$,

$$f'_+(0) = \lim_{h\to 0+} \frac{f(0 + h) - f(0)}{h} = \lim_{h\to 0+} \frac{h^2 - 0}{h} = \lim_{h\to 0+} h = 0$$

At the end point $x = 1$,

$$f'_-(1) = \lim_{h\to 0-} \frac{f(1 + h) - f(1)}{h} = \lim_{h\to 0-} \frac{(1 + h)^2 - 1}{h} = \lim_{h\to 0-} (2 + h) = 2$$

Then $f(x)$ is differentiable in $0 \leqq x \leqq 1$. We may write $f'(x) = 2x$ for any x in this interval. It is customary to write $f'_+(0) = f'(0)$ and $f'_-(1) = f'(1)$ in this case.

4.9. Find an equation for the tangent line to $y = x^2$ at the point where (a) $x = 1/3$ and (b) $x = 1$.

(a) From Problem 4.8. $f'(x_0) = 2x_0$ so that $f'(1/3) = 2/3$. Then the equation of the tangent line is

$$y - f(x_0) = f(x_0)(x - x_0) \quad \text{or} \quad y - \frac{1}{9} = \frac{2}{3}\left(x - \frac{1}{3}\right). \quad \text{i.e., } y = \frac{2}{3}x - \frac{1}{9}$$

(b) As in part (*a*), $y - f(1) = f'(1)(x - 1)$ or $y - 1 = 2(x - 1)$, i.e., $y = 2x - 1$.

Differentials

4.10. If $y = f(x) = x^3 - 6x$, find (a) Δy, (b) dy, and (c) $\Delta y - dy$.

(a) $\Delta y = f(x + \Delta x) - f(x) = \{(x + \Delta x)^3 - 6(x + \Delta x)\} - \{x^3 - 6x\}$
$= x^3 + 3x^2\,\Delta x + 3x(\Delta x)^2 + (\Delta x)^3 - 6x - 6\Delta x - x^3 + 6x$
$= (3x^2 - 6)\,\Delta x + 3x(\Delta x)^2 + (\Delta x)^3$

(b) dy = principal part of $\Delta y = (3x^2 - 6)\Delta x = (3x^2 - 6)dx$, since by definition $\Delta x = dx$.

Note that $f'(x) = 3x^2 - 6$ and $dy = (3x^2 - 6)dx$, i.e.; $dy/dx = 3x^2 - 6$. It must be emphasized that dy and dx are not necessarily small.

(c) From (a) and (b), $\Delta y - dy = 3x(\Delta x)^2 + (\Delta x)^3 = \epsilon\Delta x$, where $\epsilon = 3x\Delta x + (\Delta x)^2$.

Note that $\epsilon \to 0$ as $\Delta x \to 0$; i.e., $\dfrac{\Delta y - dy}{\Delta x} \to 0$ as $\Delta x \to 0$. Hence, $\Delta y - dy$ is an infinitesimal of higher order than Δx (see Problem 4.83).

In case Δx is small, dy and Δy are approximately equal.

4.11. Evaluate $\sqrt[3]{25}$ approximately by use of differentials.

If Δx is small, $\Delta y = f(x + \Delta x) - f(x) = f'(x)\,\Delta x$ approximately.

Let $f(x) = \sqrt[3]{x}$. Then $\sqrt[3]{x + \Delta x} - \sqrt[3]{x} \approx \frac{1}{3}x^{-2/3}\,\Delta x$ (where $\approx$ denotes *approximately equal to*).

If $x = 27$ and $\Delta x = -2$, we have

$$\sqrt[3]{27 - 2} - \sqrt[3]{27} \approx \frac{1}{3}(27)^{-2/3}(-2), \qquad \text{i.e., } \sqrt[3]{25} - 3 \approx -2/27$$

Then $\sqrt[3]{25} \approx 3 - 2/27$ or 2.926.

It is interesting to observe that $(2.926)^3 = 25.05$, so the approximation is fairly good.

Differentiation rules: differentiation of elementary functions

4.12. Prove the formula $\dfrac{d}{dx}\{f(x)\,g(x)\} = f(x)\dfrac{d}{dx}\,g(x) + g(x)\dfrac{d}{dx}\,f(x)$, , assuming f and g are differentiable.

By definition,

$$\frac{d}{dx}\{f(x)\,g(x)\} = \lim_{\Delta x \to 0} \frac{f(x + \Delta x)\,g(x + \Delta x) - f(x)\,g(x)}{\Delta x}$$

$$= \lim_{\Delta x \to 0} \frac{f(x + \Delta x)\{g(x + \Delta x) - g(x)\} + g(x)\{f(x + \Delta x) - f(x)\}}{\Delta x}$$

$$= \lim_{\Delta x \to 0} f(x + \Delta x)\left\{\frac{g(x + \Delta x) - g(x)}{\Delta x}\right\} + \lim_{\Delta x \to 0} g(x)\left\{\frac{f(x + \Delta x) - f(x)}{\Delta x}\right\}$$

$$= f(x)\frac{d}{dx}\,g(x) + g(x)\frac{d}{dx}\,f(x)$$

Another method:

Let $u = f(x)$, $\upsilon = g(x)$. Then $\Delta u = f(x + \Delta x) - f(x)$ and $\Delta\upsilon = g(x + \Delta x) - g(x)$; i.e., $f(x + \Delta x) = u + \Delta u$, $g(x + \Delta x) = \upsilon + \Delta\upsilon$. Thus,

$$\frac{d}{dx}uv = \lim_{\Delta x \to 0} \frac{(u + \Delta u)(\upsilon + \Delta\upsilon) - u\upsilon}{\Delta x} = \lim_{\Delta x \to 0} \frac{u\Delta\upsilon + \upsilon\Delta u + \Delta u\Delta\upsilon}{\Delta x}$$

$$= \lim_{\Delta x \to 0}\left(u\frac{\Delta\upsilon}{\Delta x} + \upsilon\frac{\Delta u}{\Delta x} + \frac{\Delta u}{\Delta x}\Delta\upsilon\right) = u\frac{d\upsilon}{dx} + \upsilon\frac{du}{dx}$$

where it is noted that $\Delta\upsilon \to 0$ as $\Delta x \to 0$, since υ is supposed differentiable and thus continuous.

4.13. If $y = f(u)$ where $u = g(x)$, prove that $\dfrac{dy}{dx} = \dfrac{dy}{du} \cdot \dfrac{du}{dx}$, assuming that f and g are differentiable.

Let x be given an increment $\Delta x \neq 0$. Then, as a consequence, u and y take on increments Δu and Δy, respectively, where

$$\Delta y = f(u + \Delta u) - f(u), \qquad \Delta u = g(x + \Delta x) - g(x) \tag{1}$$

Note that as $\Delta x \to 0$, $\Delta y \to 0$ and $\Delta u \to 0$.

If $\Delta u \neq 0$, let us write $\epsilon = \dfrac{\Delta y}{\Delta u} - \dfrac{dy}{du}$ so that $\epsilon \to 0$ as $\Delta u \to 0$ and

$$\Delta y = \frac{dy}{du} \Delta u + \epsilon \, \Delta u \tag{2}$$

If $\Delta u = 0$ for values of Δx, then Equation (1) shows that $\Delta y = 0$ for these values of Δx. For such cases, we define $\epsilon = 0$.

It follows that in both cases. $\Delta u \neq 0$ or $\Delta u = 0$, Equation (2) holds. Dividing Equation (2) by $\Delta x \neq 0$ and taking the limit as $\Delta x \to 0$, we have

$$\begin{aligned} \frac{dy}{dx} &= \lim_{\Delta x \to 0} \frac{\Delta y}{\Delta x} = \lim_{\Delta x \to 0} \left(\frac{dy}{du} \frac{\Delta u}{\Delta x} + \epsilon \frac{\Delta u}{\Delta x} \right) = \frac{dy}{du} \cdot \lim_{\Delta x \to 0} \frac{\Delta u}{\Delta x} + \lim_{\Delta x \to 0} \epsilon \cdot \lim_{\Delta x \to 0} \frac{\Delta u}{\Delta x} \\ &= \frac{dy}{du} \frac{du}{dx} + 0 \cdot \frac{du}{dx} \frac{dy}{du} \cdot \frac{du}{dx} \end{aligned} \tag{3}$$

4.14. Given $\dfrac{d}{dx}(\sin x) = \cos x$ and $\dfrac{d}{dx}(\cos x) = -\sin x$, derive the following formulas:

(a) $\dfrac{d}{dx}(\tan x) = \sec^2 x$ (b) $\dfrac{d}{dx}(\sin^{-1} x) = \dfrac{1}{\sqrt{1 - x^2}}$

(a) $$\frac{d}{dx}(\tan x) = \frac{d}{dx}\left(\frac{\sin x}{\cos x} \right) = \frac{\cos x \dfrac{d}{dx}(\sin x) - \sin x \dfrac{d}{dx}}{\cos^2 x}$$

$$= \frac{(\cos x)(\cos x) - (\sin x)(-\sin x)}{\cos^2 x} = \frac{1}{\cos^2 x} = {}^2 x$$

(b) If $y = \sin^{-1} x$, then $x = \sin y$. Taking the derivative with respect to x,

$$1 = \cos y \frac{dy}{dx} \quad \text{or} \quad \frac{dy}{dx} = \frac{1}{\cos y} = \frac{1}{\sqrt{1 - \sin^2 y}} = \frac{1}{\sqrt{1 - x^2}}$$

We have supposed here that the principal value $-\pi/2 \leqq \sin^{-1} x \leqq \pi/2$ is chosen so that $\cos y$ is positive, thus accounting for our writing $\cos y = \sqrt{1 - \sin^2 y}$ rather than $\cos y = \pm\sqrt{1 - \sin^2 y}$.

4.15. Derive the formula $\dfrac{d}{dx}(\log_a u) = \dfrac{\log_a e}{u} \dfrac{du}{dx} (a > 0, a \neq 1)$, where u is a differentiable function of x.

Consider $y = f(u) = \log_a u$. By definition,

$$\begin{aligned} \frac{dy}{du} &= \lim_{\Delta u \to 0} \frac{f(u + \Delta u) - f(u)}{\Delta u} = \lim_{\Delta u \to 0} \frac{\log_a (u + \Delta u) - \log_a u}{\Delta u} \\ &= \lim_{\Delta u \to 0} \frac{1}{\Delta u} \log_a \left(\frac{u + \Delta u}{u} \right) = \lim_{\Delta u \to 0} \frac{1}{u} \log_a \left(1 + \frac{\Delta u}{u} \right)^{u/\Delta u} \end{aligned}$$

Since the logarithm is a continuous function, this can be written

$$\frac{1}{u}\log_a\left\{\lim_{\Delta u\to 0}\left(1+\frac{\Delta u}{u}\right)^{u/\Delta u}\right\}=\frac{1}{u}\log_a e$$

by Problem 2.19, with $x = u/\Delta u$.

Then by Problem 4.13, $\frac{d}{dx}(\log_a u)=\frac{\log_a e}{u}\frac{du}{dx}$.

4.16. Calculate dy/dx if (a) $xy^3 - 3x^2 = xy + 5$ and (b) $e^{xy} + y \ln x = \cos 2x$.

(a) Differentiate with respect to x, considering y as a function of x. (We sometimes say that y is an *implicit function of x*, since we cannot solve explicitly for y in terms of x.) Then

$$\frac{d}{dx}(xy)^2-\frac{d}{dx}(3x^2)=\frac{d}{dx}(xy)+\frac{d}{dx}(5) \quad \text{or} \quad (x)(3y^2y')+(y^3)(1)-6x=(x)(y')+(y)(1)+0$$

where $y' = dy/dx$. Solving,

$$y' = (6x - y^3 + y)/(3xy^2 - x)$$

(b) $\frac{d}{dx}(e^{xy})+\frac{d}{dx}(u\ln)=\frac{d}{dx}(\cos 2x).\quad e^{xy}(xy'+y)+\frac{y}{x}+(\ln x)y'=-2\sin 2x.$

Solving,

$$y'=-\frac{2x\sin 2x+xye^{xy}+y}{x^2e^{xy}+x\ln x}$$

4.17. If $y = \cosh(x^2 - 3x + 1)$, find (a) dy/dx and (b) d^2y/dx^2.

(a) Let $y = \cosh u$, where $u = x^2 - 3x + 1$. Then $dy/dx = \sinh u$, $du/dx = 2x - 3$, and

$$\frac{dy}{dx}=\frac{dy}{du}\cdot\frac{du}{dx}=(\sinh u)(2x-3)=(2x-3)\sinh(x^2-3x+1)$$

(b) $$\frac{d^2y}{dx^2}=\frac{d}{dx}\left(\frac{dy}{dx}\right)=\frac{d}{dx}\left(\sinh u\frac{du}{dx}\right)=\sinh u\frac{d^2u}{dx^2}+\cosh u\left(\frac{du}{dx}\right)^2$$

$$=(\sinh u)(2)+(\cosh u)(2x-3)^2=2\sinh(x^2-3x+1)+(2x-3)^2\cosh(x^2-3x+1)$$

4.18. If $x^2y + y^3 = 2$, find (a) y' and (b) y'' at the point (1, 1).

(a) Differentiating with respect to x, $x^2y' + 2xy + 3y^2y' = 0$ and

$$y'=\frac{-2xy}{x^2+3xy^2}=-\frac{1}{2}\text{ at }(1,1)$$

(b) $$y''=\frac{d}{dx}(y')=\frac{d}{dx}\left(\frac{-2xy}{x^2+3y^2}\right)=-\frac{(x^2+3y^2)(2xy'+2y)-(2xy)(2x+6yy')}{(x^2+3y^2)^2}$$

Substituting $x = 1$, $y = 1$, and $y'=-\frac{1}{2}$, we find $y''=-\frac{3}{8}$.

Mean value theorems

4.19. Prove Rolle's theorem.

Case 1: $f(x) \equiv 0$ in $[a, b]$. Then $f'(x) = 0$ for all x in (a, b).

Case 2: $f(x) \not\equiv 0$ in $[a, b]$. Since $f(x)$ is continuous, there are points at which $f(x)$ attains its maximum and minimum values, denoted by M and m, respectively (see Problem 3.34).

Since $f(x) \not\equiv 0$, at least one of the values M, m is not zero. Suppose, for example, $M \not\equiv 0$ and that $f(\xi) = M$ (see Figure 4.9). For this case, $f(\xi + h) \leqq f(\xi)$.

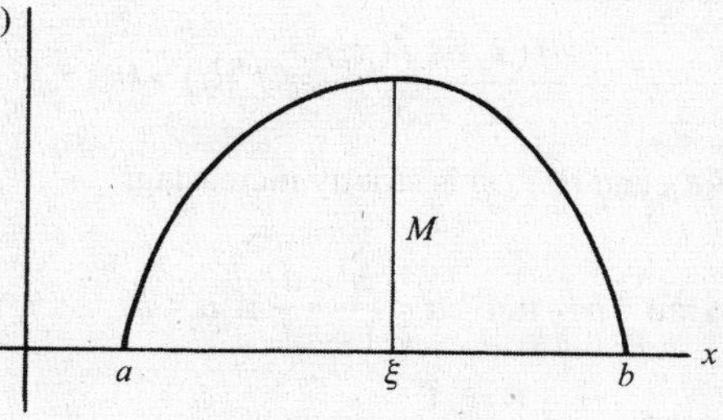

Figure 4.9

If $h > 0$, then $\dfrac{f(\xi + h) - f(\xi)}{h} \leqq 0$ and

$$\lim_{h \to 0+} \frac{f(\xi + h) - f(\xi)}{h} \leqq 0 \qquad (1)$$

If $h < 0$, then $\dfrac{f(\xi + h) - f(\xi)}{h} \geqq 0$ and

$$\lim_{h \to 0-} \frac{f(\xi + h) - f(\xi)}{h} \geqq 0 \qquad (2)$$

But, by hypothesis, $f(x)$ has a derivative at all points in (a, b). Then the right-hand derivative *(1)* must be equal to the left-hand derivative *(2)*. This can happen only if they are both equal to zero, in which case $f'(\xi) = 0$ as required.

A similar argument can be used in case $M = 0$ and $m \neq 0$.

4.20. Prove the mean value theorem.

Define $F(x) = f(x) - f(a) - (x - a)\dfrac{f(b) - f(a)}{b - a}$.

Then $f(a) = 0$ and $f(b) = 0$.

Also, if $f(x)$ satisfies the conditions on continuity and differentiability specified in Rolle's theorem, then $F(x)$ satisfies them also.

Then, applying Rolle's theorem to the function $F(x)$, we obtain

$$F'(\xi) = f'(\xi) - \frac{f(b) - f(a)}{b - a} = 0, \quad a < \xi < b \quad \text{or} \quad f'(\xi) = \frac{f(b) - f(a)}{b - a}, \quad a < \xi < b$$

4.21. Verify the mean value theorem for $f(x) = 2x^2 - 7x + 10$, $a = 2$, $b = 5$.

$f(2) = 4$, $f(5) = 25$, $f'(\xi) = 4\xi - 7$. Then the mean value theorem states that $4\xi - 7 = (25 - 4)/(5 - 2)$ or $\xi = 3.5$. Since $2 < \xi < 5$, the theorem is verified.

4.22. If $f'(x) = 0$ at all points of the interval (a, b), prove that $f(x)$ must be a constant in the interval.

Let $x_1 < x_2$ be any two different points in (a, b). By the mean value theorem for $x_1 < \xi < x_2$,

$$\frac{f(x^2) - f(x_1)}{x^2 - x^2} = f'(\xi) = 0$$

Thus $f(x_1) = f(x_2) =$ constant. From this it follows that if two functions have the same derivative at all points of (a, b), the functions can differ only by a constant.

4.23. If $f'(x) > 0$ at all points of the interval (a, b), prove that $f(x)$ is strictly increasing.

Let $x_1 < x_2$ be any two different points in (a, b). By the mean value theorem for $x_1 < \xi < x_2$,

$$\frac{f(x_2)-f(x_1)}{x_2 - x_1} = f'(\xi) > 0$$

Then $f(x_2) > f(x_1)$ for $x_2 > x_1$, and so $f(x)$ is strictly increasing.

4.24. (a) Prove that $\dfrac{b-a}{1+b^2} < \tan^{-1} b - \tan^{-1} a < \dfrac{b-a}{1+a^2}$ if $a < b$.

(b) Show that $\dfrac{\pi}{4} + \dfrac{3}{25} < \tan^{-1}\dfrac{4}{3} < \dfrac{\pi}{4} + \dfrac{1}{6}$.

(a) Let $f(x) = \tan^{-1}x$. Since $f'(x) = 1/(1+x^2)$ and $f'(\xi) = 1/(1+\xi^2)$, we have by the mean value theorem

$$\frac{\tan^{-1} b - \tan^{-1} a}{b-a} = \frac{1}{1+\xi^2} \qquad a < \xi < b$$

Since $\xi > a$, $1/(1+\xi^2) < 1/(1+a^2)$. Since $\xi < b$, $1/(1+\xi^2) > 1/(1+b^2)$. Then

$$\frac{1}{1+b^2} < \frac{\tan^{-1} b - \tan^{-1} a}{b-a} < \frac{1}{1+a^2}$$

and the required result follows on multiplying by $b - a$.

(b) Let $b = 4/3$ and $a = 1$ in the result of (a). Then, since $\tan^{+} 1 = \pi/4$, we have

$$\frac{3}{25} < \tan^{-1}\frac{4}{3} - \tan^{-1} 1 < \frac{1}{6} \quad \text{or} \quad \frac{\pi}{4} + \frac{3}{25} < \tan^{-1}\frac{4}{3} < \frac{\pi}{4} + \frac{1}{6}$$

4.25. Prove Cauchy's generalized mean value theorem.

Consider $G(x) = f(x) - f(a) - \alpha\{g(x) - g(a)\}$, where α is a constant. Then $G(x)$ satisfies the conditions of Rolle's theorem, provided $f(x)$ and $g(x)$ satisfy the continuity and differentiability conditions of Rolle's theorem and if $G(a) = G(b) = 0$. Both latter conditions are satisfied if the constant $\alpha = \dfrac{f(b)-f(a)}{g(b)-g(a)}$.

Applying Rolle's theorem, $G'(\xi) = 0$ for $a < \xi < b$, we have

$$f'(\xi) - ag'(\xi) = 0 \quad \text{or} \quad \frac{f'(\xi)}{g'(\xi)} = \frac{f(b)-f(a)}{g(b)-g(a)}, \quad a < \xi < b$$

as required.

L'Hospital's rule

4.26. Prove L'Hospital's rule for the case of the "indeterminate forms" (a) 0/0 and (b) ∞/∞.

(a) We shall suppose that $f(x)$ and $g(x)$ are differentiable in $a < x < b$ and $f(x_0) = 0$, $g(x_0) = 0$, where $a < x_0 < b$.

By Cauchy's generalized mean value theorem (Problem 4.25),

$$\frac{f(x)}{g(x)} = \frac{f(x)-f(x_0)}{g(x)-g(x_0)} = \frac{f'(\xi)}{g'(\xi)} \qquad x_0 < \xi < x$$

Then

$$\lim_{x\to x_0+} \frac{f(x)}{g(x)} = \lim_{x\to x_0+} \frac{f'(\xi)}{g'(\xi)} = \lim_{x\to x_0+} \frac{f'(x)}{g'(x)} = L$$

since as $x \to x_0+$, $\xi \to x_0+$.

Modification of this procedure can be used to establish the result if $x \to x_0-$, $x \to x_0$, $x \to \infty$, or $x \to -\infty$.

(b) We suppose that $f(x)$ and $g(x)$ are differentiable in $a < x < b$, and $\lim\limits_{x\to x_0+} f(x) = \infty$, $\lim\limits_{x\to x_0+} g(x) = \infty$ where $a < x_0 < b$.

Assume x_1 is such that $a < x_0 < x < x_1 < b$. By Cauchy's generalized mean value theorem,

$$\frac{f(x)-f(x_1)}{g(x)-g(x_1)} = \frac{f'(\xi)}{g'(\xi)} \qquad x < \xi < x_1$$

Hence,

$$\frac{f(x)-f(x_1)}{g(x)-g(x_1)} = \frac{f(x)}{g(x)} \cdot \frac{1-f(x_1)/f(x)}{1-g(x_1)/g(x)} = \frac{f'(\xi)}{g'(\xi)}$$

from which we see that

$$\frac{f(x)}{g(x)} = \frac{f'(\xi)}{g'(\xi)} \cdot \frac{1-g(x_1)/g(x)}{1-f(x_1)/f(x)} \tag{1}$$

Let us now suppose that $\lim\limits_{x \to x_0+} \dfrac{f'(x)}{g'(x)} = L$ and write Equation (1) as

$$\frac{f(x)}{g(x)} = \left(\frac{f'(\xi)}{g'(x)} - L\right)\left(\frac{1-g(x_1)/g(x)}{1-f(x_1)/f(x)}\right) + L\left(\frac{1-g(x_1)/g(x)}{1-f(x_1)/f(x)}\right) \tag{2}$$

We can choose x_1 so close to x_0 that $|f'(\xi)/g'(\xi) - L| < \epsilon$. Keeping x_1 fixed, we see that

$$\lim_{x \to x_0+}\left(\frac{1-g(x_1)/g(x)}{1-f(x_1)/f(x)}\right) = 1 \text{ since } 1 \lim_{x \to x_0+} f(x)_1 = \infty \text{ and } \lim_{x \to x_0+} g(x) = \infty$$

Then taking the limit as $x \to x_0+$ on both sides of (2), we see that, as required,

$$\lim_{x \to x_0+} \frac{f(x)}{g(x)} = L = \lim_{x \to x_0+} \frac{f'(x)}{g'(x)}$$

Appropriate modifications of this procedure establish the result if $x \to x_0-$, $x \to x_0$, $x \to \infty$, or $x \to -\infty$.

4.27. Evaluate (a) $\lim\limits_{x \to 0} \dfrac{e^{2x}-1}{x}$ and (b) $\lim\limits_{x \to 1} \dfrac{1+\cos\pi x}{x^2-2x+1}$.

All of these have the "indeterminate form" 0/0.

(a) $\lim\limits_{x \to 0} \dfrac{e^{2x}-1}{x} = \lim\limits_{x \to 0} \dfrac{2e^{2x}}{1} = 2$

(b) $\lim\limits_{x \to 1} \dfrac{1+\cos\pi x}{x^2-2x+1} = \lim\limits_{x \to 1} \dfrac{-\pi \sin \pi x}{2x-2} = \lim\limits_{x \to 1} \dfrac{-\pi^2 + \cos\pi x}{2} = \dfrac{\pi^2}{2}$

Note: Here L'Hospital's rule is applied twice, since the first application again yields the "indeterminate form" 0/0 and the conditions for L'Hospital's rule are satisfied once more.

4.28. Evaluate (a) $\lim\limits_{x \to \infty} \dfrac{3x^2-x+5}{5x^2-6x-3}$ and (b) $\lim\limits_{x \to \infty} x^2 e^{-x}$.

All of these have or can be arranged to have the "indeterminate form" ∞/∞.

(a) $\lim\limits_{x \to \infty} \dfrac{3x^2-x+5}{5x^2-6x-3} = \lim\limits_{x \to \infty} \dfrac{6x-1}{10x+6} = \lim\limits_{x \to \infty} \dfrac{6}{10} = \dfrac{3}{5}$

(b) $\lim\limits_{x \to \infty} x^2 e^{-x} = \lim\limits_{x \to \infty} \dfrac{x^2}{e^x} = \lim\limits_{x \to \infty} \dfrac{2x}{e^x} = \lim\limits_{x \to \infty} \dfrac{2}{e^x} = 0$

4.29. Evaluate $\lim\limits_{x \to 0+} x^2 \ln x$.

$$\lim_{x \to 0+} x^2 \ln x = \lim_{x \to 0+} \frac{\ln x}{1/x^2} = \lim_{x \to 0+} \frac{1/x}{-2/x^3} \lim_{x \to 0+} \frac{-x^2}{2} = 0$$

The given limit has the "indeterminate form" $0 \cdot \infty$. In the second step the form is altered so as to give the indeterminate form ∞/∞, and L'Hospital's rule is then applied.

4.30. Find $\lim_{x\to 0} (\cos x)^{1/x^2}$.

Since $\lim_{x\to 0} \cos x = 1$ and $\lim_{x\to 0} 1/x^2 = \infty$, the limit takes the "indeterminate form" 1^∞.

Let $F(x) = (\cos x)^{1/x2}$. Then $\ln F(x) = (\ln \cos x)/x^2$, to which L'Hospital's rule can be applied. We have

$$\lim_{x\to 0}\frac{\ln\cos x}{x^2} = \lim_{x\to 0}\frac{(-\sin x)/(\cos x)}{2x} = \lim_{x\to 0}\frac{-\sin x}{2x\cos x} = \lim_{x\to 0}\frac{-\cos x}{-2x\sin x + 2\cos x} = -\frac{1}{2}.$$

Thus, $\lim_{x\to 0} \ln F(x) = -\frac{1}{2}$. But since the logarithm is a continuous function, $\lim_{x\to 0} \ln F(x) = \ln(\lim_{x\to 0} F(x))$. Then

$$\ln(\lim_{x\to 0} F(x)) = -\frac{1}{2} \quad \text{or} \quad \lim_{x\to 0} F(x) = \lim_{x\to 0} (\cos x)^{1/x^2} = e^{-1/2}$$

4.31. If $F(x) = (e^{3x} - 5x)^{1/x}$, find (*a*) $\lim_{x\to 0} F(x)$ and (*b*) $\lim_{x\to 0} F(x)$.

The respective indeterminate forms in (*a*) and (*b*) are ∞^0 and 1^∞.

Let $G(x) = \ln F(x) = \dfrac{(\ln (e^{3x} - 5x)}{x}$. Then $\lim_{x\to\infty} G(x)$ and $\lim_{x\to 0} G(x)$ assume the indeterminate forms ∞/∞ and 0/0, respectively, and L'Hospital's rule applies. We have

(a) $$\lim_{x\to 0}\frac{\ln(e^{3x}-5x)}{x} = \lim_{x\to\infty}\frac{3e^{3x}-5}{e^{3x}-5x} = \lim_{x\to 0}\frac{9e^{3x}}{3e^{3x}-5} = \lim_{x\to\infty}\frac{27e^{3x}}{9e^{3x}} = 3$$

Then, as in Problem 4.30, $\lim_{x\to\infty} (e^{3x} - 5x)^{1/x} = e^3$.

(b) $$\lim_{x\to 0}\frac{\ln (e^{3x}-5x)}{x} = \lim_{x\to 0}\frac{3e^{3x}-5}{e^{3x}-5x} = -2$$

4.32. Suppose the equation of motion of a particle is $x = \sin(c_1 t + c_2)$, where c_1 and c_2 are constants (simple harmonic motion). (a) Show that the acceleration of the particle is proportional to its distance from the origin. (b) If $c_1 = 1$, $c_2 = \pi$, and $t \geq 0$, determine the velocity and acceleration at the endpoints and at the midpoint of the motion.

(a) $$\frac{dx}{dt} c_1 \cos(c_1 t + c_2), \quad \frac{d^2x}{dt^2} = c_1^2 \sin(c_1 t + c_2) = -c_1^2 x$$

This relation demonstrates the proportionality of acceleration and distance.

(b) The motion starts at 0 and moves to –1. Then it oscillates between this value and 1. The absolute value of the velocity is zero at the endpoints, and that of the acceleration is maximum there. The particle coasts through the origin (zero acceleration), while the absolute value of the velocity is maximum there.

4.33. Use Newton's method to determine $\sqrt{3}$ to three decimal points of accuracy.

$\sqrt{3}$ is a solution of $x^2 - 3 = 0$, which lies between 1 and 2. Consider $f(x) = x^2 - 3$, then $f'(x) = 2x$. The graph of f crosses the x axis between 1 and 2. Let $x_0 = 2$. Then $f(x_0) = 1$ and $f'(x_0) = 1.75$. According to the Newton formula, $x_1 = x_0 - \dfrac{f(x_0)}{f'(x_0)} = 2 - .25 = 1.75$.

Then $x_2 = x_1 - \dfrac{f(x_1)}{f'(x_1)} = 1.732$. To verify the three-decimal-point accuracy, note that $(1.732)^2 = 2.9998$ and $(1.7333)^2 = 3.0033$.

Miscellaneous problems

4.34. If $x = g(t)$ and $y = f(t)$ are twice differentiable, find (a) dy/dx and (b) d^2y/dx^2.

(a) Letting primes denote derivatives with respect to t, we have

$$\frac{dy}{dx} = \frac{dy/dt}{dx/dt} = \frac{f'(t)}{g'(t)} \quad \text{if } g'(t) \neq 0$$

(b) $$\frac{d^2y}{dx^2} = \frac{d}{dx}\left(\frac{dy}{dx}\right) = \frac{d}{dx}\left(\frac{f'(t)}{g'(t)}\right) = \frac{\frac{d}{dt}\left(\frac{f'(t)}{g'(t)}\right)}{dx/dt} = \frac{\frac{d}{dt}\left(\frac{f'(t)}{g'(t)}\right)}{g'(t)}$$

$$= \frac{1}{g'(t)}\left\{\frac{g'(t)f''(t) - f'(t)g''(t)}{[g'(t)]^2}\right\} = \frac{g'(t)f''(t) - f'(t)g''(t)}{[g'(t)]^3} \quad \text{if } g'(t) \neq 0$$

4.35. Let $f(x) = \begin{cases} e^{-1/x2}, & x \neq 0 \\ 0, & x \neq 0 \end{cases}$. Prove that (a) $f'(0) = 0$ and (b) $f''(0) = 0$.

(a) $f'_+(0) = \lim\limits_{h\to 0+} \dfrac{f(h) - f(0)}{h} = \lim\limits_{h\to 0+} \dfrac{e^{-1/h^2} - 0}{h} = \lim\limits_{h\to 0+} \dfrac{e^{-1/h^2}}{h}$

If $h = 1/u$, using L'Hospital's rule this limit equals

$$\lim_{u\to\infty} ue^{-u^2} = \lim_{u\to\infty} u/e^{u^2} = \lim_{u\to\infty} 1/2ue^{u^2} = 0$$

Similarly, replacing $h \to 0+$ by $h \to 0-$ and $u \to \infty$ by $u \to -\infty$, we find $f'_-(0) = 0$. Thus, $f'_+(0) = f'_-(0) = 0$, and so $f'(0) = 0$.

(b) $f''_+(0) = \lim\limits_{h\to 0+} \dfrac{f'(h) - f'(0)}{h} = \lim\limits_{h\to 0+} \dfrac{e^{-1/h^2} \cdot 2h^{-3} - 0}{h} = \lim\limits_{h\to 0+} \dfrac{2e^{-1/h^2}}{h} = \lim\limits_{u\to\infty} \dfrac{2u^4}{e^{u^2}} = 0$

by successive applications of L'Hospital's rule.

Similarly, $f''_-(0) = 0$ and so $f''(0) = 0$.

In general, $f^{(n)}(0) = 0$ for $n = 1, 2, 3, \ldots$

4.36. Find the length of the longest ladder which can be carried around the corner of a corridor whose dimensions are indicated in Figure 4.10, if it is assumed that the ladder is carried parallel to the floor.

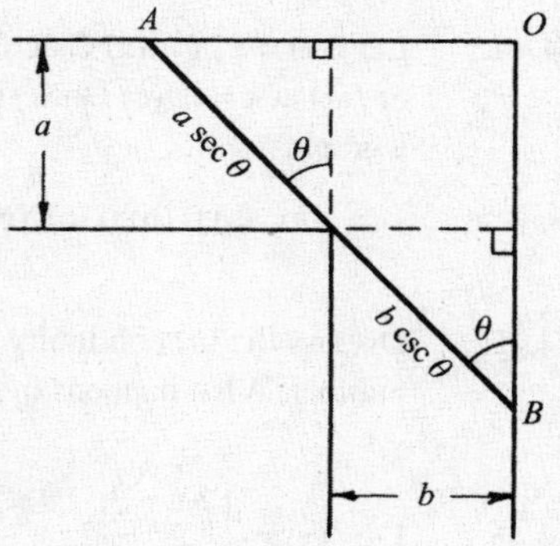

Figure 4.10

The length of the *longest* ladder is the same as the *shortest* straight-line segment AB (Figure 4.10), which touches both outer walls and the corner formed by the inner walls.

As seen from Figure 4.10, the length of the ladder AB is $L = a \sec \theta + b \csc \theta$.

L is a minimum when $dL/d\theta = a \sec \theta \tan \theta - b \csc \theta \cot \theta = 0$;

i.e., $a\sin^3 \theta = b \cos^3 \theta$ or $\tan\theta = \sqrt[3]{b/a}$. Then $\sec\theta = \dfrac{\sqrt{a^{2/3} + b^{2/3}}}{a^{1/3}}$

and $\cos\theta = \dfrac{\sqrt{a^{2/3} + b^{2/3}}}{b^{1/3}}$ so that $L = a \sec\theta + b \csc\theta = (a^{2/3} + b^{2/3})^{3/2}$.

Although it is geometrically evident that this gives the minimum length, we can prove this analytically by showing that $d^2L/d\theta^2$ for $\theta = \tan^{-1} \sqrt[3]{b/a}$ is positive (see Problem 4.78).

SUPPLEMENTARY PROBLEMS

Derivatives

4.37. Use the definition to compute the derivatives of each of the following functions at the indicated point: (a) $(3x - 4)/(2x+3)$, $x = 1$, (b) $x^3 - 3x^2 + 2x - 5$, $x = 2$, (c) $\sqrt{x}$, $x = 4$, and (d) $\sqrt[3]{6x-4}$, $x = 2$.

Ans. (a) 17/25, (b) 2, (c) $\frac{1}{4}$, (d) $\frac{1}{2}$

4.38. Show from definition that (a) $\frac{d}{dx}x^4 = 4x^3$ and (b) $\frac{d}{dx}\frac{3+x}{3-x} = \frac{6}{(3-x)^2}$, $x \neq 3$.

4.39. Let $f(x) = \begin{cases} x^3 \sin 1/x, & x \neq 0 \\ 0, & x = 0 \end{cases}$ Prove that (a) $f(x)$ is continuous at $x = 0$. (b) $f(x)$ has a derivative at $x = 0$, and (c) $f'(x)$ is continuous at $x = 0$.

4.40. Let $f(x) = \begin{cases} xe^{-1/x^2}, & x \neq 0 \\ 0, & x = 0 \end{cases}$. Determine whether $f(x)$ (a) is is continuous at $x = 0$, and (b) has a derivative at $x = 0$.

Ans. (a) Yes (b) Yes, 0

4.41. Give an alternative proof of the theorem in Problem 4.3, using "ϵ, δ" definitions.

4.42. If $f(x) = e^x$, show that $f'(x_0) = e^{x_0}$ depends on the result $\lim_{h\to 0}(e^h - 1)/h = 1$.

4.43. Use the results $\lim_{h\to 0}(\sin h)/h = 1$. $\lim_{h\to 0}(1-\cos h)/h = 0$ to prove that if $f(x) = \sin x$, $f'(x_0) = \cos x_0$.

Right-and left-hand derivatives

4.44. Let $f(x) = x|x|$. (a) Calculate the right-hand derivative of $f(x)$ at $x = 0$. (b) Calculate the left-hand derivative of $f(x)$ at $x = 0$. (c) Does $f(x)$ have a derivative at $x = 0$? (d) Illustrate the conclusions in (a), (b), and (c) from a graph.

Ans. (a) 0 (b) 0 (c) Yes, 0

4.45. Discuss the (*a*) continuity and (*b*) differentiability of $f(x) = x^p \sin 1/x$, $f(0) = 0$, where p is any positive number. What happens in case p is any real number?

4.46. Let $f(x) = \begin{cases} 2x-3, & 0 \leqq x \leqq 2 \\ x^2 - 3, & 2 < x \leqq 4 \end{cases}$. Discuss the (*a*) continuity and (*b*) differentiability of $f(x)$ in $0 \leqq x \leqq 4$.

4.47. Prove that the derivative of $f(x)$ at $x = x_0$ exists if and only if $f'_+(x_0) = f'_-(x_0)$.

4.48. (a) Prove that $f(x) = x^3 - x^2 + + 5x - 6$ is differentiable in $a \leqq x \leqq b$, where a and b are any constants. (b) Find equations for the tangent lines to the curve $y = x^3 - x^2 + 5x - 6$ at $x = 0$ and $x = 1$. Illustrate by means of a graph. (c) Determine the point of intersection of the tangent lines in (b). (d) Find $f'(x), f''(x), f'''(x), f^{(\text{IV})}(x), \ldots$

Ans. (b) $y = 5x - 6$, $y = 6x - 7$ (c) $(1, -1)$ (d) $3x^2 - 2x + 5, 6x - 2, 6, 0, 0, 0, \ldots$

4.49. If $f(x) = x^2|x|$, discuss the existence of successive derivatives of $f(x)$ at $x = 0$.

Differentials

4.50. If $y = f(x) = x + 1/x$, find (a) Δy, (b) dy, (c) $\Delta y - dy$, (d) $(\Delta y - dy)/\Delta x$, and (e) dy/dx.

Ans. (a) $\Delta x - \dfrac{\Delta x}{x(x+\Delta x)}$ (b) $\left(1 - \dfrac{1}{x^2}\right)\Delta x$ (c) $\dfrac{(\Delta x)^2}{x^2(x+\Delta x)}$ (d) $\dfrac{\Delta x}{x^2(x+\Delta x)}$ (e) $1 - \dfrac{1}{x^2}$
Note: $\Delta x = dx$.

4.51. If $f(x) = x^2 + 3x$, find (a) Δy, (b) dy, (c) $\Delta y/\Delta x$, (d) dy/dx, and (e) $(\Delta y - dy)/\Delta x$, if $x = 1$ and $\Delta x = .01$.

Ans. (a) .0501, (b) .05, (c) 5.01, (d) 5, (e) .01

4.52. Using differentials, compute approximate values for each of the following: (a) sin 31°, (b) ln(1.12), (c) $\sqrt[5]{36}$.

Ans. (a) 0.515, (b) 0.12, (c) 2.0125

4.53. If $y = \sin x$, evaluate (a) Δy and (b) dy. (c) Prove that $(\Delta y - dy)/\Delta x \to 0$ as $\Delta x \to 0$.

Differentiation rules and elementary functions

4.54. Prove the following:

(a) $\dfrac{d}{dx}\{f(x) + g(x)\} = \dfrac{d}{dx}f(x) + \dfrac{d}{dx}g(x)$

(b) $\dfrac{d}{dx}\{f(x) - g(x)\} = \dfrac{d}{dx}f(x) - \dfrac{d}{dx}g(x)$

(c) $\dfrac{d}{dx}\left\{\dfrac{f(x)}{g(x)}\right\} = \dfrac{g(x)f'(x) - f(x)g'(x)}{[g(x)]^2}, \quad g(x) \neq 0.$

4.55. Evaluate (a) $\dfrac{d}{dx}\left\{x^3 \ln\left(x^2 - 2x + 5\right)\right\}$ at $x = 1$ and (b) $\dfrac{d}{dx}\left\{\sin^2(3x + \pi/6\right\}$ at $x = 0$.

Ans. (a) $3\ln 4$ (b) $\dfrac{3}{2}\sqrt{3}$

4.56. Derive these formulas: (a) $\dfrac{d}{dx}a^u = a^u \ln a \dfrac{du}{dx}$, $a > 0$, $a \neq 1$; $\dfrac{d}{dx}\csc u = -\csc u \cot u \dfrac{du}{dx}$; and (c) $\dfrac{d}{dx}\tanh u = \text{sec h}^2 u \dfrac{du}{dx}$ where u is a differentiable function of x.

4.57. Compute (a) $\dfrac{d}{dx}\tan^{-1} x$, (b) $\dfrac{d}{dx}\csc^{-1} x$, (c) $\dfrac{d}{dx}\sinh^{-1} x$, and (d) $\dfrac{d}{dx}\coth^{-1} x$, paying attention to the use of principal values.

4.58. If $y = x^x$, compute dy/dx. (Hint: Take logarithms before differentiating.)

Ans. $x^x(1 + \ln x)$

4.59. If $y = \{\ln(3x + 2)\}^{\sin^{-1}(2x+.5)}$, find dy/dx at $x = 0$.

Ans. $\left(\dfrac{\pi}{4\ln 2} + \dfrac{2\ln \ln 2}{\sqrt{3}}\right)(\ln 2)^{\pi/6}$

4.60. If $y = f(u)$, where $u = g(\upsilon)$ and $\upsilon = h(x)$, prove that $\dfrac{dy}{dx} = \dfrac{dy}{du}\cdot\dfrac{du}{dv}\cdot\dfrac{dv}{dx}$ assuming f, g, and h are differentiable.

4.61. Calculate (*a*) dy/dx and (*b*) $d^2 y/dx^2$ if $xy - \ln y = 1$.

Ans. (a) $y^2/(1 - xy)$ (b) $(3y^3 - 2xy^4)/(1 - xy)^3$, provided $xy \neq 1$

4.62. If $y = \tan x$, prove that $y''' = 2(1 + y^2)(1 + 3y^2)$.

4.63. If $x = \sec t$ and $y = \tan t$, evaluate (a) dy/dx, (b) d^2y/dx^2, and (c) d^3y/dx^3, at $t = \pi/4$.

Ans. (a) $\sqrt{2}$ (b) -1 (c) $3\sqrt{2}$

4.64. Prove that $\dfrac{d^2y}{dx^2} = -\dfrac{d^2x}{dy^2}\bigg/\left(\dfrac{dx}{dy}\right)^3$, stating precise conditions under which it holds.

4.65. Establish formulas (a) 7 and (b) 18 on Pages 73 and 78.

Mean value theorems

4.66. Let $f(x) = 1 - (x - 1)^{2/3}$, $0 \leqq x \leqq 2$. (*a*) Construct the graph of $f(x)$. (*b*) Explain why Rolle's theorem is not applicable to this functions; i.e., there is no value ξ for which $f'(\xi) = 0$, $0 < \xi < 2$.

4.67. Verify Rolle's theorem for $f(x) = x^2(1 - x)^2$, $0 \leqq x \leqq 1$.

4.68. Prove that between any two real roots of $e^x \sin x = 1$ there is at least one real root of $e^x \cos x = -1$. (Hint: Apply Rolle's theorem to the function $e^{-x} - \sin x$.)

4.69. (a) If $0 < a < b$, prove that $(1 - a/b) < \ln b/a < (b/a - 1)$. (b) Use the result of (*a*) to show that $\dfrac{1}{6} < \ln 1.2\dfrac{1}{5}$.

4.70. Prove that $(\pi/6 + \sqrt{3}/15) < \sin^{-1}.6 < (\pi/6 + 1/8)$ by using the mean value theorem.

4.71. Show that the function $F(x)$ in Problem 4.20 represents the difference in ordinates of curve ACB and line AB at any point x in (a, b).

4.72. (a) If $f'(x) \leqq 0$ at all points of (a, b), prove that $f(x)$ is monotonic decreasing in (a, b). (b) Under what conditions is $f(x)$ strictly decreasing in (a, b)?

4.73. (a) Prove that $(\sin x)/x$ is strictly decreasing in $(0, \pi/2)$. (b) Prove that $0 \leqq \sin x \leqq 2x/\pi$ for $0 \leqq x \leqq \pi/2$.

4.74. (a) Prove that $\dfrac{\sin b - \sin a}{\cos a - \cos b} = \cot\xi$, where ξ is between a and b. (b) By placing $a = 0$ and $b = x$ in (*a*), show that $\xi = x/2$. Does the result hold if $x < 0$?

L'Hospital's Rule

4.75. Evaluate each of the following limits.

(a) $\lim_{x\to 0} \dfrac{x - \sin x}{x^3}$

(b) $\lim_{x\to 0} \dfrac{e^{2x} - 2^{ex} + 1}{\cos 3x - 2\cos 2x + \cos x}$

(c) $\lim_{x\to 1} (x^2 - 1)\tan \pi x/2$

(d) $\lim_{x\to\infty} x^3 e^{-2x}$

(e) $\lim_{x\to 0+} x^3 \ln x$

(f) $\lim_{x\to 0} (3^x - 2^x)/x$

(g) $\lim_{x\to\infty} (1 - 3/x)^{2x}$

(h) $\lim_{x\to\infty} (1 + 2x)^{1/3x}$

(i) $\lim_{x\to 0} (1/x - \csc x)$

(j) $\lim_{x\to 0} x^{\sin x}$

(k) $\lim_{x\to\infty} (1/x^2 - \cot^2 x)$

(l) $\lim_{x\to 0} \dfrac{\tan^{-1} x - \sin^{-1} x}{x(1 - \cos x)}$

(m) $\lim_{x\to\infty} x \ln\left(\dfrac{x+3}{x-3}\right)$

(n) $\lim_{x\to 0} \left(\dfrac{\sin x}{x}\right)^{1/x^2}$

(o) $\lim_{x\to\infty} (x + e^x + e^{2x})^{1/x}$

(p) $\lim_{x\to 0+} (\sin x)^{1/\ln x}$

Ans. (a) $\dfrac{1}{6}$ (b) -1 (c) $-4/\pi$ (d) 0 (e) 0 (f) $\ln 3/2$ (g) e^{-6} (h) 1 (i) 0 (j) 1 (k) $\dfrac{2}{3}$ (l) $\dfrac{1}{3}$ (m) 6 (n) $e^{-1/6}$ (o) e^2 (p) e

Miscellaneous problems

4.76. Prove that $\sqrt{\dfrac{1-x}{1+x}} < \dfrac{\ln(1+x)}{\sin^{-1} x} < 1$ if $0 < x < 1$.

4.77. If $\Delta f(x) = f(x + \Delta x) - f(x)$, (a) prove that $\Delta\{\Delta f(x)\} = \Delta^2 f(x) = f(x + 2\Delta x) - 2f(x + \Delta x) + f(x)$; (b) derive an expression for $\Delta^n f(x)$ where n is any positive integer; and (c) show that $\lim_{\Delta x\to 0} \dfrac{\Delta^n f(x)}{(\Delta x)^n} = f^{(n)}(x)$ if this limit exists.

4.78. Complete the analytic proof mentioned at the end of Problem 4.36.

4.79. Find the relative maximum and minima of $f(x) = x^2$, $x > 0$.

Ans. $f(x)$ has a relative minimum when $x = e^{-1}$.

4.80. A train moves according to the rule $x = 5t^3 + 30t$, where t and x are measured in hours and miles, respectively. (a) What is the acceleration after 1 minute? (b) What is the speed after 2 hours?

4.81. A stone thrown vertically upward has the law of motion $x = -16t^2 + 96t$. (Assume that the stone is at ground level at $t = 0$, that t is measured in seconds, and that x is measured in feet.) (*a*) What is the height of the stone at $t = 2$ seconds? (*b*) To what height does the stone rise? (*c*) What is the initial velocity, and what is the maximum speed attained?

4.82. A particle travels with constant velocities υ_1 and υ_2 in mediums I and II, respectively (see Figure 4.11). Show that in order to go from point P to point Q in the least time, it must follow path PAQ where A is such that

$$(\sin\theta_1)/(\sin\theta_2) = \upsilon_1/\upsilon_2$$

Note: This is Snell's law, a fundamental law of optics first discovered experimentally and then derived mathematically.

4.83. A variable α is called an *infinitesimal* if it has zero as a limit. Given two infinitesimals α and β, we say that α is an infinitesimal of *higher order* (or the *same order*) if $\lim \alpha/\beta = 0$ (or $\lim \alpha/\beta = l \neq 0$). Prove that as $x \to 0$, (a) $\sin^2 2x$ and $(1 - \cos 3x)$ are infinitesimals of the same order, and (b) $(x^3 - \sin^3 x)$ is an infinitesimal of higher order than $\{x - \ln(1 + x) - 1 + \cos x\}$.

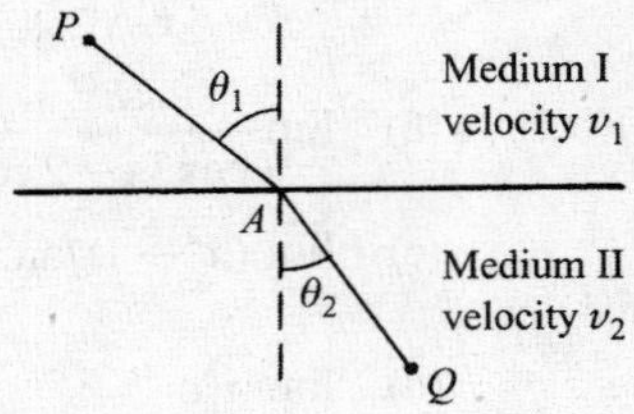

Figure 4.11

4.84. Why can we not use L'Hospital's rule to prove that $\lim\limits_{x\to 0} \dfrac{x^2 \sin 1/x}{\sin x} = 0$ (see Problem 3.91)?

4.85. Can we use L'Hospital's rule to evaluate the limit of the sequence $u_n = n^3 e^{-n^2}$, $n = 1, 2, 3, \ldots$? Explain.

4.86. (1) Determine decimal approximations with at least three places of accuracy for each of the following irrational numbers: (a) $\sqrt{2}$, (b) $\sqrt{5}$, and (c) $7^{1/3}$.

(2) The cubic equation $x^3 - 3x^2 + x - 4 = 0$ has a root between 3 and 4. Use Newton's method to determine it to at least three places of accuracy.

4.87. Using successive applications of Newton's method, obtain the positive root of (a) $x^3 - 2x^2 - 2x - 7 = 0$ and (b) $5 \sin x = 4x$ to three decimal places.

Ans. (a) 3.268 (b) 1.131

4.88. If D denotes the operator d/dx so that $Dy \equiv dy/dx$ while $D^k y \equiv d^k y/dx^k$, prove *Leibniz's formula*

$$D^n(uv) = (D^n u)v + {}_nC_1(D^{n-1}u)(Dv) + {}_nC_2(D^{n-2}u)(D^2 v) + \cdots + {}_nC_r(D^{n-r}u)D^r v + \cdots + uD^n v$$

where ${}_nC_r = \binom{n}{r}$ are the binomial coefficients (see Problem 1.95).

4.89. Prove that $\dfrac{d^n}{dx^n}(x^2 \sin x) = \{x^2 - n(n-1)\}\sin(x + n\pi/2) - 2nx\cos(x + n\pi/2)$.

4.90. If $f'(x_0) = f''(x_0) = \ldots = f^{(2n)}(x_0) = 0$ but $f^{(2n+1)}(x_0) \neq 0$, discuss the behavior of $f(x)$ in the neighborhood of $x = x_0$. The point x_0 in such case is often called a *point of inflection*. This is a generalization of the previously discussed case corresponding to $n = 1$.

4.91. Let $f(x)$ be twice differentiable in (a, b) and suppose that $f'(a) = f'(b) = 0$. Prove that there exists at least one point ξ in (a, b) such that $|f''(\xi)| \geq \dfrac{4}{(b-a)^2}\{f(b) - f(a)\}$. Give a physical interpretation involving the velocity and acceleration of a particle.

CHAPTER 5

Integrals

Introduction of the Definite Integral

The geometric problems that motivated the development of the integral calculus (determination of lengths, areas, and volumes) arose in the ancient civilizations of northern Africa. Where solutions were found, they related to concrete problems such as the measurement of a quantity of grain. Greek philosophers took a more abstract approach. In fact, Eudoxus (around 400 B.C.) and Archimedes (250 B.C.) formulated ideas of integration as we know it today.

Integral calculus developed independently and without an obvious connection to differential calculus. The calculus became a "whole" in the last part of the seventeenth century when Isaac Barrow, Isaac Newton, and Gottfried Wilhelm Leibniz (with help from others) discovered that the integral of a function could be found by asking what was differentiated to obtain that function.

The following introduction of integration is the usual one. It displays the concept geometrically and then defines the integral in the nineteenth-century language of limits. This form of definition establishes the basis for a wide variety of applications.

Consider the area of the region bound by $y = f(x)$, the x axis, and the joining vertical segments (ordinates) $x = a$ and $x = b$. (See Figure 5.1.)

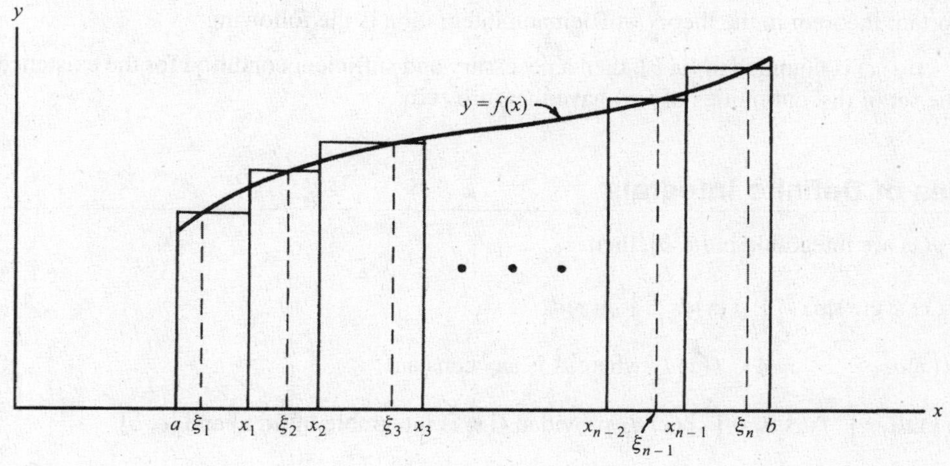

Figure 5.1

Subdivide the interval $a \leq x \leq b$ into n subintervals by means of the points $x_1, x_2, \ldots, x_{n-1}$, chosen arbitrarily. In each of the new intervals $(a, x_1), (x_1, x_2), \ldots, (x_{n-1}, b)$ choose points $\xi_1, \xi_2, \ldots, \xi_n$ arbitrarily. Form the sum

$$f(\xi_1)(x_1 - a) + f(\xi_2)(x_2 - x_1) + f(\xi_3)(x_3 - x_2) + \cdots + f(\xi_n)(b - x_{n-1}) \tag{1}$$

By writing $x_0 = a$, $x_n = b$, and $x_k - x_{k-1} = \Delta x_k$, this can be written

$$\sum_{k=1}^{n} f(\xi_k)(x_k - x_{k-1}) = \sum_{k=1}^{n} f(\xi_k)\Delta x_k \tag{2}$$

Geometrically, this sum represents the total area of all rectangles in Figure 5.1.

We now let the number of subdivisions n increase in such a way that each $\Delta x_k \to 0$. If, as a result, the sum (1) or (2) approaches a limit which does not depend on the mode of subdivision, we denote this limit by

$$\int_a^b f(x)dx = \lim_{n\to\infty} \sum_{k=1}^{n} f(\xi_k)\, \Delta\, x_k \tag{3}$$

This is called the *definite integral of $f(x)$ between a and b*. In this symbol, $f(x)\,dx$ is called the *integrand* and $[a, b]$ is called the *range of integration*. We call a and b the limits of integration, a being the lower limit of integration and b the upper limit.

The limit (3) exists whenever $f(x)$ is continuous (or piecewise continuous) in $a \leqq x \leqq b$ (see Problem 5.31). When this limit exists we say that f is *Riemann integrable* or simply *integrable* in $[a, b]$.

The definition of the definite integral as the limit of a sum was established by Cauchy around 1825. It was named for Georg Friedrich Bernhard Riemann because he made extensive use of it in this 1850 exposition of integration.

Geometrically, the value of this definite integral represents the area bounded by the curve $y = f(x)$, the x axis, and the ordinates at $x = a$ and $x = b$ only if $f(x) \geqq 0$. If $f(x)$ is sometimes positive and sometimes negative, the definite integral represents the algebraic sum of the areas above and below the x axis, treating areas above the x axis as positive and areas below the x axis as negative.

Measure Zero

A set of points on the x axis is said to have *measure zero* if the sum of the lengths of intervals enclosing all the points can be made arbitrarily small (less than any given positive number ε). We can show (see Problem 5.6) that any countable set of points on the real axis has measure zero. In particular, the set of rational numbers which is countable (see Problems 1.17 and 1.59), has measure zero.

An important theorem in the theory of Riemann integration is the following:

Theorem. If $f(x)$ is bounded in $[a, b]$, then a necessary and sufficient condition for the existence of $\int_a^b f(x)\,dx$ is that the set of discontinuities of $f(x)$ have measure zero.

Properties of Definite Integrals

If $f(x)$ and $g(x)$ are integrable in $[a, b]$, then

1. $\int_a^b \{f(x) \pm g(x)\}dx = \int_a^b f(x)dx \pm \int_a^b g(x)dx$
2. $\int_a^b Af(x)dx = A\int_a^b f(x)dx$ where A is any constant
3. $\int_a^b f(x)\,dx = \int_a^c f(x)\,dx + \int_c^b f(x)\,dx$ provided $f(x)$ is integrable in $[a, c]$ and $[c, b]$
4. $\int_a^b f(x)dx = -\int_b^a f(x)dx$
5. $\int_a^a f(x)dx = 0$
6. If in $a \leqq x \leqq b$, $m \leqq f(x) \leqq M$ where m and M are constants, then $m(b-a) \leqq \int_a^b f(x)\,dx \leqq M(b-a)$
7. If in $a \leqq x \leqq b$, $f(x) \leqq g(x)$, then $\int_a^b f(x)dx \leqq \int_a^b g(x)\,dx$
8. $\left|\int_a^b f(x)dx\right| \leqq \int_a^b |\,f(x)\,|\,dx$ if $a < b$

Mean Value Theorems for Integrals

As in differential calculus, the mean value theorems listed here are existence theorems. The first one generalizes the idea of finding an arithmetic mean (i.e., an average value of a given set of values) to a continuous function over an interval. The second mean value theorem is an extension of the first one, which defines a weighted average of a continuous function.

By analogy, consider determining the arithmetic mean (i.e., average value) of temperatures at noon for a given week. This question is resolved by recording the seven temperatures, adding them, and dividing by 7. To generalize from the notion of arithmetic mean and ask for the average temperature for the week is much more complicated because the spectrum of temperatures is now continuous. However, it is reasonable to believe that there exists a time at which the *average* temperature takes place. The manner in which the integral can be employed to resolve the question is suggested by the following example.

Let f be continuous on the closed interval $a \leq x \leq b$. Assume the function is represented by the correspondence $y = f(x)$, with $f(x) > 0$. Insert points of equal subdivision, $a = x_0, x_1, \ldots, x_n = b$. Then all $\Delta x_k = x_k - x_{k-1}$ are equal and each can be designated by Δx. Observe that $b - a = n\,\Delta x$. Let ξ_k be the midpoint of the interval Δx_k and $f(\xi_k)$ the value of f there. Then the average of these functional values is

$$\frac{f(\xi_1)+\cdots+f(\xi_n)}{n} = \frac{[f(\xi_1)+\cdots+f(\xi_n)]\,\Delta x}{b-a} = \frac{1}{b-a}\sum_{k=1}^{n} f(\xi_k)\,\Delta\xi_k$$

This sum specifies the average value of the n functions at the midpoints of the intervals. However, we may abstract the last member of the string of equalities (dropping the special conditions) and define

$$\lim_{n\to\infty}\frac{1}{b-a}\sum_{k=1}^{n} f(\xi_k)\,\Delta\xi_k = \frac{1}{b-a}\int_a^b f(x)dx$$

as the average value of f on $[a, b]$.

Of course, the question of for what value $x = \xi$ the average is attained is not answered; in fact, in general, only existence, not the value, can be demonstrated. To see that there is a point $x = \xi$ such that $f(\xi)$ represents the average value of f on $[a, b]$, recall that a continuous function on a closed interval has maximum and minimum values M and m, respectively. (Think of the integral as representing the area under the curve; see Figure 5.2.) Thus,

$$m(b-a) \leq \int_a^b f(x)\,dx \leq M(b-a)$$

or

$$m \leq \frac{1}{b-a}\int_a^b f(x)\,dx \leq M$$

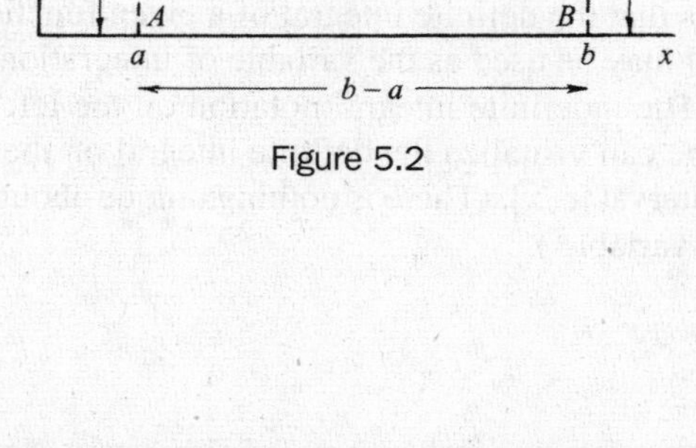

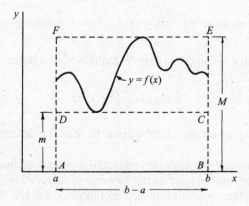

Figure 5.2

Since f is a continuous function on a closed interval, there exists a point $x = \xi$ in (a, b) intermediate to m and M such that

$$f(\xi) = \frac{1}{b-a}\int_a^b f(x)\,dx$$

While this example is not a rigorous proof of the first mean value theorem, it motivates it and provides an interpretation. (See Chapter 3, Theorem 10.)

First Mean Value Theorem If $f(x)$ is continuous in $[a, b]$, there is a point ξ in (a, b) such that

$$\int_a^b f(x)\,dx = (b-a)f(\xi) \tag{4}$$

Generalized First Mean Value Theorem If $f(x)$ and $g(x)$ are continuous in $[a, b]$, and $g(x)$ does not change sign in the interval, then there is a point ξ in (a, b) such that

$$\int_a^b f(x)g(x)dx = f(\xi)\int_a^b g(x)\,dx \tag{5}$$

This reduces to Equation (4) if $g(x) = 1$.

Connecting Integral and Differential Calculus

In the late seventeenth century the key relationship between the derivative and the integral was established. The connection which is embodied in the fundamental theorem of calculus was responsible for the creation of a whole new branch of mathematics called analysis.

Definition Any function F such that $F'(x) = f(x)$ *is called an antiderivative, primitive,* or *indefinite integral* of f.

The antiderivative of a function is not unique. This is clear from the observation that for any constant c

$$(F(x) + c)' = F'(x) = f(x)$$

The following theorem is an even stronger statement.

Theorem Any two primitives (i.e., antiderivatives) F and G of f differ at most by a constant; i.e., $F(x) - G(x) = C$.

(See the problem set for the proof of this theorem.)

EXAMPLE. If $F'(x) = x^2$, then $F(x) = \int x^2 dx = \frac{x^3}{3} + c$ is an indefinite integral (antiderivative or primitive) of x^2.

The indefinite integral (which is a function) may be expressed as a definite integral by writing

$$\int f(x)dx = \int_c^x f(t)\,dt$$

The functional character is expressed through the upper limit of the definite integral which appears on the right-hand side of the equation.

This notation also emphasizes that the definite integral of a given function depends only on the limits of integration, and thus any symbol may be used as the variable of integration. For this reason, that variable is often called a *dummy* variable. The indefinite integral notation on the left depends on continuity of f on a domain that is not described. One can visualize the definite integral on the right by thinking of the dummy variable t as ranging over a subinterval $[c, x]$. (There is nothing unique about the letter t; any other convenient letter may represent the dummy variable.)

The previous terminology and explanation set the stage for the fundamental theorem. It is stated in two parts. Part 1 states that the antiderivative of f is a new function, the integrand of which is the derivative of that function. Part 2 demonstrates how that primitive function (antiderivative) enables us to evaluate definite integrals.

The Fundamental Theorem of the Calculus

Part 1. Let f be integrable on a closed interval $[a, b]$. Let c satisfy the condition $a \leqq c \leqq b$, and define a new function

$$F(x) = \int_c^x f(t)\, dt \quad \text{if} \quad a \leqq x \leqq b$$

Then the derivative $F'(x)$ exists at each point x in the open interval (a, b), where f is continuous and $F'(x) = f(x)$. (See Problem 5.10 for proof of this theorem.)

Part 2. As in Part 1, assume that f is integrable on the closed interval $[a, b]$ and continuous in the open interval (a, b). Let F be any antiderivative so that $F'(x) = f(x)$ for each x in (a, b). If $a < c < b$, then for any x in (a, b)

$$\int_c^x f(t)\, dt = F(x) - F(c)$$

If the open interval on which f is continuous includes a and b, then we may write

$$\int_a^b f(x)\, dx = F(b) - F(a). \qquad \text{(See Problem 5.11)}$$

This is the usual form in which the theorem is used.

EXAMPLE. To evaluate $\int_1^2 x^2 dx$ we observe that $F'(x) = x^2$, $F(x) = \frac{x^3}{3} + c$, and $\int_1^2 x^2\, dx = \left(\frac{2^3}{3} + c\right) - \left(\frac{1^3}{3} + c\right) = \frac{7}{3}$. Since c subtracts out of this evaluation, it is convenient to exclude it and simply write $\frac{2^3}{3} - \frac{1^3}{3}$.

Generalization of the Limits of Integration

The upper and lower limits of integration may be variables. For example:

$$\int_{\sin x}^{\cos x} t\, dt = \left[\frac{t^2}{2}\right]_{\sin x}^{\cos x} = (\cos^2 x - \sin^2 x)/2$$

In general, if $F'(x) = f(x)$, then

$$\int_{u(x)}^{\upsilon(x)} f(t)\, dt = F[\upsilon(x)] = F[u(x)]$$

Change of Variable of Integration

If a determination of $\int f(x)\, dx$ is not immediately obvious in terms of elementary functions, useful results may be obtained by changing the variable from x to t according to the transformation $x = g(t)$. [This change of integrand that follows is suggested by the differential relation $dx = g'(t)\,dt$.] The fundamental theorem enabling us to do this is summarized in the statement

$$\int f(x)\,dx = \int f\{g(t)\}g'(t)\,dt \qquad (6)$$

where, after obtaining the indefinite integral on the right, we replace t by its value in terms of x; i.e., $t = g^{-1}(x)$. This result is analogous to the chain rule for differentiation (see Page 76).

The corresponding theorem for definite integrals is

$$\int_a^b f(x)\,dx = \int_\alpha^\beta f\{g(t)\}g'(t)\,dt \qquad (7)$$

where $g(\alpha) = a$ and $g(\beta) = b$; i.e., $\alpha = g^{-1}(a)$, $\beta = g^{-1}(b)$. This result is certainly valid if $f(x)$ is continuous in $[a, b]$ and if $g(t)$ is continuous and has a continuous derivative in $\alpha \leq t \leq \beta$.

Integrals of Elementary Functions

The following results can be demonstrated by differentiating both sides to produce an identity. In each case, an arbitrary constant c (which has been omitted here) should be added.

1. $\int u^n\,du = \dfrac{u^{n+1}}{n+1} \quad n \neq -1$
2. $\int \dfrac{du}{u} = \ln|u|$
3. $\int \sin u\,du = -\cos u$
4. $\int \cos u\,du = \sin u$
5. $\int \tan u\,du = \ln|\sec u| = -\ln|\cos u|$
6. $\int \cot u\,du = \ln|\sin u|$
7. $\int \sec u\,du = \ln|\sec u + \tan u| = \ln|\tan(u/2 + \pi/4)|$
8. $\int \csc u\,du = \ln|\csc u - \cot u| = \ln|\tan u/2|$
9. $\int \sec^2 u\,du = \tan u$
10. $\int \csc^2 u\,du = -\cot u$
11. $\int \sec u \tan u\,du = \sec u$
12. $\int \csc u \cot u\,du = -\csc u$
13. $\int a^u\,du = \dfrac{a^u}{\ln a} \quad a > 0, a \neq 1$

18. $\int \coth u\,du = \ln|\sinh u|$
19. $\int \operatorname{sech} u\,du = \tan^{-1}(\sinh u)$
20. $\int \operatorname{csch} u\,du = -\coth^{-1}(\cosh u)$
21. $\int \operatorname{sech}^2 u\,du = \tanh u$
22. $\int \operatorname{csc}h^2 u\,du = -\coth u$
23. $\int \operatorname{sech} u \tanh u\,du = -\operatorname{sech} u$
24. $\int \operatorname{csch} u \coth u\,du = -\operatorname{csch} u$
25. $\int \dfrac{du}{\sqrt{s^2 - u^2}} = \sin^{-1}\dfrac{u}{a} \quad \text{or} \quad -\cos^{-1}\dfrac{u}{a}$
26. $\int \dfrac{du}{\sqrt{u^2 \pm a^2}} = \ln|u + \sqrt{u^2 \pm a^2}|$
27. $\int \dfrac{du}{u^2 + a^2} = \dfrac{1}{a}\tan^{-1}\dfrac{u}{a} \quad \text{or} \quad -\dfrac{1}{a}\cot^{-1}\dfrac{u}{a}$
28. $\int \dfrac{du}{u^2 - a^2} = \dfrac{1}{2a}\ln\left|\dfrac{u-a}{u+a}\right|$
29. $\int \dfrac{du}{u\sqrt{a^2 \pm u^2}} = \dfrac{1}{a}\ln\left|\dfrac{u}{a + \sqrt{a^2 \pm u^2}}\right|$
30 $\int \dfrac{du}{u\sqrt{u^2 - a^2}} = \dfrac{1}{a}\cos^{-1}\dfrac{a}{u} \quad \text{or} \quad \dfrac{1}{a}\sec^{-1}\dfrac{u}{a}$

14. $\int e^u\, du = e^u$

15. $\int \sinh u\, du = \cosh u$

16. $\int \cosh u\, du = \sinh u$

17. $\int \tanh u\, du = \ln \cosh u$

31. $\int \sqrt{u^2 \pm a^2}\, du = \frac{u}{2}\sqrt{u^2 \pm a^2} \pm \frac{a^2}{2} \ln | u + \sqrt{u^2 \pm a^2} |$

32. $\int \sqrt{a^2 - u^2}\, du = \frac{u}{2}\sqrt{a^2 - u^2} + \frac{a^2}{2}\sin^{-1}\frac{u}{a}$

33. $\int e^{au} \sin bu\, du = \frac{e^{au}(a \sin bu - b \cos bu)}{a^2 + b^2}$

34. $\int e^{au} \cos bu\, du = \frac{e^{au}(a \cos bu + b \sin bu)}{a^2 + b^2}$

Special Methods of Integration

1. Integration by Parts Let u and υ be differentiable functions. According to the product rule for differentials,

$$d(u\upsilon) = u\, d\upsilon + \upsilon\, du$$

Upon taking the antiderivative of both sides of the equation, we obtain

$$u\upsilon = \int u\, d\upsilon + \int \upsilon\, du$$

This is the formula for integration by parts when written in the form

$$\int \upsilon\, dv = uv - \int \upsilon\, du \quad \text{or} \quad \int f(x)g'(x)\, dx = f(x)g(x) - \int f'(x)g(x)dx$$

where $u = f(x)$ and $\upsilon = g(x)$. The corresponding result for definite integrals over the interval $[a, b]$ is certainly valid if $f(x)$ and $g(x)$ are continuous and have continuous derivatives in $[a, b]$. See Problems 5.17 to 5.19.

2. Partial Fractions Any rational function $\frac{P(x)}{Q(x)}$ where $P(x)$ and $Q(x)$ are polynomials, with the degree of $P(x)$ less than that of $Q(x)$, can be written as the sum of rational functions having the form $\frac{A}{(ax + b)^r}, \frac{Ax + B}{(ax^2 + bx + c)^r}$ where $r = 1, 2, 3, \ldots$, which can always be integrated in terms of elementary functions.

EXAMPLE 1.

$$\frac{3x - 2}{(4x - 3)(2x - 5)^3} = \frac{A}{4x - 3} + \frac{B}{(2x + 5)^3} + \frac{C}{(2x + 5)^2} + \frac{D}{(2x + 5)}$$

EXAMPLE 2.

$$\frac{5x^2 - x + 2}{(x^2 + 2x + 4)^2(x - 1)} = \frac{Ax + B}{(x^2 + 2x + 4)^2} + \frac{Cx + D}{x^2 + 2x + 4} + \frac{E}{x - 1}$$

The constants, A, B, C, etc., can be found by clearing of fractions and equating coefficients of like powers of x on both sides of the equation or by using special methods (see Problem 5.20).

3. Rational Functions of sin x and cos x These can always be integrated in terms of elementary functions by the substitution $\tan x/2 = u$ (see Problem 5.21).

4. Special Devices Depending on the particular form of the integrand, special devices are often employed (see Problems 5.22 and 5.23).

Improper Integrals

If the range of integration $[a, b]$ is not finite or if $f(x)$ is not defined or not bounded at one or more points of $[a, b]$, then the integral of $f(x)$ over this range is called an *improper integral*. By use of appropriate limiting operations, we may define the integrals in such cases.

EXAMPLE 1

$$\int_0^{\infty} \frac{dx}{1+x^2} = \lim_{M\to\infty} \int_0^M \frac{dx}{1+x^2} = \lim_{M\to\infty} \tan^{-1} x \Big|_0^M = \lim_{M\to\infty} \tan^{-1} M = \pi/2$$

EXAMPLE 2

$$\int_0^1 \frac{dx}{\sqrt{x}} = \lim_{\epsilon\to 0+} \int_{\epsilon}^1 \frac{dx}{\sqrt{x}} = \lim_{\epsilon\to 0+} 2\sqrt{x}\Big|_{\epsilon}^1 = \lim_{\epsilon\to 0+} (2-2\sqrt{\epsilon}) = 2$$

EXAMPLE 3

$$\int_0^1 \frac{dx}{\sqrt{x}} = \lim_{\epsilon\to 0+} \int_{\epsilon}^1 \frac{dx}{x} = \lim_{\epsilon\to 0+} \ln x\Big|_{\epsilon}^1 = \lim_{\epsilon\to 0+} (-\ln \epsilon)$$

Since this limit does not exist, we say that the integral diverges (i.e., does not converge).

For further examples, see Problems 5.29 and 5.74 through 5.76. For further discussion of improper integrals, see Chapter 12.

Numerical Methods for Evaluating Definite Integrals

Numerical methods for evaluating definite integrals are available in case the integrals cannot be evaluated exactly. The following special numerical methods are based on subdividing the interval $[a, b]$ into n equal parts of length $\Delta x = (b - a)/n$. For simplicity we denote $f(a + k\Delta x) = f(x_k)$ by y_k, where $k = 0, 1, 2, \ldots, n$. The symbol $\approx$ means "approximately equal." In general, the approximation improves as n increases.

1. Rectangular Rule

$$\int_a^b f(x)dx \approx \Delta x\{y_0 + y_1 + y_2 + \cdots + y_{n-1}\} \text{ or } \Delta x\{y_1 + y_2 + y_3 + \cdots + y_n\} \tag{8}$$

The geometric interpretation is evident from Figure 5.1. When left endpoint function values $y_0, y_1, \ldots, y_{n-1}$ are used, the rule is called the *left-hand rule*. Similarly, when right endpoint evaluations are employed, it is called the *right-hand rule*.

2. Trapezoidal Rule

$$\int_a^b f(x)dx \approx \frac{\Delta x}{2}\{y_0 + 2y_1 + 2y_2 + \cdots + 2y_{n-1} + y_n\} \tag{9}$$

This is obtained by taking the mean of the approximations in Equation (8). Geometrically, this replaces the curve $y = f(x)$ by a set of approximating line segments.

3. Simpson's Rule

$$\int_a^b f(x)dx \approx \frac{\Delta x}{3}\{y_0 + 4y_1 + 2y_2 + 4y_3 + 2y_4 + 4y_5 + \cdots + 2y_{n-2} + 4y_{n-1} + y_n\} \qquad (10)$$

This formula is obtained by approximating the graph of $y = g(x)$ by a set of parabolic arcs of the form $y = ax^2 + bx + c$. The correlation of two observations lead to Equation (*10*). First,

$$\int_{-h}^{h} [ax^2 + bx + c]dx = \frac{h}{3}[2ah^2 + 6c]$$

The second observation is related to the fact that the vertical parabolas employed here are determined by three nonlinear points. In particular, consider $(-h, y_0)$, $(0, y_1)$, (h, y_2), then $y_0 = a(-h)^2 + b(-h) + c$, $y_1 = c$, $y_2 = ah^2 + bh + c$. Consequently, $y_0 + 4y_1 + y_2 = 2ah^2 + 6c$. Thus, this combination of ordinate values (corresponding to equally spaced domain values) yields the area bounded by the parabola, vertical segments, and the x axis. Now these ordinates may be interpreted as those of the function f whose integral is to be approximated. Then, as illustrated in Figure 5.3:

$$\sum_{k=1}^{n} \frac{h}{3}[y_{k-1} + 4y_k + y_{k+1}] = \frac{\Delta x}{3}[y_0 + 4y_1 + 2y_2 + 4y_3 + 2y_4 + 4y_5 + \cdots + 2y_{n-2} + 4y_{n-1} + y_n]$$

The Simpson rule is likely to give a better approximation than the others for smooth curves.

Applications

The use of the integral as a limit of a sum enables us to solve many physical and geometrical problems such as determination of areas, volumes, arc lengths, moments of intertia, and centroids.

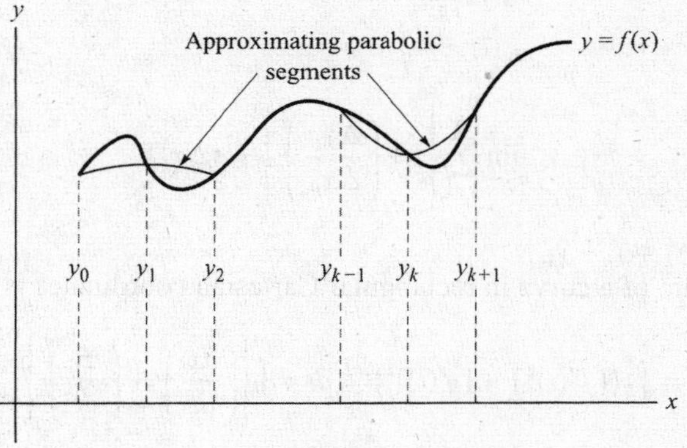

Figure 5.3

Arc Length

As you walk a twisting mountain trail, it is possible to determine the distance covered by using a pedometer. To create a geometric model of this event, it is necessary to describe the trail and a method of measuring distance along it. The trail might be referred to as a *path*, but in more exacting geometric terminology the word *curve* is appropriate. That segment to be measured is an arc of the curve. The arc is subject to the following restrictions:

1. It does not intersect itself (i.e., it is a simple arc).

2. There is a tangent line at each point.
3. The tangent line varies continuously over the arc.

These conditions are satisfied with a parametric representation $x = f(t)$, $y = g(t)$, $z = h(t)$, $a \leqq t \leqq b$, where the functions f, g, and h have continuous derivatives that do not simultaneously vanish at any point. This arc is in Euclidean three-dimensional space and is discussed in Chapter 10. In this introduction to curves and their arc length, we let $z = 0$, thereby restricting the discussion to the plane.

A careful examination of your walk would reveal movement on a sequence of straight segments, each changed in direction from the previous one. This suggests that the length of the arc of a curve is obtained as the limit of a sequence of lengths of polygonal approximations. (The polygonal approximations are characterized by the number of divisions $n \to \infty$ and no subdivision is bound from zero. (See Figure 5.4.)

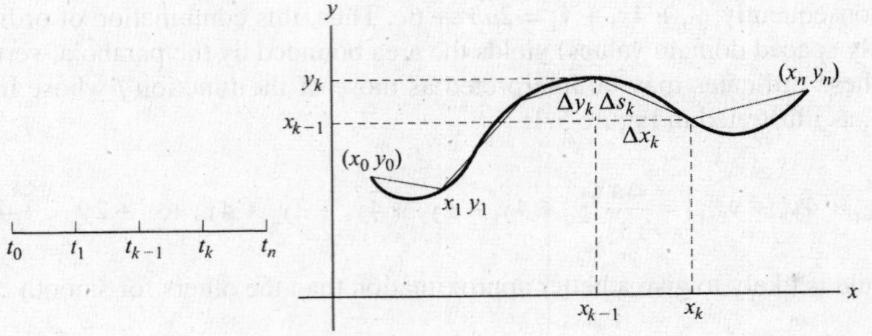

Figure 5.4

Geometrically, the measurement of the kth segment of the arc $0 \leqq t \leqq s$ is accomplished by employing the Pythagorean theorem; thus, the measure is defined by

$$\lim_{n\to\infty} \sum_{k=1}^{n} \{(\Delta x_k)^2 + (\Delta y_k)^2\}^{1/2}$$

or, equivalently,

$$\lim_{n\to\infty} \sum_{k=1}^{n} \left\{1 + \left(\frac{\Delta y_k}{\Delta x_k}\right)^2\right\}^{1/2} (\Delta x_k)$$

where $\Delta x_k = x_k - x_{k-1}$ and $\Delta y_k = y_k - y_{k-1}$.

Thus, the length of the arc of a curve in rectangular Cartesian coordinates is

$$L = \int_a^b \{[f'(t)^2] + [g'(t)]^2\}^{1/2}\, dt = \int \left\{\left(\frac{dx}{dt}\right)^2 + \left(\frac{dy}{dt}\right)^2\right\}^{1/2} dt$$

(This form may be generalized to any number of dimensions.)

Upon changing the variable of integration from t to x we obtain the planar form

$$L = \int_{f(a)}^{f(b)} \left\{1 + \left[\frac{dy}{dx}\right]^2\right\}^{1/2}$$

(This form is appropriate only in the plane.)

The generic differential formula $ds^2 = dx^2 + dy^2$ is useful, in that various representations algebraically arise from it. For example,

$$\frac{ds}{dt}$$

expresses instantaneous speed.

Area

Area was a motivating concept in introducing the integral. Since many applications of the integral are geometrically interpretable in the context of area, an extended formula is listed and illustrated here.

Let f and g be continuous functions whose graphs intersect at the graphical points corresponding to $x = a$ and $x = b$, $a < b$. If $g(x) \varepsilon f(x) \varepsilon f(x)$ on $[a, b]$, then the area bounded by $f(x)$ and $g(x)$ is

$$A = \int_a^b \{g(x) - f(x)\}dx$$

If the functions intersect in (a, b), then the integral yields an algebraic sum. For example, if $g(x) = \sin x$ and $f(x) = 0$ then

$$\int_0^{2\pi} \sin x \, dx = \cos x \Big|_0^{2\pi} = 0$$

Volumes of Revolution

Disk Method Assume that f is continuous on a closed interval $a \leq x \leq b$ and that $f(x) \varepsilon 0$. Then the solid realized through the revolution of a plane region R [bound by $f(x)$, the x axis, and $x = a$ and $x = b$] about the x axis has the volume

$$V = \pi \int_a^b [f(x)]^2 \, dx$$

This method of generating a volume is called the *disk method* because the cross sections of revolution are circular disks. See Figure 5.5(*a*).

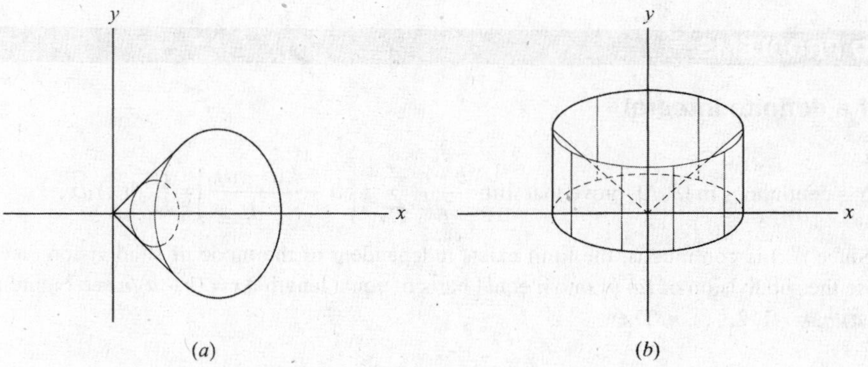

Figure 5.5

EXAMPLE. A solid cone is generated by revolving the graph of $y = kx$, $k > 0$, and $0 \leq x \leq b$ about the x axis. Its volume is

$$V = \pi \int_0^b k^2x^2 \, dx = \pi \frac{k^3x^3}{3}\Big|_0^b = \pi \frac{k^3b^3}{3}$$

Shell Method Suppose f is a continuous function on $[a, b]$, $a \varepsilon 0$, satisfying the condition $f(x) \varepsilon 0$. Let R be a plane region bounded by $f(x)$, $x = a$, $x = b$, and the x axis. The volume obtained by orbiting R about the y axis is

$$V = \int_a^b 2\pi x f(x)dx$$

This method of generating a volume is called the *shell method* because of the cylindrical nature of the vertical lines of revolution. See Figure 5.5(*b*).

EXAMPLE. If the region bounded by $y = kx$, $0 \leq x \leq b$, and $x = b$ (with the same conditions as in the previous example) is orbited about the y axis, the volume obtained is

$$V = 2\pi \int_0^b x(kx)\,dx = 2\pi k \frac{x^3}{3}\Big|_0^b = 2\pi k \frac{b^3}{3}$$

By comparing this example with that in the section on the disk method, it is clear that for the same plane region the disk method and the shell method produce different solids and, hence, different volumes.

Moment of Inertia Moment of inertia is an important physical concept that can be studied through its idealized geometric form. This form is abstracted in the following way from the physical notions of kinetic energy $K = 1/2\ m\upsilon^2$ and angular velocity $\upsilon = \omega r$ (m represents mass and υ signifies linear velocity). Upon substituting for υ,

$$K = \frac{1}{2} m\omega^2 r^2 = \frac{1}{2}(mr^2)\omega^2$$

When this form is compared to the original representation of kinetic energy, it is reasonable to identify mr^2 as rotational mass. It is this quantity, $l = mr^2$, that we call the *moment of inertia*.

Then in a purely geometric sense, we denote a plane region R described through continuous functions f and g on $[a, b]$, where $a > 0$ and $f(x)$ and $g(x)$ intersect at a and b only. For simplicity, assume $g(x)\ \varepsilon\ f(x) > 0$. Then

$$l = \int_a^b x^2[g(x) - f(x)]\,dx$$

By idealizing the plane region R as a volume with uniform density *one*, the expression $[f(x) - g(x)]\ dx$ stands in for mass and r^2 has the coordinate representation x^2. See Problem 5.25(b) for more details.

SOLVED PROBLEMS

Definition of a definite integral

5.1. If $f(x)$ is continuous in $[a, b]$, prove that $\lim\limits_{n\to\infty} \dfrac{b-a}{n} \sum\limits_{k=1}^{n} f\left(a + \dfrac{k(b-a)}{n}\right) = \int_a^b f(x)\,dx.$

Since $f(x)$ is continuous, the limit exists independent of the mode of subdivision (see Problem 5.31). Choose the subdivision of $[a, b]$ into n equal parts of equal length $\Delta x = (b - a)/n$ see Figure 5.1. Let $\xi_k = a + k(b - a)/n$, $k = 1, 2, \ldots, n$. Then

$$\lim_{n\to\infty} \sum_{k=1}^{n} f(\xi_k)\Delta x_k = \lim_{n\to\infty} \frac{b-a}{n} \sum_{k=1}^{n} f\left(a + \frac{k(b-a)}{n}\right) = \int_a^b f(x)\,dx$$

5.2. Express $\lim\limits_{n\to\infty} \dfrac{1}{n} \sum\limits_{k=1}^{n} f\left(\dfrac{k}{n}\right)$ as a definite integral.

Let $a = 0$, $b = 1$ in Problem 5.1. Then

$$\lim_{n\to\infty} \frac{1}{n} \sum_{k=1}^{n} f\left(\frac{k}{n}\right) = \int_0^1 f(x)\,dx$$

5.3. (a) Express $\int_0^1 x^2\,dx$ as a limit of a sum, and use the result to evaluate the given definite integral. (b) Interpret the result geometrically.

(a) If $f(x) = x^2$, then $f(k/n) = (k/n)^2 = k^2/n^2$. Thus, by Problem 5.2,

$$\lim_{n\to\infty}\frac{1}{n}\sum_{k=1}^{n}\frac{k^2}{n^2} = \int_0^1 x^2\,dx$$

This can be written, using Problem 1.29,

$$\begin{aligned}\int_0^1 x^2\,dx &= \lim_{n\to\infty}\frac{1}{n}\left(\frac{1^2}{n^2}+\frac{2^2}{n^2}+\cdots+\frac{n^2}{n^2}\right) = \lim_{n\to\infty}\frac{1^2+2^2+\cdots+n^2}{n^3}\\ &= \lim_{n\to\infty}\frac{n(n+1)(2n+1)}{6n^3}\\ &= \lim_{n\to\infty}\frac{(1+1/n)(2+1/n)}{6} = \frac{1}{3}\end{aligned}$$

which is the required limit.

Note: By using the fundamental theorem of the calculus, we observe that $\int_0^1 x^2\,dx = (x^3/3)\big|_0^1 = 1^3/3 - 0^3/3 = 1/3$.

(b) The area bounded by the curve $y = x^2$, the x axis, and the line $x = 1$ is equal to $\frac{1}{3}$.

5.4. Evaluate $\lim_{n\to\infty}\left\{\frac{1}{n+1}+\frac{1}{n+2}+\cdots+\frac{1}{n+n}\right\}$.

The required limit can be written

$$\lim_{n\to\infty}\frac{1}{n}\left\{\frac{1}{1+1/n}+\frac{1}{1+2/n}+\cdots+\frac{1}{1+n/n}\right\} = \lim_{n\to\infty}\frac{1}{n}\sum_{k=1}^{n}\frac{1}{1+k/n} = \int_0^1\frac{dx}{1+x} = \ln|1+x|\big|_0^1 = \ln 2$$

using Problem 5.2 and the fundamental theorem of the calculus.

5.5. Prove that $\lim_{n\to\infty}\frac{1}{n}\left\{\sin\frac{t}{n}+\sin\frac{2t}{n}+\cdots+\sin\frac{(n-1)t}{n}\right\} = \frac{1-\cos t}{t}$.

Let $a = 0$, $b = t$, $f(x) = \sin x$ in Problem 1. Then

$$\lim_{n\to\infty}\frac{1}{n}\sum_{k=1}^{n}\sin\frac{kt}{n} = \int_0^t \sin x\,dx = 1-\cos t$$

and so

$$\lim_{n\to\infty}\frac{1}{n}\sum_{k=1}^{n-1}\sin\frac{kt}{n} = \frac{1-\cos t}{t}$$

using the fact that $\lim_{n\to\infty}\frac{\sin t}{n} = 0$.

Measure zero

5.6. Prove that a countable point set has measure zero.

Let the point set be denoted by $x_1, x_2, x_3, x_4, \ldots$ and suppose that intervals of lengths less than $\varepsilon/2$, $\varepsilon/4$, $\varepsilon/8$, $\varepsilon/16, \ldots$, respectively, enclose the points, where ε is any positive number. Then the sum of the lengths of the intervals is less than $\varepsilon/2 + \varepsilon/4 + \varepsilon/8 + \ldots = \varepsilon$ [let $a = \varepsilon/2$ and $r = 1/2$ in Problem 2.25(a)], showing that the set has measure zero.

Properties of definite integrals

5.7. Prove that $\left|\int_a^b f(x)\,dx\right| \leq \int_a^b |f(x)|\,dx$ if $a < b$.

By absolute value property 2, on Page 4,

$$\left|\sum_{k=1}^{n} f(\xi_k)\Delta x_k\right| \leq \sum_{k=1}^{n} |f(\xi_k)\Delta x_k| = \sum_{k=1}^{n} |f(\xi_k)|\Delta x_k$$

Taking the limit as $n \to \infty$ and each $\Delta x_k \to 0$, we have the required result.

5.8. Prove that $\lim_{n\to\infty} \int_0^{2\pi} \frac{\sin nx}{x^2+n^2}\,dx = 0$.

$$\left|\int_0^{2\pi} \frac{\sin nx}{x^2+n^2}\,dx\right| \leq \int_0^{2\pi} \left|\frac{\sin nx}{x^2+n^2}\right| dx \leq \int_0^{2\pi} \frac{dx}{n^2} = \frac{2\pi}{n^2}$$

Then $\lim_{n\to\infty}\left|\int_0^{2\pi} \frac{\sin nx}{x^2+n^2}\,dx\right| = 0$, and so the required result follows.

Mean value theorems for integrals

5.9. Given the right triangle pictured in Figure 5.6: (a) Find the average value of h. (b) At what point does this average value occur? (c) Determine the average value of $f(x) = \sin^{-1} x$, $0 \leq x \leq \frac{1}{2}$. (Use integration by parts.) (d) Determine the average value of $f(x) = \cos^2 x$, $0 \leq x \leq \frac{\pi}{2}$.

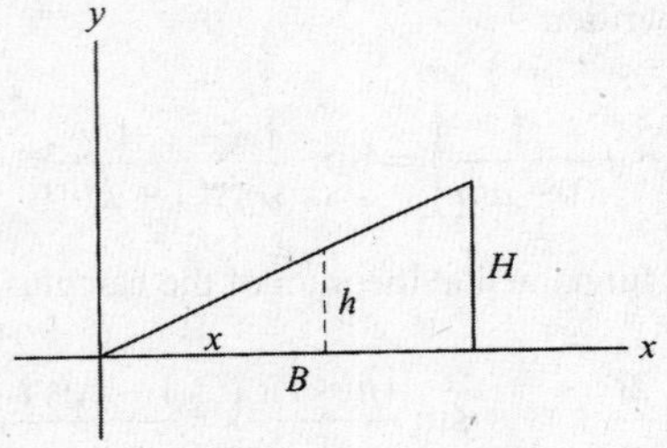

Figure 5.6

(a) $h(x) = \frac{H}{B}x$. According to the mean value theorem for integrals, the average value of the function h on the interval $[0, B]$ is

$$A = \frac{1}{B}\int_0^B \frac{H}{B}x\,dx = \frac{H}{2}$$

(b) The point ξ, at which the average value of h occurs, may be obtained by equating $f(\xi)$ with that average value, i.e., $\frac{H}{B}\xi = \frac{H}{2}$. Thus, $\xi = \frac{B}{2}$.

Fundamental theorem of the calculus

5.10. If $F(x) = \int_a^x f(t)dt$ where $f(x)$ is continuous in $[a, b]$, prove that $F'(x) = f(x)$.

$$\frac{F(x+h)-F(x)}{h} = \frac{1}{h}\left\{\int_a^{x+h} f(t)dt - \int_a^x f(t)dt\right\}$$

$$= \frac{1}{h}\int_x^{x+h} f(t)\,dt = f(\xi) \quad \xi \text{ between } x \text{ and } x+h$$

by the first mean value theorem for integrals (Page 99).

Then if x is any point interior to $[a, b]$,

$$F'(x) = \lim_{h\to 0} \frac{F(x+h) - F(x)}{h} = \lim_{h\to 0} f(\xi) = f(x)$$

since f is continuous.

If $x = a$ or $x = b$, we use right-or left-hand limits, respectively, and the result holds in these cases as well.

5.11. Prove the fundamental theorem of the calculus, Part 2 (Page 101).

By Problem 5.10, if $F(x)$ is *any* function whose derivative is $f(x)$, we can write

$$F(x) = \int_a^x f(t)\,dt + c$$

where c is any constant (see the last line of Problem 4.22).

Since $F(a) = c$, it follows that $F(x) = \int_a^b f(t)\,dt + F(a)$ or $\int_a^b f(t)\,dt = F(b) - F(a)$

5.12. If $f(x)$ is continuous in $[a, b]$, prove that $F(x) = \int_a^b f(t)\,dt$ is continuous in $[a, b]$.

If x is any point interior to $[a, b]$, then, as in Problem 5.10,

$$\lim_{h\to 0} F(x+h) - F(x) = \lim_{h\to 0} hf(\xi) = 0$$

and $F(x)$ is continuous.

If $x = a$ and $x = b$, we use right- and left-hand limits, respectively, to show that $F(x)$ is continuous at $x = a$ and $x = b$.

Another Method: By Problems 5.10 and 4.3, it follows that $F'(x)$ exists, and so $F(x)$ must be continuous.

Change of variables and special methods of integration

5.13. Prove the result in Equation (7), Page 102, for changing the variable of integration.

Let $F(x) = \int_a^x f(x)\,dx$ and $G(t) = \int_a^x f\{g(t)\}g'(t)\,dt$, where $x = g(t)$

Then $dF = f(x)dx$, $dG = f\{g(t)\}\, g'(t)dt$.

Since $dx = g'(t)\,dt$, it follows that $f(x)dx = f\{g(t)\}\, g'(t)dt$ so that $dF(x) = dG(t)$, from which $F(x) = G(t) + c$.

Now, when $x = a$, $t = \alpha$ or $F(a) = G(\alpha) + c$. But $F(a) = G(\alpha) = 0$, so that $c = 0$. Hence, $F(x) = G(t)$. Since $x = b$ when $t = \beta$, we have

$$\int_a^b f(x)\,dx = \int_\alpha^\beta f\{g(t)\}g'(t)\,dt$$

as required.

5.14. Evaluate:

(a) $\int (x+2)\sin(x^2+4x-6)\,dx$

(b) $\int \frac{\cot(\ln x)}{x}\,dx$

(c) $\int_{-1}^{1} \frac{dx}{\sqrt{(x+2)(3-x)}}$

(d) $\int 2^{-x}\tanh 2^{1-x}\,dx$

(e) $\int_0^{1/\sqrt{2}} \frac{x\sin^{-1}x^2}{\sqrt{1-x^4}}\,dx$

(f) $\int \frac{x\,dx}{\sqrt{x^2+x+1}}$

(a) **Method 1:** Let $x^2 + 4x - 6 = u$. Then $(2x + 4)\,dx = du$, $(x + 2)\,dx = \frac{1}{2}\,du$, and the integral becomes $\frac{1}{2}\int \sin u\,du = -\frac{1}{2}\cos u + c = -\frac{1}{2}\cos(x^2+4x-6) + c$.

Method 2:

$$\int (x+2)\sin(x^2+4x\_6)\,dx = \frac{1}{2}\int \sin(x^2+4x-6)\,d(x^2+4x-6)$$

$$= -\frac{1}{2}\cos(x^2+4x-6)+c$$

(b) Let $\ln x = u$. Then $(dx)/x = du$ and the integral becomes $\int \cot u\,du = \ln|\sin u| + c = \ln|\sin(\ln x)| + c$.

(c) **Method 1:** $\displaystyle\int \frac{dx}{\sqrt{(x+2)(3-x)}} = \int \frac{dx}{\sqrt{6+x-x^2}} = \int \frac{dx}{\sqrt{6+(x^2-x)}} = \int \frac{dx}{\sqrt{25/4-\left(x-\frac{1}{2}\right)^2}}$.

Letting $x-\frac{1}{2}=u$, this becomes $\displaystyle\int \frac{du}{\sqrt{25/4-u^2}} = \sin^{-1}\frac{u}{5/2}+c = \sin^{-1}\left(\frac{2x-1}{5}\right)+c$

Then

$$\int_{-1}^{1}\frac{dx}{\sqrt{(x+2)(3-x)}} = \sin^{-1}\left(\frac{2x-1}{5}\right)\Bigg|_{-1}^{1} = \sin^{-1}\left(\frac{1}{5}\right)-\sin^{-1}\left(-\frac{3}{5}\right) = \sin^{-1}.2+\sin^{-1}.6$$

Method 2: Let $x-\frac{1}{2}=u$ as in Method 1. Now, when $x=-1$, $u=-\frac{3}{2}$, and when $x=1$, $u=\frac{1}{2}$. Thus, by Formula 25, Page 102,

$$\int_{-1}^{1}\frac{dx}{\sqrt{(x+2)(3-x)}} = \int_{-1}^{1}\frac{dx}{\sqrt{25/4-\left(x-\frac{1}{2}\right)^2}} = \int_{-3/2}^{1/2}\frac{du}{\sqrt{25/4-u^2}} = \sin^{-1}\frac{u}{5/2}\Bigg|_{-3/2}^{1/2}$$

$$= \sin^{1}.2+\sin^{-1}.6$$

(d) Let $2^{1-x}=u$. Then $-2^{1-x}(\ln 2)dx = du$ and $2^{-x}\,dx = -\dfrac{du}{2\ln 2}$, so that the integral becomes

$$\frac{1}{-2\ln 2}\int \tanh u\,du = -\frac{1}{2\ln 2}\ln\cosh 2^{1-x}+c$$

(e) Let $\sin^{-1}x^2=u$. Then $du = \dfrac{1}{\sqrt{1-(x^2)^2}}2x\,dx = \dfrac{2x\,dx}{\sqrt{1-x^4}}$ and the integral becomes

$$\frac{1}{2}\int u\,du = \frac{1}{4}u^2+c = \frac{1}{4}(\sin^{-1}x^2)^2+c$$

Thus,

$$\int_0^{1/\sqrt{2}}\frac{x\sin^{-1}x^2}{\sqrt{1-x^4}}dx = \frac{1}{4}(\sin^{-1}x^2)^2\Big|_0^{1/\sqrt{2}} = \frac{1}{4}\left(\sin^{-1}\frac{1}{2}\right)^2 = \frac{\pi^2}{144}.$$

(f) $$\int\frac{x\,dx}{\sqrt{x^2+x+1}} = \frac{1}{2}\int\frac{2x+1-1}{\sqrt{x^2+x+1}}dx = \frac{1}{2}\int\frac{2x+1}{\sqrt{x^2+x+1}}dx - \frac{1}{2}\frac{dx}{\sqrt{x^2+x+1}}$$

$$= \frac{1}{2}\int(x^2+x+1)^{-1/2}\,d(x^2+x+1) - \frac{1}{2}\int\frac{dx}{\sqrt{\left(x+\frac{1}{2}\right)^2+\frac{3}{4}}}$$

$$= \sqrt{x^2+x+1} - \frac{1}{2}\ln\left|x+\frac{1}{2}+\sqrt{\left(x+\frac{1}{2}\right)^2+\frac{3}{4}}\right|+c$$

5.15. Show that $\displaystyle\int_1^2\frac{dx}{(x^2-2x+4)^{3/2}} = \frac{1}{6}$.

Write the integral as $\displaystyle\int_1^2\frac{dx}{[(x-1)^2+3]^{3/2}}$. Let $x-1=\sqrt{3}\tan u$, $dx=\sqrt{3}\sec^2 u\,du$. When $x=1$, $u=\tan^{-1}0=0$; when $x=2$, $u=\tan^{-1}1/\sqrt{3}=\pi/6$. Then the integral becomes

$$\int_0^{\pi/6} \frac{\sqrt{3}\sec^2 u\,du}{[3+3\tan^2 u]^{3/2}} = \int_0^{\pi/6} \frac{\sqrt{3}\sec^2 u\,du}{[3\sec^2 u]^{3/2}} = \frac{1}{3}\int_0^{\pi/6} \cos u\,du = \frac{1}{3}\sin u\Big|_0^{\pi/6} = \frac{1}{6}$$

5.16. Determine $\int_e^{e^2} \frac{dx}{x(\ln x)^3}$.

Let $\ln x = y$, $(dx)/x = dy$. When $x = e$, $y = 1$; when $x = e^2$, $y = 2$. Then the integral becomes

$$\int_1^2 \frac{dy}{y^3} = \frac{y^{-2}}{-2}\Big|_1^2 = \frac{3}{8}$$

5.17. Find $\int x^n \ln x\,dx$ if (a) $n \neq -1$ and if (b) $n = -1$.

(a) Use integration by parts, letting $u = \ln x$, $d\upsilon = x^n\,dx$, so that $du = (dx)/x$, $\upsilon = x^{n+1}/(n+1)$. Then

$$\int x^n \ln x\,dx = \int u\,d\upsilon = u\upsilon - \int \upsilon\,du = \frac{x^{n+1}}{n+1}\ln x - \int \frac{x^{n+1}}{n+1}\cdot\frac{dx}{x}$$

$$= \frac{x^{n+1}}{n+1}\ln x - \frac{x^{n+1}}{(n+1)^2} + c$$

(b) $\int x^{-1} \ln x\,dx = \int \ln x\,d(\ln x) = \frac{1}{2}(\ln x)^2 + c$.

5.18. Find $\int 3^{\sqrt{2x+1}}\,dx$.

Let $\sqrt{2x+1} = y$, $2x + 1 = y^2$. Then $dx = y\,dy$ and the integral becomes $\int 3^y \cdot y\,dy$. Integrate by parts, letting $u = y$, $d\upsilon = 3^y\,dy$; then $du = dy$, $\mathrm{v} = 3^y/(\ln 3)$, and we have

$$\int 3^y \cdot y\,dy = \int u\,d\upsilon = u\upsilon - \int \upsilon\,du = \frac{y\,.\,3^y}{\ln 3} - \int \frac{3^y}{\ln 3}\,dy = \frac{y\,.\,3^y}{\ln 3} - \frac{3^y}{(\ln 3)^2} + c$$

5.19. Find $\int_0^1 x\ln(x+3)\,dx$.

Let $u = \ln(x+3)$, $d\upsilon = x\,dx$. Then $du = \frac{dx}{x+3}$, $\upsilon = \frac{x^2}{2}$. Hence, on integrating by parts,

$$\int x\ln(x+3)\,dx = \frac{x^2}{2}\ln(x+3) - \frac{1}{2}\int\frac{x^2\,dx}{x+3} = \frac{x^2}{2}\ln(x+3) - \frac{1}{2}\int\left(x - 3 + \frac{9}{x+3}\right)dx$$

$$= \frac{x^2}{2}\ln(x+3) - \frac{1}{2}\left\{\frac{x^2}{2} - 3x + 9\ln(x+3)\right\} + c$$

Then

$$\int_0^1 x\ln(x+3)\,dx = \frac{5}{4} - 4\ln 4 + \frac{9}{2}\ln 3$$

5.20. Determine $\int \frac{6-x}{(x-3)(2x+5)}\,dx$.

Use the method of *partial fractions*. Let $\frac{6-x}{(x-3)(2x+5)} = \frac{A}{x-3} + \frac{B}{2x+5}$.

Method 1: To determine the constants A and B, multiply both sides by $(x-3)(2x+5)$ to obtain

$$6 - x = A(2x+5) + B(x-3) \text{ or } 6 - x = 5A - 3B + (2A+B)x \qquad (1)$$

Since this is an identity, $5A - 3B = 6$, $2A + B = -1$ and $A = 3/11$, $B = -17/11$. Then

$$\int \frac{6-x}{(x-3)(2x+5)}\,dx = \int \frac{3/11}{x-3}\,dx + \int \frac{-17/11}{2x+5}\,dx = \frac{3}{11}\ln|x-3| - \frac{17}{22}\ln|2x+5| + c$$

Method 2: Substitute suitable values for x in the identity (1). For example, letting $x = 3$ and $x = -5/2$ in (1), we find at once $A = 3/11$, $B = -17/11$.

5.21. Evaluate $\int \frac{dx}{5+3\cos x}$ by using the substitution $\tan x/2 = u$.

From Figure 5.7 we see that

$$\sin x/2 = \frac{u}{\sqrt{1+u^2}}, \qquad \cos x/2 = \frac{1}{\sqrt{1+u^2}}$$

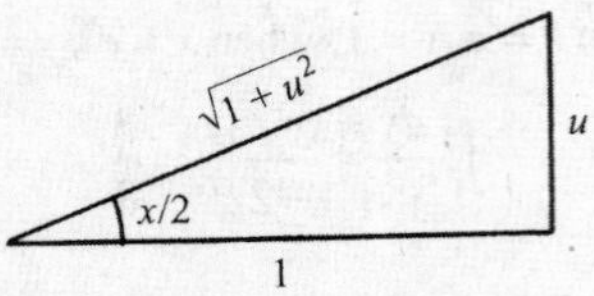

Figure 5.7

Then

$$\cos x = \cos^2 x/2 - \sin^2 x/2 = \frac{1-u^2}{1+u^2}$$

Also

$$du = \frac{1}{2}\sec^2 x/2 dx \quad \text{or} \quad dx = 2\cos^2 x/2 du = \frac{2du}{1+u^2}$$

Thus, the integral becomes

$$\int \frac{du}{u^2+4} = \frac{1}{2}\tan^{-1} u/2 + c = \frac{1}{2}\tan^{-1}\left(\frac{1}{2}\tan x/2\right) + c.$$

5.22. Evaluate $\int_0^{\pi} \frac{x\sin x}{1+\cos^2 x}\,dx$.

Let $x = \pi - y$. Then

$$I = \int_0^{\pi} \frac{x\sin x}{1+\cos^2 x}\,dx = \int_0^{\pi} \frac{(\pi - y)\sin y}{1+\cos^2 y}\,dy = \pi \int_0^{\pi} \frac{\sin y}{1+\cos^2 y}\,dy - \int_0^{\pi} \frac{y\sin y}{1+\cos^2 y}\,dy$$

$$= -\pi \int_0^{\pi} \frac{d(\cos y)}{1+\cos^2 y} - I = -\pi \tan^{-1}(\cos y)\Big|_0^{\pi} = -I = \pi^2/2 - I$$

i.e., $I = \pi^2/2 - I$ or $I = \pi^2/4$.

5.23. Prove that $\int_0^{\pi/2} \frac{\sqrt{\sin x}}{\sqrt{\sin x}+\sqrt{\cos x}}\,dx = \frac{\pi}{4}$.

Letting $x = \pi/2 - y$, we have

$$I = \int_0^{\pi/2} \frac{\sqrt{\sin x}}{\sqrt{\sin x}+\sqrt{\cos x}}\,dx = \int_0^{\pi/2} \frac{\sqrt{\cos y}}{\sqrt{\cos y}+\sqrt{\sin y}}\,dy = \int_0^{\pi/2} \frac{\sqrt{\cos x}}{\sqrt{\cos x}+\sqrt{\sin x}}\,dx$$

Then

$$I + I \int_0^{\pi/2} = \frac{\sqrt{\sin x}}{\sqrt{\sin x}+\sqrt{\cos x}}\,dx + \int_0^{\pi/2} \frac{\sqrt{\cos x}}{\sqrt{\cos x}+\sqrt{\sin x}}\,dx$$

$$= \int_0^{\pi/2} \frac{\sqrt{\sin x}+\sqrt{\cos x}}{\sqrt{\sin x}+\sqrt{\cos x}}\,dx = \int_0^{\pi/2} = \frac{\pi}{2}$$

from which $2I = \pi/2$ and $I = \pi/4$.

The same method can be used to prove that for all real values of m,

$$\int_0^{\pi/2} \frac{\sin^m x}{\sin^m x + \cos^m x}\,dx = \frac{\pi}{4}$$

(see Problem 5.89).

Note: This problem and Problem 5.22 show that some definite integrals can be evaluated without first finding the corresponding indefinite integrals.

Numerical methods for evaluating definite integrals

5.24. Evaluate $\int_0^1 \frac{dx}{1+x^2}$ approximately, using (a) the trapezodial rule, and (b) Simpson's rule, where the interval [0, 1] is divided into $n = 4$ equal parts.

Let $f(x) = 1/(1 + x^2)$. Using the notation on Page 104, we find $\Delta x = (b - a)/n = (1 - 0)/4 = 0.25$. Then, keeping four decimal places, we have $y_0 = f(0) = 1.0000$, $y_1 = f(0.25) = 0.9412$, $y_2 = f(0.50) = 0.8000$, $y_3 = f(0.75) = 0.6400$, and $y_4 = f(1) = 0.50000$.

(a) The trapezoidal rule gives

$$\frac{\Delta x}{2}\{y_0 + 2y_1 + 2y_2 + 2y_3 + y_4\} = \frac{0.25}{2}\{1.0000 + 2(0.9412) + 2(0.8000) + 2(0.6400) + 0.500\}$$
$$= 0.7828.$$

(b) Simpson's rule gives

$$\frac{\Delta x}{3}\{y_0 + 4y_1 + 2y_2 + 4y_3 + y_4\} = \frac{0.25}{2}\{1.0000 + 4(0.9412) + 2(0.8000) + 4(0.6400) + 0.500\}$$
$$= 0.7854.$$

The true value is $\pi/4 \approx 0.7854$.

Applications (area, arc length, volume, moment of inertia)

5.25. Find (a) the area and (b) the moment of inertia about the y axis of the region in the xy plane bounded by $y = 4 - x^2$ and the x axis.

(a) Subdivide the region into rectangles as in Figure 5.1. A typical rectangle is shown in Figure 5.8. Then

$$\text{Required area} = \lim_{n\to\infty}\sum_{k=1}^{n} f(\xi_k)\Delta x_k$$
$$= \lim_{n\to\infty}\sum_{k=1}^{n}(4 - \xi_k^2)\Delta x_k$$
$$= \int_{-2}^{2}(4 - x^2)\,dx = \frac{32}{3}$$

(b) Assuming unit density, the moment of inertia about the y axis of the typical rectangle shown in Figure 5.8 is $\xi_k^2 f(\xi_k)\,\Delta x_k$. Then

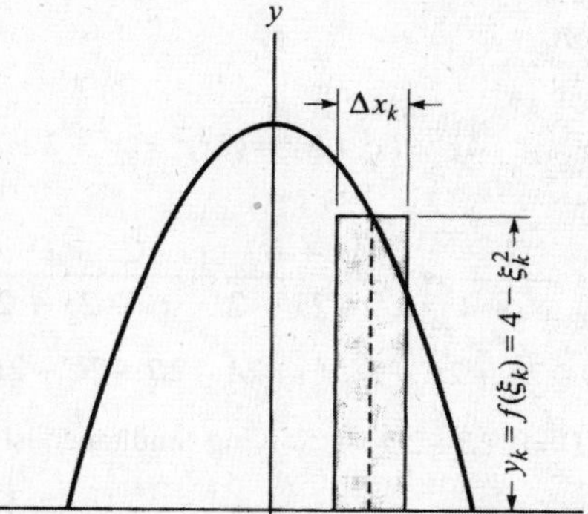

Figure 5.8

$$\text{Required moment of inertia} = \lim_{n\to\infty}\sum_{k=1}^{n}\xi_k^2 f(\xi_k)\Delta x_k = \lim_{n\to\infty}\sum_{k=1}^{n}\xi_k^2(4 - \xi_k^2)\Delta x_k$$
$$= \int_{-2}^{2} x^2(4 - x^2)\,dx = \frac{128}{15}$$

5.26. Find the length of arc of the parabola $y = x^2$ from $x = 0$ to $x = 1$.

$$\text{Required arc length} = \int_0^1 \sqrt{1+(dy/dx)^2}\,dx = \int_0^1 \sqrt{1+(2x)^2}\,dx$$

$$= \int_0^1 \sqrt{1+4x^2}\,dx = \frac{1}{2}\int_0^2 \sqrt{1+u^2}\,du$$

$$= \frac{1}{2}\left\{\frac{1}{2}u\sqrt{1+u^2} + \frac{1}{2}\ln(u+\sqrt{1+u^2})\right\}\Bigg|_0^2 = \frac{1}{2}\sqrt{5} + \frac{1}{4}\ln(2+\sqrt{5})$$

5.27. (a) (Disk method.) Find the volume generated by revolving the region of Problem 5.25 about the x axis.

$$\text{Required volume} = \lim_{n\to\infty}\sum_{k=1}^{n} \pi y_k^2 \Delta x_k = \pi\int_{-2}^{2}(4-x^2)^2\,dx = 512\pi/15.$$

(b) (Disk method.) Find the volume of the frustrum of a paraboloid obtained by revolving $f(x) = \sqrt{kx}$, $0 < a \leqq x \leqq b$ about the x axis.

$$V = \pi\int_a^b kx dx = \frac{\pi k}{2}(b^2 - a^2)$$

(c) (Shell method.) Find the volume obtained by orbiting the region of (b) about the y axis. Compare this volume with that obtained in (b).

$$V = 2\pi\int_0^b x(kx)dx = 2\pi kb^3/3$$

The solids generated by the two regions are different, as are the volumes.

Miscellaneous problems

5.28 If $f(x)$ and $g(x)$ are continuous in $[a, b]$, prove Schwarz's inequality for integrals:

$$\left(\int_a^b f(x)g(x)dx\right)^2 \leqq \int_a^b \{f(x)\}^2 dx \int_a^b \{g(x)\}^2 dx$$

We have

$$\int_a^b \{f(x)+\lambda g(x)\}^2 dx = \int_a^b \{f(x)\}^2 dx + 2\lambda\int_a^b f(x)g(x)dx + \lambda^2\int_a^b \{g(x)\}^2 dx \geqq 0$$

for all real values of λ. Hence, using Equation (1) in Problem 1.13 with

$$A^2 = \int_a^b \{g(x)\}^2 dx, \qquad B^2 = \int_a^b \{f(x)\}^2\,dx, \qquad C = \int_a^b f(x)g(x)dx$$

we find $C^2 \leqq A^2B^2$, which gives the required result.

5.29. Prove that $\lim_{M\to\infty}\int_0^M \frac{dx}{x^4+4} = \frac{\pi}{8}$.

We have $x^4 + 4 = x^4 + 4x^2 + 4 - 4x^2 = (x^2 + 2)^2 - (2x)^2 = (x^2 + 2 + 2x)(x^2 + 2 - 2x)$. According to the method of partial fractions, assume

$$\frac{1}{x^4+4} = \frac{Ax+B}{x^2+2x+2} + \frac{Cx+D}{x^2-2x+2}$$

Then $1 = (A + C)x^3 + (B - 2A + 2C + D)x^2 + (2A - 2B + 2C + 2D)x + 2B + 2D$, so that $A + C = 0$, $B - 2A + 2C + D = 0$, $2A - 2B + 2C + 2D = 0$, $2B + 2D = 1$. Solving simultaneously, $A = \frac{1}{8}$, $B = \frac{1}{4}$, $C = -\frac{1}{8}$, $D = \frac{1}{4}$. Thus,

$$\int\frac{dx}{x^4+4} = \frac{1}{8}\int\frac{x+2}{x^2+2x+2}dx - \frac{1}{8}\int\frac{x-2}{x^2-2x+2}dx$$

$$= \frac{1}{8}\int\frac{x+1}{(x+1)^2+1}dx + \frac{1}{8}\int\frac{dx}{(x+1)^2+1} - \frac{1}{8}\int\frac{x-1}{(x-1)^2+1}dx + \frac{1}{8}\int\frac{dx}{(x-1)^2+1}$$

$$= \frac{1}{16}\ln(x^2+2x+2) + \frac{1}{8}\tan^{-1}(x+1) - \frac{1}{16}\ln(x^2-2x+2) + \frac{1}{8}\tan^{-1}(x-1) + C$$

Then

$$\lim_{M\to\infty}\int_0^M \frac{dx}{x^4+4} = \lim_{M\to\infty}\left\{\frac{1}{16}\ln\left(\frac{M^2+2M+2}{M^2-2M+2}\right)+\frac{1}{8}\tan^{-1}(M+1)+\frac{1}{8}\tan^{-1}(M-1)\right\}=\frac{\pi}{8}$$

We denote this limit by $\int_0^\infty \frac{dx}{x^4+4}$, called an *improper integral of the first kind*. Such integrals are considered further in Chapter 12. See also Problem 5.74.

5.30. Evaluate $\lim_{x\to 0}\dfrac{\int_0^x \sin t^3\, dt}{x^4}$.

The conditions of L'Hospital's rule are satisfied, so that the required limit is

$$\lim_{x\to 0}\frac{\frac{d}{dx}\int_0^x \sin t^3\, dt}{\frac{d}{dx}(x^4)} = \lim_{x\to 0}\frac{\sin x^3}{4x^3} = \lim_{x\to 0}\frac{\frac{d}{dx}(\sin x^3)}{\frac{d}{dx}(4x^3)} = \lim_{x\to 0}\frac{3x^2\cos x^3}{12x^2} = \frac{1}{4}$$

5.31. Prove that if $f(x)$ is continuous in $[a, b]$, then $\int_a^b f(x)dx$ exists.

Let $\sigma = \sum_{k=1}^{n} f(\xi_k)\Delta x_k$, using the notation of Page 99. Since $f(x)$ is continuous, we can find numbers M_k and m_k, representing the l.u.b. and g.l.b. of $f(x)$ in the interval $[x_{k-1}, x_k]$, i.e., such that $m_k \leqq f(x) \leqq M_k$. We then have

$$m(b-a) \leqq s = \sum_{k=1}^{n} m_k \Delta x_k \leqq \sigma \leqq \sum_{k=1}^{n} M_k \Delta x_k = S \leqq M_k(b-a) \tag{1}$$

where m and M are the g.l.b. and l.u.b. of $f(x)$ in $[a, b]$. The sums s and S are sometimes called the *lower* and *upper sums*, respectively.

Now choose a second mode of subdivision of $[a, b]$ and consider the corresponding lower and upper sums denoted by s' and S' respectively. We have must

$$s' \leqq S \quad \text{and} \quad S' \varepsilon\, s \tag{2}$$

To prove this we choose a third mode of subdivision obtained by using the division points of both the first and second modes of subdivision and consider the corresponding lower and upper sums, denoted by t and T, respectively. By Problem 5.84, we have

$$s \leqq t \leqq T \leqq S' \quad \text{and} \quad s' \leqq t \leqq T \leqq S \tag{3}$$

which proves (2).

From (2) it is also clear that as the number of subdivisions is increased, the upper sums are monotonic decreasing and the lower sums are monotonic increasing. Since, according to Equation (1), these sums are also bounded, it follows that they have limiting values, which we shall call $\overline{s}$ and $\underline{S}$, respectively. By By Problem 5.85, $\overline{s} \leqq \underline{S}$. In order to prove that the integral exists, we must show that $\overline{s} = \underline{S}$.

Since $f(x)$ is continuous in the closed interval $[a, b]$, it is uniformly continuous. Then, given any $\varepsilon > 0$, we can take each Δx_k so small that $M_k - m_k < \varepsilon/(b-a)$. It follows that

$$S - s = \sum_{k=1}^{n}(M_k - m_k)\Delta x_k < \frac{\varepsilon}{b-a}\sum_{k=1}^{n}\Delta x_k = \varepsilon \tag{4}$$

Now $S - s = (S - \underline{S}) + (\underline{S} - \overline{s}) + (\overline{s} - s)$ and it follows that each term in parentheses is positive and so is less than ε, by Equation (4). In particular, since $\underline{S} - \overline{s}$ is a definite number, it must be zero; i.e., $\underline{S} = \overline{s}$. Thus, the limits of the upper and lower sums are equal and the proof is complete.

SUPPLEMENTARY PROBLEMS

Definition of a definite integral

5.32. (a) Express Express $\int_0^1 x^3\, dx$ as a limit of a sum. (b) Use the result of (a) to evaluate the given definite integral. (c) Interpret the result geometrically.

Ans. (b) $\dfrac{1}{4}$

5.33. Using the definition, evaluate (a) $\int_0^2 (3x+1)dx$, and (b) $\int_3^6 (x^2-4x)\, dx$.

Ans. (a) 8 (b) 9

5.34. Prove that $\lim_{n\to\infty}\left\{\dfrac{n}{n^2+1^2}+\dfrac{n}{n^2+2^2}+\cdots+\dfrac{n}{n^2+n^2}\right\}=\dfrac{\pi}{4}$.

5.35. Prove that $\lim_{n\to\infty}\left\{\dfrac{1^p+2^p+3^p+\cdots+n^p}{n^{p+1}}=\dfrac{1}{p+1}\right\}$ if $p>-1$

5.36. Using the definition, prove that $\int_a^b e^x\,dx = e^b - e^a$.

5.37. Work Problem 5.5 directly, using Problem 1.94.

5.38. Prove that $\lim_{n\to\infty}\left\{\dfrac{1}{\sqrt{n^2+1^2}}+\dfrac{1}{\sqrt{n^2+2^2}}+\cdots+\dfrac{1}{\sqrt{n^2+n^2}}\right\}=\ln(1+\sqrt{2})$.

5.38. Prove that $\lim_{n\to\infty}\sum_{k=1}^{n}\dfrac{n}{n^2+k^2x^2}=\dfrac{\tan^{-1}x}{x}$ if $x\neq 0$.

Properties of definite integrals

5.40. Prove (a) Property 2 and (b) Property 3, on Page 102.

5.41. If $f(x)$ is integrable in (a, c) and (c, b), prove that $\int_a^b f(x)dx = \int_a^c f(x)dx + \int_c^b f(x)dx$.

5.42. If $f(x)$ and $g(x)$ are integrable in $[a, b]$ and $f(x) \leqq g(x)$, prove that $\int_a^b f(x)dx \leqq \int_a^b g(x)dx$.

5.43. Prove that $1 - \cos x\ \varepsilon\ x^2/\pi$ for $0 \leqq x \leqq \pi/2$.

5.44. Prove that $\left|\int_0^1 \dfrac{\cos nx}{x+1}dx\right| \leqq \ln 2$ for all n.

5.45. Prove that $\left|\int_1^{\sqrt{3}} \dfrac{e^{-x}\sin x}{x^2+1}dx\right| \leqq \dfrac{\pi}{12e}$.

Mean value theorems for integrals

5.46. Prove result (5), Page 100. [Hint: If $m \leq f(x) \leq M$, then $mg(x) \leq f(x)g(x) \leq Mg(x)$. Now integrate and divide by $\int_a^b g(x)dx$. Then apply Theorem 9, from Chapter 3.]

5.47. Prove that there exist values ξ_1 and ξ_2 in $0 \leq x \leq 1$ such that

$$\int_0^1 \frac{\sin \pi x}{x^2+1} dx = \frac{2}{\pi(\xi_1^2+1)} = \frac{\pi}{4} \sin \pi \xi_2$$

(Hint: Apply the first mean value theorem.)

5.48. (a) Prove that there is a value ξ in $0 \leq x \leq \pi$ such that $\int_0^\pi e^{-x} \cos x\, dx = \sin \xi$. (b) Suppose a wedge in the shape of a right triangle is idealized by the region bounded by the x axis, $f(x) = x$, and $x = L$. Let the weight distribution for the wedge be defined by $W(x) = x^2 + 1$. Use the generalized mean value theorem to show that the point at which the weighted value occurs is $\frac{3L}{4} \frac{L^2+2}{L^2+3}$.

Change of variables and special methods of integration

5.49. Evaluate:

(a) $\int x^2 e^{\sin x^3} \cos x^3 dx$ (b) $\int_0^1 \frac{\tan^{-1} t}{1+t^2} dt$. (c) $\int_1^3 \frac{dx}{\sqrt{4x-x^2}}$ (d) $\int \frac{\operatorname{csch}^2 \sqrt{u}}{\sqrt{u}} du$.

Ans. (a) $\frac{1}{3} e^{\sin x^3} + c$ (b) $\pi^2/32$ (c) $\pi/3$ (d) $-2\coth\sqrt{u} + c$ (e) $\frac{1}{4} \ln 3$

5.50. Show that (a) $\int_0^1 \frac{dx}{(3+2x-x^2)^{3/2}} = \frac{\sqrt{3}}{12}$ and (b) $\int \frac{dx}{x^2\sqrt{x^2-1}} = \frac{\sqrt{x^2-1}}{x} + c$

5.51. Prove that (a) $\int \sqrt{u^2 \pm a^2}\, du = \frac{1}{2} u\sqrt{u^2 \pm a^2} \pm \frac{1}{2} a^2 \ln | u + \sqrt{u^2 \pm a^2} |$ and

(b) $\int \sqrt{a^2-u^2}\, du = \frac{1}{2} u\sqrt{a^2-u^2} + \frac{1}{2} a^2 \sin^{-1} u/a + c, \quad a > 0.$

5.52. Find $\int \frac{x\, dx}{\sqrt{x^2+2x+5}}$

Ans. $\sqrt{x^2+2x+5} - \ln | x+1+\sqrt{x^2+2x+5} | + c.$

5.53. Establish the validity of the method of integration by parts.

5.54. Evaluate (a) $\int_0^\pi x \cos 3x\, dx$ and (b) $\int x^3 e^{-2x} dx$

Ans. (a) $-2/9$ (b) $-\frac{1}{3} e^{-2x} (4x^3 + 6x^2 + 6x + 3) + c$

5.55. Show that (a) $\int_0^1 x^2 \tan^{-1} x\, dx = \frac{1}{12}\pi - \frac{1}{6} + \frac{1}{6}\ln 2$ and (b)

$$\int_{-2}^{2} \sqrt{x^2 + x + 1}\, dx = \frac{5\sqrt{7}}{4} + \frac{3\sqrt{3}}{4} + \frac{3}{8}\ln\left(\frac{5 + 2\sqrt{7}}{2\sqrt{3} - 3}\right)$$

5.56. (a) If $u = f(x)$ and $\upsilon = g(x)$ have continuous nth derivatives, prove that

$\int u\upsilon^{(n)}\, dx = u\upsilon^{(n-1)} - u'\upsilon^{(n-2)} + u''\upsilon^{(n-3)} - \cdots - (-1)^n \int u^{(n)}\, \upsilon dx$, called *generalized integration by parts.*

(b) What simplifications occur if $u^{(n)} = 0$? Discuss. (c) Use (a) to evaluate $\int_0^\pi x^4 \sin x\, dx$.

Ans. (c) $\pi^4 - 12\pi^2 + 48$

5.57. Show that $\int_0^1 \frac{x\, dx}{(x+1)^2 (x^2+1)} = \frac{\pi - 2}{8}$ [Hint: Use partial fractions; i.e., assume

$\frac{x}{(x+1)^2 (x^2+1)} = \frac{A}{(x+1)^2} + \frac{B}{x+1} + \frac{Cx + D}{x^2 + 1}$ and find A, B, C, D.]

5.58. Prove that $\int_0^\pi \frac{dx}{\alpha - \cos x} = \frac{\pi}{\sqrt{\alpha^2 - 1}}, \quad \alpha > 1.$

Numerical methods for evaluating definite integrals

5.59. Evaluate $\int_0^1 \frac{dx}{1+x}$ approximately, using (a) the trapezoidal rule and (b) Simpson's rule, taking $n = 4$. Compare with the exact value, $\ln 2 = 0.6931$.

5.60. Using (a) the trapezoidal rule and (b) Simpson's rule, evaluate $\int_0^{\pi/2} \sin^2 x\, dx$ by obtaining the values of $\sin^2 x$ at $x = 0°, 10°, \ldots, 90°$ and compare with the exact value $\pi/4$.

5.61. Prove (a) the rectangular rule and (b) the trapezoidal rule, i.e., Equations (8) and (9), Page 104.

5.62. Prove Simpson's rule.

5.63. Evaluate the following to three decimal places using numerical integration:

(a) $\int_1^2 \frac{dx}{1+x^2}$ (b) $\int_0^1 \cosh x^2 dx$

Ans. (a) 0.322 (b) 1.105

Applications

5.64. Find (a) the area and (b) the moment of inertia about the y axis of the region in the xy plane bounded by $y = \sin x$, $0 \leqq x \leqq \pi$, and the x axis, assuming unit density.

Ans. (a) 2 (b) $\pi^2 - 4$

5.65. Find the moment of inertia about the x axis of the region bounded by $y = x^2$ and $y = x$, if the density is proportional to the distance from the x axis.

Ans. $\frac{1}{8}M$, where M = mass of the region

5.66. (a) Show that the arc length of the *catenary* $y = \cosh x$ from $x = 0$ to $x = \ln 2$ is $\frac{3}{4}$. (b) Show that the length of arc of $y = x^{3/2}$, $2 \leqq x \leqq 5$ is $\frac{343}{27} - 2\sqrt{2}\,11^{3/2}$.

5.67. Show that the length of one arc of the *cycloid* $x = a(\theta - \sin\theta)$, $y = a(1 - \cos\theta)$, $(0 \leqq \tau \leqq 2\pi)$ is $8a$.

5.68. Prove that the area bounded by the ellipse $x^2/a^2 + y^2/b^2 = 1$ is πab.

5.69. (a) (Disk method.) Find the volume of the region obtained by revolving the curve $y = \sin x$, $0 \leqq x \leqq \pi$, about the x axis.

Ans. (a) $\pi^2/2$

(b) (Disk method.) Show that the volume of the frustrum of a paraboloid obtained by revolving $f(x) = \sqrt{kx}$, $0 < a \leqq x \leqq b$ about the x axis is $\pi\int_a^b kx\,dx = \frac{\pi k}{2}(b^2 - a^2)$. (c) Determine the volume obtained by rotating the region bound by $f(x) = 3$, $g(x) = 5 - x^2$ on $-\sqrt{2} \leqq x \leqq \sqrt{2}$. (d) (Shell method.) A spherical bead of radius a has a circular cylindrical hole of radius b, $b < a$, through the center. Find the volume of the remaining solid by the shell method. (e) (Shell method.) Find the volume of a solid whose outer boundary is a torus [i.e., the solid is generated by orbiting a circle $(x - a)^2 - y^2 = b^2$ about the y axis $(a > b)$].

5.70. Prove that the centroid of the region bounded by $y = \sqrt{a^2 - x^2}$, $-a \leqq x \leqq a$, and the x axis is located at $(0, 4a/3\pi)$.

5.71. (a) If $\rho = f(\phi)$ is the equation of a curve in polar coordinates, show that the area bounded by this curve and the lines $\phi = \theta$ and $\phi = \phi_2$ is $\frac{1}{2}\int_{\phi_1}^{\phi}\rho^2\,d\phi$. (b) Find the area bounded by one loop of the *lemniscate* $\rho^2 = a^2\cos 2\phi$.

Ans. (b) a^2

5.72. (a) Prove that the arc length of the curve in Problem 5.71(*a*) is $\int_{\phi_1}^{\phi_2}\sqrt{\rho^2 + (d\rho/d\phi)^2}\,d\phi$. (b) Find the length of arc of the *cardioid* $\rho = a(1 - \cos\phi)$.

Ans. (b) $8a$

Miscellaneous problems

5.73. Establish the mean value theorem for derivatives from the first mean value theorem for integrals. [Hint: Let $f(x) = F'(x)$ in (4), Page 100.]

5.74. Prove that (a) $\lim_{\epsilon\to 0+}\int_0^{4-\epsilon}\frac{dx}{\sqrt{4-x}} = 4$, (b) $\lim_{\epsilon\to 0+}\int_\epsilon^3\frac{dx}{\sqrt[3]{x}} = 6$, (c) $\lim_{\epsilon\to 0+}\int_0^{1-\epsilon}\frac{dx}{\sqrt{1-x^2}} = \frac{\pi}{2}$, and give a geometric interpretation of the results.

(These limits, denoted usually by $\int_0^4\frac{dx}{\sqrt{4-x}}$, $\int_0^3\frac{dx}{\sqrt[3]{x}}$, and $\int_0^1\frac{dx}{\sqrt{1-x^2}}$, respectively, are called *improper integrals of the second kind* (see Problem 5.29), since the integrands are not bounded in the range of integration. For further discussion of improper integrals, see Chapter 12.)

5.75. Prove that (a) $\lim_{M\to\infty}\int_0^M x^5 e^{-x}dx = 4! = 24$ and (b) $\lim_{\epsilon\to 0+}\int_1^{2-\epsilon}\frac{dx}{\sqrt{x(2-x)}} = \frac{\pi}{2}$.

5.76. Evaluate (a) $\int_0^\infty \frac{dx}{1+x^3}$, (b) $\int_0^{\pi/2}\frac{\sin 2x}{(\sin x)^{4/3}}dx$, and (c) $\int_0^\infty \frac{dx}{x+\sqrt{x^2+1}}$.

Ans. (a) $\frac{2\pi}{3\sqrt{3}}$ (b) 3 (c) does not exist

5.77. Evaluate $\lim_{x\to\pi/2}\frac{ex^2/\pi - e\pi/4 + \int_x^{\pi/2} e^{\sin t}\,dt}{1+\cos 2x}$.

Ans. $e/2\pi$

5.78. Prove: (a) $\frac{d}{dx}\int_{x^2}^{x^3}(t^2+t+1)dt = 3x^3 + x^5 - 2x^3 + 3x^2 - 2x$ and

(b) $\frac{d}{dx}\int_x^{x^2}\cos t^2\,dt = 2x\cos x^4 - \cos x^2$.

5.79. Prove that (a) $\int_0^\pi \sqrt{1+\sin x}\,dx = 4$ and (b) $\int_0^{\pi/2}\frac{dx}{\sin x + \cos x} = \sqrt{2}\ \ln(\sqrt{2}+1)$.

5.80. Explain the fallacy $I = \int_{-1}^{1}\frac{dx}{1+x^2} = -\int_{-1}^{1}\frac{dy}{1+y^2} = -I$, using the transformation $x = 1/y$. Hence, $I = 0$. But $I = \tan^{-1}(1) - \tan^{-1}(-1) = \pi/4 - (-\pi/4) = \pi/2$. Thus, $\pi/2 = 0$.

5.81. Prove that $\int_0^{1/2}\frac{\cos\pi x}{\sqrt{1+x^2}}dx \leqq \frac{1}{4}\tan^{-1}\frac{1}{2}$.

5.82. Evaluate $\lim_{n\to\infty}\left\{\frac{\sqrt{n+1}+\sqrt{n+2}+\cdots+\sqrt{2n-1}}{n^{3/2}}\right\}$.

Ans. $\frac{2}{3}(2\sqrt{2}-1)$

5.83. Prove that $f(x) = \begin{cases}1 \text{ if } x \text{ is irrational}\\ 0 \text{ if } x \text{ is rational}\end{cases}$ is not Riemann integrable in [0, 1].

[Hint: In Equation (2), Page 98, let ξ_k, $k = 1, 2, 3, \ldots, n$ be first rational and then irrational points of subdivision and examine the lower and upper sums of Problem 5.31.]

5.84. Prove the result (3) of Problem 5.31. (Hint: First consider the effect of only one additional point of subdivision.)

5.85. In Problem 5.31, prove that $\bar{s} \leq \underline{S}$. (Hint: Assume the contrary and obtain a contradiction.)

5.86. If $f(x)$ is sectionally continuous in $[a, b]$, prove that $\int_a^b f(x)dx$ exists. (Hint: Enclose each point of discontinuity in an interval, noting that the sum of the lengths of such intervals can be made arbitrarily small. Then consider the difference between the upper and lower sums.)

5.87. If $f(x) = \begin{cases}2x & 0<x<1\\ 3 & x=1\\ 6x-1 & 1<x<2\end{cases}$, find $\int_0^2 f(x)dx$. Interpret the result graphically.

Ans. 9

5.88. Evaluate $\int_0^3 \left\{x - [x] + \frac{1}{2}\right\} dx$ where $[x]$ denotes the greatest integer less than or equal to x. Interpret the result graphically.

Ans. 3

5.89. (a) Prove that $\int_0^{\pi/2} \frac{\sin^m x}{\sin^m x + \cos^m x}\, dx = \frac{\pi}{4}$ for all real values of m.

(b) Prove that $\int_0^{2\pi} \frac{dx}{1 + \tan^4 x} = \pi$.

5.90. Prove that $\int_0^{\pi/2} \frac{\sin x}{x}\, dx$ exists.

5.91. Show that $\int_0^{0.5} \frac{\tan^{-1} x}{x}\, dx = 0.4872$ approximately.

5.92. Show that $\int_0^{\pi} \frac{x\,dx}{1 + \cos^2 x} = \frac{\pi^2}{2\sqrt{2}}$.

CHAPTER 6

Partial Derivatives

Functions of Two or More Variables

The definition of a function was given in Chapter 3 (page 43). For us, the distinction for functions of two or more variables is that the domain is a set of n-tuples of numbers. The range remains one-dimensional and is referred to an interval of numbers. If $n = 2$, the domain is pictured as a two-dimensional region. The region is referred to a rectangular Cartesian coordinate system described through number pairs (x, y), and the range variable is usually denoted by z. The domain variables are independent, while the range variable is dependent.

We use the notation $f(x, y)$, $F(x, y)$, etc., to denote the value of the function at (x, y) and write $z = f(x, y)$, $z = F(x, y)$, etc. We also sometimes use the notation $z = z(x, y)$, although it should be understood that in this case z is used in two senses, namely, as a function and as a variable.

EXAMPLE. If $f(x, y) = x^2 + 2y^3$, then $f(3, -1) = (3)^2 + 2(-1)^3 = 7$.

The concept is easily extended. Thus, $w = F(x, y, z)$ denotes the value of a function at (x, y, z) (a point in three-dimensional space), etc.

EXAMPLE. If $z = \sqrt{1 - (x^2 + y^2)}$, the domain for which z is real consists of the set of points (x, y) such that $x^2 + y^2 \leqq 1$, i.e., the set of points inside and on a circle in the xy plane having center at (0, 0) and radius 1.

A three-dimensional rectangular Cartesian coordinate system is obtained by constructing three mutually perpendicular axes (the x, y, and z axes) intersecting in a point (designated by 0 and called the *origin*). This is a natural extension of the rectangular system x, y in the plane. A point in the three-dimensional Cartesian system is represented by the triple of coordinates (x, y, z). The collection of points $P(x, y, z)$, represented by the implicit equation $F(x, y, z) = 0$, is a surface. The term *surface* is used in a very broad sense and requires refinement according to the context in which it is to be used. For example, $x^2 + y^2 + z^2 = r^2$ is the algebraic representation of a surface in the large. This form might be employed in topology to indicate the property of being closed rather than open. In analysis, which is the subject of this outline of advanced calculus, the concern is with portions of a surface—that is, points and their neighborhoods. These may be obtained from implicit representations by imposing restrictions. For example,

$$z = \sqrt{r^2 - (x^2 + y^2)} \text{ with } 1x^2 + y^21 < r$$

signifies an open upper hemisphere. Problems in surface theory employ partial derivatives and relate to a point of a surface, the collection of points about it, the tangent plane at the point, and the properties of continuity and differentiability binding this structure. These concepts will be discussed in the following pages.

For functions of more than two variables such geometric interpretation fails, although the terminology is still employed. For example, (x, y, z, w) is a point in four-dimensional space, and $w = f(x, y, z)$ [or $F(x, y, z, w) = 0$] represents a *hypersurface* in four dimensions; thus, $x^2 + y^2 + z^2 + w^2 = a^2$ represents a *hypersphere* in four dimensions with radius $a > 0$ and center at (0, 0, 0, 0). $w = \sqrt{a^2 - (x^2 + y^2 + z^2)}$, $x^2 + y^2 + z^2 \leqq a^2$ describes a function generated from the hypersphere.

Neighborhoods

The set of all points (x, y) such that $|x - x_0| < \delta$, $|y - y_0| < \delta$ where $\delta > 0$ is called a *rectangular δ neighborhood* of (x_0, y_0); the set $0 < |x - x_0| < \delta$, $0 < |y - y_0| < \delta$, which excludes (x_0, y_0), is called a *rectangular deleted δ neighborhood* of (x_0, y_0). Similar remarks can be made for made for other neighborhoods; e.g., $(x - x_0)^2 + (y - y_0)^2 < \delta^2$ is a *circular δ neighborhood* of (x_0, y_0). The term *open ball* is used to designate this circular neighborhood. This terminology is appropriate for generalization to more dimensions. Whether neighborhoods are viewed as circular or square is immaterial, since the descriptions are interchangeable. Simply notice that given an open ball (circular neighborhood) of radius δ there is a centered square whose side is of length less than $\sqrt{2}\,\delta$ that is interior to the open ball, and, conversely, for a square of side δ there is an interior centered circle of radius less than δ/2. (See Figure 6.1.)

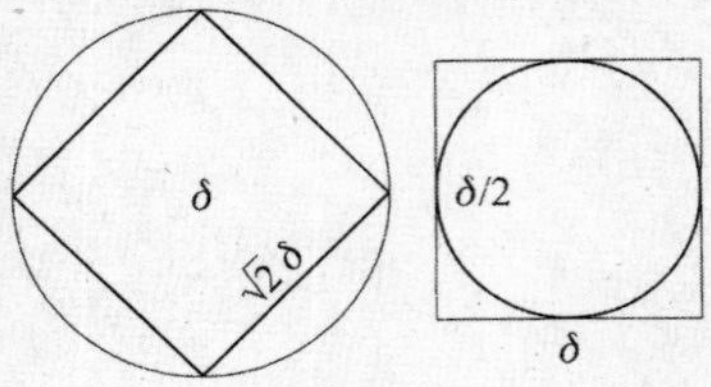

Figure 6.1

A point (x_0, y_0) is called a *limit point, accumulation point,* or *cluster point* of a point set S if every deleted δ neighborhood of (x_0, y_0) contains points of S. As in the case of one-dimensional point sets, every bounded infinite set has at least one limit point (the Bolzano-Weierstrass theorem; see Chapter 1). A set containing all its limit points is called a *closed set*.

Regions

A point P belonging to a point set S is called an *interior point* of S if there exists a deleted δ neighborhood of P all of whose points belong to S. A point P not belonging to S is called an *exterior point* of S if there exists a deleted δ neighborhood of P all of whose points do not belong to S. A point P is called a *boundary point* of S if every deleted δ neighborhood of P contains points belonging to S and also points not belonging to S.

If any two points of a set S can be joined by a path consisting of a finite number of broken line segments all of whose points belong to S, then S is called a *connected set*. A *region* is a connected set which consists of interior points or interior and boundary points. A *closed region* is a region containing all its boundary points. An *open region* consists only of interior points. The complement of a set S in the xy plane is the set of all points in the plane not belonging to S. (See Figure 6.2.)

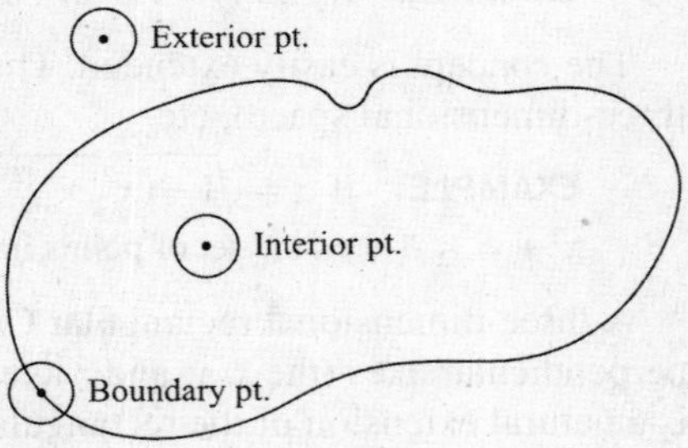

Figure 6.2

Examples of some regions are shown graphically in Figure 6.3(a), (b), and (c) . The rectangular region of Figure 6.1(a), including the boundary, represents the sets of points $a \leq x \leq b$, $c \leq y \leq d$ which is a natural extension of the closed interval $a \leq x \leq b$ for one dimension. The set $a < x < b$, $c < y < d$ corresponds to the boundary being excluded.

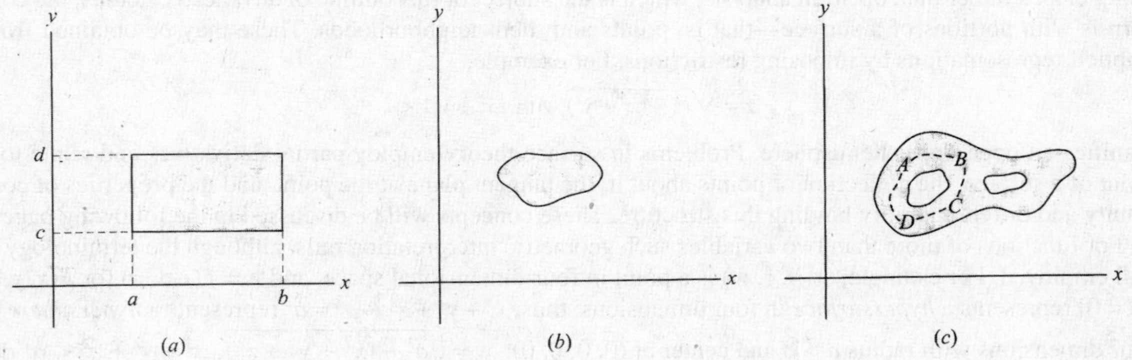

Figure 6.3

In the regions of Figure 6.3(a) and (b), any *simple closed curve* (one which does not intersect itself anywhere) lying inside the region can be shrunk to a point which also lies in the region. Such regions are called *simply connected regions*. In Figure 6.3(c), however, a simple closed curve *ABCD* surrounding one of the "holes" in the region cannot be shrunk to a point without leaving the region. Such regions are called *multiply connected regions*.

Limits

Let $f(x, y)$ be defined in a deleted δ neighborhood of (x_0, y_0) [i.e., $f(x, y)$ may be undefined at (x_0, y_0)]. We say that l is the *limit* of $f(x,y)$ as x approaches x_0 and y approaches y_0 [or (x, y) approaches (x_0, y_0)] and write $\lim\limits_{\substack{x\to x_0\\ y\to y_0}} f(x, y) = l$ [or $\lim\limits_{(x,y)\to(x_0,y_0)} f(x, y) = l$] if for any positive number δ we can find some positive number δ [depending on δ and (x_0, y_0), in general] such that $|f(x, y) - l| < \delta$ whenever $0 < |x - x_0| < \delta$ and $0 < |y - y_0| < \delta$.

If desired, we can use the deleted circular neighborhood open ball $0 < (x - x_0)^2 + (y - y_0)^2 < \delta^2$ instead of the deleted rectangular neighborhood.

EXAMPLE. Let $f(x, y) = \begin{cases} 3xy & \text{if } (x, y) \neq (1, 2) \\ 0 & \text{if } (x, y) = (1, 2) \end{cases}$: As $x \to 1$ and $y \to 2$ [or $(x, y) \to (1, 2)$], $f(x, y)$ gets closer to $3(1)(2) = 6$ and we *suspect* that $\lim\limits_{\substack{x\to 1\\ x\to 2}} f(x, y) = 6$. To *prove* this, we must show that the preceding definition of limit, with $l = 6$, is satisfied. Such a proof can be supplied by a method similar to that of Problem 6.4.

Note that $\lim\limits_{\substack{x\to 1\\ x\to 2}} f(x, y) \neq f(1, 2)$ since $f(1, 2) = 0$. The limit would, in fact, be 6 even if $f(x, y)$ were not defined at (1, 2). Thus, the existence of the limit of $f(x, y)$ as $(x, y) \to (x_0, y_0)$ is in no way dependent on the existence of a value of $f(x, y)$ at (x_0, y_0).

Note that in order for $\lim\limits_{(x,y)\to(x_0,y_0)} f(x, y)$ to exist, it must have the same value regardless of the approach of (x, y) to (x_0, y_0). It follows that if two different approaches give different values, the limit cannot exist (see Problem 6.7). This implies, as in the case of functions of one variable, that if a limit exists it is unique.

The concept of one-sided limits for functions of one variable is easily extended to functions of more than one variable.

EXAMPLE 1. $\lim\limits_{\substack{x\to 0+\\ y\to 1}} \tan^{-1}(y/x) = \pi/2$, $\lim\limits_{\substack{x\to 0-\\ y\to 1}} \tan^{-1}(y/x) = -\pi/2$.

EXAMPLE 2. $\lim\limits_{\substack{x\to 0\\ y\to 1}} \tan^{-1}(y/x)$ does not exist, as is clear from the fact that the two different approaches of Example 1 give different results.

In general, the theorems on limits, concepts of infinity, etc., for functions of one variable (see Page 25) apply as well, with appropriate modifications, to functions of two or more variables.

Iterated Limits

The *iterated limits* $\lim\limits_{x\to x_0}\left\{\lim\limits_{y\to y_0} f(x, y)\right\}$ and $\lim\limits_{y\to y_0}\left\{\lim\limits_{x\to x_0} f(x, y)\right\}$ [also denoted by $\lim\limits_{x\to x_0}\lim\limits_{y\to y_0} f(x, y)$ and $\lim\limits_{y\to y_0}\lim\limits_{x\to x_0} f(x, y)$, respectively] are not necessarily equal. Although they must be equal if $\lim\limits_{\substack{x\to x_0\\ y\to y_0}} f(x, y)$ is to exist, their equality does not guarantee the existence of this last limit.

EXAMPLE. If $(x, y) = \dfrac{x-y}{x+y}$, then $\lim\limits_{x\to 0}\left(\lim\limits_{y\to 0}\dfrac{x-y}{x+y}\right) = \lim\limits_{y\to 0}(1) = 1$ and $\lim\limits_{y\to 0}\left(\lim\limits_{x\to 0}\dfrac{x-y}{x+y}\right) = \lim\limits_{y\to 0}(-1) = -1$.
Thus, the iterated limits are not equal and so $\lim\limits_{\substack{x\to 0\\ y\to 0}} f(x, y)$ cannot exist.

Continuity

Let $f(x, y)$ be defined in a δ neighborhood of (x_0, y_0) [i.e., $f(x, y)$ must be defined *at* (x_0, y_0) as well as near it]. We say that $f(x, y)$ is *continuous* at (x_0, y_0) if for any positive number δ we can find some positive number δ [depending on δ and (x_0, y_0) in general] such that $|f(x, y) - f(x_0, y_0)| < \delta$ whenever $|x - x_0| < \delta$ and $|y - y_0| < \delta$, or, alternatively, $(x - x_0)^2 + (y - y_0)^2 < \delta^2$.

Note that three conditions must be satisfied in order for $f(x, y)$ to be continuous at (x_0, y_0):

1. $\lim\limits_{(x, y)\to(x_0, y_0)} f(x, y) = l$; i.e., the limit exists as $(x, y) \to (x_0, y_0)$.
2. $f(x_0, y_0)$ must exist; i.e., $f(x, y)$ is defined at (x_0, y_0).
3. $l = f(x_0, y_0)$.

If desired, we can write this in the suggestive form $\lim\limits_{\substack{x\to x_0\\ y\to y_0}} f(x, y) = f(\lim\limits_{x\to x_0} x, \lim\limits_{y\to y_0} y)$.

EXAMPLE. If $f(x, y) = \begin{cases} 3xy & (x, y) \neq (1, 2) \\ 0 & (x, y) = (1, 2) \end{cases}$, then $\lim\limits_{(x,y)\to(1,2)} f(x, y) = 6 \neq f(1, 2)$. Hence, $f(x, y)$ is not continuous at (1, 2). If we redefine the function so that $f(x, y) = 6$ for $(x, y) = (1, 2)$, then the function is continuous at (1, 2).

If a function is not continuous at a point (x_0, y_0), it is said to be *discontinuous* at (x_0, y_0), which is then called a *point of discontinuity*. If, as in the preceding example, it is possible to redefine the value of a function at a point of discontinuity so that the new function is continuous, we say that the point is a *removable discontinuity* of the old function. A function is said to be *continuous in a region* $\mathfrak{R}$ of the *xy* plane if it is continuous at every point of $\mathfrak{R}$.

Many of the theorems on continuity for functions of a single variable can, with suitable modification, be extended to functions of two more variables.

Uniform Continuity

In the definition of continuity of $f(x, y)$ at (x_0, y_0), δ depends on δ and also (x_0, y_0) in general. If in a region $\mathfrak{R}$ we can find a δ which depends only on δ but not on any particular point (x_0, y_0) in $\mathfrak{R}$ (i.e., the same δ will work for *all* points in $\mathfrak{R}$), then $f(x, y)$ is said to be *uniformly continuous* in $\mathfrak{R}$. As in the case of functions of one variable, it can be proved that a function which is continuous in a closed and bounded region is uniformly continuous in the region.

Partial Derivatives

The ordinary derivative of a function of several variables with respect to one of the independent variables, keeping all other independent variables constant, is called the *partial derivative* of the function with respect to the variable. Partial derivatives of $f(x, y)$ with respect to x and y are denoted by $\dfrac{\partial}{\partial x}\left[\text{or } f_x, f_x(x, y), \left.\dfrac{\partial f}{\partial x}\right|_y\right]$ and $\dfrac{\partial f}{\partial y}\left[\text{or } f_y, f_y(x, y), \left.\dfrac{\partial f}{\partial y}\right|_x\right]$, respectively, the latter notations being used when needed to emphasize which variables are held constant.

By definition,

$$\frac{\partial f}{\partial x} = \lim_{\Delta x \to 0} \frac{f(x + \Delta x, y) - f(x, y)}{\Delta x}, \qquad \frac{\partial f}{\partial y} = \lim_{\Delta y \to 0} \frac{f(x, y + \Delta y) - f(x, y)}{\Delta y} \tag{1}$$

when these limits exist. The derivatives evaluated at the particular point (x_0, y_0) are often indicated by $\left.\frac{\partial f}{\partial x}\right|_{(x_0, y_0)} = f_x(x_0, y_0)$ and $\left.\frac{\partial f}{\partial y}\right|_{(x_0, y_0)} = f_x(x_0, y_0)$, respectively.

> **EXAMPLE.** If $f(x, y) = 2x^3 + 3xy^2$, then $f_x = \partial f/\partial x = 6x^2 + 3y^2$ and $f_y = \partial f/\partial y = 6xy$. Also, $f(1, 2) = 6(1)^2 + 3(2)^2 = 18$, $f_y(1, 2) = 6(1)(2) = 12$.

If a function f has continuous partial derivatives $\partial f/\partial x$, $\partial f/\partial y$ in a region, then f must be continuous in the region. However, the existence of these partial derivatives alone is not enough to guarantee the continuity of f (see Problem 6.9).

Higher-Order Partial Derivatives

If $f(x, y)$ has partial derivatives at each point (x, y) in a region, then $\partial f/\partial x$ and $\partial f/\partial y$ are themselves functions of x and y, which may also have partial derivatives. These second derivatives are denoted by are denoted by

$$\frac{\partial}{\partial x}\left(\frac{\partial f}{\partial x}\right) = \frac{\partial^2 f}{\partial x^2} = f_{xx}, \quad \frac{\partial}{\partial y}\left(\frac{\partial f}{\partial y}\right) = \frac{\partial^2 f}{\partial y^2} = f_{yy}, \quad \frac{\partial}{\partial x}\left(\frac{\partial f}{\partial y}\right) = \frac{\partial^2 f}{\partial x\, \partial y} = f_{yx}, \quad \frac{\partial}{\partial y}\left(\frac{\partial f}{\partial x}\right) = \frac{\partial^2 f}{\partial y\, \partial x} = f_{xy} \tag{2}$$

If f_{xy} and f_{yx} are continuous, then $f_{xy} = f_{yx}$ and the order of differentiation is immaterial; otherwise they may not be equal (see Problems 6.13 and 6.41).

> **EXAMPLE.** If $f(x, y) = 2x^3 + 3xy^2$ (see preceding example), then $f_{xx} = 12x$, $f_{yy} = 6x$, and $f_{xy} = 6y = f_{yx}$. In such case $f_{xx}(1, 2) = 12$, $f_{yy}(1, 2) = 6$, and $f_{xy}(1, 2) = f_{yx}(1, 2) = 12$.

In a similar manner, higher order derivatives are defined. For example, $\frac{\partial^3 f}{\partial x^2 \partial y} = f_{yxx}$ is the derivative of f taken once with respect to y and twice with respect to x.

Differentials

(The section on differentials in Chapter 4 should be read before beginning this one.)

Let $\Delta x = dx$ and $\Delta y = dy$ be increments given to x and y, respectively. Then

$$\Delta z = f(x + \Delta x, y + \Delta y) - f(x, y) = \Delta f \tag{3}$$

is called the *increment* in $z = f(x, y)$. If $f(x, y)$ has continuous first partial derivatives in a region, then

$$\Delta z = \frac{\partial f}{\partial x}\Delta x + \frac{\partial f}{\partial y}\Delta y + \epsilon_1 \Delta x + \epsilon_2 \Delta y = \frac{\partial z}{\partial x}dx + \frac{\partial z}{\partial y}dy + \epsilon_1\, dx + \epsilon_2 dy = \Delta f \tag{4}$$

where ϵ_1 and ϵ_2 approach zero as Δx and Δy approach zero (see Problem 6.14). The expression

$$dz = \frac{\partial z}{\partial x}dx + \frac{\partial z}{\partial y}dy \quad \text{or} \quad df = \frac{\partial f}{\partial x}dx + \frac{\partial f}{\partial y}dy \tag{5}$$

is called the *total differential* or simply the *differential* of z or f, or the *principal part* of Δz or Δf. Note that $\Delta z \neq dz$ in general. However, if $\Delta x = dx$ and $\Delta y = dy$ are "small," then dz is a close approximation of Δz (see Problem 6.15). The quantities dx and dy—called *differentials* of x and y, respectively—need not be small.

The form $dz = f_x(x_0, y_0)dx + f_y(x_0, y_0)dy$ signifies a linear function with the independent variables dx and dy and the dependent range variable dz. In the one-variable case, the corresponding linear function represents the tangent line to the underlying curve. In this case, the underlying entity is a surface and the linear function generates the tangent plane at P_0. In a small enough neighborhood, this tangent plane is an approximation of the surface (i.e., the linear representation of the surface at P_0). If y is held constant, then we obtain the curve of intersection of the surface and the coordinate plane $y = y_0$. The differential form reduces to $dz = f_x(x_0, y_0)\,dx$ (i.e., the one-variable case). A similar statement follows when x is held constant. (See Figure 6.4.)

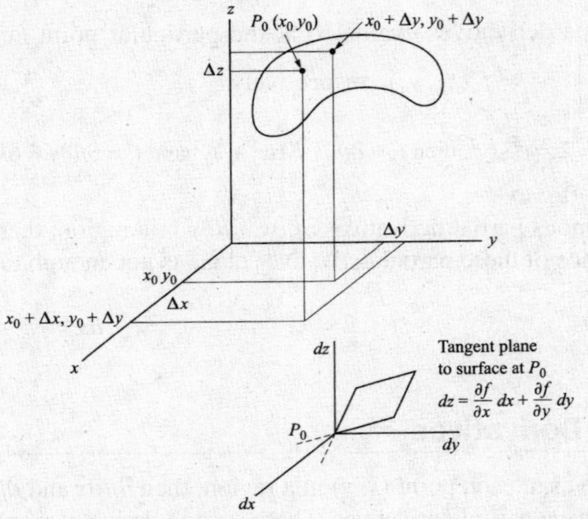

Figure 6.4

If f is such that Δf (or Δz) can be expressed in the form of Equation (4) where ϵ_1 and ϵ_2 approach zero as Δx and Δy approach zero, we call f *differentiable* at (x, y). The mere existence of f_x and f_y does not in itself guarantee differentiability; however, continuity of f_x and f_y does (although this condition happens to be slightly stronger than necessary). In case f_x and f_y are continuous in a region $\mathfrak{R}$, we say that f is *continuously differentiable* in $\mathfrak{R}$.

Theorems on Differentials

In the following, we assume that all functions have continuous first partial derivatives in a region $\mathfrak{R}$; i.e., the functions are continuously differentiable in $\mathfrak{R}$.

1. If $z = f(x_1, x_2, \ldots, x_n)$, then

$$df = \frac{\partial f}{\partial x_1}dx_1 + \frac{\partial f}{\partial x_2}dx_2 + \cdots + \frac{\partial f}{\partial x_n}dx_n \tag{6}$$

regardless of whether the variables $x_1, x_2, \ldots, x_n$ are independent or dependent on other variables (see Problem 6.20). This is a generalization of result in Equation (*5*). In Euqtion (*6*) we often use z in place of f.

2. If $f(x_1, x_2, \ldots, x_n) = c$, a constant, then $df = 0$. Note that in this case $x_1, x_2, \ldots, x_n$ cannot all be independent variables.

3. The expression $P(x, y)dx + Q(x, y)dy$, or, briefly, $P\,dx + Q\,dy$, is the differential of $f(x, y)$ if and only if $\frac{\partial P}{\partial y} = \frac{\partial Q}{\partial x}$. In such case, $P\,dx + Q\,dy$ is called an *exact differential*.

   Note: Observe that $\frac{\partial P}{\partial y} = \frac{\partial Q}{\partial x}$ implies that $\frac{\partial^2 f}{\partial y\,\partial x} = \frac{\partial^2 f}{\partial x\,\partial y}$.

4. The expression $P(x, y, z)\,dx + Q(x, y, z)dy + R(x, y, z)dz$, or, briefly, $P\,dx + Q\,dy + R\,dz$, is the differential of $f(x, y, z)$ if and only if $\frac{\partial P}{\partial y} = \frac{\partial Q}{\partial x}, \frac{\partial Q}{\partial z} = \frac{\partial R}{\partial y}, \frac{\partial R}{\partial x} = \frac{\partial P}{\partial z}$. In such case, $P\,dx + Q\,dy + R\,dz$ is called an *exact differential*.

Proofs of Theorems 3 and 4 are best supplied by methods of later chapters (see Problems 10.13 and 10.30).

Differentiation of Composite Functions

Let $z = f(x, y)$ where $x = g(r, s)$, $y = h(r, s)$ so that z is a function of r and s. Then

$$\frac{\partial z}{\partial r} = \frac{\partial z}{\partial x}\frac{\partial x}{\partial r} + \frac{\partial z}{\partial y}\frac{\partial y}{\partial r}, \qquad \frac{\partial z}{\partial s} = \frac{\partial z}{\partial x}\frac{\partial x}{\partial s} + \frac{\partial z}{\partial y}\frac{\partial y}{\partial s} \tag{7}$$

In general, if $u = F(x_1, \ldots, x_n)$ where $x_1 = f_1(r_1, \ldots, r_p), \ldots, x_n = f_n(r_1, \ldots, r_p)$, then

$$\frac{\partial u}{\partial r_k} = \frac{\partial u}{\partial x_1}\frac{\partial x_1}{\partial r_k} + \frac{\partial u}{\partial x_2}\frac{\partial x_2}{\partial r_k} + \cdots + \frac{\partial u}{\partial x_n}\frac{\partial u}{\partial r_k} \qquad k = 1, 2, \ldots, p \tag{8}$$

If, in particular, $x_1, x_2, \ldots, x_n$ depend on only one variable s, then

$$\frac{du}{ds} = \frac{\partial u}{\partial x_1}\frac{dx_1}{ds} + \frac{\partial u}{\partial x_2}\frac{dx_2}{ds} + \cdots + \frac{\partial u}{\partial x_n}\frac{dx_n}{ds} \tag{9}$$

These results, often called *chain rules*, are useful in transforming derivatives from one set of variables to another.

Higher derivatives are obtained by repeated application of the chain rules.

Euler's Theorem on Homogeneous Functions

A function represented by $F(x_1, x_2, \ldots, x_2, \ldots, x_n)$ is called *homogeneous of degree p* if, for all values of the parameter λ and some constant p, we have the identity

$$F(\lambda x_1, \lambda x_2, \ldots, \lambda x_n) = \lambda^p F(x_1, x_2, \ldots, x_n) \tag{10}$$

EXAMPLE. $F(x, y) = x^4 + 2xy^3 - 5y^4$ is homogeneous of degree 4, since

$$F(\lambda x, \lambda y) = (\lambda x)^4 + 2(\lambda x)(\lambda y)^3 - 5(\lambda y)^4 = \lambda^4(x^4 + 2xy^3 - 5y^4) = \lambda^4 F(x, y)$$

Euler's theorem on homogeneous functions states that if $F(x_1, x_2, \ldots, x_n)$ is homogeneous of degree p, then (see Problem 6.25)

$$x_1\frac{\partial F}{\partial x_1} + x_2\frac{\partial F}{\partial x_2} + \cdots + x_n\frac{\partial F}{\partial x_n} = pF \tag{11}$$

Implicit Functions

In general, an equation such as $F(x, y, z) = 0$ defines one variable—say, z—as a function of the other two variables x and y. Then z is sometimes called an *implicit function* of x and y, as distinguished from an *explicit function f*, where $z = f(x, y)$, which is such that $F[x, y, f(x, y)] \equiv 0$.

Differentiation of implicit functions requires considerable discipline in interpreting the independent and dependent character of the variables and in distinguishing the intent of one's notation. For example, suppose that in the implicit equation $F[x, y, f(x, z)] = 0$, the independent variables are x and y and that $z = f(x, y)$. In order to find $\frac{\partial f}{\partial x}$ and $\frac{\partial f}{\partial y}$, we initially write the following [observe that $F(x, t, z)$ is zero for all domain pairs (x, y); i.e., it is a constant]:

$$0 = dF = F_x\,dx + F_y\,dy + F_z\,dz$$

and then compute the partial derivatives F_x, F_y, F_z as though y, y, z constituted an independent set of variables. At this stage we invoke the dependence of z on x and y to obtain the differential form $dz = \frac{\partial f}{\partial x}dx + \frac{\partial f}{\partial y}dy$. Upon substitution and some algebra (see Problem 6.30), the following results are obtained:

$$\frac{\partial f}{\partial x} = -\frac{F_x}{F_z}, \qquad \frac{\partial f}{\partial y} = -\frac{F_y}{F_z}$$

EXAMPLE. If $0 = F(x, y, z) = x^2z + yz^2 + 2xy^2 - z^3$ and $z = f(x, y)$, then $F_x = 2xz + 2y^2$, $F_y = z^2 + 4xy$. $F_z = x^2 + 2yz - 3z^2$. Then

$$\frac{\partial f}{\partial x} = -\frac{(2xz + 2y^2)}{x^2 + 2yz - 3z^2}, \qquad \frac{\partial f}{\partial y} = -\frac{(z^2 + 4xy)}{x^2 + 2yz - 3x^2}$$

Observe that f need not be known to obtain these results. If that information is available, then (at least theoretically) the partial derivatives may be expressed through the independent variables x and y.

Jacobians

If $F(u, \upsilon)$ and $G(u, \upsilon)$ are differentiable in a region, the *Jacobian determinant*, or the *Jacobian*, of F and G with respect to u and υ is the second-order functional determinant defined by

$$\frac{\partial(F, G)}{\partial(u, \upsilon)} = \begin{vmatrix} \dfrac{\partial F}{\partial u} & \dfrac{\partial F}{\partial \upsilon} \\ \dfrac{\partial G}{\partial u} & \dfrac{\partial G}{\partial \upsilon} \end{vmatrix} = \begin{vmatrix} F_u & F_\upsilon \\ G_u & G_\upsilon \end{vmatrix} \tag{12}$$

Similarly, the third-order determinant

$$\frac{\partial(F, G, H)}{\partial(u, \upsilon, w)} = \begin{vmatrix} F_u & F_\upsilon & F_w \\ G_u & G_\upsilon & G_w \\ H_u & H_\upsilon & H_w \end{vmatrix}$$

is called the Jacobian of F, G, and H with respect to u, υ, and w. Extensions easily made.

Partial Derivatives Using Jacobians

Jacobians often prove useful in obtaining partial derivatives of implicit functions. Thus, for example, given the simultaneous equations

$$F(x, y, u, \upsilon) = 0, \qquad G(x, y, u, \upsilon) = 0$$

we may, in general, consider u and υ as functions of x and y. In this case, we have (see Problem 6.31)

$$\frac{\partial u}{\partial x} = -\frac{\dfrac{\partial(F, G)}{\partial(x, \upsilon)}}{\dfrac{\partial(F, G)}{\partial(u, \upsilon)}}, \quad \frac{\partial u}{\partial y} = -\frac{\dfrac{\partial(F, G)}{\partial(y, \upsilon)}}{\dfrac{\partial(F, G)}{\partial(u, \upsilon)}}, \quad \frac{\partial \upsilon}{\partial x} = -\frac{\dfrac{\partial(F, G)}{\partial(u, x)}}{\dfrac{\partial(F, G)}{\partial(u, \upsilon)}}, \quad \frac{\partial \upsilon}{\partial y} = -\frac{\dfrac{\partial(F, G)}{\partial(u, y)}}{\dfrac{\partial(F, G)}{\partial(u, \upsilon)}}$$

The ideas are easily extended. Thus, if we consider the simultaneous equations

$$F(u, \upsilon, w, x, y) = 0, \qquad G(u, \upsilon, w, x, y)\, y) = 0, \qquad H(u, \upsilon, w, x, y) = 0$$

we may, for example, consider u, υ, and w as functions of x and y. In this case,

$$\frac{\partial u}{\partial x} = -\frac{\dfrac{\partial(F, G, H)}{\partial(x, \upsilon, w)}}{\dfrac{\partial(F, G, H)}{\partial(u, \upsilon, w)}}, \qquad \frac{\partial w}{\partial y} = -\frac{\dfrac{\partial(F, G, H)}{\partial(u, \upsilon, y)}}{\dfrac{\partial(F, G, H)}{\partial(u, \upsilon, w)}}$$

with similar results for the remaining partial derivatives (see Problem 6.33).

Theorems on Jacobians

In the following we assume that all functions are continuously differentiable.

1. A necessary and sufficient condition that the equations $F(u, \upsilon, x, y, z) = 0$ and $G(u, \upsilon, x, y, z) = 0$ can be solved for u and υ (for example) is that $\frac{\partial(F, G)}{\partial(u, \upsilon)}$ is not identically zero in a region $\Re$.

   Similar results are valid for m equations in n variables, where $m < n$.

2. If x and y are functions of u and υ while u and υ are functions of and s, then (see Problem 6.43)

$$\frac{\partial(x, y)}{\partial(r, s)} = \frac{\partial(x, y)}{\partial(u, \upsilon)} \frac{\partial(u, \upsilon)}{\partial(r, s)} \tag{13}$$

   This is an example of a *chain rule* for Jacobians. These ideas are capable of generalization (see Problems 6.107 and 6.109, for example).

3. If $u = f(x, y)$ and $\upsilon = g(x, y)$, then a necessary and sufficient condition that a functional relation of the form $\phi(u, \upsilon) = 0$ exists between u and υ is that $\frac{\partial(u, \upsilon)}{\partial(x, y)}$ be identically zero Similar results. hold for n functions of n variables.

Further discussion of Jacobians appears in Chapter 7, where vector interpretations are employed.

Transformations

The set of equations

$$\begin{cases} x = F(u, \upsilon) \\ y = F(u, \upsilon) \end{cases} \tag{14}$$

defines, in general, a *transformation* or *mapping* which establishes a correspondence between points in the $u\upsilon$ and xy planes. If to each point in the $u\upsilon$ plane there corresponds one and only one the xy plane, and conversely, we speak of a *one-to-one* transformation or mapping. This will be so if F and G are continuously differentiable, with Jacobian not identically zero in a region. In such case (which we assume unless otherwise stated), Equations (14) are said to define a *continuously differentiable transformation* or *mapping*.

The words *transformation* and *mapping* describe the same mathematical concept in different ways. A *transformation* correlates one coordinate representation of a region of space with another. A *mapping* views this correspondence as a correlation of two distinct regions.

Under the transformation (14) a closed region $\Re$ of the xy plane is, in general, mapped into a closed region $\Re'$ of the $u\upsilon$ plane. Then if ΔA_{xy} and $\Delta A_{u\upsilon}$ denote, respectively, the areas of these regions, we can show that

$$\lim \frac{\Delta A_{xy}}{\Delta A_{u\upsilon}} = \left| \frac{\partial(x, y)}{\partial(u, \upsilon)} \right| \tag{15}$$

where lim denotes the limit as ΔA_{xy} (or $\Delta A_{u\upsilon}$) approaches zero. The Jacobian on the right of Equation (15) is often called the *Jacobian of the transformation* (14).

If we solve transformation (14) for u and υ in terms of x and y, we obtain the transformation $u = f(x, y)$, $\upsilon = g(x, y)$, often called the *inverse transformation* corresponding to (14). The Jacobians $\frac{\partial(u, \upsilon)}{\partial(x, y)}$ and $\frac{\partial(x, y)}{\partial(u, \upsilon)}$ of these transformations are reciprocals of each other (see Problem 6.43). Hence, if one Jacobian is different from zero in a region, the inverse exists and is not zero.

These ideas can be extended to transformations in three or higher dimensions. We deal further with these topics in Chapter 7, where use is made of the simplicity of vector notation and interpretation.

Curvilinear Coordinates

Rectangular Cartesian coordinates in the Euclidean plane or in three-dimensional space were mentioned at the beginning of this chapter. Other coordinate systems, the coordinate curves of which are generated from families that are not necessarily linear, are useful. These are called *curvilinear coordinates.*

EXAMPLE 1. Polar coordinates ρ, Φ are related to rectangular Cartesian coordinates through the transformation $x = \rho \cos \Phi$ and $y = \rho \sin \Phi$. The curves $\Phi = \Phi_0$ are radical lines, while $\rho = r_0$ are concentric circles. Among the convenient representations yielded by these coordinates are circles with the center at the origin. A weakness is that representations may not be defined at the origin.

EXAMPLE 2. Spherical coordinates r, Θ, Φ_0 for Euclidean three-dimensional space are generated from rectangular Cartesian coordinates by the transformation equations $x = r \sin \Theta \cos \Phi$, $y = r \sin \Theta \sin \Phi$, and $z = r \cos \Phi$. Again, certain problems lend themselves to spherical coordinates, and also uniqueness of representation can be a problem at the origin. The coordinate surfaces $r = r_0$, $\Theta = \Theta_0$, and $\Phi = \Phi_0$ are spheres, planes, and cones, respectively. The coordinate curves are the intersections of these surfaces, i.e., circles, circles, and lines.

For curvilinear coordinates in higher dimensional spaces, see Chapter 7.

Mean Value Theorem

If $f(x, y)$ is continuous in a closed region and if the first partial derivatives exist in the open region (i.e., excluding boundary points). then

$$f(x_0 + h, y_0 + k) - f(x_0, y_0) = hf_x(x_0 + \theta h, y_0 + \theta k) + kf_y(x_0 + \theta h, y_0 + \theta k) \qquad 0 < \theta < 1 \qquad (16)$$

This is sometimes written in a form in which $h = \Delta x = x - x_0$ and $k = \Delta y = y - y_0$.

SOLVED PROBLEMS

Functions and graphs

6.1. If $f(x, y) = x^3 - 2xy + 3y^2$, find: (a) $f\left(\frac{1}{x}, \frac{2}{y}\right)$; (c) $\frac{f(x, y+k) - f(x, y)}{k}$, $k \neq 0$.

(a) $f(-2,3) = (-2)^3 - 2(-2)(3) + 3(3)^2 = -8 + 12 + 27 = 31$

(b) $f\left(\frac{1}{x}, \frac{2}{y}\right) = \left(\frac{1}{x}\right)^3 - 2\left(\frac{1}{x}\right)\left(\frac{2}{y}\right) + 3\left(\frac{2}{y}\right)^2 = \frac{1}{x^3} - \frac{4}{xy} + \frac{12}{y^2}$

(c)
$$\begin{aligned}\frac{f(x, y+k) - f(x, y)}{k} &= \frac{1}{k}\{[x^3 - 2x(y+k) + 3(y+k)^2] - [x^3 - 2xy + 3y^2]\}\\ &= \frac{1}{k}(x^3 - 2xy - 2kx + 3y^2 + 6ky + 3k^2 - x^2 + 2xy - 3y^2)\\ &= \frac{1}{k}(-2kx = 6ky + 3k^2) = -2x + 6y + 3k.\end{aligned}$$

6.2. Give the domain of definition for which each of the following functions is defined and real, and indicate this domain graphically.

(a) $f(x, y) = \ln\{(16 - x^2 - y^2)(x^2 + y^2 - 4)\}$

The function is defined and real for all points (x, y) such that

$$(16 - x^2 - y^2)(x^2 + y^2 - 4) > 0, \qquad \text{i.e., } 4 < x^2 + y^2 < 16$$

which is the required domain of definition. This point set consists of all points *interior* to the circle of radius 4 with center at the origin and *exterior* to the circle of radius 2 with center at the origin, as in Figure 6.5. The corresponding region, shown shaded in Figure 6.5, is an *open region*.

(b) $f(x, y) = \sqrt{6 - (2x + 3y)}$

The function is defined and real for all points (x, y) such that $2x + 3y \leq 6$, which is the required domain of definition.

The corresponding (unbounded) region of the xy plane is shown shaded in Figure 6.6.

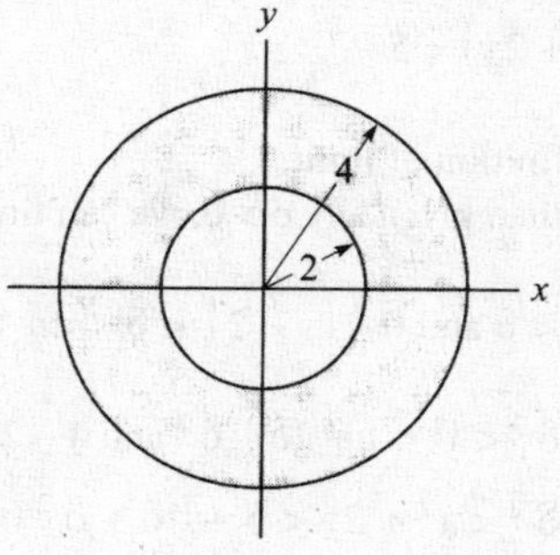

Figure 6.5

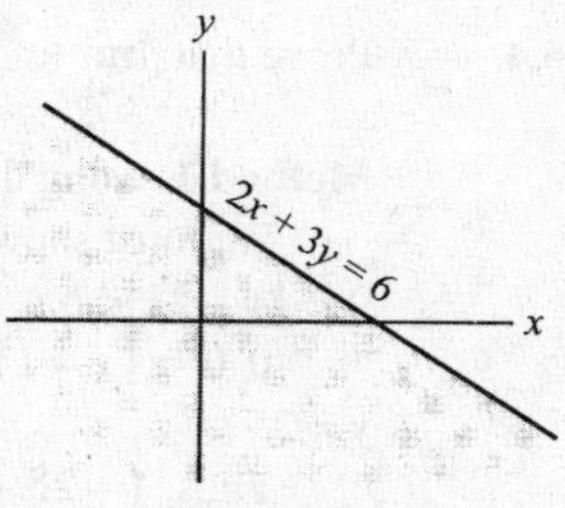

Figure 6.6

6.3. Sketch and name the surface in three-dimensional space represented by each of the following. What are the traces on the coordinate planes?

(a) $2x + 4y + 3z = 12$.

Trace on xy plane ($z = 0$) is the straight line $x + 2y = 6$, $z = 0$.
Trace on yz plane ($x = 0$) is the straight line $4y + 3z = 12$, $x = 0$.
Trace on xz plane ($y = 0$) is the straight line $2x + 3z = 12$, $y = 0$.
These are represented by AB, BC, and AC in Figure 6.7.

The surface is a plane intersecting the x, y, and z axes in the points $A(6, 0, 0)$, $B(0, 3, 0)$, and $C(0, 0, 4)$. The lengths $\overline{OA} = 6$, $\overline{OB} = 3$, and $\overline{OC} = 4$ are called the *x, y, and z intercepts*, respectively.

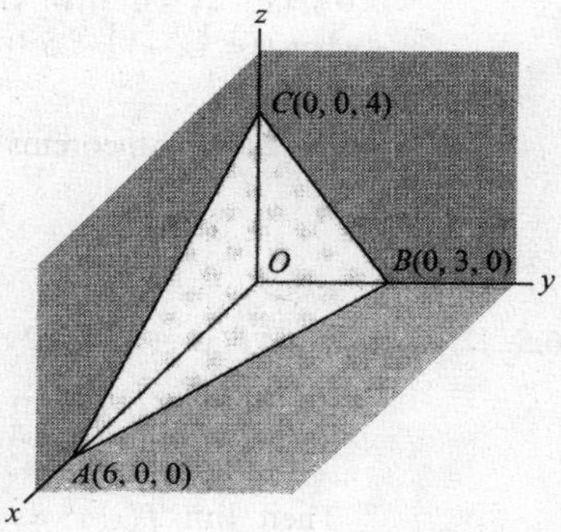

Figure 6.7

(b) $\dfrac{x^2}{a^2} + \dfrac{y^2}{b^2} - \dfrac{z^2}{c^2} = 1$

Trace on xy plane ($z = 0$) is the ellipse $\dfrac{x^2}{a^2} + \dfrac{y^2}{b^2} = 1$, $z = 0$.

Trace on yz plane ($x = 0$) is the hyperbola $\dfrac{y^2}{b^2} - \dfrac{z^2}{c^2} = 1$, $x = 0$.

Trace on xz plane ($y = 0$) is the hyperbola $\dfrac{x^2}{a^2} - \dfrac{z^2}{c^2} = 1$, $y = 0$.

Trace on any plane $z = p$ parallel to the xy plane is the ellipse $\dfrac{x^2}{a^2(1 + p^2/c^2)} + \dfrac{y^2}{b^2(1 + p^2/c^2)} = 1$.

As $|p|$ increases from zero, the elliptic cross section increases in size.
The surface is a *hyperboloid of one sheet* (see Figure 6.8).

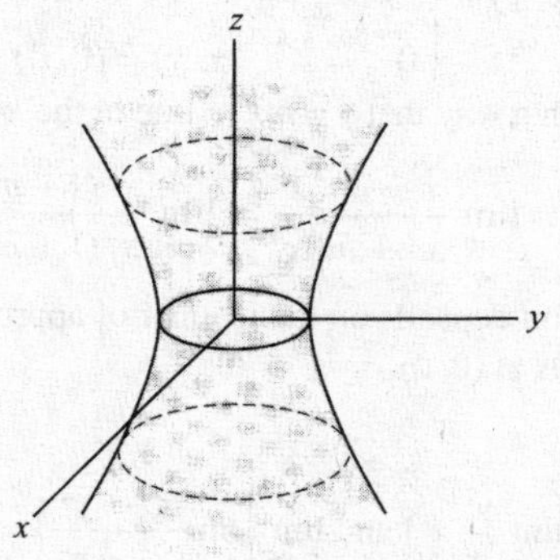

Figure 6.8

Limits and continuity

6.4. Prove that $\lim\limits_{\substack{x\to 1\\ y\to 2}} (x^2 + 2y) = 5$.

Method 1, using definition of limit.

We must show that, given any $\delta > 0$, we can find $\delta > 0$ such that $|x^2 + 2y - 5| < \delta$ when $0 < |x - 1| < \delta$, $0 < |y - 2| < \delta$.

If $0 < |x - 1| < \delta$ and $0 < |y - 2| < \delta$, then $1 - \delta\, x < 1 + < \delta$ and $2 - \delta < y < 2 + \delta$, excluding $x = 1$, $y = 2$.

Thus, $1 - 2\delta + \delta^2 < x^2 < 1 + 2\delta + \delta^2$ and $4 - 2\delta < 2y < 4 + 2\delta$. Adding

$$5 - 4\delta + \delta^2 < x^2 + 2y < 5 + 4\delta + \delta^2 \text{ or } -48 + \delta^2 < x^2 + 2y - 5 < 4\delta + \delta^2$$

Now, if $\delta \leqq 1$, it certainly follows that $-5\delta < x^2 + 2y - 5 < 5\delta$; i.e., $|x^2 + 2y - 5| < 5\delta$ whenever $0 < |x - 1| < \delta$, $0 < |y - 2| < \delta$. Then, choosing $5\delta = \epsilon$, i.e., $\delta = \epsilon/5$ (or $\delta = 1$, whichever is smaller), it follows that $|x^2 + 2y - 5| < \epsilon$ when $0 < |x - 1| < \delta$, $0 < |y - 2| < \delta$; i.e., $\lim\limits_{\substack{x\to 1\\ y\to 2}} (x^2 + 2y) = 5$.

Method 2, using theorems on limits.

$$\lim_{\substack{x\to 1\\ x\to 2}}(x^2 + 2y) = \lim_{\substack{x\to 1\\ x\to 2}} x^2 + \lim_{\substack{x\to 1\\ x\to 2}} 2y = 1 + 4 = 5$$

6.5. Prove that $f(x, y) = x^2 + 2y$ is continuous at (1, 2).

By Problem 6.4, $\lim\limits_{\substack{x\to 1\\ y\to 2}} f(x, y) = 5$. Also, $f(1, 2) = 1^2 + 2(2) = 5$.

Then $\lim\limits_{\substack{x\to 1\\ y\to 2}} f(x, y) = f(1, 2)$ and the function is continuous at (1, 2).

Alternatively, we can show, in much the same manner as in the first method of Problem 6.4, that given any $\delta > 0$ we can find $\delta > 0$ such that $|f(x, y) - f(1, 2)| < \delta$ when $|x - 1| < \delta$, $|y - 2| < \delta$.

6.6. Determine whether $f(x, y) = \begin{vmatrix} x^2 + 2y, & (x, y) \neq (1, 2) \\ 0, & (x, y) = (1, 2) \end{vmatrix}$ (a) has a limit as $x \to 1$ and $y \to 2$, and (b) is continuous at (1, 2).

(a) By Problem 6.4, it follows that $\lim\limits_{\substack{x\to 1\\ x\to 2}} f(x, y) = 5$, since the *limit* has nothing to do with the *value* at (1, 2).

(b) Since $\lim\limits_{\substack{x\to 1\\ x\to 2}} f(x, y) = 5$ and $f(1, 2) = 0$, it follows that $\lim\limits_{\substack{x\to 1\\ y\to 2}} f(x, y) \neq f(1, 2)$. Hence, the function is *discontinuous* at (1, 2).

6.7. Investigate the continuity of $f(x, y) = \begin{cases} \dfrac{x^2 - y^2}{x^2 + y^2} & (x, y) \neq (0, 0) \\ 0 & (x, y) = (0, 0) \end{cases}$. at $(0,0)$.

Let $x \to 0$ and $y \to 0$ in such a way that $y = mx$ (a line in the xy plane). Then, along this line,

$$\lim_{\substack{x\to 0\\ y\to 0}} \frac{x^2 - y^2}{x^2 + y^2} = \lim_{x\to 0} \frac{x^2 - m^2x^2}{x^2 + m^2x^2} = \lim_{x\to 0} \frac{x^2(1 - m^2)}{x^2(1 + m^2)} = \frac{1 - m^2}{1 + m^2}$$

Since the limit of the function depends on the manner of approach to (0, 0) (i.e., the slope m of the line), the function cannot be continuous at (0, 0).

Another method:

Since $\lim\limits_{x\to 0}\left\{\lim\limits_{x\to 0} \dfrac{x^2 - y^2}{x^2 + y^2}\right\} = \lim\limits_{x\to 0} \dfrac{y^2}{x^2} = 1$ and $\lim\limits_{x\to 0}\left\{\lim\limits_{x\to 0} \dfrac{x^2 - y^2}{x^2 + y^2}\right\} = -1$ are not equal, $\lim\limits_{\substack{x\to 0\\ y\to 0}} f(x, y)$ cannot exist.

Hence, $f(x, y)$ cannot be continuous at (0, 0).

Partial derivatives

6.8. If $f(x, y) = 2x^2 - xy + y^2$, find (a) $\partial f / \partial x$ and (b) $\partial f / \partial y$ at (x_0, y_0) directly from the definition.

(a)
$$\left.\frac{\partial f}{\partial x}\right|_{(x_0, y_0)} = f_x(x_0, y_0) = \lim_{h \to 0} \frac{f(x_0 + h, y_0) - f(x_0, y_0)}{h}$$
$$= \lim_{h \to 0} \frac{[2(x_0 + h)^2 - (x_0 + h)y_0 + y_0^2] = [2x_0^2 - x_0 y_0 + y_0^2]}{h}$$
$$= \lim_{h \to 0} \frac{4hx_0 + 2h^2 - hy_0}{h} = \lim_{h \to 0}(4x_0 + 2h - y_0) = 4x_0 - y_0$$

(b)
$$\left.\frac{\partial f}{\partial y}\right|_{(x_0, y_0)} = f_y(x_0, y_0) = \lim_{k \to 0} \frac{f(x_0, y_0 + k) - f(x_0, y_0)}{k}$$
$$= \lim_{k \to 0} \frac{[2x_0^2 - x_0(y_0 + k) + (y_0 + k)^2] - [2x_0^2 - x_0 y_0 + y_0^2]}{k}$$
$$= \lim_{k \to 0} \frac{-kx_0 + 2ky_0 + k^2}{k} = \lim_{k \to 0}(-x_0 + 2y_0 + k) = -x_0 + 2y_0$$

Since the limits exist for all points (x_0, y_0), we can write $f_x(x, y) = f_x = 4x - y$, $f_y(x, y) = f_y = -x + 2y$, which are themselves functions of x and y.

Note that *formally* $f_x(x_0, y_0)$ is obtained from $f(x, y)$ by differentiating with respect to x, keeping y constant and then putting $x = x_0$, $y = y_0$. Similarly, $f_y(x_0, y_0)$ is obtained by differentiating f with respect to y, keeping x constant. This procedure, while often lucrative in practice, need not always yield correct results (see Problem 6.9). It will work if the partial derivatives are continuous.

6.9. Let $f(x, y) = \begin{cases} xy/(x^2 + y^2) & (x, y) \neq (0, 0) \\ 0 & \text{otherwise} \end{cases}$. Prove that (a) both $f_x(0, 0)$ and $f_y(0, 0)$ exist but that (b) $f(x, y)$ is discontinuous at (0, 0).

(a)
$$f_x(0, 0) = \lim_{h \to 0} \frac{f(h, 0) - f(0, 0)}{h} = \lim_{h \to 0} \frac{0}{h} = 0$$
$$f_x(0, 0) = \lim_{k \to 0} \frac{f(0, 0) - f(0, 0)}{k} = \lim_{k \to 0} \frac{0}{k} = 0$$

(b) Let $(x, y) \to (0, 0)$ along the line $y = mx$ in the xy plane. Then $\lim\limits_{\substack{x \to 0 \\ y \to 0}} f(x, y) = \lim\limits_{x \to 0} \dfrac{mx^2}{x^2 + m^2x^2} = \dfrac{m}{1 + m^2}$ so that the limit depends on m and, hence, on the approach; therefore, it does not exist. Hence, $f(x, y)$ is not continuous at (0, 0).

Note that unlike the situation for functions of one variable, the existence of the first partial derivatives at a point *does not* imply continuity at the point.

Note also that if $(x, y) \neq (0, 0)$, $f_x = \dfrac{y^2 - x^2y}{(x^2 + y^2)^2}$, $f_y = \dfrac{x^3 - xy^2}{(x^2 + y^2)^2}$ and $f_x(0, 0)$, $f_y(0, 0)$ cannot be computed from them by merely letting $x = 0$ and $y = 0$. See the remark at the end of Problem 4.5(b).

6.10. If $\phi(x, y) = x^3y + e^{xy^2}$, find (a) ϕ_x, (b) $\phi_{y\,\phi xx}$, (d) ϕ_{yy}, (e) ϕ_{xy}, and (f) ϕ_{yx}.

(a)
$$\phi_x = \frac{\partial \phi}{\partial x} = \frac{\partial}{\partial x}(x^3y + e^{xy^2}) = 3x^2y + e^{xy^2} \cdot y^2 = 3x^2y + y^2e^{xy^2}$$

(b)
$$\phi_y = \frac{\partial \phi}{\partial y} = \frac{\partial}{\partial y}(x^3y + e^{xy^2}) = x^3 + e^{xy^2} \cdot 2xy = x^3 + 2xy\, e^{xy^2}$$

(c)
$$\phi_{xx} = \frac{\partial^2 \phi}{\partial x^2} = \frac{\partial}{\partial x}\left(\frac{\partial \phi}{\partial x}\right) = \frac{\partial}{\partial x}(3x^2y + y^2\, e^{xy^2}) = 6xy + y^2(e^{xy^2} \cdot y^2) = 6xy + y^4e^{xy^2}$$

(d) $$\phi_{yy} = \frac{\partial^2\phi}{\partial y^2} = \frac{\partial}{\partial y}(x^3 + 2xy\,e^{xy^2}) = 0 + 2xy\cdot\frac{\partial}{\partial y}(e^{xy^2}) + e^{xy^2}\frac{\partial}{\partial y}(2xy)$$
$$= 2xy\cdot e^{xy^2}\cdot 2xy + e^{xy^2}\cdot 2x = x^2y^2e^{xy^2} + 2x\,e^{xy^2}$$

(e) $$\phi_{xy} = \frac{\partial^2\phi}{\partial y\,\partial x} = \frac{\partial}{\partial y}\left(\frac{\partial\phi}{\partial x}\right) = \frac{\partial}{\partial y}(3x^2y + y^2\,e^{xy^2}) = 3x^2 + y^2\cdot e^{xy^2}\cdot 2xy + e^{xy^2}\cdot 2y$$
$$= 3x^2 + 2xy^3\,e^{xy^2} + 2y\,e^{xy^2}$$

(f) $$\phi_{yx} = \frac{\partial^2\phi}{\partial x\,\partial y} = \frac{\partial}{\partial x}\left(\frac{\partial\phi}{\partial y}\right) = \frac{\partial}{\partial x}(x^3 + 2xy\,e^{xy^2}) = 3x^2 + y^2\cdot e^{xy^2}\cdot 2xy + e^{xy^2}\cdot 2y$$
$$= 3x^2 + 2xy^3\,e^{xy^2} + 2y\,e^{xy^2}$$

Note that $\phi_{xy} = \phi_{yx}$ in this case. This is because the second partial derivatives exist and are continuous for all (x, y) in a region $\Re$. When this is not true, we may have $\phi_{xy} \neq \phi_{yx}$ (see Problem 6.41, for example).

6.11. Show that $U(x, y, z) = (x^2 + y^2 + z^2)^{-1/2}$ satisfies Laplace's partial differential equation $\frac{\partial^2 U}{\partial x^2} + \frac{\partial^2 U}{\partial y^2} + \frac{\partial^2 U}{\partial z^2} = 0.$

We assume here that $(x, y, z) \neq (0, 0, 0)$. Then

$$\frac{\partial U}{\partial x} = -\frac{1}{2}(x^2 + y^2 + z^2)^{-3/2}\cdot 2x = -x(x^2 + y^2 + z^2)^{-3/2}$$

$$\frac{\partial^2 U}{\partial x^2} = \frac{\partial}{\partial x}[-x(x^2 + y^2 + z^2)^{-3/2}] = (-x)[-\frac{3}{2}(x^2 + y^2 + z^2)^{-5/2}\,2x] + (x^2 + y^2 + z^2)^{-3/2}\cdot(-1)$$
$$= \frac{3x^2}{(x^2 + y^2 + z^2)^{5/2}} - \frac{(x^2 + y^2 + z^2)}{(x^2 + y^2 + z^2)^{5/2}} = \frac{2x^2 - y^2 - z^2}{(x^2 + y^2 + z^2)^{5/2}}$$

Similarly,

$$\frac{\partial^2 U}{\partial y^2} = \frac{2y^2 - x^2 - z^2}{(x^2 + y^2 + z^2)^{5/2}}, \qquad \frac{\partial^2 U}{\partial x^2} = \frac{2z^2 - x^2 - y^2}{(x^2 + y^2 + z^2)^{5/2}}$$

Adding,

$$\frac{\partial^2 U}{\partial x^2} + \frac{\partial^2 U}{\partial y^2} + \frac{\partial^2 U}{\partial z^2} = 0$$

6.12. If $z = x^2\tan^{-1}\frac{y}{x}$, find $\frac{\partial^2 z}{\partial x\,\partial y}$ at (1, 1).

$$\frac{\partial z}{\partial y} = x^2\cdot\frac{1}{1 + (y/x)^2}\frac{\partial}{\partial y}\left(\frac{y}{x}\right) = x^2\cdot\frac{x^2}{x^2 + y^2}\cdot\frac{1}{x} = \frac{x^3}{x^2 + y^2}$$

$$\frac{\partial^2 z}{\partial x\,\partial y} = \frac{\partial}{\partial x}\left(\frac{\partial z}{\partial y}\right) = \frac{\partial}{\partial x}\left(\frac{x^3}{x^2 + y^2}\right) = \frac{(x^2 + y^2)(3x^2) - (x^3)(2x)}{(x^2 + y^2)^2} = \frac{2\cdot 3 - 1\cdot 2}{2^2} = 1 \text{ at } (1, 1)$$

The result can be written $z_{xy}(1, 1) = 1$.

Note: In this calculation we are using the fact that z_{xy} is continuous at (1, 1) (see the remark at the end of Problem 6.9).

6.13. If $f(x, y)$ is defined in a region $\Re$ and if f_{xy} and f_{yx} exist and are continuous at a point of $\Re$, prove that $f_{xy} = f_{yx}$ at this point.

Let (x_0, y_0) be the point of $\Re$. Consider

$$G = f(x_0 + h, y_0 + k) - f(x_0, y_0 + k) - f(x_0 - h, y_0) + f(x_0, y_0)$$

Define

$$\phi(x, y) = f(x + h, y) - f(x, y) \tag{1}$$

$$\psi(x, y) = f(x, y + k) - f(x, y) \tag{2}$$

Then

$$G = \phi(x_0, y_0 + k) - \phi(x_0, y_0) \tag{3}$$

$$G = \psi(x_0 + h, y_0) - \psi(x_0, y_0) \tag{4}$$

Applying the mean value theorem for functions of one variable (see Page 78) to Equations (3) and (4), we have

$$G = k\phi_y(x_0, y_0 + \theta_1 k) = k\{f_y(x_0 + h, y_0 + \theta_1 - k) - f_y(x_0, y_0 + \theta_1 k)\} \quad 0 < \theta_1 < 1 \tag{5}$$

$$G = h\psi_x(x_0, + \theta_2 h, y_0) = h\{f_x(x_0 + \theta_2 h, y_0 + k) - f_x(x_0 + \theta_2 h, y_0)\} \quad 0 < \theta_2 < 1 \tag{6}$$

Applying the mean value theorem again to Equations (5) and (6), we have

$$G = hk\, f_{yx}(x_0 + \theta_3 h, y_0 + \theta_1 k) \qquad 0 < \theta_1 < 1, 0 < \theta_3 < 1 \tag{7}$$

$$G = hk\, f_{xy}(x_0 + \theta_2 h, y_0 + \theta_4 k) \qquad 0 < \theta_2 < 1, 0 < \theta_4 < 1 \tag{8}$$

From Equations (7) and (8) we have

$$f_{yx}(x_0 + \theta_3 h, y_0 + \theta_1 k) = f_{xy}(x_0 + \theta_2 h, y_0 + \theta_4 k) \tag{9}$$

Letting $h \to 0$ and $k \to 0$ in (9) we have, since f_{xy} and f_{yx} are assumed continuous at (x_0, y_0),

$$f_{yx}(x_0, y_0) = f_{xy}(d_0, y_0)$$

as required. For an example where this fails to hold, see Problem 6.41.

Differentials

6.14. Let $f(x, y)$ have continuous first partial derivatives in a region $\Re$ of the xy plane. Prove that

$$\Delta f = f(x + \Delta x, y + \Delta y) - f(x, y) = f_x \Delta x + f_y \Delta y + \delta_1 \Delta x + \delta_2 \Delta y$$

where ϵ_1 and ϵ_2 approach zero as Δx and Δy approach zero.

Applying the mean value theorem for functions of one variable (see Page 78), we have

$$\Delta f = \{f(x + \Delta x, y + \Delta y) - f(x, y + \Delta y)\} + \{f(x, y + \Delta y) - f(x, y)\} \tag{1}$$

$$= \Delta x f_x(x + \theta_1 \Delta x, y + \Delta y) + \Delta f_y(x, y + \theta_2 \Delta y) \qquad 0 < \theta_1 < 1, 0 < \theta_2 < 1$$

Since, by hypothesis, f_x and f_y are continuous, it follows that

$$f_x(x + \theta_1 \Delta x, y + \Delta y) = f_x(x, y) + \delta_1, \qquad f_y(x, y + \theta_2 \Delta y) = f_y(x, y) + \delta_2$$

where $\delta_1 \to 0$, $\delta_2 \to 0$ as $\Delta x \to 0$ and $\Delta y \to 0$.

Thus, $\Delta f = f_x \Delta x + f_y \Delta y + \delta_1 \Delta x + \delta_2 \Delta y$ as required.
Defining $\Delta x = dx$, $\Delta y = dy$, we have $\Delta f = f_x dx + f_y\, dy + \delta_1\, dx + \delta_2 dy$.
We call $df = f_x dx + f_y\, dy$ the *differential* of f (or z) or the *principal part* of Δf (or Δz).

6.15. If $z = f(x, y), = x^2 y - 3y$, find (a) Δz and (b) dz. (c) Determine Δz and dz if $x = 4$, $y = 3$, $\Delta x = -0.01$, and $\Delta y = 0.02$. (d) How might you determine $f(5.12, 6.85)$ without direct computation?

Solution:

(a) $$\begin{aligned}\Delta z &= f(x+\Delta x, y) - f(x, y)\\ &= [(x+\Delta x)^2(y+\Delta y) - 3(y+\Delta y)] - \{x^2 y - 3y\}\\ &= \underbrace{2xy\,\Delta x + (x^2-3)\,\Delta y}_{(A)} + \underbrace{(\Delta x)^2 y + 2x\,\Delta x\,\Delta y + (\Delta x)^2\,\Delta y}_{(B)}\end{aligned}$$

The sum (A) is the *principal part* of Δz and is the differential of z, i.e., dz. Thus,

(b) $dz = 2xy\,\Delta x + (x^2-3)\,\Delta y = 2xy\,dx + (x^2-3)\,dy$

Another method: $dz = \dfrac{\partial z}{\partial x}dx + \dfrac{\partial z}{\partial y}dy = 2xy\,dx + (x^2-3)\,dy$

(c) $\Delta z = f(x+\Delta x, y+\Delta y) - f(x, y) = f(4-0.01, 3+0.02) - f(4, 3)$

$= \{(3.99)^2(3.02) - 3(3.02)\} - \{(4)^2(3) - 3(3)\} = 0.018702$

$dz = 2xy\,dx + (x^2-3)\,dy = 2(4)(3)(-0.01) + (4^3-3)(0.02) = 0.02$

Note that in this case Δz and dz are approximately equal: because $\Delta x = dx$ and $\Delta y = dy$ are sufficiently small.

(d) We must find $f(x+\Delta x, y+\Delta y)$ when $x+\Delta x = 5.12$ and $y = \Delta y = 6.85$. We can accomplish this by choosing $x = 5$, $\Delta x = 0.12$, $y = 7$, $\delta y = -0.15$. Since Δx and Δy are small, we use the fact that $f(x+\Delta x, y+\Delta y) = f(x, y) + \Delta z$ is approximately equal to $f(x, y) + dz$, i.e., $z + dz$.

Now

$$z = f(x, y) = f(5, 7) = (5)^2(7) - 3(7) = 154$$

$$dz = 2xy\,dx + (x^2-3)\,dy = 2\,(5)\,(7)\,(0.12) + (5^2-3)\,(-0.15) = 5.1.$$

Then the required value is $154 + 5.1 = 159.1$ approximately. The value obtained by direct computation is 159.01864.

6.16. (a) Let $U = x^2 e^{y/x}$. Find dU. (b) Show that $(3x^2y - 2y^2)\,dx + (x^3 - 4xy + 6y^2)\,dy$ can be written as an exact differential of a function $\phi(x, y)$ and find this function.

(a) **Method 1:**

$$\frac{\partial U}{\partial x} = x^2 e^{y/x}\left(-\frac{y}{x^2}\right) + 2xe^{y/x}, \qquad \frac{\partial U}{\partial y} = x^2 e^{y/x}\left(\frac{1}{x}\right)$$

Then

$$dU = \frac{\partial U}{\partial x}dx + \frac{\partial U}{\partial y}dy = (2xe^{y/x} - ye^{y/x})\,dx + xe^{y/x}\,dy$$

Method 2:

$$\begin{aligned}dU &= x^2 d(e^{y/x}) + e^{y/x}\,d(x^2) = x^2 e^{y/x}\,d(y/x) + 2xe^{y/x}\,dx\\ &= x^2 e^{y/x}\left(\frac{x\,dy - y\,dx}{x^2}\right) + 2xe^{y/x}\,dx = (2xe^{y/x} - ye^{y/x})\,dx + xe^{y/x}\,dy\end{aligned}$$

(b) **Method 1:**

Suppose that $(3x^2y - 2y^2)\,dx + (x^3 - 4xy + 6y^2)\,dy = d\phi = \dfrac{\partial \phi}{\partial x}dx + \dfrac{\partial \phi}{\partial y}dy.$

Then

$$\frac{\partial \phi}{\partial x} = 3x^2y - 2y^2 \tag{1}$$

$$\frac{\partial \phi}{\partial y} = x^3 - 4xy + 6y^2 \tag{2}$$

From Equation (1), integrating with respect to x keeping y constant, we have

$$\phi = x^3y = 2xy^2 + f(y)$$

where $f(y)$ is the "constant" of integration. Substituting this into Equation (2) yields

$$x^3 - 4xy + F'(y) = x^3 - 4xy + 6y^2$$

from which $F'(y) = 6y^2$, i.e., $f(y) = 2y^3 + c$.

Hence, the required function is $\phi = x^3y - 2xy^2 + 2y^3 + c$, where c is an arbitrary constant.

Note that by Theorem 3, Page 130, the existence of such a function is guaranteed, since if $P = 3x^2y - 2y^2$ and $Q = x^3 - 4xy + 6y^2$, then $\partial P/\partial y = 3x^2 - 4y = \partial Q/\partial x$ identically. If $\partial P/\partial y \neq \partial Q/\partial x$, this function would not exist and the given expression would not be an exact differential.

Method 2:

$$\begin{aligned}(3x^2y - 2y^2)\,dx + (x^3 - 4xy + 6y^2)\,dy &= (3x^2y\,dx + x^3\,dy) - (2y^2\,dx + 4xy\,dy) + 6y^2\,dy \\ &= d(x^3y) - d(2xy^2) + d(2y^3) = d(x^3y - 2xy^2 + 2y^3) \\ &= d(x^3y - 2xy^2 + 2y^3 + c)\end{aligned}$$

Then the required function is $x^3y - 2xy^2 + 2y^3 + c$.

This method, called the *grouping method*, is based on our ability to recognize exact differential combinations and is less than Method 1. Naturally, before attempting to apply any method, we should determine whether the given expression is an exact differential by using Theorem 3, Page 130. See Theorem 4, Page 130.

Differentiation of composite functions

6.17. Let $z = f(x, y)$ and $x = \phi(t)$, $y = \psi(t)$ where f, ϕ, ψ are assumed differentiable. Prove

$$\frac{dz}{dt} = \frac{\partial z}{\partial x}\frac{dx}{dt} + \frac{\partial z}{\partial y}\frac{\partial y}{dt}$$

Using the results of Problem 6.14, we have

$$\frac{dz}{dt} = \lim_{\Delta t \to 0}\frac{\Delta z}{\Delta t} = \lim_{\Delta t \to 0}\left\{\frac{\partial z}{\partial x}\frac{\Delta x}{\Delta t} + \frac{\partial z}{\partial y}\frac{\Delta y}{\Delta t} + \varepsilon_1\frac{\Delta x}{\Delta t} + \varepsilon_2\frac{\Delta y}{\Delta t}\right\} = \frac{\partial z}{\partial x}\frac{dx}{dt} + \frac{\partial z}{\partial y}\frac{dy}{dt}$$

since, as $\Delta t \to 0$, we have $\Delta x \to 0$, $\Delta y \to 0$, $\varepsilon_1 \to 0$, $\dfrac{\Delta x}{\Delta t} \to \dfrac{dx}{dt}$, $\dfrac{\Delta y}{\Delta t} \to \dfrac{dy}{dt}$.

6.18. If $z = e^{xy^2}$, $x = t\cos t$, $y = t\sin t$, compute dz/dt at $t = \pi/2$.

$$\frac{dz}{dt} = \frac{\partial z}{\partial x}\frac{dx}{dt} + \frac{\partial z}{\partial y}\frac{dy}{dt} = (y^2e^{xy^2})(-t\sin t + \cos t) + (2xye^{xy^2})(t\cos t + \sin t)$$

At $t = \pi/2$, $x = 0$, $y = \pi/2$. Then $\left.\dfrac{dz}{dt}\right|_{t=\pi/2} = (\pi^2/4)(-\pi/2) + (0)(1) = -\pi^3/8$.

Another method: Substitute x and y to obtain $z = e^{t^3}\sin^2 t\cos t$ and then differentiate.

6.19. If $z = f(x, y)$ where $x = \phi(u, \upsilon)$ and $y = \psi(u, \upsilon)$, prove the following:

$$(a)\ \frac{\partial z}{\partial u} = \frac{\partial z}{\partial x}\frac{\partial x}{\partial u} + \frac{\partial z}{\partial y}\frac{\partial y}{\partial u} \qquad (b)\ \frac{\partial z}{\partial \upsilon} = \frac{\partial z}{\partial x}\frac{\partial x}{\partial \upsilon} + \frac{\partial z}{\partial y}\frac{\partial y}{\partial \upsilon}$$

(a) From Problem 6.14, assuming the differentiability of f, ϕ, ψ, we have

$$\frac{\partial z}{\partial u} = \lim_{\Delta u \to 0}\frac{\Delta z}{\Delta u} = \lim_{\Delta u \to 0}\left\{\frac{\partial z}{\partial x}\frac{\Delta x}{\Delta u} + \frac{\partial z}{\partial y}\frac{\Delta y}{\Delta u} + \varepsilon_1\frac{\Delta x}{\Delta u} + \varepsilon_2\frac{\Delta y}{\Delta u}\right\} = \frac{\partial z}{\partial x}\frac{\partial x}{\partial u} + \frac{\partial z}{\partial y}\frac{\partial y}{\partial u}$$

(b) The result is proved as in (a) by replacing Δu by $\Delta\upsilon$ and letting $\Delta\upsilon \to 0$.

6.20. Prove that $dz = \frac{\partial z}{\partial x}dx + \frac{\partial z}{\partial y}dy$ even if x and y are dependent variables.

Suppose x and y depend on three variables u, υ, w, for example. Then

$$dx = x_u\,du + x_\upsilon\,d\upsilon + x_w dw \qquad (1)$$

$$dy = y_u du + y_\upsilon\,d\upsilon + y_w\,dw \qquad (2)$$

Thus,

$$z_x dx + z_y dy = (z_x x_u + z_y y_u)\,du + (z_x x_\upsilon + z_{y}y_{\upsilon})\,d\upsilon + (z_x x_w + z_y y_w)\,dw = z_u du + z_\upsilon d\upsilon + z_w = dz$$

using obvious generalizations from Problem 6.19.

6.21. If $T = x^3 - xy + y^3$, $x = \rho\cos\phi$, and $y = \rho\sin\phi$, find (a) $\partial T/\partial\rho$, $\partial T/\partial\rho$ and (b) $\partial T/\partial\phi$.

$$\frac{\partial T}{\partial\rho} = \frac{\partial T}{\partial x}\frac{\partial x}{\partial\rho} + \frac{\partial T}{\partial y}\frac{\partial y}{\partial\rho} = (3x^2 - y)(\cos\phi) + (3y^2 - x)(\sin\phi)$$

$$\frac{\partial T}{\partial\phi} = \frac{\partial T}{\partial x}\frac{\partial x}{\partial\phi} + \frac{\partial T}{\partial y}\frac{\partial y}{\partial\phi} = (3x^2 - y)(-\rho\sin\phi) + (3y^2 - x)(\rho\cos\phi)$$

This may also be worked by direct substitution of x and y in T.

6.22. If $U = z\sin y/x$ where $x = 3r^2 + 2s$, $y = 4r - 2s^3$, and $z = 2r^2 - 3s^2$, find (a) $\partial U/\partial r$ and (b) $\partial U/\partial s$.

(a) $$\frac{\partial U}{\partial r} = \frac{\partial U}{\partial x}\frac{\partial x}{\partial r} + \frac{\partial U}{\partial y}\frac{\partial y}{\partial r} + \frac{\partial U}{\partial z}\frac{\partial z}{\partial r}$$

$$= \left\{\left(z\cos\frac{y}{x}\right)\left(-\frac{y}{x^2}\right)\right\}(6r) + \left\{\left(z\cos\frac{y}{x}\right)\left(\frac{1}{x}\right)\right\}(4) + \left(\sin\frac{y}{x}\right)(4r)$$

$$= -\frac{6ryz}{x^2}\cos\frac{y}{x} + \frac{4z}{x}\cos\frac{y}{x} + 4r\sin\frac{y}{x}$$

(b) $$\frac{\partial U}{\partial s} = \frac{\partial U}{\partial x}\frac{\partial x}{\partial s} + \frac{\partial U}{\partial y}\frac{\partial y}{\partial s} + \frac{\partial U}{\partial z}\frac{\partial z}{\partial s}$$

$$= \left\{\left(z\cos\frac{y}{x}\right)\left(-\frac{y}{x^2}\right)\right\}(2) + \left\{\left(z\cos\frac{y}{x}\right)\left(\frac{1}{x}\right)\right\}(-6s^2) + \left(\sin\frac{y}{x}\right)(-6s)$$

$$= -\frac{2yz}{x^2}\cos\frac{y}{x} - \frac{6s^2 z}{x}\cos\frac{y}{x} - 6s\sin\frac{y}{x}$$

6.23. If $x = \rho\cos\phi$, $y = \rho\sin\phi$, shown that $\left(\frac{\partial V}{\partial x}\right)^2 + \left(\frac{\partial V}{\partial y}\right)^2 = \left(\frac{\partial V}{\partial\rho}\right)^2 + \frac{1}{\rho^2}\left(\frac{\partial V}{\partial\phi}\right)^2$.

Using the subscript notation for partial derivatives, we have

$$V_\rho = V_x x_\rho + V_y y_\rho = V_x\cos\phi + V_y\sin\phi \qquad (1)$$

$$V_\phi = V_x x_\phi + V_y y_\phi = V_x(-\rho\sin\phi) + V_y(\rho\cos\phi) \qquad (2)$$

Dividing both sides of Equation (2) by ρ, we have

$$\frac{1}{\rho}V_\phi = -V_x\sin\phi + V_y\cos\phi \qquad (3)$$

Then from Equations (1) and (3), we have

$$V_\rho^2 + \frac{1}{\rho^2}V_\phi^2 = (V_x\cos\phi + V_y\sin\phi)^2 + (-V_x\sin\phi + V_y\cos\phi)^2 = V_x^2 + V_y^2$$

6.24. Show that $z = f(x^2 y)$, where f is differentiable, satisfies $x(\partial z/\partial x) = 2y\,(\partial z/\partial y)$.

Let $x^2y = u$. Then $z = f(u)$. Thus,

$$\frac{\partial z}{\partial x} = \frac{\partial z}{\partial u}\frac{\partial u}{\partial x} = f'(u)\cdot 2xy, \qquad \frac{\partial z}{\partial y} = \frac{\partial z}{\partial u}\frac{\partial u}{\partial y} = f'(u)\cdot x^2$$

Then

$$x\frac{\partial z}{\partial x} = f'(u)\cdot 2x^2y, \qquad 2y\frac{\partial z}{\partial y} = f'(u)\cdot 2x^2y \text{ and so } x\frac{\partial z}{\partial x} = 2y\frac{\partial z}{\partial y}$$

Another method: We have $dz = f'(x^2y)\,d(x^2y) = f'(x^2y)(2xy\,dx + x^2\,dy)$.

Also,

$$dz = \frac{\partial z}{\partial x}dx + \frac{\partial z}{\partial y}dy$$

Then

$$\frac{\partial z}{\partial x} = 2xy\,f'(x^2y). \qquad \frac{\partial z}{\partial y} = x^3 f'(x^2y)$$

Elimination of $f'(x^2y)$ yields $x\dfrac{\partial z}{\partial x} = 2y\dfrac{\partial z}{\partial y}$.

6.25. If for all values of the parameter λ and for some constant p, $F(\lambda x, \lambda y) = \lambda^p F(x, y)$ identically, where F is assumed differentiable, prove that $x(\partial F/\partial x) + y\,(\partial F/\partial y) = pF$.

Let $\lambda x = u$, $\lambda y = \upsilon$. Then

$$F(u, \upsilon) = \lambda^p F(x, y) \tag{1}$$

The derivative with respect to λ of the left side of Equation (*1*) is

$$\frac{\partial F}{\partial \lambda} = \frac{\partial F}{\partial u}\frac{\partial u}{\partial \lambda} + \frac{\partial F}{\partial \upsilon}\frac{d\upsilon}{\partial \lambda} = \frac{\partial F}{\partial u}x + \frac{\partial F}{\partial \upsilon}y$$

The derivative with respect to λ of the right side of Equation (1) is $p\,\lambda^{p-1}F$. Then

$$x\frac{\partial F}{\partial u} + y\frac{\partial F}{\partial \upsilon} = p\lambda^{p-1}F \tag{2}$$

Letting $\lambda = 1$ in Equation (*2*), so that $u = x$, $\upsilon = y$, we have $x(\partial F/\partial x) + y(\partial F/\partial y) = pF$.

6.26. If $F(x, y) = x^4y^2\sin^{-1} y/x$, show that $x(\partial F/\partial x) + y(\partial F/\partial y) = 6F$.

Since $F(\lambda x, \lambda y) = (\lambda x)^4(\lambda y)^2 \sin^{-1}\lambda y/\lambda x = \lambda^6 x^4y^2\sin^{-1} y/x = \lambda^6 F(x, y)$, the result follows from Problem 6.25 with $p = 6$. It can, of course, also be shown by direct differentiation.

6.27. Prove that $Y = f(x + at) + g(x - at)$ satisfies $\partial^2 Y/\partial t^2 = a^2(\partial^2 Y/\partial x^2)$, where f and g are assumed to be at least twice differentiable and a is any constant.

Let $u = x + at, \upsilon = x - at$ so that $Y = f(u) + g(\upsilon)$. Then if $f'(u) \equiv df/du,\ g'(\upsilon) \equiv dg/d\upsilon$,

$$\frac{\partial Y}{\partial t} = \frac{\partial Y}{\partial u}\frac{\partial u}{\partial t} + \frac{\partial Y}{\partial \upsilon}\frac{\partial \upsilon}{\partial t} = af'(u) - ag'(\upsilon), \qquad \frac{\partial Y}{\partial x} = \frac{\partial Y}{\partial x}\frac{\partial u}{\partial x} + \frac{\partial Y}{\partial \upsilon}\frac{\partial \upsilon}{\partial x} = f'(u) + g'(\upsilon)$$

By further differentiation, using the notation $f''(u) \equiv d^2 f/du^2$, $g''(\upsilon) \equiv d^2 g/d\upsilon^2$, we have

$$\frac{\partial^2 Y}{\partial t^2} = \frac{\partial Y_t}{\partial t} = \frac{\partial Y_t}{\partial u}\frac{\partial u}{\partial t} + \frac{\partial Y_t}{\partial \upsilon}\frac{\partial \upsilon}{\partial t} = \frac{\partial}{\partial u}\{af'(u) - ag'(\upsilon)\}(a) + \frac{\partial}{\partial \upsilon}\{af'(u) - ag'(\upsilon)\}(-a) \tag{1}$$
$$= a^2 f''(u) + a^2 g''(\upsilon)$$

$$\frac{\partial^2 Y}{\partial x^2} = \frac{\partial Y_x}{\partial x} = \frac{\partial Y_x}{\partial u}\frac{\partial u}{\partial x} + \frac{\partial Y_x}{\partial \upsilon}\frac{\partial \upsilon}{\partial x} = \frac{\partial}{\partial u}\{f'(u) + g'(\upsilon)\} + \frac{\partial}{\partial \upsilon}\{f'(u) + g')(\upsilon)\} = f''(u) + g''(\upsilon) \tag{2}$$

Then from Equations (1) and (2), $\partial^2 Y/\partial t^2 = a^2 (\partial^2 Y/\partial x^2)$.

6.28. If $x = 2r - s$ and $y = r + 2s$, find $\dfrac{\partial^2 U}{\partial y\, \partial x}$ in terms of derivatives with respect to r and s.

Solving $x = 2r - s$, $y = r + 2s$ for r and s: $r = (2x + y)/5$, $s = (2y - x)/5$.
Then $\partial r/\partial x = 2/5$, $\partial s/\partial x = -1/5$, $\partial r/\partial y = 1/5$, $\partial s/\partial y = 2/5$. Hence, we have

$$\frac{\partial U}{\partial x} = \frac{\partial U}{\partial r}\frac{\partial r}{\partial x} + \frac{\partial U}{\partial s}\frac{\partial s}{\partial x} = \frac{2}{5}\frac{\partial U}{\partial r} - \frac{1}{5}\frac{\partial U}{\partial s}$$

$$\frac{\partial^2 U}{\partial y\,\partial x} = \frac{\partial}{\partial y}\left(\frac{\partial U}{\partial x}\right) = \frac{\partial}{\partial r}\left(\frac{2}{5}\frac{\partial U}{\partial r} - \frac{1}{5}\frac{\partial U}{\partial s}\right)\frac{\partial r}{\partial y} + \frac{\partial}{\partial s}\left(\frac{2}{5}\frac{\partial U}{\partial r} - \frac{1}{5}\frac{\partial U}{\partial s}\right)\frac{\partial s}{\partial y}$$
$$= \left(\frac{2}{5}\frac{\partial^2 U}{\partial r^2} - \frac{1}{5}\frac{\partial^2 U}{\partial r\,\partial s}\right)\left(\frac{1}{5}\right) + \left(\frac{2}{5}\frac{\partial^2 U}{\partial s\,\partial r} - \frac{1}{5}\frac{\partial^2 U}{\partial s^2}\right)\left(\frac{2}{5}\right)$$
$$= \frac{1}{25}\left(2\frac{\partial^2 U}{\partial r^2} + 3\frac{\partial^2 U}{\partial r\,\partial s} - 2\frac{\partial^2 U}{\partial s^2}\right)$$

assuming U has continuous second partial derivatives.

Implicit functions and jacobians

6.29. If $U = x^3 y$, find dU/dt if

$$x^5 + y = t \text{ and} \tag{1}$$
$$x^2 + y^3 = t^2 \tag{2}$$

Equations (1) and (2) define x and y as (implicit) functions of t. Then differentiating with respect to t, we have

$$5x^4\,(dx/dt) + dy/t = 1 \tag{3}$$
$$2x\,(dx/dt) + 3y^2(dy/dt) = 2t \tag{4}$$

Solving Equations (3) and (4) simultaneously for dx/dt and dy/dt,

$$\frac{dx}{dt} = \frac{\begin{vmatrix} 1 & 1 \\ 2t & 3y^2 \end{vmatrix}}{\begin{vmatrix} 5x^4 & 1 \\ 2x & 3y^2 \end{vmatrix}} = \frac{3y^2 - 2t}{15x^4y^2 - 2x}, \quad \frac{dy}{dt} = \frac{\begin{vmatrix} 5x^4 & 1 \\ 2x & 2t \end{vmatrix}}{\begin{vmatrix} 5x^4 & 1 \\ 2x & 3y^2 \end{vmatrix}} = \frac{10x^4t - 2x}{15x^4y^2 - 2x}$$

Then $$\frac{dU}{dt} = \frac{\partial U}{\partial x}\frac{dx}{dt} + \frac{\partial U}{\partial y}\frac{dy}{dt} = (3x^2y)\left(\frac{3y^2 - 2t}{15x^4y^2 - 2x}\right) + (x^3)\left(\frac{10x^4t - 2x}{15x^4y^2 - 2x}\right).$$

6.30. If $F(x, y, z) = 0$ defines z as an implicit function of x and y in a region $\Re$ of the xy plane, prove that (a) $\partial z/\partial x = -F_x/F_z$ and (b) $\partial z/\partial y = -F_y/F_z$, where $F_z \neq 0$.

Since z is a function of x and y, $dz = \dfrac{\partial z}{\partial x}dx + \dfrac{\partial z}{\partial y}dy$.

Then $dF = \dfrac{\partial F}{\partial x}dx + \dfrac{\partial F}{\partial y}dy + \dfrac{\partial F}{\partial z}dz = \left(\dfrac{\partial F}{\partial x} + \dfrac{\partial F}{\partial z}\dfrac{\partial z}{\partial x}\right)dx + \left(\dfrac{\partial F}{\partial y} + \dfrac{\partial F}{\partial z}\dfrac{\partial z}{\partial y}\right)dy = 0.$

Since x and y are independent, we have

$$\frac{\partial F}{\partial x} + \frac{\partial F}{\partial z}\frac{\partial z}{\partial x} = 0 \tag{1}$$

$$\frac{\partial F}{\partial y} + \frac{\partial F}{\partial z}\frac{\partial z}{\partial y} = 0 \tag{2}$$

from which the required results are obtained. If desired, equations (1) and (2) can be written directly.

6.31. If $F(x, y, u, \upsilon) = 0$ and $G(x, y, u, \upsilon) = 0$, find (a) $\partial u/\partial x$, (b) $\partial u/\partial y$, (c) $\partial \upsilon/\partial x$, and (d) $\partial \upsilon/\partial y$.

The two equations in general define the dependent variables u and υ as (implicit) functions of the independent variables x and y. Using the subscript notation, we have

$$dF = F_x dx + F_y\,dy + F_u\,du + F_\upsilon d\upsilon = 0 \tag{1}$$

$$dG = G_x dx + G_y dy + G_u du + G_\upsilon\,d\upsilon = 0 \tag{2}$$

Also, since u and υ are functions of x and y,

$$du = u_x dx + u_y\,dy \tag{3}$$

$$d\upsilon = \upsilon_x dx + \upsilon_y dy \tag{4}$$

Substituting Equations (3) and (4) in (1) and (2) yields

$$dF = (F_x + F_u\,u_x + F_\upsilon\upsilon_x)dx + (F_y + F_u\,u_y + F_\upsilon\upsilon_y)dy = 0 \tag{5}$$

$$dG = (G_x + G_u u_x + G_\upsilon\upsilon_x)dx + (G_y + G_u u_y + G_\upsilon\upsilon_y)dy = 0 \tag{6}$$

Since x and y are independent, the coefficients of dx and dy in Equations (5) and (6) are zero. Hence, we obtain

$$\begin{cases} F_u\,u_x + F_\upsilon\,\upsilon_x = -F_x \\ G_u\,u_x + G_\upsilon\,\upsilon_x = -G_x \end{cases} \tag{7}$$

$$\begin{cases} F_u\,u_y + F_\upsilon\,\upsilon_y = -F_y \\ G_u\,u_y + G_\upsilon\,\upsilon_y = -G_y \end{cases} \tag{8}$$

Solving Equations (7) and (8) gives

(a) $u_x = \dfrac{\partial u}{\partial x} = \dfrac{\begin{vmatrix} -F_x & F_\upsilon \\ -G_x & G_\upsilon \end{vmatrix}}{\begin{vmatrix} F_u & F_\upsilon \\ G_u & G_\upsilon \end{vmatrix}} = -\dfrac{\dfrac{\partial(F,G)}{\partial(x,\upsilon)}}{\dfrac{\partial(F,G)}{\partial(u,\upsilon)}}$ (b) $\upsilon_x = \dfrac{\partial \upsilon}{\partial x} = \dfrac{\begin{vmatrix} F_u & -F_x \\ G_u & -G_x \end{vmatrix}}{\begin{vmatrix} F_u & F_\upsilon \\ G_u & G_\upsilon \end{vmatrix}} = -\dfrac{\dfrac{\partial(F,G)}{\partial(u,x)}}{\dfrac{\partial(F,G)}{\partial(u,\upsilon)}}$

(c) $u_y = \dfrac{\partial u}{\partial y} = \dfrac{\begin{vmatrix} -F_y & F_\upsilon \\ -G_y & G_\upsilon \end{vmatrix}}{\begin{vmatrix} F_u & F_\upsilon \\ G_u & G_\upsilon \end{vmatrix}} = -\dfrac{\dfrac{\partial(F,G)}{\partial(y,\upsilon)}}{\dfrac{\partial(F,G)}{\partial(u,\upsilon)}}$ (d) $\upsilon_y = \dfrac{\partial \upsilon}{\partial y} = \dfrac{\begin{vmatrix} F_u & -F_y \\ G_u & -G_y \end{vmatrix}}{\begin{vmatrix} F_u & F_\upsilon \\ G_u & G_\upsilon \end{vmatrix}} = -\dfrac{\dfrac{\partial(F,G)}{\partial(u,y)}}{\dfrac{\partial(F,G)}{\partial(u,\upsilon)}}$

The functional determinant $\begin{vmatrix} F_u & F_\upsilon \\ G_u & G_\upsilon \end{vmatrix}$, denoted by $\frac{\partial(F,G)}{\partial(u,\upsilon)}$ or $J\left(\frac{F,G}{u,\upsilon}\right)$, is the *Jacobian* of F and G with respect to u and υ and is supposed $\neq 0$.

Note that it is possible to devise mnemonic rules for writing at once the required partial derivatives in terms of Jacobians (see also Problem 6.33).

6.32. If $u^2 - \upsilon = 3x + y$ and $u - 2\upsilon^2 = x - 2y$, find (a) $\partial u/\partial x$, (b) $\partial \upsilon/\partial x$, (c) $\partial u/\partial y$, and (d) $\partial \upsilon/\partial y$.

Method 1: Differentiate the given equations with respect to x, considering u and υ as functions of x and y. Then

$$2u\frac{\partial u}{\partial x} - \frac{\partial \upsilon}{\partial x} = 3 \tag{1}$$

$$\frac{\partial \upsilon}{\partial x} - 4\upsilon\frac{\partial \upsilon}{\partial x} = 1 \tag{2}$$

Solving, $\frac{\partial u}{\partial x} = \frac{1-12\upsilon}{1-8u\upsilon}, \frac{\partial \upsilon}{\partial x} = \frac{2u-3}{1-8u\upsilon}$.

Differentiating with respect to y, we have

$$2u\frac{\partial u}{\partial y} - \frac{\partial \upsilon}{\partial y} = 1 \tag{3}$$

$$\frac{\partial u}{\partial y} - 4\upsilon\frac{\partial \upsilon}{\partial y} = -2 \tag{4}$$

Solving, $\frac{\partial u}{\partial y} = \frac{-2-4\upsilon}{1-8u\upsilon}, \frac{\partial \upsilon}{\partial y} = \frac{-4u-1}{1-8u\upsilon}$.

We have, of course, assumed that $1 - 8u\upsilon \neq 0$.

Method 2: The given equations are $F = u^2 - \upsilon - 3x - y = 0$, $G = u - 2\upsilon^2 - x + 2y = 0$. Then, by Problem 6.31,

$$\frac{\partial u}{\partial x} = -\frac{\dfrac{\partial(F,G)}{\partial(x,\upsilon)}}{\dfrac{\partial(F,G)}{\partial(u,\upsilon)}} = -\frac{\begin{vmatrix} F_x & F_\upsilon \\ G_x & G_\upsilon \end{vmatrix}}{\begin{vmatrix} F_u & F_\upsilon \\ G_u & G_\upsilon \end{vmatrix}} = -\frac{\begin{vmatrix} -3 & -1 \\ -1 & -4\upsilon \end{vmatrix}}{\begin{vmatrix} 2u & -1 \\ 1 & -4\upsilon \end{vmatrix}} = \frac{1-12\upsilon}{1-8u\upsilon}$$

provided $1 - 8u\upsilon \neq 0$. Similarly, the other partial derivatives are obtained.

6.33. If $F(u, \upsilon, w, x, y) = 0$, $G(u, \upsilon, w, x, y) = 0$, and $H(u, \upsilon, w, x, y) = 0$, find (a) $\left.\frac{\partial \upsilon}{\partial y}\right|_x$, (b) $\left.\frac{\partial x}{\partial \upsilon}\right|_w$, (c) $\left.\frac{\partial w}{\partial u}\right|_y$.

From three equations in five variables, we can (theoretically at least) determine three variables in terms of the remaining two. Thus, three variables are dependent and two are independent. If we were asked to determine $\partial\upsilon/\partial y$, we would know that υ is a dependent variable and y is an independent variable, but would not know the remaining independent variable. However, the particular notation $\left.\frac{\partial \upsilon}{\partial y}\right|_x$ serves to indicate that we are to obtain $\partial\upsilon/\partial y$, keeping x constant; i.e., x is the other independent variable.

(a) Differentiating the given equations with respect to y, keeping x constant, gives

$$F_u u_y + F_\upsilon \upsilon_y + F_w w_y + F_y = 0 \tag{1}$$

$$G_u u_y + G_\upsilon \upsilon_y + G_w w_y + G_y = G_y = 0 \tag{2}$$

$$H_u u_y + H_\upsilon \upsilon_y + H_w w_y + H_y = 0 \tag{3}$$

Solving simultaneously for υ_y, we have

$$\upsilon_y = \left.\frac{\partial \upsilon}{\partial y}\right|_x = -\frac{\begin{vmatrix} F_u & F_y & F_w \\ G_u & G_y & G_w \\ H_u & H_y & H_w \end{vmatrix}}{\begin{vmatrix} F_u & F_\upsilon & F_w \\ G_u & G_\upsilon & G_w \\ H_u & H_\upsilon & H_w \end{vmatrix}} = -\frac{\dfrac{\partial(F,G,H)}{\partial(u,y,w)}}{\dfrac{\partial(F,G,H)}{\partial(u,\upsilon,w)}}$$

Equations (1), (2), and (3) can also be obtained by using differentials as in Problem 6.31.

The Jacobian method is very suggestive for writing results immediately, as seen in this problem and Problem 6.31. Thus, observe that in calculating $\left.\dfrac{\partial \upsilon}{\partial y}\right|_x$ the result is the negative of the quotient of two Jacobians, the numerator containing the independent variable y and the denominator containing the dependent variable υ in the same relative positions. Using this scheme, we have

$$(b)\quad \left.\frac{\partial x}{\partial \upsilon}\right|_w = -\frac{\dfrac{\partial(F,G,H)}{\partial(\upsilon,y,u)}}{\dfrac{\partial(F,G,H)}{\partial(x,y,u)}} \qquad (c)\quad \left.\frac{\partial w}{\partial u}\right|_y = -\frac{\dfrac{\partial(F,G,H)}{\partial(u,x,\upsilon)}}{\dfrac{\partial(F,G,H)}{\partial(w,x,\upsilon)}}$$

6.34. If $z^3 - xz - y = 0$, prove that $\dfrac{\partial^2 z}{\partial x \partial y} = -\dfrac{3z^2 + x}{(3z^2 - x)^3}$.

Differentiating with respect to x, keeping y constant, and remembering that z is the dependent variable depending on the independent variables x and y, we find

$$3z^2 \frac{\partial z}{\partial x} - x\frac{\partial z}{\partial x} - z = 0$$

and

$$\frac{\partial z}{\partial x} = \frac{z}{3z^2 - x} \tag{1}$$

Differentiating with respect to y, keeping x constant, we find

$$3z^2 \frac{\partial z}{\partial y} - x\frac{\partial z}{\partial y} - 1 = 0$$

and

$$\frac{\partial z}{\partial x} = \frac{z}{3z^2 - x} \tag{2}$$

Differentiating Equation (2) with respect to x and using Equation (*1*), we have

$$\frac{\partial^2}{\partial x\,\partial y} = \frac{-1}{(3z^2 - x)^2}\left(6z\frac{\partial z}{\partial x} - 1\right) = \frac{1 - 6z\left[z/(3z^2 - x)\right]}{(3z^2 - x)^2} = -\frac{3z^2 - x}{(3z^2 - x)^3}$$

The result can also be obtained by differentiating Equation (1) with respect to y and using Equation (2).

6.35. Let $u = f(x, y)$ and $\upsilon = g(x, y)$, where f and g are continuously differentiable in some region $\Re$. Prove that a necessary and sufficient condition that there exists a functional relation between u and υ of the form $\phi(u, \upsilon) = 0$ is the vanishing of the Jacobian; i.e., $\dfrac{\partial(u,\upsilon)}{\partial(x, y)} = 0$ identically.

Necessity. We have to prove that if the functional relation $\phi(u, \upsilon) = 0$ exists, then the Jacobian $\frac{\partial(u,\upsilon)}{\partial(x,y)} = 0$ identically. To do this, we note that

$$d\phi = \phi_u du + \phi_\upsilon d\upsilon = \phi_u(u_x dx + u_y\, dy) + \phi_\upsilon(\upsilon_x dx + \upsilon_y dy)$$
$$= (\phi_u\, u_x + \phi_\upsilon\, \upsilon_x)dx + (\phi_u u_y + \phi_\upsilon \upsilon_y)dy = 0$$

Then

$$\phi_u u_x + \phi_\upsilon \upsilon_x = 0 \qquad (1)$$
$$\phi_u u_y + \phi_\upsilon \upsilon_y = 0 \qquad (2)$$

Now ϕ_u and ϕ_υ cannot be identically zero, since if they were, there would be no functional relation, contrary to hypothesis. Hence, it follows from Equations (*1*) and (2) that $\begin{vmatrix} u_x & \upsilon_x \\ u_y & \upsilon_y \end{vmatrix} = \frac{\partial(u,\upsilon)}{\partial(x,y)} = 0$ identically.

Sufficiency. We have to prove that if the Jacobian $\frac{\partial(u,\upsilon)}{\partial(x,y)} = 0$ identically, then there exists a functional relation between u and υ; i.e., $\phi(u, \upsilon) = 0$.

Let us first suppose that both $u_x = 0$ and $u_y = 0$. In this case the Jacobian is identically zero and u is a constant c_1, so that the trival functional relation $u = c_1$ is obtained.

Let us now assume that we do not have both $u_x = 0$ and $u_y = 0$; for definiteness, assume $u_x \neq 0$. We may then, according to Theorem 1, Page 133, solve for x in the equation $u = f(x, y)$ to obtain $x = F(u, y)$, from which it follows that

$$u = f\{F(u, y), y\} \qquad (1)$$
$$\upsilon = g\{F(u, y), y\} \qquad (2)$$

From these we have, respectively,

$$du = u_x dx + u_y dy = u_x(F_u du + F_y dy) + u_y dy = u_x F_u\, du + (u_x\, F_y + u_y)\, dy \qquad (3)$$
$$d\upsilon = \upsilon_x dx + \upsilon_y\, dy = \upsilon_x(F_u du + F_y\, dy) + \upsilon_y\, dy = \upsilon_x\, F_u\, du + (\upsilon_x\, F_y + \upsilon_y)\, dy \qquad (4)$$

From Equation (*3*), $u_x F_u = 1$ and $u_x F_y + u_y = 0$ or (*5*) $F_y = -u_y/u_x$. Using this, Equation (*4*) becomes

$$d\upsilon = \upsilon_x F_u\, du + \{\upsilon_x(-u_y/u_x) + \upsilon_y\}\, dy = \upsilon_x F_u\, du + \left(\frac{u_x \upsilon_y - u_y\, \upsilon_x}{u_x}\right) dy. \qquad (6)$$

But by hypothesis $\frac{\partial(u,\upsilon)}{\partial(x,y)} = \begin{vmatrix} u_x & u_y \\ \upsilon_x & \upsilon_y \end{vmatrix} = u_x\upsilon_y - u_y\upsilon_x = 0$ identically, so that Equation (*6*) becomes $d\phi = \upsilon_x F_u du$. This means essentially that, referring to Equation (*2*), $\partial\upsilon/\partial y = 0$, which means that υ is not dependent on y but depends only on u; i.e., υ is a function of u, which is the same as saying that the functional relation $\phi\,(u, \upsilon) = 0$ exists.

6.36. (a) If $u = \frac{x + y}{1 - xy}$ and $\upsilon = \tan^{-1} x + \tan^{-1} y$, find $\frac{\partial(u,\upsilon)}{\partial(x, yw)}$. (b) Are u and υ functionally related? If so, find the relationship.

(a) $$\frac{\partial(u,\upsilon)}{\partial(x,y)} = \begin{vmatrix} u_x & u_y \\ \upsilon_x & \upsilon_y \end{vmatrix} = \begin{vmatrix} \frac{1+y^2}{(1-xy)^2} & \frac{1+x^2}{(1-xy)^2} \\ \frac{1}{1+x^2} & \frac{1}{1+y^2} \end{vmatrix} = 0 \qquad \text{if } xy \neq 1$$

(b) By Problem 6.35, since the Jacobian is identically zero in a region, there must be a functional relationship between u and υ. This is seen to be $\tan \upsilon = u$; i.e., $\phi(u, \upsilon) = u - \tan \upsilon = 0$. We can show this directly by solving for x (say) in one of the equations and then substituting in the other. Thus, for example, from $\upsilon = \tan^{-1} x + \tan^{-1} y$, we find $\tan^{-1} x = \upsilon - \tan^{-1} y$ and so

$$x = \tan(\upsilon - \tan^{-1} y) = \frac{\tan \upsilon - \tan(\tan^{-1} y)}{1 + \tan \upsilon \tan(\tan^{-1} y)} = \frac{\tan \upsilon - y}{1 + y \tan \upsilon}$$

Then substituting this in $u = (x + y)/(1 - xy)$ and simplifying, we find $u = \tan \upsilon$.

6.37. (a) If $x = u - \upsilon + w$, $y = u^2 - \upsilon^2 - w^2$ and $z = u^3 + \upsilon$, evaluate the Jacobian $\dfrac{\partial(x, y, z)}{\partial(u, \upsilon, w)}$, and (b) explain the significance of the nonvanishing of this Jacobian.

(a) $$\frac{\partial(x,y,z)}{\partial(u,\upsilon,w)} = \begin{vmatrix} x_u & x_\upsilon & x_w \\ y_u & y_\upsilon & y_w \\ z_u & z_\upsilon & z_w \end{vmatrix} = \begin{vmatrix} 1 & -1 & 1 \\ 2u & -2\upsilon & -2w \\ 3u^2 & 1 & 0 \end{vmatrix} = 6wu^2 + 2u + 6u^2\upsilon + 2w$$

(b) The given equations can be solved simultaneously for u, υ, w in terms of x, y, z in a region $\mathfrak{R}$ if the Jacobian is not zero in $\mathfrak{R}$.

Transformations, curvilinear coordinates

6.38. A region $\mathfrak{R}$ in the xy plane is bounded by $x + y = 6$, $x - y = 2$, and $y = 0$. (a) Determine the region $\mathfrak{R}'$ in the $u\upsilon$ plane into which $\mathfrak{R}$ is mapped under the transformation $x = u + \upsilon$, $y = u - \upsilon$. (b) Compute $\dfrac{\partial(x, y)}{\partial(u, \upsilon)}$. (c) Compare the result of (b) with the ratio of the areas of $\mathfrak{R}$ and $\mathfrak{R}'$.

(a) The region $\mathfrak{R}$ shown shaded in Figure 6.9 (a) is a triangle bounded by the lines $x + y = 6$, $x - y = 2$, and $y = 0$, which for distinguishing purposes are shown dotted, dashed, and heavy, respectively.

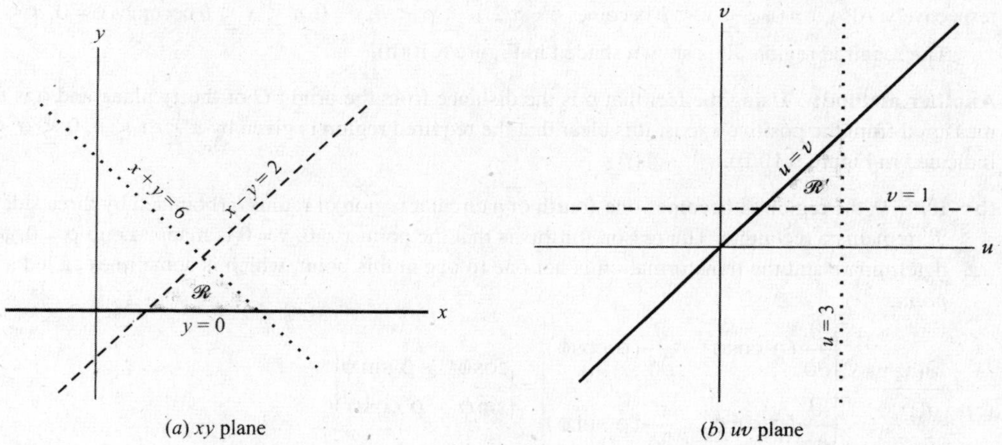

Figure 6.9

Under the given transformation, the line $x + y = 6$ is transformed into $(u + \upsilon) + (u - \upsilon) = 6$; i.e., $2u = 6$ or $u = 3$, which is a line (shown dotted) in the $u\upsilon$ plane of Figure 6.9(b).

Similarly, $x - y = 2$ becomes $(u + \upsilon) - (u - \upsilon) = 2$ or $\upsilon = 1$, which is a line (shown dashed) in the $u\upsilon$ plane. In like manner, $y = 0$ becomes $u - \upsilon = 0$ or $u = \upsilon$, which is a line shown heavy in the $u\upsilon$ plane. Then the required region is bounded by $u = 3$, $\upsilon = 1$, and $u = \upsilon$, and is shown shaded in Figure 6.9(b).

(b) $$\frac{\partial(x,y)}{\partial(u,\upsilon)} = \begin{vmatrix} \dfrac{\partial x}{\partial u} & \dfrac{\partial x}{\partial \upsilon} \\ \dfrac{\partial y}{\partial u} & \dfrac{\partial y}{\partial \upsilon} \end{vmatrix} = \begin{vmatrix} \dfrac{\partial}{\partial u}(u+\upsilon) & \dfrac{\partial}{\partial u}(u+\upsilon) \\ \dfrac{\partial}{\partial u}(u-\upsilon) & \dfrac{\partial}{\partial \upsilon}(u-\upsilon) \end{vmatrix} = \begin{vmatrix} 1 & 1 \\ 1 & -1 \end{vmatrix} = 2$$

(c) The area of triangular region $\mathfrak{R}$ is 4, whereas the area of triangular region $\mathfrak{R}'$ is 2. Hence, the ratio is 4/2 = 2, agreeing with the value of the Jacobian in (b). Since the Jacobian is constant in this case, the areas of any regions $\mathfrak{R}$ in the xy plane are twice the areas of corresponding mapped regions $\mathfrak{R}'$ in the $u\upsilon$ plane.

6.39. A region $\Re$ in the xy plane is bounded by $x^2 + y^2 = a^2$, $x^2 + y^2 = b^2$, $x = 0$, and $y = 0$, where $0 < a < b$. (a) Determine the region $\Re'$ into which $\Re$ is mapped under the transformation $x = \rho \cos \phi$, $y = \rho \sin \phi$, where $\rho > 0$, $0 \leqq \phi < 2\pi$. (b) Discuss what happens when $a = 0$. (c) compute $\dfrac{\partial(x, y)}{\partial(\rho,\phi)}$. (d) compute $\dfrac{\partial(\rho,\phi)}{\partial(x, y)}$.

(a) The region $\Re$ [shaded in Figure 6.10(a)] is bounded by $x = 0$ (dotted), $y = 0$ (dotted and dashed), $x^2 + y^2 = a^2$ (dashed), and $x^2 + y^2 = b^2$ (heavy).

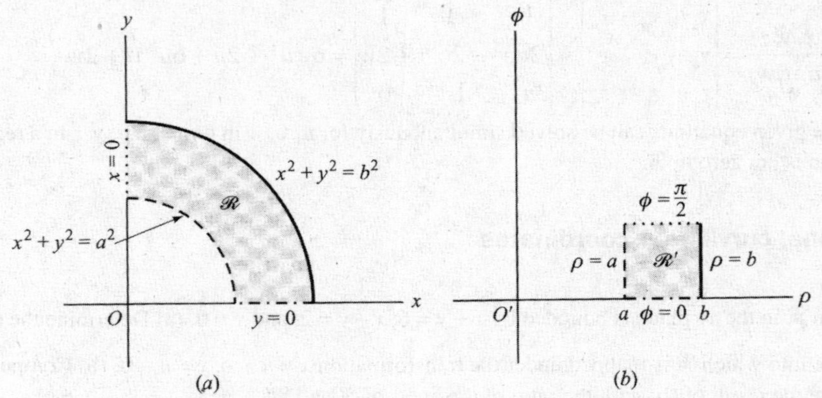

Figure 6.10

Under the given transformation, $x^2 + y^2 = a^2$ and $x^2 + y^2 = b^2$ become $\rho^2 = a^2$ and $\rho^2 = b^2$ or $\rho = a$ and $\rho = b$, respectively. Also, $x = 0$, $a \leqq y \leqq b$ becomes $\phi = \pi/2$, $a \leqq \rho \leqq b$; $y = 0$, $a \leqq x \leqq b$ becomes $\phi = 0$, $a \leqq \rho \leqq b$.

The required region $\Re'$ is shown shaded in Figure 6.10(b).

Another method: Using the fact that ρ is the distance from the origin O of the xy plane and ϕ is the angle measured from the positive x axis, it is clear that the required region is given by $a \leqq \rho \leqq b$, $0 \leqq \phi \leqq \pi/2$, as indicated in Figure 6.10(b).

(b) If $a = 0$, the region $\Re$ becomes one-fourth of a circular region of radius b (bounded by three sides), while $\Re'$ remains a rectangle. The reason for this is that the point $x = 0$, $y = 0$ is mapped into $\rho = 0$, $\phi =$ an indeterminate and the transformation is not one to one at this point, which is sometimes called a *singular point*.

(c) $$\frac{\partial(x, y)}{\partial(\rho,\phi)} = \begin{vmatrix} \dfrac{\partial}{\partial\rho}(\rho \cos\phi) & \dfrac{\partial}{\partial\phi}(\rho \cos\phi) \\ \dfrac{\partial}{\partial\rho}(\rho \sin\phi) & \dfrac{\partial}{\partial\phi}(\rho \sin\phi) \end{vmatrix} = \begin{vmatrix} \cos\phi & -\rho \sin\phi \\ \sin\phi & \rho \cos\phi \end{vmatrix}$$
$$= \rho(\cos^2 \phi + \sin^2 \phi) = \rho$$

(d) From Problem 6.43(b) we have, letting $u = \rho$, $\upsilon = \phi$, $\dfrac{\partial(x, y)}{\partial(\rho,\phi)} \dfrac{\partial(\rho,\phi)}{\partial(x, y)} = 1$ so that, thing (c), $\dfrac{\partial(\rho,\phi)}{\partial(x, y)} = \dfrac{1}{\rho}$

This can also be obtained by direct differentiation.

Note that from the Jacobians of these transformations it is clear why $\rho = 0$ (i.e., $x = 0$, $y = 0$) is a singular point.

Mean value theorem

6.40. Prove the mean value theorem for functions of two variables.

Let $f(t) = f(x_0 + ht, y_0 + kt)$. By the mean value theorem for functions of one variable,

$$F(1) = F(0) = F'(\theta)\ 0 < \theta < 1 \tag{1}$$

If $x = x_0 + ht$, $y = y_0 + kt$, then $F(t) = f(x, y)$, so that by Problem 6.17,

$$F'(t) = f_x\,(dx/dt) + f_y(dy/dt) = hf_x + kf_y \text{ and } F'(\theta) = hf_x(x_0 + \theta h, y_0 + \theta k) + kf_y(x_0 + \theta h, y_0 + \theta k)$$

where $0 < \theta < 1$. Thus, (*1*) becomes

$$f(x_0 + h, y_0 + k) - f(x_0, y_0) = hf_x(x_0 + \theta h, y_0 + \theta k) + kf_y(x_0 + \theta h, y_0 + \theta k) \tag{2}$$

where $0 < \theta < 1$ as required.

Note that Equation (2), which is analogous to Equation (1) of Problem 6.14, where $h = \Delta x$, has the advantage of being more symmetric (and also more useful), since only a single number θ is involved.

Miscellaneous problems

6.41. Let $f(x,y) = \begin{cases} xy\left(\dfrac{x^2 - y^2}{x^2 + y^2}\right) & (x,y) \neq (0,0) \\ 0 & (x,y) = (0,0) \end{cases}$.

Compute (a) $f_x(0, 0)$, (b) $f_y(0, 0)$, (c) $f_{xx}(0, 0)$ (d) $f_{yy}(0, 0)$, (e) $f_{xy}(0, 0)$, and (f) $f_{yx}(0, 0)$.

(a) $\lim\limits_{h\to 0} \dfrac{f(h,0) - f(0,0)}{h} = \lim\limits_{h\to 0} \dfrac{0}{h} = 0$

(b) $\lim\limits_{h\to o} \dfrac{f(0,k) - f(0,0)}{k} = \lim\limits_{k\to 0} \dfrac{0}{k} = 0$

If $(x, y) \neq (0, 0)$,

$$f_x(x,y) = \frac{\partial}{\partial x}\left\{xy\left(\frac{x^2 - y^2}{x^2 - y^2}\right)\right\} = xy\left(\frac{4xy^2}{(x^2 + y^2)^2}\right) + y\left(\frac{x^2 - y^2}{x^2 + y^2}\right)$$

$$f_x(x,y) = \frac{\partial}{\partial y}\left\{xy\left(\frac{x^2 - y^2}{x^2 - y^2}\right)\right\} = xy\left(\frac{-4xy^2}{(x^2 + y^2)^2}\right) + x\left(\frac{x^2 - y^2}{x^2 + y^2}\right)$$

Then

(c) $\lim\limits_{h\to 0} \dfrac{f_x(h,0) - f_x(0,0)}{h} = \lim\limits_{h\to 0} \dfrac{0}{h} = 0$

(d) $\lim\limits_{k\to 0} \dfrac{f_y(0,k) - f_y(0,0)}{k} = \lim\limits_{k\to 0} \dfrac{0}{k} = 0$

(e) $\lim\limits_{k\to 0} \dfrac{f_x(0,k) - f_x(0,0)}{k} = \lim\limits_{k\to 0} \dfrac{-k}{k} = -1$

(f) $\lim\limits_{h\to 0} \dfrac{f_y(h,0) - f_y(0,0)}{h} = \lim\limits_{h\to 0} \dfrac{h}{h} = 1$

Note that $f_{xy} \neq f_{yx}$ at (0, 0). See Problem 6.13.

6.42. Show that under the transformation $x = \rho\cos\phi$, $y = \rho\sin\phi$ the equation $\dfrac{\partial^2 V}{\partial x^2} + \dfrac{\partial^2 V}{\partial y^2} = 0$ become $\dfrac{\partial^2 V}{\partial \rho^2} + \dfrac{1}{\rho}\dfrac{\partial V}{\partial \phi} + \dfrac{1}{\rho^2}\dfrac{\partial^2 V}{\partial \phi^2} = 0.$

We have

$$\frac{\partial V}{\partial x} = \frac{\partial V}{\partial \rho}\frac{\partial \rho}{\partial x} + \frac{\partial V}{\partial \phi}\frac{\partial \phi}{\partial x} \tag{1}$$

$$\frac{\partial V}{\partial y} = \frac{\partial V}{\partial \rho}\frac{\partial \rho}{\partial y} + \frac{\partial V}{\partial \phi}\frac{\partial \phi}{\partial y} \tag{2}$$

Differentiate $x = \rho \cos\phi$, $y = \rho \sin\phi$ with respect to x, remembering that ρ and ϕ are functions of x and y

$$1 = -\rho \sin\phi \frac{\partial \phi}{\partial x} + \cos\phi \frac{\partial \rho}{\partial x}, \quad 0 = \rho \cos\phi \frac{\partial \phi}{\partial x} + \sin\phi \frac{\partial \rho}{\partial x}$$

Solving simultaneously,

$$\frac{\partial \rho}{\partial x} = \cos\phi, \quad \frac{\partial \phi}{\partial x} = -\frac{\sin\phi}{\rho} \tag{3}$$

Similarly, differentiate with respect to y. Then

$$0 = -\rho \sin\phi \frac{\partial \phi}{\partial y} + \cos\phi \frac{\partial \rho}{\partial y}, \qquad 1 = \rho \cos\phi \frac{\partial \phi}{\partial y} + \sin\phi \frac{\partial \rho}{\partial y}$$

Solving simultaneously,

$$\frac{\partial \rho}{\partial y} = \sin\phi, \quad \frac{\partial \phi}{\partial y} = \frac{\cos\phi}{\rho} \tag{4}$$

Then, from Equations (1) and (2),

$$\frac{\partial V}{\partial x} = \cos\phi \frac{\partial V}{\partial \rho} - \frac{\sin\phi}{\rho} \frac{\partial V}{\partial \phi} \tag{5}$$

$$\frac{\partial V}{\partial y} = \sin\phi \frac{\partial V}{\partial \rho} + \frac{\cos\phi}{\rho} \frac{\partial V}{\partial \phi} \tag{6}$$

Hence,

$$\begin{aligned}
\frac{\partial^2 V}{\partial x^2} &= \frac{\partial}{\partial x}\left(\frac{\partial V}{\partial x}\right) = \frac{\partial}{\partial \rho}\left(\frac{\partial V}{\partial x}\right)\frac{\partial \rho}{\partial x} + \frac{\partial}{\partial \phi}\left(\frac{\partial V}{\partial x}\right)\frac{\partial \phi}{\partial x} \\
&= \frac{\partial}{\partial \rho}\left(\cos\phi \frac{\partial V}{\partial \rho} - \frac{\sin\phi}{\rho}\frac{\partial V}{\partial \phi}\right)\frac{\partial \rho}{\partial x} + \frac{\partial}{\partial \phi}\left(\cos\phi \frac{\partial V}{\partial \rho} - \frac{\sin\phi}{\rho}\frac{\partial V}{\partial \phi}\right)\frac{\partial \phi}{\partial x} \\
&= \left(\cos\phi \frac{\partial^2 V}{\partial \rho^2} + \frac{\sin\phi}{\rho^2}\frac{\partial V}{\partial \phi} - \frac{\sin\phi}{\rho}\frac{\partial^2 V}{\partial \rho\, \partial \phi}\right)(\cos\phi) \\
&\quad + \left(-\sin\phi \frac{\partial V}{\partial \rho} + \cos\phi \frac{\partial^2 V}{\partial \rho\, \partial \phi} - \frac{\cos\phi}{\rho}\frac{\partial V}{\partial \phi} - \frac{\sin\phi}{\rho}\frac{\partial^2 V}{\partial \phi^2}\right)\left(-\frac{\sin\phi}{\rho}\right)
\end{aligned}$$

which simplifies to

$$\frac{\partial^2 V}{\partial x^2} = \cos^2\phi \frac{\partial^2 V}{\partial \rho^2} + \frac{2\sin\phi\cos\phi}{\rho^2}\frac{\partial V}{\partial \phi} - \frac{2\sin\phi\cos\phi}{\rho}\frac{\partial^2 V}{\partial \rho\, \partial \phi} + \frac{\sin^2\phi}{\rho}\frac{\partial V}{\partial \rho} + \frac{\sin^2\phi}{\rho^2}\frac{\partial^2 V}{\partial \phi^2} \tag{7}$$

Similarly,

$$\frac{\partial^2 V}{\partial y^2} = \sin^2\phi \frac{\partial^2 V}{\partial \rho^2} - \frac{2\sin\phi\cos\phi}{\rho^2}\frac{\partial V}{\partial \phi} + \frac{2\sin\phi\cos\phi}{\rho}\frac{\partial^2 V}{\partial \rho \partial \phi} + \frac{\cos^2\phi}{\rho}\frac{\partial V}{\partial \rho} + \frac{\cos^2\phi}{\rho^2}\frac{\partial^2 V}{\partial \phi^2} \tag{8}$$

Adding Equations (7) and (8), we find, as required, $\dfrac{\partial^2 V}{\partial x^2} + \dfrac{\partial^2 V}{\partial y^2} = \dfrac{\partial^2 V}{\partial \rho^2} + \dfrac{1}{\rho}\dfrac{\partial V}{\partial \rho} + \dfrac{1}{\rho^2}\dfrac{\partial^2 V}{\partial \phi^2} = 0.$

6.43. (a) If $x = f(u, \upsilon)$ and $y = g(u, \upsilon)$, where $u = \phi(r, s)$ and $\upsilon = \psi(r, s)$, prove that $\dfrac{\partial(x, y)}{\partial(r, s)} = \dfrac{\partial(x, y)}{\partial(u, \upsilon)}\dfrac{\partial(u, \upsilon)}{\partial(r, s)}$.

(b) Prove that $\dfrac{\partial(x, y)}{\partial(u, \upsilon)}\dfrac{\partial(u, \upsilon)}{\partial(x, y)} = 1$, provided $\dfrac{\partial(x, y)}{\partial(u, \upsilon)} \neq 0$, and interpret geometrically.

(a) $$\frac{\partial(x,y)}{\partial(r,s)} = \begin{vmatrix} x_r & x_s \\ y_r & y_s \end{vmatrix} = \begin{vmatrix} x_u u_r + x_\upsilon \upsilon_r & x_u u_s + x_\upsilon \upsilon_s \\ y_u u_r + y_\upsilon \upsilon_r & y_u u_s + y_\upsilon \upsilon_s \end{vmatrix}$$
$$= \begin{vmatrix} x_u & x_\upsilon \\ y_u & y_\upsilon \end{vmatrix} \begin{vmatrix} u_r & u_s \\ \upsilon_r & \upsilon_s \end{vmatrix} w = \frac{\partial(x,y)\partial(u,\upsilon)}{\partial(u,\upsilon)\partial(r,s)}$$

using a theorem on multiplication of determinants (see Problem 6.108). We have assumed here, of course, the existence of the partial derivatives involved.

(b) Place $r = x$, $s = y$ in the result of (a). Then $\dfrac{\partial(x,y)}{\partial(u,\upsilon)} \dfrac{\partial(u,\upsilon)}{\partial(x,y)} = \dfrac{\partial(x,y)}{\partial(x,y)} = 1$.

The equations $x = f(u, \upsilon)$, $y = g(u, \upsilon)$ define a transformation between points (x, y) in the xy plane and points (u, υ) in the uv plane. The inverse transformation is given by $u = \phi(x, y)$, $\upsilon = \psi(x, y)$. The result obtained states that the Jacobians of these transformations are reciprocals of each other.

6.44. Show that $F(xy, z - 2x) = 0$ satisfies, under suitable conditions, the equation $x(\partial z/\partial x) - y(\partial z/\partial y) = 2x$. What are these conditions?

Let $u = xy$, $\upsilon = z - 2x$. Then $F(u, \upsilon) = 0$ and

$$dF = F_u du + F_\upsilon d\upsilon = F_u(x\,dy + y\,dx) + F_\upsilon(dz - 2\,dx) = 0 \tag{1}$$

Taking z as dependent variable and x and y as independent variables, we have $dz = z_x\,dx + z_y\,dy$. Then substituting in Equation (*1*), we find

$$(yF_u + F_\upsilon z_x - 2)\,dx + (xF_u + F_\upsilon z_y)\,dy = 0$$

Hence, since x and y are independent, we have

$$yF_u + F_\upsilon z_x - 2 = 0 \tag{2}$$
$$xF_u + F_\upsilon z_y = 0 \tag{3}$$

Solve for F_u in Equation (*3*) and substitute in (*2*). Then we obtain the required result $xz_x - yz_y = 2x$ upon dividing by F_υ (supposed not equal to zero).

The result will certainly be valid if we assume that $F(u, \upsilon)$ is continuously differentiable and that $F_\upsilon \neq 0$.

SUPPLEMENTARY PROBLEMS

Functions and graphs

6.45. If $f(x, y) = \dfrac{2x + y}{1 - xy}$, find (a) $f(1 - 3)$, (b) $\dfrac{f(2 + h, 3) - f(2, 3)}{h}$, and (c) $f(x + y, xy)$.

Ans. (a) $-\dfrac{1}{4}$ (b) $\dfrac{11}{5(3h + 5)}$ (c) $\dfrac{2x + 2y + xy}{1 - x^2y - xy^2}$

6.46. If $g(x, y, z) = x^2 - yz + 3xy$, find (a) $g(1, -2, 2)$, (b) $g(x + 1, y - 1, z^2)$, and (c) $g(xy, xz, x + y)$.

Ans. (a) -1 (b) $x^2 - x - 2 - yz^2 + z^2 + 3xy + 3y$ (c) $x^2y^2 - x^2z - xyz + 3x^2yz$

6.47. Give the domain of definition for which each of the following functional rules is defined and real, and indicate this domain graphically: (a) $f(x, y) = \dfrac{1}{x^2 + y^2 - 1}$, (b) $f(x, y) = \ln(x + y)$, and (c) $f(x, y) = \sin^{-1}\left(\dfrac{2x - y}{x + y}\right)$.

Ans. (a) $x^2 + y^2 \neq 1$ (b) $x + y > 0$ (c) $\left|\dfrac{2x - y}{x + y}\right| \leq 1$

6.48. (a) What is the domain of definition for which $f(x, y, z) = \sqrt{\dfrac{x+y+z-1}{x^2+y^2+z^2-1}}$ is defined and real? (b) Indicate this domain graphically.

Ans. (a) $x + y + z \leqq 1, x^2 + y^2 + z^2 < 1$ and $x + y + z \geqq 1, x^2 + y^2 + z^2 > 1$

6.49. Sketch and name the surface in three-dimensional space represented by each of the following.

(a) $3x + 2z = 12$ (d) $x^2 + z^2 = y^2$ (g) $x^2 + y^2 = 2y$

(b) $4z = x^2 + y^2$ (e) $x^2 + y^2 + z^2 = 16$ (h) $z = x + y$

(c) $z = x^2 - 4y^2$ (f) $x^2 - 4y^2 - 4z^2 = 36$ (i) $y^2 = 4z$

(j) $x^2 + y^2 + z^2 - 4x + 6y + 2z - 2 = 0$

Ans. (a) plane (b) paraboloid of revolution (c) hyperbolic paraboloid (d) right circular cone (e) sphere (f) hyperboloid of two sheets (g) right circular cylinder (h) plane (i) arabolic cylinder (j) sphere, center at (2, –3, –1) and radius 4.

6.50. Construct a graph of the region bounded by $x^2 + y^2 = a^2$ and $x^2 + z^2 = a^2$, where a is a constant.

6.51. Describe graphically the set of points (x, y, z) such that (a) $x^2 + y^2 + z^2 = 1, x^2 + y^2 = z^2$ and (b) $x^2 + y^2 < z < x + y$.

6.52. The *level curves* for a function $z = f(x, y)$ are curves in the xy plane defined by $f(x, y) = c$, where c is any constant. They provide a way of representing the function graphically. Similarly, the *level surfaces* of $w = f(x, y, z)$ are the surfaces in a rectangular (xyz) coordinate system defined by $f(x, y, z) = c$, where c is any constant. Describe and graph the level curves and surfaces for each of the following functions: (a) $f(x, y) = \ln(x^2 + y^2 - 1)$, (b) $f(x, y) = 4xy$, (c) $f(x, y) = \tan^{-1} y/(x + 1)$, (d) $f(x, y) = x^{2/3} + y^{2/3}$, (e) $f(x, y, z) = x^2 + 4y^2 + 16z^2$ and (f) $\sin(x + z)/(1 - y)$.

Limits and continuity

6.53. Prove that (*a*) $\lim\limits_{\substack{x\to 4\\ y\to -1}} (3x - 2y) = 14$ and (*b*) $\lim\limits_{(x,y)\to(2,1)} (xy - 3x + 4) = 0$ by using the definition.

6.54. If $\lim f(x, y) = A$ and $\lim g(x, y) = B$, where lim denotes *limit as* $(x, y) \to (x_0, y_0)$, prove that (a) $\lim \{f(x, y) + g(x, y)\} = A + B$ and (b) $\lim \{f(x, y)\, g(x, y)\} = AB$.

6.55. Under what conditions is the limit of the quotient of two functions equal to the quotient of their limits? Prove your answer.

6.56. Evaluate each of the following limits where they exist:

(a) $\lim\limits_{\substack{x\to 1\\ y\to 2}} \dfrac{3 - x + y}{4 + x - 2y}$ (c) $\lim\limits_{\substack{x\to 4\\ y\to \pi}} x^2 \sin\dfrac{y}{x}$ (e) $\lim\limits_{\substack{x\to 0\\ y\to 1}} e^{-1/x^2 (y-1)^2}$ (g) $\lim\limits_{\substack{x\to 0+\\ y\to 1-}} \dfrac{x + y - 1}{\sqrt{x} - \sqrt{1 - y}}$

(b) $\lim\limits_{\substack{x\to 0\\ y\to 0}} \dfrac{3x - 2y}{2x - 3y}$ (d) $\lim\limits_{\substack{x\to 0\\ y\to 0}} \dfrac{x\sin(x^2 + y^2)}{x^2 + y^2}$ (f) $\lim\limits_{\substack{x\to 0\\ y\to 0}} \dfrac{2x - y}{x^2 + y^2}$ (h) $\lim\limits_{\substack{x\to 2\\ y\to 1}} \dfrac{\sin^{-1}(xy - 2)}{\tan^{-1}(3xy - 6)}$

Ans. (a) 4 (b) does not exist (c) $8\sqrt{2}$ (d) 0 (e) 0 (f) does not exist (g) 0 (h) $\dfrac{1}{3}$

6.57. Formulate a definition of limit for functions of (a) 3 and (b) n variables.

6.58. Does $\lim \dfrac{4x + y - 3z}{2x - 5y + 2z}$ as $(x, y, z) \to (0, 0, 0)$ exist? Justify your answer.

6.59. Investigate the continuity of each of the following functions at the indicated points: (a) $x^2 + y^2$, (x_0, y_0); (b) $\dfrac{x}{3x + 5y}$, (0, 0); and (c) $(x^2 + y^2)\sin\dfrac{1}{x^2 + y^2}$ if $(x, y) \neq (0, 0)$, 0 if $(x, y) = (0, 0)$, (0, 0).

Ans. (a) continuous (b) discontinuous (c) continuous

6.60. Using the definition, prove that $f(x, y) = xy + 6x$ is continuous continuous at (a) (1, 2) and (b) (x_0, y_0).

6.61. Prove that the function in Problem 6.60 is uniformly continuous in the square region defined by $0 \leq x \leq 1$, $0 \leq y \leq 1$.

Partial derivatives

6.62. If $f(x, y) = \dfrac{x - y}{x + y}$, find (a) $\partial f/\partial x$ and (b) $\partial f/\partial y$ at (2, –1) from the definition and verify your answer by differentiation rules.

Ans. (a) –2 (b) –4

6.63. If $f(x, y) = \begin{cases} (x^2 - xy)/(x + y) & \text{for } (x, y) \neq (0,0) \\ 0 & \text{for } (x, y) = (0,0) \end{cases}$ find (a) $f_x(0,0)$ and (b) $f_y(0,0)$.

Ans. (a) 1 (b) 0

6.64. Investigate $\lim\limits_{(x,y)\to(0,0)} f_x(x, y)$ for the function in the preceding problem and explain why this limit (if it exists) is or is not equal to $f_x(0, 0)$.

6.65. If $f(x, y) = (x - y)\sin(3x + 2y)$, compute (a) f_x, (b) f_y, (c) f_{xx}, (d) f_{yy}, and (e) f_{yx} at $(0, \pi/3)$.

Ans. (a) $\dfrac{1}{2}(\pi + \sqrt{3})$ (b) $\dfrac{1}{6}(2\pi - 3\sqrt{3})$ (c) $\dfrac{3}{2}(\pi\sqrt{3} - 2)$ (d) $\dfrac{2}{3}(i\sqrt{3} + 3)$ (e) $\dfrac{1}{2}(2\pi\ \sqrt{3} + 1)$

6.66. (a) Prove by direct differentiation that $z = xy\tan(y/x)$ satisfies the equation $x(\partial z/\partial x) + y(\partial z/\partial y) = 2z$ if $(x, y) \neq (0, 0)$. (b) Discuss part (a) for all other points (x, y), assuming $z = 0$ at (0, 0).

6.67. Verify that $f_{xy} = f_{yx}$ for the functions (a) $(2x - y)/(x + y)$, (b) $x \tan xy$, and (c) $\cosh(y + \cos x)$, indicating possible exceptional points, and investigate these points.

6.68. Show that $z = \ln\{(x - a)^2 + (y - b)^2\}$ satisfies $\partial^2 z/\partial x^2 + \partial^2 z/\partial y^2 = 0$ except at (a, b).

6.69. Show that $z = x\cos(y/x) + \tan(y/x)$ satisfies $x^2 z_{xx} + 2xyz_{xy} + y^2 z_{yy} = 0$ except at points for which $x = 0$.

6.70. Show that if $w = \left(\dfrac{x - y + z}{x + y - z}\right)^n$, then:

(a) $x\dfrac{\partial w}{\partial x} + y\dfrac{\partial w}{\partial y} + z\dfrac{\partial w}{\partial z} = 0$

(b) $x^2\dfrac{\partial^2 w}{\partial x^2} + y^2\dfrac{\partial^2 w}{\partial y^2} + z^2\dfrac{\partial^2 w}{\partial z^2} + 2xy\dfrac{\partial^2 w}{\partial x\,\partial y} + 2xz\dfrac{\partial^2 w}{\partial x\,\partial z} + 2yz\dfrac{\partial^2 w}{\partial y\,\partial z} = 0$

Indicate possible exceptional points.

Differentials

6.71. If $z = x^3 - xy + 3y^2$, compute (a) Δz and (b) dz, where $x = 5$, $y = 4$, $\Delta x = -0.2$, $\Delta y = 0.1$. Explain why Δz and dz are approximately equal. (c) Find Δz and dz if $x = 5$, $y = 4$, $\Delta x = -2$, $\Delta y = 1$.

Ans. (a) -11.658 (b) -12.3 (c) $\Delta z = -66$, $dz = -123$

6.72. Compute $\sqrt[5]{(3.8)^2 + 2(2.1)^3}$ approximately, using differentials.

Ans. 2.01

6.73. Find dF and dG if (a) $F(x, y) = x^3y - 4xy^2 + 8y^3$, (b) $G(x, y, z) = 8xy^2z^3 - 3x^2yz$, and (c) $F(x, y) = xy^2 \ln(y/x)$.

Ans. (a) $(3x^2y - 4y^2)\,dx + (x^3 - 8xy + 24y^2)\,dy$

(b) $(8y^2z^3 - 6xyz)\,dx + (16xyz^3 - 3x^2z)\,dy + (24xy^2z^2 - 3x^2y)\,dz$

(c) $\{y^2 \ln(y/x) - y^2\}\,dx + \{2xy \ln(y/x) + xy\}\,dy$

6.74. Prove that (a) $d(UV) = U\,dV + V\,dU$, (b) $d(U/V) = (V\,dU - U\,dV)/V^2$, (c) $d(\ln U) = (dU)/U$, and (d) $d(\tan^{-1} V) = (dV)/(1 + \upsilon^2)$, where U and V are differentiable functions of two or more variables.

6.75. Determine whether each of the following is an exact differential of a function and, if so, find the function:

(a) $(2xy^2 + 3y\cos 3x)\,dx + (2x^2y + \sin 3x)\,dy$

(b) $(6xy - y^2)\,dx + (2xe^y - x^2)\,dy$

(c) $(z^3 - 3y)\,dx + (12y^2 - 3x)\,dy + 3xz^2\,dz$

Ans. (a) $x^2y^2 + y\sin 3x + c$ (b) not exact (c) $xz^2 + 4y^3 - 3xy + c$

Differentiation of composite functions

6.76. (a) If $U(x, y, z) = 2x^2 - yz + xz^2$, $x = 2\sin t$, $y = t^2 - t + 1$, and $z = 3e^{-t}$, find dU/dt at $t = 0$. (b) If $H(x, y) = \sin(3x - y)$, $x^3 + 2y = 2t^3$, and $x - y^2 = t^2 + 3t$, find dH/dt.

Ans. (a) 24 (b) $\left(\dfrac{36t^2y + 12t + 9x^2 - 6t^2 + 6x^2t + 18}{6x^2y + 2}\right)\cos(3x - y)$

6.77. If $F(x, y) = (2x + y)/(y - 2x)$, $x = 2u - 3\upsilon$, and $y = u + 2\upsilon$, find (a) $\partial F/\partial u$, (b) $\partial F/\partial \upsilon$, (c) $\partial^2 F/\partial u^2$, (d) $\partial^2 F/\partial \upsilon^2$, and (e) $\partial^2 F/\partial u\,\partial \upsilon$, where $u = 2$, $\upsilon = 1$.

Ans. (a) 7 (b) -14 (c) 21 (d) 112 (e) -49

6.78. If $U = x^2F(y/x)$, show that, under suitable restrictions on F, $x(\partial U/\partial x) + y(\partial U/\partial y) = 2U$.

6.79. If $x = u\cos\alpha - \upsilon\sin\alpha$ and $y = u\sin\alpha + \upsilon\cos\alpha$, where α is a constant, show that $(\partial V/\partial x)^2 + (\partial V/\partial y)^2 = (\partial V/\partial u)^2 + (\partial V/\partial \upsilon)^2$.

6.80. Show that if $x = \rho\cos\phi$, $y = \rho\sin\phi$, the equation $\dfrac{\partial u}{\partial x} = \dfrac{\partial \upsilon}{\partial y}, \dfrac{\partial u}{\partial y} = -\dfrac{\partial \upsilon}{\partial x}$ becomes $\dfrac{\partial u}{\partial \rho} = \dfrac{1}{\rho}\dfrac{\partial \upsilon}{\partial \phi}, \dfrac{\partial \upsilon}{\partial \rho} = -\dfrac{1}{\rho}\dfrac{\partial u}{\partial \phi}$.

6.81. Use Problem 6.80 to show that under the transformation $x = \rho\cos\phi$, $y = \rho\sin\phi$, the equation

$$\frac{\partial^2 u}{\partial x^2} = \frac{\partial^2 u}{\partial y^2} = 0 \text{ becomes } \frac{\partial^2 u}{\partial \rho^2} = \frac{1}{\rho}\frac{\partial u}{\partial \rho} + -\frac{1}{\rho^2}\frac{\partial^2 u}{\partial \phi^2} = 0.$$

Implicit functions and jacobians

6.82. If $F(x, y) = 0$, prove that $dy/dx = -F_x/F_y$.

6.83. Find (a) dy/dx and (b) d^2y/dx^2 if $x^3 + y^3 - 3xy = 0$.

Ans. (a) $(y - x^2)/(y^2 - x)$ (b) $-2xy/(y^2 - x)^3$

6.84. If $xu^2 + \upsilon = y^3$, $2yu - x\upsilon^3 = 4x$, find (a) $\dfrac{\partial u}{\partial x}$ and (b) $\dfrac{\partial \upsilon}{\partial y}$.

Ans. (a) $\dfrac{\upsilon^3 - 3xu^2\upsilon^2 + 4}{6x^2 - u\upsilon^2 + 2y}$ (b) $\dfrac{2xu^2 + 3y^3}{3x^2u\upsilon^2 + y}$

6.85. If $u = f(x, y)$ and $\upsilon = g(x, y)$ are differentiable, prove that $\dfrac{\partial u}{\partial x}\dfrac{\partial x}{\partial u} + \dfrac{\partial \upsilon}{\partial x}\dfrac{\partial x}{\partial \upsilon} = 1$. Explain clearly which variables are considered independent in each partial derivative.

6.86. If $f(x, y, r, s) = 0$, $g(x, y, r, s) = 0$, prove that $\dfrac{\partial y}{\partial r}\dfrac{\partial r}{\partial x} + \dfrac{\partial y}{\partial s}\dfrac{\partial s}{\partial x} = 0$, explaining which variables are independent. What notation could you use to indicate the independent variables considered?

6.87. If $F(x, y) = 0$, show that $\dfrac{d^2y}{dx^2} = -\dfrac{F_{xx}F_y^2 - 2F_{xy}F_xF_y + F_{yy}F_x^2}{F_y^3}$.

6.88. Evaluate $\dfrac{\partial(F,G)}{\partial(u,\upsilon)}$ *if* $F(u,\upsilon) = 3u^2 - u\upsilon$ and $G(u,\upsilon) = 2u\upsilon^2 + \upsilon^3$.

Ans. $24u^2\upsilon + 16u\upsilon^2 - 3\upsilon^3$

6.89. If $F = x + 3y^2 - z^3$, $G = 2x^2yz$, and $H = 2z^2 - xy$, evaluate $\dfrac{\partial(F,G,H)}{\partial(x,y,z)}$ at $(1, -1, 0)$.

Ans. 10

6.90. If $u = \sin^{-1}x + \sin^{-1}y$ and $\upsilon = x\sqrt{1 - y^2} + y\sqrt{1 - x^2}$, determine whether there is a functional relationship between u and υ, and, if so, find it.

6.91. If $F = xy + yz + zx$, $G = x^2 + y^2 + z^2$, and $H = x + y + z$, determine whether there is a functional relationship connecting F, G, and H, and, if so, find it.

Ans. $H^2 - G - 2F = 0$

6.92. (a) If $x = f(u, \upsilon, w)$, $y = g(u, \upsilon, w)$, and $z = h(u, \upsilon, w)$, prove that $\dfrac{\partial(x,y,z)}{\partial(x,\upsilon,z)}\dfrac{\partial(u,\upsilon,w)}{\partial(x,y,w)} = 1$ provided $\dfrac{\partial(x,y,z)}{\partial(u,\upsilon,w)} \neq 0$. (b) Give an interpretation of the result of (a) in terms of transformations.

6.93. If $f(x, y, z) = 0$ and $g(x, y, z) = 0$, show that $\dfrac{dx}{\dfrac{\partial(f,g)}{\partial(y,z)}} = \dfrac{dy}{\dfrac{\partial(f,g)}{\partial(z,x)}} = \dfrac{dz}{\dfrac{\partial(f,g)}{\partial(x,y)}}$ giving conditions under which the result is valid.

6.94. If $x + y^2 = u$, $y + z^2 = \upsilon$, $z + x^2 = w$, find (a) $\dfrac{\partial x}{\partial u}$, (b) $\dfrac{\partial^2 x}{\partial u^2}$, (c) $\dfrac{\partial^2 x}{\partial u\,\partial \upsilon}$, assuming that the equations define x, y, and z as twice differentiable functions of u, υ, and w.

Ans. (a) $\dfrac{1}{1+8xyz}$ (b) $\dfrac{16x^2y - 8yz - 32x^2z^2}{(1+8xyz)^3}$ (c) $\dfrac{16y^2z - 8xz - 32x^2y^2}{(1+8xyz)^3}$

6.95. State and prove a theorem similar to that in Problem 6.35, for the case where $u = f(x, y, z)$, $\upsilon = g(x, y, z)$, $w = h(x, y, z)$.

Transformations, curvilinear coordinates

6.96. Given the transformation $x = 2u + \upsilon$, $y = u - 3\upsilon$, (a) sketch the region $\mathfrak{R}'$ of the uv plane into which the region $\mathfrak{R}$ of the xy plane bounded by $x = 0$, $x = 1$, $y = 0$, $y = 1$ is mapped under the transformation; (b) compute $\dfrac{\partial(x,y)}{\partial(u,\upsilon)}$; and (c) compare the result of (b) with the ratios of the areas of $\mathfrak{R}$ and $\mathfrak{R}'$.

Ans. (b) –7

6.97. (a) Prove that under a *linear transformation* $x = a_1u + a_2\upsilon$, $y = b_1u + b_2\upsilon (a_1b_2 - a_2b_1 \neq 0)$ lines and circles in the xy plane are mapped, respectively, into lines and circles in the uv plane. (b) Compute the Jacobian J of the transformation and discuss the significance of $J = 0$.

6.98. Given $x = \cos u \cosh \upsilon$, $y = \sin u \sinh \upsilon$, (a) show that, in general, the coordinate curves $u = a$ and $\upsilon = b$ in the uv plane are mapped into hyperbolas and ellipses, respectively, in the xy plane; (b) compute $\left|\dfrac{\partial(x,y)}{\partial(u,\upsilon)}\right|$; (c) compute $\left|\dfrac{\partial(u,\upsilon)}{\partial(x,y)}\right|$.

Ans. (b) $\sin^2 u \cosh^2 \upsilon + \cos^2 u \sinh^2 \upsilon$ (c) $(\sin^2 u \cosh^2 \upsilon + \cos^2 u \sinh^2 \upsilon)^{-1}$

6.99. Given the transformation $x = 2u + 3\upsilon - w$, $y = 2\upsilon + w$, $z = 2u - 2\upsilon + w$, (a) sketch the region $\mathfrak{R}'$ of the uvw space into which the region $\mathfrak{R}$ of the xyz space bounded by $x = 0$, $x = 8$, $y = 0$, $y = 4$, $z = 0$, $z = 6$ is mapped; (b) compute $\dfrac{\partial(x,y,z)}{\partial(u,\upsilon,w)}$; (c) compare the result of (b) with the ratios of the volumes of $\mathfrak{R}$ and $\mathfrak{R}'$.

Ans. (b) 1

6.100. Given the spherical coordinate transformation $x = r \sin\theta \cos\phi$, $y = r \sin\theta \sin\phi$, $z = r\cos\theta$, where r δ 0. $0 \leq \theta \leq \pi$, $0 \leq \phi < 2\pi$, describe the coordinate surfaces (a) $r = a$, (b) $\theta = b$, and (c) $\phi = c$, where a, b, c are any constants.

Ans. (a) spheres (b) cones (c) planes

6.101. (a) Verify that for the spherical coordinate transformation of Problem 6.100, $J = \dfrac{\partial(x, y, z)}{\partial(r, \theta, \phi)} = r^2 \sin\theta$.
(b) Discuss the case where $J = 0$.

Miscellaneous problems

6.102. If $F(P, V, T) = 0$, prove that (a) $\left.\dfrac{\partial P}{\partial T}\right|_v \left.\dfrac{\partial T}{\partial V}\right|_p = -\left.\dfrac{\partial P}{\partial V}\right|_T$ (b) $\left.\dfrac{\partial P}{\partial T}\right|_v \left.\dfrac{\partial T}{\partial V}\right|_p \left.\dfrac{\partial V}{\partial P}\right|_T = -1$.

These results are useful in *thermodynamics*, where *P*, *V*, and *T* correspond to pressure, volume, and temperature of a physical system.

6.103. Show that $F(x/y, z/y) = 0$ satisfies $x(\partial z/\partial x) + y(\partial z/\partial y) = z$.

6.104. Show that $F(x + y - z, x^2 + y^2) = 0$ satisfies $x(\partial z/\partial y) - y(\partial z/\partial x) = x - y$.

6.105. If $x = f(u, \upsilon)$ and $y = g(u, \upsilon)$, prove that $\dfrac{\partial \upsilon}{\partial x} = -\dfrac{1}{J}\dfrac{\partial y}{\partial u}$ where $J = \dfrac{\partial(x, y)}{\partial(u, \upsilon)}$.

6.106. If $x = f(u, \upsilon)$, $y = g(u, \upsilon)$, $z = h(u, \upsilon)$ and $F(x, y, z) = 0$, prove that $\dfrac{\partial(y, z)}{\partial(u, \upsilon)}dx + \dfrac{\partial(z, x)}{\partial(u, \upsilon)}dy + \dfrac{\partial(x, y)}{\partial(u, \upsilon)}dz = 0$

6.107. If $x = \phi(u, \upsilon, w)$, $y = \psi(u, \upsilon, w)$, $u = f(r, s)$, $\upsilon = g(r, s)$, and $w = h(r, s)$, prove that

$$\frac{\partial(x, y)}{\partial(r, s)} = \frac{\partial(x, y)}{\partial(u, \upsilon)}\frac{\partial(u, \upsilon)}{\partial(r, s)} + \frac{\partial(x, y)}{\partial(\upsilon, w)} = \frac{\partial(\upsilon, w)}{\partial(r, s)} + \frac{\partial(x, y)}{\partial(w, u)} + \frac{\partial(w, u)}{\partial(r, s)}$$

6.108. (a) Prove that $\begin{vmatrix} a & b \\ c & d \end{vmatrix} \cdot \begin{vmatrix} e & f \\ g & h \end{vmatrix} = \begin{vmatrix} ae + bg & af + bh \\ ce + dg & cf + dh \end{vmatrix}$, thus establishing the rule for the product of two second-order determinants referred to in Problem 6.43. (b) Generalize the result of (a) to determinants of 3, 4 . . .

6.109. If *x*, *y*, and *z* are functions of *u*, *υ*, and *w*, while *u*, *υ*, and *w* are functions of *r*, *s*, and *t*, prove that
$\dfrac{\partial(x, y, z)}{\partial(r, s, t)} = \dfrac{\partial(x, y, z)}{\partial(u, \upsilon, w)} \cdot \dfrac{\partial(u, \upsilon, w)}{\partial(r, s, t)}$.

6.110. Given the equation $F_j(x_1, \ldots, x_m, y_1, \ldots, y_n) = 0$ where $j = 1, 2, \ldots, n$, prove that, under suitable conditions on F_j, $\dfrac{\partial y_r}{\partial x_s} = \dfrac{\partial(F_1, F_2, \ldots, F_r, \ldots, F_n}{\partial(y_1, y_2, \ldots, x_s, \ldots, y_n} \Big/ \dfrac{\partial(F_1, F_2, \ldots, F_n}{\partial(y_1, y_2, \ldots, y_n}$

6.111. (a) If $F(x, y)$ is homogeneous of degree 2, prove that $x^2\dfrac{\partial^2 F}{\partial x^2} + 2xy\dfrac{\partial^2 F}{\partial x \partial y} + y^2\dfrac{\partial^2 F}{\partial y^2} = 2F$.

(b) Illustrate by using the special case $F(x, y) = x^2 \ln(y/x)$.

Note that the result can be written in operator form, using $D_x \equiv \partial/\partial x$ and $D_y \equiv \partial/\partial y$, as $(x D_x + y D_y)^2 F = 2F$. [Hint: Different Differentiate both sides of Problem 6.25, Equation (*1*), twice with respect to λ.]

6.112. Generalize the result of Problem 6.11 as follows. If $F(x_1, x_2, \ldots, x_n)$ is homogeneous of degree *p*, then for any positive integer *r*, if $D_{xj} \equiv \partial/\partial x_j$, $(x_1 D_{x1} + x_2 D_{x\,2} + \cdots + x_n D_{xn})^r F = p(p-1) \ldots (p - r + 1)F$.

6.113. (a) Let *x* and *y* be determined from *u* and *υ* according to $x + iy = (u + iv)^3$. Prove that under this transformation the equation $\dfrac{\partial^2\phi}{\partial x^2} + \dfrac{\partial^2\phi}{\partial y^2} = 0$ is transformed into $\dfrac{\partial^2\phi}{\partial u^2} + \dfrac{\partial^2\phi}{\partial \upsilon^2} = 0$.

(b) Is the result in (a) true if $x + iy = F(u + iv)$? Prove your statements.

CHAPTER 7

Vectors

Vectors

The foundational ideas for vector analysis were formed independently in the nineteenth century by William Rowan Hamilton and Herman Grassmann. We are indebted to the physicist John Willard Gibbs, who formulated the classical presentation of the Hamilton viewpoint in his Yale lectures, and his student E. B. Wilson, who considered the mathematical material presented in class worthy of organizing as a book (published in 1901). Hamilton was searching for a mathematical language appropriate to a comprehensive exposition of the physical knowledge of the day. His geometric presentation emphasizing magnitude and direction and compact notation for the entities of the calculus was refined in the following years to the benefit of expressing Newtonian mechanics, electromagnetic theory, and so on. Grassmann developed an algebraic and more philosophic mathematical structure which was not appreciated until it was needed for Riemanian (non-Euclidean) geometry and the special and general theories of relativity.

Our introduction to vectors is geometric. We conceive of a vector as a directed line segment $\overrightarrow{PQ}$ from one point P, called the *initial point,* to another point Q, called the *terminal point*. We denote vectors by boldfaced letters or letters with an arrow over them. Thus, $\overrightarrow{PQ}$ is denoted by **A** or $\vec{A}$, as in Figure 7.1. The *magnitude* or *length* of the vector is then denoted by $|\overrightarrow{PQ}|$, $\overline{PQ}$, $|\mathbf{A}|$ or $|\vec{A}|$.

Vectors are defined to satisfy the geometric properties discussed in the next section.

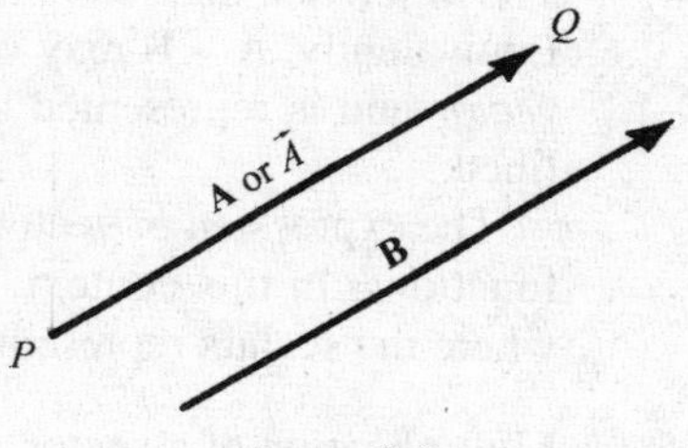

Figure 7.1

Geometric Properties of Vectors

1. Two vectors **A** and **B** are *equal* if they have the same magnitude and direction regardless of their initial points. Thus, **A** = **B** in Figure 7.1.

   In other words, a vector is geometrically represented by any one of a class of commonly directed line segments of equal magnitude. Since any one of the class of line segments may be chosen to represent it, the vector is said to be *free*. In certain circumstances (tangent vectors, forces bound to a point), the initial point is fixed; then the vector is *bound*. Unless specifically stated, the vectors in this discussion are free vectors.

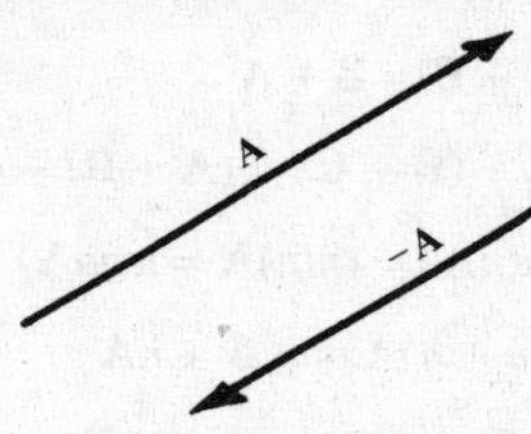

Figure 7.2

2. A vector having direction opposite to that of vector **A** but with the same magnitude is denoted by –**A** (see Figure 7.2).

3. The *sum* or *resultant* of vectors **A** and **B** of Figure 7.3(a) is a vector

C formed by placing the initial point of **B** on the terminal point of **A** and joining the initial point of **A** to the terminal point of **B** [see Figure 7.3(*b*)]. The sum **C** is written **C** = **A** + **B**. The definition here is equivalent to the *parallelogram law* for vector addition, as indicated in Figure 7.3(*c*).

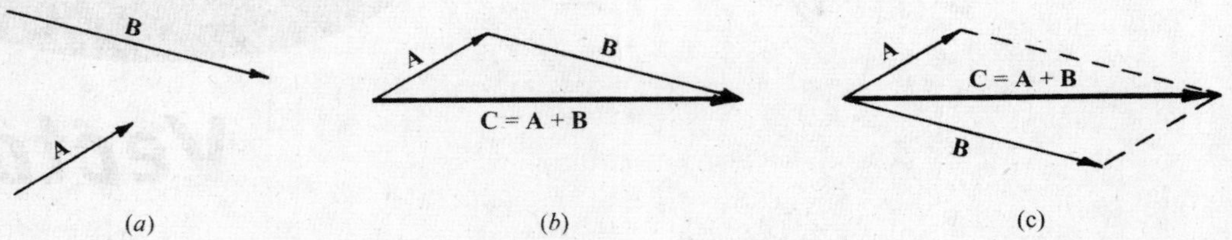

Figure 7.3

Extensions to sums of more than two vectors are immediate. For example, Figure 7.4 shows how to obtain the sum or resultant **E** of the vectors **A**, **B**, **C**, and **D**.

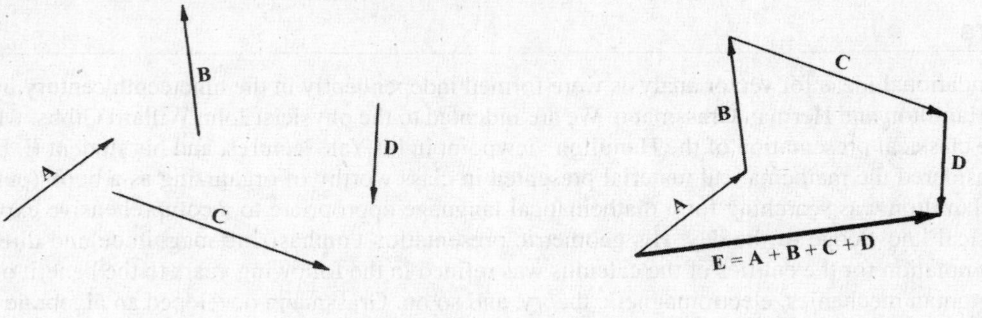

Figure 7.4

4. The *difference* of vectors **A** and **B**, represented by **A** – **B**, is that vector **C** which added to **B** gives **A**. Equivalently, **A** – **B** may be defined as **A** + (–**B**). If **A** = **B**, then **A** – **B** is defined as the *null* or *zero vector* and is represented by the symbol **0**. This has a magnitude of zero, but its direction is not defined.

   The expression of vector equations and related concepts is facilitated by the use of real numbers and functions. In this context, these are called *scalars*. This special designation arises from application where the scalars represent objects that do not have direction, such as mass, length, and temperature.

5. Multiplication of a vector **A** by a scalar m produces a vector $m\mathbf{A}$ with magnitude $|m|$ times the magnitude of **A** and direction the same as or opposite to that of **A** according as m is positive or negative. If $m = 0$, $m\mathbf{A} = 0$, the null vector.

Algebraic Properties of Vectors

The following algebraic properties are consequences of the geometric definition of a vector. (See Problems 7.1 and 7.2.)

If **A**, **B**, and **C** are vectors, and m and n are scalars, then

1.	$\mathbf{A} + \mathbf{B} = \mathbf{B} + \mathbf{A}$	Commutative law for addition
2.	$\mathbf{A} + (\mathbf{B} + \mathbf{C}) = (\mathbf{A} + \mathbf{B}) + \mathbf{C}$	Associative law for addition
3.	$m(n\mathbf{A}) = (mn)\mathbf{A} = n(m\mathbf{A})$	Associative law for multiplication
4.	$(m + n)\mathbf{A} = m\mathbf{A} + n\mathbf{A}$	Distributive law
5.	$m(\mathbf{A} + \mathbf{B}) = m\mathbf{A} + m\mathbf{B}$	Distributive law

Note that in these laws only multiplication of a vector by one or more scalars is defined. On Pages 164 and 165 we define products of vectors.

Linear Independence and Linear Dependence of a Set of Vectors

That a set of vectors $\mathbf{A}_1, \mathbf{A}_2, \ldots, \mathbf{A}_p$ is linearly independent means that $a_1\mathbf{A}_1 + a_2\mathbf{A}_2 + \cdots a_p\mathbf{A}_p = \mathbf{0}$ if and only if $a_1 = a_2 = \ldots = a_p = 0$ (i.e., the algebraic sum is zero if and only if all the coefficients are zero). The set of vectors is linearly dependent when it is not linearly independent.

Unit Vectors

Unit vectors are vectors having unit length. If $\mathbf{A}$ is any vector with magnitude $A = |\mathbf{A}| > 0$, then $\mathbf{A}/|\mathbf{A}|$ is a unit vector. If $\mathbf{a}$ is a unit vector with the same direction and sense as $\mathbf{A}$, then $\mathbf{a} = \mathbf{A}/|\mathbf{A}|$.

Rectangular (Orthogonal) Unit Vectors

The rectangular unit vectors **i**, **j**, and **k** are unit vectors having the direction of the positive x, y, and z axes of a rectangular coordinate system (see Figure 7.5). The triple **i**, **j**, **k** is said to be a *basis* of the collection of vectors. We use right-handed rectangular coordinate systems unless otherwise specified. Such systems derive their name from the fact that a right-threaded screw rotated through 90° from Ox to Oy will advance in the positive z direction. In general, three vectors **A**, **B**, and **C** which have coincident initial points and are not coplanar are said to form a *right-handed system* or *dextral system* if a right-threaded screw rotated through an angle less than 180° from **A** to **B** will advance in the direction **C** (see Figure 7.6).

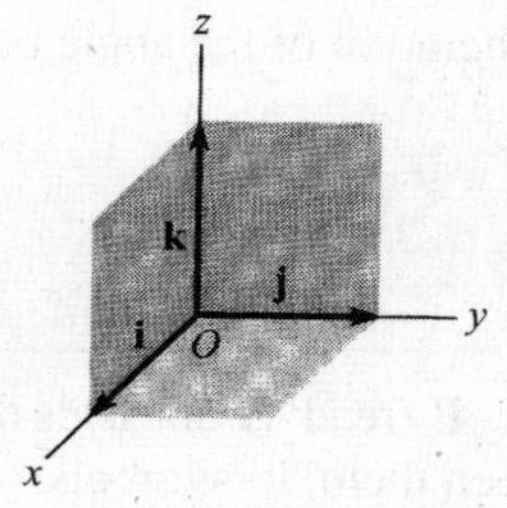

Figure 7.5

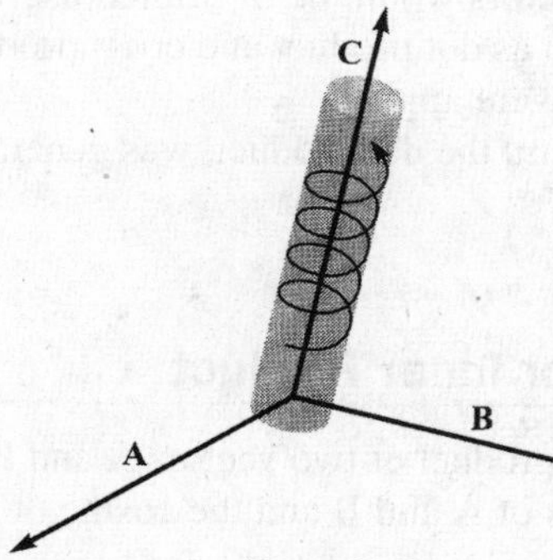

Figure 7.6

Components of a Vector

Any vector **A** in three dimensions can be represented with initial point at the origin O of a rectangular coordinate system (see Figure 7.7). Let (A_1, A_2, A_3) be the rectangular coordinates of the terminal point of vector **A** with initial point at O. The vectors $A_1\mathbf{i}$, $A_2\mathbf{j}$, and $A_3\mathbf{k}$ are called the *rectangular component vectors*, or simply *component vectors*, of **A** in the x, y, and z directions, respectively. A_1, A_2, and A_3 are called the *rectangular components*, or simply *components*, of **A** in the x, y, and z directions, respectively.

The vectors of the set $\{\mathbf{i}, \mathbf{j}, \mathbf{k}\}$ are perpendicular to one another, and they are unit vectors. The words *orthogonal* and *normal*, respectively, are used to describe these characteristics; hence, the set is what is called an *orthonormal basis*.

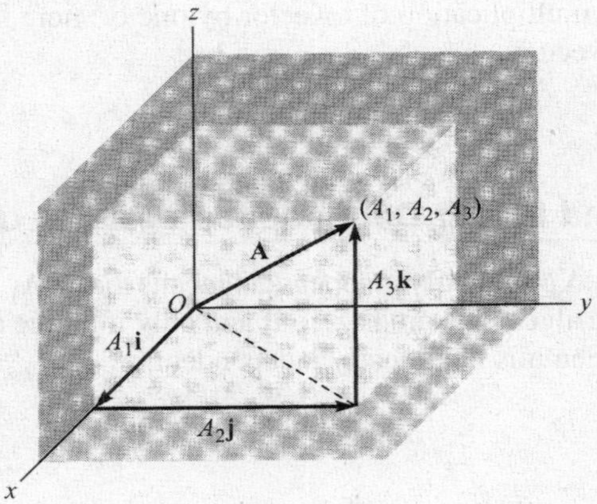

Figure 7.7

It is easily shown to be linearly independent. In an *n*-dimensional space, any set of *n* linearly independent vectors is a basis. The further characteristic of a basis is that any vector of the space can be expressed through it. It is the basis representation that provides the link between the geometric and the algebraic expressions of vectors and vector concepts.

The sum or resultant of $A_1\mathbf{i}$, $A_2\,\mathbf{j}$, and $A_3\mathbf{k}$ is the vector **A**, so that we can write

$$\mathbf{A} = A_1\mathbf{i} + A_2\mathbf{j} + A_3\mathbf{k} \tag{1}$$

The magnitude of **A** is

$$A = |\mathbf{A}| = \sqrt{A_1^2 + A_2^2 + A_3^2} \tag{2}$$

In particular, the *position vector* or *radius vector* **r** from *O* to the point (*x, y, z* is written

$$\mathbf{r} = x\mathbf{i} + y\mathbf{j} + z\mathbf{k} \tag{3}$$

and has magnitude $r = |\mathbf{r}| = \sqrt{x^2 + y^2 + z^2}$.

A theory of vectors would be of limited use without a process of multiplication. In fact, two binary processes, designated as dot product and cross product, were created to meet the geometric and physical needs to which vectors were applied.

The first of them, the dot product, was generated from consideration of the angle between two vectors.

Dot, Scalar, or Inner Product

The dot or scalar product of two vectors **A** and **B**, denoted by $\mathbf{A} \cdot \mathbf{B}$ (read: **A** dot **B**) is defined as the product of the magnitudes of **A** and **B** and the cosine of the angle between them. In symbols,

$$\mathbf{A} \cdot \mathbf{B} = AB \cos \theta, \qquad 0 \leqq \theta \leqq \pi \tag{4}$$

Assuming that neither **A** nor **B** is the zero vector, an immediate consequence of the definition is that $\mathbf{A} \cdot \mathbf{B} = 0$ if and only if **A** and **B** are perpendicular. Note that $\mathbf{A} \cdot \mathbf{B}$ is a scalar and not a vector.

The following laws are valid:

1. $\mathbf{A} \cdot \mathbf{B} = \mathbf{B} \cdot \mathbf{A}$ — Commutative law for dot products
2. $\mathbf{A} \cdot (\mathbf{B} + \mathbf{C}) = \mathbf{A} \cdot \mathbf{B} + \mathbf{A} \cdot \mathbf{C}$ — Distributive Law
3. $m(\mathbf{A} \cdot \mathbf{B}) = (m\mathbf{A}) \cdot \mathbf{B} = \mathbf{A} \cdot (m\mathbf{B}) = (\mathbf{A} \cdot \mathbf{B})m$, where *m* is a scalar
4. $\mathbf{i} \cdot \mathbf{i} = \mathbf{j} \cdot \mathbf{j} = \mathbf{k} \cdot \mathbf{k} = 1, \mathbf{i} \cdot \mathbf{j} = \mathbf{j} \cdot \mathbf{k} = \mathbf{k} \cdot \mathbf{i} = 0$
5. If $\mathbf{A} = A_1\mathbf{i} + A_2\mathbf{j} + A_3\mathbf{k}$ and $\mathbf{B} = B_1\mathbf{i} + B_2\mathbf{j} + B_3\mathbf{k}$, then $\mathbf{A} \cdot \mathbf{B} = A_1B_1 + A_2B_2 + A_3B_3$

The equivalence of this component form the dot product with the geometric definition 4 following from the law of cosines. (See Figure 7.8).

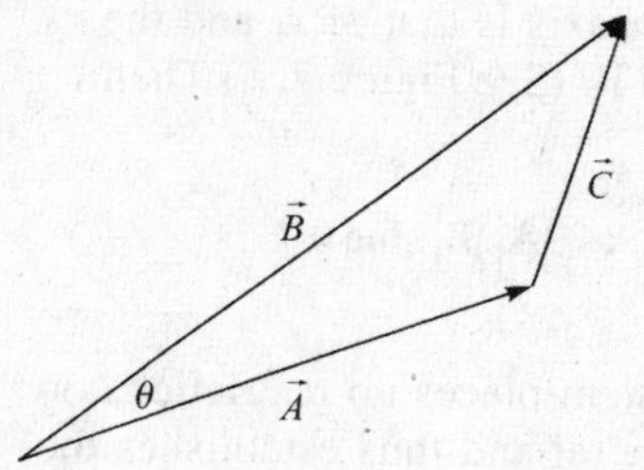

Figure 7.8

In particular,

$$|\mathbf{C}|^2 = |\mathrm{A}|^2 + |\mathbf{B}|^2 - 2|\mathbf{A}|\,|\mathbf{B}| \cos\theta$$

Since **C** = **B** – **A**, its components are $B_1 - A_1$, $B_2 - A_2$, $B_3 - A_3$ and the square of its magnitude is

$$|\mathbf{B}|^2 + |\mathbf{A}|^2 - 2(A_1B_1 + A_2B_2 + A_3B_3$$

When this representation for $|C^2|$ is placed in the original equation and cancellations are made, we obtain

$$A_1B_1 + A_2B_2 + A_3B_3 = |\mathbf{A}|\,|\mathbf{B}| \cos\theta.$$

The second form of vector multiplication—that is, the cross product—emerged from Hamilton's theory of quaternions (1844). Algebraically, the cross product is an example of a noncommutative operation. Geometrically, it generates a vector perpendicular to the initial pair of vectors, and its physical value is illustrated in electromagnetic theory, where it aids in the representation of a magnetic field perpendicular to the direction of an electric current.

Cross or Vector Product

The cross or vector product of **A** and **B** is a vector **C** = **A** × **B** (read: **A** cross **B**). The magnitude of **A** × **B** is defined as the product of the magnitudes of **A** and **B** and the sine of the angle between them. The direction of the vector **C** = **A** × **B** is perpendicular to the plane of **A** and **B**, and such that **A**, **B**, and **C** form a right-handed system. In symbols,

$$\mathbf{A} \times \mathbf{B} = AB \sin\theta\mathbf{u},\ 0 \leq \theta \leq \pi \tag{5}$$

where **u** is a unit vector indicating the direction of **A** × **B**. If **A** = **B** or if **A** is parallel to **B**, then $\sin\theta = 0$ and **A** × **B** = 0.

The following laws are valid:

1. $\mathbf{A} \times \mathbf{B} = -\mathbf{B} \times \mathbf{A}$ (Commutative law for cross products fails)
2. $\mathbf{A} \times (\mathbf{B} + \mathbf{C}) = \mathbf{A} \times \mathbf{B} + \mathbf{A} \times \mathbf{C}$ Distributive Law
3. $m(\mathbf{A} \times \mathbf{B}) = (m\mathbf{A}) \times \mathbf{B} = \mathbf{A} \times (m\mathbf{B}) = (\mathbf{A} \times \mathbf{B})m$, where m is a scalar

Also, the following consequences of the definition are important:

4. $\mathbf{i} \times \mathbf{i} = \mathbf{j} \times \mathbf{j} = \mathbf{k} \times \mathbf{k} = \mathbf{0}, \mathbf{i} \times \mathbf{j} = \mathbf{k}, \mathbf{j} \times \mathbf{k} = \mathbf{i}, \mathbf{k} \times \mathbf{i} = \mathbf{j}$
5. If $\mathbf{A} = A_1\mathbf{i} + A_2\mathbf{j} + A_3\mathbf{k}$ and $\mathbf{B} = B_1\mathbf{i} + B_1\mathbf{i} + B_2\mathbf{j} + B_3\mathbf{k}$, then

$$\mathrm{A} \times \mathrm{B} = \begin{vmatrix} \mathrm{i} & \mathrm{j} & \mathrm{k} \\ A_1 & A_2 & A_3 \\ B_1 & B_2 & B_3 \end{vmatrix}$$

The equivalence of this component representation (5) and the geometric definition may be seen as follows. Choose a coodinate system such that the direction of the x-axis is that of **A** and the xy plane is the plane of the vectors **A** and **B**. (See Figure 7.9.) Then

$$\mathbf{A}\times\mathbf{B} = \begin{vmatrix} \mathrm{i} & \mathrm{j} & \mathrm{k} \\ A_1 & 0 & 0 \\ B_1 & B_2 & 0 \end{vmatrix} = A_1B_2\mathrm{k} = |\mathbf{A}|\,|\mathbf{B}| \text{ sine } \theta\ k$$

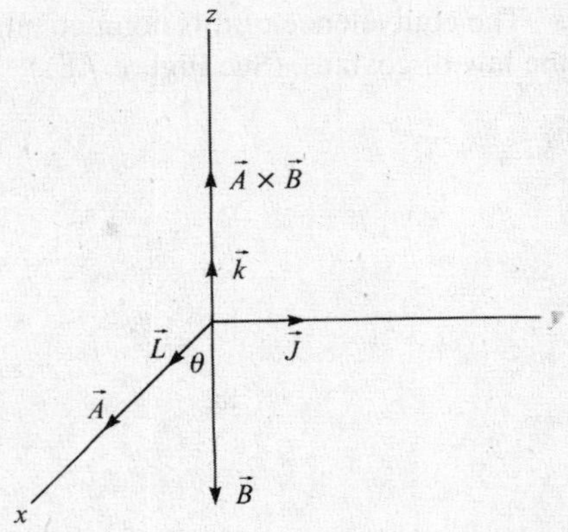

Figure 7.9

Since this choice of coordinate system places no restrictions on the vectors **A** and **B**, the result is general and thus establishes the equivalence.

6. $|\mathbf{A}\times\mathbf{B}|$ = the area of a parallelogram with sides **A** and **B**.
7. If $\mathbf{A}\times\mathbf{B} = 0$ and neither **A** nor **B** is a null vector, then **A** and **B** are parallel.

Triple Products

Dot and cross multiplication of three vectors, **A**, **B**, and **C** may produce meaningful products of the form $(\mathbf{A}\cdot\mathbf{B})\mathbf{C}$, $\mathbf{A}\cdot(\mathbf{B}\times\mathbf{C})$, and $\mathbf{A}\times(\mathbf{B}\times\mathbf{C})$. The following laws are valid:

1. $(\mathbf{A}\cdot\mathbf{B})\mathbf{C} \neq \mathbf{A}(\mathbf{B}\cdot\mathbf{C})$ in general
2. $\mathbf{A}\cdot(\mathbf{B}\times\mathbf{C}) = \mathbf{B}\cdot(\mathbf{C}\times\mathbf{A}) = \mathbf{C}\cdot(\mathbf{A}\times\mathbf{B})$ = volume of a parallelepiped having **A**, **B**, and **C** as edges, or the negative of this volume according as **A**, **B**, and **C** do or do not form a right-handed handed system. If $\mathbf{A} = A_1\mathbf{i} + A_2\mathbf{j} + A_3\mathbf{k}$, $\mathbf{B} = B_1\mathbf{i} + B_2\mathbf{j} + B_3\mathbf{k}$ and $\mathbf{C} = C_1\mathbf{i} + C_2\mathbf{j} + C_3\mathbf{k}$, then

$$\mathbf{A}\cdot(\mathbf{B}\times\mathbf{C}) = \begin{vmatrix} A_1 & A_2 & A_3 \\ B_1 & B_2 & B_3 \\ C_1 & C_2 & C_3 \end{vmatrix} \tag{6}$$

3. $\mathbf{A}\times(\mathbf{B}\times\mathbf{C}) \neq (\mathbf{A}\times\mathbf{B})\times\mathbf{C}$ (Associative law for cross products fails)
4. $\mathbf{A}\times(\mathbf{B}\times\mathbf{C}) = (\mathbf{A}\cdot\mathbf{C})\mathbf{B} - (\mathbf{A}\cdot\mathbf{B})\mathbf{C}$

   $(\mathbf{A}\times\mathbf{B})\times\mathbf{C} = (\mathbf{A}\cdot\mathbf{C})\mathbf{B} - (\mathbf{B}\cdot\mathbf{C})\mathbf{A}$

The product $\mathbf{A}\cdot(\mathbf{B}\times\mathbf{C})$ is called the *scalar triple product* or *box product* and may be denoted by [**ABC**]. The product $\mathbf{A}\times(\mathbf{B}\times\mathbf{C})$ is called the *vector triple product.*

In $\mathbf{A}\cdot(\mathbf{B}\times\mathbf{C})$ parentheses are sometimes omitted and we write $\mathbf{A}\cdot\mathbf{B}\times\mathbf{C}$. However, parentheses must be used in $\mathbf{A}\times(\mathbf{B}\times\mathbf{C})$ (see Problem 7.29). Note that $\mathbf{A}\cdot(\mathbf{B}\times\mathbf{C}) = (\mathbf{A}\times\mathbf{B})\cdot\mathbf{C}$. This is often expressed by stating that in a scalar triple product the dot and the cross can be interchanged without affecting the result (see Problem 7.26).

Axiomatic Approach to Vector Analysis

From the preceding remarks it is seen that a vector $\mathbf{r} = x\mathbf{i} + y\mathbf{j} + z\mathbf{k}$ is determined when its three components (x, y, z) relative to some coordinate system are known. In adopting an axiomatic approach, it is thus quite natural for us to make the following assumptions.

Definition A three-dimensional vector is an *ordered triplet* of real numbers with the following properties. If $\mathbf{A} = (A_1, A_2, A_3)$ and $\mathbf{B} = (B_1, B_2, B_3)$, then

1. $\mathbf{A} = \mathbf{B}$ if and only if $A_1 = B_1, A_2 = B_2, A_3 = B_3$
2. $\mathbf{A} + \mathbf{B} = (A_1 + B_1, A_2 + B_2, A_3 + B_3)$
3. $\mathbf{A} - \mathbf{B} = (A_1 - B_1, A_2 - B_2, A_3 - B_3)$
4. $\mathbf{0} = (0, 0, 0)$
5. $m\mathbf{A} = m(A_1, A_2, A_3) = (mA_1, mA_2, mA_3)$

In addition, two forms of multiplication are established.

6. $\mathbf{A} \cdot \mathbf{B} = A_1B_1 + A_2B_2 + A_3B_3$
7. Length or magnitude of $\mathrm{A} = |\mathrm{A}| = \sqrt{\mathrm{A} \cdot \mathrm{A}} = \sqrt{A_1^2 + A_2^2 + A_3^2}$
8. $\mathbf{A} \times \mathbf{B} = (A_2B_3 - A_3B_2, A_3B_1 - A_1B_3, A_1B_3, A_1B_2 - A_2B_1)$

Unit vectors are defined to be (1, 0, 0), (0, 1, 0), (0, 0, 1) and then designated by **i**, **j**, **k**, respectively, thereby identifying the components axiomatically introduced with the geometric orthonormal basis elements.

If one wishes, this axiomatic formulation (which provides a component representation for vectors) can be used to reestablish the fundamental laws previously introduced geometrically; however, the primary reason for introducing this approach was to formalize a component representation of the vectors. It is that concept that will be used in the remainder of this chapter.

Note 1 One of the advantages of component representation of vectors is the easy extension of the ideas to all dimensions. In an n-dimensional space, the component representation is

$$\mathbf{A}(A_1, A_2, \ldots, A_n)$$

An exception is the cross product which is specifically restricted to three-dimensional space. There are generalizations of the cross product to higher dimensional spaces, but there is no direct extension.)

Note 2 The geometric interpretation of a vector endows it with an absolute meaning at any point of space. The component representation (as an ordered triple of numbers) in Euclidean three space is not unique; rather, it is attached to the coordinate system employed. This follows because the components are geometrically interpreted as the projections of the arrow representation on the coordinate directions. Therefore, the projections on the axes of a second coordinate system (rotated, for example) from the first one will be different. (See Figure 7.10.) Therefore, for theories where groups of coordinate systems play a role, a more adequate component definition of a vector is as a collection of ordered triples of numbers, each one identified with a coordinate system of the group, and any two related by a coordinate transformation. This viewpoint is indispensable in Newtonian mechanics, electromagnetic theory, special relativity, and so on.

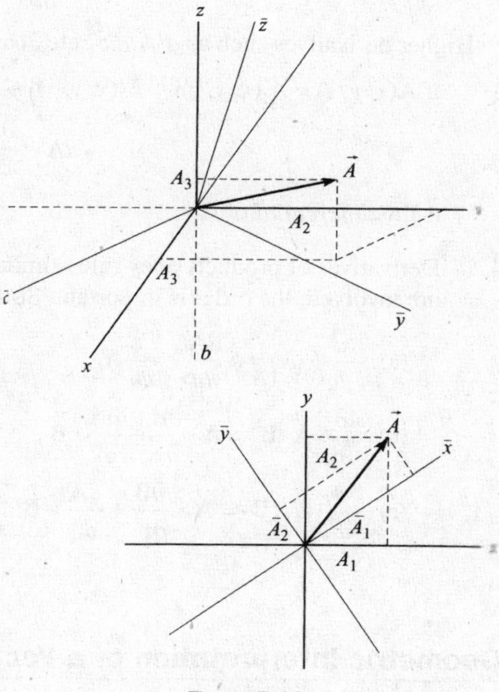

Figure 7.10

Vector Functions

If corresponding to each value of a scalar u we associate a vector $\mathbf{A}$, then $\mathbf{A}$ is called a *function* of u denoted by $\mathbf{A}(u)$. In three dimensions we can write $\mathbf{A}(u) = A_1(u)\mathbf{i} + A_2(u)\mathbf{j} + A_3(u)\mathbf{k}$.

The function concept is easily extended. Thus, if to each point (x, y, z) there corresponds a vector **A**, then **A** is a function of (x, y, z), indicated by $\mathbf{A}(x, y, z) = A_1(x, y, z)\mathbf{i} + A_2(x, y, z)\mathbf{j}, + A_3(x, y, z)\mathbf{k}$.

We sometimes say that a vector function **A** defines a *vector field* since it associates a vector with each point of a region. Similarly, $\phi(x, y, z)$ defines a *scalar field* since it associates a scalar with each point of a region.

Limits, Continuity, and Derivatives of Vector Functions

Limits, continuity, and derivatives of vector functions follow rules similar to those for scalar functions already considered. The following statements show the analogy which exists.

1. The vector function represented by $\mathbf{A}(u)$ is said to be *continuous* at u_0 if, given any positive number δ, we can find some positive number δ such that $|\mathbf{A}(u) - \mathbf{A}(u_0)| < \delta$ whenever $|u - u_0| < \delta$. This is equivalent to the statement $\lim_{u\to u_0} \mathbf{A}(u) = \mathbf{A}(u_0)$.
2. The derivative of $\mathbf{A}(u)$ is defined as

$$\frac{d\mathbf{A}}{du} = \lim_{\Delta u\to 0} \frac{\mathbf{A}(u+\Delta u) - \mathbf{A}(u)}{\Delta u}$$

provided this limit exists. In case $\mathbf{A}(u) = A_1(u)\mathbf{i} + \mathbf{A}_2\, A_2(u)\mathbf{j} + A_3(u)\mathbf{k}$; then

$$\frac{d\mathbf{A}}{du} = \frac{dA_1}{du}\mathbf{i} + \frac{dA_2}{du}\mathbf{j} + \frac{dA_3}{du}\mathbf{k}$$

Higher derivatives such as $d^2\mathbf{A}/du^2$, etc., can be similarly defined.

3. If $\mathbf{A}(x, y, z) = A_1(x, y, z)\mathbf{i} + A_2(x, y, z)\mathbf{j} + A_3(x, y, z)\mathbf{k}$; then

$$d\mathbf{A} = \frac{\partial \mathbf{A}}{\partial x}dx + \frac{\partial \mathbf{A}}{\partial y}dy + \frac{\partial \mathbf{A}}{\partial z}dz$$

is the *differential* of **A**.

4. Derivatives of products obey rules similar to those for scalar functions. However, when cross products are involved, the order is important. Some examples are

(a) $\frac{d}{du}(\phi\mathbf{A}) = \phi\frac{d\mathbf{A}}{du} + \frac{d\phi}{du}\mathbf{A}$.

(b) $\frac{\partial}{\partial y}(\mathbf{A}\cdot\mathbf{B}) = \mathbf{A}\cdot\frac{\partial \mathbf{B}}{\partial y} + \frac{\partial \mathbf{A}}{\partial y}\cdot\mathbf{B}$

(c) $\frac{\partial}{\partial z}(\mathbf{A}\times\mathbf{B}) = \mathbf{A}\times\frac{\partial \mathbf{B}}{\partial z} + \frac{\partial \mathbf{A}}{\partial z}\times\mathbf{B}$ (maintain the order of **A** and **B**)

Geometric Interpretation of a Vector Derivative

If **r** is the vector joining the origin O of a coordinate system and the point (x, y, z), then specification of the vector function $\mathbf{r}(u)$ defines x, y, and z as functions of u (**r** is called a *position vector*). As u changes, the terminal point of **r** describes a *space curve* (see Figure 7.11) having parametric equations $x = x(u)$, $y = y(u)$, $z = z(u)$. If the parameter u is the are length s measured from some fixed point on the curve, then recall from the discussion of arc length that $ds^2 = d\mathbf{r}\cdot d\mathbf{r}$. Thus,

$$\frac{d\mathbf{r}}{ds} = \mathbf{T} \tag{7}$$

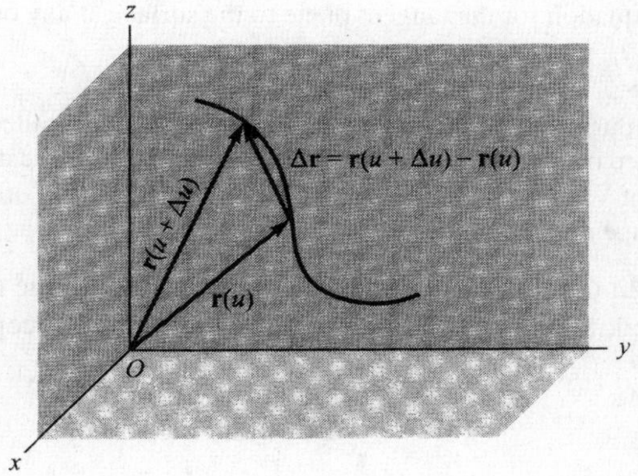

Figure 7.11

is a unit vector in the direction of the tangent to the curve and is called the *unit tangent vector*. If u is the time t, then

$$\frac{d\mathbf{r}}{dt} = \mathbf{v} \tag{8}$$

is the *velocity* with which the terminal point of **r** describes the curve. We have

$$\mathbf{v} = \frac{d\mathbf{r}}{dt} = \frac{d\mathbf{r}}{ds}\frac{ds}{dt} = \frac{ds}{dt}\mathbf{T} = \upsilon\mathbf{T} \tag{9}$$

from which we see that the magnitude of **v** is $\upsilon = ds/dt$. Similarly,

$$\frac{d^2\mathbf{r}}{dt^2} = \mathbf{a} \tag{10}$$

is the *acceleration* with which the terminal point of **r** describes the curve. These concepts have important applications in *mechanics* and *differential geometry*.

A primary objective of vector calculus is to express concepts in an intuitive and compact form. Success is nowhere more apparent than in applications involving the partial differentiation of scalar and vector fields. [Illustrations of such fields include implicit surface representation $\Phi\{x, y, z(x, y) = 0$, the electromagnetic potential function $\Phi(x, y, z)$, and the electromagnetic vector field $\mathbf{F}(x, y, z)$.] To give mathematics the capability of addressing theories involving such functions, William Rowan Hamilton and others of the nineteenth century introduced derivative concepts called *gradient, divergence*, and *curl*, and then developed an analytic structure around them.

An intuitive understanding of these entities begins with examination of the differential of a scalar field, i.e.,

$$d\Phi = \frac{\partial\Phi}{\partial x}dx + \frac{\partial\Phi}{dy}dy + \frac{\partial\Phi}{\partial z}dz$$

Now suppose the function Φ is constant on a surface S and that C; $x = f_1(t)$, $y = f_2(t)$, $z = f_3(t)$ is a curve on S. At any point of this curve, $\frac{dr}{dt} = \frac{dx}{dt}\mathbf{i} + \frac{dx}{dt}\mathbf{j} + \frac{dz}{dt}\mathbf{k}$ lies in the tangent plane to the surface. Since this statement is true for every surface curve through a given point, the differential $d\mathbf{r}$ spans the tangent plane. Thus, the triple $\frac{\partial\Phi}{\partial x}, \frac{\partial\Phi}{\partial y}, \frac{\partial\Phi}{\partial z}$ represents a vector perpendicular to S. With this special geometric characteristic in mind we define

$$\nabla\Phi = \frac{\partial\Phi}{\partial x}\mathbf{i} + \frac{\partial\Phi}{\partial y}\mathbf{j} + \frac{\partial\Phi}{\partial z}\mathbf{k}$$

to be the *gradient of the scalar field* Φ.

Furthermore, we give the symbol ∇ a special significance by naming it *del*.

EXAMPLE 1. f $\Phi(x, y, z) = 0$ is an implicitly defined surface, then, because the function always has the value zero for points on it, the condition of constancy is satisfied and $\nabla\phi$ is normal to the surface at any of its points.

This allows us to form an equation for the tangent plane to the surface at any one of its points. See Problem 7.36.

EXAMPLE 2. For certain purposes, surfaces on which Φ is constant are called *level surfaces*. In meteorology, surfaces of equal temperature or of equal atmospheric pressure fall into this category. From the previous development, opment, we see that $\nabla\Phi$ is perpendicular to the level surface at any one of its points and, hence, has the direction of maximum change at that point.

The introduction of the vector operator ∇ and the interaction of it with the multiplicative properties of dot and cross come to mind. Indeed, this line of thought does lead to new concepts called *divergence* and *curl*. A summary follows.

Gradient, Divergence, and Curl

Consider the vector operator ∇ (*del*) defined by

$$\nabla \equiv \mathbf{i}\frac{\partial}{\partial x} + \mathbf{j}\frac{\partial}{\partial y} + \mathbf{k}\frac{\partial}{\partial z} \tag{11}$$

Then if $\phi(x, y, z)$ and $\mathbf{A}(x, y, z)$ have continuous first partial derivatives in a region (a condition which is in many cases stronger than necessary), we can define the following.

1. **Gradient.** The *gradient* of ϕ is defined by

$$\begin{aligned}\text{grad}\,\phi = \nabla\phi &= \left(\mathbf{i}\frac{\partial}{\partial x} + \mathbf{j}\frac{\partial}{\partial y} + \mathbf{k}\frac{\partial}{\partial z}\right)\phi = \mathbf{i}\frac{\partial\phi}{\partial x} + \mathbf{j}\frac{\partial\phi}{\partial y} + \mathbf{k}\frac{\partial\phi}{\partial z} \\ &= \frac{\partial\phi}{\partial x}\mathbf{i} + \frac{\partial\phi}{\partial y}\mathbf{j} + \frac{\partial\phi}{\partial z}\mathbf{k}\end{aligned} \tag{12}$$

2. **Divergence.** The *divergence* of $\mathbf{A}$ is defined by

$$\begin{aligned}\text{div}\,\mathbf{A} = \nabla\cdot\mathbf{A} &= \left(\mathbf{i}\frac{\partial}{\partial x} + \mathbf{j}\frac{\partial}{\partial y} + \mathbf{k}\frac{\partial}{\partial z}\right)\cdot(A_1\mathbf{i} + A_2\mathbf{j} + A_3\mathbf{k}) \\ &= \frac{\partial A_1}{\partial x} + \frac{\partial A_2}{\partial y} + \frac{\partial A_3}{\partial z}\end{aligned} \tag{13}$$

3. **Curl.** The *curl* of $\mathbf{A}$ is defined by

$$\begin{aligned}\text{curl}\,\mathbf{A} = \nabla\times\mathbf{A} &= \left(\mathbf{i}\frac{\partial}{\partial x} + \mathbf{j}\frac{\partial}{\partial y} + \mathbf{k}\frac{\partial}{\partial z}\right)\times(A_1\mathbf{i} + A_2\mathbf{j} + A_3\mathbf{k}) \\ &= \begin{vmatrix} \mathbf{i} & \mathbf{j} & \mathbf{k} \\ \frac{\partial}{\partial x} & \frac{\partial}{\partial y} & \frac{\partial}{\partial z} \\ A_1 & A_2 & A_3 \end{vmatrix} \\ &= \mathbf{i}\begin{vmatrix} \frac{\partial}{\partial y} & \frac{\partial}{\partial z} \\ A_2 & A_3 \end{vmatrix} - \mathbf{j}\begin{vmatrix} \frac{\partial}{\partial x} & \frac{\partial}{\partial z} \\ A_1 & A_2 \end{vmatrix} + \mathbf{k}\begin{vmatrix} \frac{\partial}{\partial x} & \frac{\partial}{\partial y} \\ A_1 & A_2 \end{vmatrix} \\ &= \left(\frac{\partial A_3}{\partial y} - \frac{\partial A_2}{\partial z}\right)\mathbf{i} + \left(\frac{\partial A_1}{\partial z} - \frac{\partial A_3}{\partial x}\right)\mathbf{j} + \left(\frac{\partial A_2}{\partial x} - \frac{\partial A_1}{\partial y}\right)\mathbf{k}\end{aligned}$$

Note that in the expansion of the determinant, the operators $\partial/\partial x$, $\partial/\partial y$, $\partial/\partial z$ must precede A_1, A_2, A_3. In other words, ∇ is a vector operator, not a vector. When employing it, the laws of vector algebra either do not apply or at the very least must be validated. In particular, $\nabla \times \mathbf{A}$ is a new vector obtained by the specified partial differentiation on $\mathbf{A}$, while $\mathbf{A} \times \nabla$ is an operator waiting to act upon a vector or a scalar.

Formulas Involving ∇

If the partial derivatives of $\mathbf{A}$, $\mathbf{B}$, U, and V are assumed to exist, then

1. $\nabla(U + V) = \nabla U + \nabla V$ or grad $(U + V)$ = grad u + grad V
2. $\nabla \cdot (\mathbf{A} + \mathbf{B}) = \nabla \cdot \mathbf{A} + \nabla \cdot \mathbf{B}$ or div $(\mathbf{A} + \mathbf{B})$ + div $\mathbf{A}$ + div $\mathbf{B}$
3. $\nabla \times (\mathbf{A} + \mathbf{B}) = \nabla \times \mathbf{A} + \nabla \times \mathbf{B}$ or curl $(\mathbf{A} + \mathbf{B})$ = curl $\mathbf{A}$ + curl $\mathbf{B}$
4. $\nabla \cdot (U\mathbf{A}) = (\nabla U) \cdot \mathbf{A} + U(\nabla \cdot \mathbf{A})$
5. $\nabla \times (U\mathbf{A}) = (\nabla U) \times \mathbf{A} + U(\nabla \times \mathbf{A})$
6. $\nabla \cdot (\mathbf{A} \times \mathbf{B}) = \mathbf{B} \cdot (\nabla \times \mathbf{A}) - \mathbf{A} \cdot (\nabla \times \mathbf{B})$
7. $\nabla \times (\mathbf{A} \times \mathbf{B}) = (\mathbf{B} \cdot \nabla)\mathbf{A} - \mathbf{B}(\nabla \cdot \mathbf{A}) - (\mathbf{A} \cdot \nabla)\mathbf{B} + \mathbf{A}(\nabla \cdot \mathbf{B})$
8. $\nabla(\mathbf{A} \cdot \mathbf{B}) = (\mathbf{B} \cdot \nabla)\mathbf{A} + (\mathbf{A} \cdot \nabla)\mathbf{B} + \mathbf{B} \times (\nabla \times \mathbf{A}) + \mathbf{A} \times (\nabla \times \mathbf{B})$
9. $\nabla.(\nabla U) \equiv \nabla^2 U \equiv \dfrac{\partial^2 U}{\partial x^2} + \dfrac{\partial^2 U}{\partial y^2} + \dfrac{\partial^2 U}{\partial z^2}$ is called the *Laplacian* of U.

   and $\nabla^2 \equiv \dfrac{\partial^2}{\partial x^2} + \dfrac{\partial^2}{\partial y^2} + \dfrac{\partial^2}{\partial z^2}$ is called the *Lapacian operator*.
10. $\nabla \times (\nabla U) = 0$. The curl of the gradient of U is zero.
11. $\nabla \cdot (\nabla \times \mathbf{A}) = 0$. The divergence of the curl of $\mathbf{A}$ is zero.
12. $\nabla \times (\nabla \times \mathbf{A}) = \nabla(\nabla \cdot \mathbf{A}) - \nabla^2\mathbf{A}$

Vector Interpretation of Jacobians and Orthogonal Curvilinear Coordinates

The transformation equations

$$x = f(u_1, u_2, u_3), \qquad y = g(u_1, u_2, u_3), \qquad z = h(u_1, u_2, u_3) \tag{15}$$

(where we assume that f, g, h are continuous, have continuous partial derivatives, and have a single-valued inverse) establish a one-to-one correspondence between points in an xyz and $u_1 u_2 u_3$ rectangular coordinate system. In vector notation, the transformation (15) can be written

$$\mathbf{r} = x\mathbf{i} + y\mathbf{j} + z\mathbf{k} = f(u_1, u_2, u_3)\mathbf{i} + g(u_1, u_2, u_3)\mathbf{j} + h(u_1, u_2, u_3)\mathbf{k} \tag{16}$$

A point P in Figure 7.12 can then be defined not only by *rectangular coordinates* (x, y, z) but by coordinates (u_1, u_2, u_3) as well. We call (u_1, u_2, u_3) the *curvilinear coordinates* of the point.

If u_2 and u_3 are constant, then as u_1 varies, $\mathbf{r}$ describes a curve which we call the u_1 *coordinate curve*. Similarly, we define the u_2 and u_3 coordinate curves through P.

From Equation (16), we have

$$d\mathbf{r} = \frac{\partial \mathbf{r}}{\partial u_1} du_1 + \frac{\partial \mathbf{r}}{\partial u_2} du_2 + \frac{\partial \mathbf{r}}{\partial u_3} du_3 \tag{17}$$

The collection of vectors $\dfrac{\partial \mathbf{r}}{\partial u_1}, \dfrac{\partial \mathbf{r}}{\partial u_2}, \dfrac{\partial \mathbf{r}}{\partial u_3}$ is a basis for the vector structure associated with the curvilinear system. If the cur-

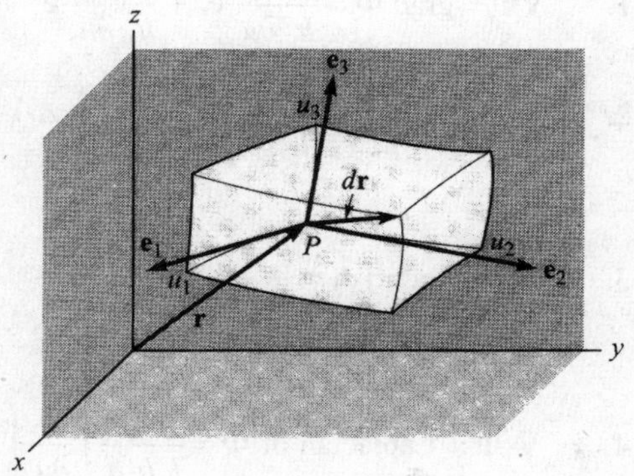

Figure 7.12

vilinear system is orthogonal, then so is this set; however, in general, the vectors are not unit vectors. he differential form for are length may be written

$$ds^2 = g_{11}(du_1)^2 + g_{22}(du_2)^2 + g_{33}(du_3)^2$$

where

$$g_{11} = \frac{\partial \mathbf{r}}{\partial u_1} \cdot \frac{\partial \mathbf{r}}{\partial u_1}, \quad g_{22} = \frac{\partial \mathbf{r}}{\partial u_2} \cdot \frac{\partial \mathbf{r}}{\partial u_2}, \quad g_{33} = \frac{\partial \mathbf{r}}{\partial u_3} \cdot \frac{\partial \mathbf{r}}{\partial u_3}$$

The vector $\partial \mathbf{r}/\partial u_1$ is tangent to the u_1 coordinate curve at P. If $\mathbf{e}_1$ is a unit vector at P in this direction, we can write $\partial \mathbf{r}/\partial u_1 = h_1 \mathbf{e}_1$ where $h_1 = |\partial \mathbf{r}/\partial u_1|$. Similarly, we can write $\partial \mathbf{r}/\partial u_2 = h_2 \mathbf{e}_2$ and $\partial \mathbf{r}/\partial u_3 = h_3 \mathbf{e}_3$, where $h_2 = |\partial \mathbf{r}/\partial u_2|$ and $h_3 = |\partial \mathbf{r}/\partial u_3|$, respectively. Then Equation (17) can be written

$$d\mathbf{r} = h_1\, du_1 \mathbf{e}_1 + h_2\, du_2\, \mathbf{e}_2 + h_3\, du_3\, \mathbf{e}_3 \tag{18}$$

The quantities h_1, h_2, h_3 are sometimes called *scale factors*.

If $\mathbf{e}_1$, $\mathbf{e}_2$, $\mathbf{e}_3$ are mutually perpendicular at any point P, the curvilinear coordinates are called *orthogonal*. Since the basis elements are unit vectors as well as orthogonal, this is an orthonormal basis. In such case the element of arc length ds is given by

$$ds^2 = d\mathbf{r} \cdot d\mathbf{r} = h^2{}_1\, du^2{}_1 + h^2{}_2\, du^2{}_2 + h^2{}_3\, du^2{}_3 \tag{19}$$

and corresponds to the square of the length of the diagonal in the preceding parallelepiped.

Also, in the case of othogonal coordinates referred to the orthonormal basis $\mathbf{e}_1$, $\mathbf{e}_2$, $\mathbf{e}_3$, the volume of the parallelepiped is given by

$$dV = |g_{jk}|\, du_1\, du_2\, du_3 = |(h_1\, du_1\, \mathbf{e}_1) \cdot (h_2\, du_2\, \mathbf{e}_2) \times (h_3\, du_3\, \mathbf{e}_3)| = h_1 h_2 h_3\, du_1\, du_2\, du_3 \tag{20}$$

which can be written as

$$dV = \left| \frac{\partial \mathbf{r}}{\partial u_1} \cdot \frac{\partial \mathbf{r}}{\partial u_2} \times \frac{\partial \mathbf{r}}{\partial u_3} \right| du_1 du_2 du_3 = \left| \frac{\partial(x, y, z)}{\partial(u_1, u_2, u_3)} \right| du_1 du_2 du_3 \tag{21}$$

where $\partial(x, y, z)/\partial(u_1, u_2, u_3)$ is the *Jacobian* of the transformation.

It is clear that when the Jacobian vanishes there is no parallelepiped and this explains geometrically the significance of the vanishing of a Jacobian as treated in Chapter 6.

Note: The further significance of the Jacobian vanishing is that the transformation degenerates at the point.

Gradient Divergence, Curl, and Laplacian in Orthogonal Curvilinear Coordinates

If Φ is a scalar function and $\mathbf{A} = A_1\, \mathbf{e}_1 + A_2\, \mathbf{e}_2 + A_3 \mathbf{e}_3$ a vector function of orthogonal curvilinear coordinates u_1, u_2, u_3, we have the following results.

1. $\nabla \Phi = \text{grad } \Phi = \dfrac{1}{h_1} \dfrac{\partial \Phi}{\partial u_1} \mathbf{e}_1 + \dfrac{1}{h_2} \dfrac{\partial \Phi}{\partial u_2} \mathbf{e}_2 + \dfrac{1}{h_3} \dfrac{\partial \Phi}{\partial u_3} \mathbf{e}_3$

2. $\nabla \cdot \mathbf{A} = \text{div } \mathbf{A} = \dfrac{1}{h_1 h_2 h_3} \left[\dfrac{\partial}{\partial u_1} (h_2, h_3 A_1) + \dfrac{\partial}{\partial u_2} (h_3 h_1 A_2) + \dfrac{\partial}{\partial u_3} (h_1 h_2 A_3) \right]$

3. $\nabla \times \mathbf{A} = \text{curl } \mathbf{A} = \dfrac{1}{h_1 h_2 h_3} \begin{vmatrix} h_1 \mathbf{e}_1 & h_2 \mathbf{e}_2 & h_3 \mathbf{e}_3 \\ \dfrac{\partial}{\partial u_1} & \dfrac{\partial}{\partial u_2} & \dfrac{\partial}{\partial u_3} \\ h_1 A_1 & h_2 A_2 & h_3 A_3 \end{vmatrix}$

4. $\nabla^2 \Phi = \text{Laplacian of } \Phi = \dfrac{1}{h_1 h_2 h_3} \left[\dfrac{\partial}{\partial u_1} \left(\dfrac{h_2 h_3}{h_1} \dfrac{\partial \Phi}{\partial u_1} \right) + \dfrac{\partial}{\partial u_2} \left(\dfrac{h_3 h_1}{h_2} \dfrac{\partial \Phi}{\partial u_2} \right) + \dfrac{\partial}{\partial u_3} \left(\dfrac{h_1 h_2}{h_3} \dfrac{\partial \Phi}{\partial u_3} \right) \right]$

These reduce to the usual expressions in rectangular coordinates if we replace (u_1, u_2, u_3) by (x, y, z), in which case $\mathbf{e}_1$, $\mathbf{e}_2$, and $\mathbf{e}_3$ are replaced by $\mathbf{i}$, $\mathbf{j}$, and $\mathbf{k}$ and $h_1 = h_2 = h_3 = 1$.

Special Curvilinear Coordinates

Cylindrical Coordinates (ρ, φ, z) See Figure 7.13.

Transformation equations:

$$x = \rho \cos \phi,\ y = \rho \sin \phi,\ z = z$$

where $\rho \geqq 0$, $0 \leqq \phi < 2\pi$, $-\infty < z < \infty$.

Scale factors: $h_1 = 1$, $h_2 = \rho$, $h_3 = 1$

Element of arc length: $ds^2 = d\rho^2 + \rho^2\, \rho^2\, d\phi^2 + dz^2$

Jacobian: $\dfrac{\partial(x, y, z)}{\partial(\rho, \phi, z)} = \rho$

Element of volume: $dV = \rho\, d\rho\, d\phi\, dz$

Laplacian:

$$\nabla^2 U = \frac{1}{\rho}\frac{\partial}{\partial \rho}\left(\rho \frac{\partial U}{\partial \rho}\right) + \frac{1}{\rho^2} + \frac{\partial^2 U}{\partial \phi^2} + \frac{\partial^2 U}{\partial z^2}$$

$$= \frac{\partial^2 U}{\partial \rho^2} + \frac{1}{\rho}\frac{\partial U}{\partial \rho} + \frac{1}{\rho^2} + \frac{\partial^2 U}{\partial \phi^2} + \frac{\partial^2 U}{\partial z^2}$$

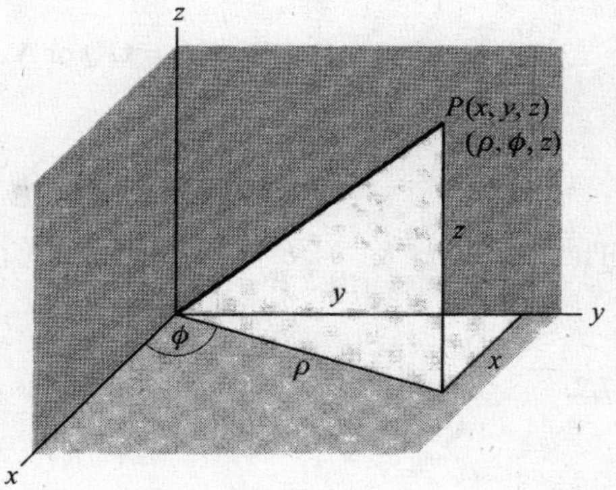

Figure 7.13

Note that corresponding results can be obtained for polar coordinates in the plane by omitting z dependence. In such case, for example, $ds^2 = d\,\rho^2 + \rho^2\, d\phi^2$, while the element of volume is replaced by the element of area, $dA = \rho\, d\,\rho\, d\,\phi$.

Spherical Coordinates (r, θ, φ) See Figure 7.14.

Transformation equations:

$$x = r \sin \theta \cos \phi,\ y = r \sin \theta \sin \phi,\ z = r \cos \theta$$

where $r \geqq 0$, $0 \leqq \theta \leqq \pi$, $0 \leqq \phi < 2\pi$.

Scale factors: $h_1 = 1$, $h_2 = r$, $h_3 = r \sin \theta$

Element of arc length: $ds^2 = dr^2 + r^2\, d\,\theta^2 + r^2 \sin^2 \theta\, d\phi^2$

Jacobian: $\dfrac{\partial(x, y, z)}{\partial(r, \theta, \phi)} = r^2 \sin\theta$

Element of volume: $dV = r^2 \sin \theta\, dr\, d\theta\, d\,\phi$

Laplacian: $\nabla^2 U = \dfrac{1}{r^2}\dfrac{\partial}{\partial r}\left(r^2 \dfrac{\partial U}{\partial r}\right)$

$$+ \frac{1}{r^2 \sin\theta}\frac{\partial}{\partial \theta}\left(\sin\theta \frac{\partial U}{\partial \theta}\right) + \frac{1}{r^2 \sin^2\theta}\frac{\partial^2 U}{\partial \phi^2}$$

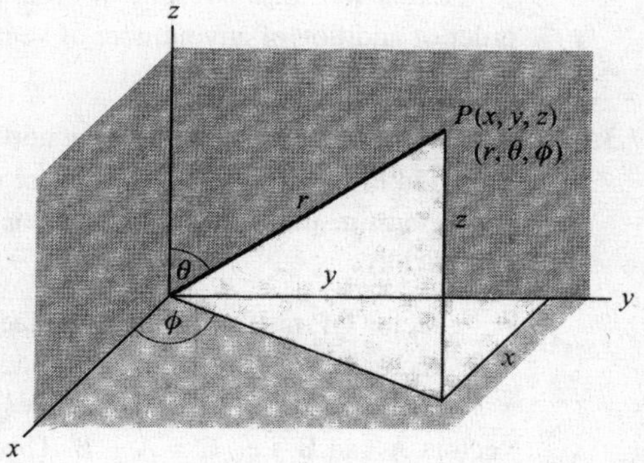

Figure 7.14

Other types of coordinate systems are possible.

SOLVED PROBLEMS

Vector algebra

7.1. Show that addition of vectors is commutative, i.e., **A** + **B** = **B** + **A**. See Figure 7.15.

$$\mathbf{OP} + \mathbf{PQ} = \mathbf{OQ} \text{ or } \mathbf{A} + \mathbf{B} = \mathbf{C}$$

and

$$\mathbf{OR} + \mathbf{RQ} = \mathbf{OQ} \text{ or } \mathbf{B} + \mathbf{A} = \mathbf{C}$$

Then **A** + **B** = **B** + **A**.

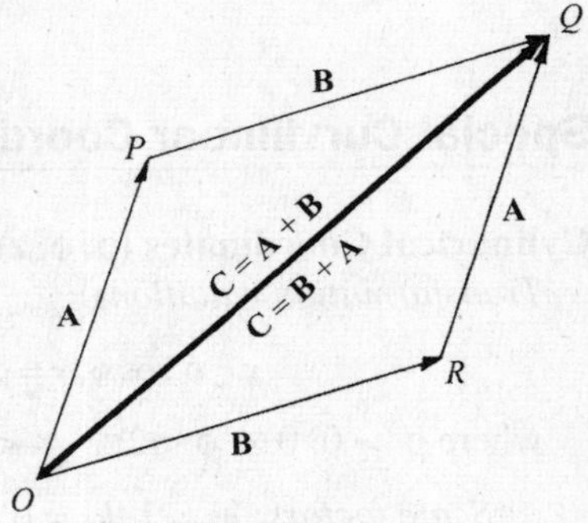

Figure 7.15

7.2. Show that the addition of vectors is associative, i.e., **A** + (**B** + **C**) = (**A** + **B**) + **C**. See Figure 7.16.

$$\mathbf{OP} + \mathbf{PQ} = \mathbf{OQ} = (\mathbf{A} + \mathbf{B}) \text{ and } \mathbf{PQ} + \mathbf{QR} = \mathbf{PR} = (\mathbf{B} + \mathbf{C})$$

Since

$$\mathbf{OP} + \mathbf{PR} = \mathbf{OR} = \mathbf{D}, \text{ i.e., } \mathbf{A} + (\mathbf{B} + \mathbf{C}) = \mathbf{D}$$

$$\mathbf{OQ} + \mathbf{QR} = \mathbf{OR} = \mathbf{D}, \text{ i.e., } (\mathbf{A} + \mathbf{B}) + \mathbf{C} = \mathbf{D}$$

we have **A** + (**B** + **C**) = (**A** + **B**) + **C**.

Extensions of the results of Problems 7.1 and 7.2 show that the order of addition of any number of vectors is immaterial.

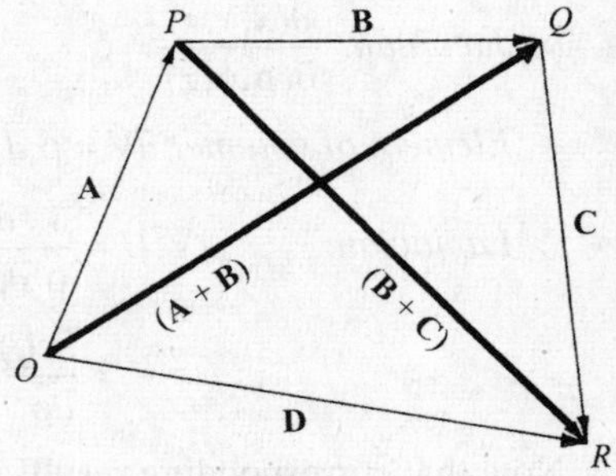

Figure 7.16

7.3. An automobile travels 3 miles due north, then 5 miles northeast as shown in Figure 7.17. Represent these displacements graphically and determine the resultant displacement (a) graphically and (b) analytically.

Vector **OP** or **A** represents displacement of 3 miles due north.
Vector **PQ** or **B** represents displacement of 5 miles northeast.
Vector **OQ** or **C** represents the resultant displacement or sum of vectors **A** and **B**, i.e., **C** = **A** + **B**. This is the *triangle law* of vector addition.

The resultant vector **OQ** can also be obtained by constructing the diagonal of the parallelogram *O P Q R* having vectors **OP** = **A** and **OR** (equal to vector **PQ** or **B**) as sides. This is the *parallelogram law* of vector addition.

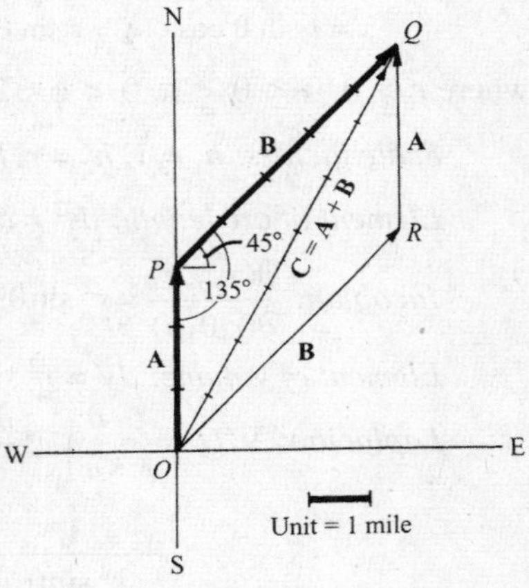

Figure 7.17

(a) *Graphical Determination of Resultant.* Lay off the 1-mile unit on vector **OQ** to find the magnitude 7.4 miles (approximately).

Angle $EOQ = 61.5°$, using a protractor. Then vector **OQ** has magnitude 7.4 miles and direction 61.5° north of east.

(b) *Analytical Determination of Resultant.* From triangle *OPQ*, denoting the magnitudes of **A, B, C** by *A, B, C*, we have by the law of cosines

$$C^2 = A^2 + B^2 - 2\,AB \cos OPQ = 3^2 + 5^2 - 2(3)(5)\cos 135° = 34 + 15\sqrt{2} = 55.21$$

and $C = 7.43$ (approximately).

By the law of sines, $\dfrac{A}{\sin \angle OQP} = \dfrac{C}{\sin \angle OPQ}$. Then

$$\sin \angle OPQ = \frac{A \sin \angle OPQ}{C} = \frac{3(0.707)}{7.43} = 0.2855 \quad \text{and} \quad \angle OQP = 16°35'$$

Thus, vector **OQ** has magnitude 7.43 miles and direction $(45° + 16°35') = 61°35'$ north of east.

7.4. Prove that if **a** and **b** are noncollinear, then $x\mathbf{a} + y\mathbf{b} = 0$ implies $x = y = 0$. Is the set $\{\mathbf{a}, \mathbf{b}\}$ linearly independent or linearly dependent?

Suppose $x \neq 0$. Then $x\mathbf{a} + y\mathbf{b} = \mathbf{0}$ implies $x\mathbf{a} = -y\mathbf{b}$ or $\mathbf{a} = -(y/x)\,\mathbf{b}$; i.e., **a** and **b** must be parallel to the same line (collinear), contrary to hypothesis. Thus, $x = 0$; then $y\mathbf{b} = \mathbf{0}$, from which $y = 0$. The set is linearly independent.

7.5. Prove that $x_1\mathbf{a} + y_1\mathbf{b} = x_2\mathbf{a} + y_2\mathbf{b}$, where **a** and **b** are noncollinear, then $x_1 = x_2$ and $y_1 = y_2$.

$x_1\mathbf{a} + y_1\mathbf{b} = x_2\mathbf{a} + y_2\mathbf{b}$ can be written

$$x_1\mathbf{a} + y_1\mathbf{b} - (x_2\mathbf{a} + y_2\mathbf{b}) = \mathbf{0} \text{ or } (x_1 - x_2)\mathbf{a} + (y_1 - y_2)\mathbf{b} = \mathbf{0}$$

Hence, by Problem 7.4, $x_1 - x_2 = 0$, $y_1 - y_2 = 0$, or $x_1 = x_2$, $y_1 = y_2$.

Extensions are possible (see Problem 7.49).

7.6. Prove that the diagonals of a parallelogram bisect each other.

Let $ABCD$ be the given parallelogram with diagonals intersecting at P, as shown in Figure 7.18.

Since $\mathbf{BD} + \mathbf{a} = \mathbf{b}$, $\mathbf{BD} = \mathbf{b} - \mathbf{a}$. Then $\mathbf{BP} = x(\mathbf{b} - \mathbf{a})$.

Since $\mathbf{AC} = \mathbf{a} + \mathbf{b}$, $\mathbf{AP} = y(\mathbf{a} + \mathbf{b})$.

But $\mathbf{AB} = \mathbf{AP} + \mathbf{PB} = \mathbf{AP} - \mathbf{BP}$; i.e., $\mathbf{a} = y(\mathbf{a} + \mathbf{b}) - x(\mathbf{b} - \mathbf{a}) = (x + y)\mathbf{a} + (y - x)\mathbf{b}$.

Since **a** and **b** are noncollinear, we have, by Problem 7.5, $x + y = 1$ and $y - x = 0$; i.e., $x = y = 1/2$ and P is the midpoint of both diagonals.

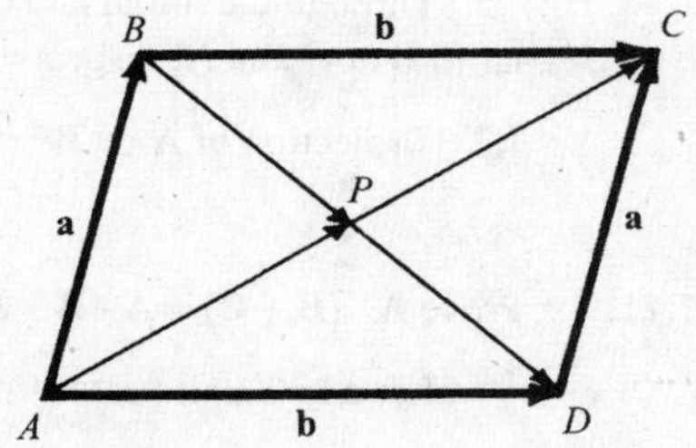

Figure 7.18

7.7. Prove that the line joining the midpoints of two sides of a triangle is parallel to the third side and has half its length.

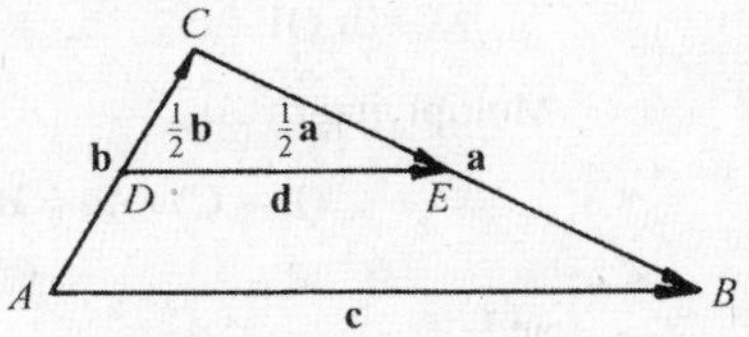

Figure 7.19

From Figure 7.19, $\mathbf{AC} + \mathbf{CB} = \mathbf{AB}$ or $\mathbf{b} + \mathbf{a} = \mathbf{c}$.

Let $\mathbf{DE} = \mathbf{d}$ be the line joining the midpoints of sides AC and CB.

Then $\mathbf{d} = \mathbf{DC} + \mathbf{CE} = \dfrac{1}{2}\mathbf{b} + \dfrac{1}{2}\mathbf{a} = \dfrac{1}{2}(\mathbf{b} + \mathbf{a}) = \dfrac{1}{2}\mathbf{c}$.

Thus, **d** is parallel to **c** and has half its length.

7.8. Prove that the magnitude A of the vector $\mathbf{A} = A_1\mathbf{i} + A_2\mathbf{j} + A_3\mathbf{k}$ is $A = \sqrt{A_1^2 + A_2^2 + A_3^2}$. See Figure 7.20.

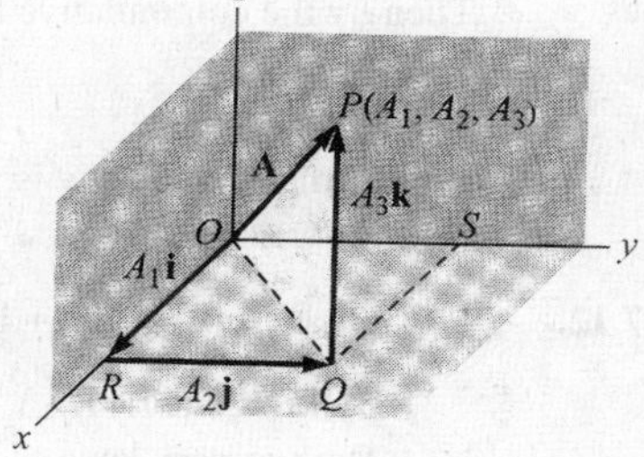

Figure 7.20

By the Pythagorean theorem,

$$(\overline{OP})^2 = (\overline{OQ})^2 + (\overline{QP})^2$$

where $(\overline{OP})$ denotes the magnitude of vector **OP**, etc. Similarly, $(\overline{OQ})^2 = (\overline{OR})^2 + (\overline{RQ})^2$.

Then $(\overline{OP})^2 = (\overline{OR})^2 + (\overline{RQ})^2 + (\overline{QP})^2$ or $A^2 = A_1^2 + A_2^2 + A_3^2$, i.e.. $A = \sqrt{A_1^2 + A_2^2 + A_3^2}$.

7.9. Determine the vector having initial point $P(x_1, y_1, z_1)$ and terminal point $Q(x_2, y_2, z_2)$ and find its magnitude. See Figure 7.21.

The position vector of P is $\mathbf{r}_1 = x_1\mathbf{i} + y_1\mathbf{j} + z_1\mathbf{k}$.
The position vector of Q is $\mathbf{r}_2 = x_2\mathbf{i} + y_2\mathbf{j} + z_2\mathbf{k}$.
$\mathbf{r}_1 = \mathbf{PQ} = \mathbf{r}_2$ or

$$\mathbf{PQ} = \mathbf{r}_2 - \mathbf{r}_1 = (x_2\mathbf{i} + y_2\mathbf{j} + z_2\mathbf{k}) - (x_1\mathbf{i} + y_1\mathbf{j} + z_1\mathbf{k})$$
$$= (x_2 - x_1)\mathbf{i} + (y_2 - y_1)\mathbf{j} + (z_2 - z_1)\mathbf{k}$$

Magnitude of $\mathbf{PQ} = \overline{PQ}$

$$= \sqrt{(x_2 - x_1)^2 + (y_2 - y_1)^2 + (z_2 - z_1)^2}.$$

Note that this is the distance between points P and Q.

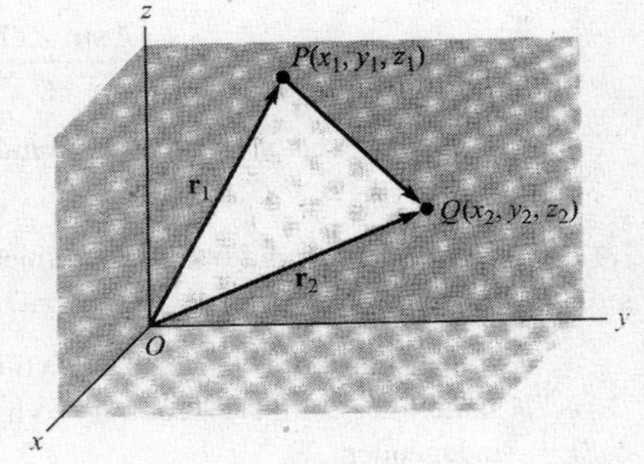

Figure 7.21

The dot or scalar product

7.10. Prove that the projection of **A** on **B** is equal to $\mathbf{A} \cdot \mathbf{b}$, where **b**, where **b** is a unit vector in the direction of **B**.

Through the initial and terminal points of **A** pass planes perpendicular to **B** at G and H, respectively, as in Figure 7.22; then

Projection of **A** on $\mathbf{B} = \overline{GH} = \overline{EF} = A \cos \theta = \mathbf{A} \cdot \mathbf{b}$

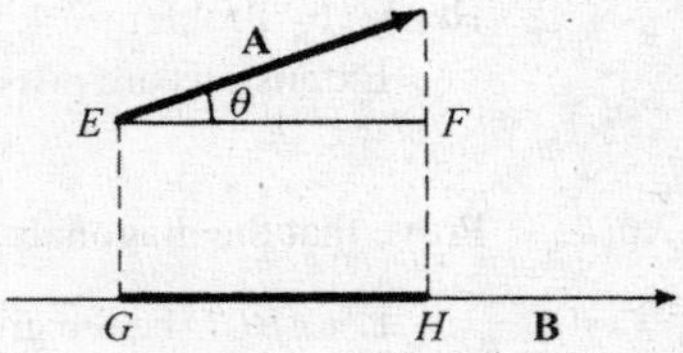

Figure 7.22

7.11. Prove $\mathbf{A} \cdot (\mathbf{B} + \mathbf{C}) = \mathbf{A} \cdot \mathbf{B} + \mathbf{A} \cdot \mathbf{C}$. See Figure 7.23.

Let **a** be a unit vector in the direction of **A**; then

Projection of $(\mathbf{B} + \mathbf{C})$ on **A** = projection of **B** on **A** + projection of **C** on **A**

$$(\mathbf{B} + \mathbf{C}) \cdot \mathbf{a} = \mathbf{B} \cdot \mathbf{a} + \mathbf{C} \cdot \mathbf{a}$$

Multiplying by A.

$$(\mathbf{B} + \mathbf{C}) \cdot A\mathbf{a} = \mathbf{B} \cdot A\mathbf{a} + \mathbf{C} \cdot A\mathbf{a}$$

and

$$(\mathbf{B} + \mathbf{C}) \cdot \mathbf{A} = \mathbf{B} \cdot \mathbf{A} + \mathbf{C} \cdot \mathbf{A}$$

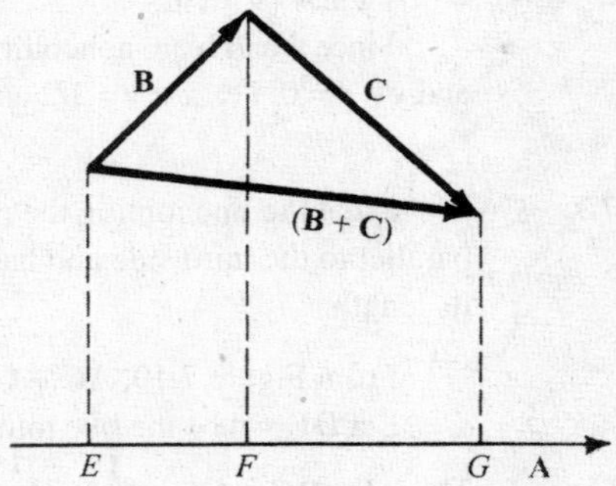

Figure 7.23

Then by the commutative law for dot products.

$$\mathbf{A} \cdot (\mathbf{B} + \mathbf{C}) = \mathbf{A} \cdot \mathbf{B} + \mathbf{A} \cdot \mathbf{C}$$

and the distributive law is valid.

7.12. Prove that $(\mathbf{A} + \mathbf{B}) \cdot (\mathbf{C} + \mathbf{D}) = \mathbf{A} \cdot \mathbf{C} + \mathbf{A} \cdot \mathbf{D} + \mathbf{B} \cdot \mathbf{C} + \mathbf{B} \cdot \mathbf{D}$.

By Problem 7.11, $(\mathbf{A} + \mathbf{B}) \cdot (\mathbf{C} + \mathbf{D}) = \mathbf{A} \cdot (\mathbf{C} + (\mathbf{C} + \mathbf{D}) + \mathbf{B} \cdot (\mathbf{C} + \mathbf{D}) = \mathbf{A} \cdot \mathbf{C} + \mathbf{A} \cdot \mathbf{D} + \mathbf{B} \cdot \mathbf{C} + \mathbf{B} \cdot \mathbf{D}$. The ordinary laws of algebra are valid for dot products where the operations are defined.

7.13. Evaluate each of the following.

(a) $\mathbf{i}\cdot\mathbf{i} = |\mathbf{i}|\ |\mathbf{i}| \cos 0° = (1)(1)(1) = 1$

(b) $\mathbf{i}\cdot\mathbf{k} = |\mathbf{i}|\ |\mathbf{k}| \cos 90° = (1)(1)(0) = 0$

(c) $\mathbf{k}\cdot\mathbf{j} = |\mathbf{k}|\ |\mathbf{j}| \cos 90° = (1)(1)(0) = 0$

(d) $\mathbf{j}\cdot(2\mathbf{i} - 3\mathbf{j} + \mathbf{k}) = 2\mathbf{j}\cdot\mathbf{i} - 3\mathbf{j}\cdot\mathbf{j} + \mathbf{j}\cdot\mathbf{k} = 0 - 3 + 0 = -3$

(e) $(2\mathbf{i} - \mathbf{j})\cdot(3\mathbf{i} + \mathbf{k}) = 2\mathbf{i}\cdot(3\mathbf{i} + \mathbf{k}) - \mathbf{j}\cdot(3\mathbf{i} + \mathbf{k}) = 6\mathbf{i}\cdot\mathbf{i} + 2\mathbf{i}\cdot\mathbf{k} - 3\mathbf{j}\cdot\mathbf{i} - \mathbf{j}\cdot\mathbf{k} = 6 + 0 - 0 - 0 = 6$

7.14. If $\mathbf{A} = A_1\mathbf{i} + A_2\mathbf{j} + A_3\mathbf{k}$ and $\mathbf{B} = B_1\mathbf{i} + B_2\mathbf{j} + B_3\mathbf{k}$, prove that $\mathbf{A}\cdot\mathbf{B} = A_1B_1 + A_2B_2 + A_3B_3$.

$$\begin{aligned}\mathbf{A}\cdot\mathbf{B} &= (A_1\mathbf{i} + A_2\mathbf{j} + A_3\mathbf{k})\cdot(B_1\mathbf{i} + B_2\mathbf{j} + B_3\mathbf{k})\\ &= A_1\mathbf{i}\cdot(B_1\mathbf{i} + B_2\mathbf{j} + B_3\mathbf{k}) + A_2\mathbf{j}\cdot(B_1\mathbf{i} + B_2\mathbf{j} + B_3\mathbf{k}) + A_3\mathbf{k}\cdot(B_1\mathbf{i} + B_2\mathbf{j} + B_3\mathbf{k})\\ &= A_1B_1\mathbf{i}\cdot\mathbf{i} + A_1B_2\mathbf{i}\cdot\mathbf{j} + A_1B_3\mathbf{i}\cdot\mathbf{k} + A_2B_1\mathbf{j}\cdot\mathbf{i} + A_2B_2\mathbf{j}\cdot\mathbf{j} + A_2B_3\mathbf{j}\cdot\mathbf{k}\\ &\quad + A_3B_1\mathbf{k}\cdot\mathbf{i} + A_3B_2\mathbf{k}\cdot\mathbf{j} + A_3B_3\mathbf{k}\cdot\mathbf{k}\\ &= A_1B_1 + A_2B_2 + A_3B_3\end{aligned}$$

since $\mathbf{i}\cdot\mathbf{j} = \mathbf{k}\cdot\mathbf{k} = 1$ and all other dot products are zero.

7.15. If $\mathbf{A} = A_1\mathbf{i} + A_2\mathbf{j} + A_3\mathbf{k}$. show that $A = \sqrt{\mathbf{A}\cdot\mathbf{A}} = \sqrt{A_1^2 + A_2^2 + A_3^2}$.

$$\mathbf{A}\cdot\mathbf{A} = (A)(A)\cos 0 = A^2. \text{ Then } A = \sqrt{\mathbf{A}\cdot\mathbf{A}}.$$

$$\text{Also, } \mathbf{A}\cdot\mathbf{A} = (A_1\mathbf{i} + A_2\mathbf{j} + A_3\mathbf{k})\cdot(A_1\mathbf{i} + A_2\mathbf{j} + A_3\mathbf{k})$$

$$= (A_1)(A_1) + (A_2)(A_2) + (A_3)(A_3) = A_1^2 + A_2^2 + A_3^2$$

By Problem 7.14, taking $\mathbf{B} = \mathbf{A}$.

Then $A = \sqrt{\mathbf{A}\cdot\mathbf{A}} = \sqrt{A_1^2 + A_2^2 + A_3^2}$. is the magnitude of **A**. Sometimes $\mathbf{A}\cdot\mathbf{A}$ is written $\mathbf{A}^2$.

The cross or vector product

7.16. Prove $\mathbf{A}\times\mathbf{B} = -\mathbf{B}\times\mathbf{A}$.

$\mathbf{A}\times\mathbf{B} = \mathbf{C}$ has magnitude $AB \sin\theta$ and direction such that **A**, **B**, and **C** form a right-handed system. See Figure 7.24(*a*).

$\mathbf{B}\times\mathbf{A} = \mathbf{D}$ has magnitude $BA \sin\theta$ and direction such that **B**, **A**, and **D** form a right-handed system. See Figure 7.24(*b*).

Then **D** has the same magnitude as **C** but is opposite in direction; i.e., $\mathbf{C} = -\mathbf{D}$ or $\mathbf{A}\times\mathbf{B} = -\mathbf{B}\times\mathbf{A}$.

The commutative law for cross products is not valid.

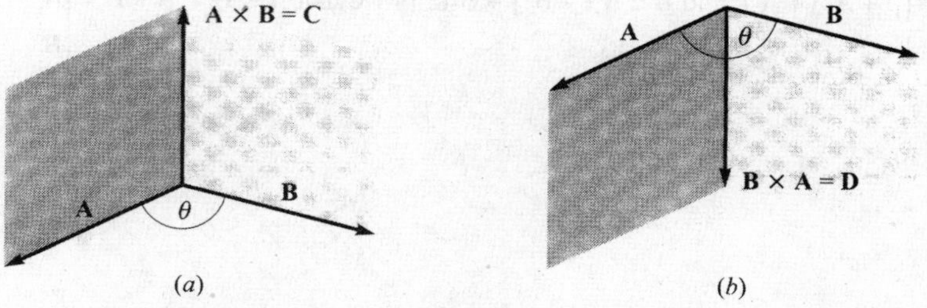

Figure 7.24

7.17. Prove that $\mathbf{A} \times (\mathbf{B} + \mathbf{C}) = \mathbf{A} \times \mathbf{B} + \mathbf{A} \times \mathbf{C}$ for the case where **A** is perpendicular to **B** and also to **C**.

Since **A** is perpendicular to **B**, $\mathbf{A} \times \mathbf{B}$ is a vector perpendicular to the plane of **A** and **B** and having magnitude $AB \sin 90° = AB$ or magnitude of **AB**. This is equivalent to multiplying vector **B** by A and rotating the resultant vector through 90° to the position shown in Figure 7.25.

Similarly, $\mathbf{A} \times \mathbf{C}$ is the vector obtained by multiplying **C** by A and rotating the resultant vector through 90° to the position shown.

In like manner, $\mathbf{A} \times (\mathbf{B} + \mathbf{C})$ is the vector obtained by multiplying $\mathbf{B} + \mathbf{C}$ by A and rotating the resultant vector through 90° to the position shown.

Since $\mathbf{A} \times (\mathbf{B} + \mathbf{C})$ is the diagonal of the parallelogram with $\mathbf{A} \times \mathbf{B}$ and $\mathbf{A} \times \mathbf{C}$ as sides, we have $\mathbf{A} \times (\mathbf{B} + \mathbf{C}) = \mathbf{A} \times \mathbf{B} + \mathbf{A} \times \mathbf{C}$.

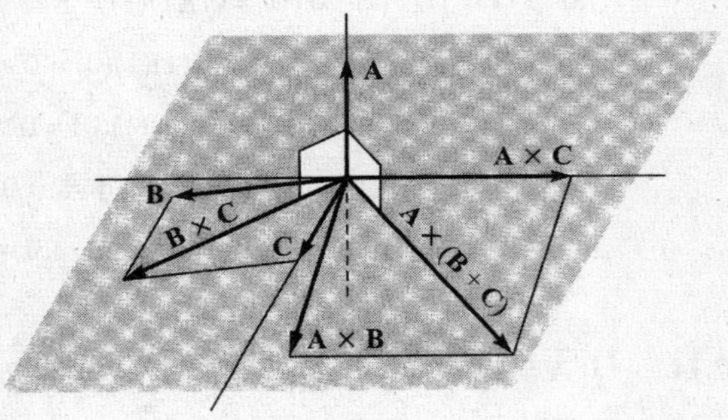

Figure 7.25

7.18. Prove that $\mathbf{A} \times (\mathbf{B} + \mathbf{C}) = \mathbf{A} \times \mathbf{B} + \mathbf{A} \times \mathbf{C}$ in the general case where **A**, **B**, and **C** are noncoplanar. See Figure 7.26.

Resolve **B** into two component vectors, one perpendicular to **A** and the other parallel to **A**, and denote them by $\mathbf{B}_\perp$ and $\mathbf{B}_\parallel$ respectively. Then $\mathbf{B} = \mathbf{B}_\perp + \mathbf{B}_\parallel$.

If θ is the angle between **A** and **B**, then $B_\perp = B \sin \theta$. Thus, the magnitude of $\mathbf{A} \times \mathbf{B}_\perp$ is $AB \sin \theta$, the same as the magnitude of $\mathbf{A} \times \mathbf{B}$. Also, the direction of $\mathbf{A} \times \mathbf{B}_\perp$ is the same as the direction of $\mathbf{A} \times \mathbf{B}$. Hence, $\mathbf{A} \times \mathbf{B}_\perp = \mathbf{A} \times \mathbf{B}$.

Similarly, if **C** is resolved into two component vectors $\mathbf{C}_\parallel$ and $\mathbf{C}_\perp$, parallel and perpendicular, respectively, to **A**, then $\mathbf{A} \times \mathbf{C}_\perp = \mathbf{A} \times \mathbf{C}$.

Also, since $\mathbf{B} + \mathbf{C} = \mathbf{B}_\perp + \mathbf{B}_\parallel + \mathbf{C}_\perp + \mathbf{C}_\parallel = (\mathbf{B}_\perp + \mathbf{C}_\perp) + (\mathbf{B}_\parallel + \mathbf{C}_\parallel)$, it follows that

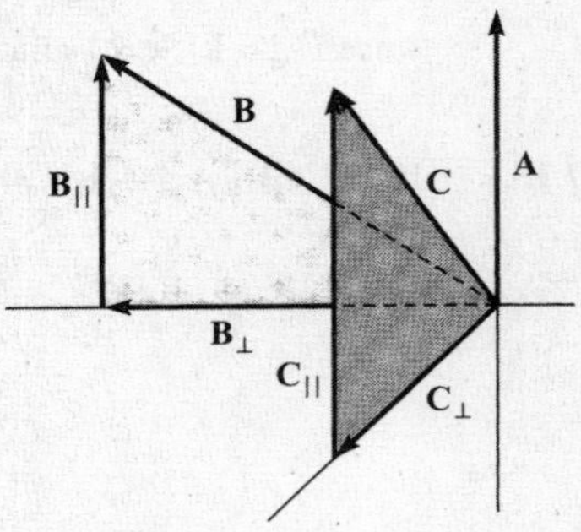

Figure 7.26

$$\mathbf{A} \times (\mathbf{B}_\perp + \mathbf{C}_\perp) = \mathbf{A} \times (\mathbf{B} + \mathbf{C})$$

Now $\mathbf{B}_\perp$ and $\mathbf{C}_\perp$ are vectors perpendicular to **A**, and so by Problem 7.17,

$$\mathbf{A} \times (\mathbf{B}_\perp + \mathbf{C}_\perp) = \mathbf{A} \times \mathbf{B}_\perp + \mathbf{A} \times \mathbf{C}_\perp$$

Then

$$\mathbf{A} \times (\mathbf{B} + \mathbf{C}) = \mathbf{A} \times \mathbf{B} + \mathbf{A} \times \mathbf{C}$$

and the distributive law holds. Multiplying by -1, using Problem 7.16, this becomes $(\mathbf{B} + \mathbf{C}) \times \mathbf{A} = \mathbf{B} \times \mathbf{A} + \mathbf{C} \times \mathbf{A}$. Note that the order of factors in cross products is important. The usual laws of algebra apply only if proper order is maintained.

7.19. (a) If $\mathbf{A} = A_1\mathbf{i} + A_2\mathbf{j} + A_3\mathbf{k}$ and $\mathbf{B} = B_1\mathbf{i} + B_2\mathbf{j} + B_3\mathbf{k}$, prove that $\mathbf{A} \times \mathbf{B} = \mathbf{A}\times\mathbf{B} = \begin{vmatrix} \mathbf{i} & \mathbf{j} & \mathbf{k} \\ A_1 & A_2 & A_3 \\ B_1 & B_2 & B_3 \end{vmatrix}$.

$$
\begin{aligned}
\mathbf{A}\times\mathbf{B} &= (A_1\mathbf{i}+A_2\mathbf{j}+A_3\mathbf{k})\times(B_1\mathbf{i}+B_2\mathbf{j}+B_3\mathbf{k})\\
&= A_1\mathbf{i}\times(B_1\mathbf{i}+B_2\mathbf{j}+B_3\mathbf{k})+A_2\mathbf{j}\times(B_1\mathbf{i}+B_2\mathbf{j}+B_3\mathbf{k})+A_3\mathbf{k}\times(B_1\mathbf{i}+B_2\mathbf{j}+B_3\mathbf{k})\\
&= A_1B_1\mathbf{i}\times\mathbf{i}+A_1B_2\mathbf{i}\times\mathbf{j}+A_1B_3\mathbf{i}\times\mathbf{k}+A_2B_1\mathbf{j}\times\mathbf{i}+A_2B_2\mathbf{j}\times\mathbf{j}+A_2B_3\mathbf{j}\times\mathbf{k}\\
&\quad +A_3B_1\mathbf{k}\times\mathbf{i}+A_3B_2\mathbf{k}\times\mathbf{j}+A_3B_3\mathbf{k}\times\mathbf{k}\\
&= (A_2B_3-A_3B_2)\mathbf{i}+(A_3B_1-A_1B_3)\mathbf{j}+(A_1B_2-A_2B_1)\mathbf{k} = \begin{vmatrix}\mathbf{i} & \mathbf{j} & \mathbf{k}\\ A_1 & A_2 & A_3\\ B_1 & B_2 & B_3\end{vmatrix}
\end{aligned}
$$

(b) Use the determinant representation to prove the result of Problem 7.18.

7.20. If $\mathbf{A} = 3\mathbf{i} - \mathbf{j} + 2\mathbf{k}$ and $\mathbf{B} = 2\mathbf{i} + 3\mathbf{j} - \mathbf{k}$, find $\mathbf{A}\times\mathbf{B}$.

$$
\begin{aligned}
\mathbf{A}\times\mathbf{B} &= \begin{vmatrix}\mathbf{i} & \mathbf{j} & \mathbf{k}\\ 3 & -1 & 2\\ 2 & 3 & -1\end{vmatrix} = \mathbf{i}\begin{vmatrix}-1 & 2\\ 3 & -1\end{vmatrix} - \mathbf{j}\begin{vmatrix}3 & 2\\ 2 & -1\end{vmatrix} + \mathbf{k}\begin{vmatrix}3 & -1\\ 2 & 3\end{vmatrix}\\
&= -5\mathbf{i}+7\mathbf{j}+11\mathbf{k}
\end{aligned}
$$

7.21. Prove that the area of a parallelogram with sides **A** and **B** is $|\mathbf{A}\times\mathbf{B}|$. See Figure 7.27.

$$
\begin{aligned}
\text{Area of parallelogram} &= h\,|\mathbf{B}|\\
&= |\mathbf{A}|\sin\theta\,|\mathbf{B}|\\
&= |\mathbf{A}\times\mathbf{B}|
\end{aligned}
$$

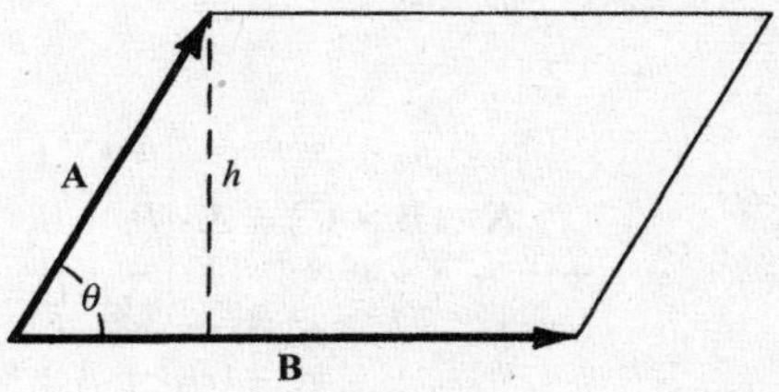

Figure 7.27

Note that the area of the triangle with sides **A** and **B** = 1/2 $|\mathbf{A}\times\mathbf{B}|$.

7.22. Find the area of the triangle with vertices at P(2, 3, 5), Q(4, 2, –1), and R(3, 6, 4).

$$\text{PQ} = (4-2)\mathbf{i} + (2-3)\mathbf{j} + (-1-5)\mathbf{k} = 2\mathbf{i} - \mathbf{j} - 6\mathbf{k}$$

$$\text{PR} = (3-2)\mathbf{i} + (6-3)\mathbf{j} + (4-5)\mathbf{k} = \mathbf{i} + 3\mathbf{j} - \mathbf{k}$$

$$
\begin{aligned}
\text{Area of triangle} &= \frac{1}{2}|\text{PQ}\times\text{PR}| = \frac{1}{2}|(2\mathbf{i}-\mathbf{j}-6\mathbf{k})\times(\mathbf{i}+3\mathbf{j}-\mathbf{k})|\\
&= \frac{1}{2}\left\|\begin{matrix}\mathbf{i} & \mathbf{j} & \mathbf{k}\\ 2 & -1 & -6\\ 1 & 3 & -1\end{matrix}\right\| = \frac{1}{2}|19\mathbf{i}-4\mathbf{j}+7\mathbf{k}|\\
&= \frac{1}{2}\sqrt{(19)^2+(-4)^2+(7)^2} = \frac{1}{2}\sqrt{426}
\end{aligned}
$$

Triple products

7.23. Show that $\mathbf{A}\cdot(\mathbf{B}\times\mathbf{C})$ is in absolute value equal to the volume of a parallelepiped with sides **A, B,** and **C.** See Figure 7.28.

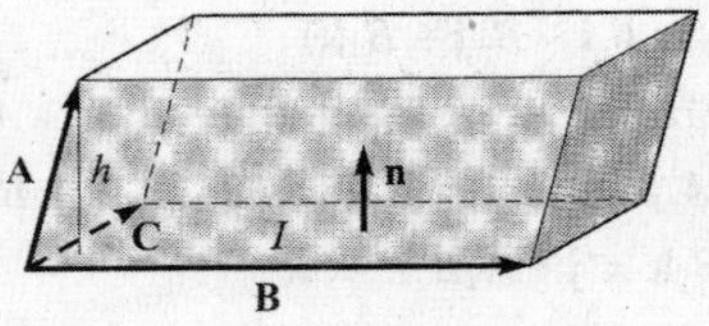

Figure 7.28

Let $\mathbf{n}$ be a unit normal to parallelogram I, having the direction of $\mathbf{B} \times \mathbf{C}$, and let h be the height of the terminal point of $\mathbf{A}$ above the parallelogram I.

$$\text{Volume of a parallelepiped} = (\text{height } h)(\text{area of parallelogram } I)$$

$$= (\mathbf{A} \cdot \mathbf{n})(|\mathbf{B} \times \mathbf{C}|)$$

$$= \mathbf{A} \cdot \{|\mathbf{B} \times \mathbf{C}|\mathbf{n}\} = \mathbf{A} \cdot (\mathbf{B} \times \mathbf{C})$$

If $\mathbf{A}$, $\mathbf{B}$ and $\mathbf{C}$ do not form a right-handed system, $\mathbf{A} \cdot \mathbf{n} < 0$ and the volume $= |\mathbf{A} \cdot (\mathbf{B} \times \mathbf{C})|$.

7.24. If $\mathbf{A} = A_1\mathbf{i} + A_2\mathbf{j} + A_3\mathbf{k}$, $\mathbf{B} = B_1\mathbf{i} + B_2\mathbf{j} + B_3\mathbf{k}$, $\mathbf{C} = C_1\mathbf{i} + C_2\mathbf{j} + C_3\mathbf{k}$ show that

$$\mathbf{A} \cdot (\mathbf{B} \times \mathbf{C}) = \begin{vmatrix} A_1 & A_2 & A_3 \\ B_1 & B_2 & B_3 \\ C_1 & C_2 & C_3 \end{vmatrix}$$

$$\mathbf{A} \cdot (\mathbf{B} \times \mathbf{C}) = \mathbf{A} \cdot \begin{vmatrix} \mathbf{i} & \mathbf{j} & \mathbf{k} \\ B_1 & B_2 & B_3 \\ C_1 & C_2 & C_3 \end{vmatrix}$$

$$= (A_1\mathrm{i} + A_2\mathrm{j} + A_3\mathrm{k}) \cdot [(B_2C_3 - B_3C_2)\mathrm{i} + (B_3C_1 - B_1C_3)\mathrm{j} + (B_1C_2 - B_2C_1)\mathrm{k}]$$

$$= A_1(B_2C_3 - B_3C_2) + A_2(B_3C_1 - B_1C_3) + A_3(B_1C_2 - B_2C_1) = \begin{vmatrix} A_1 & A_2 & A_3 \\ B_1 & B_2 & B_3 \\ C_1 & C_2 & C_3 \end{vmatrix}.$$

7.25. Find the volume of a parallelepiped with sides $\mathbf{A} = 3\mathbf{i} - \mathbf{j}$, $\mathbf{B} = \mathbf{j} + 2\mathbf{k}$, $\mathbf{C} = \mathbf{i} + 5\mathbf{j} + 4\mathbf{k}$.

By Problems 7.23 and 7.24, volume of parallelepiped $= |\mathbf{A} \cdot (\mathbf{B} \times \mathbf{C})| = \left| \begin{vmatrix} 3 & -1 & 0 \\ 0 & 1 & 2 \\ 1 & 5 & 4 \end{vmatrix} \right| = |-20| = 20.$

7.26. Prove that $\mathbf{A} \cdot (\mathbf{B} \times \mathbf{C}) = (\mathbf{A} \times \mathbf{B}) \cdot \mathbf{C}$, i.e., the dot and cross can be interchanged.

By Problem 7.24: $\mathbf{A} \cdot (\mathbf{B} \times \mathbf{C}) = \begin{vmatrix} A_1 & A_2 & A_3 \\ B_1 & B_2 & B_3 \\ C_1 & C_2 & C_3 \end{vmatrix}$, $(\mathbf{A} \times \mathbf{B}) \cdot \mathbf{C} = \mathbf{C} \cdot (\mathbf{A} \times \mathbf{B}) = \begin{vmatrix} C_1 & C_2 & C_3 \\ A_1 & A_2 & A_3 \\ B_1 & B_2 & B_3 \end{vmatrix}$

Since the two determinants are equal, the required result follows.

7.27. Let $\mathbf{r}_1 = x_1\,\mathbf{i} + y_1\mathbf{j} + z_1\mathbf{k}$, $\mathbf{r}_2 = x_2\mathbf{i} + y_2\mathbf{j} + z_2\mathbf{k}$ and $\mathbf{r}_3 = x_3\mathbf{i} + y_3\mathbf{j} + z_3\mathbf{k}$ be the position vectors of points $P_1(x_1, y_1, z_1)$, $P_2(x_2, y_x, z_2)$, and $P_3(x_3, y_3, z_3)$. Find an equation for the plane passing through P_1, P_2, and P_3. See Figure 7.29.

We assume that P_1, P_2, and P_3 do not lie in the same straight line; hence, they determine a plane.

Let $\mathbf{r} = x\mathbf{i} + y\mathbf{j} + z\mathbf{k}$ denote the position vectors of any point $P(x, y, z)$ in the plane. Consider vectors $\overrightarrow{P_1P_2} = \mathbf{r}_2 - \mathbf{r}_1$, $\overrightarrow{P_1P_3} = \mathbf{r}_3 - \mathbf{r}_1$ and $\overrightarrow{P_1P} = \mathbf{r} - \mathbf{r}_1$ which all lie in the plane. Then

$$P_1P \cdot P_1P_2 \times P_1P_3 = 0$$

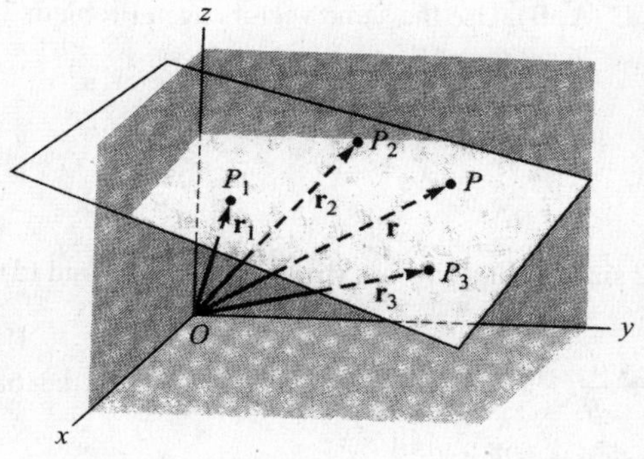

Figure 7.29

or

$$(\mathbf{r}-\mathbf{r}_1)\cdot(\mathbf{r}_2-\mathbf{r}_1)\times(\mathbf{r}_3-\mathbf{r}_1)=0$$

In terms of rectangular coordinates this becomes

$$[(x-x_1)\mathbf{i}+(y-y_1)\mathbf{j}+(z-z_1)\mathbf{k}]\cdot[(x_2-x_1)\mathbf{i}+(y_2-y_1)\mathbf{j}+(z_2-z_1)\,\mathbf{k}]$$

$$\times\,[(x_3-x_1)\mathbf{i}+(y_3-y_1)\mathbf{j}+(z_3-z_1)\mathbf{k}]=0$$

or, using Problem 7.24,

$$\begin{vmatrix} x-x_1 & y-y_1 & z-z_1 \\ x_2-x_1 & y_2-y_1 & z_2-z_1 \\ x_3-x_1 & y_3-y_1 & z_3-z_1 \end{vmatrix}=0$$

7.28. Find an equation for the plane passing through the points $P_1(3, 1, -2)$, $P_2(-1, 2, 4)$, $P_3(2, -1, 1)$.

The positions vectors of P_1, P_2, P_3 and any point $P(x, y, z)$ on the plane are respectively

$$\mathbf{r}_1=3\mathbf{i}+\mathbf{j}-2\mathbf{k},\ \mathbf{r}_2=-\mathbf{i}+2\mathbf{j}+4\mathbf{k},\ \mathbf{r}_3=2\mathbf{i}-\mathbf{j}+\mathbf{k},\ \mathbf{r}=x\mathbf{i}+y\mathbf{j}+z\mathbf{k}$$

Then $\mathbf{PP}_1=\mathbf{r}-\mathbf{r}_1$, $\mathbf{P}_2\mathbf{P}_1=\mathbf{r}_2-\mathbf{r}_1$, $\mathbf{P}_3\mathbf{P}_1=\mathbf{r}_3-\mathbf{r}_1$, all lie in the required plane and so the required equation is $(\mathbf{r}-\mathbf{r}_1)\cdot(\mathbf{r}_2-\mathbf{r}_1)\times(\mathbf{r}_3-\mathbf{r}_1)=0$, i.e.,

$$\{(x-3)\mathbf{i}+(y-1)\mathbf{j}+(z+2)\mathbf{k}\}\cdot\{-4\mathbf{i}+\mathbf{j}+6\mathbf{k}\}\times\{-\mathbf{i}-2\mathbf{j}+3\mathbf{k}\}=0$$

$$\{(x-3)\mathbf{i}+(y-1)\mathbf{j}+(z+2)\mathbf{k}\}\cdot\{15\,\mathbf{i}+6\mathbf{j}+9\mathbf{k}\}=0$$

$$15(x-3)+6(y-1)+9(z+2)=0 \qquad \text{or} \qquad 5x-2y+3z=11$$

Another method: By Problem 7.27, the required equation is

$$\begin{vmatrix} x-3 & y-1 & z+2 \\ -1-3 & 2-1 & 4+2 \\ 2-3 & -1-1 & 1+2 \end{vmatrix}=0 \quad \text{or} \quad 5x+2y+3z=11$$

7.29 (1) If $\mathbf{A}=\mathbf{i}+\mathbf{j}$, $\mathbf{B}=2\mathbf{i}-3\mathbf{j}+\mathbf{k}$, and $\mathbf{C}=4\mathbf{j}-3\mathbf{k}$, find (a) $(\mathbf{A}\times\mathbf{B})\times\mathbf{C}$, (b) $\mathbf{A}\times(\mathbf{B}\times\mathbf{C})$.

(a) $\mathbf{A}\times\mathbf{B}=\begin{vmatrix} \mathrm{i} & \mathrm{j} & \mathrm{k} \\ 1 & 1 & 0 \\ 2 & -3 & 1 \end{vmatrix}=\mathbf{i}-\mathbf{j}-5\mathbf{k}$. Then $(\mathbf{A}\times\mathbf{B})\times\mathbf{C}=\begin{vmatrix} \mathbf{i} & \mathbf{j} & \mathbf{k} \\ 1 & -1 & -5 \\ 0 & 4 & -3 \end{vmatrix}=23\mathbf{i}+3\mathbf{j}+4\mathbf{k}$.

(b) $\mathbf{B}\times\mathbf{C}=\begin{vmatrix} \mathrm{i} & \mathrm{j} & \mathrm{k} \\ 2 & -3 & 1 \\ 0 & 4 & -3 \end{vmatrix}=5\mathbf{i}+6\mathbf{j}+8\mathbf{k}$. Then $\mathbf{A}\times(\mathbf{B}\times\mathbf{C})=\begin{vmatrix} \mathbf{i} & \mathbf{j} & \mathbf{k} \\ 1 & 1 & 0 \\ 5 & 6 & 8 \end{vmatrix}=8\mathbf{i}-8\mathbf{j}+\mathbf{k}$.

It can be proved that, in general, $(\mathbf{A}\times\mathbf{B})\times\mathbf{C}\neq\mathbf{A}\times(\mathbf{B}\times\mathbf{C})$.

7.29 (2) $\mathbf{A} \times (\mathbf{B} \times \mathbf{C}) = \mathbf{B}(\mathbf{A}, \mathbf{C}) - \mathbf{C}(\mathbf{A}, \mathbf{B})$. Use the same vectors as in Problem 7.29 (1). (Note: sometimes remembered with the phrase "back to cab.")

Derivatives

7.30. If $\mathbf{r} = (t^3 + 2t)\mathbf{i} - 3e^{-2t}\mathbf{j} + 2\sin 5t\mathbf{k}$, find (a) $\frac{d\mathbf{r}}{dt}$, (b) $\left|\frac{d\mathbf{r}}{dt}\right|$, (c) $\frac{d^2\mathbf{r}}{dt^2}$, and (d) $\left|\frac{d^2\mathbf{r}}{dt^2}\right|$ at $t = 0$, and give a possible physical significance.

(a) $$\frac{d\mathbf{r}}{dt} = \frac{d}{dt}(t^3 + 2t)\mathbf{i} + \frac{d}{dt}(-3e^{-2t})\mathbf{j} + \frac{d}{dt}(2\sin 5t)\mathbf{k} = (3t^2 + 2)\mathbf{i} + 6e^{-2t}\mathbf{j} + 10\cos 5t\mathbf{k}$$

At $t = 0, d\mathbf{r}/dt = 2\mathbf{i} + 6\mathbf{j} + 10\mathbf{k}$

(b) From (*a*), $|d\mathbf{r}/dt| = \sqrt{(2)^2 + (6)^2 + (10)^2} = \sqrt{140} = 2\sqrt{35}$ at $t = 0$.

(c) $$\frac{d^2\mathbf{r}}{dt^2} = \frac{d}{dt}\left(\frac{d\mathbf{r}}{dt}\right) = \frac{d}{dt}\{(3t^2 + 2)\mathbf{i} + 6e^{-2t}\mathbf{j} + 10\cos 5t\mathbf{k}\} = 6t\mathbf{i} - 12e^{-2t}\mathbf{j} - 50\sin 5t\mathbf{k}$$

At $t = 0, d^2\mathbf{r}/dt^2 = -12\mathbf{j}$.

(d) From (c), $|d^2\mathbf{r}/dt^2| = 12$ at $t = 0$.

If t represents time, these represent, respectively, the velocity, magnitude of the velocity, acceleration, and magnitude of the acceleration at $t = 0$ of a particle moving along the space curve $x = t^3 + 2t$, $y = -3e^{-2t}$, $z = 2\sin 5t$.

7.31. Prove that $\frac{d}{du}(\mathbf{A}\cdot\mathbf{B}) = \mathbf{A}\cdot\frac{d\mathbf{B}}{du} + \frac{d\mathbf{A}}{du}\cdot\mathbf{B}$ where **A** and **B** are differentiable functions of u.

Method 1:

$$\frac{d}{du}(\mathbf{A}\cdot\mathbf{B}) = \lim_{\Delta u \to 0} \frac{(\mathbf{A} + \Delta\mathbf{A})\cdot(\mathbf{B} + \Delta\mathbf{B}) - \mathbf{A}\cdot\mathbf{B}}{\Delta u}$$
$$= \lim_{\Delta u \to 0} \frac{\mathbf{A}\cdot\Delta\mathbf{B} + \Delta\mathbf{A}\cdot\mathbf{B} + \Delta\mathbf{A}\cdot\Delta\mathbf{B}}{\Delta u}$$
$$= \lim_{\Delta u \to 0}\left(\mathbf{A}\cdot\frac{\Delta\mathbf{B}}{\Delta u} + \frac{\Delta\mathbf{A}}{\Delta u}\cdot\mathbf{B} + \frac{\Delta\mathbf{A}}{\Delta u}\cdot\Delta\mathbf{B}\right) = \mathbf{A}\cdot\frac{d\mathbf{B}}{du} + \frac{d\mathbf{A}}{du}\cdot\mathbf{B}$$

Method 2:

Let $\mathbf{A} = A_1\mathbf{i} + A_2\mathbf{j} + A_3\mathbf{k}$, $\mathbf{B} + B_1\mathbf{i} + B_2\mathbf{j} + B_3\mathbf{k}$. Then

$$\frac{d}{du}(\mathbf{A}\cdot\mathbf{B}) = \frac{d}{du}(A_1B_1 + A_2B_2 + A_3B_3)$$
$$= \left(A_1\frac{dB_1}{du} + A_2\frac{dB_2}{du} + A_3\frac{dB_3}{du}\right) + \left(\frac{dA_1}{du}B_1 + \frac{dA_2}{du}B_2 + \frac{dA_3}{du}B_3\right)$$
$$= \mathbf{A}\cdot\frac{d\mathbf{B}}{du} + \frac{d\mathbf{A}}{du}\cdot\mathbf{B}$$

7.32. If $\phi(x, y, z) = x^2yz$ and $\mathbf{A} = 3x^2y\mathbf{i} + yz^2\mathbf{j} - xz\mathbf{k}$, find $\mathbf{A} = 3x^2yi + yz^2j - xzk$, find $\frac{\partial^2}{\partial y\,\partial z}(\phi\mathbf{A})$ at the point $(1, -2, -1)$.

$$\phi\mathbf{A} = (x^2yz)(3x^2yi + yz^2\mathbf{j} - xz\mathbf{k}) = 3\mathbf{x}^4y^2z\mathbf{i} + x^2y^2z^3\mathbf{j} - x^3yz^2\mathbf{k}$$

$$\frac{\partial}{\partial z}(\phi\mathbf{A}) = \frac{\partial}{\partial z}(3\mathbf{x}^4y^2z\mathbf{i} + x^2y^2z^3\mathbf{j} - x^3y^2z^3\mathbf{k}) = 3\mathbf{x}^4y^2\mathbf{i} + 3x^2y^2z^2\mathbf{j} - 2x^3yz\mathbf{k}$$

$$\frac{\partial^2}{\partial y\,\partial z}(\phi\mathbf{A}) = \frac{\partial}{\partial z}(3\mathbf{x}^4y^2\mathrm{i} + 3x^2y^2z^2\mathbf{j} - 2x^3yz\mathbf{k}) = 6\mathbf{x}^4y\mathbf{i} + 6x^2yz^2\mathbf{j} - 2x^3z\mathbf{k}$$

If $x = 1$, $y = -2$, $z = -1$, this becomes $-12\mathbf{i} - 12\mathbf{j} + 2\mathbf{k}$.

7.33. If $\mathbf{A} = x^2 \sin y\mathbf{i} + z^2 \cos y\mathbf{j} - xy^2\mathbf{k}$, find $d\mathbf{A}$.

Method 1:

$$\frac{\partial \mathbf{A}}{\partial x} = 2x\sin y\mathbf{i} - y^2\mathbf{k}, \quad \frac{\partial \mathbf{A}}{\partial y} x^2 \cos y\mathbf{i} - z^2 \sin y\mathbf{j} - 2xy\mathbf{k}, \quad \frac{\partial \mathbf{A}}{\partial z} = 2z\cos y\mathbf{j}$$

$$\begin{aligned} d\mathbf{A} &= \frac{\partial \mathbf{A}}{\partial x}dx + \frac{\partial \mathbf{A}}{\partial y}dy + \frac{\partial \mathbf{A}}{\partial z}dz \\ &= (2x\sin y\mathbf{i} - y^2\mathbf{k})dx + (x^2\cos y\mathbf{i} - z^2\sin y\mathbf{j} - 2xy\mathbf{k})dy + (2z\cos y\mathbf{j})dz \\ &= (2x\sin y\, dx + x^2\cos y\, dy)\mathbf{i} + (2z\cos y\, dz - z^2\sin y\, dy)\mathbf{j} - (y^2dx + 2xy\, dy)\mathbf{k} \end{aligned}$$

Method 2:

$$d\mathbf{A} = d(x^2 \sin y)\mathbf{i} + d(z^2 \cos y)\mathbf{j} - d(xy^2)\mathbf{k}$$

$$= (2x \sin y\, dx + x^2 \cos y\, dy)\mathbf{i} + (2z \cos y\, dz - z^2 \sin y\, dy)\mathbf{j} - (y^2\, dx + 2xy\, dy)\mathbf{k}$$

Gradient, divergence, and curl

7.34. If $\phi = x^2yz^3$ and $\mathbf{A} = xz\mathbf{i} - y^2\mathbf{j} + 2x^2y\mathbf{k}$, find (a) $\nabla\phi$, (b) $\nabla \cdot \mathbf{A}$, (c) $\nabla \times \mathbf{A}$, (d) div $(\phi\mathbf{A})$, (e) curl $(\phi\mathbf{A})$.

(a) $$\begin{aligned}\nabla\phi &= \left(\mathbf{i}\frac{\partial}{\partial x} + \mathbf{j}\frac{\partial}{\partial y} + \mathbf{k}\frac{\partial}{\partial z}\right)\phi = \frac{\partial\phi}{\partial x}\mathbf{i} + \frac{\partial\phi}{\partial y}\mathbf{j} + \frac{\partial\phi}{\partial z}\mathbf{k} = \frac{\partial}{\partial x}(x^2yz^3)\mathbf{i} + \frac{\partial}{\partial x}(x^2yz^3)\mathbf{j} + \frac{\partial}{\partial z}(x^2yz^3)\mathbf{k} \\ &= 2xyz^3\mathbf{i} + x^2z^3\mathbf{j} + 3x^2yz^2\mathbf{k}\end{aligned}$$

(b) $$\begin{aligned}\nabla\cdot\mathbf{A} &= \left(\mathbf{i}\frac{\partial}{\partial x} + \mathbf{j}\frac{\partial}{\partial y} + \mathbf{k}\frac{\partial}{\partial z}\right)\cdot(xz\mathbf{i} - y^2\mathbf{j} + 2x^2y\mathbf{k}) \\ &= \frac{\partial}{\partial x}(xz) + \frac{\partial}{\partial y}(-y^2) + \frac{\partial}{\partial z}(2x^2y) = z - 2y\end{aligned}$$

(c) $$\begin{aligned}\nabla\times\mathbf{A} &= \left(\mathbf{i}\frac{\partial}{\partial x} + \mathbf{j}\frac{\partial}{\partial y} + \mathbf{k}\frac{\partial}{\partial z}\right)\times(xz\mathbf{i} - y^2\mathbf{j} + 2x^2y\mathbf{k}) \\ &= \begin{vmatrix}\mathbf{i} & \mathbf{j} & \mathbf{k} \\ \partial/\partial x & \partial/\partial y & \partial/\partial z \\ xz & -y^2 & 2x^2y\end{vmatrix} \\ &= \left(\frac{\partial}{\partial x}(2x^2y) - \frac{\partial}{\partial z}(-y^2)\right)\mathbf{i} + \left(\frac{\partial}{\partial z}(xz) - \frac{\partial}{\partial x}(2x^2y)\right)\mathbf{j} + \left(\frac{\partial}{\partial x}(-y^2) - \frac{\partial}{\partial y}(xz)\right)\mathbf{k} \\ &= 2x^2\mathbf{i} + (x - 4xy)\mathbf{j}\end{aligned}$$

(d) $$\begin{aligned}\text{div}\,(\phi\mathbf{A}) &= \nabla\cdot(\phi\mathbf{A}) = \nabla\cdot(x^3yz^4\mathbf{i} - x^2y^3z^3\mathbf{j} + 2x^4y^2z^3\mathbf{k}) \\ &= \frac{\partial}{\partial x}(x^3yz^4) + \frac{\partial}{\partial y}(-x^2y^3z^3) + \frac{\partial}{\partial z}(2x^4y^2z^3) \\ &= 3x^2yz^4 - 3x^2y^2z^3 + 6x^4y^2z^2\end{aligned}$$

(e) $$\begin{aligned}\text{curl}\,(\phi\mathbf{A}) &= \nabla\times(\phi\mathbf{A}) = \nabla\times(x^3yz^4\mathbf{i} - x^2y^3z^3\mathbf{j} + 2x^4y^2z^3\mathbf{k}) \\ &= \begin{vmatrix}\mathbf{i} & \mathbf{j} & \mathbf{k} \\ \partial/\partial x & \partial/\partial y & \partial/\partial z \\ x^3yz^4 & -x^2y^3z^3 & 2x^4y^2z^3\end{vmatrix} \\ &= (4x^4yz^3 - 3x^2y^3z^2)\mathbf{i} + (4x^3yz^3 - 8x^3y^2z^3)\mathbf{j} - (2xy^3z^3 + x^3z^4)\mathbf{k}\end{aligned}$$

7.35. Prove $\nabla \cdot (\phi \mathbf{A}) = (\nabla\phi) \cdot \mathbf{A} + \phi (\nabla \cdot \mathbf{A})$.

$$\begin{aligned}\nabla \cdot (\phi \mathbf{A}) &= \nabla \cdot (\phi\, A_1\mathbf{i} + \phi\, A_2\mathbf{j} + \phi\, A_3\mathbf{k}) \\ &= \frac{\partial}{\partial x}(\phi A_1) + \frac{\partial}{\partial y}(\phi A_2) + \frac{\partial}{\partial z}(\phi A_3) \\ &= \frac{\partial \phi}{\partial x}A_1 + \frac{\partial \phi}{\partial y}A_2 + \frac{\partial \phi}{\partial z}A_3 + \phi\left(\frac{\partial A_1}{\partial x} + \frac{\partial A_2}{\partial y} + \frac{\partial A_3}{\partial z}\right) \\ &= \left(\frac{\partial \phi}{\partial x}\mathbf{i} + \frac{\partial \phi}{\partial y}\mathbf{j} + \frac{\partial \phi}{\partial z}\mathbf{k}\right) \cdot (A_1\mathbf{i} + A_2\mathbf{j} + A_3\mathbf{k}) \\ &\quad + \phi\left(\frac{\partial}{\partial x}\mathbf{i} + \frac{\partial}{\partial y}\mathbf{j} + \frac{\partial}{\partial z}\mathbf{k}\right) \cdot (A_1\mathbf{i} + A_2\mathbf{j} + A_3\mathbf{k}) \\ &= (\nabla\phi) \cdot \mathbf{A} + \phi\, (\nabla \cdot \mathbf{A})\end{aligned}$$

7.36. Express a formula for the tangent plane to the surface $\phi(x, y, z) = 0$ at one of its points $P_0(x_0, y_0, z_0)$.

$$(\nabla\phi)_0 \cdot (\mathbf{r} - \mathbf{r}_0) = 0$$

7.37. Find a unit normal to the surface $2x^2 + 4yz - 5z^2 = -10$ at the point $P(3, -1, 2)$.

By Problem 7.36, a vector normal to the surface is

$$\nabla(2x^2 + 4yz - 5z^2) = 4x\mathbf{i} + 4z\mathbf{j} + (4y - 10z)\mathbf{k} = 12\mathbf{i} + 8\mathbf{j} - 24\mathbf{k} \text{ at } (3, -1, 2)$$

Them a unit normal to the surface at P is

$$\frac{12\mathbf{i} + 8\mathbf{j} - 24\mathbf{k}}{\sqrt{(12)^2 + (8)^2 + (-24)^2}} = \frac{3\mathbf{i} + 2\mathbf{j} - 6\mathbf{k}}{7}$$

Another unit normal to the surface at P is

$$-\frac{3\mathbf{i} + 2\mathbf{j} - 6\mathbf{k}}{7}$$

7.38. If $\phi = 2x^2y - xz^3$, find (a) $\nabla\phi$ and (b) $\nabla^2 \phi$.

(a) $\nabla\phi = \dfrac{\partial\phi}{\partial x}\mathbf{i} + \dfrac{\partial\phi}{\partial y}\mathbf{j} + \dfrac{\partial\phi}{\partial z}\mathbf{k} = (4xy - z^3)\mathbf{i} + 2x^2\mathbf{j} - 3xz^2\mathbf{k}$

(b) $\nabla^2\phi = \text{Laplacian of } \phi = \nabla \cdot \nabla\phi = \dfrac{\partial}{\partial x}(4xy - z^3) + \dfrac{\partial}{\partial y}(2x^2) + \dfrac{\partial}{\partial z}(-3xz^2) = 4y - 6xz$

Another method:

$$\begin{aligned}\nabla^2\phi &= \frac{\partial^2\phi}{\partial x^2} + \frac{\partial^2\phi}{\partial y^2} + \frac{\partial^2\phi}{\partial z^2} = \frac{\partial^2}{\partial x^2}(2x^2y - xz^3) + \frac{\partial^2}{\partial y^2}(2x^2y - xz^3) + \frac{\partial^2}{\partial z^2}(2x^2y - xz^3) \\ &= 4y - 6xz\end{aligned}$$

7.39. Prove div curl $\mathbf{A} = 0$.

$$\text{div curl } \mathbf{A} = \nabla \cdot (\nabla \times \mathbf{A}) = \nabla \cdot \begin{vmatrix} \mathbf{i} & \mathbf{j} & \mathbf{k} \\ \partial/\partial x & \partial/\partial y & \partial/\partial z \\ A_1 & A_2 & A_3 \end{vmatrix}$$

$$= \nabla \cdot \left[\left(\frac{\partial A_3}{\partial y} - \frac{\partial A_2}{\partial z} \right)\mathbf{i} + \left(\frac{\partial A_1}{\partial z} - \frac{\partial A_3}{\partial x} \right)\mathbf{j} + \left(\frac{\partial A_2}{\partial x} - \frac{\partial A_1}{\partial y} \right)\mathbf{k} \right.$$

$$= \frac{\partial}{\partial x}\left(\frac{\partial A_3}{\partial y} - \frac{\partial A_2}{\partial z} \right) + \frac{\partial}{\partial y}\left(\frac{\partial A_1}{\partial z} - \frac{\partial A_3}{\partial x} \right) + \frac{\partial}{\partial z}\left(\frac{\partial A_2}{\partial x} - \frac{\partial A_1}{\partial y} \right)$$

$$= \frac{\partial^2 A_3}{\partial x\, \partial y} - \frac{\partial^2 A_2}{\partial x\, \partial z} + \frac{\partial^2 A_1}{\partial y\, \partial z} - \frac{\partial^2 A_3}{\partial y\, \partial x} + \frac{\partial^2 A_2}{\partial z\, \partial x} - \frac{\partial^2 A_1}{\partial z\, \partial y}$$

$$= 0$$

assuming that $\mathbf{A}$ has continuous second partial derivatives so that the order of differentiation is immaterial.

Jacobians and curvilinear coordinates

7.40. Find ds^2 in (a) cylindrical and (b) spherical coordinates and determine the scale factors.

(a) **Method 1:**

$$x = \rho \cos \phi,\ y = \rho \sin \phi,\ = z$$

$$dx = -\rho \sin \phi d\phi + \cos \phi d\phi\rho, \qquad dy = \rho \cos \phi d\phi + \sin \phi d\rho, \qquad dz = dz$$

Then

$$ds^2 = dx^2 + dy^2 + dz^2 = (-\rho \sin\phi d\phi + \cos\phi d\rho)^2 + (\rho \cos\phi d\phi + \sin\phi d\rho)^2 + (dz)^2$$
$$= (d\rho)^2 + \rho^2 (d\phi)^2 + (dz)^2 = h_1^2 (d\rho)^2 + h_2^2 (d\phi)^2 + d_3^2 (dz)^2$$

and $h_1 = h_\rho = 1$, $h_2 = h_\phi = \rho$, $h_3 = h_z = 1$ are the scale factors.

Method 2: The position vector is $\mathbf{r} = \rho \cos \phi \mathbf{i} + \rho \sin \phi \mathbf{j} + z\mathbf{k}$. Then

$$d\mathbf{r} = \frac{\partial \mathbf{r}}{\partial \rho} d\rho + \frac{\partial \mathbf{r}}{\partial \phi} d\phi + \frac{\partial \mathbf{r}}{\partial z} dz$$
$$= (\cos\phi \mathbf{i} + \sin\phi \mathbf{j}) d\rho + (-\rho \sin\phi \mathbf{i} + \rho \cos\phi \mathbf{j}) d\phi + \mathbf{k}\, dz$$
$$= (\cos\phi d\rho - \rho \sin\phi d\phi)\mathbf{i} + (\sin\phi d\rho + \rho\cos\phi d\phi)\mathbf{j} + \mathbf{k}\, dz$$

Thus, $ds^2 = d\mathbf{r} \cdot d\mathbf{r} = (\cos \phi d\rho - \rho \sin \phi d\phi)^2 + (\sin \phi d\rho + \rho \cos \phi d\phi)^2 + (dz)^2$

$$= (d\,\rho)^2 + \rho^2 (d\,\phi)^2 + (dz)^2$$

(b) $x = r \sin \theta \cos \phi, \quad y = r \sin \theta \sin \phi, \quad z = r \cos \theta$

Then

$$dx = -r \sin \theta \sin \phi d\phi + r \cos \theta \cos \phi d\theta + \sin \theta \cos \phi dr$$

$$dy = r \sin \theta \cos \phi d\phi + r \cos \theta \sin \phi d\theta + \sin \theta \sin \phi dr$$

$$dz = -r \sin \theta d\phi + \cos \theta dr$$

and

$$(ds)^2 = (dx)^2 + (dy)^2 + (dz)^2 = (dr)^2 + r^2 (d\theta)^2 + r^2 \sin^2 \theta\, (d\phi)^2$$

The scale factors are $h_1 = h_r = 1$, $h_2 = h_\theta = r$, $h_3 = h_\phi = r \sin \theta$.

7.41. Find the volume element dV in (a) cylindrical and (b) spherical coordinates and sketch.

The volume element in orthogonal curvilinear coordinates u_1, u_2, u_3 is

$$dV = h_1 h_2 h_3\, du_1 du_2 du_3 = \left|\frac{\partial(x, y, z)}{\partial(u_1, u_2, u_3)}\right| du_1 du_2 du_3$$

(a) In cylindrical coordinates, $u_1 = \rho$, $u_2 = \phi$, $u_3 = z$, $h_1 = 1$, $h_2 = \rho$, $h_3 = 1$ [see Problem 7.40(a)]. Then

$$dV = (1)(\rho)(1)d\rho d\phi dz = \rho d\rho d\phi dz$$

This can also be observed directly from Figure 7.30(*a*).

(b) In spherical coordinates, $u_1 = r$, $u_2 = \theta$, $u_3 = \phi$, $h_1 = 1$, $h_2 = r$, $h_3 = r \sin\theta$ [see Problem 7.40(b)]. Then

$$dV = (1)(r)(r \sin\theta)\, dr d\theta d\phi = r^2 \sin\theta dr d\theta d\,\phi$$

This can also be observed directly from Figure 7.30(*b*).

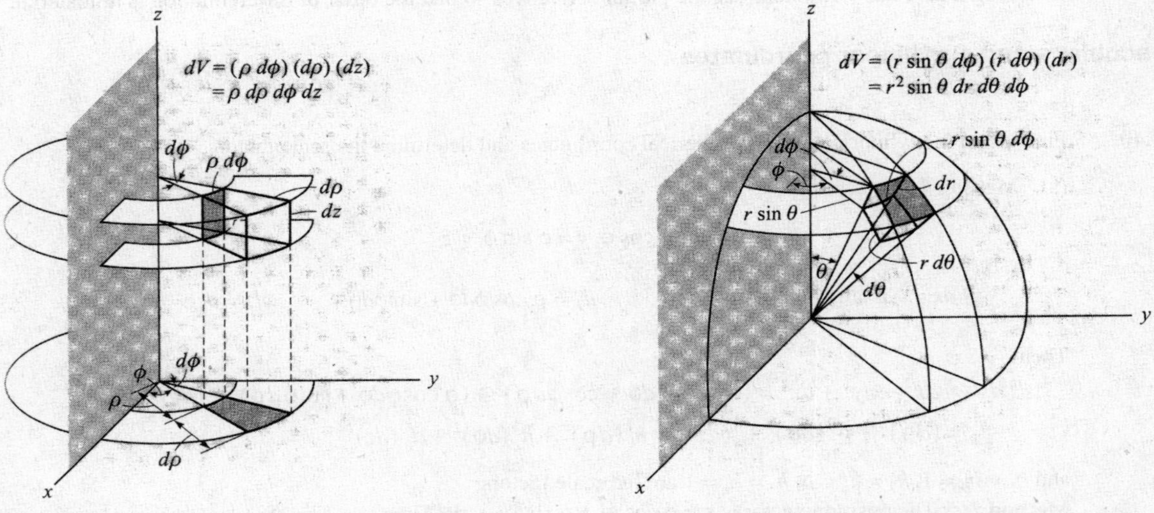

(*a*) Volume element in cylindrical coordinates.

(*b*) Volume element in spherical coordinates.

Figure 7.30

7.42. Express in cylindrical coordinates: (a) grad Φ, (b) div **A**, and (c) $\nabla^2 \Phi$.

Let $u_1 = \rho$, $u_2 = \phi$, $u_3 = z$, $h_1 = 1$, $h_2 = \rho$, $h_3 = 1$ [see Problem 7.40(a) and (b).] Then

(a) $$\text{grad}\,\Phi = \nabla\Phi = \frac{1}{1}\frac{\partial\Phi}{\partial\rho}\mathbf{e}_1 + \frac{1}{\rho}\frac{\partial\Phi}{\partial\phi}\mathbf{e}_2 + \frac{1}{1}\frac{\partial\Phi}{\partial z}\mathbf{e}_3 = \frac{\partial\Phi}{\partial\rho}\mathbf{e}_1 + \frac{1}{\rho}\frac{\partial\Phi}{\partial\phi}\mathbf{e}_2 + \frac{\partial\Phi}{\partial z}\mathbf{e}_3$$

where $\mathbf{e}_1$, $\mathbf{e}_2$, $\mathbf{e}_3$ are the unit vectors in the directions of increasing ρ, ϕ, z, respectively.

(b) $$\text{div}\,\mathbf{A} = \nabla\cdot\mathbf{A} = \frac{1}{(1)(\rho)(1)}\left[\frac{\partial}{\partial\rho}((\rho)(1)A_1) + \frac{\partial}{\partial\phi}((1)(1)A_2) + \frac{\partial}{\partial z}((1)(\rho)A_3)\right]$$

$$= \frac{1}{\rho}\left[\frac{\partial}{\partial\rho}(\rho A_1) + \frac{\partial A_2}{\partial\phi} + \frac{\partial A_3}{\partial z}\right]$$

where $\mathbf{A} = A_1\mathbf{e}_1 + A_2\mathbf{e}_2 + A_3\mathbf{e}_3$.

(c) $$\nabla^2\Phi = \frac{1}{(1)(\rho)(1)}\left[\frac{\partial}{\partial\rho}\left(\frac{(\rho)(1)}{(1)}\frac{\partial\phi}{\partial\rho}\right) + \frac{\partial}{\partial\phi}\left(\frac{(1)(1)}{(\rho)}\frac{\partial\Phi}{\partial\phi}\right) + \frac{\partial}{\partial z}\left(\frac{(1)(\rho)}{(1)}\frac{\partial\Phi}{\partial z}\right)\right]$$

$$= \frac{1}{\rho}\frac{\partial}{\partial\rho}\left(\rho\frac{\partial\Phi}{\partial\rho}\right) + \frac{1}{\rho^2}\frac{\partial^2\Phi}{\partial\phi^2} + \frac{\partial^2\Phi}{\partial z^2}$$

Miscellaneous problems

7.43. Prove that grad $f(r) = \frac{f'(r)}{r}\mathbf{r}$, where $r = \sqrt{x^2+y^2+z^2}$ and $f'(r) = df$ is df/dr is assumed to exist.

$$\begin{aligned}\text{grad } f(r) &= \nabla f(r) = \frac{\partial}{\partial x}f(r)\mathbf{i} + \frac{\partial}{\partial y}f(r)\mathbf{j} + \frac{\partial}{\partial z}f(r)\mathbf{k} \\ &= f'(r)\frac{\partial r}{\partial x}\mathbf{i} + f'(r)\frac{\partial r}{\partial y}\mathbf{j} + f'(r)\frac{\partial r}{\partial z}\mathbf{k} \\ &= f'(r)\frac{x}{r}\mathbf{i} + f'(r)\frac{y}{r}\mathbf{j} + f'(r)\frac{z}{r}\mathbf{k} = \frac{f'(r)}{r}(x\mathbf{i}+y\mathbf{j}+z\mathbf{k}) = \frac{f'(r)}{r}\mathbf{r}\end{aligned}$$

Another method: In orthogonal curvilinear coordinates u_1, u_2, u_3, we have

$$\nabla\Phi = \frac{1}{h_1}\frac{\partial\Phi}{\partial u_1}\mathbf{e}_1 + \frac{1}{h_2}\frac{\partial\Phi}{\partial u_2}\mathbf{e}_2 + \frac{1}{h_3}\frac{\partial\Phi}{\partial u_3}\mathbf{e}_3 \qquad (1)$$

If, in particular, we use spherical coordinates, we have $u_1 = r$, $u_2 = \theta$, $u_3 = \phi$. Then letting $\Phi = f(r)$, a function of r alone, the last two terms on the right of Equation *(1)* are zero. Hence, on observing that $\mathbf{e}_1 = \mathbf{r}/r$ and $h_1, = 1$, we have, the result

$$\nabla f(r) = \frac{1}{1}\frac{\partial f(r)}{\partial r}\frac{\mathbf{r}}{r} = \frac{f'(r)}{r}\mathbf{r} \qquad (2)$$

7.44. (a) Find the Laplacian of $\phi = f(r)$. (*b*) Prove that $\phi = 1/r$ is a solution of Laplace's equation $\nabla^2\phi = 0$.

(a) By Problem 7.43,

$$\nabla\phi = \nabla f(r) = \frac{f'(r)}{r}\mathbf{r}$$

By Problem 7.35, assuming that $f(r)$ has continuous second partial derivatives, we have

$$\begin{aligned}\text{Laplacian of } \phi &= \nabla^2\phi = \nabla\cdot(\nabla\phi) = \nabla\cdot\left\{\frac{f'(r)}{r}\mathbf{r}\right\} \\ &= \nabla\left\{\frac{f'(r)}{r}\right\}\cdot\mathbf{r} + \frac{f'(r)}{r}(\nabla\cdot\mathbf{r}) = \frac{1}{r}\frac{d}{dr}\left\{\frac{f'(r)}{r}\right\}\mathbf{r}\cdot\mathbf{r} + \frac{f'(r)}{r}(3) \\ &= \frac{rf''(r)+f'(r)}{r^3}r^2 + \frac{3f'(r)}{r} = f''(r) + \frac{2}{r}f'(r)\end{aligned}$$

Another method: In spherical coordinates, we have

$$\nabla^2 U = \frac{1}{r^2}\frac{\partial}{\partial r}\left(r^2\frac{\partial U}{\partial r}\right) + \frac{1}{r^2\sin\theta}\frac{\partial}{\partial\theta}\left(\sin\theta\frac{\partial U}{\partial\theta}\right) + \frac{1}{r^2\sin^2\theta}\frac{\partial^2 U}{\partial\phi^2}$$

If $U = f(r)$, the last two terms on the right are zero and we find

$$\nabla^2 f(r) = \frac{1}{r^2}\frac{d}{dr}(r^2 f'(r)) = f''(r) + \frac{2}{r}f'(r)$$

(b) From the result in (a), we have

$$\nabla^2\left(\frac{1}{r}\right) = \frac{d^2}{dr^2}\left(\frac{1}{r}\right) + \frac{2}{r}\frac{d}{dr}\left(\frac{1}{r}\right) = \frac{2}{r^3} - \frac{2}{r^3} = 0$$

showing that $1/r$ is a solution of Laplace's equation.

7.45. A particle moves along a space curve $\mathbf{r} = \mathbf{r}(t)$, where t is the time measured from some initial time. If $\upsilon = |d\mathbf{r}/dt| = ds/dt$ is the magnitude of the velocity of the particle (s is the arc length along the space curve measured from the initial position), prove that the acceleration $\mathbf{a}$ of the particle is given by

$$\mathbf{a} = \frac{dv}{dt}\mathbf{T} + \frac{v^2}{\rho}\mathbf{N}$$

where $\mathbf{T}$ and $\mathbf{N}$ are unit tangent and normal vectors to the space curve and

$$\rho = \left|\frac{d^2\mathbf{r}}{ds^2}\right|^{-1} = \left\{\left(\frac{d^2x}{ds^2}\right)^2 + \left(\frac{d^2y}{ds^2}\right)^2 + \left(\frac{d^2z}{ds^2}\right)^2\right\}^{-1/2}$$

The velocity of the particle is given by $\mathbf{v} = \upsilon\mathbf{T}$. Then the acceleration is given by

$$\mathbf{a} = \frac{d\mathbf{v}}{dt} = \frac{d}{dt}(\upsilon\mathbf{T}) = \frac{dv}{dt}\mathbf{T} + \upsilon\frac{d\mathbf{T}}{dt} = \frac{d\upsilon}{dt}\mathbf{T} + \upsilon\frac{d\mathbf{T}}{ds}\frac{ds}{dt} = \frac{d\upsilon}{dt}\mathbf{T} + \upsilon^2\frac{d\mathbf{T}}{ds} \tag{1}$$

Since $\mathbf{T}$ has a unit magnitude, we have $\mathbf{T}\cdot\mathbf{T} = 1$. Then, differentiating with respect to s,

$$\mathbf{T}\cdot\frac{d\mathbf{T}}{ds} + \frac{d\mathbf{T}}{ds}\cdot\mathbf{T} = 0, \quad 2\mathbf{T}\cdot\frac{d\mathbf{T}}{ds} = 0 \quad \text{or} \quad \mathbf{T}\cdot\frac{d\mathbf{T}}{ds} = 0$$

from which it follows that $d\mathbf{T}/ds$ is perpendicular to $\mathbf{T}$. Denoting by $\mathbf{N}$ the unit vector in the direction of $d\mathbf{T}/ds$, and called the *principal normal* to the space curve, we have

$$\frac{d\mathbf{T}}{ds} = \kappa\mathbf{N} \tag{2}$$

where k is the magnitude of $d\mathbf{T}/ds$. Now, since $\mathbf{T} = d/\mathbf{r}/ds$ [see Equation (7), Page 168], we have $d\mathbf{T}/ds = d^2\mathbf{r}/ds^2$. Hence,

$$\kappa = \left|\frac{d^2\mathbf{r}}{ds^2}\right| = \left\{\left(\frac{d^2x}{ds^2}\right)^2 + \left(\frac{d^2y}{ds^2}\right)^2 + \left(\frac{d^2}{ds}\right.\right.$$

Defining $\rho = 1/k$, Equation (2) becomes $d\mathbf{T}/ds = \mathbf{N}/\rho$. Thus, from Equation (*1*) we have, as required,

$$\mathbf{a} = \frac{d\upsilon}{dt}\mathbf{T} + \frac{\upsilon^2}{\rho}\mathbf{N}$$

The components $d\upsilon/dt$ and υ^2/ρ in the direction of $\mathbf{T}$ and $\mathbf{N}$ are called the *tangential* and *normal* components of the acceleration, the latter being sometimes called the *centripetal acceleration*. The quantities ρ and k are, respectively, the *radius of curvature* and *curvature* of the space curve.

SUPPLEMENTARY PROBLEMS

Vector algebra

7.46. Given any two vectors $\mathbf{A}$ and $\mathbf{B}$, illustrate geometrically the equality $4\mathbf{A} + 3(\mathbf{B} - \mathbf{A}) = \mathbf{A} + 3\mathbf{B}$.

7.47. A man travels 25 miles northeast, 15 miles due east, and 10 miles due south. By using an appropriate scale, determine graphically (a) how far and (b) in what direction he is from his starting position. Is it possible to determine the answer analytically?

Ans. 33.6 miles, 13.2° north of east

7.48. If $\mathbf{A}$ and $\mathbf{B}$ are any two nonzero vectors which do not have the same direction, prove that $m\mathbf{A} + n\mathbf{B}$ is a vector lying in the plane determined by $\mathbf{A}$ and $\mathbf{B}$.

7.49. If **A**, **B**, and **C** are non-coplanar vectors (vectors which do not all lie in the same plane) and $x_1\mathbf{A} + y_1\mathbf{B} + z_1\mathbf{C} = x_2\mathbf{A} + y_2\mathbf{B} + z_2\mathbf{C}$, prove that, necessarily, $x_1 = x_2$, $y_1 = y_2$, $z_1 = z_2$.

7.50. Let *ABCD* be any quadrilateral and points *P*, *Q*, *R*, and *S* the midpoints of successive sides. Prove that (a) *PQRS* is a parallelogram and (b) the perimeter of *PQRS* is equal to the sum of the lengths of the diagonals of *ABCD*.

7.51. Prove that the medians of a triangle intersect at a point which is a trisection point of each median.

7.52. Find a unit vector in the direction of the resultant of vectors $\mathbf{A} = 2\mathbf{i} - \mathbf{j} + \mathbf{k}$, $\mathbf{B} = \mathbf{i} + \mathbf{j} + 2\mathbf{k}$, $\mathbf{C} = 3\mathbf{i} - 2\mathbf{j} + 4\mathbf{k}$.

Ans. $(6\mathbf{i} - 2\mathbf{j} + 7\mathbf{k})/\sqrt{89}$

The dot or scalar product

7.53. Evaluate $|(\mathbf{A} + \mathbf{B}) \cdot (\mathbf{A} - \mathbf{B})|$ if $\mathbf{A} = 2\mathbf{i} - 3\mathbf{j} + 5\mathbf{k}$ and $\mathbf{B} = 3\mathbf{i} + \mathbf{j} - 2\mathbf{k}$.

Ans. 24

7.54. Verify the consistency of the law of cosines for a triangle. [Hint: Take the sides of **A**, **B**, **C** where $\mathbf{C} = \mathbf{A} - \mathbf{B}$. Then use $\mathbf{C} \cdot \mathbf{C} = (\mathbf{A} - \mathbf{B}) \cdot (\mathbf{A} - \mathbf{B})$.]

7.55. Find a so that $2\mathbf{i} - 3\mathbf{j} + 5\mathbf{k}$ and $3\mathbf{i} + a\mathbf{j} - 2\mathbf{k}$ are perpendicular.

Ans. $a = -4/3$

7.56. If $\mathbf{A} = 2\mathbf{i} + \mathbf{j} + \mathbf{k}$, $\mathbf{B} = \mathbf{i} - 2\mathbf{j} + 2\mathbf{k}$ and $\mathbf{C} = 3\mathbf{i} - 4\mathbf{j} + 2\mathbf{k}$, find the projection of $\mathbf{A} + \mathbf{C}$ in the direction of **B**.

Ans. 17/3

7.57. A triangle has vertices at $A(2, 3, 1)$, $B(-1, 1, 2)$, and $C(1, -2, 3)$. Find (a) the length of the median drawn from B to side AC and (b) the acute angle which this median makes with side BC.

Ans. (a) $\frac{1}{2}\sqrt{26}$ (b) $\cos^{-1}\sqrt{91/14}$

7.58. Prove that the diagonals of a rhombus are perpendicular to each other.

7.59. Prove that the vector $(AB + BA)/(A + B)$ represents the bisector of the angle between **A** and **B**.

The cross or vector product

7.60. If $\mathbf{A} = 2\mathbf{i} - \mathbf{j} + \mathbf{k}$ and $\mathbf{B} = \mathbf{i} + 2\mathbf{j} - 3\mathbf{k}$, find $|(2\mathbf{A} + \mathbf{B}) \times (\mathbf{A} - 2\mathbf{B})|$.

Ans. $5\sqrt{3}$

7.61. Find a unit vector perpendicular to the plane of the vectors $\mathbf{A} = 3\mathbf{i} - 2\mathbf{j} + 4\mathbf{k}$ and $\mathbf{B} = \mathbf{i} + \mathbf{j} - 2\mathbf{k}$.

Ans. $\pm(2\mathbf{j} + \mathbf{k})/\sqrt{5}$

7.62. If $\mathbf{A} \times \mathbf{B} = \mathbf{A} \times \mathbf{C}$, does $\mathbf{B} = \mathbf{C}$ necessarily?

7.63. Find the area of the triangle with vertices (2, –3, 1), (1, –1, 2), (–1, 2, 3).

Ans. $\frac{1}{2}\sqrt{3}$

7.64 Find the shortest distance from the point (3, 2, 1) to the plane determine by (1, 1, 0), (3, –1, 1), (–1, 0, 2).

Ans. 2

Triple products

7.65. If $\mathbf{A} = 2\mathbf{i} + \mathbf{j} - 3\mathbf{k}$, $\mathbf{B} = \mathbf{i} - 2\mathbf{j} + \mathbf{k}$, $\mathbf{C} = -\mathbf{i} + \mathbf{j} - 4$, find (a) $\mathbf{A} \cdot (\mathbf{B} \times \mathbf{C})$, (b) $\mathbf{C} \cdot (\mathbf{A} \times \mathbf{B})$, (c) $\mathbf{A} \times (\mathbf{B} \times \mathbf{C})$, and (d) $(\mathbf{A} \times \mathbf{B}) \times \mathbf{C}$.

Ans. (a) 20 (b) 20 (c) $8\mathbf{i} - 19\mathbf{j} - \mathbf{k}$ (d) $25\mathbf{i} - 15\mathbf{j} - 10\mathbf{k}$

7.66. Prove that (a) $\mathbf{A} \cdot (\mathbf{B} \times \mathbf{C}) = \mathbf{B} \cdot (\mathbf{C} \times \mathbf{A}) = \mathbf{C} \cdot (\mathbf{A} \times \mathbf{B})$ and (b) $\mathbf{A} \times (\mathbf{B} \times \mathbf{C}) = \mathbf{B}(\mathbf{A} \cdot \mathbf{C}) - \mathbf{C}(\mathbf{A} \cdot \mathbf{B})$.

7.67. Find an equation for the plane passing through (2, –1, –2), (–1, 2, –3), (4, 1, 0).

Ans. $2x + y - 3z = 9$

7.68. Find the volume of the tetrahedron with vertices at (2, 1, 1), (1, –1, 2), (0, 1, –1), (1, –2, 1).

Ans. $\frac{4}{3}$

7.69. Prove that $(\mathbf{A} \times \mathbf{B}) \cdot (\mathbf{C} \times \mathbf{D}) + (\mathbf{B} \times \mathbf{C}) \cdot (\mathbf{A} \times \mathbf{D}) + (\mathbf{C} \times \mathbf{A}) \cdot (\mathbf{B} \times \mathbf{D}) = 0$.

Derivatives

7.70 A particle moves along the space curve $\mathbf{r} = e^{-t}\cos t\mathbf{i} + e^{-t}\sin t\mathbf{j} + e^{-t}\mathbf{k}$. Find the magnitude of the (a) the velocity and (b) the acceleration at any time t.

Ans. (a) $\sqrt{3}e^{-1}$ (b) $\sqrt{5}e^{-1}$

7.71. Prove that $\frac{d}{du}(\mathbf{A} \times \mathbf{B}) = A \times \frac{d\mathbf{B}}{du} + \frac{d\mathbf{A}}{du} \times \mathbf{B}$ where **A** and **B** are differentiable functions of u.

7.72. Find a unit vector tangent to the space curve $x = t, y = t^2\ z = t^3$ at the point where $t = 1$.

Ans. $(\mathbf{i} + 2\mathbf{j} + 3\mathbf{k})/\sqrt{14}$

7.73. If $\mathbf{r} = \mathbf{a}\cos \omega t + \mathbf{b}\sin \omega t$, where **a** and **b** are any constant noncollinear vectors and ω is a constant scalar. prove that (a) $\mathrm{r} = \frac{d\mathbf{r}}{dr} = \omega(\mathrm{a} \times \mathrm{b})$ and (b) $\frac{d^2\mathbf{r}}{dt^2} + \omega^2\mathbf{r} = 0$.

7.74. If $\mathbf{A} = x^2\mathbf{i} - y\mathbf{j} + xz\mathbf{k}$, $\mathbf{B} = y\mathbf{i} + x\mathbf{j} - xyz\mathbf{k}$, and $\mathbf{C} = \mathbf{i} - y\mathbf{j} + x^3 z\mathbf{k}$, find (a) $\frac{\partial^2}{\partial x\, \partial y}(\mathbf{A} + \mathbf{B})$ and (b) $d[\mathbf{A} \cdot (\mathbf{B} \times \mathbf{C})]$ at the point (1, –1, 2).

Ans. (a) $-4\mathbf{i} + 8\mathbf{j}$ (b) $8\,dx$

7.75. If $R = x^2 y\mathbf{i} - 2y^2 z\mathbf{j} + xy^2 z^2\mathbf{k}$, find $\left|\frac{\partial^2 \mathbf{B}}{\partial x^2} \times \frac{\partial^2 \mathbf{R}}{\partial y^2}\right|$ at the point (2, 1, – 2).

Ans. $16\sqrt{5}$

Gradient, divergence, and curl

7.76. If $U, V, \mathbf{A}, \mathbf{B}$ have continuous partial derivatives, prove that (a) $\nabla(U + V) = \nabla U + \nabla V$, (b) $\nabla \cdot (\mathbf{A} + \mathbf{B}) = \nabla \cdot \mathbf{A} + \nabla \cdot \mathbf{B}$ and (c) $\nabla \times (\mathbf{A} + \mathbf{B}) = \nabla \times \mathbf{A} + \nabla \times \mathbf{B}$.

7.77. If $\phi = xy + yz + zx$ and $\mathbf{A} = x^2 y\mathbf{i} + y^2 z\mathbf{j} + z^2 x\mathbf{k}$, find (a) $\mathbf{A} \cdot \nabla\phi$, (b) $\phi\nabla \cdot \mathbf{A}$, and (c) $(\nabla\phi) \times \mathbf{A}$ at the point (3, –1, 2).

Ans. (a) 25, (b) 2, (c) $56\mathbf{i} - 30\mathbf{j} + 47\mathbf{k}$

7.78. Show that $\nabla \times (r^2\mathbf{r}) = 0$ where $\mathbf{r} = x\mathbf{i} + y\mathbf{j} + z\mathbf{k}$ and $r = |\mathbf{r}|$.

7.79. Prove that (a) $\nabla \times (U\mathbf{A}) = (\nabla U) \times \mathbf{A} + U(\nabla \times \mathbf{A})$ and (b) $\nabla \cdot (\mathbf{A} \times \mathbf{B}) = \mathbf{B} \cdot (\nabla \times \mathbf{A}) - \mathbf{A} \cdot (\nabla \times \mathbf{B})$.

7.80. Prove that curl grad $u = 0$, stating appropriate conditions on U.

7.81. Find a unit normal to the surface $x^2 y - 2xz + 2y^2 z^4 = 10$ at the point (2, 1, –1).

Ans. $\pm(3\mathbf{i} + 4\mathbf{j} - 6\mathbf{k})\ \sqrt{61}$

7.82. If $\mathbf{A} = 3xz^2\mathbf{i} - yz\mathbf{j} + (x + 2z)\mathbf{k}$, find curl $\mathbf{A}$.

Ans. $-6x\mathbf{i} + (6z - 1)\mathbf{k}$

7.83. (a) Prove that $\nabla \times (\nabla \times \mathbf{A}) = -\nabla^2 \mathbf{A} + \nabla(\nabla \cdot \mathbf{A})$. (b) Verify the result in (a) if $\mathbf{A}$ is given as in problem 7.82.

Jacobians and curvilinear coordinates

7.84. Prove that $\left|\frac{\partial(x, y, z)}{\partial(u_1, u_2, u_3)}\right| = \left|\frac{\partial \mathrm{r}}{\partial u_1} \cdot \frac{\partial \mathrm{r}}{\partial u_2} \times \frac{\partial \mathrm{r}}{\partial u_3}\right|$.

7.85. Express (a) grad Φ, (b) div $\mathbf{A}$, and (c) $\nabla^2 \Phi$ in spherical coordinates.

Ans. (a) $\frac{\partial \Phi}{\partial r}\mathrm{e}_1 + \frac{1}{r}\frac{\partial \Phi}{\partial \theta}\mathrm{e}_2 + \frac{1}{r \sin\theta}\frac{\partial \Phi}{\partial \phi}\mathrm{e}_3$

(b) $\frac{1}{r^2}\frac{\partial}{\partial r}(r^2 A_1) + \frac{1}{r \sin\theta}\frac{\partial}{\partial \theta}(\sin\theta A_2) + \frac{1}{r \sin\theta}\frac{\partial A^3}{\partial \phi}$ where $\mathbf{A} = A_1\mathrm{e}_1 + A_2\mathrm{e}_2 + A_3\mathrm{e}_3$

(c) (c) $\frac{1}{r^2}\frac{\partial}{\partial r}\left(r^2 \frac{\partial \Phi}{\partial r}\right) + \frac{1}{r^2 \sin\theta}\frac{\theta}{\partial \theta}\left(\sin\theta \frac{\partial \Phi}{\partial \theta}\right) + \frac{1}{r^2 \sin^2\theta}\frac{\partial^2 \Phi}{\partial \phi^2}$

7.86. The transformation from rectangular to *parabolic cylindrical coordinates* is defined by the equations $x = 1/2 (u^2 - \upsilon^2)$, $y = u\,\upsilon$, $z = z$. (a) Prove that the system is orthogonal. (b) Find ds^2 and the scale factors. (c) Find the Jacobian of the transformation and the volume element.

Ans. (b) $ds^2 = (u^2 + \upsilon^2)du^2 + (u)^2 + \upsilon^2)d\,\upsilon^2 + dz^2$, $h_1 = h_2 = \sqrt{u^2 + \upsilon^2}$, $h_3 = 1$

(c) $u^2 + \upsilon^2$, $(u^2 + \upsilon^2)\, du\, d\upsilon\, dz$

7.87. Write (a) $\nabla^2 \Phi$ and (b) div **A** in parabolic cylindrical coordinates.

Ans. (a) $\nabla^2\Phi = \dfrac{1}{u^2 + \upsilon^2}\left(\dfrac{\partial^2\Phi}{\partial u^2} + \dfrac{\partial^2\Phi}{\partial \upsilon^2}\right) + \dfrac{\partial^2\Phi}{\partial z^2}$

(b) $\text{div}\,\mathbf{A} = \dfrac{1}{u^2 + \upsilon^2}\left\{\dfrac{\partial}{\partial u}(\sqrt{u^2 + \upsilon^2}\, A_1) + \dfrac{\partial}{\partial \upsilon}(\sqrt{u^2 + \upsilon^2}\, A_2)\right\} + \dfrac{\partial A_3}{\partial z}$

7.88. Prove that for orthogonal curvilinear coordinates,

$$\nabla\Phi = \frac{\mathbf{e}_1}{h_1}\frac{\partial\Phi}{\partial u_1} + \frac{\mathbf{e}_2}{h_2}\frac{\partial\Phi}{\partial u_2} + \frac{\mathbf{e}_3}{h_3}\frac{\partial\Phi}{\partial u_3}$$

(Hint: Let $\nabla\Phi = a_1\mathbf{e}_1 + a_2\mathbf{e}_2 + a_3\mathbf{e}_3$ and use the fact that $d\Phi = \nabla\Phi \cdot d\mathbf{r}$ must be the same in both rectangular and curvilinear coordinates.)

7.89. Give a vector interpretation to the theorem in Problem 6.35.

Miscellaneous problems

7.90. If **A** is a differentiable function of u and $|\mathbf{A}(u)| = 1$, prove that $d\mathbf{A}/du$ is perpendicular to **A**.

7.91. Prove formulas 6, 7, and 8 on Page 171.

7.92. If ρ and ϕ are polar coordinates and A, B, n are any constants, prove that $U = \rho^n (A \cos n\phi + B \sin n\,\phi)$ satisfies Laplace's equation.

7.93. If $V = \dfrac{2\cos\theta + 3\sin^3\theta\cos\phi}{r^2}$, find $\nabla^2 V$.

Ans. $\dfrac{6\sin\theta\cos\phi(4 - 5\sin^2\theta)}{r^4}$

7.94. Find the most general function of (a) the cylindrical coordinate ρ, (b) the spherical coordinate r, and (c) the spherical coordinate θ which satisfies Laplace's equation.

Ans. (a) $A + B\ln\rho$ (b) $A + B/r$ (c) $A + B\ln(\csc\theta - \cot\theta)$ where A and B are any constants

7.95. Let **T** and **N** denote, respectively, the unit *tangent vector* and unit *principal normal* vector to a space curve $\mathbf{r} = \mathbf{r}(u)$, where $\mathbf{r}(u)$ is assumed differentiable. Define a vector $\mathbf{B} = \mathbf{T} = \mathbf{T} \times \mathbf{N}$ called the unit *binormal vector* to the space curve. Prove that

$$\frac{d\mathbf{T}}{ds} = \kappa\mathbf{N}, \quad \frac{d\mathbf{B}}{ds} = -\tau\mathbf{N}, \quad \frac{d\mathbf{N}}{ds} = \tau\mathbf{B} - \kappa\mathbf{T}$$

These are called the *Frenet-Serret* formulas and are of fundamental importance in *differential geometry*. In these formulas k is called the *curvature*, τ is called the *torsion*; and the reciprocals of these, $\rho = 1/k$ and $\sigma = 1/\tau$, are called the *radius of curvature* and *radius of torsion*, respectively.

7.96. (a) Prove that the radius of curvature at any point of the plane curve $y = f(x)$, $z = 0$ where $f(x)$ is differentiable, is given by

$$\rho = \left|\frac{(1 + y'^2)^{3/2}}{y''}\right|$$

(b) Find the radius of curvature at the point $(\pi/2, 1, 0)$ of the curve $y = \sin x$, $z = 0$.

Ans. (b) $2\sqrt{2}$

7.97. Prove that the acceleration of a particle along a space curve is given respectively in (a) cylindrical and (b) spherical coordinates by

$$(\ddot{\rho} - \rho\dot{\phi}^2)\mathbf{e}_\rho + (\rho\ddot{\phi} + 2\dot{\rho}\dot{\phi})\mathbf{e}_\phi + \ddot{z}\mathbf{e}_z$$

$$(\ddot{r} - r\dot{\theta}^2 - r\dot{\phi}^2 \sin^2\theta)\mathbf{e}_r + (r\ddot{\theta} + 2\dot{r}\dot{\theta} - r\dot{\phi}^2 \sin\theta \cos\theta)\mathbf{e}_\theta + (2\dot{r}\dot{\phi} \sin\theta + 2r\dot{\theta}\dot{\phi} \cos\theta + r\ddot{\phi} \sin\theta)e_\phi$$

where dots denote time derivatives and $\mathbf{e}_\rho$, $\mathbf{e}_\phi$, $\mathbf{e}_z$, $\mathbf{e}_r$, $\mathbf{e}_\theta$, $\mathbf{e}_\phi$ are unit vectors in the directions of increasing ρ, ϕ, z, r, θ, ϕ, respectively.

7.98. Let **E** and **H** be two vectors assumed to have continuous partial derivatives (of second order at least) with respect to position and time. Suppose further that **E** and **H** satisfy the equations

$$\nabla \cdot \mathbf{E} = 0, \quad \nabla \cdot \mathbf{H} = 0, \quad \nabla \times \mathbf{E} = -\frac{1}{c}\frac{\partial \mathbf{H}}{\partial t}, \quad \nabla \times \mathbf{H} = \frac{1}{c}\frac{\partial \mathbf{E}}{\partial t}, \tag{1}$$

prove that **E** and **H** satisfy the equation

$$\nabla^2\psi = \frac{1}{c^2}\frac{\partial^2\psi}{\partial t^2} \tag{2}$$

where ψ is a generic meaning and, in particular, can represent any component of **E** or **H**.

[The vectors **E** and **H** are called *electric* and *magnetic field vectors* in *electromagnetic theory*. Equations (*1*) are a special case of *Maxwell's equations*. The result (2) led Maxwell to the conclusion that light was an electromagnetic phenomena. The constant c is the velocity of light.]

7.99. Use the relations in Problem 7.98 to show that

$$\frac{\partial}{\partial t}\{\frac{1}{2}(E^2 + H^2)\} + c\nabla \cdot (\mathbf{E} \times \mathbf{H}) = 0$$

7.100. Let A_1, A_2, A_3 be the components of vector **A** in an xyz rectangular coordinate system with unit vectors $\mathbf{i}_1$, $\mathbf{i}_2$, $\mathbf{i}_3$ (the usual **i, j, k** vectors), and A'_1, A'_2, A'_3 the components of **A** in an $x'\, y'\, z'$ rectangular coordinate system which has the same origin as the xyz system but is rotated with respect to it and has the unit vectors $\mathbf{i}'_1$, $\mathbf{i}'_2$, $\mathbf{i}'_3$. Prove that the following relations (often called *invariance* relations) must hold:

$$A_n = l_{1n}A'_1 + l_{2n}A'_2 + l_{3n}A'_3 \qquad n = 1, 2, 3$$

where $\mathbf{i}'_m \cdot \mathbf{i}_n = l_{mn}$.

7.101. If **A** is the vector of Problem 7.100, prove that the divergence of **A** ($\nabla \cdot \mathbf{A}$) is an invariant (often called a *scalar invariant*); i.e., prove that

$$\frac{\partial A_1'}{\partial x'} + \frac{\partial A_2'}{\partial y'} + \frac{\partial A_3'}{\partial z'} = \frac{\partial A_1}{\partial x} + \frac{\partial A_2}{\partial y} + \frac{\partial A_3}{\partial z}$$

The results of this and the preceding problem express an obvious requirement that physical quantities must not depend on coordinate systems in which they are observed. Such ideas when generalized lead to an important subject called *tensor analysis*, which is basic to the *theory of relativity*.

7.102. Prove that (a) $\mathbf{A} \cdot \mathbf{B}$, (b) $\mathbf{A} \times \mathbf{B}$, and (c) $\nabla \times \mathbf{A}$ are invariant under the transformation of Problem 7.100.

7.103. If u_1, u_2, u_3 are orthogonal curvilinear coordinates, prove that

$$\text{(a)}\ \frac{\partial(u_1, u_2, u_3)}{\partial(x, y, z)} = \nabla u_1 \cdot \nabla u_2 \times \nabla u_3 \quad \text{(b)}\ \left(\frac{\partial \mathbf{r}}{\partial u_1} \cdot \frac{\partial \mathbf{r}}{\partial u_2} \times \frac{\partial \mathbf{r}}{\partial u_3}\right)(\nabla u_1 \cdot \nabla u_2 \times \nabla u_3) = 1$$

and give the significance of these in terms of Jacobians.

7.104. Use the axiomatic approach to vectors to prove relation 8 on Page 167.

7.105. A set of n vectors $\mathbf{A}_1, \mathbf{A}_2, \ldots, \mathbf{A}_n$ is called *linearly dependent* if there exists a set of scalars $c_1, c_2, \ldots, c_n$ not all zero such that $c_1\mathbf{A}_1 + c_2\mathbf{A}_2 + \cdots + c_n\mathbf{A}_n = \mathbf{0}$ identically; otherwise, the set is called *linearly independent*. (a) Prove that the vectors $\mathbf{A}_1 = 2\mathbf{i} - 3\mathbf{j} + 5\mathbf{k}$, $\mathbf{A}_2 = \mathbf{i} + \mathbf{j} - 2\mathbf{k}$, $\mathbf{A}_3 = 3\mathbf{i} - 7\mathbf{j} + 12\mathbf{k}$ are linearly dependent. (b) Prove that any four three-dimensional vectors are linearly dependent. (c) Prove that a necessary and sufficient condition that the vectors $\mathbf{A}_1 = a_1\mathbf{i} + b_1\mathbf{j} + c_1\mathbf{k}$, $\mathbf{A}_2 = a_2\mathbf{i} + b_2\mathbf{j} + c_2\mathbf{k}$, and $\mathbf{A}_3 = a_3\mathbf{i} + b_3\mathbf{j} + c_3\mathbf{k}$ be linearly independent is that $\mathbf{A}_1 \cdot \mathbf{A}_2 \times \mathbf{A}_3 \neq 0$. Give a geometrical interpretation of this.

7.106. A complex number can be defined as an ordered pair (a, b) of real numbers a and b subject to certain rules of operation for addition and multiplication. (a) What are these rules? (b) How can the rules in (a) be used to define subtraction and division? (c) Explain why complex numbers can be considered as two-dimensional vectors. (d) Describe similarities and differences between various operations involving complex numbers and the vectors considered in this chapter.

CHAPTER 8

Applications of Partial Derivatives

Applications to Geometry

The theoretical study of curves and surfaces began more than two thousand years ago when Greek philosopher-mathematicians explored the properties of conic sections, helixes, spirals, and surfaces of revolution generated from them. While applications were not on their minds, many practical consequences evolved. These included representation of the elliptical paths of planets about the sun, employment of the focal properties of paraboloids, and use of the special properties of helixes to construct the double helical model of DNA.

The analytic tool for studying functions of more than one variable is the *partial derivative*. Surfaces are a geometric starting point, since they are represented by functions of two independent variables. Vector forms of many of these concepts were introduced in Chapter 7. In this chapter, corresponding coordinate equations are exhibited.

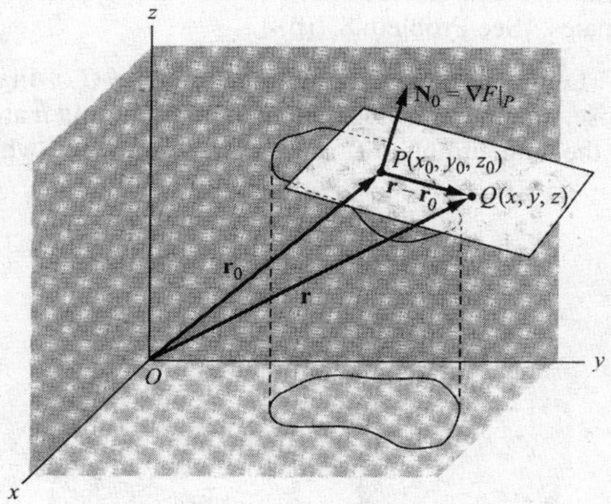

Figure 8.1

1. Tangent Plane to a Surface Let $F(x, y, z) = 0$ be the equation of a surface S such as that shown in Figure 8.1. Assume that F, and all other functions in this chapter are continuously differentiable unless otherwise indicated. Suppose we wish to find the equation of a tangent plane to S at the point $P(x_0, y_0, z_0)$. A vector normal to S at this point is $\mathbf{N}_0 = \nabla F\big|_P$, the subscript P indicating that the gradient is to be evaluated at the point $P(x_0, y_0, z_0)$.

If $\mathbf{r}_0$ and $\mathbf{r}$ are tangent, the vectors drawn, respectively, from O to $P(x_0, y_0, z_0)$ and $Q(x, y, z)$ on the tangent plane, the equation of the plane is

$$(\mathbf{r}-\mathbf{r}_0)\cdot\mathbf{N}_0=(\mathbf{r}-\mathbf{r}_0)\cdot\nabla F\big|_P=0 \tag{1}$$

since $\mathbf{r}-\mathbf{r}_0$ is perpendicular to $\mathbf{N}_0$.

In rectangular form this is

$$\left.\frac{\partial F}{\partial x}\right|_P(x-x_0)+\left.\frac{\partial F}{\partial y}\right|_P(y-y_0)+\left.\frac{\partial F}{\partial z}\right|_P(z-z_0)=0 \tag{2}$$

In case the equation of the surface is given in orthogonal curvilinear coordinates in the form $F(u_1, u_2, u_3) = 0$, the equation of the tangent plane can be obtained using the result on Page 172 for the gradient in these coordinates. See Problem 8.4.

2. Normal Line to a Surface. Suppose we require equations for the normal line to the surface S at $P(x_0, y_0, z_0)$, i.e., the line perpendicular to the tangent plane of the surface at P. If we now let $\mathbf{r}$ be the vector drawn from O in Figure 8.1 to any point (x, y, z) on the normal $\mathbf{N}_0$, we see that $\mathbf{r}-\mathbf{r}_0$ is collinear with $\mathbf{N}_0$, and so the required condition is

$$(\mathbf{r}-\mathbf{r}_0\times\mathbf{N}_0=(\mathbf{r}-\mathbf{r}_0)\times\times\nabla F\big|_P=\mathbf{0} \tag{3}$$

By expressing the cross product in the determinant form

$$\begin{vmatrix} i & j & k \\ x-x_0 & y-y_0 & z-z_0 \\ F_x|_P & F_y|_P & F_z|_P \end{vmatrix}$$

we find that

$$\frac{x-x_0}{\left.\frac{\partial F}{\partial x}\right|_P}=\frac{y-y_0}{\left.\frac{\partial F}{\partial y}\right|_P}=\frac{z-z_0}{\left.\frac{\partial F}{\partial z}\right|_P} \tag{4}$$

Setting each of these ratios equal to a parameter (such as t or u) and solving for x, y, and z yields the *parametric equations* of the normal line.

The equations for the normal line can also be written when the equation of the surface is expressed in orthogonal curvilinear coordinates. [See Problem 8.1(b).]

3. Tangent Line to a Curve Let the parametric equations of curve C of Figure 8.2 be $x = f(u)$, $y = g(u)$, $z = h(u)$, where we shall suppose, unless otherwise indicated, that f, g, and h are continuously differentiable. We wish to find equations for the tangent line to C at the point $P(x_0, y_0, z_0)$ where $u = u_0$.

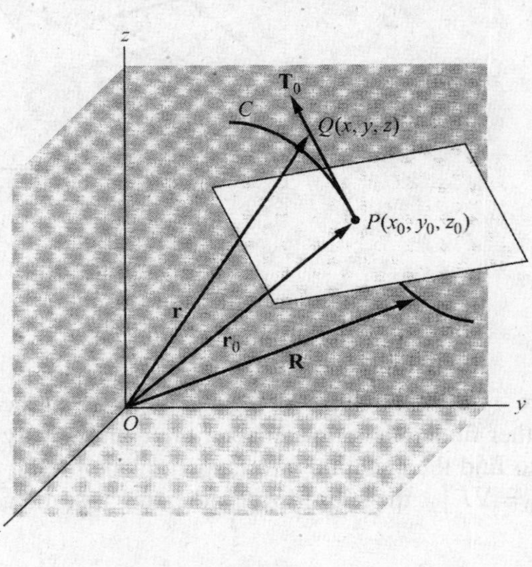

Fig. 8.2

If $\mathbf{R} = f(u)\mathbf{i} + g(u)\mathbf{j} + h(u)\mathbf{k}$, then a vector tangent to C at the point P is given by $\mathbf{T}_0 = \left.\frac{d\mathbf{R}}{du}\right|_P$. If $\mathbf{r}_0$ and $\mathbf{r}$ denote the vectors drawn respectively from O to $P(x_0, y_0, z_0)$ and $Q(x, y, z)$ on the tangent line, then $\mathbf{r} - \mathbf{r}_0$ is collinear with $\mathbf{T}_0$. Thus,

$$(\mathbf{r}-\mathbf{r}_0)\times\mathbf{T}_0 = (\mathbf{r}-\mathbf{r}_0)\times\left.\frac{d\mathbf{R}}{du}\right|_P = 0 \tag{5}$$

In rectangular form this becomes

$$\frac{x-x_0}{f'(u_0)} = \frac{y-y_0}{g'(u_0)} = \frac{z-z_0}{h'(u_0)} \tag{6}$$

The parametric form is obtained by setting each ratio equal to u.

If the curve C is given as the intersection of two surfaces with equations $F(x, y, z) = 0$ and $G(x, y, z) = 0$, observe that $\nabla F \times \nabla G$ has the direction of the line of intersection of the tangent planes; therefore, the corresponding equations of the tangent line are

$$\frac{x-x_0}{\left.\begin{vmatrix} F_y & F_z \\ G_y & G_z \end{vmatrix}\right|_P} = \frac{y-y_0}{\left.\begin{vmatrix} F_z & F_x \\ G_z & G_x \end{vmatrix}\right|_P} = \frac{z-z_0}{\left.\begin{vmatrix} F_x & F_y \\ G_x & G_y \end{vmatrix}\right|_P} \tag{7}$$

Note that the determinants in Equation (7) are Jacobians. A similar result can be found when the surfaces are given in terms of orthogonal curvilinear coordinates.

4. Normal Plane to a Curve Suppose we wish to find an equation for the normal plane to curve C at $P(x_0, y_0, z_0)$ in Figure 8.2 (i.e., the plane perpendicular to the tangent line to C at this point). Letting $\mathbf{r}$ be the vector from O to any point (x, y, z) on this plane, it follows that $\mathbf{r} - \mathbf{r}_0$ is perpendicular to $\mathbf{T}_0$. Then the required equation is

$$(\mathbf{r}-\mathbf{r}_0)\cdot\mathbf{T}_0 = (\mathbf{r}-\mathbf{r}_0)\cdot\left.\frac{d\mathbf{R}}{du}\right|_P = 0 \tag{8}$$

When the curve has parametric equations $x = f(u)$, $y = g(u)$, $z = h(u)$ this becomes

$$f'(u_0)(x-x_0) + g'(u_0)(y-y_0) + h'(u_0)(z-z_0) = 0 \tag{9}$$

Furthermore, when the curve is the intersection of the implicitly defined surfaces

$$F(x, y, z) = 0 \text{ and } G(x, y, z) = 0$$

then

$$\left.\begin{vmatrix} F_y & F_z \\ G_y & G_z \end{vmatrix}\right|_P (x-x_0) + \left.\begin{vmatrix} F_z & F_x \\ G_z & G_x \end{vmatrix}\right|_P (y-y_0) + \left.\begin{vmatrix} F_x & F_y \\ G_x & G_y \end{vmatrix}\right|_P (z-z_0) = 0 \tag{10}$$

5. Envelopes Solutions of differential equations in two variables are geometrically represented by one-parameter families of curves. Sometimes such a family characterizes a curve called an *envelope*.

For example, the family of all lines (see Problem 8.9) one unit from the origin may be represented by $x \sin\alpha - y\cos\alpha - 1 = 0$, where α is a parameter. The envelope of this family is the circle $x^2 + y^2 = 1$.

If $\phi(x, y, z) = 0$, is a one-parameter family of curves in the xy plane, there may be a curve E which is tangent at each point to some member of the family and such that each member of the family is tangent to E. If E exists, its equation can be found by solving simultaneously the equations

$$\phi(x, y, \alpha) = 0,\ \phi_\alpha(x, y, \alpha) = 0 \tag{11}$$

and E is called the *envelope* of the family.

The result can be extended to determine the envelope of a one-parameter family of surfaces $\phi(x, y, z, \alpha)$. This envelope can be found from

$$\phi(x, y, z, \alpha) = 0,\ \phi_\alpha(x, y, z, \alpha) = 0 \tag{12}$$

Extensions to two- (or more) parameter families can be made.

Directional Derivatives

Suppose $F(x, y, z)$ is defined at a point (x, y, z) on a given space curve C. Let $F(x + \Delta x, y + \Delta y, z + \Delta z)$ be the value of the function at a neighboring point on C and let Δs denote the length of arc of the curve between those points. Then

$$\lim_{\Delta s \to 0} \frac{\Delta F}{\Delta s} = \lim_{\Delta s \to 0} \frac{F(x+\Delta x,\ y+\Delta y,\ z+\Delta z) - F(x,\ y,\ z)}{\Delta s} \tag{13}$$

if it exists, is called the *directional derivative* of F at the point (x, y, z) along the curve C and is given by

$$\frac{dF}{ds} = \frac{\partial F}{\partial x}\frac{dx}{ds} + \frac{\partial F}{\partial y}\frac{dy}{ds} + \frac{\partial F}{\partial z}\frac{dz}{ds} \tag{14}$$

In vector form this can be written

$$\frac{dF}{ds} = \left(\frac{\partial F}{\partial x}\mathbf{i} + \frac{\partial F}{\partial y}\mathbf{j} + \frac{\partial F}{\partial z}\mathbf{k}\right) \cdot \left(\frac{dx}{ds}\mathbf{i} + \frac{dy}{ds}\mathbf{j} + \frac{dz}{ds}\mathbf{k}\right) = \nabla F \cdot \frac{dr}{ds} = \nabla F \cdot \mathbf{T} \tag{15}$$

from which it follows that the directional derivative is given by the component of ∇F in the direction of the tangent to C.

Thus, the maximum value of the directional derivative is given by $|\nabla F|$ and these maxima occur in directions normal to the surfaces $F(x, y, z) = c$ is (where c is any constant), which are sometimes called *equipotential surfaces* or *level surfaces*.

Differentiation Under the Integral Sign

Let

$$\phi(\alpha) = \int_{u_1}^{u_2} f(x, \alpha)\, dx \qquad a \leqq \alpha \leqq b \tag{16}$$

where u_1 and u_2 may depend on the parameter α. Then

$$\frac{d\phi}{d\alpha} = \int_{u_2}^{u_1} \frac{\partial f}{\partial \alpha} dx + f(u_2, \alpha)\frac{du_2}{d\alpha} - f(u_1, \alpha)\frac{du_1}{d\alpha} \tag{17}$$

for $a \leqq \alpha \leqq b$, if $f(x, \alpha)$ and $\partial f/\partial \alpha$ are continuous in both x and α in some region of the $x\alpha$ plane including $u_1 \leqq x \leqq u_2$, $a \leqq \alpha \leqq b$ and if u_1 and u_2 are continuous and have continuous derivatives for $a \leqq \alpha \leqq b$.

In case u_1 and u_2 are constants, the last two terms of Equation (*17*) are zero.

The result (*17*), called *Leibniz's rule*, is often useful in evaluating definite integrals (see Problems 8.15 and 8.29).

Integration Under the Integral Sign

If $\phi(\alpha)$ is defined by Equation (*16*) and $f(x, \alpha)$ is continuous in x and α in a region including $u_1 \leqq x \leqq u_2$, $a \leqq x \leqq b$, then if u_1 and u_2 are constants,

$$\int_a^b \phi(\alpha)\, d\alpha = \int_a^b \left\{ \int_{u_1}^{u_2} f(x,\alpha)dx \right\} d\alpha = \int_{u_1}^{u_2} \left\{ \int_a^b f(x,\alpha)d\alpha \right\} dx \tag{18}$$

The result is known as *interchange of the order of integration* or *integration under the integral sign*. (See Problem 8.18.)

Maxima and Minima

In Chapter 4 we briefly examined relative extrema for functions of one variable. The general idea was that for points of the graph of $y = g(x)$ that were locally highest or lowest, the condition $g'(x) = 0$ was necessary. Such points $P_0(x_0)$ were called *critical points*. [See Figure 8.5(*a* and *b*).] The condition $g'(x) = 0$ was useful in searching for relative maxima and minima, but it was not decisive. [See Figure 8.3(*c*).]

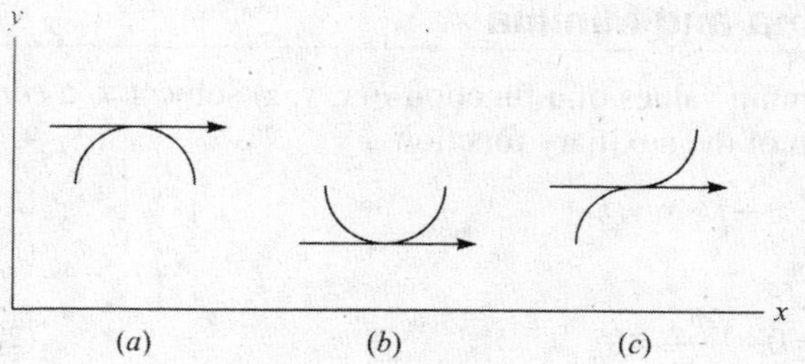

Figure 8.3

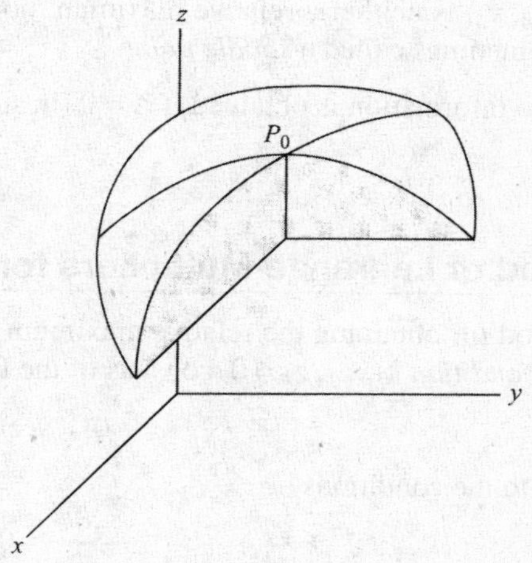

Figure 8.4

To determine the exact nature of the function at a critical point, P_0, $g''(x_0)$) had to be examined.

	> 0		counterclockwise rotation (relative minimum)
$g''(x_0)$	< 0	implied	a clockwise rotation (relative maximum)
	$= 0$		need for further investigation

This section describes the n ecessary and sufficient conditions for relative extrema of functions of two variables. Geometrically, we think of surfaces S represented by $z = f(x, y)$. If at a point $P_0(x_0, y_0)$ then $f_x(x, y_0) = 0$, means that the curve of intersection of S and the plane $y = y_0$ has a tangent parallel to the x axis. Similarly, $f_y(x_0, y_0) = 0$ indicates that the curve of intersection of S and the cross section $x = x_0$ has a tangent parallel the y axis. (See Problem 8.20.)

Thus,

$$f_x(x, y_0) = 0, f_y(x_0, y) = 0$$

are necessary conditions for a relative extrema of $z = f(x, y)$ at P_0; however, they are not sufficient because there are directions associated with a rotation through 360° that have not been examined. Of course, no differentiation between relative maxima and relative minima has been made. (See Figure 8.4.)

A very special form $f_{xy} - f_x f_y$, invariant under plane rotation and capable of characterizing the roots of a quadratic equation $Ax^2 + 2Bx + C = 0$, allows us to form sufficient conditions for relative extrema. (See Problem 8.21.)

A point (x_0, y_0) is called a *relative maximum point* or *relative minimum point* of $f(x, y)$ respectively according as $f(x_0 + h, y_0 + k) < f(x_0, y_0)$ or $f(x_0 + h, y_0 + k) > f(x_0, y_0)$ for all h and k such that $0 < |h| < \delta$, $0 < |k| < \delta$ where δ is a sufficiently small positive number.

A necessary condition that a differentiable function $f(x, y)$ have a relative maximum or minimum is

$$\frac{\partial f}{\partial x} = 0, \qquad \frac{\partial f}{\partial y} = 0 \tag{19}$$

If (x_0, y_0) is a point (called a *critical point*) satisfying Equations (19) and if Δ is defined by

$$\Delta = \left.\left\{\left(\frac{\partial^2 f}{\partial x^2}\right)\left(\frac{\partial^2 f}{\partial y^2}\right) - \left(\frac{\partial^2 f}{\partial x\, \partial y}\right)^2\right\}\right|_{(x_0, y_0)} \tag{20}$$

then

1. (x_0, y_0) is a relative maximum point if $\Delta > 0$ and $\left.\frac{\partial^2 f}{\partial x^2}\right|_{(x_0, y_0)} < 0 \left(\text{or} \left.\frac{\partial^2 f}{\partial y^2}\right|_{(x_0, y_0)} < 0 \right)$.

2. (x_0, y_0) is a relative minimum point of $\Delta > 0$ and $\left.\frac{\partial^2 f}{\partial x^2}\right|_{(x_0, y_0)} > 0 \left(\text{or} \left.\frac{\partial^2 f}{\partial y^2}\right|_{(x_0, y_0)} > 0 \right)$.

3. (x_0, y_0) is neither a relative maximum nor a relative minimum point if $\Delta < 0$. If $\Delta < 0$, < 0, (x_0, y_0) is sometimes called a *saddle point*.

4. No information is obtained if $\Delta = 0$ (in such case further investigation is necessary).

Method of Lagrange Multipliers for Maxima and Minima

A method for obtaining the relative maximum or minimum values of a function $F(x, y, z)$ subject to a *constraint condition* $\phi(x, y, z) = 0$, consists of the formation of the auxiliary function

$$G(x, y, z) \equiv F(x, y, z) + \lambda\phi(x, y, z) \tag{21}$$

subject to the conditions

$$\frac{\partial G}{\partial x} = 0, \quad \frac{\partial G}{\partial y} = 0, \quad \frac{\partial G}{\partial z} = 0 \tag{22}$$

which are necessary conditions for a relative maximum or minimum. The parameter λ, which is independent of x, y, z, is called a *Lagrange multiplier*.

The conditions in Equation (22) are equivalent to $\nabla G = 0$, and, hence, $0 = \nabla F + \lambda \nabla \phi$.

Geometrically, this means that ∇F and $\nabla \phi$ are parallel. This fact gives rise to the method of Lagrange multipliers in the following way.

Let the maximum value of F on $\phi(x, y, z) = 0$ be A and suppose it occurs at $P_0(x_0, y_0, z_0)$. (A similar argument can be made for a minimum value of F.) Now consider a family of surfaces $F(x, y, z) = C$.

The member $F(x, y, z) = A$ passes through P_0, while those surface $F(x, y, z) = B$ with $B < A$ do not. (This choice of a surface, i.e., $f(x, y, z) = A$, geometrically imposes the condition $\phi(x, y, z) = 0$ on F.) Since at P_0 the condition $0 = \nabla F + \lambda \nabla \phi$ tells us that the gradients of $F(x, y, z) = A$ and $\phi(x, y, z)$ are parallel, we know that the surfaces have a common tangent plane at a point that is maximum for F. Thus, $\nabla G = 0$ is a necessary condition for a relative maximum of F at P_0. Of course, the condition is not sufficient. The critical point so determined may not be unique and it may not produce a relative extremum.

The method can be generalized. If we wish to find the relative maximum or minimum values of a function $F(x_1, x_2, x_3, \ldots, x_n)$ subject to the *constraint conditions* $\phi(x_1, \ldots, x_n) = 0$, $\phi_2(x_1, \ldots, x_n) = 0, \ldots, \phi_k(x_1, \ldots, x_n) = 0$, we form the we form the auxiliary function

$$G(x_1, x_2, \ldots, x_n) \equiv F + \lambda_1\phi_1 + \lambda_2\phi_2 + \cdots + \lambda_k\phi_k \tag{23}$$

subject to the (necessary) conditions

$$\frac{\partial G}{\partial x_1} = 0, \quad \frac{\partial G}{\partial x_2} = 0, \ \ldots\ldots\ldots, \ \frac{\partial G}{\partial x_n} \equiv 0 \tag{24}$$

where $\lambda_1, \lambda_2, \ldots, \lambda_k$, which are independent of $x_1, x_2, \ldots, x_n$, are the *Lagrange multipliers*.

Applications to Errors

The theory of differentials can be applied to obtain errors in a function of x, y, z, etc., when the errors in x, y, z, etc., are known. See Problem 8.28.

SOLVED PROBLEMS

Tangent Plane And Normal Line To A Surface

8.1. Find equations for the (a) tangent plane and (b) normal line to the surface $x^2yz + 3y^2 = 2xz^2 - 8z$ at the point $(1, 2, -1)$.

(a) The equation of the surface is $F = x^2yz + 3y^2 - 2xz^2 + 8z = 0$. A normal line to the surface at $(1, 2, -1)$ is

$$\mathbf{N}_0 = \nabla F\big|_{(1,2,-1)} = (2xyz - 2z^2)\mathbf{i} + (x^2z + 6y)\mathbf{j} + (xy - 4xz + 8)\mathbf{k}\big|_{(1,2,-1)}$$

$$= -6\mathbf{i} + 11\mathbf{j} + 14\mathbf{k}$$

Referring to Figure 8.1:
The vector from O to any point (x, y, z) on the tangent plane is $\mathbf{r} = x\mathbf{i} + y\mathbf{j} + z\mathbf{k}$.
The vector from O to the point $(1, 2, -1)$ on the tangent plane is $\mathbf{r}_0 = \mathbf{i} + 2\mathbf{j} - \mathbf{k}$.
The vector $\mathbf{r} - \mathbf{r}_0 = (x - 1)\mathbf{i} + (y - 2)\mathbf{j} + (z + 1)\mathbf{k}$ lies in the tangent plane and is thus perpendicular to $\mathbf{N}_0$.

Then the required equation is

$$(\mathbf{r} - \mathbf{r}_0) \cdot \mathbf{N}_0 = 0 \quad \text{i.e.,} \quad \{(x-1)\mathbf{i} + (y-2)\mathbf{j} + (z+1)\mathbf{k}\} \cdot \{-6\mathbf{i} + 11\mathbf{j} + 14\mathbf{k}\} = 0$$

$$-6(x - 1) + 11(y - 2) + 14(z + 1) = 0 \text{ or } 6x - 11y - 14z + 2 = 0$$

(b) Let $\mathbf{r} = x\mathbf{i} + y\mathbf{j} + z\mathbf{k}$ be the vector from O to any point (x, y, z) of the normal $\mathbf{N}_0$. The vector from O to the point $(1, 2, -1)$ on the normal is $\mathbf{r}_0 = 2\mathbf{i} + 2\mathbf{j} - \mathbf{k}$. The vector $\mathbf{r} - \mathbf{r}_0 = (x - 1)\mathbf{i} + (y - 2)\mathbf{j} + (z + 1)\mathbf{k}$ is collinear with $\mathbf{N}_0$. Then

$$(r - r_0) \times N_0 = 0 \qquad \text{i.e.,} \qquad \begin{vmatrix} \mathbf{i} & \mathbf{j} & \mathbf{k} \\ x-1 & y-2 & x+1 \\ -6 & 11 & 14 \end{vmatrix} = 0$$

which is equivalent to the equations

$$11(x - 1) = -6(y - 2), \; 14(y - 2) = 11(z + 1), \; 14(x - 1) = -6(z + 1)$$

These can be written as

$$\frac{x-1}{-6} = \frac{y-2}{11} = \frac{z+1}{14}$$

often called the *standard form* for the equations of a line. By setting each of these ratios equal to the parameter t, we have

$$x = 1 - 6t, \quad y = 2 + 11t, \quad z = 14t - 1$$

called the *parametric equations* for the line.

8.2. In what point does the normal line of Problem 8.1(b) meet the plane $x + 3y - 2z = 10$?

Substituting the parametric equations of Problem 8.1(b), we have

$$1 - 6t + 3(2 + 11t) - 2(14t - 1) = 10 \text{ or } t = -1$$

Then $x = 1 - 6t = 7$, $y = 2 + 11t = -9$, $z = 14t - 1 = -15$ and the required point is $(7, -9, -15)$.

8.3. Show that the surface $x^2 - 2yz + y^3 = 4$ is perpendicular to any member of the family of surfaces $x^2 + 1 = (2 - 4a)y^2 + az^2$ at the point of intersection $(1, -1, 2)$.

Let the equations of the two surfaces be written in the form

$$F = x^2 - 2yz + y^3 - 4 = 0 \text{ and } G = x^2 + 1 - (2 - 4a)y^2 - az^2 = 0$$

Then

$$\nabla F = 2x\mathbf{i} + (3y^2 - 2z)\mathbf{j} - 2y\mathbf{k}, \qquad \nabla G = 2x\mathbf{i} - 2(2 - 4a)y\mathbf{j} - 2az\mathbf{k}$$

Thus, the normals to the two surfaces at (1, –1, 2) are given by

$$\mathbf{N}_1 = 2\mathbf{i} - \mathbf{j} + 2\mathbf{k},\ \mathbf{N}_2 = 2\mathbf{i} + 2(2 - 4a)\mathbf{j} - 4a\mathbf{k}$$

Since $\mathbf{N}_1 \cdot \mathbf{N}_2 = (2)(2) - 2(2 - 4a) - (2)(4a) \equiv 0$, it it follows that $\mathbf{N}_1$ and $\mathbf{N}_2$ are perpendicular for all a, and so the required result follows.

8.4. The equation of a surface is given in spherical coordinates by $F(r, \theta, \phi) = 0$, where we suppose that F is continuously differentiable. (a) Find an equation for the tangent plane to the surface at the point (r_0, θ_0, ϕ_0). (b) Find an equation for the tangent plane to the surface $r = 4 \cos \theta$ at the point $(2\sqrt{2},\ \pi/4, 3\pi/4)$. (c) Find a set of equations for the normal line to the surface in ( ) at the indicated point.

(a) The gradient of Φ in orthogonal curvilinear coordinates is

$$\nabla\Phi = \frac{1}{h_1}\frac{\partial\Phi}{\partial u_1}\mathbf{e}_1 + \frac{1}{h_2}\frac{\partial\Phi}{\partial u_2}\mathbf{e}_2 + \frac{1}{h_3}\frac{\partial\Phi}{\partial u_3}\mathbf{e}_3$$

where
$$\mathbf{e}_1 = \frac{1}{h_1}\frac{\partial\mathbf{r}}{\partial u_1}, \quad \mathbf{e}_2 = \frac{1}{h_2}\frac{\partial\mathbf{r}}{\partial u_2}, \quad \mathbf{e}_3 = \frac{1}{h_3}\frac{\partial\mathbf{r}}{\partial u_3}$$

(see Pages 170 and 172).

In spherical coordinates $u_1 = r$, $u_2 = \theta$, $u_3 = \phi$, $h_1 = 1$, $h_2 = r$, $h_3 = r \sin\theta$ and $\mathbf{r} = x\mathbf{i} + y\mathbf{j} + z\mathbf{k} = r \sin\theta \cos\phi\mathbf{i} + r \sin\theta \sin\phi\mathbf{j} + r\cos\theta\mathbf{k}$.

Then

$$\begin{cases} \mathbf{e}_1 = \sin\theta\ \cos\theta\,\mathbf{i} + \sin\theta\ \sin\theta\,\mathbf{j} + \cos\theta\,\mathbf{k} \\ \mathbf{e}_2 = \cos\theta\ \cos\phi\,\mathbf{i} + \cos\theta\ \sin\phi\,\mathbf{j} - \sin\theta\,\mathbf{k} \\ \mathbf{e}_3 = -\sin\phi\,\mathbf{i} + \cos\phi\,\mathbf{j} \end{cases} \tag{1}$$

and

$$\nabla F = \frac{\partial F}{\partial r}\mathbf{e}_1 + \frac{1}{r}\frac{\partial F}{\partial\theta}\mathbf{e}_2 + \frac{1}{r\sin\theta}\frac{\partial F}{\partial\phi}\mathbf{e}_3 \tag{2}$$

As on Page 196, the required equation is $(\mathbf{r} - \mathbf{r}_0) \cdot \nabla F|_P = 0$.
Now, substituting Equations (1) and (2), we have

$$\begin{aligned} \nabla F|_P = & \left\{\left.\frac{\partial F}{\partial r}\right|_P \sin\theta_0\ \cos\phi_0 + \frac{1}{r_0}\left.\frac{\partial F}{\partial\theta}\right|_P \cos\theta_0\ \cos\phi_0 - \frac{\sin\phi_0}{r_0\sin\theta_0}\left.\frac{\partial F}{\partial\phi}\right|_P\right\}\mathbf{i} \\ & + \left\{\left.\frac{\partial F}{\partial r}\right|_P \sin\theta_0\ \sin\phi_0 + \frac{1}{r_0}\left.\frac{\partial F}{\partial\theta}\right|_P \cos\theta_0\ \sin\phi_0 + \frac{\cos\phi_0}{r_0\sin\theta_0}\left.\frac{\partial F}{\partial\phi}\right|_P\right\}\mathbf{j} \\ & + \left\{\left.\frac{\partial F}{\partial r}\right|_P \cos\theta_0 - \frac{1}{r_0}\left.\frac{\partial F}{\partial\theta}\right|_P \sin\theta_0\right\}\mathbf{k} \end{aligned}$$

Denoting the expressions in braces by A, B, C, respectively, so that $\nabla F|_p = A\mathbf{i} + B\mathbf{j} + C\mathbf{k}$, we see that the required equation is $A(x - x_0) + B(y - y_0) + C(z - z_0) = 0$. This can be written in spherical coordinates by using the transformation equations for x, y, and z in these coordinates.

(b) We have $F = r - 4\cos\theta = 0$. Then $\partial F/\partial r = 1$, $\partial F/\partial\theta = 4\sin\theta$, $\partial F/\partial\phi = 0$.

Since $r_0 = 2\sqrt{2}$, $\theta_0 = \pi/4$, $\phi_0 = 3\pi/4$, we have from (a), $\nabla F|_p = A\mathbf{i} + B\mathbf{j} + C\mathbf{k} = -\mathbf{i} + \mathbf{j}$.

From the transformation equations, the given point has rectangular coordinates $(-\sqrt{2}, \sqrt{2}, 2)$, and so $r - r_0 = (x + \sqrt{2})\mathbf{i} + (y - \sqrt{2})\mathbf{j} + (z - 2)\mathbf{k}$.

The required equation of the plane is thus $-(x + \sqrt{2}) + (y - \sqrt{2}) = 0$ or $y - x = 2\sqrt{2}$. In spherical coordinates this becomes $r \sin\theta \sin\phi - r \sin\theta \cos\phi = 2\sqrt{2}$.

In rectangular coordinates the equation $r = 4\cos\theta$ becomes $x^2 + y^2 + (z-2)^2) = 4$ and the tangent plane can be determined from this as in Problem 8.1. In other cases, however, it may not be so easy to obtain the equation in rectangular form, and in such cases the method of part *(a)* is simpler to use.

(c) The equations of the normal line can be represented by

$$\frac{x+\sqrt{2}}{-1} = \frac{y-\sqrt{2}}{1} = z = 2$$

Tangent Line and Normal Plane to a Curve

8.5. Find equations for (a) the tangent line and (b) the normal plane to the curve $x = t - \cos t$, $y = 3 + \sin 2t$, $z = 1 + \cos 3t$ at the point where $t = 1/2\pi$.

(a) The vector from origin O (see Figure 8.2) to any point of curve C is $\mathbf{R} = (t - \cos t)\mathbf{i} + (3 + \sin 2t)\mathbf{j} + (1 + \cos 3t)\mathbf{k}$. Then a vector tangent to C at the point where $t = \frac{1}{2}\pi$ is

$$\mathbf{T}_0 = \left.\frac{d\mathbf{R}}{dt}\right|_{t=1/2\pi} = (1+\sin t)\, i + 2\cos 2t\,\mathbf{j} - 3\sin 3t\,\mathbf{k}\,|_{t=1/2\pi} = 2\mathbf{i} - 2\mathbf{j} + 3\mathbf{k}$$

The vector from O to the point where $t = 1/2\pi$ is $\mathbf{r}_0 = 1/2\pi\mathbf{i} + 3\mathbf{j} + \mathbf{k}$.

The vector from O to any point (x, y, z) on the tangent line is $\mathbf{r} = x\mathbf{i} + y\mathbf{j} + z\mathbf{k}$.

Then $\mathbf{r} - \mathbf{r}_0 = (x - \frac{1}{2}\pi)\mathbf{i} + y - 3)\mathbf{j} + (z - 1)\,\mathbf{k}$ is collinear with $\mathbf{T}_0$, so that the required equation is

$$(\mathbf{r} - \mathbf{r}_0) \times \mathbf{T}_0 = \mathbf{0}, \quad \text{i.e.,} \quad \begin{vmatrix} \mathbf{i} & \mathbf{j} & \mathbf{k} \\ x - \frac{1}{2}\pi & y-3 & z-1 \\ 2 & -2 & 3 \end{vmatrix} = 0$$

and the required equations are $\dfrac{x - \frac{1}{2}\pi}{2} = \dfrac{y-3}{-2} = \dfrac{z-1}{3}$ or, in parametric form, $x = 2t + \frac{1}{2}\pi$, $y = 3 - 2t$, $z = 3t + 1$.

(b) Let $r = x\mathbf{i} + y\mathbf{j} + z\mathbf{k}$ be the vector from O to any point (x, y, z) of the normal plane. The vector from O to the point where $t = \frac{1}{2}\pi$ is $\mathbf{r}_0 = \frac{1}{2}\pi\mathbf{i} + 3\mathbf{j} + \mathbf{k}$. The vector $\mathbf{r} - \mathbf{r}_0 = (x - \frac{1}{2}\pi)\mathbf{i} + (y - 3)\mathbf{j} + (z - 1)\mathbf{k}$ lies in the normal plane and, hence, is perpendicular to $\mathbf{T}_0$. Then the required equation is $(\mathbf{r} - \mathbf{r}_0) \cdot \mathbf{T}_0 = 0$ or $2(x - \frac{1}{2}\pi) - 2(y - 3) + 3(z - 1) = 0$.

8.6. Find equations for (a) the tangent line and (b) the normal plane to the curve $3x^2y + y^2z = -2$, $2xz - x^2y = 3$ at the point $(1, -1, 1)$.

(a) The equations of the surfaces intersecting in the curve are

$$F = 3x^2y + y^2z + 2 = 0,\; G = 2xz - x^2y - 3 = 0$$

The normals to each surface at the point $P(1, -1, 1)$ are, respectively,

$$\mathbf{N}_1 = \nabla F|_P = 6xy\,\mathbf{i} + (3x^2 + 2yz)\mathbf{j} + y^2\mathbf{k} = -6 + \mathbf{j} + \mathbf{k}$$

$$\mathbf{N}_2 = \nabla G|_P = (2z - 2xy)\mathbf{i} - x^2\mathbf{j} + 2x\mathbf{k} = 4\mathbf{i} - \mathbf{j} + 2\mathbf{k}$$

Then a tangent vector to the curve at P is

$$\mathbf{T}_0 = \mathbf{N}_1 \times \mathbf{N}_2 = (-6\mathbf{i} + \mathbf{j} + \mathbf{k}) \times (4 - \mathbf{j} + 2\mathbf{k}) = 3\mathbf{i} + 16\mathbf{j} + 2\mathbf{k}$$

Thus, as in Problem 8.5(a), the tangent line is given by

$$(\mathbf{r} - \mathbf{r}_0) \times \mathbf{T}_0 = 0 \text{ or } \{(x-1)\mathbf{i} + (y+1)\mathbf{j} + (z-1)\mathbf{k}\} \times \{3\mathbf{i} + 16\mathbf{j} + 2\mathbf{k}\} = \mathbf{0}$$

i.e.,

$$\frac{x-1}{3} = \frac{y+1}{16} = \frac{z-1}{2} \quad \text{or} \quad x = 1 + 3t, \quad y = 16t - 1, \quad z = 2t + 1$$

(b) As in Problem 8.5(b), the normal plane is given by

$$(r - \mathbf{r}_0) \cdot \mathbf{T}_0 = 0 \text{ or } \{(x-1)\mathbf{i} + (y+1)\mathbf{j} + (z-1)\mathbf{k}\} \cdot \{3\mathbf{i} + 16\mathbf{j} + 2\mathbf{k}\} = 0$$

i.e.,

$$3(x-1) + 16(y+1) + 2(z-1) = 0 \text{ or } 3x + 16y + 2z = -11$$

The results in (a) and (b) can also be obtained by using Equations (7) and (10), respectively, from Page 197.

8.7. Establish equation (10), from Page 197.

Suppose the curve is defined by the intersection of two surfaces whose equations are $F(x, y, z) = 0$, $G(x, y, z) = 0$, where we assume F and G to be continuously differentiable.

The normals to each surface at point P are given, respectively, by $\mathbf{N}_1 = \nabla F\,|_P$ and $\mathbf{N}_2 = \nabla G\,|_P$. Then a tangent vector to the curve at P is $\mathbf{T}_0 = \mathbf{N}_1 \times \mathbf{N}_2 = \nabla F\,|_P \times \nabla G\,|_P$. Thus, the equation of the normal plane is $(\mathbf{r} - \mathbf{r}_0) \cdot \mathbf{T}_0 = 0$. Now

$$\mathbf{T}_0 = \nabla F\,|_P \times \nabla G\,|_P = \{(F_x i + F_y j + F_z k) \times (G_x i + G_y j + G_z k)\}\,|_P$$

$$= \begin{vmatrix} \mathbf{i} & \mathbf{j} & \mathbf{k} \\ F_x & F_y & F_z \\ G_x & G_y & G_z \end{vmatrix}_P = \begin{vmatrix} F_y & F_z \\ G_y & G_z \end{vmatrix}_P \mathbf{i} + \begin{vmatrix} F_z & F_x \\ G_z & G_x \end{vmatrix}_P \mathbf{j} + \begin{vmatrix} F_x & F_y \\ G_x & G_y \end{vmatrix}_P \mathbf{k}$$

and so the required equation is

$$(r - r_0)\,.\,\nabla F\,|_P = 0 \quad or \quad \begin{vmatrix} F_y & F_z \\ G_y & G_z \end{vmatrix}_P (x - x_0) + \begin{vmatrix} F_z & F_x \\ G_z & G_x \end{vmatrix}_P (y - y_0) + \begin{vmatrix} F_x & F_y \\ G_x & G_y \end{vmatrix}_P (z - z_0) = 0$$

Envelopes

8.8. Prove that the envelope of the family $\phi(x, y, \alpha) = 0$, if it exists, can be obtained by solving simultaneously the equations $\phi = 0$ and $\phi_\alpha = 0$.

Assume parametric equations of the envelope to be $x = f(\alpha)$, $y = g(\alpha)$. Then $\phi(f(\alpha), g(\alpha), \alpha) = 0$ identically, so, upon differentiating with respect to α (assuming that ϕ, f, and g have continuous derivatives), we have

$$\phi_x f'(\alpha) + \phi_y g'(\alpha) + \phi_\alpha = 0 \tag{1}$$

The slope of any member of the family $\phi(x, y, \alpha) = 0$ at (x, y) is given by $\phi_x\, dx + \phi_y\, dy = 0$ or $\dfrac{dy}{dx} = -\dfrac{\phi_x}{\phi_y}$.

The slope of the envelope at (x, y) is $\dfrac{dy}{dx} = \dfrac{dy/d\alpha}{dx/d\alpha} = \dfrac{g'(\alpha)}{f'(\alpha)}$. Then at any point where the envelope and a member of the family are tangent, we must have

$$-\frac{\phi_x}{\phi_y} = \frac{g'(\alpha)}{f'(\alpha)} \quad \text{or} \quad \phi_x f'(\alpha) + \phi_y g'(\alpha) = 0 \tag{2}$$

Comparing Equations (2) with (1), we see that $\phi_\alpha = 0$, and the required result follows.

8.9. (a) Find the envelope of the family $x \sin \alpha + y \cos \alpha = 1$. (b) Illustrate the results geometrically.

(a) By Problem 8.8 the envelope, if it exists, is obtained by solving simultaneously the equations $\phi(x, y, \alpha) = x \sin \alpha + y \cos\alpha - 1 = 0$ and $\phi_\alpha(x, y, \alpha) = x \cos \alpha - y \cos \alpha = 0$. From these equations we find $x = \sin \alpha$, $y = \cos \alpha$ or $x^2 + y^2 = 1$.

(b) The given family is a family of straight lines, some members of which are indicated in Figure 8.5. The envelope is the circle $x^2 + y^2 = 1$.

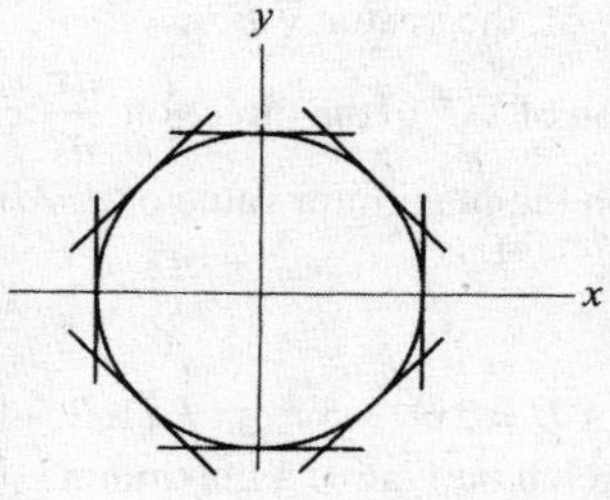

Fig. 8.5

8.10. Find the envelope of the family of surfaces $z = 2\alpha x - a^2 y$.

By a generalization of Problem 8.8, the required envelope, if it exists, is obtained by solving simultaneously the equations

$$\phi = 2\alpha x - \alpha^2 y - z = 0 \tag{1}$$

and

$$\phi_\alpha = 2x - 2\alpha y = 0 \tag{2}$$

From Equation (2), $\alpha = x/y$. Then substitution in Equation (1) yields $x^2 = yz$, the required envelope.

8.11. Find the envelope of the two-parameter family of surfaces $z = \alpha x + \beta y - \alpha\beta$.

The envelope of the family $F(x, y, z, \alpha, \beta) = 0$, if it exists, is obtained by eliminating α and β between the equations $F = 0$, $F_\alpha = 0$, $F_\beta = 0$ (see Problem 8.43). Now

$$F = z - \alpha x - \beta y + \alpha\beta = 0,\ F_\alpha = -x + \beta = 0,\ F_\beta = -y + \alpha = 0$$

Then $\beta = x$, $\alpha = y$, and we have $z = xy$.

Directional derivatives

8.12. Find the directional derivative of $F = x^2yz^3$ along the curve $x = e^{-u}$, $y = 2 \sin u + 1$, $z = u - \cos u$ at the point P where $u = 0$.

The point P corresponding to $u = 0$ is $(1, 1, -1)$. Then

$$\nabla F = 2xyz^3\mathbf{i} + x^2z^3\mathbf{j} + 3x^2yz^2\mathbf{k} = -2\mathbf{i} - \mathbf{j} + 3\mathbf{k} \text{ at } P$$

A tangent vector to the curve is

$$\begin{aligned}\frac{dr}{du} &= \frac{d}{du}\{e^{-u}\mathbf{i} + (2\sin u + 1)\mathbf{j} + (u - \cos u)k\}\\ &= -e^{-u}\mathbf{i} + 2\cos u\mathbf{j} + (1 + \sin u)\mathbf{k} = -\mathbf{i} + 2\mathbf{j} + \mathbf{k} \ at\ P\end{aligned}$$

and the unit tangent vector in this direction is

$$\mathbf{T}_0 = \frac{-\mathbf{i} + 2\mathbf{j} + \mathbf{k}}{\sqrt{6}}$$

Then

$$\text{Directional derivative} = \nabla F \cdot \mathbf{T}_0 = (-2\mathbf{i} - \mathbf{j} + 3\mathbf{k}) \cdot \left(\frac{-\mathbf{i} + 2\mathbf{j} + \mathbf{k}}{\sqrt{6}}\right) = \frac{3}{\sqrt{6}} = \frac{1}{2}\sqrt{6}$$

Since this is positive, F is increasing in this direction.

8.13. Prove that the greatest rate of change of F, i.e., the maximum directional derivative, takes place in the direction of, and has the magnitude of, the vector ∇F.

$\frac{dF}{ds} = \nabla F \cdot \frac{d\mathbf{r}}{ds}$ is the projection of ∇F in the direction $\frac{d\mathbf{r}}{ds}$. This projection is a maximum when ∇F and $d\mathbf{r}/ds$ have the same direction. Then the maximum value of dF/ds takes place in the direction of ∇F, and the magnitude is $|\nabla F|$.

8.14. (a) Find the directional derivative of $U = 2x^3y - 3y^2z$ at $P(1, 2, -1)$ in a direction toward $Q(3, -1, 5)$. (b) In what direction from P is the directional derivative a maximum? (c) What is the magnitude of the maximum directional derivative?

(a) $\nabla U = 6x^2y\mathbf{i} + (2x^3 - 6yz)\mathbf{j} - 3y^2\mathbf{k} = 12\mathbf{i} + 14\mathbf{j} - 12\mathbf{k}$ at P.

The vector from P to $Q = (3-1)\mathbf{i} + (-1-2)\mathbf{j} + [5-(-1)]\mathbf{k} = 2\mathbf{i} - 3\mathbf{j} + 6\mathbf{k}$

$$\text{The unit vector from } P \text{ to } Q = \mathbf{T} = \frac{2\mathbf{i} - 3\mathbf{j} + 6\mathbf{k}}{\sqrt{(2)^2 + (-3)^2 + (6)^2}} = \frac{2\mathbf{i} - 3\mathbf{j} + 6\mathbf{k}}{7}$$

Then

$$\text{Directional derivative at } P = (12\mathbf{i} + 14\mathbf{j} - 12\mathbf{k}) \cdot \left(\frac{2\mathbf{i} - 3\mathbf{j} + 6\mathbf{k}}{7}\right) = -\frac{90}{7}$$

i.e., U is decreasing in this direction.

(b) From Problem 8.13, the directional derivative is a maximum in the direction $12\mathbf{i} + 14\mathbf{j} - 12\mathbf{k}$.

(c) From Problem 8.13, the value of the maximum directional derivative is $|12\mathbf{i} + 14\mathbf{j} - 12\mathbf{k}| = \sqrt{144 + 196 + 144} = 22$.

Differentiation under the integral sign

8.15. Prove Leibnitz's rule for differentiating under the integral sign.

Let $\phi(\alpha) = \int_{u_1(\alpha)}^{u_2(\alpha)} f(x, \alpha)\,dx.$ Then

$$\Delta\phi = \phi(\alpha + \Delta\alpha) - \phi(\alpha) = \int_{u_1(\alpha+\Delta\alpha)}^{u_2(\alpha+\Delta\alpha)} f(x, \alpha + \Delta\alpha)\,dx - \int_{u_1(\alpha)}^{u_2(\alpha)} f(x, \alpha)\,dx$$

$$= \int_{u_1(\alpha+\Delta\alpha)}^{u_1(\alpha)} f(x, \alpha + \Delta\alpha)\,dx + \int_{u_1(\alpha)}^{u_2(\alpha)} f(x, \alpha + \Delta\alpha) + \int_{u_1(\alpha)}^{u_2(\alpha+\Delta\alpha)} f(x, \alpha + \Delta\alpha)\,dx - \int_{u_1(\alpha)}^{u_2(\alpha)} f(x, \alpha)\,dx$$

$$= \int_{u_1(\alpha)}^{u_2(\alpha)} [f(x, \alpha + \Delta\alpha) - f(x, \alpha)]\,dx + \int_{u_2(\alpha)}^{u_2(\alpha+\Delta\alpha)} f(x, \alpha + \Delta\alpha)\,dx - \int_{u_1(\alpha)}^{u_1(\alpha+\Delta\alpha)} f(x, \alpha + \Delta\alpha)\,dx$$

By the mean value theorem for integrals, we have

$$\int_{u_1(\alpha)}^{u_2(\alpha)} [f(x, \alpha + \Delta\alpha) - f(x, \alpha)]\,dx = \Delta\alpha \int_{u_1(\alpha)}^{u_2(\alpha)} f(x, \xi)\,dx \tag{1}$$

$$\int_{u_1(\alpha)}^{u_1(\alpha+\Delta\alpha)} f(x, \alpha + \Delta\alpha)\,dx = f(\xi_1, \alpha + \Delta\alpha)\,[u_1(\alpha + \Delta\alpha) - u_1(\alpha)] \tag{2}$$

$$\int_{u_2(\alpha)}^{u_2(\alpha+\Delta\alpha)} f(x,\alpha+\Delta\alpha)\,dx = f(\xi_2,\alpha+\Delta\alpha)\,[u_1(\alpha+\Delta\alpha)-u_1(\alpha)] \qquad (3)$$

where ξ is between $\alpha + \Delta\alpha$, ξ_1 is between $u_1(\alpha)$ and $u_1(\alpha+\Delta\alpha)$ and ξ_2 is between $u_2(\alpha)$ and $u_2(\alpha+\Delta\alpha)$. Then

$$\frac{\Delta\phi}{\Delta\phi} = \int_{u_1(\alpha)}^{u_2(\alpha)} f_\alpha(x,\xi)\,dx + f(\xi_2,\alpha+\Delta\alpha)\frac{\Delta u_2}{\Delta\alpha} - f(\xi_1,\alpha+\Delta\alpha)\frac{\Delta u_1}{\Delta\alpha}$$

Taking the limit as $\Delta\alpha \to 0$, making use of the fact that the functions are assumed to have continuous derivatives, we obtain

$$\frac{d\phi}{d\phi} = \int_{u_1(\alpha)}^{u_2(\alpha)} f_\alpha(x,\xi)\,dx + f(\xi_2,\alpha+\Delta\alpha)\frac{\Delta u_2}{\Delta\alpha} - f(\xi_1,\alpha+\Delta\alpha)\frac{\Delta u_1}{\Delta\alpha}$$

8.16. If $\phi(\alpha) = \int_{\alpha}^{\alpha^2} \frac{\sin\alpha x}{x}\,dx$, find $\phi'(\alpha)$ where $\alpha \neq 0$.

By Leibniz's rule,

$$\begin{aligned}\phi'(\alpha) &= \int_{\alpha}^{\alpha^2} \frac{\partial}{\partial a}\left(\frac{\sin\alpha x}{x}\right)dx + \frac{\sin(\alpha\cdot\alpha^2)}{\alpha^2}\frac{d}{d\alpha}(\alpha^2) - \frac{\sin(\alpha\cdot\alpha)}{\alpha}\frac{d}{d\alpha}(\alpha)\\ &= \int_{\alpha}^{\alpha^2}\cos\alpha x\,dx + \frac{2\sin\alpha^3}{\alpha} - \frac{\sin\alpha^2}{\alpha}\\ &= \left.\frac{\sin\alpha x}{\alpha}\right|_{\alpha}^{a^2} + \frac{2\sin\alpha^3}{\alpha} - \frac{\sin\alpha^2}{\alpha} = \frac{3\sin\alpha^3 - 2\sin\alpha^2}{\alpha}\end{aligned}$$

8.17. If $\int_0^{\pi} \frac{dx}{\alpha - \cos x} = \frac{\pi}{\sqrt{\alpha^2-1}} \cdot \alpha > 1$, find $\int_0^{\pi} \frac{dx}{(2-\cos x)^2}$. (See Problem 5.58.)

By Leibniz's rule, if $\phi(\alpha) = \int_0^{\pi} \frac{dx}{(\alpha-\cos x)} = \pi(\alpha^2-1)^{-1/2}$, then

$$\phi(\alpha) = -\int_0^{\pi}\frac{dx}{(\alpha-\cos x)} = \frac{1}{2}\pi(\alpha^2-1)^{-3/2}2\alpha = \frac{-\pi\alpha}{(\alpha^2-1)^{3/2}}$$

Thus, $\int_0^{\pi}\frac{dx}{(\alpha-\cos x)} = \frac{-\pi\alpha}{(\alpha^2-1)^{3/2}}$, from which $\int_0^{\pi}\frac{dx}{(2-\cos x)^2} = \frac{2\pi}{3\sqrt{3}}$.

Integration under the integral sign

8.18. Prove the result (18), on Page 198, for integration under the integral sign.

Consider:

$$\psi(\alpha) = \int_{u_1}^{u_2}\left\{\int_{\alpha}^{\alpha} f(x,\alpha)\,dx\right\}dx \qquad (2)$$

By Leibniz's rule,

$$\psi'(\alpha) = \int_{u_1}^{u_2}\frac{\partial}{\partial\alpha}\left\{\int_{\alpha}^{\alpha} f(x,\alpha)\,dx\right\}dx = \int_{u_1}^{u_2} f(x,\alpha)\,dx = \phi(\alpha)$$

Then, by integration,

$$\psi(\alpha) = \int_{\alpha}^{\alpha}\phi(\alpha)\,d\alpha + c \qquad (2)$$

Since $\psi(a) = 0$ from Equation (1), we have $c = 0$ in (2). Thus from Equation (1) and (2) with $c = 0$, we find

$$\int_{u_1}^{u_2} \left\{ \int_{\alpha}^{\alpha} f(x, \alpha)\, dx \right\} dx = \int_{\alpha}^{\alpha} \left\{ \int_{u_1}^{u_2} f(x, \alpha)\, dx \right\} dx$$

Putting $\alpha = b$, the required result follows.

8.19. Prove that $\int_0^{\pi} \ln \left(\frac{b - \cos x}{a - \cos x} \right) dx = \pi \ln \left(\frac{b + \sqrt{b^2 - 1}}{a + \sqrt{a^2 - 1}} \right)$ if $a, b > 1$.

From Problem 5.58, $\int_0^{\pi} \frac{dx}{\alpha - \cos x} = \frac{\pi}{\sqrt{\alpha^2 - 1}} \cdot \alpha > 1.$

Integrating the left side with respect to α from a to b yields

$$\int_0^{\pi} \left\{ \int_a^b \frac{dx}{\alpha - \cos x} \right\} dx = \int_0^{\pi} \text{In}\, (\alpha - \cos x) \Big|_a^b \, dx \int_0^{\pi} \text{In} \left(\frac{b - \cos x}{a - \cos x} \right) dx$$

Integrating the right side with respect to α to b yields

$$\int_0^{\pi} \frac{\pi\, d\alpha}{\sqrt{\alpha^2 - 1}} = \pi \ln\, (\alpha + \sqrt{a^2 - 1}) \Big|_a^b = \pi \ln \left(\frac{b + \sqrt{b^2 - 1}}{a + \sqrt{a^2 - 1}} \right)$$

and the required result follows.

Maxima and minima

8.20. Prove that a necessary condition for $f(x, y)$ to have a relative extremum (maximum or minimum) at (x_0, y_0) is that $f_x(x_0, y_0) = 0, f_y(x_0, y_0) = 0$.

If $f(x_0, y_0)$ is to be an extreme value for $f(x, y)$, then it must be an extreme value for both $f(x, y_0)$ and $f(x_0, y)$. But a necessary condition that these have extreme values at $x_x = 0$ and $y = y_0$, respectively, is $f_x(x_0, y_0) = 0$, $f_y(x_0, y_0) = 0$ (using results for functions of one variable).

8.21. Let f be continuous and have continuous partial derivatives of order two, at least, in a region R with the critical point $P_0(x_0, y_0)$ an interior point. Determine the sufficient conditions for relative extrema at P_0.

In the case of one variable, sufficient conditions for a relative extrema were formulated through the second derivative [if positive then a relative minimum, if negative then a relative maximum, if zero a possible point of inflection but more investigation is necessary]. In the case of $z = f(x, y)$ that is before us we can expect the second partial derivatives to supply information. (See Figure 8.6.)

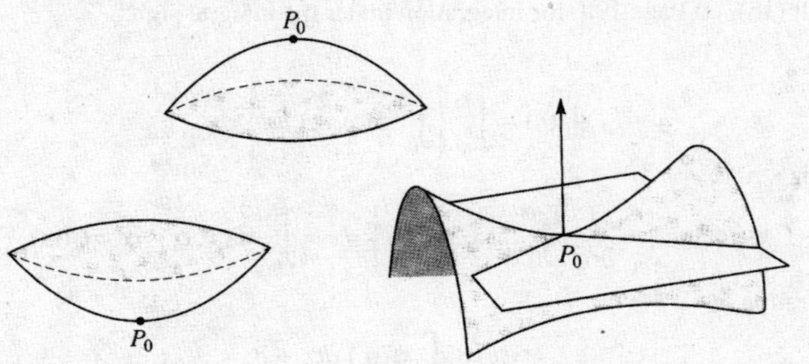

Fig. 8.6

First observe that solutions of the quadratic equation $At^2 + 2Bt + C = 0$ are $t = \dfrac{-2B \pm \sqrt{4B^2 - 4AC}}{2A}$.

Further observe that the nature of these solutions is determined by $B^2 - AC$. If the quantity is positive, the solutions are real and distinct; if negative, they are complex conjugate; and if zero, the two solutions are coincident.

The expression $B^2 - AC$ also has the property of invariance with respect to plane rotations

$$x = \bar{x}\cos\theta - \bar{y}\sin\theta$$

$$y = \bar{x}\cos\theta - \bar{y}\sin\theta$$

It has been discovered that with the identifications $A = f_{xx}$, $B = f_{xy}$, $C = f_{yy}$, we have the partial derivative form $f^2_{xy} - f_{xx}f_{yy}$ that characterizes relative extrema.

The demonstration of invariance of this form can be found in analytic geometric books. However, if you would like to put the problem in the context of the second partial derivative, observe that

$$f_{\bar{x}} = f_x\frac{\partial x}{\partial \bar{x}} + f_x\frac{\partial y}{\partial \bar{x}} = f_x\cos\theta + f_y\sin\theta$$

$$f_{\bar{y}} = f_x\frac{\partial x}{\partial \bar{x}} + f_y\frac{\partial y}{\partial \bar{y}} = -f_x\sin\theta + f_y\cos\theta$$

Then, using the chain rule to compute the second partial derivatives and proceeding by straightforward but tedious calculation, we show that.

$$f^2_{xy} = f_{xx}f_{yy} = f^2_{\bar{x}\bar{y}} - f_{\bar{x}\bar{x}}f_{\bar{y}\bar{y}}$$

The following equivalences are a consequence of this invariant form (independently of direction in the tangent plane at P_0):

$$f^2_{xy} = f_{xx}f_{yy} < 0 \quad \text{and} \quad f_{xx}f_{yy} > 0 \tag{1}$$

$$f^2_{xy} = f_{xx}f_{yy} < 0 \quad \text{and} \quad f_{xx}f_{yy} < 0 \tag{2}$$

The key relation is (1) because in order that this equivalence hold, both terms $f_x f_y$ must have the same sign. We can look to the one-variable case (make the same argument for each coordinate direction) and conclude that there is a relative minimum at P_0 if both partial derivatives are positive and a relative maximum if both are negative. We can make this argument for any pair of coordinate directions because of the invariance under rotation that was established.

If relation (2) holds, then the point is called a *saddle point*. If the quadratic form is zero, no information results.

Observe that this situation is analogous to the one-variable extreme value theory in which the nature of f at x, and with $f'(x) = 0$, is undecided if $f''(x) = 0$.

8.22. Find the relative maxima and minima of $f(x, y) = x^3 + y^3 - 3x - 12y + 20$.

$f_x = 3x^2 - 3 = 0$ when $x = \pm 1$, $f_y = 3y^2 - 12 = 0$ when $y = \pm 2$. Then critical points are $P(1, 2)$, $Q(-1, 2)$, $R(1, -2)$, $S(-1, -2)$.

$f_{xx} = 6x$, $f_{yy} = 0$. Then $\Delta = f_{xx}f_{yy} - f^2_{xy} = 36xy$.

At $P(1, 2)$, $\Delta > 0$ and f_{xx} (or f_{yy}) > 0; hence P is a relative minimum point.

At $Q(-1, 2)$, $\Delta < 0$ and Q is neither a relative maximum or minimum point.

At $R(1, -2)$, $\Delta < 0$ and R is neither a relative maximum or minimum point.

At $S(-1, -2)$, $\Delta > 0$ and f_{xx} (or f_{yy}) < 0 so S is a relative maximum point.

Thus, the relative minimum value of $f(x, y)$ occurring at P is 2, while the relative maximum value occurring at S is 38. Points Q and R are *saddle points*.

8.23. A rectangular box, open at the top, is to have a volume of 32 cubic feet. What must be the dimensions so that the total surface is a minimum?

If x, y, and z are the edges (see Fig. 8.7), then

$$\text{Volume of box} = V = xyz = 32 \tag{1}$$

$$\text{Surface area of box} = S = xy + 2yz + 2xz \tag{2}$$

or, since $z = 32/xy$ from Equation (1),

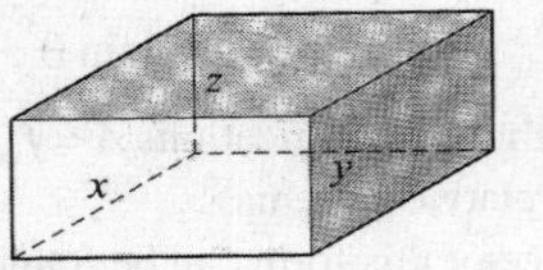

Fig. 8.7

$$\frac{\partial S}{\partial x} = y - \frac{64}{x^2} = 0 \text{ when } x^2 y = 64 \tag{3}$$

$$\frac{\partial S}{\partial y} = x - \frac{64}{y^2} = 0 \text{ when } xy^2 = 64 \tag{4}$$

Dividing Equations (3) and (4), we find $y = x$ so that $x^3 = 64$ or $x = y = 4$ and $z = 2$.

For $x = y = 4$, $\Delta = S_{xx}S_{yy} - S_{xy}^2 = \left(\frac{128}{x^3}\right)\left(\frac{128}{y^3}\right) - 1 > 0$ and $s_{xx} = \frac{128}{x^3} > 0$. Hence, it follows that the dimensions 4 feet × 4 feet × 2 feet give the minimum surface.

Lagrange multipliers for maxima and minima

8.24. Consider $F(x, y, z)$ subject to the constraint condition $G(x, y, z) = 0$. Prove that a necessary condition that $F(x, y, z)$ have an extreme value is that $F_xG_y - F_yG_x = 0$.

Since $G(x, y, z) = 0$, we can consider z as a function of x and y—say, $z = f(x, y)$. A necessary condition that $F[x, y, f(x, y)]$ have an extreme value is that the partial derivatives with respect to x and y be zero. This gives

$$F_x + F_z z_x = 0 \tag{1}$$

$$F_y + F_z Z_y = 0 \tag{2}$$

Since $G(x, y, z) = 0$, we also have

$$G_x + G_x z_x = 0 \tag{3}$$

$$G_y + G_z z_y = 0 \tag{4}$$

From Equations (1) and (3) we have

$$F_x G_x - F_x G_x = 0 \tag{5}$$

and from Equations (2) and (4) we have

$$F_y G_z - F_z G_y = 0 \tag{6}$$

Then from Equations (5) and (6) we find $F_xG_y - F_yG_x = 0$.

These results hold only if $F_z \neq 0$, $G_z \neq 0$.

8.25. Referring to Problem 8.24, show that the stated condition is equivalent to the conditions $\phi_x = 0$, $\phi_y = 0$ where $\phi = F + \lambda G$ and λ is a constant.

If $\phi_x = 0$, $F_x + \lambda G_x = 0$. If $\phi_y = 0$, $F_y + \lambda G_y = 0$. Elimination of λ between these equations yields $F_xG_y - F_y G_x = 0$.

The multiplier λ is the *Lagrange multiplier*. If desired, we can consider equivalently $\phi = \lambda F + G$ where $\phi_x = 0$, $\phi_y = 0$.

8.26. Find the shortest distance from the origin to the hyperbola $x^2 + 8xy + 7y^2 = 225$, $z = 0$.

We must find the minimum value of $x^2 + y^2$ (the square of the distance from the origin to any point in the xy plane) subject to the constraint $x^2 + 8xy + 7y^2 = 225$.

According to the method of Lagrange multipliers, we consider $\phi = x^2 + 8xy + 7y^2 - 225 + \lambda(x^2 + y^2)$. Then

$$\phi_x = 2x + 8y + 2\lambda x = 0 \quad \text{or} \quad (\lambda + 1)\,x + 4y = 0 \tag{1}$$

$$\phi_y = 8x + 14y + 2\lambda y = 0 \quad \text{or} \quad 4x + (\lambda + 7)y = 0 \tag{2}$$

From Equations (1) and (2), since $(x, y) \neq (0, 0)$, we must have

$$\begin{vmatrix} \lambda + 1 & 4 \\ 4 & \lambda + 7 \end{vmatrix} = 0, \text{ i.e., } \lambda^2 + 8\lambda - 9 = \text{ or } \lambda = 1, -9$$

Case 1: $\lambda = 1$. From Equation (*1*) or (2), $x = -2y$, and substitution in $x^2 + 8xy + 7y^2 = 225$ yields $-5y^2 = 225$, for which no real solution exists.

Case 2: $\lambda = -9$. From Equation (*1*) or (2), $y = 2x$, and substitution in $x^2 + 8xy + 7y^2 = 225$ yields $45x^2 = 225$. Then $x^2 = 5$, $y^2 = 4x^2 = 20$ and so $x^2 + y^2 = 25$. Thus, the required shortest distance is $\sqrt{25} = 5$.

8.27. (a) Find the maximum and minimum values of $x^2 + y^2 + z^2$ subject to the constraint conditions $x^2/4 + y^2/5 + z^2/25 = 1$ and $z = x + y$. (b) Give a geometric interpretation of the result in (*a*).

(a) We must find the extrema of $F = x^2 + y^2 + z^2$ subject to the constraint conditions $\phi_1 = \frac{x^2}{4} + \frac{x^2}{5} + \frac{z^2}{25} - 1 = 0$ and $\phi_2 = x + y - z = 0$. In this case we use two Lagrange multipliers λ_1, λ_2 and consider the function

$$G = F + \lambda_1\phi_1 + \lambda_2\phi_2 = x^2 + y^2 + z^2 + \lambda_1\left(\frac{x^2}{4} + \frac{y^2}{5} + \frac{z^2}{25} - 1\right) + \lambda_2(x + y - z)$$

Taking the partial derivatives of G with respect to x, y, z and setting them equal to zero, we find

$$G_x = 2x + \frac{\lambda_1 x}{2} + \lambda_2 = 0, \quad G_y = 2y + \frac{2\lambda_1 y}{5} + \lambda_2 = 0, \quad G_x = 2z + \frac{2\lambda_1 z}{25} - \lambda_2 = 0 \tag{1}$$

Solving these equations for x, y, z, we find

$$x = \frac{-2\lambda_2}{\lambda_1 + 4}, \quad y = \frac{-5\lambda_2}{2\lambda_1 + 10}, \quad z = \frac{25\lambda_2}{2\lambda_1 + 50} \tag{2}$$

From the second constraint condition, $x + y - z = 0$, we obtain, on division by λ_2, assumed different from zero (this is justified, since otherwise we would have $x = 0$, $y = 0$, $z = 0$, which would not satisfy the first constraint condition), the result

$$\frac{2}{\lambda_1 + 4} + \frac{5}{2\lambda_1 + 10} + \frac{25}{2\lambda_1 + 50} = 0$$

Multiplying both sides by $2(\lambda_1 + 4)(\lambda_1 + 5)(\lambda_1 + 5)(\lambda_1 + 25)$ and simplifying yields

$$17\lambda_1^2 + 245\lambda_1 + 750 = 0 \text{ or } (\lambda_1 + 10)(17\lambda_1 + 75) = 0$$

from which $\lambda_1 = -10$ or $-75/17$.

Case 1: $\lambda_1 = -10$.

From (2), $x = \frac{1}{3}\lambda_2$, $y = \frac{1}{2}\lambda_2$, $z = 5/6\lambda_2$. Substituting in the first constraint condition, $x^2/4 + y^2/5 + z^2/25 = 1$, yields $\lambda_2^2 = 180/19$ or $\lambda_2 = \pm 6\sqrt{5/19}$. This gives the two critical points

$$\left(2\sqrt{5/19}, 3\sqrt{5/19}, 5\sqrt{5/19}\right), \left(-2\sqrt{5/19}, -3\sqrt{5/19}, -5\sqrt{5/19}\right)$$

The value of $x^2 + y^2 + z^2$ corresponding to these critical points is $(20 + 45 + 125)/19 = 10$.

Case 2: $\lambda_1 = -75/17$.

From (2), $x = 34/7\,\lambda_2$, $y = -17/4\lambda_2$, $z = 17/28\lambda_2$. Substituting in the first constraint condition, $x^2/4 + y^2/5 + z^2/25 = 1$, yields $\lambda_2 = \pm 140/(17\sqrt{646})$ which give the critical points

$$\left(40/\sqrt{646}, -35\sqrt{646}, 5/\sqrt{646}\right), \left(-40/\sqrt{646}, -35\sqrt{646}, -5/\sqrt{646}\right)$$

The value of $x^2 + y^2 + z^2$ corresponding to these is $(1600 + 1225 + 25)/646 = 75/17$.

Thus, the required maximum value is 10 and the minimum value is 75/17.

(b) Since $x^2 + y^2 + z^2$ represents the square of the distance of (x, y, z) from the origin (0, 0, 0), the problem is equivalent to determining the largest and smallest distances from the origin to the curve of intersection of the ellipsoid $x^2/4 + y^2/5 + z^2/25 = 1$ and the plane $z = x + y$. Since this curve is an ellipse, we have the interpretation that $\sqrt{10}$ and $\sqrt{75/17}$ are the lengths of the semimajor and semiminor axes of this ellipse.

The fact that the maximum and minimum values happen to be given by $-\lambda_1$ in both Case 1 and Case 2 is more than a coincidence. It follows, in fact, on multiplying Equations (*1*) by x, y, and z in succession and adding, for we then obtain

$$2x^2 + \frac{\lambda_1 x^2}{2} + \lambda_2 x + 2y^2 + \frac{2\lambda_1 y^2}{5} + \lambda_2 y + 2z^2 + \frac{2\lambda_1 z^2}{25} - \lambda_2 z = 0$$

i.e.,

$$x^2 + y^2 + z^2 + \lambda_1\left(\frac{x^2}{4} + \frac{y^2}{5} + \frac{z^2}{25}\right) + \lambda_2\,(x + y - z) = 0$$

Then, using the constraint conditions, we find $x^2 + y^2 + z^2 = -\lambda_1$.

For a generalization of this problem, see Problem 8.76.

Applications to errors

8.28. The period T of a simple pendulum of length l is given by $T = 2\sqrt{l/g}$. Find (a) the error and (b) the percent error made in computing T by using $l = 2$ m and $g = 9.75$ m/ sec^2, if the true values are $l = 19.5$m and $g = 9.81$ m/sec^2.

(a) $T = 2\pi l^{1/2} g^{-1/2}$. Then

$$dT = (2\pi\, g^{-1/2}(\frac{1}{2}l^{-1/2}\,dl) + (2\pi\, l^{1/2})(-\frac{1}{2}g^{-3/2}\,dg) = \frac{\pi}{\sqrt{lg}}\,dl - \pi\sqrt{\frac{1}{g^3}}\,dg \qquad (1)$$

Error in $g = \Delta g = dg = +0.06$; error in $l = \Delta l = dl = -0.5$

The error in T is actually ΔT, which is in this case approximately equal to dT. Thus, we have from Equation (1),

$$\text{Error in } T = dT = \frac{\pi}{\sqrt{(2)(9.75)}}(-0.05) - \pi\sqrt{\frac{2}{(9.75)^3}}(+0.06) = -0.0444 \text{ sec (approx.)}$$

The value of T for $l = 2$, $g = 9.75$ is $T = 2\pi\sqrt{\frac{2}{9.75}} = 2.846$ sec (approx.)

(b) Percent error (or relative error) in $T = \dfrac{dT}{T} = \dfrac{-0.0444}{2.846} = -1.56\%$.

Another method: Since $\ln T = \ln 2\pi + \dfrac{1}{2}\ln l - \dfrac{1}{2}\ln g$,

$$\frac{dT}{T} = \frac{1}{2}\frac{dl}{l} - \frac{1}{2}\frac{dg}{g} = \frac{1}{2}\left(\frac{-0.05}{2}\right) - \frac{1}{2}\left(\frac{+0.06}{9.75}\right) = -1.56\% \quad (2)$$

as before. Note that Equation (2) can be written

$$\text{Percent error in } T = \frac{1}{2} \text{ Percent error in } l - \frac{1}{2} \text{ Percent error in } g$$

Miscellaneous problems

8.29. Evaluate $\displaystyle\int_0^1 \frac{x-1}{\ln x}\,dx$.

In order to evaluate this integral, we resort to the following device. Define

$$\phi(\alpha) = \int_0^1 \frac{x^\alpha - 1}{\ln x}\,dx \qquad \alpha > 0$$

Then by Leibniz's rule

$$\phi'(\alpha) = \int_0^1 \frac{\partial}{\partial\alpha}\left(\frac{x^\alpha - 1}{\ln x}\right)dx = \int_0^1 \frac{x^\alpha \ln x}{\ln x}\,dx = \int_0^1 dx = \frac{1}{\alpha + 1}$$

Integrating with respect to α, $\phi(\alpha) = \ln(\alpha + 1) + c$. But since $\phi(0) = 0$, $c = 0$, and so $\phi(\alpha) = \ln(\alpha + 1)$. Then the value of the required integral is $\phi(1) = \ln 2$.

The applicability of Leibniz's rule can be justified here, since if we define $F(x, \alpha) = (x^\alpha - 1)/\ln x$, $0 < x < 1$, $F(0, \alpha) = 0$, $F(1, \alpha) = \alpha$, then $F(x, \alpha)$ is continuous in both x and α for $0 \leqq x \leqq 1$ and all finite $\alpha > 0$.

8.30. Find constants a and b for which $F(a, b) = \displaystyle\int_0^\pi \{\sin x - (ax^2 + bx)\}^2\,dx$ is a minimum.

The necessary conditions for a minimum are $\partial F/\partial a = 0$. Performing these differentiations, we obtain

$$\frac{\partial F}{\partial a} = \int_0^\pi \frac{\partial}{\partial a}\{\sin x - (ax^2 + bx)\}^2\,dx = -2\int_0^\pi x^2\{\sin x - (ax^2 + bx)\}\,dx = 0$$

$$\frac{\partial F}{\partial b} = \int_0^\pi \frac{\partial}{\partial b}\{\sin x - (ax^2 + bx)\}^2\,dx = -2\int_0^\pi x\{\sin x - (ax^2 + bx)\}\,dx = 0$$

From these we find

$$\begin{cases} \alpha \displaystyle\int_0^\pi x^4\,dx + b\int_0^\pi x^3\,dx = \int_0^\pi x^2 \sin x\,dx \\ \alpha \displaystyle\int_0^\pi x^3\,dx + b\int_0^\pi x^2\,dx = \int_0^\pi x \sin x\,dx \end{cases}$$

or

$$\begin{cases} \dfrac{\pi^5 a}{5} + \dfrac{\pi^4 b}{4} = \pi^2 - 4 \\ \dfrac{\pi^4 a}{4} + \dfrac{\pi^3 b}{3} = \pi \end{cases}$$

Solving for a and b, we find

$$a = \frac{20}{\pi^3} - \frac{320}{\pi^5} \approx -0.40065, \qquad b = \frac{240}{\pi^4} - \frac{12}{\pi^2} \approx 1.24798$$

We can show that for these values, $F(a, b)$ is indeed a minimum using the sufficiency conditions on Page 200.

The polynomial $ax^2 + bx$ is said to be a *least square approximation* of sin x over the interval $(0, \pi)$. The ideas involved here are of importance in many branches of mathematics and their applications.

SUPPLEMENTARY PROBLEMS

Tangent plane and normal line to a surface

8.31. Find the equations of (a) the tangent plane and (b) the normal line to the surface $x^2 + y^2 = 4z$ at $(2, -4, 5)$.

Ans. (a) $x - 2y - z = 5$ (b) $\dfrac{x-2}{1} = \dfrac{y+4}{-2} = \dfrac{z-5}{-1}$

8.32. If $z = f(x, y)$, prove that the equations for the tangent plane and normal line at point $P(x_0, y_0, z_0)$ are given, respectively, by (a) $z - z_0 = f_x|_P (x - x_0) + f_y|_P (y - y_0)$ and (b) $\dfrac{x - x_0}{f_x|_P} = \dfrac{y - y_0}{f_x|_P} = \dfrac{z - z_0}{-1}$.

8.33. Prove that the acute angle γ between the z axis and the normal line to the surface $F(x, y, z) = 0$ at any point is given by $\sec\gamma = \sqrt{F_x^2 + F_y^2\ F_z^2}\ /|F_z|$.

8.34. The equation of a surface is given in cylindrical coordinates by $F(\rho, \phi, z) = 0$, where F is continuously differentiable. Prove that the equations of (a) the tangent plane and (*b*) the normal line at the point $P(\rho_0, \phi_0, z_0)$ are given, respectively, by $A(x - x_0) + B(y - y_0) + C(z - z_0) = 0$ and $\dfrac{x - x_0}{A} = \dfrac{y - y_0}{B} = \dfrac{z - z_0}{C}$ where $x_0 = \rho_0 \cos \phi_0$, $y_0 = \rho_0 \sin \phi_0$ and $A = F_\rho|p \cos\phi_0 - \dfrac{1}{\rho} F_\phi|p \sin\phi_0$, $B = F_\rho|p \sin\phi_0 + \dfrac{1}{\rho} F_\phi|p \cos\phi_0$, and $C = F_z|_P$.

8.35. Use Problem 8.34 to find the equation of the tangent plane to the surface $\pi z = \rho\phi$ at the point where $\rho = 2$, $\phi = \pi/2$, $z = 1$. To check your answer, work the problem using rectangular coordinates.

Ans. $2x - \pi y + 2\pi z = 0$

Tangent line and normal plane to a curve

8.36. Find the equations of (a) the tangent line and (b) the normal plane to the space curve $x = 6 \sin t$, $y = 4 \cos 3t$. $z = 2 \sin 5t$ at the point where $t = \pi/4$.

Ans. (a) $\dfrac{x - 3\sqrt{2}}{3} = \dfrac{y + 2\sqrt{2}}{-6} = \dfrac{z + \sqrt{2}}{-5}$ (b) $3x - 6y - 5z = 26\sqrt{2}$

8.37. The surfaces $x + y + z = 3$ and $x^2 - y^2 + 2z^2 = 2$ intersect in a space curve. Find the equations of (a) the tangent line and (b) the normal plane to this space curve at the point $(1, 1, 1)$.

Ans. (a) $\dfrac{x-1}{-3} = \dfrac{y-1}{1} = \dfrac{z-1}{2}$ (b) $3x - y - 2z = 0$

Envelopes

8.38. Find the envelope of each of the following families of curves in the *xy* plane: (a) $y = ax - \alpha^2$ and (b) $\frac{x^2}{\alpha} + \frac{y^2}{1-\alpha} = 1$. In each case construct a graph.

Ans. (a) $x^2 = 4y$ (b) $x + y = \pm 1$, $x - y = \pm 1$

8.39. Find the envelope of a family of lines having the property that the length intercepted between the *x* and *y* axes is a constant *a*.

Ans. $x^{2/3} + y^{2/3} = a^{2/3}$

8.40. Find the envelope of the family of circles having centers on the parabola $y = x^2$ and passing through its vertex. [Hint: Let (α, α^2) be any point on the parabola.]

Ans. $x^2 = -y^3/(2y + 1)$

8.41. Find the envelope of the normals (called an *evolute*) to the parabola $y = \frac{1}{2}x^2$ and construct a graph.

Ans. $8(y - 1)^3 = 27x^2$

8.42. Find the envelope of the following families of surfaces: (a) $\alpha(x - y) - \alpha^2 z = 1$ and (b) $(x - \alpha)^2 + y^2 = 2\alpha z$.

Ans. (a) $4z = (x - y)^2$, (b) $y^2 = z^2 + 2xz$

8.43. Prove that the envelope of the two-parameter family of surfaces $F(x, y, z, \alpha, \beta) = 0$, if it exists, is obtained by eliminating α and β in the equations $F = 0$, $F_\alpha = 0$, and $F_\beta = 0$.

8.44. Find the envelope of the two-parameter families (a) $z = \alpha x + \beta y - \alpha^2 - \beta^2$ and (b) $x \cos\alpha + y\cos\beta + z\cos\gamma = a$ where $\cos^2\alpha + \cos^2\beta + \cos^2\gamma = 1$ and *a* is a constant.

Ans. (a) $4z = x^2 + y^2$ (b) $x^2 + y^2 + z^2 = a^2$

Directional derivatives

8.45. (a) Find the directional derivative of $U = 2xy - z^2$ at $(2, -1, 1)$ in a direction toward $(3, 1, -1)$. (b) In what direction is the directional derivative a maximum? (c) What is the value of this maximum?

Ans. (a) 10/3 (b) $-2\mathbf{i} + 4\mathbf{j} - 2\mathbf{k}$ (c) $2\sqrt{6}$

8.46. The temperature at any point (x, y) in the *xy* plane is given by $T = 100xy/(x^2 + y^2)$. (a) Find the directional derivative at the point (2, 1) in a direction making an angle of 60° with the positive *x* axis. (b) In what direction from (2, 1) would the derivative be a maximum? (c) What is the value of this maximum?

Ans. (a) $12\sqrt{3} - 6$ (b) in a direction making an angle of $\pi - \tan^{-1} 2$ with the positive *x* axis. or in the direction $-\mathbf{i} + 2\mathbf{j}$ (c) $12\sqrt{5}$

8.47. Prove that if $F(\rho, \phi, z)$ is continuously differentiable, the maximum directional derivative of *F* at any point is given by $\sqrt{\left(\frac{\partial F}{\partial \rho}\right)^2 + \frac{1}{\rho^2}\left(\frac{\partial F}{\partial \phi}\right)^2 + \left(\frac{\partial F}{\partial z}\right)^2}$.

Differentiation under the integral sign

8.48. If $\phi(\alpha) = \int_{\sqrt{\alpha}}^{1/\alpha} \cos\alpha x^2\, dx$, find $\dfrac{d\phi}{d\alpha}$

Ans. $\int_{\sqrt{\alpha}}^{1/\alpha} x^2 \sin\alpha x^2\, dx - \dfrac{1}{\alpha^2}\cos\dfrac{1}{\alpha} - \dfrac{1}{2\sqrt{\alpha}}\cos\alpha^2$

8.49. (a) If $F(\alpha) = \int_0^{\alpha^2} \tan^{-1}\dfrac{x}{\alpha}\, dx$, find $\dfrac{dF}{d\alpha}$ by Leibniz's rule. (b) Check the result in (a) by direct integration.

Ans. (a) $2\alpha \tan^{-1}\alpha - \dfrac{1}{2}\,\text{In}\,(\alpha^2 + 1)$

8.50. Given $\int_0^1 x^p\, dx = \dfrac{1}{p+1}$, $p > -1$, prove that $\int_0^1 x^p (\text{In}\, x)^m\, dx = \dfrac{(-1)^m\, m!}{(p+1)^{m+1}}$

8.51. Prove that $\int_0^{\pi} \text{In}\,(1 + \alpha\cos x)\, dx = \pi\ \text{In}\left(\dfrac{1+\sqrt{1-\alpha^2}}{2}\right), |\alpha| < 1.$

8.52. Prove that $\int_0^{\pi} \text{In}\,(1 - 2\alpha\cos x + \alpha^2)\, dx = \begin{cases} \pi\ \text{In}\,\alpha^2, & |\alpha| < 1 \\ 0, & |\alpha| > 1 \end{cases}$. Discuss the case $|\alpha| = 1$.

8.53. Show that $\int_0^{\pi} \dfrac{dx}{(5 - 3\cos x)^3} = \dfrac{59\pi}{2048}$.

Integration under the integral sign

8.54. Verify that $\int_0^1 \left\{ \int_1^2 (\alpha^2 - x^2)\, dx \right\} d\alpha = \int_1^2 \left\{ \int_0^1 (\alpha^2 - x^2)\, d\alpha = \int_1^2 \right\} dx$

8.55. Starting with the result $\int_0^{2\pi} (\alpha - \sin x)\, dx = 2\pi\alpha$, prove that for all constants a and b, x) $dx = 2\pi\alpha$, prove that for all constants a and b,

$$\int_0^{2\pi} \{(b - \sin x)^2 - (a - \sin x)^2\}\, dx = 2\pi(b^2 - a^2)$$

8.56. Use the result $\int_0^{2\pi} \dfrac{dx}{\alpha + \sin x} = \dfrac{2\pi}{\sqrt{\alpha^2 - 1}}$, $\alpha > 1$ to prove that $\int_0^{2\pi} \text{In}\left(\dfrac{5 + 3\sin x}{5 + 4\sin x}\right) dx = 2\pi\ \text{In}\left(\dfrac{9}{8}\right)$

8.57. (a) Use the result $\int_0^{\pi/2} \dfrac{dx}{1 + \alpha\cos x} = \dfrac{\cos^{-1}\alpha}{\sqrt{1-\alpha^2}}$, $0 \leq \alpha < 1$ to show that for $0 \leq a < 1$, $0 \leq b < 1$,

$$\int_0^{\pi/2} \sec x\ \text{In}\left(\frac{1 + b\cos x}{1 + a\cos x}\right) dx = \frac{1}{2}\{(\cos^{-1} a)^2 - (\cos^{-1} b)^2\}$$

(b) Show that $\int_0^{\pi/2} \sec x\ \text{In}\,(1 + \dfrac{1}{2}\cos x)\, dx = \dfrac{5\pi^2}{72}$.

Maxima and minima, lagrange multipliers

8.58. Find the maxima and minima of $F(x, y, z) = xy^2\ z) = xy^2z^3$ subject to the conditions $x + y + z = 6$, $x > 0$, $y > 0$, $z > 0$.

Ans. maximum value = 108 at $x = 1$, $y = 2$, $z = 3$

8.59. What is the volume of the largest rectangular parallelepiped which can be inscribed in the ellipsoid $x^2/9 + y^2/16 + z^2/36 = 1$?

Ans. $64\sqrt{3}$

8.60. (a) Find the maximum and minimum values of $x^2 + y^2$ subject to the condition $3x^2 + 4xy + 6y^2 = 140$. (b) Give a geometrical interpretation of the results in (a).

Ans. maximum value = 70, minimum value = 20

8.61. Solve Problem 8.23 using Lagrange multipliers.

8.62. Prove that in any triangle ABC there is a point P such that $\overline{PA}^2 + \overline{PB}^2 + \overline{PC}^2$ is a minimum and that P is the intersection of the medians.

8.63. (a) Prove that the maximum and minimum values of $f(x, y) = x^2 + xy + y^2$ in the unit square $0 \leqq x \leqq 1, 0 \leqq y \leqq 1$ are 3 and 0, respectively. (b) Can the result of (a) be obtained by setting the partial derivatives of $f(x, y)$ with respect to x and y equal to zero. Explain.

8.64. Find the extreme values of z on the surface $2x^2 + 3y^2 + z^2 - 12xy + 4xz = 35$.

Ans. maximum = 5, minimum = –5

8.65. Establish the method of Lagrange multipliers in the case where we wish to find the extreme values of $F(x, y, z)$ subject to the two constraint conditions $G(x, y, z) = 0$, $H(x, y, z) = 0$.

8.66. Prove that the shortest distance from the origin to the curve of intersection of the surfaces $xyz = a$ and $y = bx$, where $a > 0$, $b > 0$, is $3\sqrt{a(b^2 + 1)/2b}$.

8.67. Find the volume of the ellipsoid $11x^2 + 9y^2 + 15z^2 - 4xy + 10yz - 20xz = 80$.

Ans. $64\pi\sqrt{2}/3$

Applications to errors

8.68. The diameter of a right circular cylinder is measured as 6.0 ± 0.03 inches, while its height is measured as 4.0 ± 0.02 inches. What is the largest possible (a) error and (b) percent error made in computing the volume?

Ans. (a) 1.70 in^3 (b) 1.5 percent

8.69. The sides of a triangle are measured to be 12.0 and 15.0 feet, and the included angle is 60.0°. If the lengths can be measured to within 1 percent accuracy, while the angle can be measured to within 2 percent accuracy, find the maximum error and percent error in determining the (a) area and (b) the opposite side of the triangle.

Ans. (a) 2.501 ft^2, 3.21 percent (b) 0.287 ft, 2.08 percent

Miscellaneous problems

8.70. If ρ and ϕ are cylindrical coordinates, a and b are any positive constants, and n is a positive integer, prove that the surface $\rho^n \sin n\phi = a$ and $\rho^n \cos n\phi = b$ are mutually perpendicular along their curves of intersection.

8.71. Find an equation for (a) the tangent plane and (b) the normal line to the surface $8r\theta\phi = \pi^2$ at the point where $r = 1, \theta = \pi/4, \phi = \pi/2$, (r, θ, ϕ) being spherical coordinates.

Ans. (a) $4x - (\pi^2 + 4\pi)y + (4\pi - \pi^2)z = -\pi^2\sqrt{2}$ (b) $\dfrac{x}{-4} = \dfrac{y - \sqrt{2}/2}{\pi^2 + 4\pi} = \dfrac{z - \sqrt{2}/2}{\pi^2 - 4\pi}$

8.72. (a) Prove that the shortest distance from the point (a, b, c) to the plane $Ax + By + Cz + D = 0$ is

$$\left|\frac{Aa + Bb + Cc + D}{\sqrt{A^2 + B^2 + C^2}}\right|$$

(b) Find the shortest distance from $(1, 2, -3)$ to the plane $2x - 3y + 6z = 20$.

Ans. (b) 6

8.73. The potential V due to a charge distribution is given in spherical coordinates (r, θ, ϕ) by

$$V = \frac{p\cos\theta}{r^2}$$

where p is a constant. Prove that the maximum directional derivative at any point is

$$\frac{p\sqrt{\sin^2\theta + 4\cos^2\theta}}{r^3}$$

8.74. Prove that $\displaystyle\int_0^1 \frac{x^m - x^n}{\ln x}\,dx = \ln\left(\frac{m+1}{n+1}\right)$ if $m > 0, n > 0$. Can you extend the result to the case $m > -1, n > -1$?

8.75. (a) If $b^2 - 4ac < 0$ and $a > 0, c > 0$, prove that the area of the ellipse $ax^2 + bxy + cy^2 = 1$ is $2\pi/\sqrt{4ac - b^2}$. (Hint: Find the maximum and minimum values of $x^2 + y^2$ subject to the constraint $ax^2 + bxy + cy^2 = 1$.)

8.76. Prove that the maximum and minimum distances from the origin to the curve of intersection defined by $x^2/a^2 + y^2/b^2 + z^2/c^2 = 1$ and $Ax + By + Cz = 0$ can be obtained by solving for d the equation

$$\frac{A^2a^2}{a^2 - d^2} + \frac{B^2b^2}{b^2 - d^2} + \frac{C^2c^2}{c^2 - d^2} = 0$$

8.77. Prove that the last equation in the preceding problem always has two real solutions d_1^2 and d_2^2 for any real nonzero constants a, b, c and any real constants A, B, C (not all zero). Discuss the geometrical significance of this.

8.78. (a) Prove that $\displaystyle I_M = \int_0^M \frac{dx}{(x^2 + \alpha^2)^2} = \frac{1}{2\alpha^3}\tan^{-1}\frac{M}{\alpha} + \frac{M}{2\alpha^2(\alpha^2 + M^2)}$.

(b) Find $\lim\limits_{M\to\infty} I_M$. This can be denoted by $\displaystyle\int_0^\infty \frac{dx}{(x^2 + \alpha^2)^2}$.

(c) Is $\displaystyle\lim_{M\to\infty} \frac{d}{d\alpha}\int_0^M \frac{dx}{(x^2 + \alpha^2)^2} = \frac{d}{d\alpha}\lim_{M\to\infty}\int_0^M \frac{dx}{(x^2 + \alpha^2)^2}$?

8.79. Find the point on the paraboloid $z = x^2 + y^2$ which is closest to the point (3, –6, 4).

Ans. (1, –2, 5)

8.80. Investigate the maxima and minima of $f(x, y) = (x^2 - 2x + 4y^2 - 8y)^2$.

Ans. Minimum value = 0

8.81. (*a*) Prove that $\displaystyle\int_0^{\pi/2} \frac{\cos x\, dx}{\alpha \cos x + \sin x} = \frac{\alpha\pi}{2(\alpha^2+1)} - \frac{\ln \alpha}{\alpha^2+1}$.

(b) Use (*a*) to Prove that $\displaystyle\int_0^{\pi/2} \frac{\cos^2 x\, dx}{(2\cos x + \sin x)^2} = \frac{3\pi + 5 - 8\ln 2}{50}$.

8.82. (*a*) Find sufficient conditions for a relative maximum or minimum of $w = f(x, y, z)$.

(b) Examine $w = x^2 + y^2 + z^2 - 6xy + 8xz - 10yz$ for maxima and minima.

[Hint: For (*a*) use the fact that the *quadratic form* $A\alpha^2 + B\beta^2 + C\gamma^2 + 2D\alpha\beta + 2E\alpha\gamma + 2F\beta\gamma > 0$ (i.e., is *positive definite*) if

$$A > 0, \qquad \begin{vmatrix} A & D \\ D & B \end{vmatrix} > 0. \qquad \begin{vmatrix} A & D & F \\ D & B & E \\ F & E & C \end{vmatrix} > 0$$

CHAPTER 9

Multiple Integrals

Much of the procedure for double and triple integrals may be thought of as a reversal of partial differentiation and otherwise is analogous to that for single integrals. However, one complexity that must be addressed relates to the domain of definition. With single integrals, the functions of one variable were defined on intervals of real numbers. Thus, the integrals only depended on the properties of the functions. The integrands of double and triple integrals are functions of two and three variables, respectively, and as such are defined on two- and three-dimensional regions. These regions have a flexibility in shape not possible in the single-variable cases. For example, with functions of two variables, and the corresponding double integrals, rectangular regions $a \leqq x \leqq b, c \leqq y \leqq d$ are common. However, in many problems the domains are regions bounded above and below by segments of plane curves. In the case of functions of three variables, and the corresponding triple integrals other than the regions $a \leqq x \leqq b, c \leqq y \leqq d, e \leqq z \leqq f$, there are those bounded above and below by portions of surfaces. In very special cases, double and triple integrals can be directly evaluated. However, the systematic technique of *iterated integration* is the usual procedure. It is here that the reversal of partial differentiation comes into play.

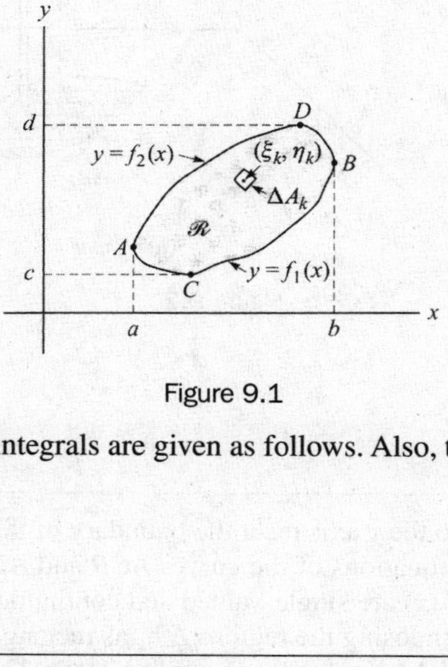

Figure 9.1

Definitions of double and triple integrals are given as follows. Also, the method of iterated integration is described.

Double Integrals

Let $F(x, y)$ be defined in a closed region $\Re$ of the xy plane (see Figure 9.1). Subdivide $\Re$ into n subregions $\Delta\Re_k$ of area ΔA_k, $k = 1, 2, \ldots, n$. Let (ξ_k, η_k) be some point of ΔA_k. Form the sum

$$\sum_{k=1}^{n} F(\xi_k, \eta_k)\,\Delta A_k \qquad (1)$$

Consider

$$\lim_{n\to\infty} \sum_{k=1}^{n} F(\xi_k, \eta_k)\,\Delta A_k \qquad (2)$$

where the limit is taken so that the number n of subdivisions increases without limit and such that the largest linear dimension of each ΔA_k approaches zero. See Figure 9.2(a). If this limit exists, it is denoted by

$$\iint_{\Re} F(x, y)\, dA \tag{3}$$

and is called the *double integral* of $F(x, y)$ over the region $\Re$.

It can be proved that the limit does exist if $F(x, y)$ is continuous (or sectionally continuous) in $\Re$.

The double integral has a great variety of interpretations with any individual one dependent on the form of the integrand. For example, if $F(x, y) = \rho(x, y)$ represents the variable density of a flat iron plate, then the double integral $\int_A \rho\, dA$ of this function over a same-shaped plane region A is the mass of the plate. In Figure 9.2(b) we assume that $F(x, y)$ is a height function [established by a portion of a surface $z = F(x, y)$] for a cylindrically shaped object. In this case the double integral represents a volume.

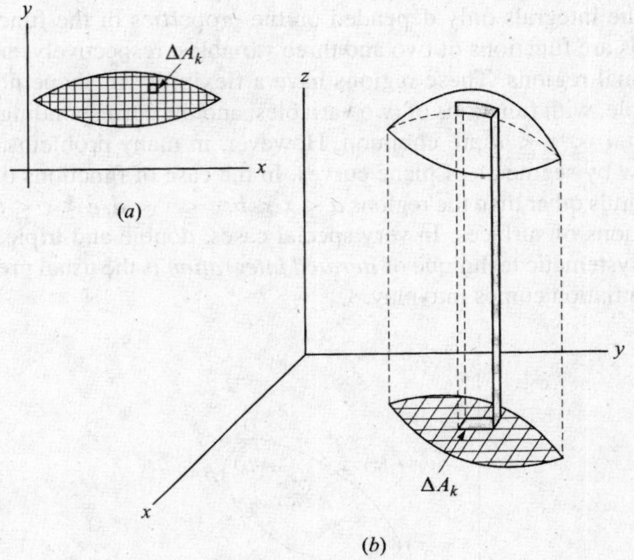

Figure 9.2

Iterated Integrals

If $\Re$ is such that any lines parallel to the y axis meet the boundary of $\Re$ in, at most, two points (as is true in Figure 9.1), then we can write the equations of the curves ACB and ADB bounding $\Re$ as $y = f_1(x)$ and $y = f_2(x)$, respectively, where $f_1(x)$ and $f_2(x)$ are single-valued and continuous in $a \leq x \leq b$. In this case we can evaluate the double integral (3) by choosing the regions $\Delta\Re_k$ as rectangles formed by constructing a grid of lines parallel to the x and y axes and ΔA_k as the corresponding areas. Then Equation (3) can be written

$$\iint_{\Re} F(x, y)\, dx\, dy = \int_{x=a}^{b} \int_{y=f_1(x)}^{f_2(x)} F(x, y)\, dy\, dx$$
$$= \int_{x=a}^{b} \left\{ \int_{y=f_1(x)}^{f_2(x)} F(x, y)\, dy \right\} dx \tag{4}$$

where the integral in braces is to be evaluated first (keeping x constant) and finally integrating with respect to x from a to b. The result (4) indicates how a double integral can be evaluated by expressing it in terms of two single integrals called *iterated integrals*.

The process of iterated integration is visually illustrated in Figure 9.3(*a*) and (*b*) and further illustrated as follows.

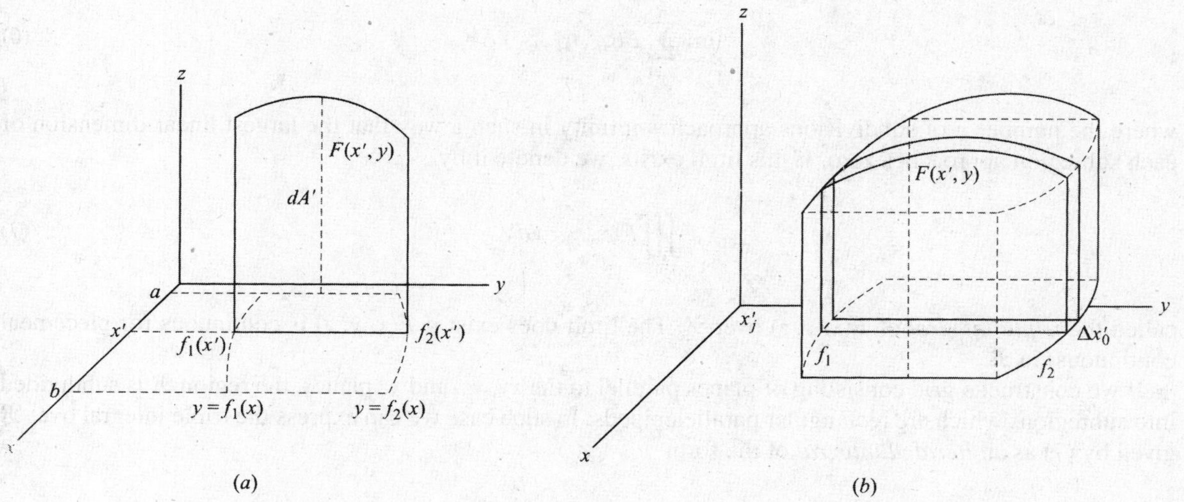

Figure 9.3

The general idea, as demonstrated with respect to a given three-space region, is to establish a plane section, integrate to determine its area, and then add up all the plane sections through an integration with respect to the remaining variable. For example, choose a value of x (say, $x = x'$). The intersection of the plane $x = x'$ with the solid establishes the plane section. In it, $z = F(x', y)$ is the height function, and if $y = f_1(x)$ and $y = f_2(x)$ for all z) are the bounding cylindrical surfaces of the solid, then the width is $f_2(x') - f_1(x')$, i.e., $y_2 - y_1$. Thus, the area of the section is $A = \int_{y_1}^{y_2} F(x', y)\, dy$. Now establish slabs $A_j \Delta x_j$, where, for each interval $\Delta x_j = x_j - x_{j-r}$, there is an intermediate value x_j'. Then sum these to get an approximation to the target volume. Adding the slabs and taking the limit yields

$$V = \lim_{x \to \infty} \sum_{j=1}^{n} A_j \Delta x_j = \int_a^b \left(\int_{y_1}^{y_2} F(x, y)\, dx \right) dx$$

In some cases the order of integration is dictated by the geometry. For example, if $\Re$ is such that any lines parallel to the x axis meet the boundary of $\Re$ in, at most, two points (as in Figure 9.1), then the equations of curves *CAD* and *CBD* can be written $x = g_1(y)$ and $x = g_2(y)$, respectively, and we find, similarly,

$$\iint_{\Re} F(x, y)\, dx\, dy = \int_{y=c}^{d} \int_{x=g_1(y)}^{g_2(y)} F(x, y)\, dx\, dy$$

$$= \int_{y=c}^{d} \left\{ \int_{x=g_1(y)}^{g_2(y)} F(x, y)\, dx \right\} dy \qquad (5)$$

If the double integral exists, Equations (4) and (5) yield the same value. (See, however, Problem 9.21.) In writing a double integral, either of the forms (4) or (5), whichever is appropriate, may be used. We call one form an *interchange of the order of integration* with respect to the other form.

In case $\Re$ is not of the type shown in Figure 9.3, it can generally be subdivided into regions $\Re_1, \Re_2, \ldots$, which are of this type. Then the double integral over $\Re$ is found by taking the sum of the double integrals over $\Re_1, \Re_2, \ldots$.

Triple Integrals

These results are easily generalized to closed regions in three dimensions. For example, consider a function $F(x, y, z)$ defined in a closed three-dimensional region $\Re$. Subdivide the region into n subregions of volume ΔV_k, $k = 1, 2, \ldots, n$. Letting (ξ_k, η_k, ζ_k) be some point in each subregion, we form

$$\lim_{x\to\infty} \sum_{k=1}^{n} F(\xi_k, \eta_k, \xi_k)\, \Delta V_k \tag{6}$$

where the number n of subdivisions approaches infinity in such a way that the largest linear dimension of each subregion approaches zero. If this limit exists, we denote it by

$$\iiint_{\Re} F(x, y, z)\, dV \tag{7}$$

called the *triple integral* of $F(x, y, z)$ over $\Re$. The limit does exist if $F(x, y, z)$ is continuous (or piecemeal continuous) in $\Re$.

If we construct a grid consisting of planes parallel to the xy, yz, and xz planes, the region $\Re$ is subdivided into subregions which are rectangular parallelepipeds. In such case we can express the triple integral over $\Re$ given by (7) as an *iterated integral* of the form

$$\int_{x=a}^{b} \int_{y=g_1(x)}^{g_2(a)} \int_{z=f_1(x,y)}^{f_2(x,y)} F(x, y, z)dx\, dy\, dz = \int_{x=a}^{b} \left[\int_{y=g_1(x)}^{g_2(x)} \left\{ \int_{z=f_1(x,y)}^{f_2(x,Y)} F(x, y, z)\, dz \right\} dy \right] dx \tag{8}$$

(where the innermost integral is to be evaluated first) or the sum of such integrals. The integration can also be performed in any other order to give an equivalent result.

The interated triple integral is a sequence of integrations, first from surface portion to surface portion, then from curve segment to curve segment, and finally from point to point. (See Figure 9.4.)

Extensions to higher dimensions are also possible.

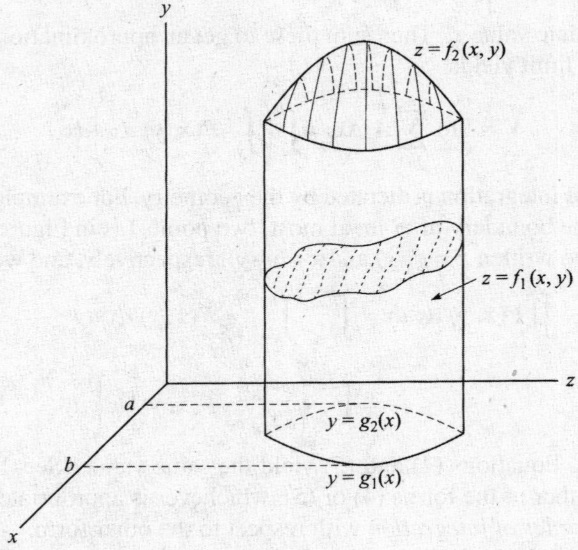

Figure 9.4

Transformations of Multiple Integrals

In evaluating a multiple integral over a region $\Re$, it is often convenient to use coordinates other than rectangular, such as the curvilinear coordinates considered in Chapters 6 and 7.

If we let (u, υ) be curvilinear coordinates of points in a plane, there will be a set of transformation equations $x = f(u, \upsilon)$, $y = g(u, \upsilon)$ mapping points (x, y) of the xy plane into points (u, υ) of the $u\upsilon$ plane.

In such case the region $\Re$ of the xy plane is mapped into a region $\Re'$ of the uv plane. We then have

$$\iint F(x, y)\, dx\, dy = \iint G(u, \upsilon) \left|\frac{\partial(x, y)}{\partial(u, \upsilon)}\right| du\, d\upsilon \tag{9}$$

where $G(u, \upsilon), \equiv F\{f(u, \upsilon), g(u, \upsilon)\}$ and

$$\frac{\partial(x, y)}{\partial(u, \upsilon)} \equiv \begin{vmatrix} \dfrac{\partial x}{\partial u} & \dfrac{\partial x}{\partial \upsilon} \\ \dfrac{\partial y}{\partial u} & \dfrac{\partial y}{\partial \upsilon} \end{vmatrix} \tag{10}$$

is the *Jacobian* of x and y with respect to u and υ (see Chapter 6).

Similarly, if (u, υ, w) are curvilinear coordinates in three dimensions, there will be a set of transformation equations $x = f(u, \upsilon, w)$, $y = g(u, \upsilon, w)$, $z = h(u, \upsilon, w)$ and we can write

$$\iiint\limits_{\Re} F(x, y, z)\, dx\, dy\, dz = \iiint\limits_{\Re} G(u, \upsilon, w) \left|\frac{\partial(x, y, z)}{\partial(u, \upsilon, w)}\right| du\, dv\, dw \tag{11}$$

where $G(u, \upsilon, w) \equiv F\{(f(u, \upsilon, w), g(u, \upsilon, w), h(u, \upsilon, w)\}$ and

$$\frac{\partial(x, y, z)}{\partial(u, \upsilon, w)} \equiv \begin{vmatrix} \dfrac{\partial x}{\partial u} & \dfrac{\partial x}{\partial \upsilon} & \dfrac{\partial x}{\partial w} \\ \dfrac{\partial y}{\partial u} & \dfrac{\partial y}{\partial \upsilon} & \dfrac{\partial y}{\partial w} \\ \dfrac{\partial z}{\partial u} & \dfrac{\partial z}{\partial \upsilon} & \dfrac{\partial z}{\partial w} \end{vmatrix} \tag{12}$$

is the Jacobian of x, y, and z with respect to u, υ, and w.

The results (9) and (11) correspond to change of variables for double and triple integrals. Generalizations to higher dimensions are easily made.

The Differential Element of Area in Polar Coordinates, Differential Elements of Area in Cylindral and Spherical Coordinates

Of special interest is the differential element of area dA for polar coordinates in the plane, and the differential elements of volume dV for cylindrical and spherical coordinates in three-space. With these in hand, the double and triple integrals as expressed in these systems are seen to take the following forms. (See Figure 9.5.)

The transformation equations relating cylindrical coordinates to rectangular Cartesian ones appear in Chapter 7, in particular,

$$x = \rho \cos \phi,\ y = \rho \sin \phi,\ z = z$$

The coordinate surfaces are circular cylinders, planes, and planes. (See Figure 9.5.)

At any point of the space (other than the origin), the set of vectors $\left\{\dfrac{\partial \mathbf{r}}{\partial \rho}, \dfrac{\partial \mathbf{r}}{\partial \phi}, \dfrac{\partial \mathbf{r}}{\partial z}\right\}$ constitutes an orthogonal basis.

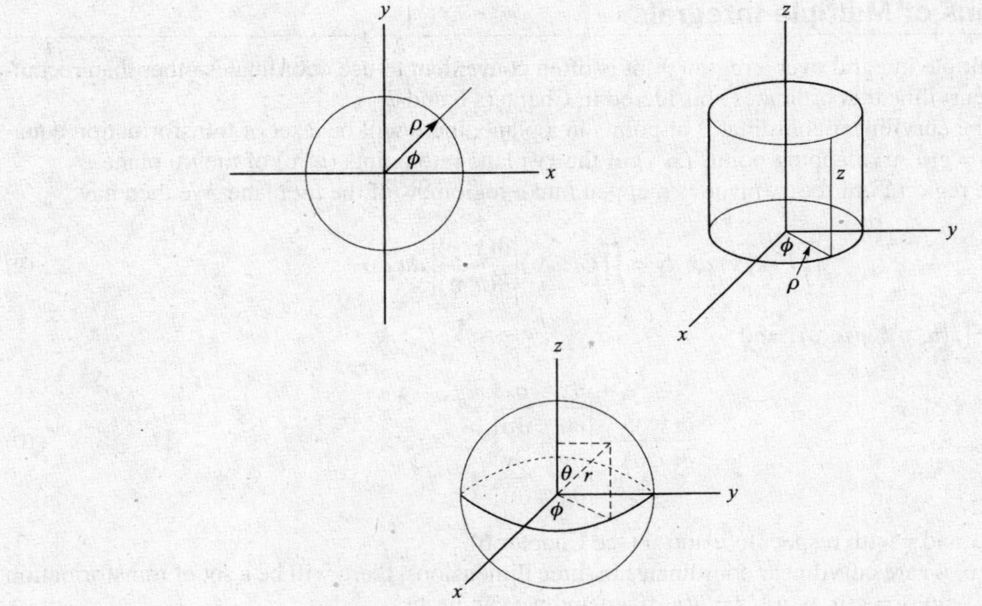

Figure 9.5

In the cylindrical case, $\mathbf{r} = \rho\cos\phi\mathbf{i} + \rho\sin\phi\mathbf{j} + z\mathbf{k}$ and the set is

$$\frac{\partial \mathbf{r}}{\partial \rho} = \cos\phi \mathrm{i} + \sin\phi \,\mathrm{j}, \quad \frac{\partial \mathbf{r}}{\partial \rho} = -\rho\sin\phi \mathrm{i} + \rho\cos\phi \mathbf{j}, \quad \frac{\partial \mathbf{r}}{\partial z} = \mathbf{k}$$

Therefore, $\dfrac{\partial \mathbf{r}}{\partial \rho} \cdot \dfrac{\partial \mathbf{r}}{\partial \phi} \times \dfrac{\partial \mathbf{r}}{\partial z} = \rho.$

That the geometric interpretation of $\dfrac{\partial \mathbf{r}}{\partial \rho} \cdot \dfrac{\partial \mathbf{r}}{\partial \phi} \times \dfrac{\partial \mathbf{r}}{\partial z}\, d\rho\, d\phi\, dz$ is an infinitesimal rectangular parallelepiped suggests that the differential element of volume in cylindrical coordinates is

$$dV = \rho\, d\rho\, d\phi\, dz$$

Thus, for an integrable but otherwise arbitrary function $F(\rho, \phi, z)$ of cylindrical coordinates, the iterated triple integral takes the form

$$\int_{z_1}^{z_2} \int_{g_1(z)}^{g_2(z)} \int_{f_1(\phi,z)}^{f_2(\phi,z)} F(\rho,\phi,z)\rho\, d\rho\, d\phi\, dz$$

The differential element of area for polar coordinates in the plane results by suppressing the z coordinate. It is

$$dA = \left| \frac{\partial \mathbf{r}}{\partial \rho} \times \frac{\partial \mathbf{r}}{\partial \phi} \right| d\rho\, d\phi$$

and the iterated form of the double integral is

$$\int_{\rho_1}^{\rho_2} \int_{\phi_1(\rho)}^{\phi_2(\rho)} F(\rho,\phi)\rho\, d\rho\, d\phi$$

The transformation equations relating spherical and rectangular Cartesian coordinates are

$$x = r\sin\theta\cos\phi,\ y = r\sin\theta\sin\phi,\ z = r\cos\theta$$

In this case the coordinate surfaces are spheres, cones, and planes. (See Figure 9.5.)

Following the same pattern as with cylindrical coordinates we discover that

$$dV = r^2 \sin\theta\, dr\, d\theta\, d\phi$$

and the iterated triple integral of $F(r, \theta, \phi)$ has the spherical representation

$$\int_{r_1}^{r_2}\int_{\theta_1(\phi)}^{\theta_2(\phi)}\int_{\phi_1(r,\theta)}^{\phi_2(r,\theta)} F(r,\theta,\phi)\, r^2 \sin\theta \, dr\, d\theta\, d\phi$$

Of course, the order of these integrations may be adapted to the geometry.

The coordinate surfaces in spherical coordinates are spheres, cones, and planes. If r is held constant—say, $r = a$—then we obtain the differential element of surface area

$$dA = a^2 \sin\theta \, d\theta\, d\phi$$

The first octant surface area of a sphere of radius a is

$$\int_0^{\pi/2}\int_0^{\pi/2} a^2 \sin\theta\, d\theta\, d\phi = \int_0^{\pi/2} a^2 (-\cos\theta)\Big|_0^{\frac{\pi}{2}} d\phi = \int_0^{\pi/2} a^2\, d\phi = a^2\frac{\pi}{2}$$

Thus, the surface area of the sphere is $4\pi a^2$.

SOLVED PROBLEMS

Double integrals

9.1. (a) Sketch the region $\Re$ in the xy plane bounded by $y = x^2$, $x = 2$, $y = 1$. (b) Give a physical interpretation to $\iint_{\Re} (x^2 + y^2)\, dx\, dy$. (c) Evaluate the double integral in (b).

(a) The required region $\Re$ is shown shaded in Figure 9.6.

(b) Since $x^2 + y^2$ is the square of the distance from any point (x, y) to $(0, 0)$, we can consider the double integral as representing the *polar moment of inertia* (i.e., moment of intertia with respect to the origin) of the region $\Re$ (assuming unit density).

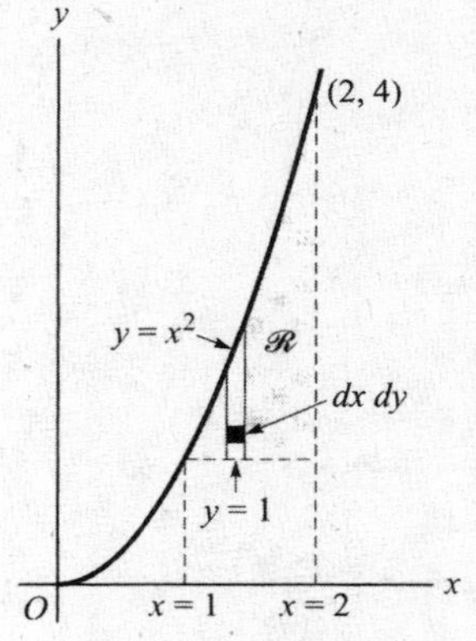

Figure 9.6

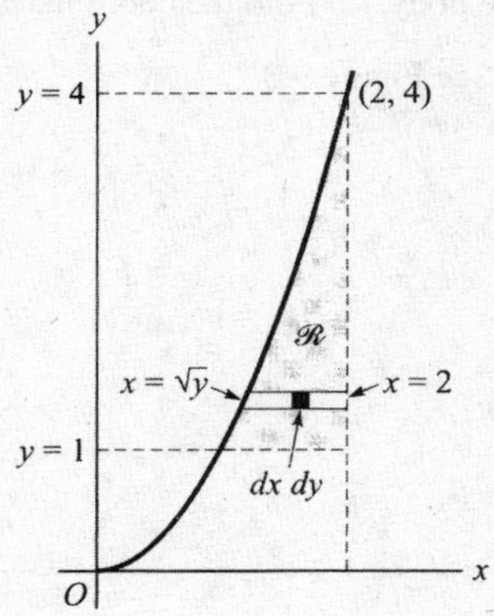

Figure 9.7

We can also consider the double integral as representing the *mass* of the region $\Re$, assuming a density varying as $x^2 + y^2$.

(c) **Method 1:** The double integral can be expressed as the iterated integral

$$\int_{x=1}^{2}\int_{y=1}^{x^2} (x^2 + y^2)\, dy\, dx = \int_{x=1}^{2}\left\{\int_{y=1}^{x^2} (x^2 + y^2)\, dy\right\} dx = \int_{x=1}^{2} x^2 y + \frac{y^2}{3}\Bigg|_{y=1}^{x^2}$$

$$dx = \int_{x=1}^{2}\left(x^4 + \frac{x^6}{3} - x^2 - \frac{1}{3}\right) dx = \frac{1006}{105}$$

The integration with respect to y (keeping x constant) from $y = 1$ to $y = x^2$ corresponds formally to summing in a vertical column (see Figure 9.6). The subsequent integration with respect to x from $x = 1$ to $x = 2$ corresponds to addition of contributions from all such vertical columns between $x = 1$ and $x = 2$.

Method 2: The double integral can also be expressed as the iterated integral

$$\int_{y=1}^{4}\int_{x=\sqrt{y}}^{2}(x^2+y^2)\,dx\,dy = \int_{y=1}^{4}\left\{\int_{x=\sqrt{y}}^{2}(x^2+y^2)\,dx\right\}dy = \int_{y=1}^{4}\frac{x^3}{3}+xy^2\Big|_{x=\sqrt{y}}^{2}\,dy$$

$$= \int_{x=1}^{2}\left(x^4+\frac{x^6}{3}-x^2-\frac{1}{3}\right)dx = \frac{1006}{105}$$

In this case the vertical column of region $\mathfrak{R}$ in Figure 9.6 is replaced by a horizontal column, as in Figure 9.7. Then the integration with respect to x (keeping y constant) from $x = \sqrt{y}$ to $x = 2$ corresponds to summing in this horizontal column. Subsequent integration with respect to y from $y = 1$ to $y = 4$ corresponds to addition of contributions for all such horizontal columns between $y = 1$ and $y = 4$.

9.2. Find the volume of the region bounded by the elliptic paraboloid $z = 4 - x^2 - \frac{1}{4}y^2$ and the plane $z = 0$.

Because of the symmetry of the elliptic paraboloid, the result can be obtained by multiplying the first octant volume by 4.

Letting $z = 0$ yields $4x^2 + y^2 = 16$. The limits of integration are determined from this equation. The required volume is

$$4\int_0^2\int_0^{2\sqrt{4-x^2}}\left(4-x^2-\frac{1}{4}y^2\right)dy\,dx = 4\int_0^2\left(4y-x^2y-\frac{1}{4}\frac{y^3}{3}\right)\Bigg|_0^{2\sqrt{4-x^2}}dx = 16\Pi$$

Hint: Use trigonometric substitutions to complete the integrations.

9.3. The geometric model of a material body is a plane region R bounded by $y = x^2$ and $y = \sqrt{2-x^2}$ on the interval $0 \leq x \leq 1$, and with a density function $\rho = xy$. (a) Draw the graph of the region. (b) Find the mass of the body. (c) Find the coordinates of the center of mass.

(a) See Figure 9.8.

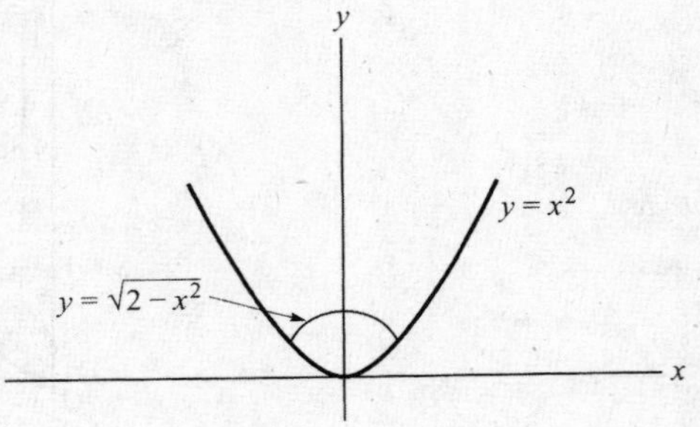

Figure 9.8

(b) $$M = \int_a^b\int_{f_1}^{f_2}\rho\,dy\,dx = \int_0^1\int_{x^2}^{\sqrt{2-x^2}} yx\;dy\,dx = \int_0^1\left[\frac{y^2}{2}\right]_{x^2}^{\sqrt{2-x^2}} x\,dx$$

$$= \int_0^1\frac{1}{2}x(2-x^2-x^4)\,dx = \left[\frac{x^2}{2}-\frac{x^4}{8}-\frac{x^6}{12}\right]_0^1 = \frac{7}{24}$$

(c) The coordinates of the center of mass are defined to be

$$\overline{x} = \frac{1}{M}\int_a^b\int_{f_1(x)}^{f_2(x)} x\,\rho\,dy\,dx \quad\text{and}\quad \overline{y} = \frac{1}{M}\int_a^b\int_{f_1(x)}^{f_2(x)} y\,\rho\,dy\,dx$$

where

$$M = \int_a^b \int_{f_1(x)}^{f_2(x)} \rho \, dy \, dx$$

Thus,

$$M\,\bar{x} = \int_0^1 \int_{x^2}^{\sqrt{2x-x^2}} x \, xy \, dy \, dx = \int_0^1 x^2 \left[\frac{y^2}{2}\right]_{x^2}^{\sqrt{2-x^2}} dx = \int_0^1 x^2 \frac{1}{2}[2 - x^2 - x^4] dx$$

$$= \left[\frac{x^3}{3} - \frac{x^5}{10} - \frac{x^7}{14}\right]_0^1 = \frac{1}{3} - \frac{1}{10} \frac{1}{14} = \frac{17}{105}$$

$$M\bar{y} = \int_0^1 \int_{x^2}^{\sqrt{2x-5}} yx \, dy \, dx = -\frac{13}{120} + 4\frac{\sqrt{2}}{15}$$

9.4. Find the volume of the region common to the intersecting cylinders $x^2 + y^2 = a^2$ and $x^2 + z^2 = a^2$.

Required volume = 8 times volume of region shown in Figure 9.9

$$= 8 \int_{x=0}^{a} \int_{y=0}^{\sqrt{a^2-x^2}} z \, dy \, dx$$

$$= 8 \int_{x=0}^{a} \int_{y=0}^{\sqrt{a^2-x^2}} = \sqrt{a^2 - x^2} \, dy \, dx$$

$$= 8 \int_{x=0}^{a} (a^2 - x^2) \, dx = \frac{16a^3}{3}$$

As an aid in setting up this integral, note that $z \, dy \, dx$ corresponds to the volume of a column such as shown darkly shaded in Figure 9.9. Keeping x constant and integrating with respect to y from $y = 0$ to $y = \sqrt{a^2 - x^2}$ corresponds to adding the volumes of all such columns in a slab parallel to the yz plane, thus giving the volume of this slab. Finally, integrating with respect to x from $x = 0$ to $x = a$ corresponds to adding the volumes of all such slabs in the region, thus giving the required volume.

9.5. Find the volume of the region bounded by $z = x + y$, $z = 6$, $x = 0$, $y = 0$, $z = 0$.

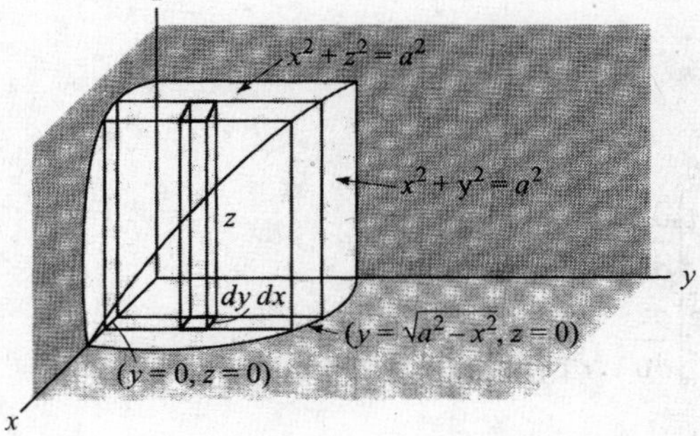

Figure 9.9

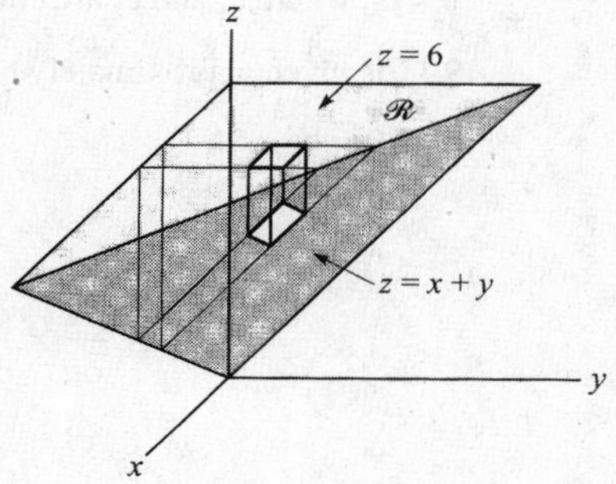

Figure 9.10

Required volume = volume of region shown in Figure 9.10

$$= \int_{x=0}^{6} \int_{y=0}^{6-x} \{6 - (x + y)\} dy \, dx$$

$$= \int_{x=0}^{6} (6 - x)\, y - \frac{1}{2} y^2 \Big|_{y=0}^{6-x} dx$$

$$= \int_{x=0}^{6} \frac{1}{2} (6 - x)^2 \, dx = 36$$

In this case the volume of a typical column (shown darkly shaded) corresponds to $\{6 - (x + y)\}\, dy\, dx$. The limits of integration are then obtained by integrating over the region $\Re$ of Figure 9.10. Keeping x constant and integrating with respect to y from $y = 0$ to $y = 6 - x$ (obtained from $z = 6$ and $z = x + y$) corresponds to summing all columns in a slab parallel to the yz plane. Finally, integrating with respect to x from $x = 0$ to $x = 6$ corresponds to adding the volumes of all such slabs and gives the required volume.

Transformation of double integrals

9.6. Justify Equation (9), Page 225, for changing variables in a double integral.

In rectangular coordinates, the double integral of $F(x, y)$ over the region $\Re$ (shaded in Figure 9.11) is $\iint_{\Re} (F(x, y)\, dx\, dy$. We can also evaluate this double integral by considering a grid formed by a family of u and υ curvilinear coordinate curves constructed on the region $\Re$, as shown in Figure 9.11.

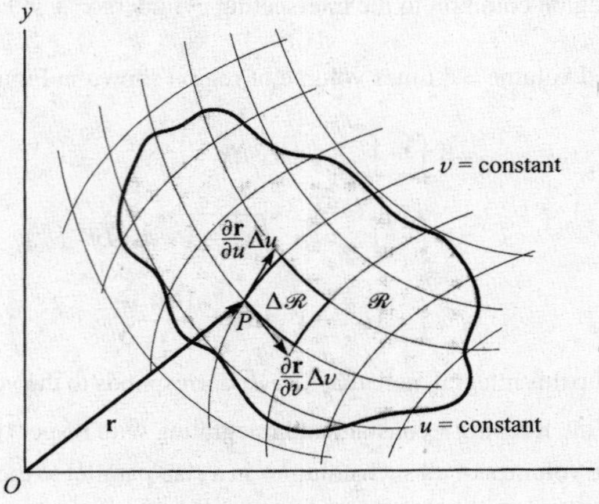

Figure 9.11

Let P be any point with coordinates (x, y) or (u, υ), where $x = f(u, \upsilon)$ and $y = g(u, \upsilon)$. Then the vector $\mathbf{r}$ from O to p is given by $\mathbf{r} = x\mathbf{i} + y\mathbf{j} = f(u, \upsilon)\mathbf{i} + g(u, \upsilon)\mathbf{j}$. The tangent vectors to the coordinate curves $u = c_1$ and $\upsilon = c_2$, where c_1 and c_2 are constants, are $\partial\mathbf{r}/\partial\upsilon$ and $\partial\mathbf{r}/\partial u$, respectively. Then the area of region $\Delta\Re$ of Figure 9.11 is given approximately by $\left|\frac{\partial\mathbf{r}}{\partial u} \times \frac{\partial\mathbf{r}}{\partial \upsilon}\right| \Delta u\, \Delta\upsilon$.

But

$$\frac{\partial\mathbf{r}}{\partial u} \times \frac{\partial\mathbf{r}}{\partial \upsilon} = \begin{vmatrix} \mathbf{i} & \mathbf{j} & \mathbf{k} \\ \frac{\partial x}{\partial u} & \frac{\partial y}{\partial u} & 0 \\ \frac{\partial x}{\partial \upsilon} & \frac{\partial y}{\partial \upsilon} & 0 \end{vmatrix} = \begin{vmatrix} \frac{\partial x}{\partial u} & \frac{\partial y}{\partial u} \\ \frac{\partial x}{\partial \upsilon} & \frac{\partial y}{\partial \upsilon} \end{vmatrix} \mathbf{k} = \frac{\partial(x, y)}{\partial(u, \upsilon)} \mathbf{k}$$

so that

$$\left|\frac{\partial\mathbf{r}}{\partial u} \times \frac{\partial\mathbf{r}}{\partial \upsilon}\right| \Delta u\, \Delta\upsilon = \left|\frac{\partial(x, y)}{\partial(u, \upsilon)}\right| \Delta u\, \Delta\upsilon$$

The double integral is the limit of the sum

$$\sum F\{f(u,\upsilon), g(u,\upsilon)\} \left|\frac{\partial(x, y)}{\partial(u, \upsilon)}\right| \Delta u\, \Delta\upsilon$$

taken over the entire region $\Re$. An investigation reveals that this limit is

$$\iint_{\Re} F\{f(u,\upsilon), g(u,\upsilon)\} = \left|\frac{\partial(x, y)}{\partial(u, \upsilon)}\right| du\, d\upsilon$$

where $\Re'$ is the region in the $u\upsilon$ plane into which the region $\Re$ is mapped under the transformation $x = f(u, \upsilon)$, $y = g(u, \upsilon)$.

Another method of justifying this method of change of variables makes use of line integrals and Green's theorem in the plane (see Problem 10.32).

9.7. If $u = x^2 - y^2$ and $\upsilon = 2xy$, find $\partial(u, \upsilon)$ in terms of u and υ.

$$\frac{\partial(u,\upsilon)}{\partial(x,y)} = \begin{vmatrix} u_x & u_y \\ \upsilon_x & \upsilon_y \end{vmatrix} = \begin{vmatrix} 2x & -2y \\ 2y & 2x \end{vmatrix} = 4(x^2 + y^2)$$

From the identity $(x^2 + y^2)^2 = (x^2 - y^2)^2 + (2xy)^2$, we have

$$(x^2 + y^2)^2 = u^2 + \upsilon^2 \qquad \text{and} \qquad x^2 + y^2 = \sqrt{u^2 + \upsilon^2}$$

Then, by Problem 6.43,

$$\frac{\partial(x,y)}{\partial(u,\upsilon)} = \frac{1}{\partial(u,\upsilon)/\partial(x,y)} = \frac{1}{4(x^2 + y^2)} = \frac{1}{4\sqrt{u^2 + \upsilon^2}}$$

Another method: Solve the given equations for x and y in terms of u and υ and find the Jacobian directly.

9.8. Find the polar moment of inertia of the region in the xy plane bounded by $x^2 - y^2 = 1$, $x^2 - y^2 = 9$, $xy = 2$, $xy = 4$, assuming unit density.

Under the transformation $x^2 - y^2 = u$, $2xy = \upsilon$, the required region $\Re$ in the xy plane, shaded in Figure 9.12(a), is mapped into region $\Re'$ of the $u\upsilon$ plane, shaded in Figure 9.12(b). Then:

$$\text{Required polar moment of inertia} = \iint_{\Re} (x^2 + y^2)\, dx\, dy = \iint_{\Re'} (x^2 + y^2) \left|\frac{\partial(x,y)}{\partial(u,\upsilon)}\right| du\, d\upsilon$$

$$= \iint_{\Re'} \sqrt{u^2 + \upsilon^2} \frac{du\, d\upsilon}{4\sqrt{u^2 + \upsilon^2}} = \frac{1}{4}\int_{u=1}^{9}\int_{\upsilon=4}^{8} du\, d\upsilon = 8$$

where we have used the results of Problem 9.7.

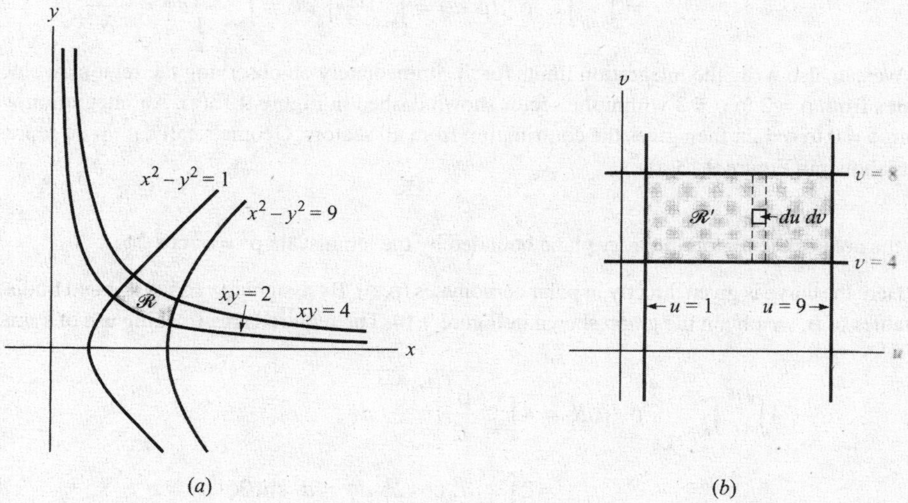

Figure 9.12

Note that the limits of integration for the region $\Re'$ can be constructed directly from the region $\Re$ in the xy plane without actually constructing the region $\Re'$. In such case we use a grid, as in Problem 9.6. The coordinates (u, υ) are curvilinear coordinates, in this case called *hyperbolic coordinates*.

9.9 Evaluate $\iint_{\Re} \sqrt{x^2+y^2}\,dx\,dy$, where $\Re$ is the region in the xy plane bounded by $x^2+y^2=4$ and $x^2+y^2=9$.

The presence of x^2+y^2 suggests the use of polar coordinates (ρ, ϕ), where $x = \rho\cos\phi$, $y = \rho\sin\phi$ (see Problem 6.39). Under this transformation the region $\Re$ [Figure 9.13(*a*)] is mapped into the region $\Re'$ [Figure 9.13(*b*)].

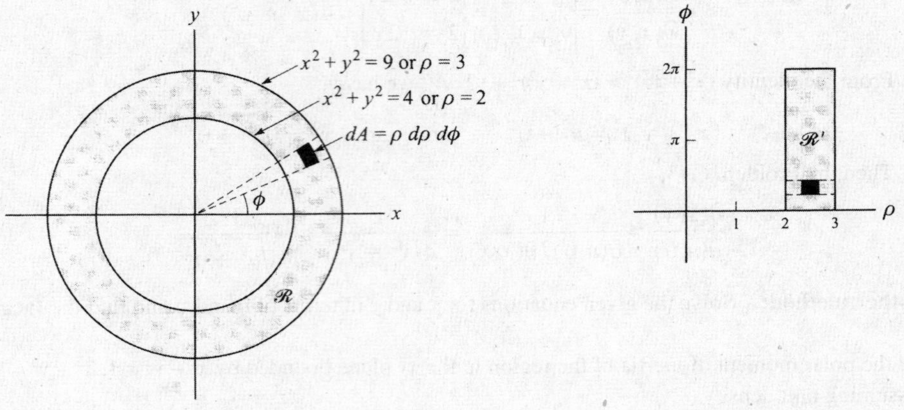

Figure 9.13

Since $\dfrac{\partial(x, y)}{\partial(\rho, \phi)} = \rho$, it follows that

$$\iint_{\Re} \sqrt{x^2+y^2}\;dx\,dy = \iint_{\Re'} \sqrt{x^2+y^2}\left|\frac{\partial(x,y)}{\partial(\rho,\phi)}\right| d\rho\,d\phi = \iint_{\Re'} \rho \cdot \rho\, d\rho\,d\phi$$

$$= \int_{\phi=0}^{2\pi}\int_{\rho=2}^{3} \rho^2 d\rho\,d\phi = \int_{\phi=0}^{2\pi} \frac{\rho^3}{3}\Big|_2^3 d\phi = \int_{\phi=0}^{2\pi} \frac{19}{3} d\phi = \frac{38\pi}{3}$$

We can also write the integration limits for $\Re'$ immediately on observing the region $\Re$, since for fixed ϕ. ρ varies from $\rho = 2$ to $\rho = 3$ within the sector shown dashed in Figure 9.13(a). An integration with respect to ϕ from $\phi = 0$ to $\phi = 2\pi$ then gives the contribution from all sectors. Geometrically, $\rho\,d\rho\,d\phi$ represents the area dA, as shown in Figure 9.13(a).

9.10. Find the area of the region in the xy plane bounded by the lemniscate $\rho^2 = a^2\cos 2\phi$.

Here the curve is given directly in polar coordinates (ρ, ϕ). By assigning various to ϕ and finding corresponding values of ρ, we obtain the graph shown in Figure 9.14. The required area (making use of symmetry) is

$$4\int_{\phi=0}^{\pi/4}\int_{\rho=0}^{a\sqrt{\cos 2\phi}} \rho\,d\rho\,d\phi = 4\int_{\phi=0}^{\pi/4} \frac{\rho^3}{2}\Bigg|_{\rho=0}^{a\sqrt{\cos 2\phi}} d\phi$$

$$= 2\int_{\phi=0}^{\pi/4} a^2\cos 2\phi\,d\phi = a^2\sin 2\phi\Big|_{\phi=0}^{\pi/4} = a^2$$

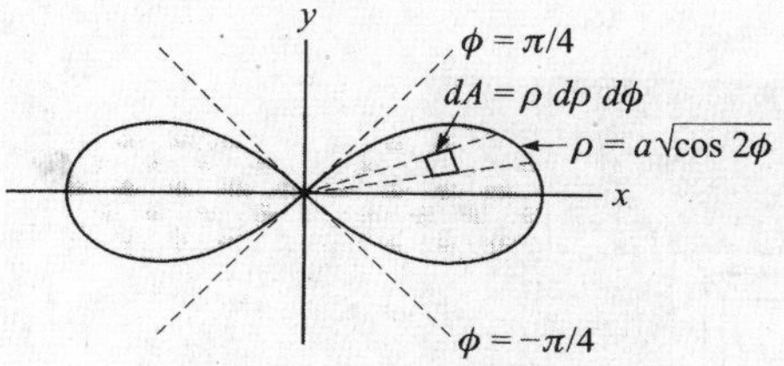

Figure 9.14

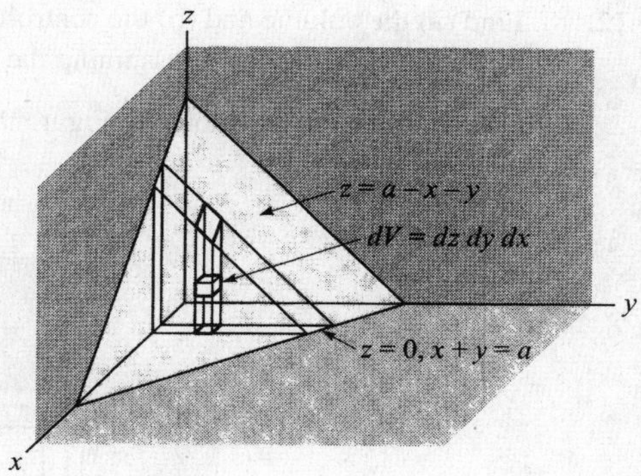

Figure 9.15

Triple integrals

9.11. (a) Sketch the three-dimensional region $\mathfrak{R}$ bounded by $x + y + z = a$ $(a > 0)$, $x = 0$, $y = 0$, $z = 0$. (b) Give a physical interpretation to

$$\iiint_{\mathfrak{R}} (x^2 + y^2 + z^2)\,dx\,dy\,dz$$

(c) Evaluate the triple integral in (b).

(a) The required region $\mathfrak{R}$ is shown in Figure 9.15.

(b) Since $x^2 + y^2 + z^2$ is the square of the distance from any point (x, y, z) to $(0, 0, 0)$, we can consider the triple integral as representing the *polar moment of inertia* (i.e., moment of inertia with respect to the origin) of the region $\mathfrak{R}$ (assuming unit density).

We can also consider the triple integral as representing the *mass* of the region if the density varies as $x^2 + y^2 + z^2$.

(c) The triple integral can be expressed as the iterated integral

$$\int_{x=0}^{a}\int_{y=0}^{a-x}\int_{z=0}^{a-x-y}(x^2 + y^2 + z^2)\,dz\,dy\,dx$$

$$= \int_{x=0}^{a}\int_{y=0}^{a-x} x^2z + y^2z + \frac{z^3}{3}\bigg|_{z=0}^{a-x-y} dy\,dx$$

$$= \int_{x=0}^{a}\int_{y=0}^{a-x}\left\{x^2(a-x) - x^2y + (a-x)y^2 - y^3 + \frac{(a-x-y)^3}{3}\right\} dy\,dx$$

$$= \int_{x=0}^{a} x^2(a-x)y - \frac{x^2y^2}{2} + \frac{(a-x)y^3}{3} - \frac{y^4}{4} - \frac{(a-x-y)^4}{12}\bigg|_{y=0}^{a-x} dx$$

$$= \int_0^a\left\{x^2(a-x)^2 - \frac{x^2(a-x)^2}{2} + \frac{(a-x)^4}{3} - \frac{(a-x)^4}{4} + \frac{(a-x)^4}{12}\right\} dx$$

$$= \int_0^a\left\{\frac{x^2(a-x)^2}{2} + \frac{(a-x)^4}{6}\right\} dx = \frac{a^5}{20}$$

The integration with respect to z (keeping x and y constant) from $z = 0$ to $z = a - x - y$ corresponds to summing the polar moments of inertia (or masses) corresponding to each cube in a vertical column. The subsequent integration with respect to y from $y = 0$ to $y = a - x$ (keeping x constant) corresponds to addition of contributions from all vertical columns contained in a slab parallel to the yz plane. Finally, integration with respect to x from $x = 0$ to $x = a$ adds up contributions from all slabs parallel to the yz plane.

Although this integration has been accomplished in the order z, y, x, any other order is is clearly possible and the final answer should be the same.

9.12. Find (a) the volume and (b) the centroid of the region $\Re$ bounded by the parabolic cylinder $z = 4 - x^2$ and the planes $x = 0$, $y = 6$, $z = 0$, assuming the density to be a constant σ.

The region $\Re$ is shown in Figure 9.16.

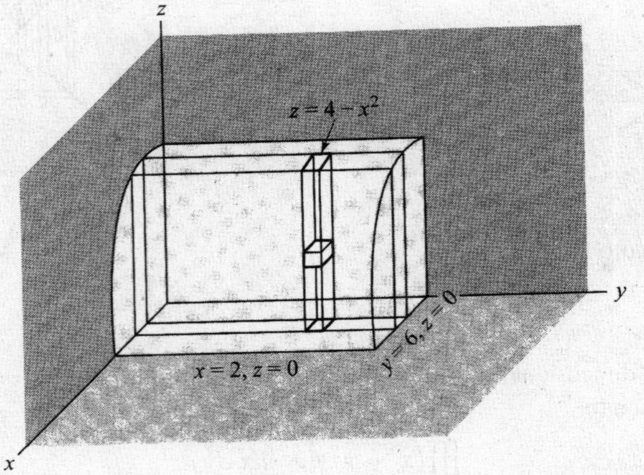

Figure 9.16

(a) Required volume $= \iiint dx\, dy\, dz$

$$= \int_{x=0}^{2}\int_{y=0}^{6}\int_{z=0}^{4-x^2} dz\, dy\, dx$$

$$= \int_{x=0}^{2}\int_{y=0}^{6} (4-x^2)dy\, dx$$

$$= \int_{x=0}^{2} (4-x^2)y\Big|_{y=0}^{6} dx$$

$$= \int_{x=0}^{2} (2-6x^2)\, dx = 32$$

(b) Total mass $= \int_{x=0}^{2}\int_{y=0}^{6}\int_{z=0}^{4-x^2} \sigma\, dz\, dy\, dx = 32\sigma$ by (a), since σ is constant. Then

$$\bar{x} = \frac{\text{Total moment about } yz \text{ plane}}{\text{Total mass}} = \frac{\int_{x=0}^{2}\int_{y=0}^{6}\int_{z=0}^{4-x^2} \sigma\, x\, dz\, dy\, dx}{\text{Total mass}} = \frac{24}{32\sigma} = \frac{3}{4}$$

$$\bar{y} = \frac{\text{Total moment about } xz \text{ plane}}{\text{Total mass}} = \frac{\int_{x=0}^{2}\int_{y=0}^{6}\int_{z=0}^{4-x^2} \sigma\, y\, dz\, dy\, dx}{\text{Total mass}} = \frac{96\sigma}{32\sigma} = 3$$

$$\bar{z} = \frac{\text{Total moment about } xy \text{ plane}}{\text{Total mass}} = \frac{\int_{x=0}^{2}\int_{y=0}^{6}\int_{z=0}^{4-x^2} \sigma\, z\, dz\, dy\, dx}{\text{Total mass}} = \frac{256\sigma/5}{32\sigma} = \frac{8}{5}$$

Thus, the centroid has coordinates (3/4, 3, 8/5).

Note that the value for $\bar{y}$ could have been predicted because of symmetry.

Transformation of triple integrals

9.13. Justify Equation (11), Page 225, for changing variables in a triple integral.

By analogy with Problem 9.6, we construct a grid of curvilinear coordinate surfaces which subdivide the region $\Re$ into subregions, a typical one of which is $\Delta\Re$ (see Figure 9.17).

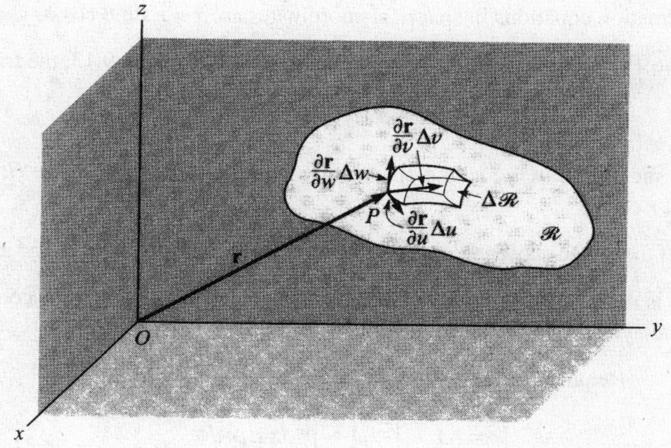

Figure 9.17

The vector **r** from the origin O to point P is

$$r = x\mathbf{i} + y\mathbf{j} + z\mathbf{k} = f(u, \upsilon, w)\mathbf{i} + g(u, \upsilon, w)\mathbf{j} + h(u, \upsilon, w)\mathbf{k}$$

assuming that the transformation equations are $x = f(u, \upsilon, w)$, $y = g(u, \upsilon, w)$, and $z = h(u, \upsilon, w)$.

Tangent vectors to the coordinate curves corresponding to the intersection of pairs of coordinate surfaces are given by $\partial\mathbf{r}/\partial u$, $\partial\mathbf{r}/\partial\upsilon$, $\partial\mathbf{r}/\partial w$. Then the volume of the region $\Delta\Re$ of Figure 9.17 is given approximately by

$$\left|\frac{\partial\mathbf{r}}{\partial u}\cdot\frac{\partial\mathbf{r}}{\partial \upsilon}\times\frac{\partial\mathbf{r}}{\partial w}\right|\Delta u\ \Delta \upsilon\ \Delta w = \left|\frac{\partial(x,y,z)}{\partial(u,\upsilon,w)}\right|\Delta u\ \Delta\upsilon\ \Delta w$$

The triple integral of $F(x, y, z)$ over the region is the limit of the sum

$$\sum F\{f(u,\upsilon,w), g(u,\upsilon,w), h(u,\upsilon,w)\}\left|\frac{\partial(x,y,z)}{\partial(u,\upsilon,w)}\right|\Delta u\Delta\upsilon\Delta w$$

An investigation reveals that this limit is

$$\iiint F\{f(u,\upsilon,w), g(u,\upsilon,w), h(u,\upsilon,w)\}\left|\frac{\partial(x,y,z)}{\partial(u,\upsilon,w)}\right| du\, d\upsilon\, dw$$

where $\Re'$ is the region in the $u\upsilon w$ space into which the region $\Re$ is mapped under the transformation.

Another method for justifying this change of variables in triple integrals makes use of Stokes's theorem (see Problem 10.84).

9.14. What is the mass of a circular cylindrical body represented by the region $0 \leqq \rho \leqq c$, $0 \leqq \phi \leqq 2\pi$, $0 \leqq z \leqq h$, and with the density function $\mu = z \sin^2 \phi$?

$$M = \int_0^h \int_0^{2\pi} \int_0^c z\sin^2\phi\rho\ d\rho\ d\phi\ dz = \pi$$

9.15. Use spherical coordinates to calculate the volume of a sphere of radius a.

$$V = 8\int_0^a \int_0^{\pi/2} \int_0^{\pi/2} a^2 \sin\theta\, dr\ d\theta\ d\phi = \frac{4}{3}\pi a^3$$

9.16. Express $\iiint\limits_{\Re'} F(x, y, z)\, dx\, dy\, dz$ in (a) cylindrical and (b) spherical coordinates.

(a) The transformation equations in cylindrical coordinates are $x = \rho\cos\phi$, $y = \rho\sin\phi$, $z = z$.

As in Problem 6.39, $\partial(x, y, z)/\partial(\rho, \phi, z) = \rho$. Then, by Problem 9.13, the triple integral becomes

$$\iiint\limits_{\Re'} G(\rho,\phi,z)\rho\, d\rho\ d\phi\ dz$$

where $\Re'$ is the region in the ρ, ϕ, z space corresponding to $\Re$ and where $G(\rho, \phi, z \equiv F(\rho\cos\phi, \rho\sin\phi, z)$.

(b) The transformation equations in spherical coordinates are $x = r \sin\theta \cos\phi$, $y = r \sin\theta \sin\phi$, $z = r\cos\theta$.

By Problem 6.101, $\partial(x, y, z)/\partial(r, \theta, \phi) = r^2 \sin\theta$. Then, by Problem 9.13, the triple integral becomes

$$\iiint_{\mathfrak{R}'} H(r,\theta,\phi) r^2 \sin\theta \, dr \, d\theta \, d\phi$$

where $\mathfrak{R}'$ is the region in the r, θ, ϕ space corresponding to $\mathfrak{R}$, and where $H(r, \theta, \phi) \equiv F(r \sin\theta \cos\phi, r \sin\theta \sin\phi, r\cos\theta)$.

9.17. Find the volume of the region above the xy plane bounded by the paraboloid $z = x^2 + y^2$ and the cylinder $x^2 + y^2 = a^2$.

The volume is most easily found by using cylindrical coordinates. In these coordinates the equations for the paraboloid and cylinder are, respectively, $z = \rho^2$ and $\rho = a$. Then

$$\text{Required volume} = 4 \text{ times volume shown in Figure 9.18}$$

$$= 4\int_{\phi=0}^{\pi/2}\int_{\rho=0}^{a}\int_{z=0}^{\rho^2} \rho \, dz \, d\rho \, d\phi$$

$$= 4\int_{\phi=0}^{\pi/2}\int_{\rho=0}^{a} \rho^3 d\rho \, d\phi$$

$$= 4\int_{hi=0}^{\pi/2} \left.\frac{\rho^4}{4}\right|_{=0}^{a} d\phi = \frac{\pi}{2}a^4$$

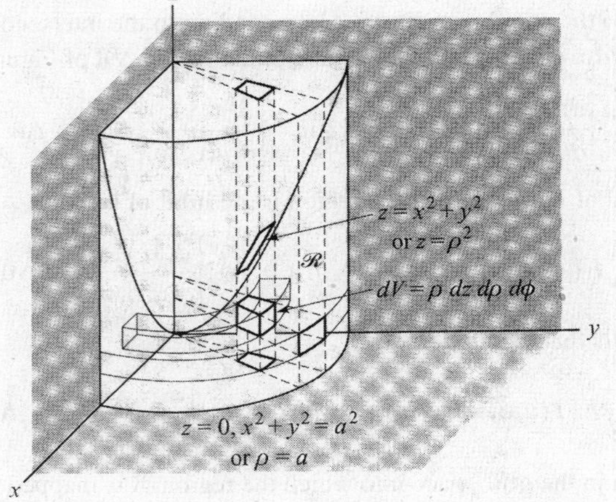

Figure 9.18

The integration with respect to z (keeping ρ and ϕ constant) from $z = 0$ to $z = \rho^2$ corresponds to summing the cubical volumes (indicated by dV) in a vertical column extending from the xy plane to the paraboloid. The subsequent integration with respect to ρ (keeping ϕ constant) from $\rho = 0$ to $\rho = a$ corresponds to addition of volumes of all columns in the wedge-shaped region. Finally, integration with respect to ϕ corresponds to adding volumes of all such wedge-shaped regions.

The integration can also be performed in other orders to yield the same result.

We can also set up the integral by determining the region $\mathfrak{R}'$ in ρ, ϕ, z space into which $\mathfrak{R}$ is mapped by the cylindrical coordinate transformation.

9.18. (a) Find the moment of inertia about the z axis of the region in Problem 9.17, assuming that the density is the constant σ. (b) Find the radius of gyration.

(a) The moment of inertia about the z axis is

$$I_z = 4\int_{\phi_0}^{\pi/2}\int_{\rho=0}^{a}\int_{z=0}^{\rho^2} \rho^2 \sigma\rho \, dz \, d\rho \, d\phi$$

$$= 4\sigma\int_{\phi=0}^{\pi/2}\int_{\rho=0}^{a} \rho^5 d\rho \, d\phi = 4\sigma\int_{\phi=0}^{\pi/2} \left.\frac{\rho^6}{6}\right|_{\rho=0}^{a} d\phi = \frac{\pi a^6 \sigma}{3}$$

The result can be expressed in terms of the mass M of the region, since, by Problem 9.17,

$$M = \text{volume} \times \text{desnity} = \frac{\pi}{2} a^4 \sigma \quad \text{so that} \quad I_z = \frac{\pi a^6 \sigma}{3} = \frac{\pi a^6}{3} \cdot \frac{2M}{\pi a^4} = \frac{2}{3} M a^2$$

Note that in setting up the integral for I_z we can think of $\sigma\, \rho\, dz\, d\rho\, d\phi\; dz\, d\rho\, d\phi$ as being the mass of the cubical volume element, $\rho^2 \sigma\, \rho\, dz\, d\rho\, d\phi$ as the moment of inertia of this mass with respect to the z axis, and $\iiint_{\mathfrak{R}'} \rho^2 \sigma \rho\, dz\, d\rho\, d\phi$ as the total moment of inertia about the z axis. The limits of integration are determined as in Problem 9.17.

(b) The radius of gyration is the value K such that $MK^2 = \frac{2}{3} Ma^2$; i.e., $K^2 = \frac{2}{3} a^2$ or $K = a\sqrt{2/3}$.

The physical significance of K is that if all the mass M were concentrated in a thin cylindrical shell of radius K, then the moment of inertia of this shell about the axis of the cylinder would be I_z.

9.19. (a) Find the volume of the region bounded above by the sphere $x^2 + y^2 + z^2 = a^2$ and below by the cone $z^2 \sin^2 \alpha = (x^2 + y^2) \cos^2 \alpha$, where α is a constant such that $0 \leq \alpha \leq \pi$. (b) From the result in (a), find the volume of a sphere of radius a.

In spherical coordinates the equation of the sphere is $r = a$ and that of the cone is $\theta = \alpha$. This can be seen directly or by using the transformation equations $x = r \sin \theta \cos \phi$, $y = r \sin \theta \sin \phi$, $z = r \cos \theta$. For example, $z^2 \sin^2 \alpha = (x^2 + y^2) \cos^2 \alpha$ becomes, on using these equations, $r^2 \cos^2 \theta \sin^2 \alpha = (r^2 \sin^2 \theta \cos^2 \phi + r^2 \sin^2 \theta \sin^2 \phi) \cos^2 \alpha$, i.e., $r^2 \cos^2 \theta \sin^2 \alpha = r^2 \sin^2 \theta \cos^2 \alpha$, from which $\tan \theta = \pm \tan \alpha$ and so $\theta = \alpha$ or $\theta = \pi - \alpha$. It is sufficient to consider one of these—say, $\theta = \alpha$.

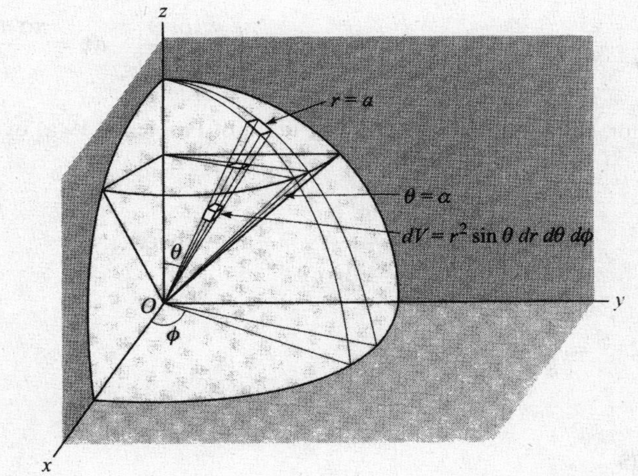

Figure 9.19

(a) Required volume = 4 times volume (shaded) in Figure 9.19

$$= 4 \int_{\phi=0}^{\pi/2} \int_{\theta=0}^{\alpha} \int_{r=0}^{\rho^2} r^2 \sin\theta \; dr\; d\theta\; d\phi$$

$$= 4 \int_{\phi=0}^{\pi/2} \int_{\theta=0}^{\alpha} \frac{r^3}{3} \sin\theta \Big|_{r=0}^{\alpha} d\theta\; d\phi$$

$$= \frac{4a^3}{3} \int_{\phi=0}^{\pi/2} \int_{\theta=0}^{\alpha} \sin\theta\; d\theta\; d\phi$$

$$= \frac{4a^3}{3} \int_{\phi=0}^{\pi/2} -\cos\theta \Big|_{\theta=0}^{\alpha} d\phi$$

$$= \frac{2\pi a^3}{3} (1 - \cos\alpha)$$

The integration with respect to r (keeping θ and ϕ constant) from $r = 0$ to $r = a$ corresponds to summing the volumes of all cubical elements (such as indicated by dV) in a column extending from $r = 0$ to $r = a$. The subsequent integration with respect to θ (keeping ϕ constant) from $\theta = 0$ to $\theta = \pi/4$ corresponds to summing the volumes of all columns in the wedge-shaped region. Finally, integration with respect to ϕ corresponds to adding volumes of all such wedge-shaped regions.

(b) Letting $\alpha = \pi$, the volume of the sphere thus obtained is

$$\frac{2\pi a^3}{3}(1-\cos\pi) = \frac{4}{3}\pi a^3$$

9.20. (a) Find the centroid of the region in Problem 9.19. (b) Use the result in (a) to find the centroid of a hemisphere.

(a) The centroid ($\bar{x}$, $\bar{y}$, $\bar{z}$) is, due to symmetry, given by $\bar{x} = \bar{y} = 0$ and

$$\bar{z} = \frac{\text{Total moment about } xy \text{ plane}}{\text{Total mass}} = \frac{\iiint z\sigma\, dV}{\iiint \sigma\, dV}$$

Since $z = r\cos\theta$ and σ is constant, the numerator is

$$4\sigma\int_{\phi=0}^{\pi/2}\int_{\theta=0}^{\alpha}\int_{r=0}^{\rho^2} r\cos\theta\cdot r^2\sin\theta\, dr\, d\theta\, d\phi = 4\sigma\int_{\phi=0}^{\pi/2}\int_{\theta=0}^{\alpha}\left.\frac{r^4}{4}\right|_{r=0}^{a}\sin\theta\cos\theta\, d\theta\, d\phi$$

$$= \sigma a^4\int_{\phi=0}^{\pi/2}\int_{\theta=0}^{\alpha}\sin\theta\cos\theta\, d\theta\, d\phi$$

$$= \sigma a^4\int_{\phi=0}^{\pi/2}\left.\frac{sin^2\theta}{2}\right|_{\theta=0}^{a} d\phi = \frac{\pi\sigma a^4\sin^2\alpha}{4}$$

The denominator, obtained by multiplying the result of Problem 9.19(a) by σ, is $\frac{2}{3}\pi\sigma a^3$ $(1 - \cos\alpha)$. Then

$$\bar{z} = \frac{\frac{1}{4}\pi\sigma a^4\sin^2\alpha}{\frac{2}{3}\pi\sigma a^3(1-\cos\alpha)} = \frac{3}{8}a(1+\cos\alpha).$$

(b) Letting $\alpha = \pi/2$, $\bar{z} = \frac{3}{8}$a.

Miscellaneous problems

9.21. Prove that (a) $\int_0^1\left\{\int_0^1\frac{x-y}{(x+y)^3}dy\right\}dx = \frac{1}{2}$ and (b) $\int_0^1\left\{\int_0^1\frac{x-y}{(x+y)^3}dx\right\}dy = -\frac{1}{2}$,

(a) $$\int_0^1\left\{\int_0^1\frac{x-y}{(x+y)^3}dy\right\}dx = \int_0^1\left\{\int_0^1\frac{2x-(x+y)}{(x+y)^3}dy\right\}dx$$

$$= \int_0^1\left\{\int_0^1\left(\frac{2x}{(x+y)^3} - \frac{1}{(x+y)^2}\right)dy\right\}dx$$

$$= \int_0^1\left.\left(\frac{-x}{(x+y)^2} - \frac{1}{x+y}\right)\right|_{y=0}^{1} dx$$

$$= \int_0^1\frac{dx}{(x+y)^2} - \left.\frac{-1}{x+1}\right|_0^1 = \frac{1}{2}$$

(b) This follows at once on formally interchanging x and y in (a) to obtain

$$\iint_{\Re} \frac{x-y}{(x+y)^3}\,dx\,dy, \int_0^1 \left\{ \int_0^1 \frac{x-y}{(x+y)^3}\,dx \right\} dy = -\frac{1}{2}$$ and then multiplying both sides by –1.

This example shows that interchange in order of integration may not always produce equal results. A sufficient condition under which the order may be interchanged is that the double integral over the corresponding region exists. In this case $\iint_{\Re} \frac{x-y}{(x+y)^3}\,dx\,dy$, where $\Re$ is the region $0 \leq x \leq 1$, $0 \leq y \leq 1$, fails to exist because of the discontinuity of the integrand at the origin. The integral is actually an *improper* double integral (see Chapter 12).

9.22. Prove that $\int_0^x \left\{ \int_0^t F(u)du \right\} dt = \int_0^x (x-u)F(u)du$

Let $I(x) = \int_0^x \left\{ \int_0^t F(u)du \right\} dt, \quad J(x) = \int_0^z (x-u)F(u)du.$ Then

$$I'(x) = \int_0^z F(u)du, \quad J'(x) = \int_0^z F(u)du$$

using Leibniz's rule, Page 198. Thus, $I'(x) = J'(x)$, and so $I(x) = J(x) = c$, where c is a constant. Since $I(0) = J(0) = 0$, $c = 0$, and so $I(x) = J(x)$.

The result is sometimes written in the form

$$\int_0^x \int_0^x F(x)dx^2 = \int_0^x (x-u)F(u)du$$

The result can be generalized to give (see Problem 9.58)

$$\int_0^x \int_0^x \cdots \int_0^x F(x)dx^n = \frac{1}{(n-1)!} \int_0^x (x-u)^{n-1} F(u)du$$

SUPPLEMENTARY PROBLEMS

Double integrals

9.23. (a) Sketch the region $\Re$ in the xy plane bounded by $y^2 = 2x$ and $y = x$. (b) Find the area of $\Re$. (c) Find the polar moment of inertia of $\Re$, assuming constant density σ.

Ans. (b) $\frac{2}{3}$ (c) $48\sigma/35 = 72M/35$, where M is the mass of $\Re$

9.24. Find the centroid of the region in problem 9.23.

Ans. $\bar{x} = \frac{4}{5}, \bar{y} = 1$

9.25. Given $\int_{y=0}^{3} \int_{x=1}^{\sqrt{4-y}} (x+y)dx\,dy$, (a) sketch the region and give a possible physical interpretation of the double integral, (b) interchange the order of integration, and (c) evaluate the double integral.

Ans. (b) $\int_{x=1}^{2} \int_{y=1}^{4-x^2} (x+y)dy\,dx$ (c) 241/60

9.26. Show that $\int_{x=1}^{2} \int_{y=\sqrt{x}}^{x} \sin \frac{\pi x}{2y}\,dx + \int_{x=2}^{4} \int_{y=\sqrt{x}}^{2} \sin \frac{\pi x}{2y}\,dy\,dx = \frac{4(\pi+2)}{\pi^3}$.

9.27. Find the volume of the tetrahedron bounded by $x/a + y/b + z/c = 1$ and the coordinate planes.

Ans. $abc/6$

9.28. Find the volume of the region bounded by $z = x^3 + y^2$, $z = 0$, $x = -a$, $x = -a$, $y = -a$, $y = a$.

Ans. $8a^4/3$

9.29. Find (a) the moment of inertia about the z axis and (b) the centroid of the region in Problem 9.28, assuming a constant density σ.

Ans. (a) $\frac{112}{45}a^6\sigma = \frac{14}{15}Ma^2$, where M = mass (b) $\bar{x} = \bar{y} = 0, \bar{z} = \frac{7}{15}a^2$

Transformation of double integrals

9.30. Evaluate $\iint_{\Re} \sqrt{x^2 + y^2}\, dx\, dy$, where $\Re$ is the region $x^2 + y^2 \leq a^2$.

Ans. $\frac{2}{3}\pi a^3$

9.31. If $\Re$ is the region of Problem 9.30, evaluate $\iint_{\Re} e^{-(x^2+y^2)} dx\, dy$.

Ans. $\pi(1 - e^{-a^2})$

9.32. By using the transformation $x + y = u$, $y = uv$, show that $\int_{x=0}^{1}\int_{y=0}^{1-x} e^{y/(x+y)} dy\, dx = \frac{e-1}{2}$.

9.33. Find the area of the region bounded by $xy = 4$, $xy = 8$, $xy^3 = 5$, $xy^3 = 15$. (Hint: Let $xy = u$, $xy^3 = v$.)

Ans. 2 ln 3

9.34. Show that the volume generated by revolving the region in the first quadrant bounded by the parabolas $y^2 = x$, $y^2 = 8x$, $x^2 = y$, $x^2 = 8y$ about the x axis is $279\pi/2$. (Hint: Let $y^2 = ux$, $x^2 = vy$.)

9.35. Find the area of the region in the first quadrant bounded by $y = x^3$, $y = 4x^3$, $x = y^3$, $x = 4y^3$.

Ans. $\frac{1}{8}$

9.36. Let $\Re$ be the region bounded by $x + y = 1$, $x = 0$, $y = 0$. Show that $\iint_{\Re} \cos\left(\frac{x-y}{x+y}\right) dx\, dy = \frac{\sin 1}{2}$. (Hint: Let $x - y = u$, $x + y = v$.)

Triple integrals

9.37. (a) Evaluate $\int_{x=0}^{1}\int_{y=0}^{1}\int_{z=\sqrt{x^2+y^2}}^{2} xyz\, dz\, dy\, dx$. (b) Give a physical interpretation to the integral in (a).

Ans. (a) $\frac{3}{8}$

9.38. Find (a) the volume and (b) the centroid of the region in the first octant bounded by $x/a + y/b + z/c = 1$, where a, b, c are positive.

Ans. (a) $abc/6$ (b) $\bar{x} = a/4$, $\bar{y} = b/4$, $\bar{z} = c/4$

9.39. Find (a) the moment of inertia and (b) the radius of gyration about the z axis of the region in Problem 9.38.

Ans. (a) $M(a2 + b2)/10$ (b) $\sqrt{(a2 + b2)/10}$

9.40. Find the mass of the region corresponding to $x2 + y2 + z2 \leq 4$, $x \in 0$, $y \in 0$, $z \in 0$, if the density is equal to xyz.

Ans. 4/3

9.41. Find the volume of the region bounded by $z = x2 + y2$ and $z = 2x$.

Ans. $\pi/2$

Transformation of triple integrals

9.42. Find the volume of the region bounded by $z = 4 - x^2 - y^2$ and the xy plane.

Ans. 8π

9.43. Find the centroid of the region in Problem 9.42, assuming constant density σ.

Ans. $\bar{x} = \bar{y} = 0, \bar{z} = \dfrac{4}{3}$

9.44. (a) Evaluate $\iiint_{\Re} \sqrt{x^2 + y^2 + z^2}\, dx\, dy\, dz$, where $\Re$ is the region bounded by the plane $z = 3$ and the cone $z = z = \sqrt{x^2 + y^2}$. (b) Give a physical interpretation of the integral in (a). (Hint: Perform the integration in cylindrical coordinates in the order ρ, z, ϕ.)

Ans. $27\pi(2\sqrt{2} - 1)/2$

9.45. Show that the volume of the region bonded by the cone $z = \sqrt{x^2 + y^2}$ and the paraboloid $z = x^2 + y^2$ is $\pi/6$.

9.46. Find the moment of inertia of a right circular cylinder of radius a and height b, about its axis if the density is proportional to the distance from the axis.

Ans. $\dfrac{3}{5} Ma^2$

9.47. (a) Evaluate $\iiint_{\Re} \dfrac{dx\, dy\, dz}{(x^2 + y^2 + z^2)^{3\ 2}}$, where $\Re$ is the region bounded by the spheres $x^2 + y^2 + z^2 = a^2$ and $x^2 + y^2 + z^2 = b^2$ where $a > b > 0$. (b) Give a physical interpretation of the integral in (a).

Ans. (a) $4\pi \ln(a/b)$

9.48. (a) Find the volume of the region bounded above by the sphere $r = 2a \cos\theta$ and below by the cone $\phi = \alpha$, where $0 < \alpha < \pi/2$. (b) Discuss the case $\alpha = + \pi/2$.

Ans. $\dfrac{4}{3}\pi a^3(1 - \cos^4 \alpha)$

9.49. Find the centroid of a hemispherical shell having outer radius a and inner radius b if the density (a) is constant and (b) varies as the square of the distance from the base. Discuss the case $a = b$.

Ans. Taking the z axis as the axis of symmetry: (a) $\bar{x} = \bar{y} = 0$, $\bar{z} = \dfrac{3}{8}(a^4 - b^4)/(a^3 - b^3)$ (b) $\bar{x} = \bar{y} = 0$, $\bar{z} = \dfrac{5}{8}(a^6 - b^6)/(a^5 - b^5)$

Miscellaneous problems

9.50. Find the mass of a right circular cylinder of radius a and height b if the density varies as the square of the distance from a point on the circumference of the base.

Ans. $\dfrac{1}{6}\pi a^2 bk(9a^2 + 2b^2)$, where k = constant of proportionality

9.51. Find (a) the volume and (b) the centroid of the region bounded above by the sphere $x^2 + y^2 + z^2 = a^2$ and below by the plane $z = b$ where $a > b > 0$, assuming constant density.

Ans. (a) $\dfrac{1}{3}\pi(2a^3 - 3a^2b + b^3)$ (b) $\bar{x} = \bar{y} = 0$, $\bar{z} = \dfrac{3}{4}(a + b)^2/(2a + b)$

9.52. A sphere of radius a has a cylindrical hole of radius b bored from it, the axis of the cylinder coinciding with a diameter of the sphere. Show that the volume of the sphere which remains is $\dfrac{4}{3}\pi[a^3 - (a^2 - b^2)^{3/2}]$.

9.53. A simple closed curve in a plane is revolved about an axis in the plane which does not intersect the curve. Prove that the volume generated is equal to the area bounded by the curve multiplied by the distance traveled by the centroid of the area (Pappus's theorem).

9.54. Use Problem 9.53 to find the volume generated by revolving the circle $x^2 + (y - b)^2 = a^2$, $b > a > 0$ about the x axis.

Ans. $2\pi^2 a^2 b$

9.55. Find the volume of the region bounded by the hyperbolic cylinders $xy = 1$, $xy = 9$, $xz = 4$, $xz = 36$, $yz = 25$, $yz = 49$. (Hint: Let $xy = u$, $xz = \upsilon$, $yz = w$.)

Ans. 64

9.56 Evaluate $\iiint_{\Re} \sqrt{1-(x^2/a^2+y^2/b^2+z^2/c^2)}\, dx\, dy\, dz$ where $\Re$ is the region interior to the ellipsoid $x^2/a^2 + y^2/b^2 + z^2/c^2 = 1$. (Hint: Let $x = au$, $y = b\upsilon$, $z = cw$. Then use spherical coordinates.)

Ans. $\frac{1}{4}\pi^2 abc$

9.57. If $\Re$ is the region $x^2 + xy + y^2 \leqq 1$, prove that $\iint_{\Re} e^{-(x^2+xy+y^2)} dx\, dy = \dfrac{2\pi}{e\sqrt{3}}(e-1)$. (Hint: Let $x = u \cos\alpha - \upsilon \sin\alpha$, $y = u \sin\alpha + \upsilon \cos\alpha$ and choose α so as to eliminate the xy term in the integrand. Then let $u = a\rho \cos\phi$, $\upsilon = b\rho \sin\phi$ where a and b are appropriately chosen.)

9.58. Prove that $\int_0^x \int_0^x \cdots \int_0^x F(x)dx^n = \dfrac{1}{(n-1)!}\int_0^x (x-u)^{n-1} F(u)du$ for $n = 1, 2, 3, \ldots$ (see problem 9.22).

CHAPTER 10

Line Integrals, Surface Integrals, and Integral Theorems

Construction of mathematical models of physical phenomena requires functional domains of greater complexity than the previously employed line segments and plane regions. This section makes progress in meeting that need by enriching integral theory with the introduction of segments of curves and portions of surfaces as domains. Thus, single integrals as functions defined on curve segments take on new meaning and are then called *line integrals*. Stokes's theorem exhibits a striking relation between the line integral of a function on a closed curve and the double integral of the surface portion that is enclosed. The divergence theorem relates the triple integral of a function on a three-dimensional region of space to its double integral on the bounding surface. The elegant language of vectors best describes these concepts; therefore, it would be useful to reread the introduction to Chapter 7, where the importance of vectors is emphasized. (The integral theorems also are expressed in coordinate form.)

Line Integrals

The objective of this section is to geometrically view the domain of a vector or scalar function as a segment of a curve. Since the curve is defined on an interval of real numbers, it is possible to refer the function to this primitive domain, but to do so would suppress much geometric insight.

A curve C in three-dimensional space may be represented by parametric equations:

$$x = f_1(t),\ y = f_2(t),\ z = f_3(t),\ a \leqq t \leqq b \tag{1}$$

or in vector notation:

$$\mathbf{x} = \mathbf{r}(\mathrm{t}) \tag{2}$$

where

$$\mathbf{r}(t) = x\mathbf{i} + y\mathbf{j} + z\mathbf{k}$$

(see Figure 10.1).

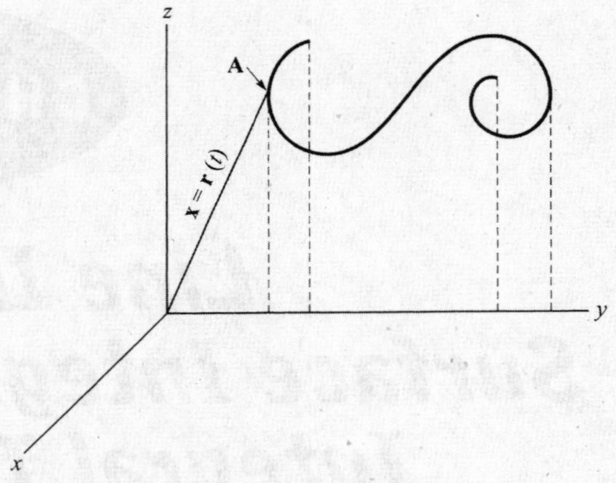

Figure 10.1

For this discussion it is assumed that **r** is continuously differentiable. While (as we are doing) it is convenient to refer the Euclidean space to a rectangular Cartesian coordinate system, it is not necessary. (For example, cylindrical and spherical coordinates sometimes are more useful.) In fact, one of the objectives of the vector language is to free us from any particular frame of reference. Then, a vector $\mathbf{A}[x(t), y(t, z(t)]$ or a scalar Θ is pictured on the domain C, which, according to the parametric representation, is referred to the real number interval $a \leq t \leq b$.

The integral

$$\int_C \mathbf{A} \cdot d\mathbf{r} \tag{3}$$

of a vector field **A** defined on a curve segment C is called a *line integral*. The integrand has the representation

$$A_1\, dx + A_2\, dy + A_3\, dz$$

obtained by expanding the dot product.

The scalar and vector integrals

$$\int_C \Theta(t)dt = \lim_{n\to\infty} \sum_{k=1}^{n} \Theta(\xi_k, \eta_k, \zeta_k)\Delta t_k \tag{4}$$

$$\int_C \mathbf{A}(t)dt = \lim_{n\to\infty} \sum_{k=1}^{n} \mathbf{A}(\xi_k, \eta_k, \zeta_k)\Delta t_k \tag{5}$$

can be interpreted as line integrals; however, they do not play a major role [except for the fact that the scalar integral (3) takes the form (4)].

The following three basic ways are used to evaluate the line integral (3):

1. The parametric equations are used to express the integrand through the parameter t. Then

$$\int_C \mathbf{A} \cdot d\mathbf{r} = \int_{t_1}^{t_2} \mathbf{A} \cdot \frac{d\mathbf{r}}{dt} dt$$

2. If the curve C is a plane curve (for example, in the xy plane) and has one of the representations $y = f(x)$ or $x = g(y)$, then the two integrals that arise are evaluated with respect to x or y, whichever is more convenient.

3. If the integrand is a perfect differential, then it may be evaluated through knowledge of the endpoints (that is, without reference to any particular joining curve). (See the section on independence of path on Page 246; also see Page 251.)

These techniques are further illustrated for plane curves in the next section and for three-space in the problems.

Evaluation of Line Integrals for Plane Curves

If the equation of a curve C in the plane $z = 0$ is given as $y = f(x)$, the line integral (2) is evaluated by placing $y = f(x)$, $dy = f'(x)\,dx$ in the integrand to obtain the definite integral

$$\int_{a_1}^{a_2} P\{x, f(x)\}dx + Q\{x, f(x)\}f'(x)dx \tag{6}$$

which is then evaluated in the usual manner.

Similarly, if C is given as $x = g,(y)$, then $dx = g'\ (y)\ dy$ and the line integral becomes

$$\int_{b_1}^{b_2} P\{g(y), y\}g'(y)dy + Q\{g(y), y\}dy \tag{7}$$

If C is given in parametric form $x = \phi(t)$, $y = \psi(t)$, the line integral becomes

$$\int_{t_1}^{t_2} P\{\phi(t), \psi(t)\}\phi'(t)dt + Q\{\phi(t), \psi(t)\}, \psi'(t)dt \tag{8}$$

where t_1 and t_2 denote the values of t corresponding to points A and B, respectively.

Combinations of these methods may be used in the evaluation. If the integrand $\mathbf{A} \cdot d\mathbf{r}$ is a perfect differential $d\Theta$, then

$$\int_C \mathbf{A} \cdot d\mathbf{r} = \int_{(a,b)}^{(c,d)} d\Theta = \Theta(c, d) - \Theta(a, b) \tag{9}$$

Similar methods are used for evaluating line integrals along space curves.

Properties of Line Integrals Expressed for Plane Curves

Line integrals have properties which are analogous to those of ordinary integrals. For example:

1. $\int_C P(x, y)dx + Q(x, y)dy = \int_C P(x, y)dx + \int_C Q(x, y)dy$

2. $\int_{(a_1, b_1)}^{(a_2, b_2)} P\,dx + Q\,dy = -\int_{(a_2, b_2)}^{(a_1, b_1)} P\,dx + q\,dy$

Thus, reversal of the path of integration changes the sign of the line integral.

3. $\int_{(a_1, b_1)}^{(a_2, b_2)} P\,dx + Q\,dy = \int_{(a_1, b_1)}^{(a_3, b_3)} P\,dx + Q\,dy + \int_{(a_3, b_3)}^{(a_2, b_2)} P\,dx + Q\,dy$

where (a_3, b_3) is another point on C.

Similar properties hold for line integrals in space.

Simple Closed Curves, Simply and Multiply Connected Regions

A *simple closed curve* is a closed curve which does not intersect itself anywhere. Mathematically, a curve in the xy plane is defined by the parametric equations $x = \phi(t)$, $y = \psi(t)$ where ϕ and ψ are single-valued and continuous in an interval $t_1 \leq t \leq t_2$. If $\phi(t_1) = \phi(t_2)$ and $\psi(t_1) = \psi(t_2)$, the curve is said to be *closed.* If $\phi(u) = \phi(\upsilon)$ and $\psi(u) = \psi(\upsilon)$ only when $u = \upsilon$ (except in the special case where $u = t_1$ and $\upsilon = t_2$), the curve is closed and does not intersect itself, and so is a simple closed curve. We shall also assume, unless otherwise stated, that ϕ and ψ are piecewise differentiable in $t_1 \leq t \leq t_2$.

If a plane region has the property that any closed curve in it can be continuously shrunk to a point without leaving the region, then the region is called *simple connected*; otherwise, it is called *multiply connected* (see Figure 10.2 and Page 127).

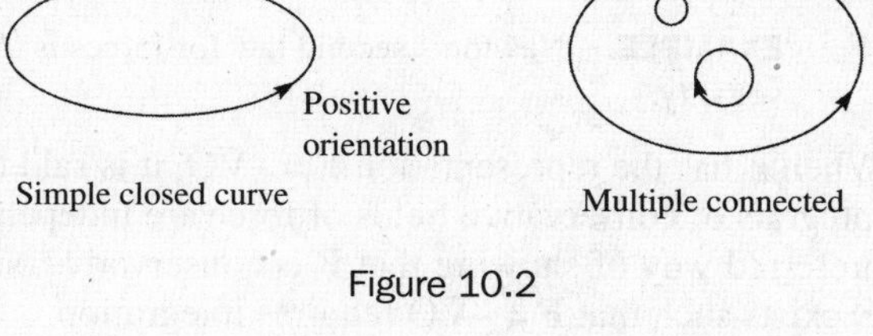

Figure 10.2

As the parameter t varies from t_1 to t_2, the plane curve is described in a certain sense or direction. For curves in the xy plane, we arbitrarily describe this direction as *positive* or *negative* according as a person traversing the curve in this direction with his head pointing in the positive z direction has the region enclosed by the curve always toward his left or right, respectively. If we look down upon a simple closed curve in the xy plane, this amounts to saying that traversal of the curve in the counterclockwise direction is taken as positive, while traversal in the clockwise direction is taken as negative.

Green's Theorem in the Plane

This theorem is needed to prove Stokes's theorem (Page 251). Then it becomes a special case of that theorem.

Let P, Q, $\partial P/\partial y$, $\partial Q/\partial x$ be single-valued and continuous in a simple connected region $\Re$ bounded by a simple closed curve C. Then

$$\oint_C P\,dx + Q\,dy = \iint_{\Re} \left(\frac{\partial Q}{\partial x} - \frac{\partial P}{\partial y} \right) dx\,dy \tag{10}$$

where $\oint_C$ is used to emphasize that C is closed and that it is described in the positive direction.

This theorem is also true for regions bounded by two or more closed curves (i.e., multiply connected regions). See Problem 10.10.

Conditions for a Line Integral to Be Independent of the Path

The line integral of a vector field **A** is independent of path if its value is the same regardless of the (allowable) path from initial to terminal point. (Thus, the integral is evaluated from knowledge of the coordinates of these two points.)

For example, the integral of the vector field $\mathbf{A} = y\mathbf{i} + x\mathbf{j}$ is independent of path since

$$\int_C \mathbf{A} \cdot d\mathbf{r} = \int_C y\,dx + x\,dy = \int_{x_1\,y_1}^{x_2\,y_2} d(xy) = x_2 y_2 - x_1 y_1$$

Thus, the value of the integral is obtained without reference to the curve joining P_1 and P_2.

This notion of the independence of path of line integrals of certain vector fields, important to theory and application, is characterized by the following three theorems.

Theorem 1 A necessary and sufficient condition that $\int_C \mathbf{A} \cdot d\mathbf{r}$ be independent of path is that there exists a scalar function Θ such that $\mathbf{A} = \nabla\Theta$.

Theorem 2 A necessary and sufficient condition that the line integral $\int_C \mathbf{A} \cdot d\mathbf{r}$ be independent of path is that $\nabla \times \mathbf{A} = \mathbf{0}$.

Theorem 3 If $\nabla \times \mathbf{A} = \mathbf{0}$, then the line integral of **A** over an allowable closed path is 0; i.e., $\oint_C \mathbf{A} \cdot d\mathbf{r} = 0$.

If C is a plane curve, then Theorem 3 follows immediately from Green's theorem, since in the plane case $\nabla \times \mathbf{A}$ reduces to

$$\frac{\partial A_1}{\partial y} = \frac{\partial A_2}{\partial x}$$

EXAMPLE. Newton's second law for forces is $\mathbf{F} = \dfrac{d(m\mathbf{v})}{dt}$, where m is the mass of an object and **v** is its velocity.

When **F** has the representation $\mathbf{F} = -\nabla\Theta$, it is said to be conservative. The previous theorems tell us that the integrals of conservative fields of force are independent of path. Furthermore, showing that $\nabla \times \mathbf{F} = \mathbf{0}$ is the preferred way of showing that **F** is conservative, since it involves differentiation, while demonstrating that Θ exists such that $\mathbf{F} = -\nabla\Theta$ requires integration.

Surface Integrals

Our previous double integrals have been related to a very special surface, the plane. Now we consider other surfaces. yet, the approach is quite similar. Surfaces can be viewed intrinsically, i.e., as non-Euclidean spaces: however, we do not do that. Rather, the surface is thought of as embedded in a three-dimensional Euclidean space and expressed through a two-parameter vector representation:

$$\mathbf{x} = \mathbf{r}(\upsilon_1, \upsilon_2)$$

While the purpose of the vector representation is to be general (that is, interpretable through any allowable three-space coordinate system), it is convenient to initially think in terms of rectangular Cartesian coordinates: therefore, assume

$$\mathbf{r} = x\mathbf{i} + y\mathbf{j} + z\mathbf{k}$$

and that there is a parametric representation

$$x = r(\upsilon_1, \upsilon_2),\ y = r(\upsilon_1, \upsilon_2),\ z = r(\upsilon_1, \upsilon_2) \tag{11}$$

The functions are assumed to be continuously differentiable.

The parameter curves υ_2 = const and υ_1 const establish a coordinate system on the surface (just as y = const and x = const form such a system in the plane). The key to establishing the surface integral of a function is the differential element of surface area. (For the plane, that element is $dA = dx, dy$.)

At any point P of the surface

$$d\mathbf{x} = \frac{\partial \mathbf{r}}{\partial \upsilon_1} d\upsilon_1 + \frac{\partial \mathbf{r}}{\partial \upsilon_2} d\upsilon_2$$

spans the tangent plane to the surface. In particular, the directions of the coordinate curves υ_2 = const and υ_1 = const are designated by $d\mathbf{x}_1 = \frac{\partial \mathbf{r}}{\partial \upsilon_1} d\upsilon_1$ and $d\mathbf{x}_2 \frac{\partial \mathbf{r}}{\partial \upsilon_2} d\upsilon_2$, respectively (see Figure 10.3).

The cross product

$$d\mathbf{x}_1\ x\ d\mathbf{x}_2 = \frac{\partial \mathbf{r}}{\partial \upsilon_1} \times \frac{\partial \mathbf{r}}{\partial \upsilon_2} d\upsilon_1\, d\upsilon_2$$

is normal to the tangent plane at P, and its magnitude $\left| \frac{\partial \mathbf{r}}{\partial \upsilon_1} \times \frac{\partial \mathbf{r}}{\partial \upsilon_2} \right|$ is the area of a differential coordinate parallelogram.

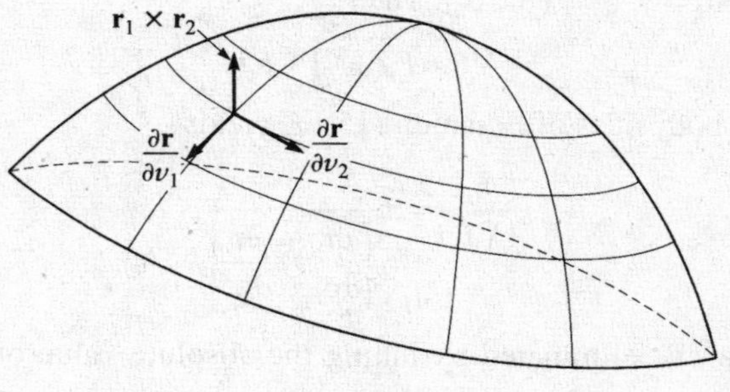

Figure 10.3

(This is the usual geometric interpreation of the cross product abstracted to the differential level.) This strongly suggests the following definition:

Definition The differential element of surface area is

$$dS = \left| \frac{\partial \mathbf{r}}{\partial \upsilon_1} \times \frac{\partial \mathbf{r}}{\partial \upsilon_2} \right| d\upsilon_1\, d\upsilon_2 \tag{12}$$

For a function $\Theta(\upsilon_1, \upsilon_2)$ that is everywhere integrable on S,

$$\iint_S \Theta\, dS = \iint_S \Theta(\upsilon_1, \upsilon_2)\left|\frac{\partial \mathbf{r}}{\partial \upsilon_1}\times\frac{\partial \mathbf{r}}{\partial \upsilon_2}\right| d\upsilon_1\, d\upsilon_2 \tag{13}$$

is the surface integral of the function Θ.

In general, the surface integral must be referred to three-space coordinates to be evaluated. If the surface has the Cartesian representation $z = f(x, y)$ and the identifications

$$\upsilon_1 = x,\ \upsilon_2 = y,\ z = f(\upsilon, \upsilon_2)$$

are made, then

$$\frac{\partial \mathbf{r}}{\partial \upsilon_1} = \mathbf{i} + \frac{\partial z}{\partial x}\mathbf{k},\quad \frac{\partial \mathbf{r}}{\partial \upsilon_2} = \mathbf{j} + \frac{\partial z}{\partial y}\mathbf{k}$$

and

$$\frac{\partial \mathbf{r}}{\partial \upsilon_2}\times\frac{\partial \mathbf{r}}{\partial \upsilon_2} = \mathbf{k} - \frac{\partial z}{\partial x}\mathbf{j} - \frac{\partial z}{\partial x}\mathbf{i}$$

Therefore,

$$\left|\frac{\partial \mathbf{r}}{\partial \upsilon_1}\times\frac{\partial \mathbf{r}}{\partial \upsilon_2}\right| = \left[1 + \left(\frac{\partial z}{\partial x}\right)^2 + \left(\frac{\partial z}{\partial y}\right)^2\right]^{1/2}$$

Thus, the surface integral of Θ has the special representation

$$S = \iint_S \Theta(x, y, z)\left[1 + \left(\frac{\partial z}{\partial x}\right)^2 + \left(\frac{\partial z}{\partial y}\right)^2\right]^{1/2} dx\, dy \tag{14}$$

If the surface is given in the implicit form $F(x, y, z) = 0$, then the gradient may be employed to obtain another representation. To establish it, recall that, at any surface point P, the gradient ∇F is perpendicular (normal) to the tangent plane (and, hence, to S).

Therefore, the following equality of the unit vectors holds (up to sign):

$$\frac{\nabla F}{|\nabla F|} = \pm\left(\frac{\partial \mathbf{r}}{\partial x}\times\frac{\partial \mathbf{r}}{\partial y}\right)\Big/\left|\frac{\partial \mathbf{r}}{\partial \upsilon_1}\times\frac{\partial \mathbf{r}}{\partial \upsilon_2}\right| \tag{15}$$

[A conclusion of the theory of implicit functions is that from $F(x, y, z) = 0$ (and under appropriate conditions) there can be produced an explicit representation $z = f(x, y)$ of a portion of the surface. This is an existence statement. The theorem does not say that this representation can be explicitly produced.] With this fact in hand, we again let $\upsilon_1 = x$, $\upsilon_2 = y$, $z = f(\upsilon_1, \upsilon_2)$. Then

$$\nabla F = F_x\mathbf{i} + f_y\mathbf{j} + F_z\mathbf{k}$$

Taking the dot product of both sides of Equation (15), K yields

$$\frac{F_z}{|\nabla F|} = \pm\frac{1}{\left|\frac{\partial \mathbf{r}}{\partial \upsilon_1}\times\frac{\partial \mathbf{r}}{\partial \upsilon_2}\right|}$$

The ambiguity of sign can be eliminated by taking the absolute value of both sides of the equation. Then

$$\left|\frac{\partial \mathbf{r}}{\partial \upsilon_1}\times\frac{\partial \mathbf{r}}{\partial \upsilon_2}\right| = \frac{|\nabla F|}{|F_z|} = \frac{[(F_x)^2 + (F_y)^2 + (F_z)^2]^{1/2}}{|F_z|}$$

and the surface integral of Θ takes the form

$$\iint_S \frac{[(F_x)^2 + (F_y)^2 + (F_z)^2]^{1/2}}{|F_z|}\, dx\, dy \tag{16}$$

The formulas (14) and (16) also can be introduced in the following nonvectorial manner.

Let S be a two-sided surface having projection $\mathfrak{R}$ on the xy plane, as in Figure 10.4. Assume that an equation for S is $z = f(x, y)$, where f is single-valued and continuous for all x and y in $\mathfrak{R}$. Divide $\mathfrak{R}$ into n subregions of area ΔA_p, $p = 1, 2, \ldots, n$, and erect a vertical column on each of these subregions to intersect S in an area ΔS_p.

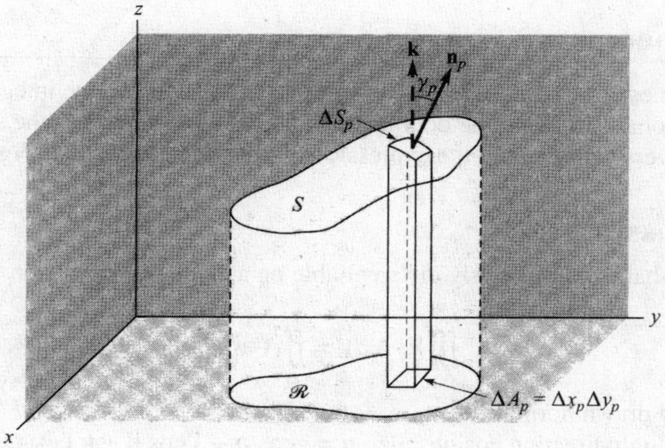

Figure 10.4

Let $\phi(x, y, z)$ be single-valued and continuous at all points of S. Form the sum

$$\sum_{p=1}^{n} \phi(\xi_p, \eta_p, \zeta_p)\Delta S_p \tag{17}$$

where (ξ_p, η_p, ζ_p) is some point of ΔS_p. If the limit of this sum as $n \to \infty$ in such a way that each $\Delta S_p \to 0$ exists, the resulting limit is called the *surface integral* of $\phi(x, y, z)$ over S and is designated by

$$\iint_S \phi(x, y, z)\, dS \tag{18}$$

Since $\Delta S_p = |\sec \gamma_p|\, \Delta A_p$ approximately, where γ_p is the angle between the normal line to S and the positive z axis, the limit of the sum (17) can be written

$$\iint_{\mathfrak{R}} \phi(x, y, z)\, |\sec\gamma|\, dA \tag{19}$$

The quantity $|\sec \gamma|$ is given by

$$|\sec\gamma| = \frac{1}{|\mathbf{n}_p \cdot \mathbf{k}|} = \sqrt{1 + \left(\frac{\partial z}{\partial x}\right)^2 + \left(\frac{\partial z}{\partial y}\right)^2} \tag{20}$$

Then, assuming that $z = f(x, y)$ has continuous (or sectionally continuous) derivatives in $\mathfrak{R}$, (19) can be written in rectangular form as

$$\iint_{\mathfrak{R}} \phi(x, y, z) \sqrt{1 + \left(\frac{\partial z}{\partial x}\right)^2 + \left(\frac{\partial z}{\partial y}\right)^2}\, dx\, dy \tag{21}$$

In case the equation for S is given as $F(x, y, z) = 0$, (21) can also be written

$$\iint_{\mathfrak{R}} \phi(x, y, z) \frac{\sqrt{(F_x)^2 + (F_y)^2 + (F_z)^2}}{|F_z|}\, dx\, dy \tag{22}$$

The results (21) or (22) can be used to evaluate (18).

In the preceding we have assumed that S is such that any line parallel to the z axis intersects S in only one point. In case S is not of this type, we can usually subdivide S into surfaces $S_1, S_2, \ldots$ which are of this type. Then the surface integral over S is defined as the sum of the surface integrals over $S_1, S_2, \ldots$.

The results stated hold when S is projected onto a region $\mathfrak{R}$ on the xy plane. In some cases it is better to project S onto the yz or xz planes. For such cases, (18) can be evaluated by appropriately modifying (21) and (22).

The Divergence Theorem

The divergence theorem establishes equality between a triple integral (volume integral) of a function over a region of three-dimensional space and the double integral of the function over the surface that bounds that region. This relation is very important in the expression of physical theory. (See Figure 10.5.)

Divergence (or Gauss) Theorem

Let **A** be a vector field that is continuously differentiable on a closed-space region V bounded by a smooth surface S. Then

$$\iiint_V \nabla \cdot \mathbf{A}\, dV = \iint_S \mathbf{A} \cdot \mathbf{n}\, dS \tag{23}$$

where **n** is an outwardly drawn normal.

If **n** is expressed through direction cosines, i.e., $\mathbf{n} = \mathbf{i}\cos\alpha + \mathbf{j}\cos\beta + \mathbf{k}\cos\gamma$, then Equation (23) may be written

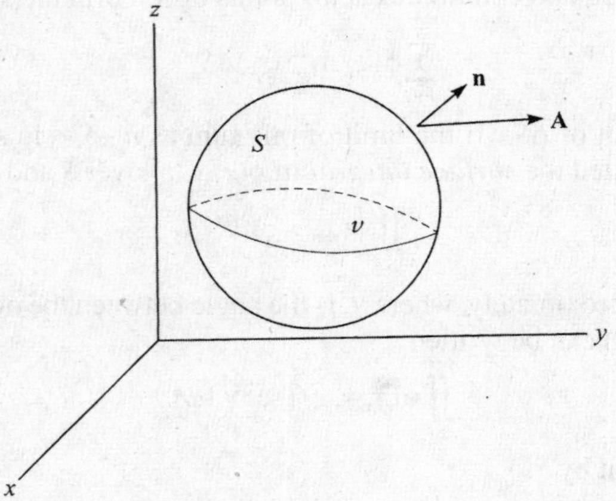

Figure 10.5

$$\iiint \left(\frac{\partial A_1}{\partial x} + \frac{\partial A_2}{\partial y} + \frac{\partial A_3}{\partial z} \right) dV = \iint_S (A_1 \cos\alpha + A_2 \cos\beta + A_3 \cos\gamma)\, dS \tag{24}$$

The rectangular Cartesian component form of Equation (23) is

$$\iiint_V \left(\frac{\partial A_1}{\partial x} + \frac{\partial A_2}{\partial y} + \frac{\partial A_3}{\partial z} \right) dV = \iint_S (A_1\, dy\, dz + A_2\, dz\, dx + A_3\, dx\, dy) \tag{25}$$

EXAMPLE. If **B** is the magnetic field vector, then one of Maxwell's equations of electromagnetic theory is $\nabla \cdot \mathbf{B} = 0$. When this equation is substituted into the left member of Equation (23), the right member tells us that the magnetic flux through a closed surface containing a magnetic field is zero. A simple interpretation of this fact results by thinking of a magnet enclosed in a ball. All magnetic lines of force that flow out of the ball must return (so that the total flux is zero). Thus, the lines of force flow from one pole to the other, and there is no dispersion.

Stokes's Theorem

Stokes's theorem establishes the equality of the double integral of a vector field over a portion of a surface and the line integral of the field over a simple closed curve bounding the surface portion. (See Figure 10.6.)

Suppose a closed curve C bounds a smooth surface portion S. If the component functions of $\mathbf{x} = \mathbf{r}(\upsilon_1, \upsilon_2)$ have continuous mixed partial derivatives, then for a vector field $\mathbf{A}$ with continuous partial derivatives on S

$$\oint_C \mathbf{A}\cdot d\mathbf{r} = \iint_S \mathbf{n}\cdot\nabla\times\mathbf{A}\, dS \tag{26}$$

where $\mathbf{n} = \cos\alpha\mathbf{i} + \cos\beta\mathbf{j} + \cos\gamma\mathbf{k}$ with α, β, and γ represeting the angles made by the outward normal $\mathbf{n}$ and $\mathbf{i}$, $\mathbf{j}$, and $\mathbf{k}$, respectively.

Then the component form of Equation (26) is

$$\oint_C (A_1dx + A_2dy + A_3dz) = \iint_S \left[\left(\frac{\partial A_3}{\partial y} - \frac{\partial A_2}{\partial z}\right)\cos\alpha + \left(\frac{\partial A_1}{\partial z} - \frac{\partial A_3}{\partial x}\right)\cos\beta + \left(\frac{\partial A_2}{\partial x} - \frac{\partial A_1}{\partial y}\right)\cos\gamma\right] dS \tag{27}$$

If $\nabla \times \mathbf{A} = \mathbf{0}$, Stokes's theorem tells us that $\oint_C \mathbf{A}\cdot d\mathbf{r} = 0$. This is Theorem 3 on Page 233.

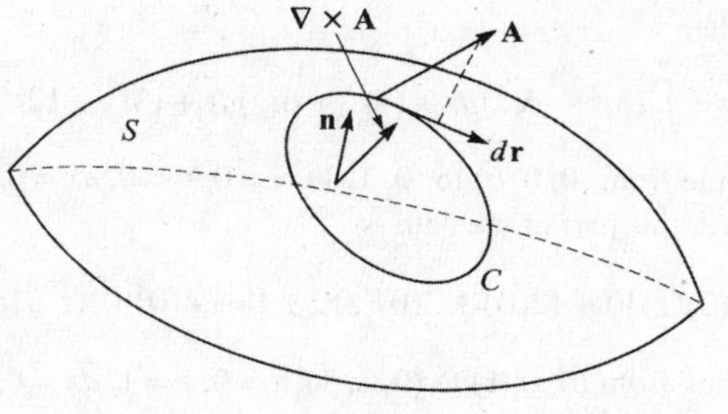

Figure 10.6

SOLVED PROBLEMS

Line integrals

10.1. Evaluate $\int_{(0,1)}^{(1,2)} (x^2 - y)\,dx + (y^2 + x)\,dy$ along (a) a straight line from (0, 1) to (1, 2), (b) a straight lines from (0, 1) to (1, 1) and then from (1, 1) to (1, 2), and (c) the parabola $x = t$, $y = t^2 + 1$.

(a) An equation for the line joining (0, 1) and (1, 2) in the xy plane is $y = x + 1$. Then $dy = dx$ and the line integral equals

$$\int_{x=0}^{1} \{x^2 - (x+1)\}\,dx + \{(x+1)^2 + x\}\,dx = \int_0^1 (2x^2 + 2x)\,dx = 5/3$$

(b) Along the straight line from (0, 1) to (1, 1), $y = 1$, $dy = 0$ and the line integral equals

$$\int_{x=0}^{1} (x^2 - 1)\,dx + (1+x)(0) = \int_0^1 (x^2 - 1)\,dx = -2/3$$

Along the straight line from (1, 1) to (1, 2), $x = 1$, $dx = 0$ and the line integral equals

$$\int_{y=1}^{2} (1-y)(0) + (y^2+1)\,dy = \int_t^2 (y^2 + 1)\,dy = 10/3$$

Then the required value $= -2/3 + 10/3 = 8.3$.

(c) Since $t = 0$ at (0, 1) and $t = 1$ at (1, 2), the line integral equals

$$\int_{t=0}^{1} \{t^2 - (t^2+1)\,dt + \{(t^2+1)^2 + t\}2t\,dt = \int_0^1 (2t^5 + 4t^2 + 2t^2 + 2t - 1)\,dt = 2$$

10.2. If $\mathbf{A} = (3x^2 - 6yz)\mathbf{i} + (2y + 3xz)\mathbf{j} + (1 - 4xyz^2)\mathbf{k}$, evaluate $\int_C \mathbf{A} \cdot d\mathbf{r}$ from (0,1,1) to (1,1,1) along the following paths C:

(a) $x = t,\ y = t^2,\ z = t^3$

(b) The straight lines from (0, 0, 0) to (0, 0, 1), then to (0, 1, 1), and then to (1, 1, 1)

(c) The straight line joining (0, 0, 0) and (1, 1, 1)

$$\int_C \mathbf{A}\cdot d\mathbf{r} = \int_C \{(3x^2 - 6yz)\mathbf{i} + (2y + 3xz)\}\mathbf{j} + (1 - 4xyz^2)\mathbf{k}\}\cdot(dx\mathbf{i} + dy\mathbf{j} + dz\mathbf{k})$$
$$= \int_C \{(3x^2 - 6yz)dx + (2y + 3xz)dy + (1 - 4xyz^2)dz$$

(a) If $x = t$, $y = t^2$, $z = t^3$, points (0, 0, 0) and (1, 1, 1) correspond to $t = 0$ and $t = 1$, respectively. Then

$$\int_C \mathbf{A}\cdot d\mathbf{r} = \int_{t=0}^{1} \{3t^2 - 6(t^2)(t^3)\}dt + \{2t^2 + 3(t)(t^3)\}d(t^2) + \{1 - 4(t)(t^2)(t^3)^2\}d(t^3)$$
$$= \int_{t=0}^{1} \{3t^2 - 6t^5\}dt + (4t^3 + 6t^5)\}dt + (3t^2 - 12t^{11})dt = 2$$

Another method: Along C, $\mathbf{A} = (3t^2 - 6t^5)\,\mathbf{i} + (2t^2 + 3t^4)\mathbf{j} + (1 - 4t^9)\mathbf{k}$ and $\mathbf{r} = x\mathbf{i} + y\mathbf{j} + z\mathbf{k} = t\mathbf{i} + t^2\mathbf{j} + t^3\mathbf{k}$, $d\mathbf{r} = (\mathbf{i} + 2t\mathbf{j} + 3t^2\mathbf{k})dt$. Then

$$\int_C \mathbf{A}\cdot d\mathbf{r} = \int_0^1 (3t^2 - 6t^5)dt + (4t^3 + 6t^5)dt + (3t^2 - 12t^{11})dt \;= 2$$

(b) Along the straight line from (0, 0, 0) to (0, 1, 1), $x = 0$, $y = 0$, $dx = 0$, $dy = 0$, while z varies from 0 to 1. Then the integral over this part of the path is

$$\int_{z=0}^{1} \{3(0)^2 - 6(0)(z)\}0 + \{2(0) + 3(0)(z)0 + \{1 - 4(0)(0)(z^2)\}dz = \int_{z=0}^{1} dz = 1$$

Along the straight line from (0, 0, 1) to (0, 1, 1), $x = 0$, $z = 1$, $dx = 0$, $dz = 0$, while y varies from 0 to 1. Then the integral over this part of the path is

$$\int_{y=0}^{1} \{3(0)^2 - 6(y)(1)\}0 + \{2y + 3(0)(1)\}dy + \{1 - 4(0)(y)(1)^2\}0 = \int_{y=0}^{1} 2y\,dy = 1$$

Along the straight line from (0, 1, 1), to (1, 1, 1), $v = 1$, $z = 1$, $dy = 0$, $dz = 0$, while x varies from 0 to 1. Then the integral over this part of the path is

$$\int_{x=0}^{1} \{3x^2 - 6(1)(1)\}dx + \{2(1) + 3x(1)\}0 + \{1 - 4x(1)(1)^2\}0 = \int_{x=0}^{1} (3x^2 - 6)dx = -5$$

Adding,

$$\int_C \mathbf{A}\cdot dx = 1 + 1 - 5 = -3$$

(c) The straight line joining (0, 0, 0) and (1, 1, 1) is given in parametric form by $x = t$, $y = t$, $z = t$. Then

$$\int_C \mathbf{A}\cdot d\mathbf{r} = \int_{t=0}^{1} (3t^2 - 6t^2)\,dt + (2t + 3t^2)\,dt + (1 - 4t^4)\,dt = 6/5$$

10.3. Find the work done in moving a particle once around an ellipse C in the xy plane, if the ellipse has its center at the origin with semimajor and semiminor axes 4 and 3, respectively, as indicated in Figure 10.7, and if the force field is given by

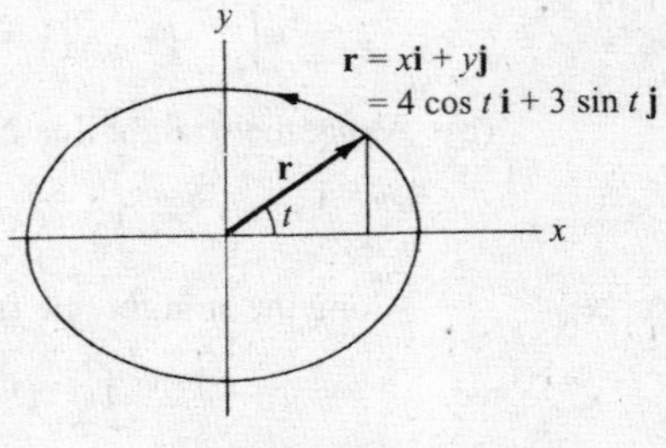

Figure 10.7

$$\mathbf{F} = (3x - 4y + 2z)\mathbf{i} + (4x + 2y - 3z^2)\mathbf{j} + (2xz - 4y^2 + z^3)\mathbf{k}$$

In the plane $z = 0$, $\mathbf{F} = (3x - 4y)\mathbf{i} + (4x + 2y)\mathbf{j} - 4y^2\mathbf{k}$, and $d\mathbf{r} = dx\mathbf{i} + dy\mathbf{j}$, so that the work done is

$$\oint_C \mathbf{F}\cdot d\mathbf{r} = \int_C \{(3x - 4y)\mathbf{i} + (4x + 2y)\mathbf{j} - 4y^2\mathbf{k}\}\cdot(dx\mathbf{i} + dy\mathbf{j})$$
$$= \oint_C (3x - 4y)\,dx + (4x + 2y)\,dy$$

Choose the parametric equations of the ellipse as $x = 4\cos t$, $y = 3\sin t$, where t varies from 0 to 2π (see Figure 10.7). Then the line integral equals

$$\int_{t=0}^{2\pi} \{3(4\cos t) - 4(3\sin t)\}\{-4\sin t\}dt + \{4(4\cos t) + 2(3\sin t)\}\{3\cos t\}dt$$

$$= \int_{t=0}^{2\pi} (48 - 30\sin t\cos t)dt = (48t - 15\sin^2 t)\Big|_0^{2\pi} = 96\pi$$

In traversing C we have chosen the counterclockwise direction indicated in Figure 10.7. We call this the *positive* direction or say that C has been traversed in the *positive sense*. If C were traversed in the clockwise (negative) direction, the value of the integral would be -96π.

10.4. Evaluate $\int_C y\,ds$ along the curve C given by $y = 2\sqrt{x}$ from $x = 3$ to $x = 24$.

Since $ds = \sqrt{dx^2 + dy^2} = \sqrt{1 + (y')^2}\,dx = \sqrt{1 + 1/x\,dx}$, we have

$$\int_C y\,ds = \int_2^{24} 2\sqrt{x}\sqrt{1 + 1/x}\,dx = 2\int_3^{24} \sqrt{x+1}\,dx = \frac{4}{3}(x+1)^{3/2}\Big|_3^{24} = 156$$

Green's theorem in the plane

10.5. Prove Green's theorem in the plane if C is a closed curve which has the property that any straight line parallel to the coordinate axes cuts C in, at most, two points.

Let the equations of the curves AEB and AFB (see Figure 10.8) be $y = Y_1(x)$ and $y = Y_2(x)$, respectively. If $\Re$ is the region bounded by C, we have

$$\iint_{\Re} \frac{\partial P}{\partial y}dx\,dy = \int_{x=a}^{b}\left[\int_{y=Y_1(x)}^{Y_2(x)} \frac{\partial P}{\partial y}dy\right]dx$$

$$= \int_{x=a}^{b} P(x, y)\Big|_{y=Y_1(x)}^{Y_2(x)} dx = \int_a^b [P(x, Y_2) - P(x, Y_1)]dx$$

$$= -\int_a^b P(x, Y_1)dx - \int_b^a P(x, Y_2)dx = -\oint_C P dx$$

Then

$$\oint_C P dx = -\iint_{\Re} \frac{\partial P}{\partial y}dx\,dy \tag{1}$$

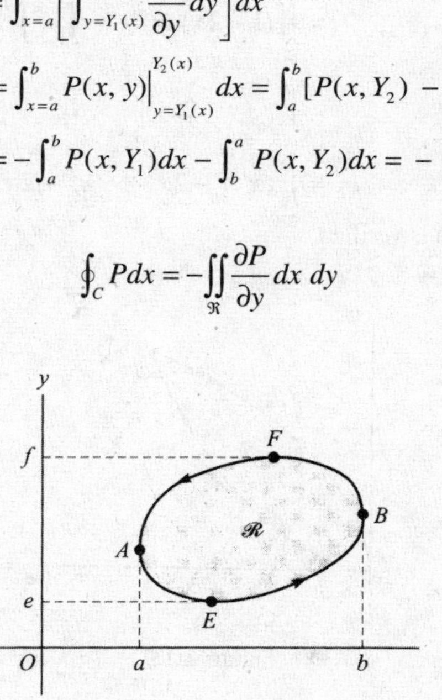

Figure 10.8

Similarly, let the equations of curves EAF and EBF be $x = X_1(y)$ and $x = X_2(y)$, respectively. Then

$$\iint_{\Re} \frac{\partial Q}{\partial x}dx\,dy = \int_{y=c}^{f}\left[\int_{x=X_1(y)}^{X_2(y)} \frac{\partial Q}{\partial x}dx\right]dy = \int_c^f [Q(X_2, y) - Q(X_1, y)]dy$$

$$= \int_f^c Q(X_1, y)dy + \int_c^f Q(X_2, y)dy = \oint_C Q\,dy$$

Then

$$\oint_C Q\,dy = \iint_{\Re} \frac{\partial Q}{\partial x}\,dx\,dy \tag{2}$$

Adding Equations (1) and (2),

$$\oint_C P\,dx + Q\,dy = \iint_{\Re} \left(\frac{\partial Q}{\partial x} - \frac{\partial P}{\partial y}\right) dx\,dy$$

10.6. Verify Green's theorem in the plane for

$$\oint_C (2xy - x^2)dx + (x + y^2)dy$$

where C is the closed curve of the region bounded by $y = x^2$ and $y^2 = x$.

The plane curve $y = x^2$ and $y^2 = x$ intersect at (0, 0) and (1, 1). The positive direction in traversing C is as shown in Figure 10.9.

Along $y = x^2$, the line integral equals

$$\int_{x=0}^{1} \{(2x)(x^2) - x^2\}dx + \{x + (x^2)^2\}d(x^2) = \int_0^1 (2x^3 + x^2 + 2x^5)dx = 7/6$$

Along $y^2 = x$, the line integral equals

$$\int_{y=1}^{0} \{(2)(y^2)(y) - (y^2)^2\}\,d(y^2) + \{y^2 + y^2\}\,dy = \int_1^0 (4y^4 - 2y^5 + 2y^2)dy = -17/15$$

Then the required line integral = 7/6 – 17/15 = 1/30.

$$\begin{aligned}\iint_{\Re}\left(\frac{\partial Q}{\partial x} - \frac{\partial P}{\partial y}\right)dx\,dy &= \iint_{\Re}\left\{\frac{\partial}{\partial x}(x + y^2) - \frac{\partial}{\partial y}(2xy - x^2)\right\} dx\,dy \\ &= \iint_{\Re}(1 - 2x)dx\,dy = \int_{x=0}^{1}\int_{y=x^2}^{\sqrt{x}} (1 - 2x)dy\,dx \\ &= \int_{x=0}^{1} (y - 2xy)\Big|_{y=x^2}^{\sqrt{x}} dx \\ &= \int_0^1 (x^{1/2} - 2x^{3/2} - x^2 + 2x^3)\,dx = 1/30\end{aligned}$$

Hence, Green's theorem is verified.

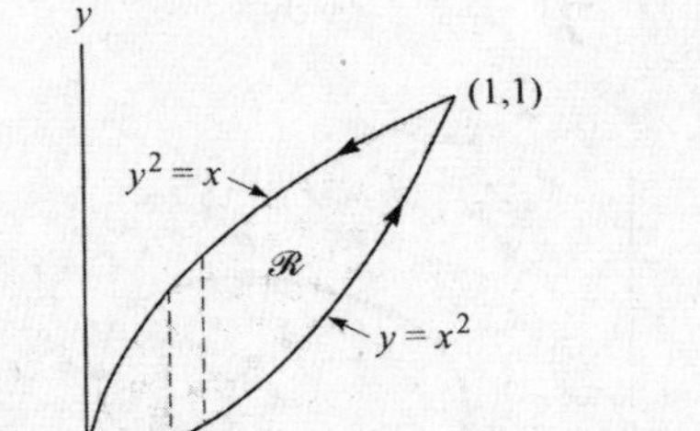

Figure 10.9

10.7. Extend the proof of Green's theorem in the plane given in Problem 10.5 to the curves C for which lines parallel to the coordinate axes may cut C in more than two points.

Consider a closed curve C such as is shown in Figure 10.10, in which lines parallel to the axes may meet C in more than two points. By constructing line ST, the region is divided into two regions $\Re_1$ and $\Re_2$, which are of the type considered in Problem 10.5 and for which Green's theorem applies, i.e.,

$$\int_{STUS} P\,dx + Q\,dy = \iint_{\Re_1}\left(\frac{\partial Q}{\partial x} - \frac{\partial P}{\partial y}\right)dx\,dy \tag{1}$$

$$\int_{SVTS} P\,dx + Q\,dy = \iint_{\Re_1} \left(\frac{\partial Q}{\partial x} - \frac{\partial P}{\partial y} \right) dx\,dy \tag{2}$$

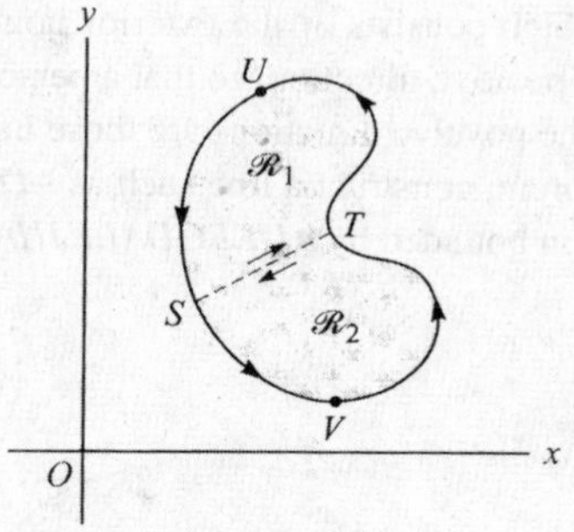

Figure 10.10

Adding the left-hand sides of Equations (1) and (2), and omitting the integrand $P\,dx + Q\,dy$ in each case, we have

$$\int_{STUS} + \int_{SVTS} = \int_{ST} + \int_{TUS} + \int_{SVT} + \int_{TS} = \int_{TUS} + \int_{SVT} = \int_{TUSVT}$$

using the fact that $\int_{ST} = -\int_{TS}$.

Adding the right-hand sides of Equations (1) and (2), omitting the integrand, $\iint_{\Re_1} + \iint_{\Re_2} = \iint_{\Re}$, where $\Re$ consists of regions $\Re_1$ and $\Re_2$.

Then $\int_{TUSVT} P\,dx + Q\,dy = \iint_{\Re} \left(\frac{\partial Q}{\partial x} - \frac{\partial P}{\partial y} \right) dx\,dy$, and the theorem is proved.

A region $\Re$ such as is considered here and in Problem 10.5, for which any closed lying in $\Re$ can be continuously shrunk to a point without leaving $\Re$, is called a *simply connected region*. A region which is not simply connected is called *multiply connected*. We have shown here that Green's theorem in the plane applies to simply connected regions bounded by closed curves. In Problem 10.10 the theorem is extended to multiply connected regions.

For more complicated simply connected regions, it may be necessary to construct more lines, such as ST, to establish the theorem.

10.8. Show that the area bounded by a simple closed curve C is given by $\frac{1}{2}\oint_C x\,dy - y\,dx$.

In Green's theorem, put $P = -y$, $Q = x$. Then

$$\oint_C x\,dy - y\,dx = \iint_{\Re} \left(\frac{\partial}{\partial x}(x) - \frac{\partial}{\partial y}(-y) \right) dx\,dy = 2\iint_{\Re} dx\,dy = 2A$$

where A is the required area. Thus, $A = \frac{1}{2}\oint_C x\,dy - y\,dx$.

10.9. Find the area of the ellipse $x = a\cos\theta$, $y = b\sin\theta$.

$$\text{Area} = \frac{1}{2}\oint_C x\,dy - y\,dx = \frac{1}{2}\int_0^{2\pi} (a\cos\theta)(b\cos\theta)d\theta - (b\sin\theta)(-a\sin\theta)d\theta$$

$$= \frac{1}{2}\int_0^{2\pi} ab(\cos^2\theta + \sin^2\theta)d\theta = \frac{1}{2}\int_0^{2\pi} ab\,d\theta = \pi ab$$

10.10. Show that Green's theorem in the plane is also valid for a multiply connected region $\Re$ such as is shown in Figure 10.11.

The shaded region $\Re$, shown in Figure 10.11, is multiply connected, since not every closed curve lying in $\Re$ can be shrunk to a point without leaving $\Re$, as is observed by considering a curve surrounding *DEFGD*, for example. The boundary of $\Re$, which consists of the exterior boundary *AHJKLA* and the interior boundary *DEFGD*, is to be traversed in the positive direction, so that a person traveling in this direction always has the region on his left. It is seen that the positive directions are those indicated Figure 10.11.

In order to establish the theorem, construct a line such as *AD*, called a *crosscut*, connecting the exterior and interior boundaries. The region bounded by *ADEFGDALKJHA* is simply connected, and so Green's theorem is valid. Then

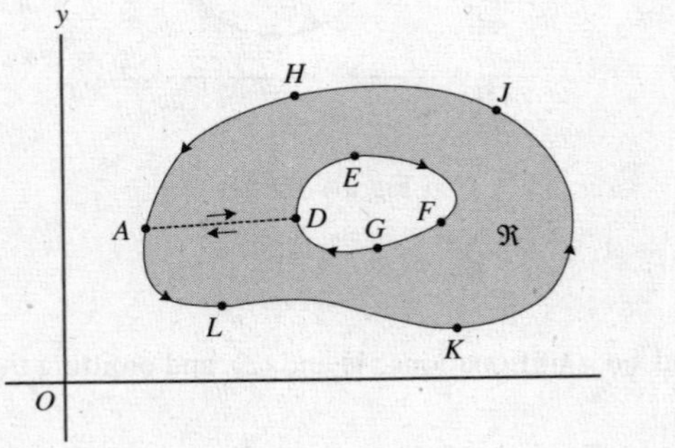

Figure 10.11

$$\oint_{ADEFGDALKJHA} P\,dx + Q\,dy = \iint_{\Re}\left(\frac{\partial Q}{\partial x} - \frac{\partial P}{\partial y}\right)dx\,dy$$

But the integral on the left, leaving out the integrand, is equal to

$$\int_{AD} + \int_{DEFGD} + \int_{DA} + \int_{ALKJHA} = \int_{DEFGD} + \int_{ALKJHA}$$

since $\int_{AD} = -\int_{DA}$. Thus, if C_1 is the curve *ALKJHA*, C_2 is the curve *DEFGD*, and C is the boundary of $\Re$ consisting of C_1 and C_2 (traversed in the positive directions), then $\int_{C_1} + \int_{C_2} = \int_C$ and so

$$\oint_C P\,dx + Q\,dy = \iint_{\Re}\left(\frac{\partial Q}{\partial x} - \frac{\partial P}{\partial y}\right)dx\,dy$$

Independence of the path

10.11. Let $P(x, y)$ and $Q(x, y)$ be continuous and have continuous first partial derivatives at each point of a simply connected region $\Re$. Prove that a necessary and sufficient condition that $\oint_C P\,dx + Q\,dy = 0$ around every closed path C in $\Re$ is that $\partial P/\partial y = \partial Q/\partial x$ identically in $\Re$.

Sufficiency. Suppose $\partial P/\partial y = \partial Q/\partial x$. Then, by Green's theorem,

$$\oint_C P\,dx + Q\,dy = \iint_{\Re}\left(\frac{\partial Q}{\partial x} - \frac{\partial P}{\partial y}\right)dx\,dy = 0$$

where $\Re$ is the region bounded by C.

Necessity. Suppose $\oint_C P\,dx + Q\,dy = 0$ around every closed path C in $\Re$ and that $\partial P/\partial y \neq \partial Q/\partial x$ at some point of $\Re$. In particular, suppose $\partial P/\partial y - \partial Q/\partial x > 0$ at the point (x_0, y_0).

By hypothesis, $\partial P/\partial y$ and ∂Q are continuous in $\Re$, so that there must be some region τ containing (x_0, y_0) as an interior point for which $\partial P/\partial y - \partial Q/\partial x > 0$. If Γ is the boundary of τ, then by Green's theorem,

$$\oint_C P\,dx + Q\,dy = \iint_{\tau}\left(\frac{\partial Q}{\partial x} - \frac{\partial P}{\partial y}\right)dx\,dy > 0$$

contradicting the hypothesis that $\oint_\Gamma P dx + Q\, dy = \iint_\tau \left(\frac{\partial Q}{\partial x} - \frac{\partial P}{\partial y} \right) dx\, dy > 0$ for *all* closed curves in in $\mathfrak{R}$. Thus, $\partial Q/\partial x - \partial P/\partial y$ cannot be positive.

Similarly, we can show that $\partial Q/\partial x - \partial P/\partial y$ cannot be negative, and it follows that it must be identically zero; i.e., $\partial P/\partial y = \partial Q/\partial x$ identically in $\mathfrak{R}$.

10.12 Let P and Q be defined as in Problem 10.11. Prove that a necessary and sufficient condition that $\int_A^B P\, dx + Q\, dy$ be independent of the path in $\mathfrak{R}$ joining points A and B is that $\partial P/\partial y = \partial\, Q/\partial x$ identically in $\mathfrak{R}$.

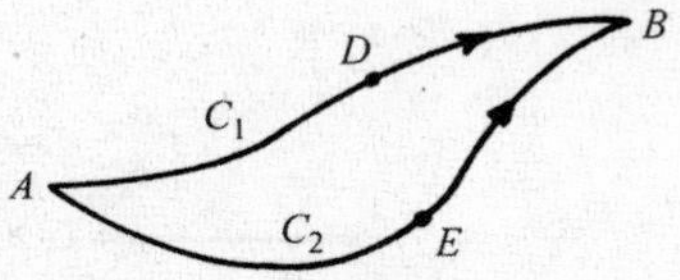

Figure 10.12

Sufficiency. If $\partial P/\partial y = \partial Q/\partial x$, then by Problem 10.11,

$$\int_{ADBEA} P\, dx + Q\, dy = 0$$

(See Figure 10.12.) From this, omitting for brevity the integrand $P\, dx + Q\, dy$, we have

$$\int_{ADB} + \int_{BEA} = 0, \quad \int_{ADB} = -\int_{BEA} = \int_{AEB} \quad \text{and so} \quad \int_{C_1} = \int_{C_2}$$

i.e., the integral is independent of the path.

Necessity. If the integral is independent of the path, then for all paths C_1 and C_2 in $\mathfrak{R}$ we have

$$\int_{C_1} = \int_{C_2}, \quad \int_{ADB} = \int_{AEB} \quad \text{and} \quad \int_{ADBEA} = 0$$

From this it follows that the line integral around any closed path in $\mathfrak{R}$ is zero, and, hence, by Problem 10.11 that $\partial P/\partial y = \partial Q/\partial x$.

10.13. Let P and Q be as in Problem 10.11. (a) Prove that a necessary and sufficient condition that $P\, dx + Q\, dy$ be an exact differential of a function $\varphi(x, y)$ is that $\partial P/\partial y = \partial Q/\partial x$. (b) Show that in such case $\int_A^B P\, dx + Q\, dy = \int_A^B d\phi = \phi(B) - \phi(A)$ where A and B are any two points.

(a) **Necessity.** If $P\, dx + Q\, dy = d\,\phi = \frac{\partial \phi}{\partial x} dx + \frac{\partial \phi}{\partial y} dy$, an exact differential, then

$$\partial\phi/\partial x = P \tag{1}$$
$$\partial\phi/\partial y = 0 \tag{2}$$

Thus, by differentiating Equations (1) and (2) with respect to y and x, respectively, $\partial P/\partial y = \partial Q/\partial x$, since we are assuming continuity of the partial derivatives.

Sufficiency. By Problem 10.12, if $\partial P/\partial y = \partial Q/\partial x$, then $\int P\, dx + Q\, dy$ is independent of the path joining two points. In particular, let the two points be (a, b) and (x, y) and define

$$\phi(x, y) = \int_{(a,b)}^{(x,y)} P\, dx + Q\, dy$$

Then

$$\phi(x + \Delta x, y) - \phi(x, y) = \int_{(a,b)}^{(x+\Delta x, y)} P\, dx + Q\, dy - \int_{(a,b)}^{(x,y)} P\, dx + Q\, dy$$
$$= \int_{(x,y)}^{(x+\Delta x, y)} P\, dx + Q\, dy$$

Since the last integral is independent of the path joining (x, y) and (x + Δx, y), we can choose the path to be a straight line joining these points (see Figure 10.13) so that $dy = 0$. Then, by the mean value theorem for integrals,

$$\frac{\phi(x+\Delta x, y) - \phi(x, y)}{\Delta x} = \frac{1}{\Delta x}\int_{(x, y)}^{(x+\Delta x, y)} P\,dx = P(x + \theta\,\Delta x, y) \qquad 0 < \theta < 1$$

Taking the limit as $\Delta x \to 0$, we have $\partial\phi/\partial x = P$.

Similarly, we can show that $\partial\phi/\partial y = Q$.

Thus, it follows that $P\,dx + Q\,dy = \dfrac{\partial\phi}{\partial x}dx + \dfrac{\partial\phi}{\partial y}dy = d\phi$.

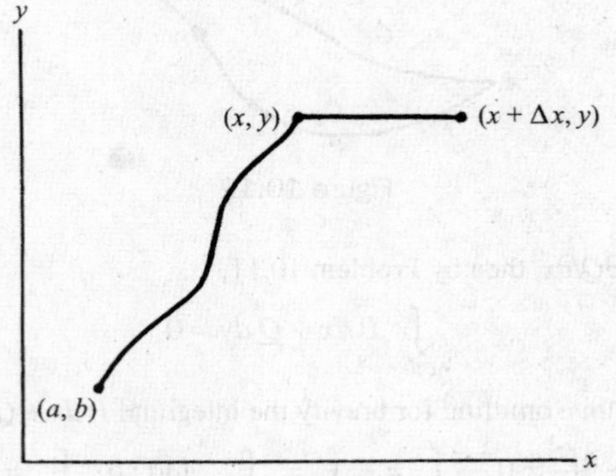

Figure 10.13

(b) Let $A = (x_1, y_1)$ and $B = (x_2, y_2)$. From (a),

$$\phi(x, y) = \int_{(a, b)}^{(x, y)} P\,dx + Q\,dy.$$

Then, omitting the integrand $P\,dx + Q\,dy$, we have

$$\int_A^B = \int_{(x_1, y_1)}^{(x_2, y_2)} = \int_{(a, b)}^{(x_2, y_2)} - \int_{(a, b)}^{(x_1, y_1)} = \phi(x_2, y_2) - \phi(x_1, y_1) = \phi(B) - \phi(A)$$

10.14. (a) Prove that $\int_{(1,2)}^{(3,4)} (6xy^2 - y^3)\,dx + (6x^2y - 3xy^2)\,dy\ dy$ is independent of the path joining (1, 2) and (3, 4). (b) Evaluate the integral in (a).

(a) P = 6xy2 – y3, Q = 6x2 y – 3xy2. Then ∂P/∂y = 12xy – 3y2 = ∂Q/∂x and, by Problem 10.12, the line integral is independent of the path.

(b) **Method 1:** Since the line integral is independent of the path, choose any path joining (1, 2) and (3, 4), for example, that consisting of lines from (1, 2) to (3, 2) (along which $y = 2$, $dy = 0$) and then (3, 2) to (3, 4) (along which $x = 3$, $dx = 0$). Then the required integral equals

$$\int_{x=1}^{3} (24x - 8)dx + \int_{y=2}^{4} (54y - 9y^2)dy = 80 + 156 = 236$$

Method 2: Since $\dfrac{\partial P}{\partial y} = \dfrac{\partial Q}{\partial x}$, we must have

$$\frac{\partial\phi}{\partial y} = 6x^2y - 3xy^2 \qquad (1)$$

$$\frac{\partial\phi}{\partial y} = 6x^2y - 3xy^2 \qquad (2)$$

From Equation (1), $\phi = 3x^2y^2 - xy^3 + f(y)$. From Equation (2), $\phi = 3x^2y^2 - xy^3 + g(x)$. The only way in which these two expressions for ϕ are equal is if $f(y) = g(x) = c$, a constant. Hence, $\phi = 3x^2y^2 - xy^3 + c$. Then, by Problem 10.13.

$$\int_{(1,2)}^{(3,4)} (6xy^2 - y^3)dx + (6x^2y - 3xy^2)dy = \int_{(1,2)}^{(3,4)} d(3x^2y^2 - xy^3 + c)$$
$$= 3x^2y^2 - xy^3 + c\Big|_{(1,2)}^{(3,4)} = 236$$

Note that in this evaluation the arbitrary constant c can be omitted. See also Problem 6.16.

We could also have noted by inspection that

$$(6xy^2 - y^3)dx + (6x^2y - 3xy^2)dy = (6xy^2dx + 6x^2y\,dy) - (y^3dx + 3xy^2dy)$$
$$= d(3x^2y^2) - d(xy^3) = d(3x^2y^2 - xy^3)$$

from which it is clear that $\phi = 3x^2y^2 - xy^3 + c$.

10.15. Evaluate $\oint (x^2y\cos x + 2xy\sin x - y^2e^x)dx + (x^2\sin x - 2ye^x)dy$ around the hypocycloid $x^{2/3} + y^{2/3} = a^{2/3}$.

$$P = x^2y\cos x + 2xy\sin x - y^2e^x,\ Q = x^2\sin x - 2ye^x$$

Then $\partial P/\partial y = x^2\cos x + 2x\sin x - 2ye^x = \partial Q/\partial x$, so that, by Problem 10.11, the line integral around any closed path—in particular, $x^{2/3} + y^{2/3} = a^{2/3}$—is zero.

Surface integrals

10.16. If γ is the angle between the normal line to any point (x, y, z) of a surface S and the positive z axis, prove that

$$|\sec\gamma| = \sqrt{1 + z_x^2 + z_y^2} = \frac{\sqrt{F_x^2 + F_y^2 + F_z^2}}{|F_z|}$$

according as the equation for S is $z = f(x, y)$ or $F(x, y, z) = 0$.

If the equation for S is $F(x, y, z) = 0$, a normal to S at (x, y, z) is $\nabla F = F_x\mathbf{i} + F_y\mathbf{j} + F_z\mathbf{k}$. Then

$$\nabla F \cdot \mathbf{K} = |\nabla F||\mathbf{k}|\cos\gamma \quad \text{or} \quad F_z = \sqrt{F_x^2 + F_y^2 + F_z^2}\cos\gamma$$

from which $|\sec\gamma| = \dfrac{\sqrt{F_x^2 + F_y^2 + F_z^2}}{|F_z|}$ as required. as required.

In case the equation is $z = f(x, y)$, we can write $F(x, y) = 0$, from which $F_x = -z_x$, $F_y - z_y$, $F_z = 1$ and we find $|\sec\gamma| = \sqrt{1 + z_x^2 + z_y^2}$.

10.17. Evaluate $\iint_S U(x, y, z)dS$, where S is the surface of the paraboloid $z = 2 - (x^2 + y^2)$ above the xy plane and $U(x, y, z)$ is equal to (a) 1, (b) $x^2 + y^2$, y^2, and (c) $3z$. Give a physical interpretation in each case. (See Figure 10.14.)

The required integral is equal to

$$\iint_{\Re} U(x, y, z)\sqrt{1 + z_x^2 + z_y^2}\,dx\,dy. \qquad (1)$$

where $\Re$ is the projection of S on the xy plane given by $x^2 + y^2 = 2$, $z = 0$.

Since $z_x = -2x$, $z_y = -2y$, (1) can be written

$$\iint_{\Re} U(x, y, z)\sqrt{1 + 4x^2 + 4y^2}\,dx\,dy \qquad (2)$$

(a) If U(x, y, z) = 1, (2) becomes

$$\iint_{\Re} \sqrt{1 + 4x^2 + 4y^2}\,dx\,dy$$

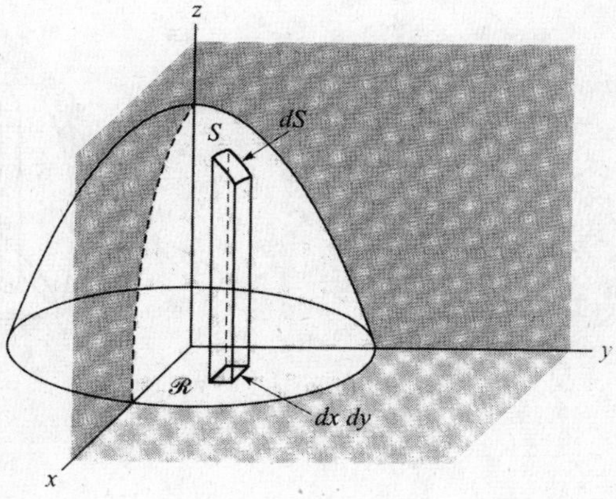

Figure 10.14

To evaluate this, transform to polar coordinates (ρ, ϕ). Then the integral becomes

$$\int_{\phi=0}^{2\pi}\int_{\rho=0}^{\sqrt{2}}\sqrt{1+4\rho^2}\;\rho\,d\rho\,d\phi=\int_{\phi=0}^{2\pi}\frac{1}{12}(1+4\rho^2)^{3/2}\Big|_{\rho=0}^{\sqrt{2}}\,d\phi=\frac{13\pi}{3}.$$

Physically, this could represent the surface area of *S* or the mass of *S* assuming unit density.

(b) If U(x, y, z) = x2 + y2, (2) becomes $\iint_{\Re}(x^2+y^2)\sqrt{1+4x^2+4y^2}\,dx\,dy$ or, in polar coordinates,

$$\int_{\phi=0}^{2\pi}\int_{\rho=0}^{\sqrt{2}}\rho^3\sqrt{1+4\rho^2}\;d\rho\,d\phi=\frac{149\pi}{30}$$

where the integration with respect to ρ is accomplished by the substitution $\sqrt{1+4\rho^2}=u$.

Physically, this could represent the moment of inertia of *S* about the *z* axis assuming unit density, or the mass of *S* assuming a density = $x^2 + y^2$.

(c) If U(x, y, z) = 3z, (2) becomes

$$\iint_{\Re}3z\sqrt{1+4x^2+4y^2}\,dx\,dy=\iint_{\Re}3\{2-(x^2+y^2)\}\sqrt{1+4x^2+4y^2}\,dx\,dy$$

or, in polar coordinates,

$$\int_{\phi=0}^{2\pi}\int_{\rho=0}^{\sqrt{2}}3\rho(2-\rho^2)\sqrt{1+4\rho^2}\,d\rho\,d\phi=\frac{111\pi}{10}$$

Physically, this could represent the mass of *S* assuming a density = 3*z*, or three times the first moment of *S* about the *xy* plane.

10.18. Find the surface area of a hemisphere of radius *a* cut off by a cylinder having this radius as diameter.

Equations for the hemisphere and cylinder (see Figure 10.15) are given, respectively, by $x^2 + y^2 + z^2 = a^2$ (or $z\ \sqrt{a^2-x^2-y^2}$) and $(x - a/2)^2 + y^2 = a^2/4$ (or $x^2 + y^2 = ax$).

Since

$$z_x=\frac{-x}{\sqrt{a^2-x^2-y^2}}\quad\text{and}\quad z_y=\frac{-y}{\sqrt{a^2-x^2-y^2}}$$

we have

$$\text{Required surface area}=2\iint_{\Re}\sqrt{1+z_x^2+z_y^2}\,dx\,dy=2\iint_{\Re}\frac{a}{\sqrt{a^2-x^2-y^2}}\,dx\,dy$$

Two methods of evaluation are possible.

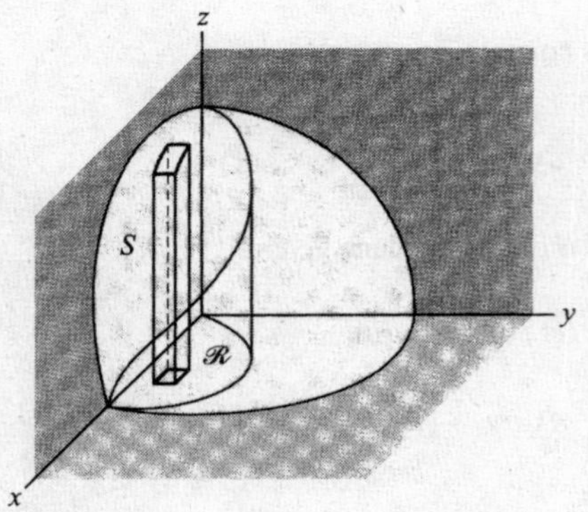

Figure 10.15

Method 1: Using polar coordinates.

Since $x^2 + y^2 = ax$ in polar coordinates is $\rho = a\cos\phi$, the integral becomes

$$2\int_{\phi=0}^{\pi/2}\int_{\rho=0}^{a\cos\phi}\frac{a}{\sqrt{a^2-\rho^2}}\rho\,d\rho\,d\phi = 2a\int_{\phi=0}^{\pi/2}-\sqrt{a^2-\rho^2}\Big|_{\rho=0}^{a\cos\phi}d\phi$$

$$= 2a^2\int_0^{\pi/2}(1-\sin\phi)d\phi = (\pi-2)a^2$$

Method 2: The integral is equal to

$$2\int_{x=0}^{a}\int_{y=0}^{\sqrt{ax-x^2}}\frac{a}{\sqrt{a^2-x^2-y^2}}dx\,dy = 2a\int_{x=0}^{a}\sin^{-1}\frac{y}{\sqrt{ax-x^2}}\Bigg|_{y=0}^{\sqrt{ax-x^2}}dx$$

$$= 2a\int_0^a \sin^{-1}\sqrt{\frac{x}{a+x}}dx$$

Letting $x = a\tan^2\theta$, this integral becomes

$$4a^2\int_0^{\pi/4}\theta\tan\theta\sec^2\theta d\theta = 4a^2\left\{\frac{1}{2}\theta\tan^2\theta\,|_0^{\pi/4} - \frac{1}{2}\int_0^{\pi/4}\tan^2\theta d\theta\right\}$$

$$= 2a^2\left\{\theta\tan^2\theta\,|_0^{\pi/4} - \int_0^{\pi/4}(\sec^2\theta-1)d\theta\right\}$$

$$= 2a^2\left\{\pi/4 - (\tan\theta-\theta)\,|_0^{\pi/4}\right\} = (\pi-2)a^2$$

Note that these integrals are actually *improper* and should be treated by appropriate limiting procedures (see Problem 5.74 and Chapter 12).

10.19. Find the centroid of the surface in Problem 10.17.

By symmetry, $\bar{x} = \bar{y} = 0$ and $\bar{z} = \dfrac{\iint_S z\,dS}{\iint_S z\,dS} = \dfrac{\iint_{\Re} z\sqrt{1+4x^2+4y^2}\,dx\,dy}{\iint_{\Re}\sqrt{1+4x^2+4y^2}\,dx\,dy}$

The numerator and denominator can be obtained from the results of Problems 10.17(c) and 10.17(a), respectively, and we thus have $\bar{z} = \dfrac{37\pi/10}{13\pi/3} = \dfrac{111}{130}$.

10.20. Evalute $\iint_S \mathbf{A}\cdot\mathbf{n}\,dS$, where $\mathbf{A} = xy\mathbf{i} - x^2\mathbf{j} + (x+z)\mathbf{k}$, S is that portion of the plane $2x + 2y + z = 6$ included in the first octant, and $\mathbf{n}$ is a unit normal to S. (See Figure 10.16.)

A normal to S is $\nabla(2x+2y+z-6) = 2\mathbf{i}+2\mathbf{j}+\mathbf{k}$, and so

$$\mathbf{n} = \frac{2\mathbf{i}+2\mathbf{j}+\mathbf{k}}{\sqrt{2^2+2^2+1^2}} = \frac{2\mathbf{i}+2\mathbf{j}+\mathbf{k}}{3}$$

Then

$$\mathbf{A}\cdot\mathbf{n} = \{xy\mathbf{i} - x^2\mathbf{j} + (x+z)\mathbf{k}\}\cdot\left(\frac{2\mathbf{i}+2\mathbf{j}+\mathbf{k}}{3}\right)$$

$$= \frac{2xy - 2x^2 + (x+z)}{3}$$

$$= \frac{2xy - 2x^2 + (x+6-2x-2y)}{3}$$

$$= \frac{2xy - 2x^2 - x - 2y + 6}{3}$$

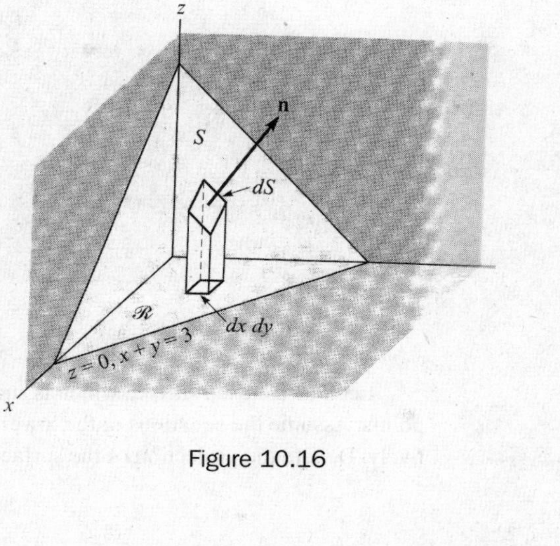

Figure 10.16

The required surface integral is, therefore,

$$\iint_S \left(\frac{2xy - 2x^2 - x - 2y + 6}{3}\right) dS = \iint_{\Re} \left(\frac{2xy - 2x^2 - x - 2y + 6}{3}\right)\sqrt{1 + z_x^2 + z_y^2}\,dx\,dy$$

$$= \iint_{\Re} \left(\frac{2xy - 2x^2 - x - 2y + 6}{3}\right)\sqrt{1^2 + 2^2 + 2^2}\,dx\,dy$$

$$= \int_{x=0}^{3}\int_{y=0}^{3-x} (2xy - 2x^2 - x - 2y + 6)\,dy\,dx$$

$$= \int_{x=0}^{3} (xy^2 - 2x^2y - xy - y^2 + 6y)\Big|_0^{3-x} dx = 27/4$$

10.21. In dealing with surface integrals we have restricted ourselves to surface which are two-sided. Give an example of a surface which is not two-sided.

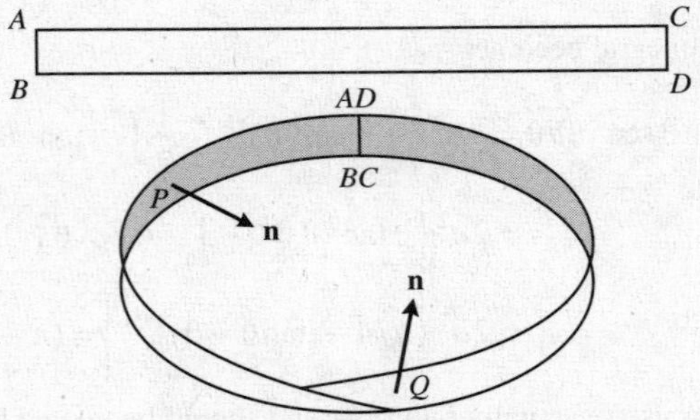

Figure 10.17

Take a strip of paper such as *ABCD*, as shown in Figure 10.17. Twist the strip so that points *A* and *B* fall on *D* and *C*, respectively, as in the figure. If **n** is the positive normal at point *P* of the surface, we find that as **n** moves around the surface, it reverses its original direction when it reaches *P* again. If we tried to color only one side of the surface, we would find the whole thing colored. This surface, called a *Möbius strip*, is an example of a one-sided surface. This is sometimes called a *nonorientable* surface. A two-sided surface is *orientable*.

The divergence theorem

10.22. Prove the divergence theorem. (See Figure 10.18.)

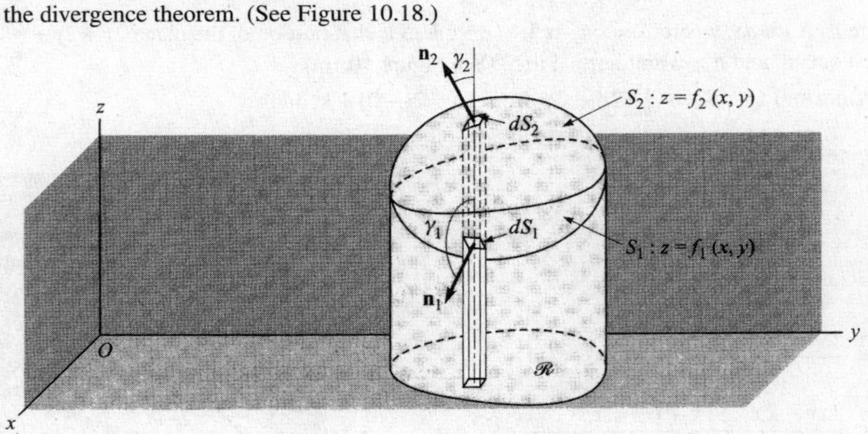

Figure 10.18

Let S be a closed surface which is such that any line parallel to the coordinate axes cuts S in, at most, two points. Assume the equations of the lower and upper portions S_1 and S_2 to be $z = f_1(x, y)$ and $z = f_2(x, y)$ respectively. Denote the projection of the surface on the xy plane by $\Re$. Consider

$$\iiint_V \frac{\partial A_3}{\partial z} dV = \iiint_V \frac{\partial A_3}{\partial z} dz\, dy\, dx = \iint_{\Re} \left[\int_{z=f_1(x,y)}^{f_2(x,y)} \frac{\partial A_3}{\partial z} dz \right] dy\, dx$$

$$= \iint_{\Re} A_3(x,y,z)\Big|_{z=f_1}^{f_2} dy\, dx = \iint_{\Re} [A_3(x,y,f_2) - A_3(x,y,f_1)]\, dy\, dx$$

For the upper portion S_2, $dy\, dx = \cos \gamma_2\, dS_2 = \mathbf{k} \cdot \mathbf{n}_2\, dS_2$ since the normal $\mathbf{n}_2$ to S_2 makes an acute angle γ_2 with $\mathbf{k}$.

For the lower portion S_1, $dy\, dx = -\cos \gamma_1\, dS_1 = -\mathbf{k} \cdot \mathbf{n}_1\, dS_1$ since the normal $\mathbf{n}_1$ to S_1 makes an obtuse angle γ_1 with $\mathbf{k}$.

Then

$$\iint_{\Re} A_3(x,y,f_2)\, dy\, dx = \iint_{S_2} A_3 \mathbf{k} \cdot \mathbf{n}_2\, dS_2$$

$$\iint_{\Re} A_3(x,y,f_1)\, dy\, dx = -\iint_{S_1} A_3 \mathbf{k} \cdot \mathbf{n}_1\, dS_1$$

and

$$\iint_{\Re} A_3(x,y,f_2)\, dy\, dx - \iint_{\Re} A_3(x,y,f_1)\, dy\, dx = \iint_{S_2} A_3 \mathbf{k} \cdot \mathbf{n}_2 dS_2 + \iint_{S_1} A_3 \mathbf{k} \cdot \mathbf{n}_1 dS_1$$

$$= \iint_S A_3 \mathbf{k} \cdot \mathbf{n}\, dS$$

so that

$$\iiint_V \frac{\partial A_3}{\partial z} dV = \iint_S A_3 \mathbf{k} \cdot \mathbf{n}\, dS \tag{1}$$

Similarly, by projecting S on the other coordinate planes,

$$\iiint_V \frac{\partial A_1}{\partial x} dV = \iint_S A_3 \mathbf{i} \cdot \mathbf{n}\, dS \tag{2}$$

$$\iiint_V \frac{\partial A_2}{\partial y} dV = \iint_S A_3 \mathbf{j} \cdot \mathbf{n}\, dS \tag{3}$$

Adding Equations (1), (2), and (3),

$$\iiint_V \left(\frac{\partial A_1}{\partial x} + \frac{\partial A_2}{\partial y} + \frac{\partial A_3}{\partial z} \right) dV = \iint_S (A_1 \mathbf{i} + A_2 \mathbf{j} + A_3 \mathbf{k})\, \mathbf{n}\, dS$$

or

$$\iiint_V \nabla \cdot \mathbf{A}\, dV = \iint_S \mathbf{A} \cdot \mathbf{n}\, dS$$

The theorem can be extended to surfaces which are such that lines parallel to the coordinate axes meet them in more than two points. To establish this extension, subdivide the region bounded by S into subregions whose surfaces do satisfy this condition. The procedure is analogous to that used in Green's theorem for the plane.

10.23. Verify the divergence theorem for $\mathbf{A} = (2x - z)\mathbf{i} + x^2 y \mathbf{j} - xz^2 \mathbf{k}$ taken over the region bounded by $x = 0$, $x = 1$, $y = 0$, $y = 1$, $z = 0$, $z = 1$.

We first evaluate $\iint_S \mathbf{A} \cdot \mathbf{n} dS$, where S is the surface of the cube in Figure 10.19.

Face DEFG: $\mathbf{n} = \mathbf{i}$, $x = 1$. Then

$$\iint_{DEFG} \mathbf{A} \cdot \mathbf{n}\, dS = \int_0^1 \int_0^1 \{(2 - z)\mathbf{i} + \mathbf{j} - z^2 \mathbf{k}\} \cdot \mathbf{i}\, dy\, dz$$

$$= \int_0^1 \int_0^1 (2 - z)\, dy\, dz = 3/2$$

Face ABCO: $\mathbf{n} = -\mathbf{i}$, $x = 0$. Then

$$\iint_{ABCO} \mathbf{A}\cdot\mathbf{n}\, dS = \int_0^1\int_0^1 (-z\mathbf{i})\cdot(-\mathbf{i})\, dy\, dz$$

$$= \int_0^1\int_0^1 z\, dy\, dz = 1/2$$

Face ABEF: $\mathbf{n} = \mathbf{j}$, $y = 1$. Then

$$\iint_{ABEF} \mathbf{A}\cdot\mathbf{n}\, dS = \int_0^1\int_0^1 \{(2-z)\mathbf{i} + x^2\mathbf{j} - xz^2\mathbf{k}\}\cdot\mathbf{j}\, dx\, dz = \int_0^1\int_0^1 x^2\, dx\, dz = 1/3$$

Face OGDC: $\mathbf{n} = -\mathbf{j}$, $y = 0$. Then

$$\iint_{OGDC} \mathbf{A}\cdot\mathbf{n}\, dS = \int_0^1\int_0^1 \{(2x-z)\mathbf{i} - xz^2\mathbf{k}\}\cdot(-\mathbf{j})\, dx\, dz = 0$$

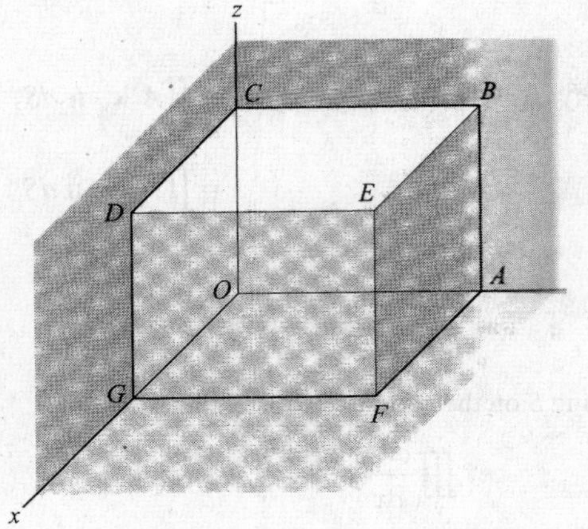

Figure 10.19

Face BCDE: $\mathbf{n} = \mathbf{k}$, $z = 1$. Then

$$\iint_{BCDE} \mathbf{A}\cdot\mathbf{n}\, dS = \int_0^1\int_0^1 \{(2x-1)\mathbf{i} + x^2y\mathbf{j} - x\mathbf{k}\}\cdot\mathbf{k}\, dx\, dy = \int_0^1\int_0^1 -x\, dx\, dy - 1/2$$

Face AFGO: $\mathbf{n} = -\mathbf{k}$, $z = 0$. Then

$$\iint_{AFGO} \mathbf{A}\cdot\mathbf{n}\, dS = \int_0^1\int_0^1 \{2x\mathbf{i} - x^2y\mathbf{j}\}\cdot(-\mathbf{k})\, dx\, dy = 0$$

Adding, $\iint_S \mathbf{A}\cdot\mathbf{n}\, dS = \frac{3}{2} + \frac{1}{2} + \frac{1}{3} + 0 - \frac{1}{2} + 0 = \frac{11}{6}$. Since

$$\iiint_V \nabla\cdot\mathbf{A}\, dV = \int_0^1\int_0^1\int_0^1 (2 + x^2 - 2xz)dx\, dy\, dz = \frac{11}{6}$$

the divergence theorem is verified in this case.

10.24. Evaluate $\iint_S \mathbf{A}\cdot\mathbf{n}\, ds$, where S is a closed surface.

By the divergence theorem,

$$\iint_S \mathbf{r}\cdot\mathbf{n}\, dS = \iiint_V \nabla\cdot\mathbf{r}\, dV$$

$$= \iiint_V \left(\frac{\partial}{\partial x}\mathbf{i} + \frac{\partial}{\partial y}\mathbf{j} + \frac{\partial}{\partial z}\mathbf{k}\right)\cdot(x\mathbf{i} + y\mathbf{j} + z\mathbf{k})dV$$

$$= \iiint_V \left(\frac{\partial x}{\partial x}\mathbf{i} + \frac{\partial y}{\partial y}\mathbf{j} + \frac{\partial z}{\partial z}\mathbf{k}\right) dV = 3\iiint_V dV = 3V$$

where V is the volume enclosed by S.

10.25. Evaluate $\iint_S xz^2 dy\,dz + (x^2y - z^3)dz\,dx + (2xy + y^2z)dx\,dy$, where S is the entire surface of the hemispherical region bounded by $z = \sqrt{a^2 - x^2 - y^2}$ and $z = 0$ (a) by the divergence theorem (Green's theorem in space) and (b) directly.

(a) Since $dy\,dz = dS\cos\alpha$, $dz\,dx = dS\,dS\cos\beta$, and $dx\,dy = dS\cos\gamma$, the integral can be written

$$\iint_S \{xz^2 \cos\alpha + (x^2y - z^3)\cos\beta + (2xy + y^2z)\cos\gamma\}dS = \iint_S \mathbf{A}\cdot\mathbf{n}\,dS$$

where $\mathbf{A} = xz^2\mathbf{i} + (x^2y - z^3)\,\mathbf{j} + (2xy + y^2z)\mathbf{k}$ and $\mathbf{n} = \cos\alpha\mathbf{i} + \cos\beta\mathbf{j} + \cos\gamma\mathbf{k}$, the outward drawn unit normal. Then, by the divergence theorem the integral equals

$$\iiint_V \nabla\cdot\mathbf{A}\,dV = \iiint_V \left\{\frac{\partial}{\partial x}(xz^2) + \frac{\partial}{\partial y}(x^2y - z^3) + \frac{\partial}{\partial z}(2xy + y^2z)\right\}dV = \iiint_V (x^2 + y^2 + z^2)dV$$

where V is the region bounded by the hemisphere and the xy plane.

By use of spherical coordinates, as in Problem 9.19, this integral is equal to

$$4\int_{\phi=0}^{\pi/2}\int_{\theta=0}^{\pi/2}\int_{r=0}^{a} r^2\cdot r^2 \sin\theta\,dr\,d\theta\,d\phi = \frac{2\pi a^5}{5}$$

(b) If S_1 is the convex surface of the hemispherical region and S_2 is the base ($z = 0$), then

$$\iint_{S_1} xz^2\,dy\,dz = \int_{y=-a}^{a}\int_{z=0}^{\sqrt{a^2-y^2}} z^2\sqrt{a^2 - y^2 - z^2}\,dz\,dy - \int_{y=-a}^{a}\int_{z=0}^{\sqrt{a^2-x^2}} -z^2\sqrt{a^2 - y^2 - z^2}\,dz\,dx$$

$$\iint_{S_1} (x^2y - z^3)\,dy\,dx = \int_{x=-a}^{a}\int_{x=0}^{\sqrt{a^2-x^2}} \{x^2\sqrt{a^2 - y^2 - z^2} - z^3\}\,dz\,dx$$

$$-\int_{x=-a}^{a}\int_{z=0}^{\sqrt{a^2-x^2}} \{-x^2\sqrt{a^2 - x^2 - z^2} - z^3\}\,dz\,dx$$

$$\iint_{S_1} (2xy - y^2z)\,dx\,dy = \int_{x=-a}^{a}\int_{y=-\sqrt{a^2-x^2}}^{\sqrt{a^2-x^2}} \{2xy + y^2\sqrt{a^2 - y^2 - z^2}\}\,dy\,dx$$

$$\iint_{S_2} xz^2 dy\,dz = 0, \qquad \iint_{S_2} (x^2y - z^3)\,dz\,dx = 0,$$

$$\iint_{S_2} (2xy - y^2z)\,dx\,dy = \iint_{S_2} \{2xy - y^2(0)\}\,dx\,dy = \int_{x=-a}^{a}\int_{y=-\sqrt{a^2-x^2}}^{\sqrt{a^2-x^2}} 2xy\,dy\,dx = 0$$

By addition of the preceding, we obtain

$$4\int_{y=0}^{a}\int_{x=0}^{\sqrt{a^2-y^2}} z^2\sqrt{a^2 - y^2 - z^2}\,dz\,dy + 4\int_{x=0}^{a}\int_{z=0}^{\sqrt{a^2-x^2}} x^2\sqrt{a^2 - x^2 - z^2}\,dz\,dx$$

$$+4\int_{x=0}^{a}\int_{y=0}^{\sqrt{a^2-x^2}} y^2\sqrt{a^2 - x^2 - y^2}\,dy\,dx$$

Since by symmetry all these integrals are equal, the result, on using polar coordinates, is

$$12\int_{x=0}^{a}\int_{y=0}^{\sqrt{a^2-x^2}} y^2\sqrt{a^2 - x^2 - y^2}\,dy\,dx = 12\int_{\phi=0}^{\pi/2}\int_{\rho=0}^{a} \rho^2\sin^2\phi\sqrt{a^2 - \rho^2}\,\rho\,d\rho\,d\phi = \frac{2\pi a^5}{5}$$

Stokes's theorem

10.26. Prove Stokes's theorem.

Let S be a surface which is such that its projections on the xy, yz, and xz planes are are regions bounded by simple closed curves, as indicated in Figure 10.20. Assume S to have representation $z = f(x, y)$ or $x = g(y, z)$ or $y = h(x, z)$, where f, g, and h are single-valued, continuous, and differentiable functions. We must show that

$$\iint_S (\nabla \times \mathbf{A}) \cdot \mathbf{n}\, dS = \iint_S [\nabla \times (A_1\mathbf{i} + A_2\mathbf{j} + A_3\mathbf{k})] \cdot \mathbf{n}\, dS$$
$$= \int_C A \cdot d\mathbf{r}$$

where C is the boundary of S.

Consider first $\iint_S [\nabla \times (A_1\mathbf{i})] \cdot \mathbf{n}\, dS$.

Since $\nabla \times (A_1\mathbf{i}) = \begin{vmatrix} \mathbf{i} & \mathbf{j} & \mathbf{k} \\ \frac{\partial}{\partial x} & \frac{\partial}{\partial y} & \frac{\partial}{\partial z} \\ A_1 & 0 & 0 \end{vmatrix} = \frac{\partial A_1}{\partial z}\mathbf{j} - \frac{\partial A_1}{\partial y}\mathbf{k},$

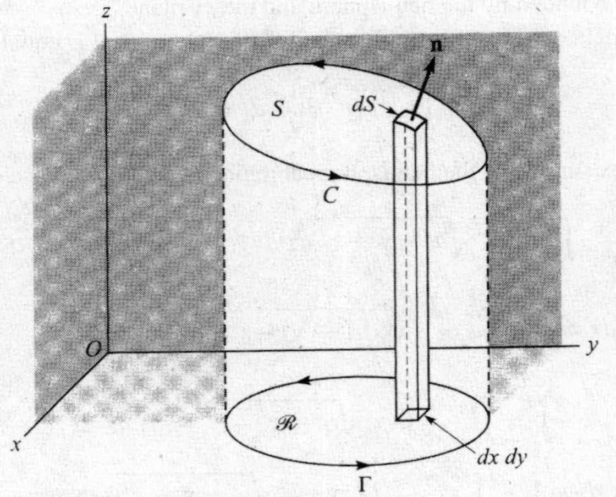

Figure 10.20

$$[\nabla \times (A_1\mathbf{i})] \cdot \mathbf{n}\, dS = \left(\frac{\partial A_1}{\partial z}\mathbf{n} \cdot \mathbf{j} - \frac{\partial A_1}{\partial y}\mathbf{n} \cdot \mathbf{k}\right) dS \tag{1}$$

If $z = f(x, y)$ is taken as the equation of S, then the position vector to any point of S is $\mathbf{r} = x\mathbf{i} + y\mathbf{j} + z\mathbf{k} = x\mathbf{i} + y\mathbf{j} + f(x, y)\mathbf{k}$ so that $\frac{\partial \mathbf{r}}{\partial y} = \mathbf{j} + \frac{\partial z}{\partial y}\mathbf{k} = \mathbf{j} + \frac{\partial f}{\partial y}\mathbf{k}$. But $\frac{\partial \mathbf{r}}{\partial y}$ is a vector tangent to S and thus perpendicular to $\mathbf{n}$, so that

$$\mathbf{n} \cdot \frac{\partial \mathbf{r}}{\partial y} = \mathbf{n} \cdot \mathbf{j} + \frac{\partial z}{\partial y}\mathbf{n} \cdot \mathbf{k} = 0 \quad \text{or} \quad \mathbf{n} \cdot \mathbf{j} = -\frac{\partial z}{\partial y}\mathbf{n} \cdot \mathbf{k}$$

Substitute in Equation (1) to obtain

$$\left(\frac{\partial A_1}{\partial z}\mathbf{n} \cdot \mathbf{j} - \frac{\partial A_1}{\partial y}\mathbf{n} \cdot \mathbf{k}\right) dS = \left(\frac{\partial A_1}{\partial z}\frac{\partial z}{\partial y}\mathbf{n} \cdot \mathbf{k} - \frac{\partial A_1}{\partial y}\mathbf{n} \cdot \mathbf{k}\right) dS$$

or

$$[\nabla \times (A_1\mathbf{i})] \cdot \mathbf{n}\, dS = -\left(\frac{\partial A_1}{\partial z} + \frac{\partial A_1}{\partial y}\frac{\partial z}{\partial y}\right)\mathbf{n} \cdot \mathbf{k}\, dS \tag{2}$$

Now on S, $A_1(x, y, z) = A_1[x, y, f(x, y)] = F(x, y)$; hence, $\frac{\partial A_1}{\partial y} + \frac{\partial A_1}{\partial z}\frac{\partial z}{\partial y} = \frac{\partial F}{\partial y}$ and Equation (2) becomes

$$[\nabla \times (A_1\mathbf{i})] \cdot \mathbf{n}\, dS = -\frac{\partial F}{\partial y}\mathbf{n} \cdot \mathbf{k}\, dS = -\frac{\partial F}{\partial y}dx\, dy$$

Then

$$\iint_S [\nabla \times (A_1\mathbf{i})] \cdot \mathbf{n}\, dS = \iint_{\Re} -\frac{\partial F}{\partial y} dx\, dy$$

where $\Re$ is the projection of S on the xy plane. By Green's theorem for the plane, the last integral equals $\oint_\Gamma F\, dx$ where Γ is the boundary of $\Re$. Since at each point (x, y) of Γ the value of F is the same as the value of A_1 at each point (x, y, z) of C, and since dx is the same for both curves, we must have

$$\oint_\Gamma F\, dx = \oint_C A_1\, dx$$

or

$$\iint_S [\nabla \times (A_1\mathbf{i})] \cdot \mathbf{n}\, dS = \oint_C A_1\, dx$$

Similarly, by projections on the other coordinate planes,

$$\iint_S [\nabla \times (\mathbf{A_2 j})] \cdot \mathbf{n}\, dS = \oint_C A_2\, dy, \quad \iint_S [\nabla \times (A_3\mathbf{k})] \cdot \mathbf{n}\, dS = \oint_C A_3\, dz$$

Thus, by addition,

$$\iint_S (\nabla \times \mathbf{A}) \cdot \mathbf{n}\, dS = \oint_C \mathbf{A} \cdot d\mathbf{r}$$

The theorem is also valid for surfaces S which may not satisfy these imposed restrictions. Assume that S can be subdivided into surfaces $S_1, S_2, \ldots S_k$ with boundaries $C_1, C_2, \ldots, C_k$, which do satisfy the restrictions. Then Stokes's theorem holds for each such surface. Adding these surface integrals, the total surface integral over S is obtained. Adding the corresponding line integrals over $C_1, C_2 \ldots, C_k$, the line integral over C is obtained.

10.27. Verify Stokes's theorem for $\mathbf{A} = 3y\,\mathbf{i} - xz\mathbf{j} + yz^2\mathbf{k}$, where S is the surface of the paraboloid $2z = x^2 + y^2$ bounded by $z = 2$ and C is its boundary. See Figure 10.21.

The boundary C of S is a circle with equations $x^2 + y^2 = 4$, $z = 2$ and parametric equations $x = 2\cos t$, $y = 2\sin t$, $z = 2$, where $0 \leqq t < 2\pi$. Then

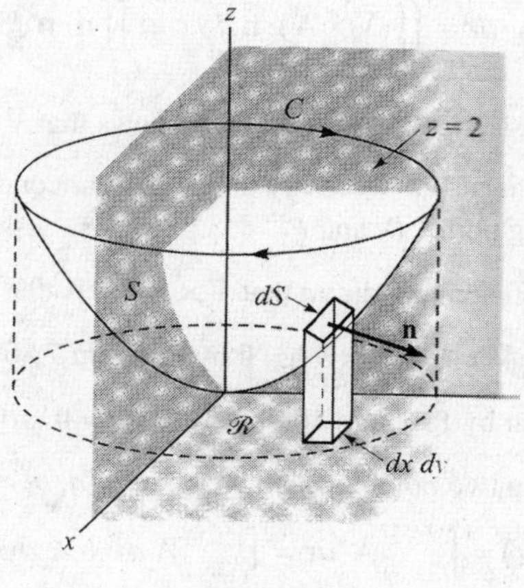

Figure 10.21

$$\oint_C \mathbf{A} \cdot d\mathbf{r} = \oint_C 3y\, dx - xz\, dy + yz^2 dz$$
$$= \int_{2\pi}^{0} 3(2\sin t)(-2\sin t)dt - (2\cos t)(2)(2\cos t)dt$$
$$= \int_0^{2\pi} (12\sin^2 t + 8\cos^2 t)dt = 20\pi$$

Also,

$$\nabla \times \mathbf{A} = \begin{vmatrix} \mathbf{i} & \mathbf{j} & \mathbf{k} \\ \dfrac{\partial}{\partial x} & \dfrac{\partial}{\partial y} & \dfrac{\partial}{\partial z} \\ 3y & -xz & yz^2 \end{vmatrix} = (z^2 + x)\mathbf{i} - (z+3)\mathbf{k}$$

and

$$\mathbf{n} = \frac{\nabla(x^2 + y^2 - 2z)}{|\nabla(x^2 + y^2 - 2z)|} = \frac{x\mathbf{i} + y\mathbf{j} - \mathbf{k}}{\sqrt{x^2 + y^2 + 1}}.$$

Then

$$\iint_S (\nabla \times \mathbf{A}) \cdot \mathbf{n}\, dS = \iint_{\mathfrak{R}} (\nabla \cdot A) \cdot \mathbf{n} \frac{dx\, dy}{|\mathbf{n} \cdot \mathbf{k}|} = \iint_{\mathfrak{R}} (xz^2 + x^2 + z + 3)\, dx\, d$$

$$= \iint_{\mathfrak{R}} \left\{ x\left(\frac{x^2 + y^2}{2}\right)^2 + x^2 + \frac{x^2 + y^2}{2} + 3 \right\} dx\, dy$$

In polar coordinates this becomes

$$\int_{\phi=0}^{2\pi} \int_{\rho=0}^{2} \{(\rho \cos\phi)(\rho^4/2) + \rho^2 \cos^2\phi + \rho^2/2 + 3\}\rho\, d\rho\, d\phi = 20\pi$$

10.28. Prove that a necessary and sufficient condition that $\oint_C \mathbf{A} \cdot d\mathbf{r} = 0$ for every closed curve C is that $\nabla \times \mathbf{A} = 0$ identically.

Sufficiency. Suppose $\nabla \times \mathbf{A} = \mathbf{0}$. Then, by Stokes's theorem,

$$\oint_C \mathbf{A} \cdot d\mathbf{r} = \iint_S (\nabla \times \mathbf{A}) \cdot \mathbf{n}\, dS = 0$$

Necessity. Suppose $\oint_C \mathbf{A} \cdot d\mathbf{r} = 0$ around every closed path C, and assume $\nabla \times \mathbf{A} \neq \mathbf{0}$ at some point P. Then, assuming $\nabla \times \mathbf{A}$ is continuous, there will be a region with P as an interior point, where $\nabla \times \mathbf{A} \neq \mathbf{0}$. Let S be a surface contained in this region whose normal $\mathbf{n}$ at each point has the same direction as $\nabla \times \mathbf{A}$; i.e., $\nabla \times \mathbf{A} = \alpha\mathbf{n}$ where α is a positive constant. Let C be the boundary of S. Then, by Stokes's theorem,

$$\oint_C \mathbf{A} \cdot d\mathbf{r} = \iint_S (\nabla \times \mathbf{A}) \cdot \mathbf{n}\, dS = \alpha \iint_S \mathbf{n} \cdot \mathbf{n}\, dS > 0$$

which contradicts the hypothesis that $\oint_C \mathbf{A} \cdot d\mathbf{r} = 0$ and shows that $\nabla \times \mathbf{A} = 0$.

It follows that $\nabla \times \mathbf{A} = \mathbf{0}$ is also a necessary and sufficient condition for a line integral $\int_{P_1}^{P_2} \mathbf{A} \cdot d\mathbf{r}$ to be independent of the path joining points P_1 and P_2.

10.29. Prove that a necessary and sufficient condition that $\nabla \times \mathbf{A} = \mathbf{0}$ is that $\mathbf{A} = \nabla\phi$.

Sufficiency. If $\mathbf{A} = \nabla\phi$, then $\nabla \times \mathbf{A} = \nabla \times \nabla\phi = \mathbf{0}$ by Problem 7.80.

Necessity. If $\nabla \times \mathbf{A} = \mathbf{0}$, then by Problem 10.28, $\oint_C \mathbf{A} \cdot d\mathbf{r} = \mathbf{0}$ around every closed path and $\int_{P_1}^{P_2} \mathbf{A} \cdot d\mathbf{r}$ is independent of the path joining two points, which we take as (a, b, c) and (x, y, z). Let us define

$$\phi(x, y, z) = \int_{(a,b,c)}^{(x,y,z)} \mathbf{A} \cdot d\mathbf{r} = \int_{(a,b,c)}^{(x,y,z)} A_1 dx + A_2 dy + A_3 dz$$

Then

$$\phi(x + \Delta x, y, z) - \phi(x, y, z) = \int_{(x,y,z)}^{(x+\Delta x,y,z)} A_1 dx + A_2 dy + A_3 dz$$

Since the last integral is independent of the path joining (x, y, z) and $(x + \Delta x, y, z)$, we can choose the path to be a straight line joining these points so that dy and dz are zero. Then

$$\frac{\phi\,(x + \Delta x, y, z) - \phi(x, y, z)}{\Delta x} = \frac{1}{\Delta x} \int_{(x,y,z)}^{(x+\Delta x,y,z)} A_1 dx = A_1(x + \theta\Delta x, y, z) \quad 0 < \theta < 1$$

where we have applied the law of the mean for integrals.

Taking the limit of both sides as $\Delta x \to 0$ gives $\partial\phi/\partial x = A_1$.

Similarly, we can show that $\partial\phi/\partial y = A_2$, $\partial\phi/\partial z = A_3$. Thus,

$$\mathbf{A} = A_1\mathbf{i} + A_2\mathbf{j} + A_3\mathbf{k} = \frac{\partial\phi}{\partial x}\mathbf{i} + \frac{\partial\phi}{\partial y}\mathbf{j} + \frac{\partial\phi}{\partial z}\mathbf{k} = \nabla\phi.$$

10.30. (a) Prove that a necessary and sufficient condition that $A_1\,dx + A_2\,dy + A_3\,dz = d\phi$, an exact differential, is that $\nabla \times \mathbf{A} = \mathbf{0}$ where $\mathbf{A} = A_1\mathbf{i} + A_2\mathbf{j} + A_3\mathbf{k}$. (b) Show that in such case,

$$\int_{(x_1,y_1,z_1)}^{(x_2,y_2,z_2)} A_1dx + A_2dy + A_3dz = \int_{(x_1,y_1,z_1)}^{(x_2,y_2,z_2)} d\phi = \phi(x_2,y_2,z_2) - \phi(x_1,y_1,z_1)$$

(a) **Necessity.** If $A_1dx + A_2dy + A_3dz = d\phi = \frac{\partial\phi}{\partial x}dx + \frac{\partial\phi}{\partial y}dy + \frac{\partial\phi}{\partial z}dz$, then

$$\frac{\partial\phi}{\partial x} = A_1 \quad (1)$$

$$\frac{\partial\phi}{\partial y} = A_2 \quad (2)$$

$$\frac{\partial\phi}{\partial z} = A_3 \quad (3)$$

Then, by differentiating, and assuming continuity of the partial derivatives, we have

$$\frac{\partial A_1}{\partial y} = \frac{\partial A_2}{\partial x}, \quad \frac{\partial A_2}{\partial z} = \frac{\partial A_3}{\partial y}, \quad \frac{\partial A_1}{\partial z} = \frac{\partial A_3}{\partial x}$$

which is precisely the condition $\nabla \times \mathbf{A} = \mathbf{0}$.

Another method: If $A_1\,dx + A_2\,dy + A_3\,dz = d\phi$, then

$$\mathbf{A} = A_1\mathbf{i} + A_2\mathbf{j} + A_3\mathbf{k} = \frac{\partial\phi}{\partial x}\mathbf{i} + \frac{\partial\phi}{\partial y}\mathbf{j} + \frac{\partial\phi}{\partial z}\mathbf{k} = \nabla\phi.$$

from which $\nabla \times \mathbf{A} = \nabla \times \nabla\phi = \mathbf{0}$.

Sufficiency. If $\nabla \times \mathbf{A} = \mathbf{0}$, then by Problem 10.29, $\mathbf{A} = \nabla\phi$ and

$$A_1dx + A_2dy + A_3dz = \mathbf{A}\cdot d\mathbf{r} = \nabla\phi\cdot d\mathbf{r} = \frac{\partial\phi}{\partial x}dx + \frac{\partial\phi}{\partial y}dy + \frac{\partial\phi}{\partial z}dz = d\phi$$

(b) From (a),

$$\phi(x,y,z) = \int_{(a,b,c)}^{(x,y,z)} A_1dx + A_2dy + A_3dz$$

Then, omitting the integrand $A_1\,dx + A_2\,dy + A_3\,dz$, we have

$$\int_{(x_1,y_1,z_1)}^{(x_2,y_2,z_2)} = \int_{(a,b,c)}^{(x_2,y_2,z_2)} - \int_{(a,b,c)}^{(x_1,y_1,z_1)} = \phi(x_2,y_2,z_2) - \phi(x_1,y_1,z_1)$$

10.31. (a) Prove that $\mathbf{F} = (2xz^3 + 6y)\mathbf{i} + (6x - 2yz)\mathbf{j} + (3x^2z^2 - y^2)\mathbf{k}$ is a conservative force field. (b) Evaluate $\int_C \mathbf{F}\cdot d\mathbf{r}$ where C is any path from $(1, -1, 1)$ to $(2, 1, -1)$. (c) Give a physical interpretation of the results.

(a) A force field $\mathbf{F}$ is conservative if the line integral $\int_C \mathbf{F}\cdot d\mathbf{r}$ is independent of the path C joining any two points. A necessary and sufficient condition that F be conservative is that $\nabla \times \mathbf{F} = \mathbf{0}$.

Since here $\nabla\times\mathbf{F} = \begin{vmatrix} \mathbf{i} & \mathbf{j} & \mathbf{k} \\ \frac{\partial}{\partial x} & \frac{\partial}{\partial y} & \frac{\partial}{\partial z} \\ 2xz^3 + 6y & 6x - 2yz & 3x^2z^2 - y^2 \end{vmatrix} = \mathbf{0}$, **F is conservative**

(b) **Method 1:** By Problem 10.30, $\mathbf{F}\cdot d\mathbf{r} = (2xz^3 + 6y)dx + (6x - 2yz)dy + (3x^2z^2 - y^2)dz$ is an exact differential $d\phi$, where ϕ is such that

$$\frac{\partial\phi}{\partial x} = 2xz^3 + 6y \tag{1}$$

$$\frac{\partial\phi}{\partial y} = 6x - 2yz \tag{2}$$

$$\frac{\partial\phi}{\partial z} = 3x^2z^2 - y^2 \tag{3}$$

From these we obtain, respectively,

$$\phi = x^2z^3 + 6xy + f_1(y, z)$$

$$\phi = 6xy - y^2z + f_2(x, z)$$

$$\phi = x^2y^2 - y^2z + f_3(x, y)$$

These are consistent if $f_1(y, z) = -y^2z + c$, $f_2(x, z) = x^2z^3 + c$, and $f_3(x, y) = 6xy + c$, in which case $\phi = x^2z^3 + 6xy - y^2z + c$. Thus, by Problem 10.30,

$$\int_{(1,-1,1)}^{(2,1,-1)} \mathbf{F}\cdot d\mathbf{r} = x^2z^3 + 6xy - y^2z + c\,\Big|_{(1,-1,1)}^{(2,1,-1)} = 15$$

Alternatively, we may notice by inspection that

$$\mathbf{F}\cdot d\mathbf{r} = (2xz^3\,dx + 3x^2z^2\,dz) + (6y\,dx + 6x\,dy) - (2yz\,dy + y^2\,dz)$$

$$= d(x^2z^3) + d(6xy) - d(y^2z) = d(x^2z^3 + 6xy - y^2z + c) \text{ from which } \phi \text{ is determined.}$$

Method 2: Since the integral is independent of the path, we can choose any path to evaluate it; in particular, we can choose the path consisting of straight lines from (1, –1, 1) to (2, –1, 1), then to (2, 1, 1) and then to (2, 1, –1). The result is

$$\int_{x=1}^{2} (2x - 6)\,dx + \int_{y=-1}^{1} (12 - 2y)dy + \int_{z=1}^{-1} (12z^2 - 1)dz = 15$$

where the first integral is obtained from the line integral by placing $y = -1$, $z = 1$, $dy = 0$, $dz = 0$; the second integral, by placing $x = 2$, $z = 1$, $dx = 0$, $dz = 0$; and the third integral, by placing $x = 2$, $y = 1$, $dx = 0$, $dy = 0$.

(c) Physically, $\int_C \mathbf{F}\cdot d\mathbf{r}$ represents the work done in moving an object from (1, –1, 1) to (2, 1, –1) along C. In a conservative force field, the work done is independent of the path C joining these points.

Miscellaneous problems

10.32. (a) If $x = f(u, \upsilon)$, $y = g(u, \upsilon)$ defines a transformation which maps a region $\Re$ of the xy plane into a region $\Re'$ of the $u\upsilon$ plane, prove that

$$\iint_{\Re} dx\,dy = \iint_{\Re'} \left|\frac{\partial(x, y)}{\partial(u,\upsilon)}\right| du\,d\upsilon$$

(b) Interpret geometrically the result in (a).

(a) If C (assumed to be a simple closed curve) is the boundary of $\Re$, then by Problem 10.8,

$$\iint_{\Re} dx\,dy = \frac{1}{2}\oint_C x\,dy - y\,dx \tag{1}$$

Under the given transformation, the integral on the right of Equation (1) becomes

$$\frac{1}{2}\oint_{C'} x\left(\frac{\partial y}{\partial u}du + \frac{\partial y}{\partial \upsilon}d\upsilon\right) - y\left(\frac{\partial y}{\partial u}du + \frac{\partial x}{\partial \upsilon}d\upsilon\right) = \frac{1}{2}\oint_{C'}\left(x\frac{\partial y}{\partial u} - y\frac{\partial x}{\partial u}\right)du + \left(x\frac{\partial y}{\partial \upsilon} - y\frac{\partial x}{\partial \upsilon}\right)d\upsilon \tag{2}$$

where C' is the mapping of C in the $u\upsilon$ plane (we suppose the mapping to be such that C' is a simple closed curve also).

By Green's theorem, if $\Re'$ is the region in the $u\upsilon$ plane bounded by C', the right side of Equation (2) equals

$$\frac{1}{2}\iint_{\Re'}\left|\frac{\partial}{\partial u}\left(x\frac{\partial y}{\partial \upsilon}-y\frac{\partial x}{\partial \upsilon}\right)-\frac{\partial}{\partial \upsilon}\left(x\frac{\partial y}{\partial u}-y\frac{\partial x}{\partial u}\right)\right| du\,d\upsilon=\iint_{\Re'}\left|\frac{\partial x}{\partial u}\frac{\partial y}{\partial \upsilon}-\frac{\partial x}{\partial \upsilon}\frac{\partial y}{\partial u}\right| du\,d\upsilon$$

$$=\iint_{\Re'}\left|\frac{\partial(x,y)}{\partial(u,\upsilon)}\right| du\,d\upsilon$$

where we have inserted absolute value signs so as to ensure that the result is nonnegative, as is $\int\int_{\Re} dx\,dy$.

In general, we can show (see Problem 10.83) that

$$\iint_{\Re} F(x,y)\,dx\,dy=\iint_{\Re'} F\{f(u,\upsilon),g(u,\upsilon)\}\left|\frac{\partial(x,y)}{\partial(u,\upsilon)}\right| du\,d\upsilon \tag{3}$$

(b) $\iint_{\Re} dx\,dy$ and $\iint_{\Re'}\left|\frac{\partial(x,y)}{\partial(u,\upsilon)}\right| du\,d\upsilon$ represent the area of region R, the first expressed in rectangular coordinates, the second in curvilinear coordinates. See Page 225, and the introduction of the differential element of surface area for an alternative to (a).

10.33. Let $\mathbf{F}=\dfrac{-y\mathbf{i}+x\mathbf{j}}{x^2+y^2}$. (a) Calculate $\nabla\times\mathbf{F}$. (b) Evaluate $\oint\mathbf{F}\cdot d\mathbf{r}$ around any closed path and explain the results.

(a) $$\nabla\times\mathbf{F}=\begin{vmatrix}\mathbf{i} & \mathbf{j} & \mathbf{k}\\ \dfrac{\partial}{\partial x} & \dfrac{\partial}{\partial y} & \dfrac{\partial}{\partial z}\\ \dfrac{-y}{x^2+y^2} & \dfrac{x}{x^2+y^2} & 0\end{vmatrix}=\mathbf{0}\text{ in any region excluding }(0,0).$$

(b) $\oint\mathbf{F}\cdot d\mathbf{r}=\oint_C\dfrac{-y\,dx=x\,dy}{x^2+y^2}$. Let x = ρ cos ϕ, y = ρ sin ϕ, where (ρ, ϕ) are polar coordinates. Then

$$dx=-\rho\sin\phi\,d\phi+d\rho\cos\phi,\quad dy=\rho\cos\phi\,d\phi+d\rho\sin\phi$$

and so

$$\frac{-y\,dx=x\,dy}{x^2+y^2}=d\phi=d\left(\text{are tan}\frac{y}{x}\right)$$

For a closed curve *ABCDA* [see Figure 10.22 (*a*)] surrounding the origin, ϕ = 0 at *A* and ϕ = 2π after a complete circuit back to *A*. In this case the line integral equals $\int_0^{2\pi} d\phi=2\pi$.

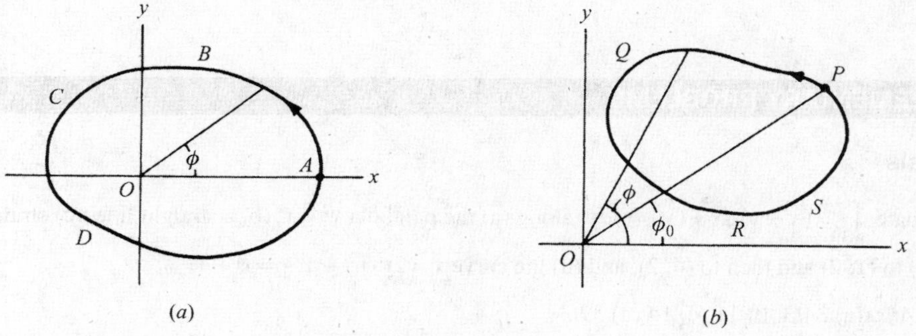

Figure 10.22

For a closed curve *PQRSP* [see Figure 10.22(*b*)] not surrounding the origin, $\phi = \phi_0$ at *P* and $\phi = \phi_0$ after a complete circuit back to *P*. In this case the line integral equals $\int_{\phi_0}^{\phi_0} d\phi = 0$.

Since $\mathbf{F} = P\mathbf{i} + Q\mathbf{j}$, $\nabla \times \mathbf{F} = \mathbf{0}$ is equivalent to $\partial P/\partial y = \partial Q/\partial x$, the results would seem to contradict those of Problem 10.11. However, no contradiction exists, since $P = \dfrac{-y}{x^2 + y^2}$ and $Q = \dfrac{x}{x^2 + y^2}$ do not have continuous derivatives throughout any region including (0, 0), and this was assumed in Problem 10.11.

10.34. If div **A** denotes the divergence of a vector field **A** at a point *P*, show that

$$\operatorname{div} \mathbf{A} = \lim_{\Delta V \to 0} \frac{\iint\limits_{\Delta S} \mathbf{A} \cdot \mathbf{n} dS}{\Delta V}$$

where ΔV is the volume enclosed by the surface ΔS and the limit is obtained by shrinking ΔV to the point *P*.

By the divergence theorem,

$$\iiint\limits_{\Delta V} \operatorname{div} \mathbf{A} \, dV = \iint\limits_{\Delta S} \mathbf{A} \cdot \mathbf{n} \, dS$$

By the mean value theorem for integrals, the left side can be written

$$\overline{\operatorname{div} \mathbf{A}} \iiint\limits_{\Delta V} dV = \overline{\operatorname{div} \mathbf{A}} \, \Delta V$$

where $\overline{\operatorname{div} \mathbf{A}}$ is some value intermediate between the maximum and minimum of div **A** throughout ΔV. Then

$$\overline{\operatorname{div} \mathbf{A}} = \frac{\iint\limits_{\Delta S} \mathbf{A} \cdot \mathbf{n} dS}{\Delta V}$$

Taking the limit as $\Delta V \to 0$ such that *P* is always interior to ΔV, $\overline{\operatorname{div} \mathbf{A}}$ approaches the value div **A** at point *P*: hence,

$$\operatorname{div} \mathrm{A} = \lim_{\Delta V \to 0} \frac{\iint\limits_{\Delta S} \mathrm{A} \cdot \mathrm{n} dS}{\Delta V}$$

This result can be taken as a starting point for defining the divergence of **A**, and from it all the properties may be derived, including proof of the divergence theorem. We can also use this to extend the concept of divergence to coordinate systems other than rectangular (see Page 170).

Physically, $\left(\iiint\limits_{\Delta S} \mathrm{A} \cdot \mathrm{n} dS \right) / \Delta V$ represents the flux or net outflow per unit volume of the vector **A** from the surface ΔS. If div **A** is positive in the neighborhood of a point *P*, it means that the outflow from *P* is positive, and we call *P* a *source*. Similarly, if div **A** is negative in the neighborhood of *P*, the outflow is really an inflow, and *P* is called a *sink*. If in a region there are no sources or sinks, then div **A** = 0, and we call **A** a *solenoidal* vector field.

SUPPLEMENTARY PROBLEMS

Line Integrals

10.35. Evaluate $\int_{(1,1)}^{(4,2)} (x + y)dx + (y - x)dy$ along (a) the parabola $y^2 = x$, (b) a straight line, (c) straight lines from (1, 1) to (1, 2) and then to (4, 2), and (d) the curve $x = 2t^2 + t + 1$, $y = t^2 + 1$.

Ans. (a) 34/3 (b) 11 (c) 14 (d) 32/3

10.36. Evaluate $\oint(2x - y + 4)dx + (5y + 3x - 6)dy$ around a triangle in the xy plane with vertices at (0, 0), (3, 0), (3, 2) traversed in a counterclockwise direction.

Ans. 12

10.37. Evaluate the line integral in Problem 10.36 around a circle of radius 4 with center at (0, 0).

Ans. 64π

10.38. (a) If $\mathbf{F} = (x^2 - y^2)\mathbf{i} + 2xy\mathbf{j}$, evaluate $\int_C \mathbf{F} \cdot d\mathbf{r}$ along the curve C in the xy plane given by $y = x^2 - x$ from the point (1, 0) to (2, 2). (b) Interpret physically the result obtained.

Ans. (a) 124/15

10.39. Evaluate $\int_C (2x + y)ds$, where C is the curve in the xy plane given by $x^2 + y^2 = 25$ and s is the are length parameter, from the point (3, 4) to (4, 3) along the shortest path.

Ans. 15

10.40. If $\mathbf{F} = (3x - 2y)\mathbf{i} + (y + 2z)\mathbf{j} - x^2\mathbf{k}$, evaluate $\int_C \mathbf{F} \cdot d\mathbf{r}$ from (0, 0, 0) to (1, 1, 1), where C is a path consisting of (a) the curve $x = t$, $y = t^2$, $z = t^3$, (b) a straight line joining these points, (c) the straight lines from (0, 0, 0) to (0, 1, 0), then to (0, 1, 1) and then to (1, 1, 1), and (d) the curve $x = z^2$, $z = y^2$.

Ans. (a) 23/15 (b) 5/3 (c) 0 (d) 13/15

10.41. If $\mathbf{T}$ is the unit tangent vector to a curve C (plane or space curve) and $\mathbf{F}$ is a given force field, prove that under appropriate conditions $\int_C \mathbf{F} \cdot d\mathbf{r} = \int_C \mathbf{F} \cdot \mathbf{T}\, ds$, where s is the arc length parameter. Interpret the result physically and geometrically.

Green's theorem in the plane, independence of the path

10.42. Verify Green's theorem in the plane for $\oint_C (x^2 - xy^3)dx + (y^2 - 2xy)dy$, where C is a square with vertices at (0, 0), (2, 0), (2, 2), (0, 2) and counterclockwise orientation.

Ans. Common value = 8

10.43. Evaluate the line integrals of (a) Problem 10.36 and (b) Problem 10.37 by Green's theorem.

10.44. (a) Let C be any simple closed curve bounding a region having area A. Prove that if $a_1, a_2, a_3, b_1, b_2, b_3$ are constants, $\oint_C (a_1x + a_2y + a_3)dx + (b_1x + b_2y + b_3)dy = (b_1 - a_2)A$ (b) Under what conditions will the line integral around any path C be zero?

Ans. (b) $a_2 = b_1$

10.45. Find the area bounded by the hypocycloid $x^{2/3} + y^{2/3} = a^{2/3}$. (Hint: Parametric equations are $x = a\cos^3 t$, $y = a\sin^3 t$, $0 \leq t \leq 2\pi$.)

Ans. $3\pi a^2/8$

10.46. If $x = \rho\cos\phi$, $y = \rho\sin\phi$, prove that $\frac{1}{2}\oint x\,dy - y\,dx = \frac{1}{2}\int \rho^2 d\phi$ and interpret.

10.47. Verify Green's theorem in the plane for $\oint_C (x^3 - x^2y)dx + xy^2dy$, where C is the boundary of the region enclosed by the circles $x^2 + y^2 = 4$ and $x^2 + y^2 = 16$.

Ans. Common value = 120π

10.48. (a) Prove that $\int_{(1,0)}^{(2,1)} (2xy - y^4 + 3)\,dx + (x^2 - 4xy^3)dy$ is independent of the path joining (1, 0) and (2, 1).
(b) Evaluate the integral in (a).

Ans. (b) 5

10.49. Evaluate $\int_C (2xy^3 - y^2 \cos x)\,dx + (1 - 2y \sin x + 3x^2y^2)dy$ along the parabola $2x = \pi y^2$ from (0, 0) to ($\pi/2$, 1).

Ans. $\pi^2/4$

10.50. Evaluate the line integral in Problem 10.49 around a parallelogram with vertices at (0, 0), (3, 0), (5, 2), (2, 2).

Ans. 0

10.51. (a) Prove that $G = (2x^2 + xy - 2y^2)\,dx + (3x^2 + 2xy)\,dy$ is not an exact differential. (b) Prove that $e^{y/x}\,G/x$ is an exact differential of ϕ and find ϕ. (c) Find a solution of the differential equation $(2x^2 + xy - 2y^2)\,dx + (3x^2 + 2xy)\,dy = 0$.

Ans. (b) $\phi = e^{y/x}(x^2 + 2xy) + c$ (c) $x^2 + 2xy + ce^{-y/x} = 0$

Surface integrals

10.52. (a) Evaluate $\iint_S (x^2 + y^2)\,dS$, where S is the surface of the cone $z^2 = 3(x^2 + y^2)$ bounded by $z = 0$ and $z = 3$.
(b) Interpret physically the result in (a).

Ans. (a) 9π

10.53. Determine the surface area of the plane $2x + y + 2z = 16$ cut off by (a) $x = 0, y = 0, x = 2, y = 3$ and (b) $x = 0, y = 0$, and $x^2 + y^2 = 64$.

Ans. (a) 9 (b) 24π

10.54. Find the surface of the paraboloid $2z = x^2 + y^2$ which is outside the cone $z = \sqrt{x^2 + y^2}$.

Ans. $\dfrac{2}{3}\pi(5\sqrt{5} - 1)$

10.55. Find the area of the surface of the cone $z^2 = 3(x^2 + y^2)$ cut out by the paraboloid $z = x^2 + y\,y^2$.

Ans. 6π

10.56. Find the surface area of the region common to the intersecting cylinders $x^2 + y^2 = a^2$ and $x^2 + z^2 = a^2$.

Ans. $16a^2$

10.57. (a) Obtain the surface area of the sphere $x^2 + y^2 + z^2 = a^2$ contained within the cone $z \tan \alpha = \sqrt{x^2 + y^2}$, $0 < \alpha < \pi/2$. (b) Use the result in (a) to find the surface area of a hemisphere. (c) Explain why formally placing $\alpha = \pi$ in the result of (a) yields the total surface area of a sphere.

Ans. (a) $2\pi a^2(1 - \cos \alpha)$ (b) $2\pi a^2$ (consider the limit as $\alpha \to \pi/2$)

10.58. Determine the moment of inertia of the surface of a sphere of radius a about a point on the surface. Assume a constant density σ.

Ans. $2Ma^2$, where mass $M = 4\pi a^2\,\sigma$

10.59. (a) Find the centroid of the surface of the sphere $x^2 + y^2 + z^2 = a^2$ contained within the cone $z \tan \alpha = \sqrt{x^2 + y^2}$, $0 < \alpha < \pi/2$. (b) From the result in (a) obtain the centroid of the surface of a hemisphere.

Ans. (a) $1/2a(1 + \cos\alpha)$ (b) $a/2$

The divergence theorem

10.60. Verify the divergence theorem for $\mathbf{A} = (2xy + z)\mathbf{i} + y^2\mathbf{j} - (x + 3y)\mathbf{k}$ taken over the region bounded by $2x + 2y + z = 6$, $x = 0$, $y = 0$, $z = 0$.

Ans. common value = 27

10.61. Evaluate $\iint_S \mathrm{F} \cdot n\, dS$. where $\mathbf{F} = (z^2 - x)\mathbf{i} - xy\mathbf{j} + 3z\mathbf{k}$ and S is the surface of the region bounded by $z = 4 - y^2$, $x = 0$, $x = 3$ and the xy plane.

Ans. 16

10.62. Evaluate $\iint_S \mathrm{A} \cdot n\, dS$. where $\mathbf{A} = (2x + 3z)\mathbf{i} - (xz + y)\mathbf{j} + (y^2 + 2z)\mathbf{k}$ and S is the surface of the sphere having center at $(3, -1, 2)$ and radius 3.

Ans. 108π

10.63. Determine the value of $\iint_S x\, dy\, dz + y\, dz\, dx + z\, dx\, dy$, where S is the surface of the region bounded by the cylinder $x^2 + y^2 = 9$ and the planes $z = 0$ and $z = 3$, (a) by using the divergence theorem and (b) directly.

Ans. 81π

10.64. Evaluate $\iint_S 4xz\, dy\, dz - y^2 dz\, dx + yz\, dx\, dy$, where S is the surface of the cube bounded by $x = 0$, $y = 0$, $z = 0$, $x = 1$, $y = 1$, $z = 1$, (a) directly and (b) By Green's theorem in space (divergence theorem).

Ans. 3/2

10.65. Prove that $\iint_S (\nabla \times \mathbf{A}) \cdot \mathbf{n}\, dS$ for any closed surface S.

10.66. Prove that $\iint_S \mathbf{n}\, dS = 0$. where n is the outward drawn normal to any closed surface S. (Hint: Let $\mathbf{A} = \Theta\mathbf{c}$, where $\mathbf{c}$ is an arbitrary vector constant.) Express the divergence theorem in this special case. Use the arbitrary property of $\mathbf{c}$.

10.67. If $\mathbf{n}$ is the unit outward drawn normal to any closed surface S bounding the region V, prove that

$$\iiint_V \operatorname{div} \mathbf{n}\, dV = S$$

Stokes's theorem

10.68. Verify Stokes's theorem for $\mathbf{A} = 2y\mathbf{i} + 3x\mathbf{j} - z^2\mathbf{k}$, where S is the upper half surface of the sphere $x^2 + y^2 + z^2 = 9$ and C is its boundary.

Ans. Common value = 9π

10.69. Verify Stokes's theorem for $\mathbf{A} = (y + z)\mathbf{i} - xz\mathbf{j} + y^2\mathbf{k}$, where S is the surface of the region in the first octant bounded by $2x + z = 6$ and $y = 2$ which is not included in the (a) xy plane, (b) plane $y = 2$, and (c) plane $2x + z = 6$ and C is the corresponding boundary.

Ans. The common value is (a) –6, (b) –9, and (c) –18.

10.70. Evaluate $\iint_S (\nabla \times \mathbf{A}) \cdot \mathbf{n}\, dS$. where $\mathbf{A} = (x - z)\mathbf{i} + (x^3 + yz)\mathbf{j} - 3xy^2\mathbf{k}$ and S is the surface of the cone $z = 2 - \sqrt{x^2 + y^2}$ above the xy plane.

Ans. 12π

10.71. If V is a region bounded by a closed surface S and $\mathbf{B} = \nabla \times \mathbf{A}$, prove that $\iint_S \mathbf{B} \cdot \mathbf{n}\, dS = 0$

10.72. (a) Prove that $\mathbf{F} = (2xy + 3)\mathbf{i} + (x^2 - 4z)\mathbf{j} - 4y\mathbf{k}$ is a conservative force field. (b) Find ϕ such that $\mathbf{F} = \nabla\phi$. (c) Evaluate $\int_C \mathbf{F} \cdot d\mathbf{r}$, where C is any path from (3, –1, 2) to (2, 1, –1).

Ans. (b) $\phi = x^2y - 4yz + 3x$ + constant (c) 6

10.73. Let C be any path joining any point on the sphere $x^2 + y^2 + z^2 = a^2$ to any point on the sphere $x^2 + y^2 + z^2 = b^2$. Show that if $\mathbf{F} = 5r^3\mathbf{r}$, where $\mathbf{r} = x\mathbf{i} + y\mathbf{j} + z\mathbf{k}$, then $\int_C \mathbf{F} \cdot d\mathbf{r} = b^5 - a^5$.

10.74. In Problem 10.73 evaluate $\int_C \mathbf{F} \cdot d\mathbf{r}$ is $\mathbf{F} = f(r)\mathbf{r}$, where $f(r)$ is assumed to be continuous.

Ans. $\int_a^b r f(r)\,dr$

10.75. Determine whether there is a function ϕ such that $\mathbf{F} = \nabla\phi$, where (a) $\mathbf{F} = (xz - y)\mathbf{i} + (x^2y + z^3)\mathbf{j} + (3xz^2 - xy)\mathbf{k}$, and (b) $\mathbf{F} = 2xe^{-y}\mathbf{i} + (\cos z - x^2 e^{-y})\mathbf{j} - y \sin z\mathbf{k}$. If so, find it.

Ans. (a) ϕ does not exist (b) $\phi = x^2e^{-y} + y \cos z$ + constant

10.76. Solve the differential equation $(z^3 - 4xy)\, dx + (6y - 2x^2)\, dy + (3xz^2 + 1)\, dz = 0$.

Ans. $xz^3 - 2x^2y + 3y^2 + z$ = constant

Miscellaneous problems

10.77. Prove that a necessary and sufficient condition that $\oint_C \frac{\partial U}{\partial x} dy - \frac{\partial U}{\partial y} dx$ be zero around every simple closed path C in a region $\Re$ (where U is continuous and has continuous partial derivatives of order two, at least) is that $\frac{\partial^2 U}{\partial x^2} + \frac{\partial^2 U}{\partial y^2} = 0$.

10.78. Verify Green's theorem for a multiply connected region containing two "holes" (see Problem 10.10).

10.79. If $P\,dx + Q\,dy$ is not an exact differential but $\mu(P\,dx + Q\,dy)$ is an exact differential where μ is some function of x and y, then μ is called an *integrating factor*. (a) Prove that if F and G are functions of x alone then $(Fy + G)\,dx + dy$ has an integrating factor μ which is a function of x alone, and find μ. What must be assumed about F and G? (b) Use (a) to find solutions of the differential equation $xy' = 2x + 3y$.

Ans. (a) $\mu = e^{\int F(x)\,dx}$ (b) $y = cx^3 - x$, where c is any constant

10.80. Find the surface area of the sphere $x^2 + y^2 + (z - a)^2 = a^2$ contained within the paraboloid $z = x^2 + y^2$.

Ans. $2\pi a$

10.81. If $f(r)$ is a continuously differentiable function of $r = \sqrt{x^2 + y^2 + z^2}$. prove that

$$\iint_S f(r)\mathbf{n}dS = \iiint_V \frac{f'(r)}{r}\mathbf{r}dV$$

10.82. Prove that $\iint \nabla \times (\phi \mathbf{n}) dS = \mathbf{0}$, where ϕ is any continuously differentiable scalar function of position and $\mathbf{n}$ is a unit outward drawn normal to a closed surface S. (See Problem 10.66.)

10.83. Establish Equation (3). Problem 10.32, by using Green's theorem in the plane. [Hint: Let the closed region $\Re$ in the xy plane have boundary C and suppose that under the transformation $x = f(u, \upsilon)$, $y = g(u, \upsilon)$, these are transformed into $\Re'$ and C' in the $u\upsilon$ plane, respectively.] First prove that $\iint_\Re F(x, y)dxdy = \int_C Q(x, y)dy$

where $\partial Q / \partial y = F(x, y)$. Then show that, apart from sign, this last integral is equal to

$\int_C Q[f(u,\upsilon), g(u,\upsilon)]\left|\frac{\partial g}{\partial u}du + \frac{\partial g}{\partial \upsilon}d\upsilon\right|$. Finally, use Green's theorem to transform this into

$\iint_{\Re'} F[f(u,\upsilon), g(u,\upsilon)]\left|\frac{\partial(x, y)}{\partial(u,\upsilon)}\right| du d\upsilon.$

10.84. If $x = f(u, \upsilon, w)$, $y = g(u, \upsilon, w)$, $z = h(u, \upsilon, w)$ defines a transformation which maps a region $\Re$ of xyz space into a region $\Re'$ of $u\upsilon w$ space, prove, using Stokes's theorem, that

$$\iiint_\Re F(x, y, z)dxdydz = \iiint_{\Re'} G(u,\upsilon, w)\left|\frac{\partial(x, y, z)}{\partial(u,\upsilon, w)}\right| du d\upsilon dw$$

where $G(u, \upsilon, w) \equiv F[f(u, \upsilon, w), g(u, \upsilon, w), h(u, \upsilon, w)]$. State sufficient conditions under which the result is valid. See Problem 10.83. Alternatively, employ the differential element of volume $dV = \frac{\partial \mathbf{r}}{\partial u} \cdot \frac{\partial \mathbf{r}}{\partial \upsilon} \times \frac{\partial \mathbf{r}}{\partial w} du\, d\upsilon\, dw$ (recall the geometric meaning).

10.85. (a) Show that, in general, the equation $\mathbf{r} = \mathbf{r}(u, \upsilon)$ geometrically represents a surface. (b) Discuss the geometric significance of $u = c_1$, $\upsilon = c_2$, where c_1 and c_2 are constants. (c) Prove that the element of arc length on this surface is given by

$$ds^2 = E\, du^2 + 2F\, du\, d\upsilon + G\, d\upsilon^2$$

where $E = \frac{\partial \mathbf{r}}{\partial u} \cdot \frac{\partial \mathbf{r}}{\partial u}$, $F = \frac{\partial \mathbf{r}}{\partial u} \cdot \frac{\partial \mathbf{r}}{\partial \upsilon}$, and $G = \frac{\partial \mathbf{r}}{\partial \upsilon} \cdot \frac{\partial \mathbf{r}}{\partial \upsilon}$.

10.86. (a) Referring to Problem 10.85, show that the element of surface area is given by $dS = \sqrt{EG - F^2}\, du\, d\upsilon$. (b) Deduce from (a) that the area of a surface r = r(u, υ) is $\iint_S \sqrt{EG - F^2}\, du d\upsilon$. [Hint: Use the fact that

$\left|\frac{\partial \mathrm{r}}{\partial u} \times \frac{\partial \mathrm{r}}{\partial \upsilon}\right| = \sqrt{\left(\frac{\partial \mathrm{r}}{\partial u} \times \frac{\partial \mathrm{r}}{\partial \upsilon}\right) \cdot \left(\frac{\partial \mathrm{r}}{\partial u}\right) \times \left(\frac{\partial \mathrm{r}}{\partial \upsilon}\right)}$ and then use the identity $(\mathbf{A} \times \mathbf{B}) \cdot (\mathbf{C} \times \mathbf{D}) = (\mathbf{A} \cdot \mathbf{C})(\mathbf{B} \cdot \mathbf{D}) - (\mathbf{A} \cdot \mathbf{D})(\mathbf{B} \cdot \mathbf{C})$.]

10.87. (a) Prove that $\mathbf{r} = a \sin u \cos \upsilon \mathbf{i} + a \sin u \sin \upsilon \mathbf{j} + a \cos u$, $0 \leqq u \leqq \pi$, $0 \leqq \upsilon < 2\pi$ represents a sphere of radius a. (b) Use Problem 10.86 to show that the surface area of this sphere is $4\pi a^2$.

10.88. Use the result of Problem 10.34 to obtain div $\mathbf{A}$ in (a) cylindrical and (b) spherical coordinates. See Page 173.

CHAPTER 11

Infinite Series

The early developers of the calculus, including Newton and Leibniz, were well aware of the importance of infinite series. The values of many functions such as sine and cosine were geometrically obtainable only in special cases. Infinite series provided a way of developing extensive tables of values for them.

This chapter begins with a statement of what is meant by infinite series, then the question of when these sums can be assigned values is addressed. Much information can be obtained by exploring infinite sums of constant terms; however, the eventual objective in analysis is to introduce series that depend on variables. This presents the possibility of representing functions by series. Afterward, the question of how continuity, differentiability, and integrability play a role can be examined.

The question of dividing a line segment into infinitesimal parts has stimulated the imaginations of philosophers for a very long time. In a corruption of a paradox introduced by Zeno of Elea (in the fifth century B.C.) a dimensionless frog sits on the end of a one-dimensional log of unit length. The frog jumps halfway, and then halfway and halfway ad infinitum. The question is whether the frog ever reaches the other end. Mathematically, an unending sum,

$$\frac{1}{2}+\frac{1}{4}+\cdots+\frac{1}{2^n}+\cdots$$

is suggested. Common sense tells us that the sum must approach 1 even though that value is never attained. We can form sequences of partial sums

$$S_1=\frac{1}{2},\ S_2=\frac{1}{2}+\frac{1}{4},\ \ldots,\ S_n=\frac{1}{2}+\frac{1}{4}+\cdots+\frac{1}{2^n}S_{n+1}\cdots$$

and then examine the limit. This returns us to Chapter 2 and the modern manner of thinking about the infinitesimal.

In this chapter, consideration of such sums launches us on the road to the theory of infinite series.

Definitions of Infinite Series and Their Convergence and Divergence

Definition The sum

$$S=\sum_{n=1}^{\infty}u_n=u_1+u_2+\cdots+u_n+\cdots \tag{1}$$

is an *infinite series*. Its value, if one exists, is the limit of the sequence of partial sums $\{S_n\}$

$$S=\lim_{n\to\infty}S_n \tag{2}$$

If there is a unique value, the series is said to *converge* to that *sum S*. If there is not a unique sum, the series is said to *diverge*.

Sometimes the character of a series is obvious. For example, the series $\sum_{n=1}^{\infty}\frac{1}{2^n}$ generated by the frog on the log surely converges, while $\sum_{n=1}^{\infty}n$ is divergent. On the other hand, the variable series $1-x+x^2-x^3+x^4-x^5+\cdots$ raises questions.

This series may be obtained by carrying out the division $\frac{1}{1-x}$. If $-1 < x < 1$, the sum S_n yields an approximation to $\frac{1}{1-x}$ and Equation (2) is the exact value. The indecision arises for $x = -1$. Some very great mathematicians, including Leonhard Euler, thought that S should be equal to $\frac{1}{2}$, as is obtained by substituting -1 into $\frac{1}{1-x}$. The problem with this conclusion arises with examination of $1 - 1 + 1 - 1 + 1 - 1 + \cdots$ and observation that appropriate associations can produce values of 1 or 0. Imposition of the condition of uniqueness for convergence puts this series in the category of divergent and eliminates such possibility of ambiguity in other cases.

Fundamental Facts Concerning Infinite Series

1. If Σu_n converges, then $\lim_{n\to\infty} u_n = 0$ (see Problem 2.26). The converse, however, is not necessarily true; i.e., if $\lim_{n\to\infty} u_n = 0$, Σu_n may or may not converge. It follows that if the nth term of a series does *not* approach zero, the series is divergent.
2. Multiplication of each term of a series by a constant different from zero does not affect the convergence or divergence.
3. Removal (or addition) of a *finite* number of terms from (or to) a series does not affect the convergence or divergence.

Special Series

1. **Geometric series** $\sum_{n=1}^{\infty} ar^{n-1} = a + ar + ar^2 + \cdots$, where a and r are constants, converges to $S = \frac{a}{1-r}$ if $|r| < 1$ and diverges if $|r| \geqq 1$. The sum of the first n terms is $S_n = \frac{a(1-r^n)}{1-r}$ (see Problem 2.25).
2. **The *p* series** $\sum_{n=1}^{\infty} \frac{1}{n^p} = \frac{1}{1^p} + \frac{1}{2^p} + \frac{1}{3^p} + \cdots$, where p is a constant, converges for $p > 1$ and diverges for $p \leqq 1$. The series with $p = 1$ is called the *harmonic series*.

Tests for Convergence and Divergence of Series of Constants

More often than not, exact values of infinite series cannot be obtained. Thus, the search turns toward information about the series. In particular, its convergence or divergence comes into question. The following tests aid in discovering this information.

1. **The comparison test** for series of nonnegative terms.
   (a) *Convergence.* Let $\upsilon_n \geqq 0$ for all $n > N$ and suppose that $\Sigma \upsilon_n$ converges. Then if $0 \leqq u_n \leqq \upsilon_n$ for all $n > N$, Σu_n also converges. Note that $n > N$ means *from some term onward.* Often, $N = 1$.

EXAMPLE. Since $\frac{1}{2^n + 1} \leqq \frac{1}{2^n}$ and $\sum \frac{1}{2^n}$ converges, $\sum \frac{1}{2^{n+1}}$ also converges.

(b) *Divergence.* Let $\upsilon_n \geqq 0$ for all $n > N$ and suppose that $\Sigma\upsilon_n$ diverges. Then if $u_n \geqq \upsilon_n$ for all $n > N$, Σu_n also diverges.

EXAMPLE. Since $\dfrac{1}{\ln n} > \dfrac{1}{n}$ and $\displaystyle\sum_{n=2}^{\infty}\frac{1}{n}$ diverges, $\displaystyle\sum_{n=2}^{\infty}\frac{1}{\ln n}$ also diverges.

2. **The limit-comparison** or **quotient test** for series of nonnegative terms.

   (a) If $u_n \geqq 0$ and $\upsilon_n \geqq 0$ and if $\displaystyle\lim_{n\to\infty}\frac{u_n}{\upsilon_n} = A \neq 0$ or ∞, then Σu_n and $\Sigma\upsilon_n$ either both converge or both diverge.

   (b) If $A = 0$ in (a) and $\Sigma\upsilon_n$ converges, then Σu_n converges.

   (c) If $A = \infty$ in (a) and $\Sigma\upsilon_n$ diverges, then Σu_n diverges.

This test is related to the comparison test and is often a very useful alternative to it. In particular, taking $\upsilon_n = 1/n^p$, we have the following theorems from known facts about the p series.

Theorem 1 Let $\displaystyle\lim_{n\to\infty} n^p u_n = A$. Then

(i) Σu_n conyerges if $p > 1$ and A is finite.

(ii) Σu_n diverges if $p \leqq 1$ and $A \neq 0$ (A may be infinite).

EXAMPLES 1. $\displaystyle\sum\frac{n}{4n^3-2}$ converges since $\displaystyle\lim_{n\to\infty} n^2\cdot\frac{n}{4n^3-2} = \frac{1}{4}$.

2. $\displaystyle\sum\frac{\ln n}{\sqrt{n+1}}$ diverges since $\displaystyle\lim_{n\to\infty} n^{1/2}\cdot\frac{\ln n}{(n+1)^{1/2}} = \infty$.

3. **Integral test** for series of non-negative terms.

   If $f(x)$ is positive, continuous, and monotonic decreasing for $x \geqq N$ and is such that $f(n) = u_n$, $n = N, N+1, N+2, \ldots$, then Σu_n converges or diverges according as $\displaystyle\int_N^{\infty} f(x)\,dx = \lim_{M\to\infty}\int_n^M f(x)\,dx$ converges or diverges. In particular, we may have $N = 1$, as is often true in practice.

   This theorem borrows from Chapter 12, since the integral has an unbounded upper limit. (It is an improper integral. The convergence or divergence of these integrals is defined in much the same way as for infinite series.)

EXAMPLE: $\displaystyle\sum_{n=1}^{\infty}\frac{1}{n^2}$ converges since $\displaystyle\lim_{M\to\infty}\int_1^M\frac{dx}{x^2} = \lim_{M\to\infty}\left(1-\frac{1}{M}\right)$ exists.

4. **Alternating series test.** An *alternating series* is one whose successive terms are alternately positive and negative.

An alternating series converges if the following two conditions are satisfied (see Problem 11.15).

   (a) $|u_{n+1}| \leqq |u_n|$ for $n \geqq N$. (Since a fixed number of terms does not affect the convergence or divergence of a series, N may be any positive integer. Frequently it is chosen to be 1.)

   (b) $\displaystyle\lim_{n\to\infty} u_n = 0\left(\text{or}\ \lim_{n\to\infty}|u_n| = 0\right)$

EXAMPLE. For the series $1-\dfrac{1}{2}+\dfrac{1}{3}-\dfrac{1}{4}+\dfrac{1}{5}-\cdots = \displaystyle\sum_{n=1}^{\infty}\frac{(-1)^{n-1}}{n}$, we have $u_n = \dfrac{(-1)^{n-1}}{n}$, $|u_n| = \dfrac{1}{n}$, $|u_{n+1}| = \dfrac{1}{n+1}$. Then for $n \geqq 1$, $|u_{n+1}| \leqq |u_n|$. Also $\displaystyle\lim_{n\to\infty}|u_n| = 0$. Hence, the series converges.

Theorem 2 The numerical error made in stopping at any particular term of a convergent alternating series which satisfies conditions (a) and (b) is less than the absolute value of the next term.

EXAMPLE. If we stop at the fourth term of the series $1-\dfrac{1}{2}+\dfrac{1}{3}-\dfrac{1}{4}+\dfrac{1}{5}-\cdots$, the error made is less than $\dfrac{1}{5} = 0.2$.

5. **Absolute and conditional convergence.** The series Σu_n is called *absolutely convergent* if $\Sigma|u_n|$ converges. If Σu_n converges but $\Sigma|u_n|$ diverges, then Σu_n is called *conditionally convergent.*

Theorem 3 If $\Sigma|u_n|$ converges, then Σu_n converges. In words, an absolutely convergent series is convergent (see Problem 11.17).

EXAMPLE 1. $\frac{1}{1^2}+\frac{1}{2^2}-\frac{1}{3^2}-\frac{1}{4^2}+\frac{1}{5^2}+\frac{1}{6^2}-\cdots$ is absolutely convergent and thus convergent, since the series of absolute values $\frac{1}{1^2}+\frac{1}{2^2}+\frac{1}{3^2}+\frac{1}{4^2}+\cdots$ converges.

EXAMPLE 2. $1-\frac{1}{2}+\frac{1}{3}-\frac{1}{4}+\cdots$ converges, but $1+\frac{1}{2}+\frac{1}{3}+\frac{1}{4}+\cdots$ diverges. Thus, $1-\frac{1}{2}+\frac{1}{3}-\frac{1}{4}+\cdots$ is conditionally convergent.

Any of the tests used for series with nonnegative terms can be used to test for absolute convergence. Also, tests that compare successive terms are common. Tests 6, 8, and 9 are of this type.

6. **Ratio test.** Let $\lim\limits_{n\to\infty}\left|\frac{u_{n-1}}{u_n}\right| = L$. Then the series Σu_n

   (a) converges (absolutely) if $L < 1$.

   (b) diverges if $L > 1$.

   If $L = 1$ the test fails.

7. **The *n*th root test.** Let $\lim\limits_{n\to\infty}\sqrt[n]{|u_n|} = L$. Then the series Σu_n

   (a) converges (absolutely) if $L < 1$

   (b) diverges if $L > 1$.

   If $L = 1$ the test fails.

8. **Raabe's test.** Let $\lim\limits_{n\to\infty} n\left(1-\left|\frac{u_n+1}{u_n}\right|\right) = L$. Then the series Σu_n

   (a) converges (absolutely) if $L > 1$.

   (b) diverges or converges conditionally if $L < 1$.

   If $L = 1$ the test fails.

   This test is often used when the ratio tests fails.

9. **Gauss's test.** If $\left|\frac{u_n+1}{u_n}\right| = 1-\frac{L}{n}+\frac{c_n}{n^q}$, where $|c_n| < P$ for all $n > N$ the sequence c_n is bounded, then the series Σu_n

   (a) converges (absolutely) if $L > 1$.

   (b) diverges or converges conditionally if $L \leq 1$.

   This test is often used when Raabe's test fails.

Theorems on Absolutely Convergent Series

Theorem 4 (Rearrangement of terms.) The terms of an absolutely convergent series can be rearranged in any order, and all such rearranged series will converge to the same sum. However, if the terms of a conditionally convergent series are suitably rearranged, the resulting series may diverge or converge to *any* desired sum (see Problem 11.80).

Theorem 5 (Sums, differences, and products.) The sum, difference, and product of two absolutely convergent series is absolutely convergent. The operations can be performed as for finite series.

Infinite Sequences and Series of Functions, Uniform Convergence

We opened this chapter with the thought that functions could be expressed in series form. Such representation is illustrated by

$$\sin x = x - \frac{x^3}{3!} + \frac{x^3}{5!} - + \cdots + (-1)^{n-1} \frac{x^{2n-1}}{(2n-)!} + \cdots$$

where

$$\sin x = \lim_{n\to\infty} S_n. \quad \text{with} \quad S_1 = x, S_2 = x - \frac{x^3}{3!}, \cdots S_n = \sum_{k=1}^{n} (-1)^{k-1} \frac{x^{2k-1}}{(2k-1)!}$$

Observe that until this section the sequences and series depended on one element, n. Now there is variation with respect to x as well. This complexity requires the introduction of a new concept called *uniform convergence*, which, in turn, is fundamental in exploring the continuity, differentiation, and integrability of series.

Let $\{u_n(x)\}$, $n = 1, 2, 3, \ldots$ be a sequence of functions defined in $[a, b]$. The sequence is said to converge to $F(x)$, or to have the limit $F(x)$ in $[a, b]$, if for each $\epsilon > 0$ and each x in $[a, b]$ we can find $N > 0$ such that $|u_n(x) - F(x)| < \epsilon$ for all $n > N$. In such case we write $\lim_{n\to\infty} u_n(x) = F(x)$. The number N may depend on x as well as ϵ. If it depends *only* on ϵ and not on x, the sequence is said to converge to $F(x)$ *uniformly* in $[a, b]$ or to be *uniformly convergent* in $[a, b]$.

The infinite series of functions

$$\sum_{n}^{\infty} u_n(x) = u_1(x) + u_2(x) + u_3(x) + \cdots \tag{3}$$

is said to be convergent in $[a, b]$ if the sequence of partial sums $\{S_n(x)\}$, $n = 1, 2, 3, \ldots$, where $S_n(x) = u_1(x) + u_2(x) + \cdots + u_n(x)$, is convergent in $[a, b]$. In such case we write $\lim_{n\to\infty} S_n(x) = S(x)$ and call $S(x)$ the *sum* of the series.

It follows that $\Sigma u_n(x)$ converges to $S(x)$ in $[a, b]$ if for each $\epsilon > 0$ and each x in $[a, b]$ we can find $N > 0$ such that $|S_n(x) - S(x)| < \epsilon$ for all $n > N$. If N depends *only* on ϵ and not on x, the series is called *uniformly convergent* in $[a, b]$.

Since $S(x) - S_n(x) = R_n(x)$, the remainder after n terms, we can equivalently say that $\Sigma u_n(x)$ is uniformly convergent in $[a, b]$ if for each $\epsilon > 0$ we can find N depending on ϵ but not on x such that $|R_n(x)| < \epsilon$ for all $n > N$ and all x in $[a, b]$.

These definitions can be modified to include other intervals besides $a \leqq x \leqq b$, such as $a < x < b$, and so on.

The domain of convergence (absolute or uniform) of a series is the set of values of x for which the series of functions converges (absolutely or uniformly).

EXAMPLE 1. Suppose $u_n = x^n/n$ and $-1/2 \leqq x \leqq 1$. Now think of the constant function $F(x) = 0$ on this interval. For any $\epsilon > 0$ and any x in the interval, there is N such that for all $n > N$ $|u_n - F(x)| < \epsilon$, i.e., $|x^n/n| < \epsilon$. Since the limit does not depend on x, the sequence is uniformly convergent.

EXAMPLE 2. If $u_n = x^n$ and $0 \leqq x \leqq 1$, the sequence is not uniformly convergent because [think of the function $F(x) = 0$, $0 \leqq x < 1$, $F(1) = 1$]

$$|x^n - 0| < \varepsilon \text{ when } x^n < \varepsilon$$

thus

$$n \ln \text{x} < \ln \varepsilon$$

On the interval $0 \leqq x < 1$, and for $0 < \epsilon < 1$, both members of the inequality are negative; therefore, $n > \frac{\ln \varepsilon}{\ln x}$. Since $\frac{\ln \varepsilon}{\ln x} = \frac{\ln 1 - \ln \varepsilon}{\ln 1 - \ln x} = \frac{\ln (/\varepsilon)}{\ln (1/x)}$, it follows that we must choose N such that

$$n > N > \frac{\ln 1/\varepsilon}{\ln 1/x}$$

From this expression we see that $\epsilon \to 0$, then $\ln \ln \frac{1}{\varepsilon} \to \infty$, and also as $x \to 1$ from the left $\ln \frac{1}{x} \to 0$ from the right; thus, in either case, N must increase without bound. This dependency on both ϵ and x demonstrates that the sequence is not uniformly convergent. For a pictorial view of this example, see Figure 11.1.

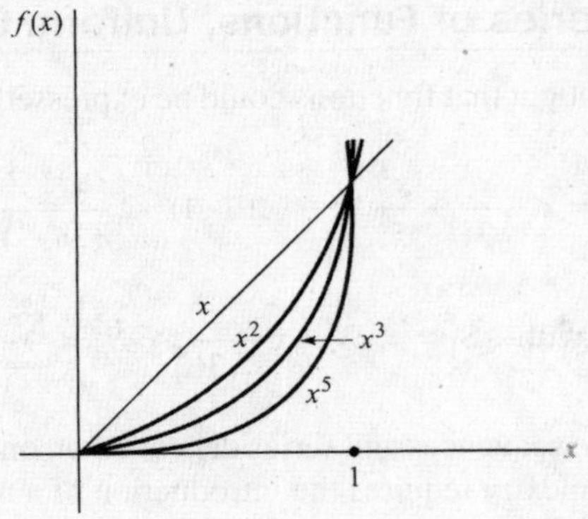

Figure 11.1

Special Tests for Uniform Convergence of Series

1. **Weierstrass M test.** If a sequence of positive constants $M_1, M_2, M_3, \ldots$ can be found such that in some interval
   (a) $|u_n(x)| \leqq M_n \qquad n = 1, 2, 3, \ldots$
   (b) ΣM_n converges

   then $\Sigma u_n(x)$ is uniformly and absolutely convergent in the interval.

   EXAMPLE. $\sum_{n=1}^{\infty} \frac{\cos nx}{n^2}$ is uniformly and absolutely convergent in $[0, 2\pi]$ since $\left|\frac{\cos nx}{n^2}\right| \leqq \frac{1}{n^2}$ and $\sum \frac{1}{n^2}$ converges.

   This test supplies a sufficient but not a necessary condition for uniform convergence, i.e., a series may be uniformly convergent even when the test cannot be made to apply.

   Because of this test, we may be led to believe that uniformly convergent series must be absolutely convergent, and conversely. However, the two properties are independent; i.e., a series can be uniformly convergent without being absolutely convergent, and conversely. See Problems 11.30 and 11.127.

2. **Dirichlet's test.** Suppose that
   (a) the sequence $\{a_n\}$ is a monotonic decreasing sequence of positive constants having limit zero.
   (b) there exists a constant P such that for $a \leqq x \leqq b$ $|u_1(x) + u_2(x) + \cdots + u_n(x)| < P, \qquad$ for all $n > N$.

   Then the series $a_1u_1(x) + a_2u_2(x) + \cdots = \sum_{n=1}^{\infty} a_nu_n(x)$ is uniformly convergent in $a \leqq x \leqq b$.

Theorems on Uniformly Convergent Series

If an infinite series of functions is uniformly convergent, it has many of the properties possessed by sums of finite series of functions, as indicated in the following theorems.

Theorem 6 If $\{u_n(x)\}, n = 1, 2, 3, \ldots$ are continuous in $[a, b]$ and if $\Sigma u_n(x)$ converges uniformly to the sum $S(x)$ in $[a, b]$, then $S(x)$ is continuous in $[a, b]$.

Briefly, this states that a uniformly convergent series of continuous functions is a continuous function. This result is often used to demonstrate that a given series is not uniformly convergent by showing that the sum function $S(x)$ is discontinuous at some point (see Problem 11.30).

In particular, if x_0 is in $[a, b]$, then the theorem states that

$$\lim_{x\to x_0}\sum_{n=1}^{\infty} u_n(x) = \sum_{n=1}^{\infty}\lim_{x\to x_0} u_n(x) = \sum_{n=1}^{\infty} u_n(x_0)$$

where we use right- or left-hand limits in case x_0 is an endpoint of $[a, b]$.

Theorem 7 If $\{u_n(x)\}$, $n = 1, 2, 3, \ldots$ are continuous in $[a, b]$ and if $\Sigma u_n(x)$ converges uniformly to the sum $S(x)$ in $[a, b]$, then

$$\int_a^b S(x)\,dx = \sum_{n=1}^{\infty}\int_a^b u_n(x)\,dx \tag{4}$$

or

$$\int_a^b \left\{\sum_{n=1}^{\infty} u_n(x)\right\} dx = \sum_{n=1}^{\infty}\int_a^b u_n(x)\,dx \tag{5}$$

Briefly, a uniformly convergent series of continuous functions can be integrated term by term.

Theorem 8 If $\{u_n(x)\}$, $n = 1, 2, 3, \ldots$ are continuous and have continuous derivatives in $[a, b]$ and if $\Sigma u_n(x)$ converges to $S(x)$ while $\Sigma u_n'(x)$ is uniformly convergent in $[a, b]$, then in $[a, b]$

$$S'(x) = \sum_{n=1}^{\infty} u_n'(x) \tag{6}$$

or

$$\frac{d}{dx}\left\{\sum_{n=1}^{\infty} u_n(x)\right\} = \sum_{n=1}^{\infty}\frac{d}{dx}u_n(x) \tag{7}$$

This shows conditions under which a series can be differentiated term by term.

Theorems similar to these can be formulated for sequences. For example, if $\{u_n(x)\}$, $n = 1, 2, 3, \ldots$ is uniformly convergent in $[a, b]$, then

$$\lim_{n\to\infty}\int_a^b u_n(x)\,dx = \int_a^b \lim_{n\to\infty} u_n(x)\,dx \tag{8}$$

which is the analog of Theorem 7.

Power Series

A series having the form

$$a_0 + a_1x + a_2x^2 + \cdots = \sum_{n=0}^{\infty} a_nx^n \tag{9}$$

where $a_0, a_1, a_2, \ldots$ are constants, is called a *power series* in x. It is often convenient to abbreviate the series (9) as Σa_nx^n.

In general, a power series converges for $|x| < R$ and diverges for $|x| > R$, where the constant R is called the *radius of convergence* of the series. For $|x| = R$, the series may or may not converge.

The interval $|x| < R$ or $-R < x < R$, with possible inclusion of endpoints, is called the *interval of convergence* of the series. Although the ratio test is often successful in obtaining this interval, it may fail, and in such cases, other tests may be used (see Problem 11.22).

The two special cases $R = 0$ and $R = \infty$ can arise. In the first case the series converges only for $x = 0$; in the second case it converges for all x, sometimes written $-\infty < x < \infty$ (see Problem 11.25). When we speak of a convergent power series, we shall assume, unless otherwise indicated, that $R > 0$.

Similar remarks hold for a power series of the form (9), where x is replaced by $(x - a)$.

Theorems on Power Series

Theorem 9 A power series converges uniformly and absolutely in any interval which lies *entirely within* its interval of convergence.

Theorem 10 A power series can be differentiated or integrated term by term over any interval lying entirely within the interval of convergence. Also, the sum of a convergent power series is continuous in any interval lying entirely within its interval of convergence.

This follows at once from Theorem 9 and the theorem on uniformly convergent series on Pages 284 and 285. The results can be extended to include endpoints of the interval of convergence by the following theorems.

Theorem 11 *Abel's theorem.* When a power series converges up to and including an endpoint of its interval of convergence, the interval of uniform convergence also extends so far as to include this endpoint. See Problem 11.42.

Theorem 12 *Abel's limit theorem.* If $\sum_{n=0}^{\infty} a_n x^n$ converges at $x = x_0$, which may be an interior point or an endpoint of the interval of convergence, then

$$\lim_{x \to x_0} \left\{ \sum_{n=0}^{\infty} a_n x^n \right\} = \sum_{n=0}^{\infty} \left\{ \lim_{x \to x_0} a_n x^n \right\} = \sum_{n=0}^{\infty} a_n x_0^n \tag{10}$$

If x_0 is an endpoint, we must use $x \to x_0 +$ or $x \to x_0 -$ in Equation (10) according as x_0 is a left- or a right-hand endpoint.

This follows at once from Theorem 11 and Theorem 6 on the continuity of sums of uniformly convergent series.

Operations With Power Series

In the following theorems we assume that all power series are convergent in some interval.

Theorem 13 Two power series can be added or subtracted term by term for each value of x common to their intervals of convergence.

Theorem 14 Two power series, for example, $\sum_{n=0}^{\infty} a_n x^n$ and $\sum_{n=0}^{\infty} b_n x^n$, can be multiplied to obtain $\sum_{n=0}^{\infty} c_n x^n$ where

$$c_n = a_0 b_n + a_1 b_{n-1} + a_2 b_{n-2} + \cdots + a_n b_0 \tag{11}$$

the result being valid for each x within the common interval of convergence.

Theorem 15 If the power series $\sum_{n=0}^{\infty} a_n x^n$ is divided by the power series $\Sigma b_n x^n$ where $b_0 \neq 0$, the quotient can be written as a power series which converges for sufficiently small values of x.

Theorem 16 If $y = \sum_{n=0}^{\infty} a_n x^n$, then by substituting $x = \sum_{n=0}^{\infty} b_n y^n$, we can obtain the coefficients b_n in terms of a_n. This process is often called *reversion of series*.

Expansion of Functions in Power Series

This section gets at the heart of the use of infinite series in analysis. Functions are represented through them. Certain forms bear the names of mathematicians of the eighteenth and early nineteenth centuries who did so much to develop these ideas.

A simple way (and one often used to gain information in mathematics) to explore series representation of functions is to assume such a representation exists and then discover the details. Of course, whatever is found must be confirmed in a rigorous manner. Therefore, assume

$$f(x) = A_0 + A_1(x-c) + A_2(x-c)^2 + \cdots + A_n(x-c)^n + \cdots$$

Notice that the coefficients A_n can be identified with derivatives of f. In particular,

$$A_0 = f(c), A_1 = f'(c), A_2 = \frac{1}{2!}f''(c), \ldots, A_n = \frac{1}{n!}f^{(n)}(c), \ldots$$

This suggests that a series representation of f is

$$f(x) = f(c) + f'(c)(x-c) + \frac{1}{2!}f''(x-c)^2 + \cdots + \frac{1}{n!}f^{(n)}(c)(x-c)^n \cdots$$

A first step in formalizing series representation of a function f, for which the first n derivatives exist, is accomplished by introducing *Taylor polynomials* of the function.

$$P_0(x) = f(c) \qquad P_1(x) = f(c) + f'(c)(x-c),$$

$$P_2(x) = f(c) + f'(c)(x-c) + \frac{1}{2!}f''(c)(x-c)^2,$$

$$P_n(x) = f(c) + f'(c)(x-c) + \cdots \frac{1}{n!}f^{(n)}(c)(x-c)^n \tag{12}$$

Taylor's Theorem

Let f and its derivatives $f', f'', \ldots, f^{(n)}$ exist and be continuous in a closed interval $a \le x \le b$ and suppose that $f^{(n+1)}$ exists in the open interval $a < x < b$. Then for c in $[a, b]$,

$$f(x) = P_n(x) + R_n(x)$$

where the remainder $R_n(x)$ may be represented in any of the three following ways.

For each n there exists ξ such that

$$R_n(x) = \frac{1}{(n+1)!}f^{(n+1)}(\xi)(x-c)^{n+1} \qquad \text{(Lagrange form)} \tag{13}$$

(ξ is between c and x.)

(The theorem with this remainder is a mean value theorem. Also, it is called Taylor's formula.)

For each n there exists ξ such that

$$R_n(x) = \frac{1}{n!}f^{(n+1)}(\xi)(x-\xi)^n(x-c) \qquad \text{(Cauchy form)} \tag{14}$$

$$R_n(x) = \frac{1}{n!}\int_c^x (x-t)^n f^{(n+1)}(t)\,dt \qquad \text{(Integral form)} \tag{15}$$

If all the derivatives of f exist, then the following form, without remainder, may be explored:

$$f(x) = \sum_{n=0}^{\infty} \frac{1}{n!}f^{(n)}(c)(x-c)^n \tag{16}$$

This infinite series is called a Taylor series, although when $c = 0$, it can also be referred to as a MacLaurin series or expansion.

We might be tempted to believe that if all derivatives of $f(x)$ exist at $x = c$, the expansion shown here would be valid. This, however, is not necessarily the case, for although one can then *formally* obtain the series on the right of the expansion, the resulting series may not converge to $f(x)$. For an example of this see Problem 11.108.

Precise conditions under which the series converges to $f(x)$ are best obtained by means of the theory of functions of a complex variable. (See Chapter 16.)

The determination of values of functions at desired arguments is conveniently approached through Taylor polynomials.

EXAMPLE. The value of sin x may be determined geometrically for $0, \frac{\pi}{6}$, and an infinite number of other arguments. To obtain values for other real number arguments, a Taylor series may be expanded about any of these points. For example, let $c = 0$ and evaluate several derivatives there; i.e., $f(0) = \sin 0 = 0, f'(0) = \cos 0 = 1, f''(0) = -\sin 0 = 0, f'''(0) = -\cos 0 = -1, f^{lv}(0) = \sin 0 = 0, f^{v}(0) = \cos 0 = 1$.

Thus, the MacLaurin expansion to five terms is

$$\sin x = 0 + x - 0 - \frac{1}{3!}x^3 + 0 - \frac{1}{51}x^5 + \cdots$$

Since the fourth term is 0, the Taylor polynomials P_3 and P_4 are equal, i.e.,

$$P_3(x) = p_4(x) = x - \frac{x^3}{3!}$$

and the Lagrange remainder is

$$R_4(x) = \frac{1}{5!}\cos\xi\ x^5$$

Suppose an approximation of the value of sin .3 is required. Then

$$P_4(.3) = .3 - \frac{1}{6}(.3)^3 \approx .2945.$$

The accuracy of this approximation can be determined from examination of the remainder. In particular (remember $|\cos \xi| \le 1$),

$$|R_4| = \left|\frac{1}{5!}\cos\xi\ (.3)^5\right| \le \frac{1}{120}\frac{243}{10^5} < .000021$$

Thus, the approximation $P_4(.3)$ for sin .3 is correct to four decimal places.

Additional insight into the process of approximation of functional values results by constructing a graph of $P_4(x)$ and comparing it to $y = \sin x$. (See Figure 11.2.)

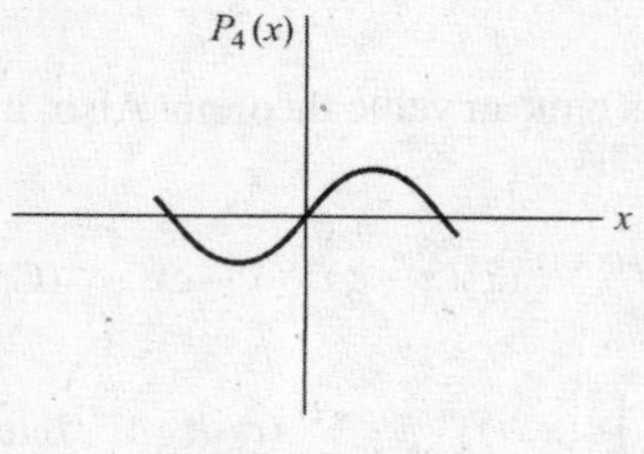

Figure 11.2

The roots of the equation are $0, \pm\sqrt{6}$. Examination of the first and second derivatives reveals a relative maximum at $x = \sqrt{2}$ and a relative minimum at $x = -\sqrt{2}$. The graph is a local approximation of the sin curve. The reader can show that $P_6(x)$ produces an even better approximation.

(For an example of series approximation of an integral, see the example that follows.)

Some Important Power Series

The following series, convergent to the given function in the indicated intervals, are frequently employed in practice:

1. $\sin x \quad = x - \frac{x^3}{3!} + \frac{x^5}{5!} - \frac{x^7}{7!} + \cdots (-1)^{n-1} \frac{x^{2n-1}}{(2n-1)!} + \cdots \quad -\infty < x < \infty$

2. $\cos x \quad = 1 - \frac{x^2}{2!} + \frac{x^4}{4!} - \frac{x^6}{6!} + \cdots (-1)^{n-1} \frac{x^{2n-2}}{(2n-2)!} + \cdots \quad -\infty < x < \infty$

3. $e^x \quad = 1 + x + \frac{x^2}{2!} + \frac{x^3}{3!} + \cdots + \frac{x^{n-1}}{(n-1)!} + \cdots \quad -\infty < x < \infty$

4. $\ln|1+x| \quad = x - \frac{x^2}{2} + \frac{x^3}{3} - \frac{x^4}{4} + \cdots (-1)^{n-1} \frac{x^n}{n} + \cdots \quad -1 < x \leqq 1$

5. $\frac{1}{2}\ln\frac{|1+x|}{1-x} \quad = x + \frac{x^3}{3} + \frac{x^5}{5} + \frac{x^7}{7} + \cdots + \frac{x^{2n-1}}{2n-1} + \cdots \quad -1 < x < 1$

6. $\tan^{-1} x \quad = x - \frac{x^3}{3} + \frac{x^5}{5} - \frac{x^7}{7} + \cdots (-1)^{n-1} \frac{x^{2n-1}}{2n-1} + \cdots \quad -1 \leqq x \leqq 1$

7. $(1+x)^p \quad = 1 + px + \frac{P(p-1)}{2!} x^2 + \cdots + \frac{P(p-1)\ldots(p-n+1)}{n!} x^n + \cdots$

This is the *binomial series*.

(a) If p is a positive integer or zero, the series terminates.

(b) If $p > 0$ but is not an integer, the series converges (absolutely) for $-1 \leqq x \leqq 1$.

(c) If $-1 < p < 0$, the series converges for $-1 < x \leqq 1$.

(d) If $p \leqq -1$, the series converges for $-1 < x < 1$.

For all p, the series certainly converges if $-1 < x < 1$.

EXAMPLE. Taylor's theorem applied to the series for e^x enables us to estimate the value of the integral $\int_0^1 e^{x^2}\, dx$. Substituting x^2 for x, we obtain

$$\int_0^1 e^{x^2}\, dx = \int_0^1 \left(1 + x + \frac{x^4}{2!} + \frac{x^6}{3!} + \frac{x^8}{4!} + \frac{e^\xi}{5!} x^{10}\right) dx$$

where

$$P_4(x) = 1 + x + \frac{1}{2!}x^4 + \frac{1}{3!}x^6 + \frac{1}{4!}x^8$$

and

$$R_4(x) = \frac{e^\xi}{5!} x^{10}, \qquad 0 < \xi < x$$

Then

$$\int_0^1 P_4(x)\, dx = 1 + \frac{1}{3} + \frac{1}{5(2!)} + \frac{1}{7(3!)} + \frac{1}{9(4!)} \approx 1.4618$$

$$\left|\int_0^1 R_4(x)\, dx\right| \le \int_0^1 \left|\frac{e^\xi}{5!} x^{10}\right| dx \le e \int_0^1 \frac{x^{10}}{5!}\, dx = \frac{e}{11.5} < .0021$$

Thus, the maximum error is less than .0021 and the value of the integral is accurate to two decimal places.

Special Topics

1. **Functions defined by series** are often useful in applications and frequently arise as solutions of differential equations. For example, the function defined by

$$J_p(x) = \frac{x^p}{2^p p!}\left\{1 - \frac{2}{2(2p+2)} + \frac{x^4}{2.4(2p+2)(2p+4)} - \cdots\right\}$$
$$= \sum_{n=0}^{\infty} \frac{(-1)^n (x/2)^{p+2n}}{n!(n+p)!} \tag{16}$$

is a solution of *Bessel's differential equation* $x^2y'' + xy' + (x^2 - p^2)y = 0$ and is thus called a *Bessel function of order p*. See Problems 11.46 and 11.110 through 11.113.

Similarly, the *hypergeometric function*

$$F(a, b; c; x) = 1 + \frac{a \cdot b}{1 \cdot c}x + \frac{a(a+1)\,b\,(b+1)}{1 \cdot 2 \cdot c(c+1)}x^2 + \cdots \tag{17}$$

is a solution of *Gauss's differential equation* $x(1 - x)y'' + \{c - (a + b + 1)x\}y' - aby = 0$.

These functions have many important properties.

2. **Infinite series of complex terms**, in particular, power series of the form $\sum_{n=0}^{\infty} a_n z^n$, where $z = x + iy$ and a_n may be complex and can be handled in a manner similar to real series.

Such power series converge for $|z| < R$; i.e., interior to a *circle of convergence* $x^2 + y^2 = R^2$, where R is the *radius of convergence* (if the series converges only for $z = 0$, we say that the radius of convergence R is zero; if it converges for all z, we say that the radius of convergence is infinite). On the boundary of this circle, i.e., $|z| = R$, the series may or may not converge, depending on the particular z.

Note that for $y = 0$ the circle of convergence reduces to the interval of convergence for real power series. Greater insight into the behavior of power series is obtained by use of the theory of functions of a complex variable (see Chapter 16).

3. **Infinite series of functions of two (or more) variables,** such as $\sum_{n=0}^{\infty} u_n(x, y)$, can be treated in a manner analogous to series in one variable. In particular, we can discuss power series in x and y having the form

$$a_{00} + (a_{10}x + a_{01}y) + (a_{20}x^2 + a_{11}xy + a_{02}y^2) + \cdots$$

using double subscripts for the constants. As for one variable, we can expand suitable functions of x and y in such power series. In particular, the Taylor theroem may be extended as follows.

Taylor's Theorem (For Two Variables)

Let f be a function of two variables x and y. If all partial derivatives of order n are continuous in a closed region and if all the $(n + 1)$ partial derivatives exist in the open region, then

$$f(x_0 + h, y_0 + k) = f(x_0, y_0) + \left(h\frac{\partial}{\partial x} + k\frac{\partial}{\partial y}\right)f(x_0, y_0) + \frac{1}{2!}\left(h\frac{\partial}{\partial x} + k\frac{\partial}{\partial y}\right)^2 f(x_0, y_0) + \cdots$$
$$+ \frac{1}{n!}\left(h\frac{\partial}{\partial x} + k\frac{\partial}{\partial y}\right)^n f(x_0, y_0) + R_n \tag{18}$$

where

$$R_n = \frac{1}{(n+1)!}\left(h\frac{\partial}{\partial x} + k\frac{\partial}{\partial y}\right)^{n+1} f(x_0 + \theta h, y_0 + \theta k), \qquad 0 < \theta < 1$$

and where the meaning of the operator notation is as follows:

$$\left(h\frac{\partial}{\partial x}+k\frac{\partial}{\partial y}\right)f = hf_x + kf_y,$$

$$\left(h\frac{\partial}{\partial x}+k\frac{\partial}{\partial y}\right)^2 = h^2 f_{xx} + 2hkf_{xy} + k^2 f_{yy}$$

and we formally expand $\left(h\frac{\partial}{\partial x}+k\frac{\partial}{\partial y}\right)^n$ by the binomial theorem.

Note: In alternate notation $h = \Delta x = x - x_0$, $k = \Delta y = y - y_0$.

If $R_n \to 0$ as $n \to \infty$ then an unending continuation of terms produces the *Taylor series for f(x, y)*. Multivariable Taylor series have a similar pattern.

4. **Double series.** Consider the array of numbers (or functions)

$$\begin{pmatrix} u_{11} & u_{12} & u_{13} & \cdots \\ u_{21} & u_{22} & u_{23} & \cdots \\ u_{31} & u_{32} & u_{33} & \cdots \\ \vdots & \vdots & \vdots & \end{pmatrix}$$

Let $S_{mn} = \sum_{p=1}^{m}\sum_{q=1}^{n} u_{pq}$ be the sum of the numbers in the first m rows and first n columns of this array. If there exists a number S such that $\lim_{\substack{m\to\infty \\ n\to\infty}} S_{mn} = S$, we say that the double series $\sum_{p=1}^{\infty}\sum_{q=1}^{\infty} u_{pq}$ *converges* to the *sum S*; otherwise, it *diverges*.

Definitions and theorems for double series are very similar to those for series already considered.

5. **Infinite products.** Let $P_n = (1 + u_1)(1 + u_1)(1 + u_2)(1 + u_3) \ldots (1 + u_n)$ denoted by $\prod_{k=1}^{n}(1+u_k)$, where we suppose that $u_k \neq -1$, $k = 1, 2, 3, \ldots$. If there exists a number $P \neq 0$ such that $\lim_{n\to\infty} P_n = P$, we say that the *infinite product* $((1+u_1)(1+u_2)(1+u_3)\ldots = \prod_{k=1}^{\infty}(1+u_k)$, or, briefly, $\Pi(1 + u_k)$, converges to P; otherwise, it diverges.

If $\Pi(1 + |u_k|)$ converges, we call the infinite product $\Pi(1 + u_k)$ *absolutely convergent*. It can be shown that an absolutely convergent infinite product converges and that factors can in such cases be rearranged without affecting the result.

Theorems about infinite products can (by taking logarithms) often be made to depend on theorems for infinite series. Thus, for example, we have the following theorem.

Theorem A necessary and sufficient condition that $\Pi(1 + u_k)$ converge absolutely is that Σu_k converge absolutely.

6. **Summability.** Let $S_1, S_2, S_3, \ldots$ be the partial sums of a divergent series Σu_n. If the sequence $S_1, \frac{S_1+S_2}{2}, \frac{S_1+S_2+S_3}{3}, \ldots$ (formed by taking arithmetic means of the first n terms of $S_1, S_2, S_3, \ldots$) converges to S, we say that the series Σu_n is *summable* in the *Césaro sense*, or *C*-1 *summable* to S (see Problem 11.51).

If Σu_n converges to S, the Césaro method also yields the result S. For this reason, the Césaro method is said to be a *regular* method of summability.

In case the Césaro limit does not exist, we can apply the same technique to the sequence $S_1, \frac{S_1+S_2}{3}, \frac{S_1+S_2+S_3}{3}, \cdots$ If the *C*-1 limit for this sequence exists and equals S, we say that Σu_k converges to S in the *C*-2 sense. The process can be continued indefinitely.

SOLVED PROBLEMS

Convergence And Divergence Of Series Of Constants

11.1. (a) Prove that $\dfrac{1}{1\cdot 3}+\dfrac{1}{3\cdot 5}+\dfrac{1}{5\cdot 7}+\cdots=\displaystyle\sum_{n=1}^{\infty}\frac{1}{(2n-1)(2n+1)}$ converges and (b) find its sum.

$$u_n=\frac{1}{(2n-1)(2n+1)}=\frac{1}{2}\left(\frac{1}{2n-1}-\frac{1}{2n+1}\right)$$

Then

$$S_n=u_1+u_2+\cdots+u_n=\frac{1}{2}\left(\frac{1}{1}-\frac{1}{3}\right)+\frac{1}{2}\left(\frac{1}{3}-\frac{1}{5}\right)+\cdots+\frac{1}{2}\left(\frac{1}{2n-1}-\frac{1}{2n+1}\right)$$

$$=\frac{1}{2}\left(\frac{1}{1}-\frac{1}{3}+\frac{1}{3}-\frac{1}{5}+\frac{1}{5}-\cdots+\frac{1}{2n-1}-\frac{1}{2n+1}\right)=\frac{1}{2}\left(1-\frac{1}{2n+1}\right)$$

Since $\lim_{n\to\infty} S_n=\lim_{n\to\infty}\frac{1}{2}\left(1-\frac{1}{2n+1}\right)=\frac{1}{2}$, the series converges and its sum is 1/2.

The series is sometimes called a *telescoping series*, since the terms of S_n, other than the first and last, cancel out in pairs.

11.2. (a) Prove that $\dfrac{2}{3}+\left(\dfrac{2}{3}\right)^2+\left(\dfrac{2}{3}\right)^3+\cdots=\displaystyle\sum_{n=1}^{\infty}\left(\frac{2}{3}\right)^n$ converges and (b) find its sum.

This is a geometric series; therefore, the partial sums are of the form $S_n=\dfrac{a(1-r^n)}{1-r}$. Since $|r|<1$ $S=\lim_{n\to\infty}S_n=\dfrac{a}{1-r}$ and in particular with $r=\dfrac{2}{3}$ and $a=\dfrac{2}{3}$, we obtain $S=2$.

11.3. Prove that the series $\dfrac{1}{2}+\dfrac{2}{3}+\dfrac{3}{4}+\dfrac{4}{5}+\cdots=\displaystyle\sum_{n=1}^{\infty}\frac{n}{n+1}$ diverges.

$\lim_{n\to\infty}u_n=\lim_{n\to\infty}\dfrac{n}{n+1}=1$. Hence, by Problem 2.26, the series is divergent.

11.4. Show that the series whose *n*th term is $u_n=\sqrt{n+1}-\sqrt{n}$ diverges although $\lim_{x\to\infty} u_n=0$.

It is a fact that $\lim_{x\to\infty} u_n=0$ follows from Problem 2.14(c).

Now $S_n=u_1+u_2+\cdots+u_n=(\sqrt{2}-\sqrt{1})+(\sqrt{3}-\sqrt{2})+\cdots+((\sqrt{n+1}-\sqrt{n})=\sqrt{n+1}-\sqrt{1}$.

The S_n increases without bound and the series diverges.

This problem shows that $\lim_{x\to\infty}=0$ is a *necessary* but not *sufficient* condition for the convergence of Σu_n. See also Problem 11.6.

Comparison test and quotient test

11.5. If $0\leq u_n\leq \upsilon_n$, $n=1,2,3,\ldots$ and if $\Sigma\upsilon_n$ converges, prove that Σu_n also converges (i.e., establish the comparison test for convergence).

Let $S_n=u_1+u_2+\cdots+u_n$, $T_n=\upsilon_1+\upsilon_2+\cdots+\upsilon_n$.

Since $\Sigma\upsilon_n$ converges, $\lim_{n\to\infty}T_n$ exists and equals T, say. Also, since $\upsilon_n\geq 0$, $T_n\leq T$.

Then $S_n=u_1+u_2+\cdots+u_n\leq \upsilon_1+\upsilon_2+\cdots+\upsilon_n\leq T$ or $0\leq S_n\leq T$.

Thus S_n is a bounded monotonic increasing sequence and must have a limit (see Chapter 2); i.e., Σu_n converges.

11.6. Using the comparison test, prove that $1 + \frac{1}{2} + \frac{1}{3} + \cdots = \sum_{n=1}^{\infty} \frac{1}{n}$ diverges.

We have

$$1 \geqq \frac{1}{2}$$

$$\frac{1}{2} + \frac{1}{3} \geqq \frac{1}{4} + \frac{1}{4} = \frac{1}{2}$$

$$\frac{1}{4} + \frac{1}{5} + \frac{1}{6} + \frac{1}{7} \geqq \frac{1}{8} + \frac{1}{8} + \frac{1}{8} + \frac{1}{8} = \frac{1}{2}$$

$$\frac{1}{8} + \frac{1}{9} + \frac{1}{10} + \cdots + \frac{1}{15} \geqq \frac{1}{16} + \frac{1}{16} + \frac{1}{16} + \cdots + \frac{1}{16} \text{(8 terms)} = \frac{1}{2}$$

and soon. Thus, to any desired number of terms.

$$1 + \left(\frac{1}{2} + \frac{1}{3}\right) + \left(\frac{1}{4} + \frac{1}{5} + \frac{1}{6} + \frac{1}{7}\right) + \cdots \geqq \frac{1}{2} + \frac{1}{2} + \frac{1}{2} + \cdots$$

Since the right-hand side can be made larger than any positive number by choosing enough terms, the given series diverges.

By methods analogous to that used here, we can show that $\sum_{n=1}^{\infty} \frac{1}{n^p}$, where p is a constant, diverges if $p \leq 1$ and converges if $p > 1$. This can also be shown in other ways [see Problem 11.13(a)].

11.7. Test for convergence or divergence $\sum_{n=1}^{\infty} \frac{\ln n}{2n^3 - 1}$.

Since In $n < n$ and $\frac{1}{2n^3 - 1} \leqq \frac{1}{n^3}$, we have $\frac{\ln n}{2n^3 - 1} \leqq \frac{n}{n^3} = \frac{1}{n^2}$.

Then the given series converges, since $\sum_{n=1}^{\infty} \frac{1}{n^2}$ converges.

11.8. Let u_n and υ_n be positive. If $\lim_{n\to\infty} \frac{u_n}{\upsilon_n}$ = constant $A \neq 0$, prove that Σu_n converges or diverges according as $\Sigma \upsilon_n$ converges or diverges.

By hypothesis, given $\epsilon > 0$, we can choose an integer N such that $\left|\frac{u_n}{\upsilon_n} - A\right| < \varepsilon$ for all $n > N$. Then for $n = N + 1, N + 2, \ldots$

$$-\varepsilon < f\frac{u_n}{\upsilon_n} - A < \varepsilon \qquad \text{or} \qquad (A - \varepsilon)\upsilon_n < u_n \, (A + \varepsilon)\upsilon_n \tag{1}$$

Summing from $N + 1$ to ∞ (more precisely, from $N + 1$ to M and then letting $M \to \infty$),

$$(A - \varepsilon) \sum_{N+1}^{\infty} \upsilon_n \leqq \sum_{N+1}^{\infty} u_n \leqq (A + \varepsilon) \sum_{N+1}^{\infty} \upsilon_n \tag{2}$$

There is no loss in generality in assuming $A - \epsilon > 0$. Then from the right-hand inequality of Equation (2), Σu_n converges when $\Sigma \upsilon_n$ does. From the left-hand inequality of Equation (2), Σu_n diverges when $\Sigma \upsilon_n$ does. For the cases $A = 0$ or $A = \infty$, see Problem 11.66.

11.9. Test for convergence: (a) $\sum_{n=1}^{\infty} \frac{4n^2 - n + 3}{n^3 + 2n}$, (b) $\sum_{n=1}^{\infty} \frac{n + \sqrt{n}}{2n^3 - 1}$, and (c) $\sum_{n=1}^{\infty} \frac{\ln n}{n^2 + 3}$.

(a) For large n, $\frac{4n^2 - n + 3}{n^3 + 2n}$ is approximately $\frac{4n^2}{n^3} = \frac{4}{n}$. Taking $u_n = \frac{4n^2 - n + 3}{n^3 + 2n}$ and $\upsilon_n = \frac{4}{n}$, we have $\lim_{n\to\infty} = \frac{u_n}{\upsilon_n} = 1$.

Since $\Sigma \upsilon_n = 4\Sigma 1/n$ diverges, Σu_n also diverges, by Problem 11.8.

Note that the purpose of considering the behavior of u_n for large n is to obtain an appropriate comparison series υ_n. In this case we could just as well have taken $\upsilon_n = 1/n$.

Another method: $\lim_{n\to\infty} n\left(\dfrac{4n^2 - n + 3}{n^3 + 2n}\right) = 4$. Then by Theorem 1, Page 281, the series converges.

(b) For large n, $u_n = \dfrac{n+\sqrt{n}}{2n^3 - 1}$ is approximately $\upsilon_n = \dfrac{n}{2n^3} = \dfrac{1}{2n^2}$.

Since $\lim_{n\to\infty} \dfrac{u_n}{\upsilon_n} = 1$ and $\sum \upsilon_n = \dfrac{1}{2}\sum \dfrac{1}{n^2}$ converges (p series with $p = 2$), the given series converges.

Another method: $\lim_{n\to\infty} n^2\left(\dfrac{n+\sqrt{n}}{2n^3-1}\right) = \dfrac{1}{2}$. Then by Theorem 1, Page 281, the series converges.

(c) $\lim_{n\to\infty} n^{3/2}\left(\dfrac{\ln n}{n^2+3}\right) \leq \lim_{n\to\infty} n^{3/2}\left(\dfrac{\ln n}{n^2}\right) = \lim_{n\to\infty} \dfrac{\ln n}{\sqrt{n}} = 0$ (by L'Hospital's rule or otherwise). Then by Theorem 1 with p = 3/2. the series converges.

Note that the method of Problem 11.6(a) yields $\dfrac{\ln n}{n^2+3} < \dfrac{n}{n^2} = \dfrac{1}{n}$, but nothing can be deduced, since $\Sigma 1/n$ diverges.

11.10. Examine for convergence: (a) $\sum_{n=1}^{\infty} e^{-n^2}$ and (b) $\sum_{n=1}^{\infty} \sin^3\left(\dfrac{1}{n}\right)$.

(a) $\lim_{n\to\infty} n^2 e^{-n^2} = 0$ (by L'Hospital's rule or otherwise). Then by Theorem 1 with p = 2, the series converges.

(b) For large n, sin(1/n) is approximately 1/n. This leads to consideration of

$$\lim_{n\to\infty} n^3 \sin^3\left(\frac{1}{n}\right) = \lim_{n\to\infty}\left\{\frac{\sin(1/n)}{1/n}\right\}^3 = 1$$

from which we deduce, by Theorem 1 with $p = 3$, that the given series converges.

Integral test

11.11. Establish the integral test (see Page 281).

We perform the proof taking $N = 1$. Modifications are easily made if $N > 1$.

From the monotonicity of $f(x)$, we have

$$u_{n+1} = f(n+1) \leq f(x) \leq f(n) = u_n \qquad n = 1, 2, 3, \ldots$$

Integrating from $x = n$ to $x = n + 1$, using Property 7, Page 98,

$$u_{n+1} \leq \int_n^{n+1} f(x)\,dx \leq u_n \qquad n = 1, 2, 3 \ldots$$

Summing from $n = 1$ to $M - 1$,

$$u_2 + u_3 + \cdots + u_M \leq \int_1^M f(x)\,dx \leq u_1 + u_2 + \cdots u_{M-1} \tag{1}$$

If $f(x)$ is strictly decreasing, the equality signs in Equation (1) can be omitted.

If $\lim_{M\to\infty}\int_1^M f(x)\,dx$ exists and is equal to S, we see from the left-hand inequality in Equation (1) that $u_2 + u_3 + \cdots + u_M$ is monotonic increasing and bounded above by S, so that Σu_n converges.

If $\lim_{M\to\infty}\int_1^M f(x)\,dx$ is unbounded, we see from right-hand inequality in Equation (1) that σu_n diverges.

Thus, the proof is complete.

11.12. Illustrate geometrically the proof in Problem 11.11.

Geometrically, $u_2 + u_3 + \cdots + u_M$ is the total area of the rectangles shown shaded in Figure 11.3, while $u_1 + u_2 + \cdots + u_{M-1}$ is the total area of the rectangles which are shaded and nonshaded.

The area under the curve $y = f(x)$ from $x = 1$ to $x = M$ is intermediate in value between the two areas given above, thus illustrating the result (1) of Problem 11.11.

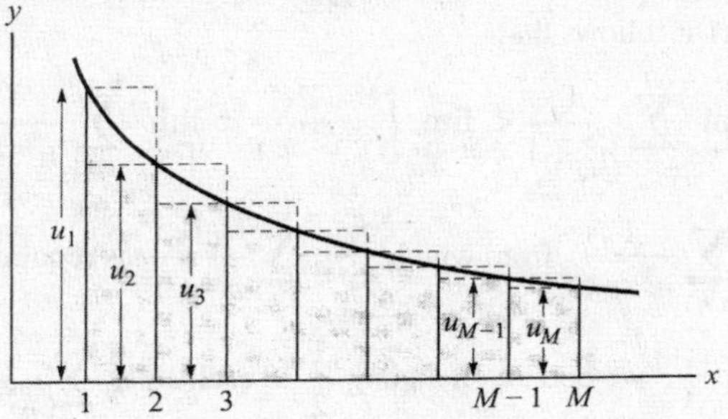

Figure 11.3

11.13. Test for convergence:

(a) $\sum_1^\infty \frac{1}{n^p}$, $p =$ constant

(b) $\sum_1^\infty \frac{n}{n^2+1}$

(c) $\sum_2^\infty \frac{1}{n \ln n}$

(d) $\sum_1^\infty n e^{-n^2}$

(a) Consider $\int_1^M \frac{dx}{x^p} = \int_1^M x^{-p}\,dx = \left.\frac{x^{1-p}}{1-p}\right|_1^M = \frac{M^{1-p}}{1-p}$, where $p \neq 1$.

If $p < 1$, $\lim_{M\to\infty} \frac{M^{1-p}-1}{1-p} = \infty$, so that integral, and thus the series, diverges.

If $p > 1$, $\lim_{M\to\infty} \frac{M^{1-p}-1}{1-p} = \frac{1}{p-1}$, so that the integral, and thus the series, converges.

If $p = 1$, $\int_1^M \frac{dx}{x^p} = \int_1^M \frac{dx}{x} = \ln M$ and $\lim_{M\to\infty} \ln M = \infty$, so that the integral, and thus the series, diverges. Thus, the series converges if $p > 1$ and diverges if $p \leq 1$.

(b) $\lim_{M\to\infty} \int_1^M \frac{x\,dx}{x^2+1} = \lim_{M\to\infty} \frac{1}{2}\ln(x^2+1)\Big|_1^M = \lim_{M\to\infty}\left\{\frac{1}{2}\ln(M^2+1) - \frac{1}{2}\right\} = \infty$ and the series diverges.

(c) $\lim_{M\to\infty} \int_1^M \frac{dx}{x\ln x} = \lim_{M\to\infty} \ln(\ln x)\Big|_2^M = \lim_{M\to\infty}\{\ln(\ln M) - \ln(\ln 2)\} = \infty$ and the series diverges.

(d) $\lim_{M\to\infty} \int_1^M xe^{-x^2}\,dx = \lim_{M\to\infty} -\frac{1}{2}e^{-x2}\Big|_1^M = \lim_{M\to\infty}\left\{\frac{1}{2}e^{-1} - \frac{1}{2}e^{-M^2}\right\} = \frac{1}{2}e^{-1}$ and the series converges.

Note that when the series converges, the value of the corresponding integral is not (in general) the same as the sum of the series. However, the approximate sum of a series can often be obtained quite accurately by using integrals. See Problem 11.74.

11.14. Prove that Prove that $\frac{\pi}{4} < \sum_{n=1}^{\infty}\frac{1}{n^2+1} < \frac{1}{2}+\frac{\pi}{4}$.

From Problem 11.11 it follows that

$$\lim_{M\to\infty}\sum_{n=2}^{M}\frac{1}{n^2+1} < \lim_{M\to\infty}\int_1^M\frac{dx}{x^2+1} < \lim_{M\to\infty}\sum_{n=1}^{M-1}\frac{1}{n^2+1}$$

i.e., $\sum_{n=2}^{\infty}\frac{1}{n^2+1} < \frac{\pi}{4} < \sum_{n=1}^{\infty}\frac{1}{n^2+1}$, from which $\frac{\pi}{4} < \sum_{n=1}^{\infty}\frac{1}{n^2+1}$ as required.

Since $\sum_{n=2}^{\infty}\frac{1}{n^2+1} < \frac{\pi}{4}$, we obtain, on adding $\frac{1}{2}$ to each side, $\sum_{n=1}^{\infty}\frac{1}{n^2+1} < \frac{1}{2}+\frac{\pi}{4}$.

The required result is therefore proved.

Alternating series

11.15. Given the alternating series $a_1 - a_2 + a_3 - a_4 + \cdots$, where $0 \leqq a_{n+1} \leqq a_n$ and where $\lim_{n\to\infty} a_n = 0$, prove that (a) the series converges and (b) the error made in stopping at any term is not greater than the absolute value of the next term.

(a) The sum of the series to 2M terms is

$$S_{2M} = (a_1 - a_2) + (a_3 - (a_4) + \cdots + (a_{2M-1} - a_{2M})$$

$$= a_1 - (a_2 - a_3) - (a_4 - a_5) - \cdots - (a_{2M-2} - a_{2M-1}) - a_{2M}$$

Since the quantities in parentheses are nonnegative, we have

$$S_{2M} \geqq 0, \qquad S_2 \leqq S_4 \leqq S_6 \leqq S_8 \leqq \cdots \leqq S_{2M} \leqq a_1$$

Therefore, $\{S_{2M}\}$ is a bounded monotonic increasing sequence and thus has limit S.

Also, $S_{2M+1} = S_{2M} + a_{2M+1}$. Since $\lim_{M\to\infty} S_{2M} = S$ and $\lim_{M\to\infty} a_{2M+1} = 0$ (for, by hypothesis, $\lim_{M\to\infty} a_n = 0$), it follows that $\lim_{M\to\infty} S_{2M+1} = \lim_{M\to\infty} S_{2M} + \lim_{M\to\infty} a_{2M+1} = S + 0 = S$.

Thus, the partial sums of the series approach the limit S and the series converges.

(b) The error made in stopping after 2M terms is

$$(a_{2M+1} - a_{2M+2}) + (a_{2M+3} - a_{2M+4}) + \cdots = a_{2M+1} - (a_{2M+2} - a_{2M+3}) - \cdots$$

and is thus nonnegative and less than or equal to a_{2M+1}, the first term which is omitted.

Similarly, the error made in stopping after $2M + 1$ terms is

$$-a_{2M+2} + (a_{2M+3} - a_{2M+4}) + \cdots = -(a_{2M+2} - a_{2M+3}) - (a_{2M+4} - a_{2M+5}) - \cdots$$

which is nonpositive and greater than $-a_{2M+2}$.

11.16. (a) Prove that the series $\sum_{n=1}^{\infty} \frac{(-1)^{n+1}}{2n-1}$. converges. (b) Find the maximum error made in approximating the sum by the first eight terms and the first nine terms of the series. (c)How many terms of the series are needed in order to obtain an error which does not exceed .001 in absolute value?

(a) The series is $1 - \frac{1}{3} + \frac{1}{5} - \frac{1}{7} + \frac{1}{9} - \cdots$. If $u_n = \frac{(-1)^{n+1}}{2n-1}$, then $a_n = |u_n| = \frac{1}{2n-1}$, $a_{n+1} = |u_{n+1}| = \frac{1}{2n+1}$. Since $\frac{1}{2n+1} \leq \frac{1}{2n-1}$ and since $\lim_{n\to\infty} \frac{1}{2n-1} 0$, it follows by Problem 11.5(a) that the series converges.

(b) Use the results of Problem 11.15(b). Then the first eight terms give $1 - \frac{1}{3} + \frac{1}{5} - \frac{1}{7} + \frac{1}{9} - \frac{1}{11} + \frac{1}{13} - \frac{1}{15}$, and the error is positive and does not exceed $\frac{1}{17}$.

Similarly, the first nine terms are $1 - \frac{1}{3} + \frac{1}{5} - \frac{1}{7} + \frac{1}{9} - \frac{1}{11} + \frac{1}{13} - \frac{1}{15} + \frac{1}{17}$ and the error is negative and greater than or equal to $-\frac{1}{19}$; i.e., the error does not exceed $\frac{1}{19}$ in absolute value.

(c) The absolute value of the error made in stopping after M terms is less than 1/(2M + 1). To obtain the desired accuracy, we must have 1/(2M + 1) $\leq$ .001, form which M $\geq$ 499.5. Thus, at least 500 terms are needed.

Absolute and conditional convergence

11.17. Prove that an absolutely convergent series is convergent.

Given that $\Sigma |u_n|$ converges, we must show that Σu_n converges.
Let $S_M = u_1 + u_2 + \cdots + u_M$ and $T_M = |u_1| + |u_2| + \cdots + |u_M|$. Then

$$S_M + T_M = (u_1 + |u_1|) + (u_2 + |u_2|) + \cdots + (u_M + |u_M|)$$

$$\leq 2|u_1| + 2|u_2| + \cdots + 2|u_M|$$

Since $\Sigma |u_n|$ converges and since $u_n + |u_n| \geq 0$, for $n = 1, 2, 3, \ldots$, it follows that $S_M + T_M$ is a bounded monotonic increasing sequence, and so $\lim_{M\to\infty} (S_M + T_M)$ exists.

Also, since $\lim_{M\to\infty} T_M$ exists (since the series is absolutely convergent by hypothesis),

$$\lim_{M\to\infty} S_M = \lim_{M\to\infty} (S_M + T_M - T_M) = \lim_{M\to\infty} (S_M + T_M) - \lim_{M\to\infty} T_M$$

must also exist, and the result is proved.

11.18. Investigate the convergence of the series $\frac{\sin\sqrt{1}}{1^{3/2}} - \frac{\sin\sqrt{2}}{2^{3/2}} + \frac{\sin\sqrt{3}}{3^{3/2}} - \cdots$.

Since each term is, in absolute value, less than or equal to the corresponding term of the series $\frac{1}{1^{3/2}} + \frac{1}{2^{3/2}} + \frac{1}{3^{3/2}} + \cdots$ which converges, it follows that the given series is absolutely convergent and, hence, convergent by Problem 11.17.

11.19. Examine for convergence and absolute convergence:

(a) $\sum_{n=1}^{\infty} \frac{(-1)^{n-1}}{n^2+1}$

(b) $\sum_{n=2}^{\infty} \frac{(-1)^{n-1}}{n \ln^2 n}$

(c) $\sum_{n=1}^{\infty} \frac{(-1)^{n-1} 2^n}{n^2}$

(a) The series of absolute values is $\sum_{n=1}^{\infty} \frac{n^2}{n^2+1}$, which is divergent by Problem 11.13(b). Hence, the given series is not absolutely convergent

However, if $a_n = |u_n| = \frac{n}{n^2+1}$ and $a_{n+1} = |u_{n+1}| = \frac{n+1}{(n+1)^2+1}$, then $a_{n+1} \leqq a_n$ for all $n \geqq 1$, and also $\lim_{n\to\infty} a_n = \lim_{n\to\infty} \frac{n}{n^2+1} = 0$. Hence, by Problem 11.15 the series converges.

Since the series converges but is not absolutely convergent, it is *conditionally convergent.*

(b) The series of absolute values is $\sum_{n=2}^{\infty} \frac{1}{n \ln^2 n}$.

By the integral test, this series converges or diverges according as $\lim_{M\to\infty} \int_2^M \frac{dx}{x \ln^2 x}$ exists or does not exist.

If $u = \ln x$, $\int \frac{dx}{x \ln^2 x} = \int \frac{du}{u^2} = -\frac{1}{u} + c = -\frac{1}{\ln x} + c.$

Hence, $\lim_{M\to\infty} \int_2^M \frac{dx}{x \ln^2 x} = \lim_{M\to\infty} \left(\frac{1}{\ln 2} - \frac{1}{\ln M} \right) = \frac{1}{\ln 2}$, and the integral exists. Thus, the series converges.

Then $\sum_{n=2}^{\infty} \frac{(-1)^{n-1}}{n \ln^2 n}$ converges absolutely and thus converges.

Another method: Since $\frac{1}{(n+1)\ln^2(n+1)} \leqq \frac{1}{n \ln^2 n}$ and $\lim_{n\to\infty} \frac{1}{n \ln^2 n} = 0$, it follows by Problem 11.15(a), that the given alternating series converges. To examine its absolute convergence, we must proceed as before.

(c) Since $\lim_{n\to\infty} u_n \neq 0$ where $u_n = \frac{(-1)^{n-1} 2^n}{n^2}$, the given series cannot be convergent. To show that $\lim_{n\to\infty} u_n \neq 0$, it suffices to show that $\lim_{n\to\infty} |u_n| = \lim_{n\to\infty} \frac{2n}{n^2} \neq 0$. This can be accomplished by L'Hospital's rule or other methods [see Problem 11.21(b)].

Ratio test

11.20. Establish the ratio test for convergence.

Consider first the series $u_1 + u_2 + u_3 + \cdots$ where each term is nonnegative. We must prove that if $\lim_{n\to\infty} \frac{u_{n+1}}{u_n} L < 1$, then necessarily Σu_n converges.

By hypothesis, we can choose an integer N so large that for all $n \geqq N$, $(u_{n+1}/u_n) < r$ where $L < r < 1$. Then

$$u_{N+1} < r\, u_N$$
$$u_{N+2} < r\, u_{N+1} < r^2 u_N$$
$$u_{N+3} < r\, u_{N+2} < r^3 u_N$$

and so on. By addition,

$$u_{N+1} + u_{N+2} + \cdots < u_N(r + r^2 + r^3 + \cdots)$$

and so the given series converges by the comparison test, since $0 < r < 1$.

In case the series has terms with mixed signs, we consider $|u_1| + |u_2|\ |u_3| + \cdots$. By the preceding proof and Problem 11.17, it follows that if $\lim_{n\to\infty}\left|\frac{u_{n+1}}{u_n}\right| = L < 1$, then Σu_n converges (absolutely).

Similarly, we can prove that if $\lim_{n\to\infty}\left|\frac{u_{n+1}}{u_n}\right| = L > 1$ the series Σu_n diverges, while if $\lim_{n\to\infty}\left|\frac{u_{n+1}}{u_n}\right| = L = 1$ the ratio test fails [see Problem 11.21(c)].

11.21. Investigate the convergence of

(a) $\sum_{n=1}^{\infty} n^4 e^{-n^2}$

(b) $\sum_{n=1}^{\infty} \frac{(-1)^{n-1}\, 2^n}{n^2}$

(c) $\sum_{n=1}^{\infty} \frac{(-1)^{n-1}\, n}{n^2+1}$

(a) Here $u_n = e^{-n^2}$. Then

$$\lim_{n\to\infty}\left|\frac{u_{n+1}}{u_n}\right| = \lim_{n\to\infty}\left|\frac{(n+1)^4\, e^{-(n+1)^2}}{n^4\, e^{-n^2}}\right| = \lim_{n\to\infty}\frac{(n+1)^4\, e^{-(n^2+2n+1)}}{n^4\, e^{-n^2}}$$

$$= \lim_{n\to\infty}\left(\frac{n+1}{n}\right)^4 e^{-2n-1} = \lim_{n\to\infty}\left(\frac{n+1}{n}\right)^4 \lim_{n\to\infty} e^{-2n-1} = 1\cdot 0 = 0$$

Since $0 < 1$, the series converges.

(b) Here $u_n = \frac{(-1)^{n-1}\, 2^n}{n^2}$. Then

$$\lim_{n\to\infty}\left|\frac{u_{n+1}}{u_n}\right| = \lim_{n\to\infty}\left|\frac{(-1)^n 2^{n+1}}{(n+1)^2}\cdot\frac{n^2}{(-1)^{n-1}\, 2^n}\right| = \lim_{n\to\infty}\frac{2n^2}{(n+1)^2} = 2$$

Since $s > 1$, the series diverges. Compare Problem 11.19(c).

(c) Here $u_n = \frac{(-1)^{n-1}\, n}{n^2+1}$. Then

$$\lim_{n\to\infty}\left|\frac{u_{n+1}}{u_n}\right| = \lim_{n\to\infty}\left|\frac{(-1)^n\,(n+1)}{(n+1)^2+1}\cdot\frac{n^2+1}{(-1)^{n-1}\, n}\right| = \lim_{n\to\infty}\frac{(n+1)\,(n^2+1)}{(n^2+2n+2)^n} = 1$$

and the ratio test fails. By using other tests [see Problem 11.19(a)], the series is seen to be convergent.

Miscellaneous tests

11.22. Test for convergence $1 + 2r + r^2 + 2r^3 + r^4 + 2r^5 + \cdots$ where (a) $r = 2/3$, (b) $r = -2/3$, (c) $r = 4/3$.

Here the ratio test is inapplicable, since $\left|\frac{u_{n+1}}{u_n}\right| = 2|r|$ or $\frac{1}{2}|r|$, depending on whether n is odd or even.

However, using the nth root test, we have

$$\sqrt[n]{|u_n|} = \begin{cases} \sqrt[n]{2|r^n|} = \sqrt[n]{2}\,|r| & \text{if } n \text{ is odd} \\ \sqrt[n]{|r^n|} = |r| & \text{if } n \text{ is even} \end{cases}$$

Then $\lim_{n\to\infty} \sqrt[n]{|u_n|} = |r|$ (since $\lim_{n\to\infty} 2^{1/n} = 1$).

Thus, if $|r| < 1$ the series converges, and if $|r| > 1$ the series diverges.
Hence, the series converges for cases (a) and (b), and diverges in case (c).

11.23. Test for convergence $\left(\frac{1}{3}\right)^2 + \left(\frac{1\cdot 4}{3\cdot 6}\right)^2 + \left(\frac{1\cdot 4\cdot 7}{3\cdot 6\cdot 9}\right)^2 + \cdots + \left(\frac{1\cdot 4\cdot 7\ldots(3n-2)}{3\cdot 6\cdot 9\ldots(3n)}\right)^2 + \cdots$.

The ratio test fails, since $\lim_{n\to\infty}\left|\frac{u_{n+1}}{u_n}\right| = \lim_{n\to\infty}\left(\frac{3n+1}{3n+3}\right)^2 = 1$. However, by Raabe's test,

$$\lim_{n\to\infty} n\left(1 - \left|\frac{u_{n+1}}{u_n}\right|\right) = \lim_{n\to\infty} n\left\{1 - \left(\frac{3n+1}{3n+3}\right)^2\right\} = \frac{4}{3} > 1$$

and so the series converges.

11.24. Test for convergence $\left(\frac{1}{2}\right)^2 + \left(\frac{1\cdot 3}{2\cdot 4}\right)^2 + \left(\frac{1\cdot 3\cdot 5}{24t}\right)^2 + \cdots + \left(\frac{1\cdot 3\cdot 5\ldots(2n-1)^2}{2\cdot 4\cdot 6\ldots(2n)}\right) + \cdots$.

The ratio test fails, since $\lim_{n\to\infty}\left|\frac{u_{n+1}}{u_n}\right| = \lim_{n\to\infty}\left(\frac{2n+1}{2n+2}\right)^2 = 1$. Also, Raabe's test fails since

$$\lim_{n\to\infty} n\left(1 - \left|\frac{u_{n+1}}{u_n}\right|\right) = \lim_{n\to\infty} n\left\{1 - \left(\frac{2n+1}{2n+2}\right)^2\right\} = 1$$

However, using long division,

$$\left|\frac{u_{n+1}}{u_n}\right| = \left(\frac{2n+1}{2n+2}\right)^2 = 1 - \frac{1}{n} + \frac{5 - 4/n}{4n^2 + 8n + 4} = 1 - \frac{1}{n} + \frac{c_n}{n^2} \quad \text{where } |c_n| < P$$

so that the series diverges by Gauss's test.

Series of functions

11.25. For what values of x do the following series converge?

(a) $\sum_{n=1}^{\infty} \frac{x^{n-1}}{n\cdot 3^n}$

(b) $\sum_{n=1}^{\infty} \frac{(-1)^{n-1}\, x^{2n-1}}{(2n-1)!}$

(c) $\sum_{n=1}^{\infty} n!\,(x-a)^n$

(d) $\sum_{n=1}^{\infty} \frac{n\,(x-1)^n}{2^n\,(3n-1)}$

(a) $u_n = \frac{x^{n-1}}{n\cdot 3^n}$. Assuming x ≠ 0 (if x = 0 the series converges), we have

$$\lim_{n\to\infty}\left|\frac{u_{n+1}}{u_n}\right| = \lim_{n\to\infty}\left|\frac{x^n}{(n+1)\cdot 3^{n+1}}\cdot\frac{n\cdot 3^n}{x^{n-1}}\right| = \lim_{n\to\infty}\frac{n}{3\,(n+1)}|x| = \frac{|x|}{3}$$

Then the series converges if $\frac{|x|}{3} < 1$, and diverges if $\frac{|x|}{3} > 1$. If $\frac{|x|}{3} = 1$. i.e., $x = \pm 3$, the test fails.

If $x = 3$ the series becomes $\sum_{n=1}^{\infty} \frac{1}{3n} = \frac{1}{3}\sum_{n=1}^{\infty}\frac{1}{n}$, which diverges.

If $x = -3$ the series becomes $\sum_{n=1}^{\infty} \frac{(-1)^{n-1}}{3n} = \frac{1}{3}\sum_{n=1}^{\infty}\frac{(-1)^{n-1}}{n}$, which converges.

Then the *interval of convergence* is $-3 \leqq x < 3$. The series diverges outside this interval.

Note that the series converges absolutely for $-3 < x < 3$. At $x = -3$ the series converges conditionally.

(b) Proceed as in (a) with $u_n = \frac{(-1)^{n-1}\,x^{2n-1}}{(2n-1)!}$. Then

$$\lim_{n\to\infty}\left|\frac{u_{n+1}}{u_n}\right| = \lim_{n\to\infty}\left|\frac{(-1)^n\,x^{2n+1}}{(2n+1)!}\cdot\frac{(2n-1)!}{(-1)^{n-1}\,x^{2n-1}}\right| = \lim_{n\to\infty}\frac{(2n-1)!}{(2n+1)!}x^2$$

$$= \lim_{n\to\infty}\frac{(2n-1)!}{(2n+1)\,(2n)\,(2n-1)!}x^2 = \lim_{n\to\infty}\frac{x^2}{(2n+1)\,(2n)} = 0$$

Then the series converges (absolutely) for all x, i.e., the interval of (absolute) convergence is $-\infty < x < \infty$.

(c) $u_n = n!\,(x-a)^n$, $\lim_{n\to\infty}\left|\frac{u_{n+1}}{u_n}\right| = \lim_{n\to\infty}\left|\frac{(n+1)!\,(x-a)^{n+1}}{n!\,(x-a)^n}\right| = \lim_{n\to\infty}(n+1)\,|x-a|$.

This limit is infinite if $x \neq a$. Then the series converges only for $x = a$.

(d) $u_n = \frac{n(x-1)^n}{2^n\,(3n-1)}$, $u_{n+1} = \frac{(n+1)\,(x-1)^{n+1}}{2^{n+1}\,(3n+2)}$. Then

$$\lim_{n\to\infty}\left|\frac{u_{n+1}}{u_n}\right| = \lim_{n\to\infty}\left|\frac{(n+1)\,(3n-1)\,(x-1)}{2n\,(3n+2)}\right| = \left|\frac{x-1}{2}\right| = \frac{|x-1|}{2}$$

Thus, the series converges for $|x-1| < 2$ and diverges for $|x-1| > 2$.

The test fails for $|x-1| = 2$; i.e., $x - 1 = \pm 2$ or $x = 3$ and $x = -1$.

For $x = 3$ the series becomes $\sum_{n=1}^{\infty}\frac{n}{3n-1}$, which diverges, since the nth term does not approach zero.

For $x = -1$ the series becomes $\sum_{n=1}^{\infty}\frac{(-1)^n\,n}{3n-1}$, which also diverges, since the $n$the term does not approach zero.

Then the series converges only for $|x-1| < 2$; i.e., $-2 < x - 1 < 2$ or $-1 < x < 3$.

11.26. For what values of x does (a) $\sum_{n=1}^{\infty}\frac{1}{2n-1}\left(\frac{x+2}{x-1}\right)^n$ and (b) $\sum_{n=1}^{\infty}\frac{1}{(x+n)\,(x+n-1)}$ converge?

(a) $u_n = \frac{1}{2n-1}\left(\frac{x+2}{x-1}\right)^n$. Then $\lim_{n\to\infty}\left|\frac{u_{n+1}}{u_n}\right| = \lim_{n\to\infty}\frac{2n-1}{2n+1}\left|\frac{x+2}{x-1}\right| = \left|\frac{x+2}{x-1}\right|$ if $x \neq 1, -2$.

Then the series converges if $\left|\frac{x+2}{x-1}\right| < 1$, it diverges if $\left|\frac{x+2}{x-1}\right| > 1$, and the test fails if $\left|\frac{x+2}{x-1}\right| = 1$, i.e., $x = -\frac{1}{2}$.

If $x = 1$, the series diverges.

If $x = -2$, the series converges.

If $x - \frac{1}{2}$, the series is $\sum_{n=1}^{\infty}\frac{(-1)^n}{2n-1}$, which converges.

Thus, the series converges for $\left|\frac{x+2}{x-1}\right| < 1$, $x = -\frac{1}{2}$, and $x = -2$, i.e., for $x \leqq -\frac{1}{2}$.

(b) The ratio test fails, since $\lim_{n\to\infty}\left|\frac{u_{n+1}}{u_n}\right| = 1$, where $u_n = \frac{1}{(x+n)(x+n-1)}$. However, noting that

$$\frac{1}{(x+n)(x+n-1)} = \frac{1}{x+n-1} - \frac{1}{x+n}$$

we see that if $x \neq 0, -1, -2, \ldots, -n$,

$$S_n = u_1 + u_2 + \cdots + u_n = \left(\frac{1}{x} - \frac{1}{x+1}\right) + \left(\frac{1}{x+1} - \frac{1}{x+2}\right) + \cdots + \left(\frac{1}{x+n-1} - \frac{1}{x+n}\right)$$

$$= \frac{1}{x} - \frac{1}{x+n}$$

and $\lim_{n\to\infty} S_n = 1/x$, provided $x \neq 0, -1, -2, -3, \ldots$.

Then the series converges for all x except $x = 0, -1, -2, -3, \ldots$, and its sum is $1/x$.

Uniform convergence

11.27. Find the domain of convergence of $(1-x) + x(1-x) + x^2(1-x) + \cdots$.

Method 1:

$$\text{Sum of first } n \text{ terms } S_n(x) = (1-x) + x(1-x) + x^2(1-x) + \cdots + x^{n-1}(1-x)$$

$$= 1 - x + x - x^2 + x^2 + \cdots + x^{n-1} - x^n$$

$$= 1 - x^n$$

If $|x| < 1$, $\lim_{n\to\infty} S_n(x) = \lim_{n\to\infty}(1 - x^n) = 1$.

If $|x| > 1$. $\lim_{n\to\infty} S_n(x)$ does not exist.

If $x = 1$, $S_n(x) = 0$ and $\lim_{n\to\infty} S_n(x) = 0$.

If $x = -1$, $S_n(x) = 1 - (-1)^n$ and $\lim_{n\to\infty} S_n(x)$ does not exist.

Thus, the series converges for $|x| < 1$ and $x = 1$, i.e., for $-1 < x \leqq 1$.

Method 2, using the ratio test: The series converges if $x = 1$. If $x \neq 1$ and $u_n = x^{n-1}(1-x)$, then $\lim_{n\to\infty}\left|\frac{u_{n+1}}{u_n}\right| = \lim_{n\to\infty}|x|$.

Thus, the series converges if $|x| < 1$ and diverges if $|x| > 1$. The test fails if $|x| = 1$. If $x = 1$, the series converges; if $x = -1$, the series diverges. Then the series converges for $-1 < x \leqq 1$.

11.28. Investigate the uniform convergence of the series of Problem 11.27 in the interval (a) $-\frac{1}{2} < x < \frac{1}{2}$, (b) $-\frac{1}{2} \leqq x \leqq \frac{1}{2}$, (c) $-.99 \leqq x \leqq .99$, (d) $-1 < x < 1$, and (e) $0 \leqq x < 2$.

(a) By Problem 11.27, $S_n(x) = 1 - x^n$, $S(x) = \lim_{n\to\infty} S_n(x) = 1$ if $-\frac{1}{2} < x < \frac{1}{2}$; thus, the series converges in this interval. We have

$$\text{Remainder after } n \text{ terms} = R_n(x) = S(x) = S_n(x) = 1 - (1 - x^n) = x^n$$

The series is *uniformly convergent* in the interval if given any $\epsilon > 0$ we can find N dependent on ϵ, *but not on x*, such that $|R_n(x)| < \epsilon$ for all $n > N$. Now

$$|R_n(x)| = |x^n| = |x|^n < \varepsilon \quad \text{when} \quad n \ln|x| < \ln \varepsilon \quad \text{or} \quad n > \frac{\ln \varepsilon}{\ln|x|}$$

since division by $\ln|x|$ (which is negative, since $|x| < \frac{1}{2}$) reverses the sense of the inequality.

But if $|x| < \frac{1}{2}$, $\ln|x| < \ln\left(\frac{1}{2}\right)$, and $n > \frac{\ln \varepsilon}{\ln|x|} > \frac{\ln \varepsilon}{\ln\left(\frac{1}{2}\right)} = N$. Thus, since N is independent of x, the series is uniformly convergent in the interval.

(b) In this case $|x| \leqq \frac{1}{2} \cdot \ln|x| \leqq \ln\left(\frac{1}{2}\right)$, and $n > \frac{\ln \varepsilon}{\ln|x|} \geqq \frac{\ln \varepsilon}{\ln\left(\frac{1}{2}\right)} = N$, so that the series is also uniformly convergent in $-\frac{1}{2} \leqq x \leqq \frac{1}{2}$.

(c) Reasoning similarly, with $\frac{1}{2}$ replaced by .99, shows that the series is uniformly convergent in $-.99 \leqq x \leqq .99$.

(d) The arguments used here break down in this case, since $\frac{\ln \varepsilon}{\ln|x|}$ can be made larger than any positive number by choosing $|x|$ sufficiently close to 1. Thus, no N exists and it follows that the series is not uniformly convergent in $-1 < x < 1$.

(e) Since the series does not even converge at all points in this interval, it cannot converge uniformly in the interval.

11.29. Discuss the continuity of the sum function $S(x) = \lim_{n\to\infty} S_n(x)$ of Problem 11.27 for the interval $0 \leqq x \leqq 1$.

If $0 \leqq x < 1$, $S(x) = \lim_{n\to\infty} S_n(x) = \lim_{n\to\infty}(1 - x^n) = 1$.

If $x = 1$, $S_n(x) = 0$ and $S(x) = 0$.

Thus, $S(x) = \begin{cases} 1 & \text{if } 0 \leqq x < 1 \\ 0 & \text{if } x = 1 \end{cases}$ and $S(x)$ is discontinuous at $x = 1$ but continuous at all other points in $0 \leqq x < 1$.

In Problem 11.34 it is shown that if a series is uniformly convergent in an interval, the sum function $S(x)$ must be continuous in the interval. It follows that if the sum function is not continuous in an interval, the series cannot be uniformly convergent. This fact is often used to demonstrate the nonuniform convergence of a series (or sequence).

11.30. Investigate the uniform convergence of $x^2 + \frac{x^2}{1+x^2} + \frac{x^2}{(1+x^2)^2} + \cdots + \frac{x^2}{(1+x^2)^n} + \cdots$.

Suppose $x \neq 0$. Then the series is a geometric series with ratio $1/(1 + x^2)$ whose sum is (see Problem 2.25).

$$S(x) = \frac{x^2}{1 - 1/(1+x^2)} = 1 + x^2$$

If $x = 0$, the sum of the first n terms is $S_n(0) = 0$; hence, $S(0) = \lim_{n\to\infty} S_n(0) = 0$.

Since $\lim_{x\to 0} S(x) = 1 \neq S(0)$, $S(x)$ is discontinuous at $x = 0$. Then, by Problem 11.34, the series cannot be *uniformly convergent* in any interval which includes $x = 0$, although it is (absolutely) *convergent* in any interval. However, it is uniformly convergent in any interval which excludes $x = 0$.

This can also be shown directly (see Problem 11.93).

Weierstrass *M* test

11.31. Prove the Weierstrass M test; i.e., if $|u_n(x)| \leqq M_n$, $n = 1, 2, 3, \ldots$, where M_n are positive constants such that ΣM_n converges, then $\Sigma u_n(x)$ is uniformly (and absolutely) convergent.

The remainder of the series $\Sigma u_n(x)$ after n terms is $R_n(x) = u_{n+1}(x) + u_{n+2}\, x) + \cdots$. Now

$$|R_n(x)| = |u_{n+1}(x) + u_{n+2}(x) + \cdots| \leqq |u_{n+1}(x)| + |u_{n+2}(x)| + \cdots \leqq M_{n+1} + M_{n+2} + \cdots$$

But $M_{n+1} + M_{n+2} + \cdots$ can be made less than ϵ by choosing $n > N$, since ΣM_n converges. Since N is clearly independent of x, we have $|R_n(x)| < \epsilon$ for $n > N$, and the series is uniformly convergent. The absolute convergence follows at once from the comparison test.

11.32. Test for uniform convergence: (a) $\sum_{n=1}^{\infty} \frac{\cos nx}{n^4}$, (b) $\sum_{n=1}^{\infty} \frac{x^n}{n^{3/2}}$, (c) $\sum_{n=1}^{\infty} \frac{\sin x^n}{n}$, and (d) $\sum_{n=1}^{\infty} \frac{1}{n^2 + x^2}$,

(a) $\left|\frac{\cos nx}{n^4}\right| \leqq \frac{1}{n^4} = M_n$. Then, since ΣMn converges (p series with p = 4 > 1), the series is uniformly (and absolutely) convergent for all x by the M test.

(b) By the ratio test, the series converges in the interval $-1 \leqq x \leqq 1$; i.e., $|x| \leqq 1$.

For all x in this interval, $\left|\frac{x^n}{n^{3/2}}\right| = \frac{|x|^n}{n^{3/2}} \leqq \frac{1}{n^{3/2}}$. Choosing $M_n = \frac{1}{n^{3/2}}$, we see that ΣM_n converges. Thus, the given series converges uniformly for $-1 \leqq x \leqq 1$ by the M test.

(c) $\left|\frac{\sin nx}{n^4}\right| \leqq \frac{1}{n}$. However, ΣM_n where $M_n = \frac{1}{n}$ does not converge. The M test cannot be used in this case and we cannot conclude anything about the uniform convergence by this test (see, however, Problem 11.125).

(d) $\left|\frac{1}{n^2 + x^2}\right| \leqq \frac{1}{n^2}$, and $\sum \frac{1}{n^2}$ converges. Then, by the M test, the given series converges uniformly for all x.

11.33. If a power series $\Sigma a_n x^n$ converges for $x = x_0$, Prove that it converges. (a) absolutely in the interval $|x| \leqq |x_0|$ and (b) uniformly in the interval $|x| \leqq |x_1|$, where $|x_1| < |x_0|$.

(a) Since $\Sigma a_n x^n{}_0$ converges, $\lim_{n\to\infty} a_n x_0^n = 0$, and so we can make $|a_n x_0^n| < 1$ by choosing n large enough; i.e., $|a_n| < |a_n| < \frac{1}{|x_0|^n}$ for $n > N$. Then

$$\sum_{N+1}^{\infty} |a_n x^n| = \sum_{N+1}^{\infty} |a_n||x|^n < \sum_{N+1}^{\infty} \frac{|x|^n}{|x_0|^n} \tag{1}$$

Since the last series in Equation (1) converges for $|x| < |x_0|$, it follows by the comparison test that the first series converges; i.e., the given series is absolutely convergent.

(b) Let $M_n = \frac{|x_1|^n}{|x_0|^n}$. Then ΣM_n converges, since $|x_1| < |x_0|$. As in (a), $|a_n x^n| < M_n$ for $|x| \leqq |x_1|$, so that by the Weierstrass M test, $\Sigma a_n x^n$ is uniformly convergent.

It follows that a power series is uniformly convergent in any interval *within* its interval of convergence.

Theorems on uniform convergence

11.34. Prove Theorem 6, Page 284.

We must show that $S(x)$ is continuous in $[a, b]$.

Now $S(x) = S_n(x) + R_n(x)$, so that $S(x + h) = S_n(x + h) + R_n(x + h)$ and thus,

$$S(x + h) - S(x) = S_n(x) + h) - S_n(x) + R_n(x + h) - R_n(x) \tag{1}$$

where we choose h so that both x and $x + h$ lie in $[a, b]$ (if $x = b$, for example, this will require $h < 0$).

Since $S_n(x)$ is a sum of a finite number of continuous functions, it must also be continuous. Then, given $\epsilon > 0$, we can find δ so that

$$|S_n(x+h) - S_n(x)| < \epsilon/3 \text{ whenever } |h| < \delta \tag{2}$$

Since the series, by hypothesis, is uniformly convergent, we can choose N so that

$$|R_n(x)| < \varepsilon/3 \text{ and } |\,|R_n(x+h)| < \varepsilon/3 \text{ for } n > N \tag{3}$$

Then from Equations (*1*), (*2*), and (*3*),

$$|S(x+h) - S(x)| \leq |S_n(x)+h) - S_n(x)| + |R_n(x+h)| + |R_n(x)| < \epsilon$$

for $|h| < \delta$, and so the continuity is established.

11.35. Prove Theorem 7, Page 285.

If a function is continuous in $[a, b]$, its integral exists. Then, since $S(x)$, $S_n(x)$ and $R_n(x)$ are continuous,

$$\int_a^a S(x) \int_a^b S_n(x)\,dx + \int_a^b R_n(x)\,dx$$

To prove the theorem we must show that

$$\left|\int_a^b S(x)\,dx - \int_a^b S_n(x)\,dx\right| = \left|\int_a^b R_n(x)\,dx\right|$$

can be made arbitrarily small by choosing n large enough. This, however, follows at once, since by the uniform convergence of the series we can make $|R_n(x)| < \epsilon/(b-a)$ for $n > N$ independent of x in $[a, b]$, and so

$$\left|\int_a^b R_n(x)\,dx\right| \leq \int_a^b |R_n(x)|\,dx < \int_a^b \frac{\varepsilon}{b-a}\,dx = \varepsilon$$

This is equivalent to the statements

$$\int_a^b S(x)\,dx = \lim_{n\to\infty} \int_a^b S_n(x)\,dx \qquad \text{or} \qquad \lim_{n\to\infty} \int_a^b S_n(x)\,dx = \int_a^b \left\{\lim_{n\to\infty} S_n(x)\right\} dx$$

11.36. Prove Theorem 8, Page 285.

Let $g(x) = \sum_{n=1}^{\infty} u_n'(x)$. Since, by hypothesis, this series converges uniformly in $[a, b]$, we can integrate term by term (by Problem 11.35) to obtain

$$\int_a^x g(x)\,dx = \sum_{n=1}^{\infty} \int_a^x u_n'(x)\,dx = \sum_{n=1}^{\infty} \{u_n(x) - u_n(a)\}$$

$$= \sum_{n=1}^{\infty} u_n(x) - \sum_{n=1}^{\infty} u_n(a) = S(x) - S(a)$$

because, by hypothesis, $\sum_{n=1}^{\infty} u_n(x)$ converges to $S(x)$ in $[a, b]$.

Differentiating both sides of $\int_0^x g(x)\,dx = S(x) - S(a)$ then shows that $g(x) = S'(x)$, which proves the theorem.

11.37. Let $S_n(x) = nxe^{-nx^2}$, $n = 1, 2, 3, \ldots$, $0 \leq x \leq 1$. (a) Determine whether $\lim_{n\to\infty} \int_0^1 S_n(x)\,dx = \int_0^1 \lim_{n\to\infty} S_n(x)\,dx$. (b) Explain the result in (a).

(a) $\int_0^1 S_n(x)\,dx = \int_0^1 nxe^{-nx^2}\,dx = -\frac{1}{2}e^{-nx^2}\Big|_0^1 = \frac{1}{2}(1 - e^{-n})$. Then

$$\lim_{n\to\infty} \int_0^1 S_n(x)\,dx = \lim_{n\to\infty} \frac{1}{2}(1 - e^{-n}) = \frac{1}{2}$$

$S(x) = \lim_{n\to\infty} S_n(x) = \lim_{n\to\infty} nxe^{-nx^2} = 0.$ whether $x = 0$ or $< x \leqq 1$. Then.

$$\int_0^1 S(x)\,dx = 0$$

It follows that $\lim_{n\to\infty} \int_0^1 S_n(x)\,dx \neq \int_0^1 \lim_{n\to\infty} S_n(x)\,dx$; i.e., the limit cannot be taken under the integral sign.

(b) The reason for the result in (a) is that although the sequence Sn(x) converges to 0. it does not converge uniformly to 0. To show this, observe that the function nxe^{-nx^2} has a maximum at $x = 1/\sqrt{2n}$ (by the usual rules of elementary calculus), the value of this maximum being $\sqrt{\frac{1}{2}}\, n\, e^{-1/2}$. Hence, as n → ∞, Sn(x) cannot be made arbitrarily small for all x and so cannot converge uniformly to 0.

11.38. Let $f(x) = \sum_{n=1}^{\infty} \frac{\sin nx}{n^3}$. Prove that $\int_0^{\pi} f(x)\,dx = 2\sum_{n=1}^{\infty} \frac{1}{(2n-1)^4}$.

We have $\left|\frac{\sin nx}{n^3}\right| \leqq \frac{1}{n^3}$. Then, by the Weierstrass *M* test, the series is uniformly convergent for all *x*, in particular $0 \leqq x \leqq \pi$, and can be integrated term by term. Thus,

$$\int_0^{\pi} f(x)\,dx = \int_0^{\pi}\left(\sum_{n=1}^{\infty} \frac{\sin nx}{n^3}\right)dx \sum_{n=1}^{\infty}\int_0^{\pi} \frac{\sin nx}{n^3}\,dx$$

$$= \sum_{n=1}^{\infty} \frac{1-\cos n\pi}{n^4} = 2\left(\frac{1}{1^4} + \frac{1}{3^4} + \frac{1}{5^4} + \cdots\right) = 2\sum_{n=1}^{\infty} \frac{1}{(2n-1)^4}$$

Power series

11.39. Prove that both the power series $\sum_{n=0}^{\infty} a_n x^n$ and the corresponding series of derivatives $\sum_{n=0}^{\infty} na_n x^{n-1}$ have the same radius of convergence.

Let $R > 0$ be the radius of convergence of $\Sigma a_n x^n$. Let $0 < |x_0| < R$. Then, as in Problem 11.33, we can choose *N* as that $| |a_n| < \frac{1}{|x_0|^n}$ for $n > N$.

Thus, the terms of the series $\Sigma |na_n x^{n-1}| = \Sigma n |a_n|\, |x|^{n-1}$ can for $n > N$ be made less than corresponding terms of the series $\sum n \frac{|x|^{n-1}}{|x_0|^n}$, which converges, by the ratio test, for $|x| < |x_0| < R$.

Hence, $\Sigma na_n x^{n-1}$ converges absolutely for all points x_0 (no matter how close $|x_0|$ is to *R*).
If, however, $|x| > R$, $\lim_{n\to\infty} a_n x^n \neq 0$ and thus $\lim_{n\to\infty} na_n x^{n-1} \neq 0$, so that $\Sigma na_n x^{n-1}$ does not converge.

Thus, *R* is the radius of convergence of $\Sigma na_n x^{n-1}$.
Note that the series of derivatives may or may not converge for values of *x* such that $|x| = R$.

11.40. Illustrate Problem 11.39 by using the series $\sum_{n=1}^{\infty} \frac{x^n}{n^2 \cdot 3^n}$.

$$\lim_{n\to\infty}\left|\frac{u_{n+1}}{u_n}\right| = \lim_{n\to\infty}\left|\frac{x^{n+1}}{(n+1)^2 \cdot 3^{n+1}} \cdot \frac{n^2 \cdot 3n}{x^n}\right| = \lim_{n\to\infty} \frac{n^2}{3(n+1)^2}|x| = \frac{|x|}{3}$$

so that the series converges for $|x| < 3$. At $x = \pm 3$ the series also converges, so that the interval of convergence is $-3 \leqq x \leqq 3$.

The series of derivatives is

$$\sum_{n=1}^{\infty} \frac{nx^{n-1}}{n^2 \cdot 3n} = \sum_{n=1}^{\infty} \frac{x^{n-1}}{n \cdot 3n}$$

By Problem 11.25(a), this has the interval of convergence $-3 \leqq x < 3$.

The two series have the same radius of convergence, i.e., $R = 3$, although they do not have the same interval of convergence.

Note that the result of Problem 11.39 can also be proved by the ratio test *if* this test is applicable. The proof given there, however, applies even when the test is not applicable, as in the series in Problem 11.22.

11.41. Prove that in any interval *within* its interval of convergence, a power series (a) represents a continuous function, say, f(x); (b) can be integrated term by term to yield the integral of f(x); and (c) can be differentiated term by term to yield the derivative of f(x).

We consider the power series $\Sigma a_n x^n$, although analogous results hold for $\Sigma a_n(x-a)^n$.

(a) This follows from Problem 11.33 and 11.34, and the fact that each term anxn of the series is continuous.

(b) This follows from Problems 11.33 and 11.35, and the fact that each term anxn of the series is continuous and thus integrable.

(c) From Problem 11.39, the series of derivatives of a power series always converges within the interval of convergence of the original power series and therefore is uniformly convergent within this interval. Thus, the required result follows from Problems 11.33 and 11.36.

If a power series converges at one (or both) endpoints of the interval of convergence, it is possible to establish (a) and (b) to include the endpoint (or endpoints). See Problem 11.42.

11.42. Prove Abel's theorem that if a power series converges at an endpoint of its interval of convergence, then the interval of uniform convergence includes this endpoint.

For simplicity in the proof, we assume the power series to be $\sum_{k=0}^{\infty} a_k x^k$ with the endpoint of its interval of convergence at $x = 1$, so that the series surely converges for $0 \leqq x \leqq 1$. Then we must show that the series converges uniformly in this interval.

Let

$$R_n(x) = a_n x^n + a_{n+1} x^{n+1} + a_{n+2} x^{n+2} + \cdots, \qquad R_n = a_n + a\, a_{n+2} + \cdots$$

To prove the required result we must show that given any $\epsilon > 0$, we can find N such that $|R_{n \text{ all } n > N,}$ where N is independent of the particular x in $0 \leqq x \leqq 1$.

Now

$$\begin{aligned} R_n(x) &= (R_n - R_{n+1})\, x^n + (R_{n+1} - R_{n+2})\, x^{n+1} + (R_{n+2} - R_{n+3})\, x^{n+2} + \cdots \\ &= R_n x^n + R_{n+1}(x^{n+1} - x^n) + R_{n+2}(x^{n+2} - x^{n+1}) + \cdots \\ &= x^n \{R_n - (1-x)(R_{n+1} + R_{n+2}\, x + R_{n+3}\, x^2 + \cdots)\} \end{aligned}$$

Hence, for $0 \leqq x < 1$,

$$|R_n(x)| \leqq |R_n| + (1-x)(|R_{n+1}| + |R_{n+2}|x + |R_{n\,|Rn+3}|\, x^2 + \cdots) \tag{1}$$

Since Σa_k converges by hypothesis, it follows that given $\epsilon > 0$, we can choose N such that $|R_k| < \epsilon/2$ for all $k \geqq n$. Then for $n > N$ we have, from Equation (1),

$$|R_n(x)| \leq \frac{\epsilon}{2} + (1-x)\left(\frac{\epsilon}{2} + \frac{\epsilon}{2} x + \frac{\epsilon}{2} x^2 + \cdots\right) = \frac{\epsilon}{2} + \frac{\epsilon}{2} = \epsilon \tag{2}$$

since $(1-x)(1 + x + x^2 + x^3 + \ldots) = 1$ (if $0 \leqq x < 1$).

Also, for $x = 1$, $|R_n(x)| = |R_n| < \epsilon$ for $n > N$.

Thus, $|R_n(x)| < \epsilon$ for all $n > N$, where N is independent of the value of x in $0 \leqq x \leqq 1$, and the required result follows.

Extensions to other power series are easily made.

11.43. Prove Abel's limit theorem (see Page 286).

As in Problem 11.42, assume the power series to be $\sum_{k=1}^{\infty} a_k x^k$, convergent for $0 \leqq x \leqq 1$.

Then we must show that $\lim_{x\to 1-} \sum_{k=0}^{\infty} a_k x^k = \sum_{k=0}^{\infty} a_k$.

This follows at once from Problem 11.42, which shows that $\Sigma a_k x^k$ is uniformly convergent for $0 \leqq x \leqq 1$, and from Problem 11.34, which shows that $\Sigma a_k x^k$ is continuous at $x = 1$.

Extensions to other power series are easily made.

11.44. (a) Prove that $\tan^{-1} x = x - \frac{x^3}{3} + \frac{x^5}{5} - \frac{x^7}{7} + \cdots$ where the series is uniformly convergent in $-1 \leqq x \leqq 1$.

(b) Prove that $\frac{\pi}{4} = 1 - \frac{1}{3} + \frac{1}{5} - \frac{1}{7} + \cdots$

(a) By Problem 2.25, with r = – x2 and a = 1, we have

$$\frac{1}{1+x^2} = 1 - x^2 + x^4 - x^6 + \cdots \quad -1 < x < 1 \tag{1}$$

Integrating from 0 to x, where $-1 < x < 1$, yields

$$\int_0^x \frac{dx}{1+x^2} = \tan^{-1} x = x - \frac{x^3}{3} + \frac{x^5}{5} - \frac{x^7}{7} + \cdots \tag{2}$$

using Problems 11.33 and 11.35.

Since the series on the right of Equation (2) converges for $x = \pm 1$, it follows by Problem 11.42 that the series is uniformly convergent in $-1 \leqq x \leqq 1$ and represents $\tan^{-1} x$ in this interval

(b) By Problem 11.43 and (a), we have

$$\lim_{x\to 1-} \tan^{-1} x = \lim_{x\to 1-} \left(x - \frac{x^3}{3} + \frac{x^5}{5} - \frac{x^7}{7} + \cdots \right) \quad \text{or} \quad \frac{\pi}{4} = 1 - \frac{1}{3} + \frac{1}{5} - \frac{1}{7} + \cdots$$

11.45. Evaluate $\int_0^1 \frac{1-e^{-x2}}{x^2}\,dx$ to three-decimal-place accuracy.

We have $e^u\ 1 + u + \frac{u^2}{2!} + \frac{u^3}{3!} + \frac{u^4}{4!} + \frac{u^5}{5!} + \cdots, \quad -\infty < u < \infty.$

Then, if $u = -x^2$, $e^{-x^2} = 1 - x^2 + \frac{x^4}{2!} - \frac{x^6}{3!} + \frac{x^8}{3!} = \frac{x^{10}}{5!} + \cdots, \quad -\infty < x < \infty.$

Thus, $\frac{1-e^{-x^2}}{x^2} = 1 - \frac{x^2}{2!} + \frac{x^4}{3!} - \frac{x^6}{4!} + \frac{x^8}{5!} - \cdots.$

Since the series converges for all x and so, in particular, converges uniformly for $0 \leqq x \leqq 1$, we can integrate term by term to obtain

$$\int_0^1 \frac{1-e^{-x^2}}{x^2}\,dx = x - \frac{x^3}{3\cdot 2!} + \frac{x^5}{5\cdot 3!} - \frac{x^7}{7\cdot 4!} + \frac{x^9}{9\cdot 5!} - \cdots \Bigg|_0^1$$

$$= 1 - \frac{1}{3\cdot 2!} + \frac{1}{5\cdot 3!} - \frac{1}{7\cdot 4!} + \frac{1}{9\cdot 5!} - \cdots$$

$$= 1 - 0.16666 + 0.03333 - 0.00595 + 0.00092 - \cdots = 0.862$$

Note that the error made in adding the first four terms of the alternating series is less than the fifth term, i.e., less than 0.001 (see Problem 11.15).

Miscellaneous problems

11.46. Prove that $y = J_p(x)$ defined by Equation (16), Page 287, satisfies Bessel's differential equation:

$$x^2y \cdot + xy' + (x^2 - p^2)\,y = 0.$$

The series for $J_p(x)$ converges for all x [see Problem 11.10 (a)]. Since a power series can be differentiated term by term within its interval of convergence, we have for all x,

$$y = \sum_{n=0}^{\infty} \frac{(-1)^n\, x^{p+2n}}{2^{p+2n}\, n!(n+p)!}$$

$$y' = \sum_{n=0}^{\infty} \frac{(-1)^n\,(p+2n)\, x^{p+2n-1}}{2^{p+2n}\, n!(n+p)!}$$

$$y'' = \sum_{n=0}^{\infty} \frac{(-1)^n\,(p+2n)(p+2n-1)\, x^{p+2n-2}}{2^{p+2n}\, n!(n+p)!}$$

Then,

$$(x^2 - p^2)\,y = \sum_{n=0}^{\infty} \frac{(-1)^n\, x^{p+2n+2}}{2^{p+2n}\, n!(n+p)!} - \sum_{n=0}^{\infty} \frac{(-1)^n\, p^2\, x^{p+2n}}{2^{p+2n}\, n!(n+p)!}$$

$$xy' = \sum_{n=0}^{\infty} \frac{(-1)^n\,(p+2n)\, x^{p+2n}}{2^{p+2n}\, n!(n+p)!}$$

$$y'' = \sum_{n=0}^{\infty} \frac{(-1)^n\,(p+2n)(p+2n-1)\, x^{p+2n}}{2^{p+2n}\, n!(n+p)!}$$

Adding,

$$\begin{aligned}
x^2y'' + xy' + (x^2 - p^2)\,y &= \sum_{n=0}^{\infty} \frac{(-1)^n\, x^{p+2n+2}}{2^{p+2n}\, n!(n+p)!} \\
&\quad + \sum_{n=0}^{\infty} \frac{(-1)^n\,[-p^2 + (p+2n) + (p+2n)(p+2n-1)]\, x^{p+2n}}{2^{p+2n}\, n!\,(n+p)!} \\
&= \sum_{n=0}^{\infty} \frac{(-1)^n\, x^{p+2n+2}}{2^{p+2n}\, n!\,(n+p)!} + \sum_{n=0}^{\infty} \frac{(-1)^n\,[4n(n+p)]\, x^{p+2n}}{2^{p+2n}\, n!\,(n+p)!} \\
&= \sum_{n=1}^{\infty} \frac{(-1)^n\, 4x^{p+2n}}{2^{p+2n-2}\,(n-1)!\,(n-1+p)!} + \sum_{n=1}^{\infty} \frac{(-1)^n\, 4x^{p+2n}}{2^{p+2n}\,(n-1)!\,(n+p-1)!} \\
&= -\sum_{n=1}^{\infty} \frac{(-1)^n\, 4x^{p+2n}}{2^{p+2n}\,(n-1)!\,(n+1-p)!} + \sum_{n=1}^{\infty} \frac{(-1)^n\, 4x^{p+2n}}{2^{p+2n}\,(n-1)!\,(n+p-1)!} \\
&= 0
\end{aligned}$$

11.47. Test for convergence the complex power series $\displaystyle\sum_{n=1}^{\infty} \frac{z^{n-1}}{n^3 \cdot 3^{n-1}}$.

Since $\displaystyle\lim_{n\to\infty}\left|\frac{u_{n+1}}{u_n}\right| = \lim_{n\to\infty}\left|\frac{z_n}{(n+1)^3 \cdot 3^n} \cdot \frac{n^3 \cdot 3^{n-1}}{z^{n-1}}\right| = \lim_{n\to\infty} \frac{n^3}{3(n+1)^3}|z| = \frac{|z|}{3}$, the series converges for $\dfrac{|z|}{3} < 1$, i.e., $|z| < 3$, and diverges for $|z| > 3$.

For $|z| = 3$, the series of absolute values is $\sum_{n=1}^{\infty} \frac{|z|^{n-1}}{n^3 \cdot 3^{n-1}} = \sum_{n=1}^{\infty} \frac{1}{n^3}$, so that the series is absolutely convergent and thus convergent for $|z| = 3$.

Thus, the series converges within and on the circle $|z| = 3$.

11.48. Assuming the power series for e^x holds for complex numbers, show that $e^{ix} = \cos x + i \sin x$.

Letting $z = ix$ in $e^2 = 1 + z + \frac{z^2}{2!} + \frac{z^3}{3!} + \cdots$, we have

$$e^{ix} = 1 + ix + \frac{i^2x^2}{2!} + \frac{i^3x^3}{3!} + \cdots = \left(1 - \frac{x^2}{2!} + \frac{x^4}{4!} - \cdots\right) + i\left(x - \frac{x^3}{3!} + \frac{x^5}{5!} - \cdots\right)x$$
$$= \cos x + i \sin x$$

Similarly, $e^{-ix} = \cos x - i \sin x$. The results are called *Euler's identities*.

11.49. Prove that $\lim_{n\to\infty}\left(1 + \frac{1}{2} + \frac{1}{3} + \frac{1}{4} + \cdots + \frac{1}{n} - \text{In } n\right)$ exists.

Letting $f(x) = 1/x$ in Equation (1), Problem 11.11, we find

$$\frac{1}{2} + \frac{1}{3} + \frac{1}{4} + \cdots + \frac{1}{M} \leqq \ln M \leqq 1 + \frac{1}{2} + \frac{1}{3} + \frac{1}{4} + \cdots + \frac{1}{M-1}$$

from which, on replacing M by n, we have

$$\frac{1}{n} \leqq 1 + \frac{1}{2} + \frac{1}{3} + \frac{1}{4} + \cdots + \frac{1}{n} - \text{In } n \leqq 1$$

Thus, the sequence $S_n = 1 + \frac{1}{2} + \frac{1}{3} + \frac{1}{4} + \cdots + \frac{1}{n} - \text{In } n$ is bounded by 0 and 1.

Consider $S_{n+1} - S_n = \frac{1}{n+1} - \ln\left(\frac{n+1}{n}\right)$. By integrating the inequality $\frac{1}{n+1} \leqq \frac{1}{x} \leqq \frac{1}{n}$ with respect to x from n to $n + 1$, we have

$$\frac{1}{n+1} \leqq \ln\left(\frac{n+1}{n}\right) \leqq \frac{1}{n} \quad \text{or} \quad \frac{1}{n+1} - \frac{1}{n} \leqq \frac{1}{n+1} - \ln\left(\frac{n+1}{n}\right) \leqq 0$$

i.e., $S_{n+1} - S_n \leqq 0$, so that S_n is monotonic decreasing.

Since S_n is bounded and monotonic decreasing, it has a limit. This limit, denoted by γ, is equal to 0.577215 . . . and is called *Euler's constant*. It is not yet known whether γ is rational or not.

11.50. Prove that the infinite product $\prod_{k=1}^{\infty}(1 + u_k)$, converges $\sum_{k=1}^{\infty} u_k$ converges.

According to the Taylor series for e^x (Page 289), $1 + x \leqq e^x$ for $x > 0$, so that

$$P_n = \prod_{k=1}^{n}(1 + u_k) = (1 + u_1)(1 + u_2)\cdots(1 + u_n) \leqq e^{u_1} . e^{u_2} \cdots e^{u_n} = e^{u_1 + u_2 + \cdots u_n}$$

Since $u_1 + u_2 + \cdots$ converges, it follows that P_n is a bounded monotonic increasing sequence and so has a limit, thus proving the required result.

11.51. Prove that the series $1 - 1 + 1 - 1 + 1 - 1 + \cdots$ is $C - 1$ summable to 1/2.

The sequence of partial sums is 1, 0, 1, 0, 1, 0, . . .

Then $S_1 = 1, \frac{S_1 + S_2}{2} = \frac{1+0}{2} = \frac{1}{2}, \frac{S_1 + S_2 + S_3}{3} = \frac{1+0+1}{3} = \frac{2}{3}, \ldots$.

Continuing in this manner, we obtain the sequence $1. \frac{1}{2}, \frac{2}{3}, \frac{1}{2}, \frac{3}{5}, \frac{1}{2}, \ldots$, the *n*th term being $T_n = \begin{cases} 1/2 & \text{if } n \text{ is even} \\ n/(2n-1) & \text{if } n \text{ is odd} \end{cases}$. Thus, $\lim_{n\to\infty} T_n = \frac{1}{2}$ and the required result follows.

11.52. (a) If $f^{(n+1)}(x)$ is continuous in $[a, b]$ prove that for c in $[a, b]$, $f(x) = f(c) + f'(c)(x-c) + \frac{1}{2!} f''(c)(x-c)^2 + \cdots + \frac{1}{n!} f^{(n)}(c)(x-c)^n + \frac{1}{n!}\int_c^x (x-1)^n f^{(n+1)}(t)\,dt$. (b) Obtain the Lagrange and Cauchy forms of the remainder in Taylor's formula. (See Page 290.)

The proof of (a) is made using mathematical induction. (See Chapter 1.) The result holds for $n = 0$, since

$$f(x) = f(c) + \int_c^x f'(t)\,dt = f(c) + f(x) - f(c)$$

We make the induction assumption that it holds for $n = k$ and then use integration by parts with

$$d\upsilon = \frac{(x-t)^k}{k!}\,dt \text{ and } u = f^{k+1}(t)$$

Then

$$\upsilon = -\frac{(x-t)^{k+1}}{(k+1)!} \quad \text{and } du = f^{k+2}(t)\,dt$$

Thus,

$$\frac{1}{k!}\int_C^x (x-t)^k f^{(k+1)}(t)\,dt = -\frac{f^{k+1}(x-t)^{k+1}}{(k+1)!}\Big|_C^X + \frac{1}{(k+1)!}\int_C^x (x-t)^{k+1} f^{(k+2)}(t)\,dt$$

$$= -\frac{f^{k+1}(x-t)^{k+1}}{(k+1)!} + \frac{1}{(k+1)!}\int_C^x (x-t)^{k+1} f^{(k+2)}(t)\,dt$$

Having demonstrated that the result holds for $k + 1$, we conclude that it holds for all positive integers.

To obtain the Lagrange form of the remainder R_n, consider the form

$$f(x) = f(c) + f'(c)(x-c) + \frac{1}{2!} f''(c)(x-c)^2 + \cdots + \frac{K}{n!}(x-c)^n$$

This is the Taylor polynomial $P_{n-1}(x)$ plus $\frac{K}{n!}(x-c)^n$. Also, it could be looked upon as P_n except that in the last term, $f^{(n)}(c)$ is replaced by a number K such that for fixed c and x the representation of $f(x)$ is exact. Now define a new function

$$\Phi(t) = f(t) - f(x) + \sum_{j=1}^{n-1} f^{(j)}(t)\frac{(x-t)^j}{j!} + \frac{K(x-t)^n}{n!}$$

The function Φ satisfies the hypothesis of Rolle's Theorem in that $\Phi(c) = \Phi(x) = 0$, the function is continuous on the interval bound by c and x, and Φ' exists at each point of the interval. Therefore, there exists ξ in the interval such that $\Phi'(\xi) = 0$. We proceed to compute Φ' and set it equal to zero.

$$\Phi'(t) = f'(t) + \sum_{j=1}^{n-1} f^{(j+1)}(t)\frac{(x-t)^j}{j!} - \sum_{j=1}^{n-1} f^{(j)}(t)\frac{(x-t)^{j-1}}{(j-1)!} - \frac{K(x-t)^{n-1}}{(n-1)!}$$

This reduces to

$$\Phi'(t) = \frac{f^{(n)}(t)}{(n-1)!}(x-t)^{n-1} - \frac{K}{(n-1)!}(x-t)^{n-1}$$

According to hypothesis, for each n there is ξ_n such that

$$\Phi(\xi_n) = 0$$

Thus,

$$K = f^{(n)}(\xi_n)$$

and the Lagrange remainder is

$$R_{n-1} = \frac{f^{(n)}(\xi_n)}{n!}(x-c)^n$$

or, equivalently,

$$R_n = \frac{1}{(n+1)!} f^{(n+1)}(\xi_{n+1})(x-c)^{n+1}$$

The Cauchy form of the remainder follows immediately by applying the mean value theorem for integrals. (See Page 287.)

11.53. Extend Taylor's theorem to functions of two variables x and y.

Define $F(t) = f(x_0 + ht, y_0 + kt)$; then, applying Taylor's theorem for one variable (about $t = 0$),

$$F(t) = F(0) + F'(0) + \frac{1}{2!}F''(0)\,t^2 + \cdots + \frac{1}{n!}F^{(n)}(0)t^n + \frac{1}{(n+1)!}F^{(n+1)}(\theta)t^{n+1},\ 0 < \theta < t$$

Now let $t = 1$

$$F(1) = f(x_0 + h, y_0 + k) = F(0) + F'(0) + \frac{1}{2!}F''(0) + \cdots + \frac{1}{n!}F^{(n)}(0) + \frac{1}{(n+1)!}F^{(n+1)}(\theta)$$

When the derivatives $F'(t), \ldots, F^{(n)}(t), F^{(n+1)}(\theta)$ are computed and substituted into the previous expression, the two-variable version of Taylor's formula results. (See Page 290, where this form and notational details can be found.)

11.54. Expand $x^2 + 3y - 2$ in powers of $x - 1$ and $y + 2$. Use Taylor's formula. with $h = x - x_0$, $k = y - y_0$, where $x_0 = 1$ and $y_0 = -2$.

$x^2 + 3y - 2 = -10 - 4(x-1) + 4(y+2) - 2(x-1)^2 + 2(x-1)(y+2) + (x-1)^2(y+2)$ (Check this algebraically.)

11.55. Prove that $\ln \frac{x+y}{2} = \frac{x+y-2}{2+\theta(x+y-2)}$, $0 < \theta\ 1$, $x > 0$, $y > 0$. (Hint: Use Taylor's formula with the linear term as the remainder.)

11.56. Expand $f(x, y) = \sin xy$ in powers of $x - 1$ and $y - \frac{\pi}{2}$ to second-degree terms.

$$1 - \frac{1}{8}\pi^2(x-1)^2 - \frac{\pi}{2}(x-1)\left(y - \frac{\pi}{2}\right) - \left(y - \frac{\pi}{2}\right)^2$$

SUPPLEMENTARY PROBLEMS

Convergence And Divergence Of Series Of Constants

11.57. (a) Prove that the series $\frac{1}{3 \cdot 7} + \frac{1}{7 \cdot 11} + \frac{1}{11 \cdot 15} + \cdots = \sum_{n=1}^{\infty} \frac{1}{(4n-1)(4n+3)}$ converges and (b) find its sum.

Ans. (b) 1/12

11.58. Prove that the convergence or divergence of a series is not affected by (a) multiplying each term by the same nonzero constant and (b) removing (or adding) a finite number of terms.

11.59. If Σu_n and Σv_n converge to A and B, respectively, prove that $\Sigma (u_n + v_n)$ converges to $A + B$.

11.60. Prove that the series $\frac{3}{2} + \left(\frac{3}{2}\right)^2 + \left(\frac{3}{2}\right)^3 + \cdots = \sum \left(\frac{3}{2}\right)^n$ diverges.

11.61. Find the fallacy: Let $S = 1 - 1 + 1 - 1 + 1 - 1 + \cdots$. Then $S = 1\ 1 - (1 - 1) - (1 - 1) - \cdots = 1$ and $S = (1 - 1) + (1 - 1) + (1 - 1) + \cdots = 0$. Hence, $1 = 0$.

Comparison test and quotient test

11.62. Test for convergence: (a) $\sum_{n=1}^{\infty} \frac{1}{n^2 + 1}$, (b) $\sum_{n=1}^{\infty} \frac{n}{4n^2 - 3}$, (c) $\sum_{n=1}^{\infty} \frac{n+2}{(n+1)\sqrt{n+3}}$, (d) $\sum_{n=1}^{\infty} \frac{3^n}{n \cdot 5^n}$,

(e) $\sum_{n=1}^{\infty} \frac{1}{5n - 3}$, and (f) $\sum_{n=1}^{\infty} \frac{2n - 1}{(3n+2)n^{4/3}}$.

Ans. (a) convergence (b) divergence (c) divergence (d) convergence (e) divergence (f) convergence

11.63. Investigate the convergence of (a) $\sum_{n=1}^{\infty} \frac{4n^2 + 5n - 2}{n(n^2 + 1)^{3/2}}$ and (b) $\sum_{n=1}^{\infty} \sqrt{\frac{n - \ln n}{n^2 + 10n^3}}$.

Ans. (a) convergence (b) divergence

11.64. Establish the comparison test for divergence (see Page 280).

11.65. Use the comparison test to prove that (a) $\sum_{n=1}^{\infty} \leq \frac{1}{n^p}$ converges if $p > 1$ and diverges if $p \leq 1$,

(b) $\sum_{n=1}^{\infty} \frac{\tan^{-1} n}{n}$ diverges, and (c) $\sum_{n=1}^{\infty} \frac{n^2}{2n}$ converges.

11.66. Establish the results (b) and (c) of the quotient test, Page 280.

11.67. Test for convergence: (a) $\sum_{n=1}^{\infty} \frac{(\ln n)^2}{n^2}$, (b) $\sum_{n=1}^{\infty} \sqrt{n \tan^{-1}(1/n^3)}$, (c) $\sum_{n=1}^{\infty} \frac{3 + \sin n}{n(1 + e^{-n})}$, and

(d) $\sum_{n=1}^{\infty} n \sin^2(1/n)$.

Ans. (a) convergence (b) divergence (c) divergence (d) divergence

11.68. If Σu_n converges, where $u_n \geq 0$ for $n > N$, and if $\lim_{n\to\infty} nu_n$ exists, prove that $\lim_{n\to\infty} nu_n = 0$.

11.69. (a) Test for convergence $\sum_{n=1}^{\infty} \frac{1}{n^{1+1/n}}$. (b) Does your answer to (a) contradict the statement about the p series made on Page 266 that $\Sigma 1/n^p$ converges for $p > 1$?

Ans. (a) divergence

Integral test

11.70. Test for convergence: (a) $\sum_{n=1}^{\infty}\frac{n^2}{2n^3-1}$, (b) $\sum_{n=2}^{\infty}\frac{1}{n(\ln n)^3}$, (c) $\sum_{n=1}^{\infty}\frac{n}{2n}$, (d) $\sum_{n=1}^{\infty}\frac{e^{-\sqrt{n}}}{\sqrt{n}}$, (e) $\sum_{n=2}^{\infty}\frac{\ln n}{n}$, and (f) $\sum_{n=10}^{\infty}\frac{2^{\ln(\ln n)}}{n\ln n}$.

Ans. (a) divergence (b) convergence (c) convergence (d) convergence (e) divergence (f) divergence

11.71. Prove that $\sum_{n=2}^{\infty}\frac{1}{n(\ln n)^p}$, where p is a constant, (a) converges if $p > 1$ and (b) diverges if $p \leq 1$.

11.72. Prove that $\frac{9}{8} < \sum_{n=1}^{\infty}\frac{1}{n^3} < \frac{5}{4}$.

11.73. Investigate the convergence of $\sum_{n=1}^{\infty}\frac{e^{\tan^{-1} n}}{n^2+1}$.

Ans. convergence

11.74. (a) Prove that $\frac{2}{3}n^{3/2}+\frac{1}{3} \leq \sqrt{1}+\sqrt{2}+\sqrt{3}+\cdots+\sqrt{n} \leq \frac{2}{3}n^{3/2}+n^{1/2}-\frac{2}{3}$. (b) Use (a) to estimate the value of $\sqrt{1}+\sqrt{2}+\sqrt{3}+\cdots+\sqrt{100}$, giving the maximum error. (c) Show how the accuracy in (b) can be improved by estimating, for example, $\sqrt{10}+\sqrt{11}+\cdots+\sqrt{100}$ and adding on the value of $\sqrt{1}+\sqrt{2}+\cdots+\sqrt{9}$ computed to some desired degree of accuracy.

Ans. (b) 671.5 ± 4.5

Alternating series

11.75. Test for convergence: (a) $\sum_{n=1}^{\infty}\frac{(-1)^{n+1}}{2n}$, (b) $\sum_{n=1}^{\infty}\frac{(-1)^n}{n^2+2n+2}$, (c) $\sum_{n=1}^{\infty}\frac{(-1)^{n+1}\,n}{3n-1}$, (d) $\sum_{n=1}^{\infty}(-1)^n\sin^{-1}\frac{1}{n}$, and (e) $\sum_{n=2}^{\infty}\frac{(-1)^n\sqrt{n}}{\ln n}$.

Ans. (a) convergence, (b) convergence, (c) divergence, (d) convergence, (e) divergence

11.76. (a) What is the largest absolute error made in approximating the sum of the series $\sum_{n=1}^{\infty}\frac{(-1)^n}{2^n(n+1)}$ by the sum of the first five terms? (b) What is the least number of terms which must be taken in order that three-decimal-place accuracy will result?

Ans. (a) 1/192 (b) eight terms

11.77. (a) Prove that $S\frac{1}{1^3}+\frac{1}{2^3}+\frac{1}{3^3}+\cdots=\frac{4}{3}\left(\frac{1}{1^3}+\frac{1}{2^3}+\frac{1}{3^3}-\cdots\right)$. (b) How many terms of the series on the right are needed in order to calculate S to six-decimal-place accuracy?

Ans. (b) at least 100 terms

Absolute and conditional convergence

11.78. Test for absolute or conditional convergence:

(a) $\sum_{n=1}^{\infty} \frac{(-1)^{n-1}}{n^2+1}$ (c) $\sum_{n=2}^{\infty} \frac{(-1)^n}{n \ln n}$ (e) $\sum_{n=1}^{\infty} \frac{(-1)^n}{2n-1} \sin \frac{1}{\sqrt{n}}$

(b) $\sum_{n=1}^{\infty} \frac{(-1)^{n-1} n}{n^2+1}$ (d) $\sum_{n=1}^{\infty} \frac{(-1)^n n^3}{(n^2+1)^{4/3}}$ (f) $\sum_{n=1}^{\infty} \frac{(-1)^{n-1} n^3}{2n-1}$

Ans. (a) absolute convergence, (b) conditional convergence, (c) conditional convergence, (d) divergence, (e) absolute convergence, (f) absolute convergence

11.79. Prove that $\sum_{n=1}^{\infty} \frac{\cos n\pi a}{x^2+n^2}$ converges absolutely for all real x and a.

11.80. If $1 - \frac{1}{2} + \frac{1}{3} + \frac{1}{4} + \cdots$ converges to S, prove that the rearranged series $1 + \frac{1}{3} - \frac{1}{2} + \frac{1}{5} + \frac{1}{7} - \frac{1}{4} + \frac{1}{9} + \frac{1}{11} - \frac{1}{6} + \cdots = \frac{3}{2} S$. Explain. (Hint: Take $\frac{1}{2}$ of the first series and write it as $1 + \frac{1}{2} + 0 - \frac{1}{4} + 0 + \frac{1}{6} + \cdots$; then add term by term to the first series. Note that $S = \ln 2$, as shown in Problem 11.100.)

11.81. Prove that the terms of an absolutely convergent series can always be rearranged without altering the sum.

Ratio test

11.82. Test for convergence: (a) $\sum_{n=1}^{\infty} \frac{(-1)^n n}{(n+1)e^n}$ (b) $\sum_{n=1}^{\infty} \frac{10^n n}{(2n+1)!}$, (c) $\sum_{n=1}^{\infty} \frac{3^n}{n^3}$, (d) $\sum_{n=1}^{\infty} \frac{(-1)^n 2^{3n}}{3^{2n}}$, and (e) $\sum_{n=1}^{\infty} \frac{(\sqrt{5}-1)^n}{n^2+1}$.

Ans. (a) convergence (absolute) (b) convergence (c) divergence (d) convergence (absolute) (e) divergence

11.83. Show that the ratio test cannot be used to establish the conditional convergence of a series.

11.84. Prove that (a) $\sum_{n=1}^{\infty} \frac{n!}{n^n}$ converges and (b) $\lim_{n\to\infty} \frac{n!}{n^n} = 0$.

Miscellaneous tests

11.85. Establish the validity of the nth root test on Page 282.

11.86. Apply the nth root test to work Problems 11.82(a), (c), (d), and (e).

11.87. Prove that $\frac{1}{3} + \left(\frac{2}{3}\right)^2 + \left(\frac{1}{3}\right)^3 + \left(\frac{2}{3}\right)^4 + \left(\frac{1}{3}\right)^5 + \left(\frac{2}{3}\right)^6 + \cdots$ converges.

11.88. Test for convergence: (a) $\frac{1}{3} + \frac{1\cdot 4}{3\cdot 6} + \frac{1\cdot 4\cdot 7}{3\cdot 6\cdot 9} + \cdots$ and (b) $\frac{2}{9} + \frac{2\cdot 5}{9\cdot 12} + \frac{2\cdot 5\cdot 8}{9\cdot 12\cdot 15} + \cdots$.

Ans. (a) divergence (b) convergence

11.89. If a, b, and d are positive numbers and $b > a$, prove that $\frac{a}{b} + \frac{a(a+d)}{b(b+d)} + \frac{a(a+d)(a+2d)}{b(b+d)(b+2d)} + \cdots$ converges if $b - a > d$, and diverges if $b - a \leqq d$.

Series of functions

11.90. Find the domain of convergence of the series (a) $\sum_{n=1}^{\infty} \frac{x^n}{n^3}$, (b) $\sum_{n=1}^{\infty} \frac{(-1)^n (x-1)^n}{2^n (3n-1)}$, (c) $\sum_{n=1}^{\infty} \frac{1}{n(1+x^2)^n}$, (d) $\sum_{n=1}^{\infty} n^2 \left(\frac{1-x}{1+x}\right)^n$, and (e) $\sum_{n=1}^{\infty} \frac{e^{nx}}{n^2 - n + 1}$

Ans. (a) $-1 \leqq x \leqq 1$ (b) $-1 < x \leqq 3$ (c) all $x \neq 0$ (d) $x > 0$ (e) $x \leqq 0$

11.91. Prove that $\sum_{n=1}^{\infty} \frac{1 \cdot 3 \cdot 5 \cdots (2n-1)}{2 \cdot 4 \cdot 6 \cdots (2n)} x^n$ converges for $-1 \leqq x < 1$.

Uniform convergence

11.92. By use of the definition, investigate the uniform convergence of the series

$$\sum_{n=1}^{\infty} \frac{x}{[1+(n-1)x][1+nx]}$$

$\left[\text{Hint: Resolve the } n\text{th term into partial fractions and show that the } n\text{th partial sum is } S_n(x) = 1 - \frac{1}{1+nx}.\right]$

Ans. Not uniformly convergent in any interval which includes $x = 0$; uniformly convergent in any other interval.

11.93. Work Problem 11.30 directly by first obtaining $S_n(x)$.

11.94. Investigate by any method the convergence and uniform convergence of the series (a) $\sum_{n=1}^{\infty} \left(\frac{x}{3}\right)^n$, (b) $\sum_{n=1}^{\infty} \frac{\sin^2 nx}{2n-1}$, and (c) $\sum_{n=1}^{\infty} \frac{x}{(1+x)^n}$, $x \geqq 0$.

Ans. (a) convergence for $|x| < 3$; uniform convergence for $|x| \leqq r < 3$ (b) uniform convergence for all x (c) convergence for $x \geqq 0$; not uniform convergence for $x \geqq 0$, but uniform convergence for $x \geqq r > 0$

11.95. If $F(x) = \sum_{n=1}^{\infty} \frac{\sin nx}{n^3}$, prove that (a) $F(x)$ is continuous for all x, (b) $\lim_{x \to 0} F(x) = 0$, and (c) $F'(x) = \sum_{n=1}^{\infty} \frac{\cos nx}{n^2}$, is continuous everywhere.

11.96. Prove that $\int_0^{\pi} \left(\frac{\cos 2x}{1 \cdot 3} + \frac{\cos 4x}{3 \cdot 5} + \frac{\cos 6x}{5 \cdot 7} + \cdots \right) dx = 0.$

11.97. Prove that $F(x) = \sum_{n=1}^{\infty} \frac{\sin nx}{\sinh n\pi}$ has derivatives of all orders for any real x.

11.98. Examine the sequence $u_n(x) = \frac{1}{1+x^{2n}}$, $n = 1, 2, 3, \ldots$, for uniform convergence.

11.99. Prove that $\lim_{n\to\infty}\int_0^1 \frac{dx}{(1+x/n)^n} = 1-e^{-1}$.

Power series

11.100. (a) Prove that In $(1+x) = x - \frac{x^2}{x} + \frac{x^3}{3} - \frac{x^4}{4} + \cdots$ (b) Prove that In $2 = 1 - \frac{1}{2} + \frac{1}{3} - \frac{1}{4} + \cdots$. (Hint: Use the fact that $\frac{1}{1+x} = 1 - x + x^2 - x^3 + \cdots$ and integrate.)

11.101. Prove that $\sin^{-1} x = x + \frac{1}{2}\frac{x^3}{3} + \frac{1\cdot 3}{2\cdot 4}\frac{x^5}{5} + \frac{1\cdot 3\cdot 5}{2\cdot 4\cdot 6}\frac{x^7}{7} + \cdots, -1 \leqq x \leqq 1$.

11.102. Evaluate (a) $\int_0^{1/2} e^{-x^2}\,dx$ and (b) $\int_0^1 \frac{1-\cos x}{x}\,dx$ to three decimal places, justifying all steps.

Ans. (a) 0.461 (b) 0.486

11.103. Evaluate (a) sin 40°, (b) cos 65°, and (c) tan 12° correctly to three decimal places.

Ans. (a) 0.643 (b) 0.423 (c) 0.213

11.104. Verify the expansions 4, 5, and 6 on Page 289.

11.105. By multiplying the series for sin x and cos x, verify that 2 sin x cos x = sin $2x$.

11.106. Show that $e^{\cos x} = e\left(1 - \frac{x^2}{2!} + \frac{4x^4}{4!} - \frac{31x^6}{6!} + \cdots\right), \quad -\infty < x < \infty$.

11.107. Obtain the expansions

(a) $\tanh^{-1} x = x + \frac{x^3}{3} + \frac{x^5}{5} + \frac{x^7}{7} + \cdots \qquad -1 < x < 1$

(b) $\ln(x + \sqrt{x^2+1}) = x - \frac{1}{2}\frac{x^3}{3} + \frac{1\cdot 3}{2\cdot 4}\frac{x^5}{5} - \frac{1\cdot 3\cdot 5}{2\cdot 4\cdot 6}\frac{x^7}{7} + \cdots \qquad -1 < x \leqq 1$

11.108. Let $f(x) = \begin{cases} e^{-1/x^2} & x \neq 0 \\ 0 & x = 0 \end{cases}$. Prove that the formal Taylor series about $x = 0$ corresponding to $f(x)$ exists but that it does not converge to the given function for any $x \neq 0$.

11.109. Prove that

(a) $\frac{\ln(1+x)}{1+x} = x - \left(1+\frac{1}{2}\right)x^2 + \left(1+\frac{1}{2}+\frac{1}{3}\right)x^3 - \cdots \qquad$ for $-1 < x < 1$

(b) $\{\ln(1+x)^2\} = x^2 - \left(1+\frac{1}{2}\right)\frac{2x^3}{3} + \left(1+\frac{1}{2}+\frac{1}{3}\right)\frac{2x^4}{4} - \cdots \qquad$ for $-1 < x \leqq 1$

Miscellaneous problems

11.110. Prove that the series for $J_p(x)$ converges (a) for all x and (b) absolutely and uniformly in any finite interval.

11.111. Prove that (a) $\frac{d}{dx}\{J_0(x)\} = -J_1(x)$, (b) $\frac{d}{dx}\{x^p J_p(x)\} = x^p J_{p-1}(x)$, and

(c) $J_{p+1}(x) = \frac{2p}{x} J_p(x) - J_{p-1}(x)$.

11.112. Assuming that the result of Problem 11.111(c) holds for $p = 0, -1, -2, \ldots$, prove that (a) $J_{-1}(x) = -J_1(x)$, (b) $J_{-2}(x) = J_2(x)$, and (c) $J_{-n}(x) = (-1)^n J_n(x)$, $n = 1, 2, 3, \ldots$.

11.113. Prove that $e^{1/2x(t-1/t)} = \sum_{p=-\infty}^{\infty} J_p(x)t^p$ (Hint: Write the left side as $e^{xt/2}\, e^{-x/2t}$, expand, and use Problem 11.112.)

11.114. Prove that $\sum_{n=1}^{\infty} \frac{(n+1)z^n}{n(n+2)^2}$ is absolutely and uniformly convergent at all points within and on the circle $|z| = 1$.

11.115. (a) If $\sum_{n=1}^{\infty} a_n x^n = \sum_{n=1}^{\infty} b_n x^n$ for all x in the common interval of convergence $|x| < R$ where $R > 0$, prove that $a_n = b_n$ for $n = 0, 1, 2, \ldots$ (b) Use (a) to show that the Taylor expansion of a function exists and the expansion is unique.

11.116. Suppose that $\overline{\lim}\, \sqrt[n]{|u_n|} = L$. Prove that Σu_n converges or diverges according as $L < 1$ or $L > 1$. If $L = 1$, the test fails.

11.117. Prove that the radius of convergence of the series $\Sigma a_n x^n$ can be determined by the following limits, when they exist, and give examples: (a) $\lim_{n\to\infty} \left|\frac{a_n}{a_{n+1}}\right|$, (b) $\lim_{n\to\infty} \frac{1}{\sqrt[n]{|a_n|}}$, and (c) $\overline{\lim_{n\to\infty}}\, \frac{1}{\sqrt[n]{|a_n|}}$.

11.118. Use Problem 11.117 to find the radius of convergence of the series in Problem 11.22.

11.119. (a) Prove that a necessary and sufficient condition that the series Σu_n converge is that, given any $\epsilon > 0$, we can find $N > 0$ depending on ϵ such that $|S_p - S_q| < \epsilon$ whenever $p > N$ and $q > N$, where $S_k = u_1 + u_2 + \cdots + u_k$. (b) Use (a) to prove that the series $\sum_{n=1}^{\infty} \frac{n}{(n+1)3^n}$ converges. (c) How could you use (a) to prove that the series $\sum_{n=1}^{\infty} \frac{1}{n}$ diverges? (Hint: Use the Cauchy convergence criterion, Page 27.)

11.120. Prove that the hypergeometric series (Page 276) (a) is absolutely convergent for $|x| < 1$, (b) is divergent for $|x| > 1$, (c) is absolutely divergent for $|x| = 1$ if $a + b - c < 0$, and (d) satisfies the differential equation $x(1-x)y'' + \{c - (a+b+1)x\}y' - aby = 0$.

11.121. If $F(a, b; c; x)$ is the hypergeometric function defined by the series on Page 290, prove that (a) $F(-p, 1; 1; -x) = (1+x)^p$, (b) $x F(1, 1; 2; -x) = \text{In}(1+x)$, and (c) $F\left(\frac{1}{2}, \frac{1}{2}; \frac{3}{2}; x^2\right) = (\sin^{-1} x)/x$.

11.122. Find the sum of the series $S(x) = x + \frac{x^3}{1\cdot 3} + \frac{x^5}{1\cdot 3\cdot 5} + \cdots$. (Hint: Show that $S'(x) - 1 + xS(x)$ and solve.)

Ans. $e^{x^2/2} \int_0^x e^{-x^2/2}\, dx$

11.123. Prove that $1 + \dfrac{1}{1 \cdot 3} + \dfrac{1}{1 \cdot 3 \cdot 5} + \dfrac{1}{1 \cdot 3 \cdot 5 \cdot 7} + \cdots = \sqrt{e}\left(1 - \dfrac{1}{2 \cdot 3} + \dfrac{1}{2^2 \cdot 2! \cdot 5} - \dfrac{1}{2^3 \cdot 3! \cdot 7} + \dfrac{1}{2^4 \cdot 4! \cdot 9} - \cdots\right)$.

11.124. Establish Dirichlet's test on Page 284.

11.125. Prove that $\displaystyle\sum_{n=1}^{\infty} \frac{\sin nx}{n}$ is uniformly convergent in any interval which does not include $0, \pm\pi, \pm 2\pi, \ldots$ (Hint: use the Dirichlet test, Page 284, and Problem 1.94.)

11.126. Establish the results on Page 289 concerning the binomial series. (Hint: Examine the Lagrange and Cauchy forms of the remainder in Taylor's theorem.)

11.127. Prove that $\displaystyle\sum_{n=1}^{\infty} \frac{(-1)^{n-1}}{n + x^2}$ converges uniformly for all x, but not absolutely.

11.128. Prove that $1 - \dfrac{1}{4} + \dfrac{1}{7} - \dfrac{1}{10} + \cdots = \dfrac{\pi}{3\sqrt{3}} + \dfrac{1}{3} \ln 2$.

11.129. If $x = ye^y$, prove that $y = \displaystyle\sum_{n=1}^{\infty} \frac{(-1)^{n-1}\, n^{n-1}}{n!} x^n$ for $-1/e < x \leq 1/e$.

11.130. Prove that the equation $e^{-\lambda} = \lambda - 1$ has only one real root and show that it is given by

$$\lambda = 1 + \sum_{n=1}^{\infty} \frac{(-1)^{n-1}\, n^{n-1}\, e^{-n}}{n!}$$

11.131. Let $\dfrac{x}{e^x - 1} = 1 + B_1 x + \dfrac{B_2 x^2}{2!} + \dfrac{B_3 x^3}{3!} + \cdots$. (a) Show that the numbers B_n, called the *Bernoulli numbers*, satisfy the recursion formula $(B + 1)^n - B^n = 0$ where B^k is formally replaced by B_k after expanding.

(b) Using (a) or otherwise, determine $B_1, \ldots, B_6$.

Ans. (b) $B_1 = -\dfrac{1}{2},\ B_2 = \dfrac{1}{6},\ B_3 = 0,\ B_4 = -\dfrac{1}{30},\ B_5 = 0,\ B_6 = \dfrac{1}{42}$

11.132. (a) Prove that $\dfrac{x}{e^x - 1} = \dfrac{x}{2}\left(\coth \dfrac{x}{2} - 1\right)$. (b) Use Problem 11.127 and (a) to show that $B_{2k+1} = 0$ if $k = 1, 2, 3, \ldots$

11.133. Derive the series expansions:

(a) $\coth x = \dfrac{1}{x} + \dfrac{x}{3} - \dfrac{x^3}{45} + \cdots + \dfrac{B_{2n}(2x)^{2n}}{(2n)!\,x} + \cdots$

(b) $\coth x = \dfrac{1}{x} + \dfrac{x}{3} - \dfrac{x^3}{45} + \cdots (-1)^n \dfrac{B_{2n}(2x)^{2n}}{(2n)!\,x} + \cdots$

(c) $\tan x = x + \dfrac{x^3}{3} + \dfrac{2x^5}{15} + \cdots (-1)^{n-1} \dfrac{2(2^{2n-1} - 1)\, B_{2n}(2x)^{2n-1}}{(2n)!} + \cdots$

(d) $\csc x = \dfrac{1}{x} + \dfrac{x}{6} + \dfrac{7}{360}x^3 + \cdots (-1)^{n-1} \dfrac{2(2^{2n-1} - 1)\, B_{2n} x^{2n-1}}{(2n)!} + \cdots$

[Hint: For (a) use Problem 11.132; for (b) replace x by ix in (a); for (c) use $\tan x = \cot x - 2 \cot 2x$; for (d) use $\csc x = \cot x + \tan x/2$.]

11.134. Prove that $\prod_{n=1}^{\infty}\left(1+\frac{1}{n^3}\right)$ converges.

11.135. Use the definition to prove that $\prod_{n=1}^{\infty}\left(1+\frac{1}{n}\right)$ diverges.

11.136. (a) Prove that $\prod_{n=1}^{\infty}(1-u_n)$, where $0 < u_n < 1$, converges if and only if Σu_n converges.

11.137. (a)Prove that $\prod_{n=2}^{\infty}\left(1-\frac{1}{n^2}\right)$ converges to $\frac{1}{2}$. (b) Evaluate the infinite product in (a) to two decimal places and compare with the true value.

11.138. Prove that the series $1 + 0 - 1 + 1 + 0 - 1 + 1 + 0 - 1 + \cdots$ is the $C - 1$ summable to zero.

11.139. Prove that the Cesaro method of summability is regular. (Hint: See Page 291.)

11.140. Prove that the series $1 + 2x + 3x^2 + 4x^3 + \cdots + nx^{n-1} + \cdots$ converges to $1/(1-x)^2$ for $|x| < 1$.

11.141. A series $\sum_{n=0}^{\infty} a$ is called *Abel summable* to S if $S = \lim_{x\to 1-}\sum_{n=0}^{\infty} a_n X^n$ exists. Prove that

(a) $\sum_{n=0}^{\infty}(-1)^n(n+1)$ $(n+1)$ is Abel summable to $\frac{1}{4}$

(b) $\sum_{n=0}^{\infty}\frac{(-1)^n(n+1)(n+2)}{2}$ is Abel summable to $\frac{1}{8}$

11.142. Prove that the double series $\sum_{m=0}^{\infty}\sum_{n=0}^{\infty}\frac{1}{(m^2+n^2)^p}$, where p is a constant, converges or diverges according as $p > 1$ or $p \leqq 1$, respectively.

11.143. (a) Prove that $\int_x^{\infty}\frac{e^{x-u}}{u}\,du = \frac{1}{x} - \frac{1}{x^2} + \frac{2!}{x^3} - \frac{3!}{x^4} + \cdots \frac{(-1)^{n-1}(n-1)!}{x^n} + (-1)^n n!\int_x^{\infty}\frac{e^{x-u}}{u^{n+1}}\,du.$

(b) Use (a) to prove that $\int_x^{\infty}\frac{e^{x-u}}{u}\,du \sim \frac{1}{x} - \frac{1}{x^2} + \frac{2!}{x^3} - \frac{3!}{x^4} + \cdots.$

CHAPTER 12

Improper Integrals

Definition of an Improper Integral

The functions that generate the Riemann integrals of Chapter 5 are continuous on closed intervals. Thus, the functions are bounded and the intervals are finite. Integrals of functions with these characteristics are called *proper integrals*. When one or more of these restrictions are relaxed, the integrals are said to be *improper*. Categories of improper integrals are established as follows.

The integral $\int_a^b f(x)\,dx$ is called an *improper integral* if

1. $a = -\infty$ or $b = \infty$ or both; i.e., one or both integration limits is infinite.
2. $f(x)$ is unbounded at one or more points of $a \leqq x \leqq b$. Such points are called *singularities* of $f(x)$.

Integrals corresponding to (1) and (2) are called *improper integrals of the first and second kinds*, respectively. Integrals with both conditions (1) and (2) are called *improper integrals of the third kind*

EXAMPLE 1. $\int_0^\infty \sin x^2\,dx$ is an improper integral of the first kind.

EXAMPLE 2. $\int_0^4 \frac{dx}{x-3}$ is an improper integral of the second kind.

EXAMPLE 3. $\int_0^\infty \frac{e^{-x}}{\sqrt{x}}dx$ is an improper integral of the third kind.

EXAMPLE 4. $\int_0^1 \frac{\sin x}{x}dx$ is a *proper integral*, since $\lim\limits_{x\to 0+} \frac{\sin x}{x} = 1$.

Improper Integrals of the First Kind (Unbounded Intervals)

If f is integrable on the appropriate domains, then the indefinite integrals $\int_a^x f(t)\,dt$ and $\int_x^a f(t)\,dt$ (with variable upper and lower limits, respectively) are functions. Through them we define three forms of the improper integral of the first kind.

Definition

(a) If f is integrable on $a \leqq x < \infty$, then $\int_a^\infty f(x)dx = \lim\limits_{x\to\infty} \int_a^x f(t)dt$.

(b) If f is integrable on $-\infty < x \leqq a$, then $\int_{-\infty}^a f(x)dx = \lim\limits_{x\to-\infty} \int_x^a f(t)\,dt$.

(c) If f is integrable on $-\infty < x < \infty$, then

$$\int_{-\infty}^\infty f(x)\,dx = \int_{-\infty}^\alpha f(x)\,dx + \int_\alpha^\infty f(x)\,dx$$

$$= \lim_{x\to-\infty} \int_x^\alpha f(t)\,dt + \lim_{x\to\infty} \int_\alpha^x f(t)\,dt$$

In (c) it is important to observe that

$$\lim_{x\to-\infty}\int_x^{\alpha} f(t)dt + \lim_{x\to\infty}\int_{\alpha}^{x} f(t)dt$$

and

$$\lim_{x\to\infty}\left[\int_{-x}^{\alpha} f(t)dt + \int_{\alpha}^{x} f(t)dt\right]$$

are not necessarily equal.

This can be illustrated with $f(x) = xe^{x^2}$. The first expression is not defined, since neither of the improper integrals (i.e., limits) is defined, while the second form yields the value 0.

EXAMPLE. The function $F(x) = \frac{1}{\sqrt{2\pi}}e^{-(x^2/2)}$ is called the *normal density function* and has numerous applications in probability and statistics. In particular (see the bell-shaped curve in Figure 12.1),

$$\int_{-\infty}^{\infty}\frac{1}{\sqrt{2\pi}}e - \frac{x^2}{2} : dx = 1$$

(See Problem 12.31 for the trick of making this evaluation.)

Perhaps at some point in your academic career you were "graded on the curve." The infinite region under the curve with the limiting area of 1 corresponds to the assurance of getting a grade. C's are assigned to those whose grades fall in a designated central section, and so on. (Of course, this grading procedure is not valid for a small number of students, but as the number increases it takes on statistical meaning.)

In this chapter we formulate tests for convergence or divergence of improper integrals. It will be found that such tests and proofs of theorems bear close analogy to convergence and divergence tests and corresponding theorems for infinite series (see Chapter 11).

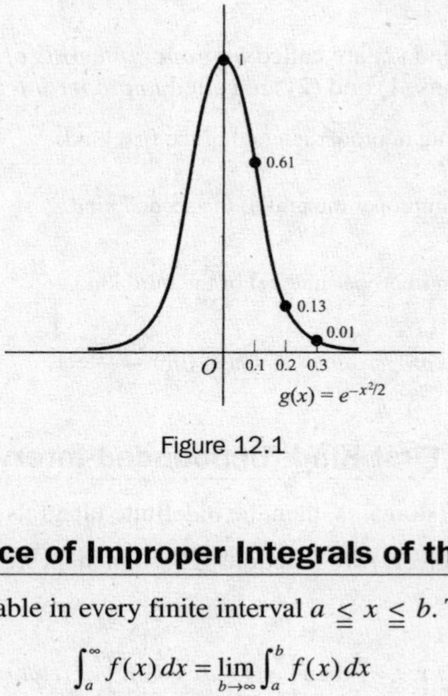

Figure 12.1

Convergence or Divergence of Improper Integrals of the First Kind

Let $f(x)$ be bounded and integrable in every finite interval $a \leq x \leq b$. Then we define

$$\int_a^{\infty} f(x)\,dx = \lim_{b\to\infty}\int_a^b f(x)\,dx \tag{1}$$

where b is a variable on the positive real numbers.

The integral on the left is called *convergent* or *divergent* according as the limit on the right does or does not exist. Note that $\int_a^{\infty} f(x)\,dx$ bears close analogy to the infinite series $\sum_{n=1}^{\infty} u_n$, where $u_n = f(n)$, while $\int_a^b f(x)\,dx$ corresponds to the partial sums of such infinite series. We often write M in place of b in Equation (1).

Similarly, we define

$$\int_{-\infty}^{b} f(x)\,dx = \lim_{a\to-\infty}\int_{a}^{b} f(x)\,dx \tag{2}$$

where a is a variable on the negative real numbers. And we call the integral on the left convergent or divergent according as the limit on the right does or does not exist.

EXAMPLE 1. $\int_1^{\infty}\frac{dx}{x^2} = \lim_{b\to\infty}\int_1^b \frac{dx}{x^2} = \lim_{b\to\infty}\left(1-\frac{1}{b}\right) = 1$ so that $\int_t^{\infty}\frac{dx}{x^2}$ converges to 1.

EXAMPLE 2. $\int_{-\infty}^{a}\cos x\,dx = \lim_{a\to\infty}\int_a^u \cos x\,dx = \lim_{a\to\infty}(\sin u - \sin a)$. Since this limit does not exit, $\int_{-\infty}^{a}\cos x\,dx$ is divergent.

In like manner, we define

$$\int_{-\infty}^{\infty} f(x)\,dx = \int_{-\infty}^{x_0} f(x)\,dx + \int_{x_0}^{\infty} f(x)\,dx \tag{3}$$

where x_0 is a real number, and we call the integral convergent or divergent according as the integrals on the right converge or not, as in definitions (1) and (2). [See the previous remarks in part (c) of the definition of improper integrals of the first kind.]

Special Improper Integrals of the First Kind

1. **Geometric or exponential integral** $\int_a^{\infty} e^{t-1x}\,dx$, where t is a constant, converges if $t > 0$ and diverges if $t \leqq 0$. Note the analogy with the geometric series if $r = e^{-t}$ so that $e^{-tx} = r^x$.
2. **The p integral of the first kind** $\int_a^{\infty}\frac{dx}{x^p}$, where p is a constant and $a > 0$, converges if $p > 1$ and diverges if $p \leqq 1$. Compare with the p series.

Convergence Tests for Improper Integrals of the First Kind

The following tests are given for cases where an integration limit is ∞. Similar tests exist where an integration limit is $-\infty$ (a change of variable $x = -y$ then makes the integration limit ∞). Unless otherwise specified, we assume that $f(x)$ is continuous and thus integrable in every finite interval $a \leqq x \leqq b$.

1. **Comparison test** for integrals with nonnegative integrands.

   (a) *Convergence.* Let $g(x) \geqq 0$ for all $x \geqq a$, and suppose that $\int_a^{\infty} g(x)\,dx$ converges. Then if $0 \leqq f(x) \leqq g(x)$ for all $x \geqq a$, $\int_a^{\infty} f(x)\,dx$ also converges.

   EXAMPLE. Since $\frac{1}{e^x+1} \leqq \frac{1}{e^x} = e^{-x}$ and $\int_a^{\infty} e^{-x}\,dx$ converges, $\int_0^{\infty}\frac{dx}{e^x+1}$ also converges.

   (b) *Divergence.* Let $g(x) \geqq 0$ for all $x \geqq a$, and suppose that $\int_a^{\infty} g(x)\,dx$ diverges. Then if $f(x) \leqq g(x)$ for all $x \geqq a$, $\int_a^{\infty} f(x)\,dx$ also diverges.

   EXAMPLE. Since $\frac{1}{\ln x} > \frac{1}{x}$ for $x \geqq 2$ and $\int_2^{\infty}\frac{dx}{x}$ diverges (p integral with $p = 1$), $\int_2^{\infty}\frac{dx}{\ln x}$ also diverges.

2. Quotient test for integrals with nonnegative integrands.

   (a) If $f(x) \geqq$ and $g(x) \geqq 0$, and if $\lim_{x\to\infty}\frac{f(x)}{g(x)} = A \neq 0$ or ∞, then $\int_a^{\infty} f(x)\,dx$ and $\int_a^{\infty} g(x)\,dx$ either both converge or both diverge.

   (b) If $A = 0$ *in* (a) and $\int_a^{\infty} g(x)\,dx$ converges, then $\int_a^{\infty} f(x)\,dx$ converges.

   (c) If $A = \infty$ in (a) and $\int_a^{\infty} g(x)\,dx$ diverges, then $\int_a^{\infty} f(x)\,dx$ diverges.

This test is related to the comparison test and is often a very useful alternative to it. In particular, taking $g(x) = 1/x^p$, we have, from known facts about the p integral, the following theorem.

Theorem 1 Let $\lim_{x\to\infty} x^p f(x) = A$. Then

(i) $\int_a^\infty f(x)\,dx$ converges if $p > 1$ and A is finite.

(ii) $\int_a^\infty f(x)\,dx$ diverges if $p \leqq 1$ and $A \neq 0$ (A may be infinite).

EXAMPLE 1. $\int_0^\infty \frac{x^2\,dx}{4x^4+25}$ converges since $\lim_{x\to\infty} x^2 \cdot \frac{x^2}{4x^4+25} = \frac{1}{4}$.

EXAMPLE 2. $\int_0^\infty \frac{x\,dx}{\sqrt{x^4+x^2+1}}$ diverges since $\lim_{x\to\infty} x \cdot \frac{x}{\sqrt{x^4+x^2+1}} = 1$.

A similar test can be devised using $g(x) = e^{-tx}$.

3. **Series test** for integrals with nonnegative integrands. $\int_a^\infty f(x)dx$ converges or diverges according as Σu_n where $u_n = f(n)$, converges or diverges.

4. **Absolute and conditional convergence.** $\int_a^\infty f(x)dx$ is called *absolutely convergent* if $\int_a^\infty |f(x)|\,dx$ converges. If $\int_a^\infty f(x)\,dx$ converges but $\int_a^\infty |f(x)|\,dx$ diverges, then $\int_a^\infty f(x)\,dx$ is called *conditionally convergent*.

Theorem 2 If $\int_a^\infty |f(x)\,dx$ converges, then $\int_a^\infty f(x)\,dx$ converges. In words, an absolutely convergent integral converges.

EXAMPLE 1. $\int_a^\infty \frac{\cos x}{x^2+1}dx$ is absolutely convergent and thus convergent, since

$$\int_0^\infty \left|\frac{\cos x}{x^2+1}\right| dx \leqq \int_0^\infty \frac{dx}{x^2+1} \text{ and } \int_0^\infty \frac{dx}{x^2+1} \text{ converges.}$$

EXAMPLE 2. $\int_0^\infty \frac{\sin x}{x}dx$ converges (see Problem 12.11), but $\int_0^\infty \left|\frac{\sin x}{x}\right| dx$ does not converge (see Problem 12.12). Thus, $\int_0^\infty \frac{\sin x}{x}dx\ dx$ is conditionally convergent.

Any of the tests used for integrals with nonnegative integrands can be used to test for absolute convergence.

Improper Integrals of the Second Kind

If $f(x)$ becomes unbounded only at the endpoint $x = a$ of the interval $a \leqq x \leqq b$, then we define

$$\int_a^b f(x)\,dx = \lim_{\in\to 0+} \int_{a+\in}^b f(x)\,dx \tag{4}$$

and define it to be an improper integral of the second kind. If the limit on the right of Equation (4) exists, we call the integral on the left *convergent*; otherwise, it is *divergent*.

Similarly, if $f(x)$ becomes unbounded only at the endpoint $x = b$ of the interval $a \leqq x \leqq b$, then we extend the category of improper integrals of the second kind.

$$\int_a^b f(x)\,dx = \lim_{\in\to 0+} \int_a^{b-\in} f(x)\,dx \tag{5}$$

Note: Be alert to the word *unbounded*. This is distinct from *undefined*. For example, $\int_a^1 \frac{\sin x}{x}dx = \lim_{\in\to 0}\int_\in^1 \frac{\sin x}{x}dx$ is a proper integral, since $\lim_{x\to 0}\frac{\sin x}{x} = 1$ and, hence, is bounded as $x \to 0$ even though the function is undefined at $x = 0$. In such case the integral on the left of Equation (5) is called convergent or divergent according as the limit on the right exists or does not exist.

Finally, the category of improper integrals of the second kind also includes the case where $f(x)$ becomes unbounded only at an interior point $x = x_0$ of the interval $a \leqq x \leqq b$; then we define

$$\int_a^b f(x)\,dx = \lim_{\epsilon_1 \to 0+} \int_a^{x_0-\epsilon_1} f(x)dx + \lim_{\epsilon_2 \to 0+} \int_{x_0+\epsilon_2}^{b} f(x)dx \tag{6}$$

The integral on the left of Equation (6) converges or diverges according as the limits on the right exist or do not exist.

Extensions of these definitions can be made in case $f(x)$ becomes unbounded at two or more points of the interval $a \leqq x \leqq b$.

Cauchy Principal Value

It may happen that the limits on the right of Equation (6) do not exist when ϵ_1 and ϵ_2 aproach zero independently. In such case it is possible that by choosing $\epsilon_1 = \epsilon_2 = \epsilon$ in (6), i.e., writing

$$\int_a^b f(x)dx = \lim_{\epsilon \to 0+} \left\{ \int_a^{x_0-\epsilon} f(x)dx + \int_{x_0+\epsilon}^{b} f(x)dx \right\} \tag{7}$$

the limit does exist. If the limit on the right of Equation (7) does exist, we call this limiting value the *Cauchy principal value* of the integral on the left. See Problem 12.14.

EXAMPLE. The natural logarithm (i.e., base e) may be defined as follows:

$$\ln x = \int_t^x \frac{dt}{t}, \qquad 0 < x < \infty$$

Since $f(x) = \dfrac{1}{x}$ is unbounded as $x \to 0$, this is an improper integral of the second kind (see Figure 12.2).

Also, $\int_0^\infty \dfrac{dt}{t}$ is an integral of the third kind, since the interval to the right is unbounded.

Now $\lim_{\epsilon \to 0} \int_\epsilon^1 \dfrac{dt}{t} = \lim_{\epsilon \to 0} [\ln 1 - \ln \epsilon] \to -\infty$ as $\epsilon \to 0$; therefore, this improper integral of the second kind is divergent. Also, $\int_t^\infty \dfrac{dt}{t} = \lim_{x \to \infty} \int_1^x \dfrac{dt}{t} = \lim_{x \to \infty} [\ln x - \ln 1] \to \infty$; this integral (which is of the first kind) also diverges.

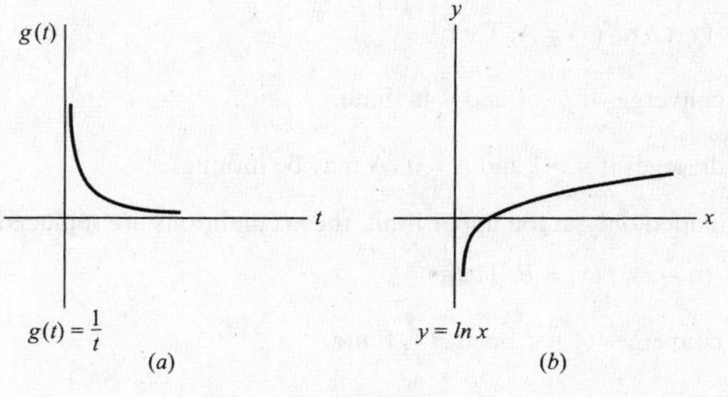

Figure 12.2

Special Improper Integrals of the Second Kind

1. $\int_a^b \dfrac{dx}{(x-a)^p}$ converges if $p < 1$ and diverges if $p \geqq 1$.

2. $\int_a^b \dfrac{dx}{(b-x)^p}$ converes if $p < 1$ and diverges if $p \geqq 1$.

These can be called *p integrals of the second kind*. Note that when $p \leqq 0$ the integrals are proper.

Convergence Tests for Improper Integrals of the Second Kind

The following tests are given for the case where $f(x)$ is unbounded only at $x = a$ in the interval $a \leqq x \leqq b$. Similar tests are available if $f(x)$ is unbounded at $x = b$ or at $x = x_0$ where $a < x_0 < b$.

1. **Comparison test** for integrals with nonnegative integrands.

   (a) *Convergence.* Let $g(x)\ \varepsilon\ 0$ for $a < x \leqq b$, and suppose that $\int_a^b g(x)\,dx$ converges. Then if $0 \leqq f(x) \leqq g(x)$ for $a < x \leqq b$, $\int_a^b f(x)\,dx$ also converges.

EXAMPLE. $\dfrac{1}{\sqrt{x^4-1}} < \dfrac{1}{\sqrt{x-1}}$ for $x > 1$. Then since $\int_1^5 \dfrac{dx}{\sqrt{x-1}}$ converges (p integral with $a = 1$, $p = \frac{1}{2}$), $\int_1^5 \dfrac{dx}{\sqrt{x^4-1}}$ also converges.

   (b) *Divergence.* Let $g(x) \geqq 0$ for $a < x \leqq b$, and suppose that $\int_a^b g(x)\,dx$ diverges. Then if $f(x) \geqq g(x)$ for $a < x \geq b$, $\int_a^b f(x)dx$ also diverges.

EXAMPLE. $\dfrac{\ln x}{(x-3)^4} > \dfrac{1}{(x-3)^4}$ for $x > 3$. Then since $\int_3^b \dfrac{dx}{(x-3)^4}$ diverges (p integral with $a = 3$, $p = 4$), $\int_3^b \dfrac{\ln x}{(x-3)^4}dx$ also diverges.

2. **Quotient test** for integrals with nonnegative integrands.

   (a) If $f(x) \geqq 0$ and $g(x) \geqq 0$ for $a < x \leqq b$, and if $\lim_{x\to a} \dfrac{f(x)}{g(x)} = A \neq 0$ or ∞, then $\int_a^b f(x)\,dx$ and $\int_a^b g(x)\,dx$ either both converge or both diverge.

   (b) If $A = 0$ in (a), and $\int_a^b g(x)dx$ converges, then $\int_a^b f(x)dx$ converges.

   (c) If $A = \infty$ in (a), and $\int_a^b g(x)dx$ diverges, then $\int_a^b f(x)dx$ diverges.

This test is related to the comparison test and is a very useful alternative to it. In particular, taking $g(x) = 1/(x-a)^p$ we have, from known facts about the p integral, the following theorems.

Theorem 3. Let $\lim_{x\to a+} (x-a)^p f(x) = A$. Then

(i) $\int_a^b f(x)dx$ converges if $p < 1$ and A is finite.

(ii) $\int_a^b f(x)dx$ diverges if $p \geqq 1$ and $A \neq 0$ (A may be infinite).

If $f(x)$ becomes unbounded only at the upper limit, these conditions are replaced by those in Theorem 4.

Theorem 4. Let $\lim_{x\to b-} (b-x)^p f(x) = B$. Then

(i) $\int_a^b f(x)dx$ converges of $p < 1$ and B is finite.

(ii) $\int_a^b f(x)dx$ diverges if $p \geqq 1$ and $B \neq 0$ (B may be infinite).

EXAMPLE 1. $\int_1^5 \dfrac{dx}{\sqrt{x^4-1}}$ converges, since $\lim_{x\to 1+} (x-1)^{1/2} \cdot \dfrac{1}{(x^4-1)^{1/2}} = \lim_{x\to 1+} \sqrt{\dfrac{x-1}{x^4-1}} = \dfrac{1}{2}$.

EXAMPLE 2. $\int_0^3 \dfrac{dx}{(3-x)\sqrt{x^2+1}}$ diverges, since $\lim_{x\to 3-} (3-x) \cdot \dfrac{1}{(3-x)\sqrt{x^2+1}} = \dfrac{1}{\sqrt{10}}$.

3. **Absolute and conditional convergence.** $\int_a^b f(x)\,dx$ is called *absolute convergent* if $\int_a^b |f(x)|\,dx$ converges. If $\int_a^b f(x)\,dx$ converges but $\int_a^b |f(x)|\,dx$ diverges, then $\int_a^b f(x)\,dx$ is called *conditionally convergent*.

Theorem 5. If $\int_a^b |f(x)|\,dx$ converges, then $\int_a^b f(x)\,dx$ converges. In words, an absolutely convergent integral converges.

EXAMPLE. $\left|\dfrac{\sin x}{\sqrt[3]{x-\pi}}\right| \leqq \dfrac{1}{\sqrt[3]{x-\pi}}$ and $\int_\pi^{4\pi} \dfrac{dx}{\sqrt[3]{x-\pi}}$ converges $\left(p \text{ integral with } a = \pi, p = \dfrac{1}{3}\right)$, it follows that $\int_\pi^{4\pi} \left|\dfrac{\sin x}{\sqrt[3]{x-\pi}}\right| dx$ converges and thus $\int_\pi^{4\pi} \dfrac{\sin x}{\sqrt[3]{x-\pi}}\,dx$ converges (absolutely).

Any of the tests used for integrals with nonnegative integrands can be used to test for absolute convergence.

Improper Integrals of the Third Kind

Improper integrals of the third kind can be expressed in terms of improper integrals of the first and second kinds, and, hence, the question of their convergence or divergence is answered by using results already established.

Improper Integrals Containing a Parameter, Uniform Convergence

Let

$$\phi(\alpha) = \int_a^\infty f(x,\alpha)\,dx \tag{8}$$

This integral is analogous to an infinite series of functions. In seeking conditions under which we may differentiate or integrate $\phi(\alpha)$ with respect to α, it is convenient to introduce the concept of convergence for integrals by analogy with infinite series.

We shall suppose that the integral (8) converges for $\alpha_1 \leqq \alpha \leqq \alpha_2$, or, briefly, $[\alpha_1, \alpha_2]$.

Definition. The integral (8) is said to be *uniformly convergent* in $[\alpha_1, \alpha_2]$ if for each $\epsilon > 0$ we can find a number N depending on ϵ but not on α, such that

$$\left|\phi(\alpha) - \int_a^u f(x,\alpha)\,dx\right| < \epsilon \qquad \text{for all } u > N \text{ and all } \alpha \text{ in } [\alpha_1, \alpha_2]$$

This can be restated by noting that $\left|\phi(\alpha) - \int_a^u f(x,\alpha)\,dx\right| = \left|\int_u^\infty f(x,\alpha)\,dx\right|$, which is analogous in an infinite series to the absolute value of the remainder after N terms.

This definition and the properties of uniform convergence to be developed are formulated in terms of improper integrals of the first kind. However, analogous results can be given for improper integrals of the second and third kinds.

Special Tests for Uniform Convergence of Integrals

1. **Weierstrass M test.** If we can find a function $M(x)\ \varepsilon\ 0$ such that
   (a) $|f(x,\alpha)| \leqq M(x)\ \alpha_1 \leqq \alpha \leqq \alpha_2, x > a$

   (b) $\int_a^\infty M(x)\,dx$ converges

   then $\int_a^\infty f(x,\alpha)\,dx$ is uniformly and absolutely convergent in $\alpha_1 \leqq \alpha \leqq \alpha_2$.

   EXAMPLE. Since $\left|\dfrac{\cos\alpha x}{x^2+1}\right| \leqq \dfrac{1}{x^2+1}$ and $\int_0^\infty \dfrac{dx}{x^2+1}$ converges, it follows that $\int_0^\infty \dfrac{\cos\alpha x}{x^2+1}\,dx$ is uniformly and absolutely convergent for all real values of α.

As in the case of infinite series, it is possible for integrals to be uniformly convergent without being absolutely convergent, and conversely.

2. **Dirichlet's test.** Suppose that
   (a) $\psi(x)$ is a positive monotonic decreasing function which approches zero as $x \to \infty$.
   (b) $\left|\int_a^u f(x,\alpha)\,dx\right| < P$ for all $u > a$ and $\alpha_1 \leqq \alpha \leqq \alpha_2$.

   Then the integral $\int_a^\infty f(x,\alpha)\psi(x)\,dx$ is uniformly convergent for $\alpha_1 \leqq \alpha \leqq \alpha_2$.

Theorems on Uniformly Convergent Integrals

Theorem 6. If $f(x, \alpha)$ is continuous for $x \,\varepsilon\, a$ and $\alpha_1 \leqq \alpha \leqq \alpha_2$, and if $\int_a^\infty f(x,\alpha)\,dx$ is uniformly convergent for $\alpha_1 \leqq \alpha \leqq \alpha_2$, then $\phi(\alpha) = \int_a^\infty f(x,\alpha)\,dx$ is continuous in $\alpha_1 \leqq \alpha \leqq \alpha_2$. In particular, if α_0 is any point of $\alpha_1 \leqq \alpha \leqq \alpha_2$, we can write

$$\lim_{\alpha\to\alpha_a} \phi(\alpha)\,d\alpha = \lim_{\alpha\to\alpha_u} \int_a^\infty f(x,\alpha) = \int_a^\infty \lim_{\alpha\to\alpha_v} f(x,\alpha)\,dx \tag{9}$$

If α_0 is one of the endpoints, we use right- or left-hand limits.

Theorem 7. Under the conditions of Theorem 6, we can integrate $\phi(\alpha)$ with respect to α_1 to α_2 to obtain

$$\int_{\alpha_1}^{\alpha_2} \phi(\alpha)d\alpha = \int_{\alpha_1}^{\alpha_2}\left\{\int_a^\infty f(x,\alpha)\,dx\right\}d\alpha \int_a^\infty\left\{\int_{\alpha_1}^{\alpha_2} f(x,\alpha)\,d\alpha\right\}dx \tag{10}$$

which corresponds to a change of the order of integration.

Theorem 8. If $f(x, \alpha)$ is continuous and has a continuous partial derivative with respect to α for $x \,\varepsilon\, a$ and $\alpha_1 \leqq \alpha \leqq \alpha_2$, and if $\int_a^\infty \frac{\partial f}{\partial \alpha}dx$ converges uniformly in $\alpha_1 \leqq \alpha \leqq \alpha_2$, then if a does not depend on α,

$$\frac{d\phi}{d\alpha} = \int_a^\infty \frac{\partial f}{\partial \alpha}dx \tag{11}$$

If a depends on α, this result is easily modified (see Leibniz's rule, Page 198).

Evaluation of Definite Integrals

Evaluation of definite integrals which are improper can be achieved by a variety of techniques. One useful device consists of introducing an appropriately placed parameter in the integral and then differentiating or integrating with respect to the parameter, employing the aforementioned properties of uniform convergence.

Laplace Transforms

Operators that transform one set of objects into another are common in mathematics. Both the derivative and the indefinite integral are examples. Logarithms provide an immediate arithmetic advantage by replacing multiplication, division, and powers, respectively, by the relatively simpler processes of addition, subtraction, and multiplication. After obtaining a result with logarithms, an antilogarithm procedure is necessary to find its image in the original framework. The Laplace transform has a role similar to that of logarithms but in the more sophisticated world of differential equations. (See Problems 12.34 and 12.36.)

The Laplace transform of a function $F(x)$ is defined as

$$f(s) = L\{F(x)\} = \int_0^{\infty} e^{-5x} F(x)\,dx \tag{12}$$

and is analogous to power series as seen by replacing e^{-s} by t so that $e^{-sx} = t^x$. Many properties of power series also apply to Laplace transforms. Table 12.1, showing Laplace transforms, is useful. In each case, a is a real constant.

TABLE 12.1

$F(x)$	$\mathcal{L}\{F(x)\}$	
a	$\dfrac{a}{s}$	$s > 0$
e^{ax}	$\dfrac{1}{s-a}$	$s > a$
$\sin ax$	$\dfrac{a}{s^2+a^2}$	$s > 0$
$\cos ax$	$\dfrac{8}{s^2+a^2}$	$s > 0$
$x^n \; n = 1, 2, 3, \ldots$	$\dfrac{n!}{s^{n+1}}$	$s > 0$
$Y'(x)$	$s\mathcal{L}\{Y(x)\} - Y(0)$	
$Y''(x)$	$s^2\mathcal{L}\{Y(x)\} - sY(0) - Y'(0)$	

Linearity

The Laplace transform is a linear operator; i.e., $\zeta\{F(x) + G(x)\} = \zeta\{F(x)\} + \zeta\{G(x)\}$.

This property is essential for returning to the solution after having calculated in the setting of the transforms. (See the example that follows and the previously cited problems.)

Convergence

The exponential e^{-st} contributes to the convergence of the improper integral. What is required is that $F(x)$ does not approach infinity too rapidly as $x \to \infty$. This is formally stated as follows: If there is some constant a such that $|F(x)| \leqq e^{ax}$ for all sufficiently large values of x, then $f(s) = \int_0^{\infty} e^{-5s} F(x)\,dx$ converges when $s > a$ and f has derivatives of all orders. (The differentiations of f can occur under the integral sign.)

Application

The feature of the Laplace transform that (when combined with linearity) establishes it as a tool for solving differential equations is revealed by applying integration by parts to $f(s) = \int_0^{x} e^{-st} F(t)\,dt$. By letting $u = F(t)$ and $dv = e^{-st}\,dt$, we obtain, after letting $x \to \infty$,

$$\int_0^{x} e^{-st} F(t)dt = \frac{1}{s}F(0) + \frac{1}{s}\int_0^{\infty} e^{-st} F'(t)\,dt$$

Conditions must be satisfied that guarantee the convergence of the integrals (for example, $e^{-st} F(t) \to 0$ as $t \to \infty$).

This result of integration by parts may be put in the form

(a) $\zeta\{F'(t)\} = s\zeta\{F(t)\} - F(0)$.

Repetition of the procedure combined with a little algebra yields

(b) $\zeta\{F''(t)\} = s^2\zeta\{F(t)\} - sF(0) - F'(0)$.

The Laplace representation of derivatives of the order needed can be obtained by repeating the process.

To illustrate application, consider the differential equation

$$\frac{d^2y}{dt^2} + 4y = 3\sin t$$

where $y = F(t)$ and $F(0) = 1$, $F'(0) = 0$. We use

$$\zeta(\sin at) = \frac{\alpha}{s^2 + a^2},\ \zeta\{\cos at\} = \frac{5}{s^2 + a^2}$$

and recall that

$$f(s) = \zeta\{F(t)\}\ \zeta\{F''(t)\} + 4\zeta\{F(t)\} = 3\zeta\{\sin t\}$$

Using (b), we obtain

$$s^2 f(s) - s + 4f(s) = \frac{3}{s^2 + 1}$$

Solving for $f(s)$ yields

$$f(s) = \frac{3}{(s^2 + 4)(s^2 + 1)} + \frac{s}{s^2 + 4} = \frac{j}{s^2 + 1} - \frac{1}{s^2 + 4} + \frac{s}{s^2 + 4}$$

(Partial fractions were employed.)

Referring to the table of Laplace transforms, we see that this last expression may be written

$$f(s) = \zeta\{\sin t\} - \frac{1}{2}\zeta\{\sin 2t) + \zeta\{\cos 2t\}$$

then, using the linearity of the Laplace transform,

$$f(s) = \zeta\left\{\sin t - \frac{1}{2}\sin 2t + \cos 2t\right\}$$

We find that

$$F(t) = \sin t - \frac{1}{2}\sin 2t + \cos 2t$$

satisfies the differential equation.

Improper Multiple Integrals

The definitions and results for improper single integrals can be extended to improper multiple integrals.

SOLVED PROBLEMS

Improper integrals

12.1. Classify according to the type of improper integral:

(a) $\displaystyle\int_{-1}^{1} \frac{dx}{\sqrt[3]{x}(x+1)}$ (c) $\displaystyle\int_{3}^{10} \frac{x\,dx}{(x-2)^2}$ (e) $\displaystyle\int_{0}^{\pi} \frac{1-\cos x}{x^2}\,dx$

(b) $\int_0^{\infty} \frac{dx}{1+\tan x}$ (d) $\int_{-\infty}^{\infty} \frac{x^2 dx}{x^4+x^2+1}$

(a) Second kind (integrand is unbounded at $x = 0$ and $x = -1$).

(b) Third kind (integration limit is infinite and integrand is unbounded where $\tan x = -1$).

(c) This is a proper integral (integrand becomes unbounded at $x = 2$, but this is outside the range of integration $3 \leqq x \leqq 10$).

(d) First kind (integration limits are infinite but integrand is bounded).

(e) This is a proper integral (since $\lim_{x\to 0+} \frac{1-\cos x}{x^2} = \frac{1}{2}$ by applying L' Hospital's rule).

12.2. Show how to transform the improper integral of the second kind, $\int_t^2 \frac{dx}{\sqrt{x(2-x)}}$, into (a) an improper integral of the first kind, (b) a proper integral.

(a) Consider $\int_t^{2-\epsilon} \frac{dx}{\sqrt{x(2-x)}}$, where $0 < \epsilon < 1$, say. Let $2-x = \frac{1}{y}$. Then the integral becomes $\int_1^{1/\epsilon} \frac{dy}{y\sqrt{2y-1}}$. As $\epsilon \to 0+$, we see that consideration of the given integral is equivalent to consideration of

$$\int_1^{\infty} \frac{dy}{y\sqrt{2y-1}},$$

which is an improper integral of the first kind.

(b) Letting $2 - x = \upsilon^2$ in the integral of (a), it becomes $2\int_{\sqrt{\epsilon}}^1 \frac{d\upsilon}{\sqrt{\upsilon^2+2}}$. We are thus led to consideration of $2\int_0^1 \frac{d\upsilon}{\sqrt{\upsilon^2+1}}$, which is a proper integral.

From this, we see that an improper integral of the first kind *may* be transformed into an improper integral of the second kind, and conversely (actually this can *always* be done).

We also see that an improper integral may be transformed into a proper integral (this can only *sometimes* be done).

Improper integrals of the first kind

12.3. Prove the comparison test (Page 326) for convergence of improper integrals of the first kind.

Since $0 \leqq f(x) \leqq g(x)$ for $x \geqq a$, we have, using Property 7, Page 98,

$$0 \leqq \int_a^b f(x)\,dx \leqq \int_a^b g(x)\,dx \leqq \int_a^{\infty} g(x)\,dx$$

But, by hypothesis, the last integral exists. Thus, $\lim_{b\to\infty} \int_a^b f(x)\,dx$ exists, and, hence, $\int_a^{\infty} f(x)\,dx$ converges.

12.4. Prove the quotient test (a) on Page 326.

By hypothesis, $\lim_{x\to\infty} \frac{f(x)}{g(x)} = A > 0$. Then, given any $\epsilon > 0$, we can find N such that $\left|\frac{f(x)}{g(x)} - A\right| < \epsilon$ when $x \geqq N$. Thus, for $x \geqq N$, we have

$$A-\epsilon \leqq \frac{f(x)}{g(x)} \leqq A+\epsilon \qquad or \qquad (A-\epsilon)g(x) \leqq f(x) \leqq (A+\epsilon)g(x)$$

Then

$$(A-\epsilon)\int_N^b g(x)\,dx \leqq \int_N^b f(x)\,dx \leqq (A+\epsilon)\int_N^b g(x)\,dx \tag{1}$$

There is no loss of generality in choosing $A - \epsilon > 0$.

If $\int_a^{\infty} g(x)\,dx$ converges, then by the inequality on the right of Equation (1),

$\lim_{b\to\infty} \int_N^b f(x)\,dx$ exists, and so $\int_a^{\infty} f(x)\,dx$ converges

If $\int_a^\infty g(x)\,dx$ diverges, then by the inequality on the left of Equation (1), $\lim_{b\to\infty}\int_N^b f(x)\,dx = \infty$, and so $\int_a^\infty f(x)\,dx$ diverges.

For the cases where $A = 0$ and $A = \infty$, see Problem 12.41.

As seen in this and Problem 12.3, there is, in general, a marked similarity between proofs for infinite series and improper integrals.

12.5. Test for convergence: (*a*) $\int_1^\infty \frac{x\,dx}{3x^4+5x^2+1}$ and (*b*) $\int_2^\infty \frac{x^2-1}{\sqrt{x^6+16}}\,dx$.

(a) **Method 1:** For large x, the integrand is approximately x/3x4 = 1/3x3.

Since $\frac{x}{3x^4+5x^2+1} \leq \frac{1}{3x^3}$ and $\frac{1}{3}\int_1^\infty \frac{dx}{x^3}$ converges (p integral with $p = 3$), it follows by the comparison test that $\int_1^\infty \frac{x\,dx}{3x^4+5x^2+1}$ also converges.

Note that the purpose of examining the integrand for large x is to obtain a suitable comparison integral.

Method 2: *Let* $f(x) = \frac{x}{3x^4+5x^2+1}$, $g(x) = \frac{1}{x^3}$. Since $\lim_{x\to\infty}\frac{f(x)}{3}$, and $\int_1^\infty g(x)\,dx$ converges, $\int_1^\infty f(x)\,dx$ also converges by the quotient test.

Note that in the comparison function $g(x)$, we have discarded the factor $\frac{1}{3}$. However, it could just as well have been included.

Method 3: $\lim_{x\to\infty} x^3\left(\frac{x}{3x^4+5x^2+1}\right) = \frac{1}{3}$. Hence, by Theorem 1, Page 324, the required integral converges.

(b) **Method 1:** For large x, the integrand is approximately $x^2/x^6 = 1/x$.

For $x \geqq 2$, $\frac{x^2-1}{\sqrt{x^6+1}} \geqq \frac{1}{2}\cdot\frac{1}{x}$. Since $\frac{1}{2}\int_2^\infty \frac{dx}{x}$ diverges, $\int_2^\infty \frac{x^2-1}{\sqrt{x^6+16}}\,dx$ also diverges.

Method 2: Let $f(x) = \frac{x^2-1}{\sqrt{x^6-16}}$, $g(x) = \frac{1}{x}$. Then, since $\lim_{x\to\infty}\frac{f(x)}{g(x)} = 1$ and $\int_2^\infty g(x)\,dx$ diverges, $\int_2^\infty f(x)\,dx$ also diverges.

Method 3: Since $\lim_{x\to\infty} x\left(\frac{x^2-1}{\sqrt{x^6+16}}\right) = 1$, the required integral diverges by Theorem 1, Page 324.

Note that Method 1 may (and often does) require us to obtain a suitable inequality factor (in this case, $\frac{1}{2}$ or any positive constant less than $\frac{1}{2}$) before the comparison test can be applied. Methods 2 and 3, however, do not require this.

12.6. Prove that $\int_0^\infty e^{-x^2}\,dx$ converges.

$\lim_{x\to\infty} x^2 e^{-x^2} = 0$ (by L'Hospital's rule or otherwise). Then, by Theorem 1, with $A = 0$, $p = 2$, the given integral converges. Compare Problem 11.10(a).

12.7. Examine for convergence: (a) $\int_1^\infty \frac{\ln x}{x+a}\,dx$, where a is a positive constant and (b) $\int_0^\infty \frac{1-\cos x}{x^2}\,dx$.

(a) $\lim_{x\to\infty} x.\frac{\ln x}{x+a} = \infty$. Hence, by Theorem 1, Page 324, with $A = \infty$, $p = 1$, the given integral diverges.

(b) $\int_0^\infty \frac{1-\cos x}{x^2}\,dx = \int_0^\pi \frac{1-\cos x}{x^2}\,dx + \int_\pi^\infty \frac{1-\cos x}{x^2}\,dx.$

The first integral on the right converges [see Problem 12.1(e)].

Since $\lim_{x\to\infty} x^{3/2}\left(\frac{1-\cos x}{x^2}\right)=0$, the second integral on the right converges by Theorem 1, Page 324, with $A = 0$ and $p = 3/2$.

Thus, the given integral converges.

12.8. Test for convergence: (*a*) $\int_{-\infty}^{-1}\frac{e^x}{x}dx$ and (*b*) $\int_{-\infty}^{\infty}\frac{x^3+x^2}{x^6+1}dx$.

(a) Let x = – y. Then the integral becomes $-\int_1^{\infty}\frac{e^{-y}}{y}dy$.

Method 1: $\frac{e^{-y}}{y} \leqq e^{-y}$ for $\leqq 1$. Then, since $\int_1^{\infty} e^{-y}dy$ converges, $\int_1^{\infty}\frac{e^{-y}}{y}dy$ converges; hence, the given integral converges.

Method 2: $\lim_{x\to\infty} y^2\left(\frac{e^{-y}}{y}\right)=\lim_{y\to\infty} ye^{-y}=0$. Then the given integral converges by Theorem 1, Page 324, with $A = 0$ and $p = 2$.

(b) Write the given integral as $\int_{-\infty}^{0}\frac{x^3+x^2}{x^6+1}dx+\int_0^{\infty}\frac{x^3+x^2}{x^6+1}dx$. Letting $x = -y$ in the first integral, it becomes $-\int_0^{\infty}\frac{y^3-y^2}{y^6+1}dy$. Since $\lim_{y\to\infty} y^3\left(\frac{y^3-y^2}{y^6+1}\right)=1$, this integral converges.

Since $\lim_{x\to\infty} x^3\left(\frac{x^3+x^2}{x^6+1}\right)=1$, the second integral converges.

Thus, the given integral converges.

Absolute and conditional convergence for improper integrals of the first kind

12.9. Prove that $\int_{\alpha}^{\infty} f(x)\,dx$ converges if $\int_0^{\infty} |\,f(x)\,|\,dx$ converges; i.e., an absolutely convergent integral is convergent.

We have $-|f(x)| \leqq f(x) \leqq |f(x)|$; i.e., $0 \leqq f(x) + |f(x)| 2|f(x)|$. Then

$$0 \leqq \int_a^b [f(x) \leqq +|\,f(x)]\,dx \leqq 2\int_a^b |\,f(x)\,|\,dx$$

If $\int_{\alpha}^{\infty} |f(x)|\,dx$ converges, it follows that $\int_{\alpha}^{\infty} |f(x) + |f(x)|]\,dx$ converges. Hence, by subtracting $\int_{\alpha}^{\infty} |f(x)|\,dx$, which converges, we see that $\int_{\alpha}^{\infty} f(x)\,dx$ converges.

12.10. Prove that $\int_t^{\infty}\frac{\cos x}{x^2}dx$ converges.

Method 1: $\left|\frac{\cos x}{x^2}\right| \leqq \frac{1}{x^2}$ for $x \geqq 1$. Then by the comparison test, since $\int_t^{\infty}\frac{dx}{x^2}$ converges, it follows that $\int_t^{\infty}\left|\frac{\cos x}{x^2}\right|d$ x converges; $\int_t^{\infty}\left|\frac{\cos x}{x^2}\right|d$ x converges, i.e., $\int_t^{\infty}\frac{\cos x}{x^2}$ converges absolutely, and so converges by Problem 12.9.

Method 2: Since $\lim_{x\to\infty} x^{3/2}\left|\frac{\cos x}{e^2}\right| = \lim_{x\to\infty}\left|\frac{\cos x}{x^{1/2}}\right| = 0$ it follows from Theorem 1, Page 324, with $A = 0$ and $p = 3/2$. that $\int_t^{\infty}\left|\frac{\cos x}{x^2}\right|dx$ converges, and, hence, $\int_t^{\infty}\frac{\cos x}{x^2}dx$ converges (absolutely).

12.11. Prove that $\int_t^{\infty} \frac{\sin x}{x}\,dx$ converges.

Since $\int_0^1 \frac{\sin x}{x}\,dx$ converges $\left(\text{because } \frac{\sin x}{x} \text{ is continuous in } 0 < x \leqq 1 \text{ and } \lim_{x\to 0+} \frac{\sin x}{x} = 1\right)$, we need only show that $\int_1^{\infty} \frac{\sin x}{x}\,dx$ converges.

Method 1: Integration by parts yields

$$\int_t^M \frac{\sin x}{x}\,dx = -\frac{\cos x}{x}\bigg|_1^M + \int_t^M \frac{\cos x}{x^2}\,dx = \cos 1 - \frac{\cos M}{M} + \int_t^M \frac{\cos x}{x^2}\,dx \qquad (1)$$

or, on taking the limit on both sides of Equation (1) as $M \to \infty$ and using the fact that $\lim_{M\to\infty} \frac{\cos M}{M} = 0$,

$$\int_1^{\infty} \frac{\sin x}{x}\,dx = \cos 1 + \int_1^{\infty} \frac{\cos x}{x^2}\,dx \qquad (2)$$

Since the integral on the right of Equation (2) converges by Problem 12.10, the required result follows. The technique of integration by parts to establish convergence is often useful in practice.

Method 2:

$$\int_0^{\infty} \frac{\sin x}{x}\,dx = \int_0^{\pi} \frac{\sin x}{x}\,dx + \int_{\pi}^{2\pi} \frac{\sin x}{x}\,dx + \ldots + \int_{n\pi}^{(n+1)\pi} \frac{\sin x}{x}\,dx + \ldots$$
$$= \sum_{n=0}^{\infty} \int_{n\pi}^{(n+1)\pi} \frac{\sin x}{x}\,dx$$

Letting $x = \upsilon + n\pi$, the summation becomes

$$\sum_{n=0}^{\infty} (-1)^n \int_0^{\pi} \frac{\sin \upsilon}{n + n\pi}\,d\upsilon \int_0^{\pi} \frac{\sin \upsilon}{\upsilon}\,d\upsilon + \int_0^{\pi} \frac{\sin \upsilon}{\upsilon + \pi}\,d\upsilon + \int_0^{\pi} \frac{\sin \upsilon}{\upsilon + 2\pi}\,d\upsilon - \ldots$$

This is an alternating series. Since $\frac{1}{\upsilon + n\pi} \leqq \frac{1}{\upsilon + (n+1)\pi}$ and $\sin \upsilon \geqq 0$ in $[0, \pi]$, it follows that

$$\int_0^{\pi} \frac{\sin \upsilon}{\upsilon + n\pi}\,d\upsilon \leqq \int_0^{\pi} \frac{\sin \upsilon}{\upsilon + (n+1)\pi}\,d\upsilon$$

Also,

$$\lim_{n\to\infty} \int_0^{\pi} \frac{\sin \upsilon}{\upsilon + n\pi}\,d\upsilon \leqq \lim_{n\to\infty} \int_0^{\pi} \frac{d\upsilon}{n\pi} = 0$$

Thus, each term of the alternating series is, in absolute value, less than or equal to the preceding term, and the nth term approaches zero as $n \to \infty$. Hence, by the alternating series test (Page 281), the series and, thus, the integral converge.

12.12. Prove that $\int_0^{\infty} \frac{\sin x}{x}\,dx$ converges conditionally.

Since, by Problem 12.11, the given integral converges, we must show that it is not absolutely convergent; i.e., $\int_0^{\infty} \left|\frac{\sin x}{x}\right| dx$ diverges. diverges.

As in Problem 12.11, Method 2, we have

$$\int_0^{\infty} \left|\frac{\sin x}{x}\right| dx = \sum_{n=0}^{\infty} \int_{n\pi}^{(n+1)\pi} \left|\frac{\sin x}{x}\right| dx = \sum_{n=0}^{\infty} \int_0^{\pi} \frac{\sin \upsilon}{\upsilon + n\pi}\,d\upsilon \qquad (1)$$

Now $\frac{1}{\upsilon + n\pi} \geqq \frac{1}{(n+1)\pi}$ for $0 \leqq \upsilon \leqq \pi$. Hence,

$$\int_0^{\pi} \frac{\sin \upsilon}{\upsilon + n\pi}\,d\upsilon \geqq \frac{1}{(n+1)\pi} \int_9^{\pi} \sin \upsilon\,d\upsilon = \frac{2}{(n+1)\pi} \qquad (2)$$

Since $\sum_{n=0}^{\infty} \frac{2}{(n+1)\pi}$ diverges, the series on the right of Equation (1) diverges by the comparison test.

Hence, $\int_0^{\infty} \left|\frac{\sin x}{x}\right| dx$ diverges and the required result follows.

Improper integrals of the second kind, cauchy principal value

12.13. (a) Prove that $\int_{-1}^{7} \frac{dx}{\sqrt[3]{x+1}}$ converges and (b) find its value.

The integrand is unbounded at $x = -1$. Then we define the integral as

$$\lim_{\epsilon \to 0+} \int_{-1+\epsilon}^{7} \frac{dx}{\sqrt[3]{x+1}} = \lim_{\epsilon \to 0+} \left.\frac{(x+1)^{2/3}}{2/3}\right|_{-1+\epsilon}^{7} = \lim_{\epsilon \to 0+} \left(6 - \frac{3}{2}\epsilon^{2/3}\right) = 6$$

This shows that the integral converges to 6.

12.14. Determine whether $\int_{-1}^{5} \frac{dx}{(x-1)^3}$ converges (a) in the usual sense and (b) in the Cauchy principal value sense.

(a) By definition,

$$\int_{-1}^{5} \frac{dx}{(x-1)^3} = \lim_{\epsilon_1 \to 0+} \int_{-1}^{1-\epsilon_1} \frac{dx}{(x-1)^3} + \lim_{\epsilon_2 \to 0+} \int_{1+\epsilon_2}^{5} \frac{dx}{(x-1)^3}$$

$$= \lim_{\epsilon_1 \to 0+} \left(\frac{1}{8} - \frac{1}{2\epsilon_1^2}\right) + \lim_{\epsilon_2 \to 0+} \left(\frac{1}{2\epsilon_2^2} - \frac{1}{32}\right)$$

and, since the limits do not exist, the integral does not converge in the usual sense.

(b) Since

$$\lim_{\epsilon \to 0+} \left\{\int_{-1}^{1-\epsilon} \frac{dx}{(x-1)^3} + \int_{1+\epsilon}^{5} \frac{dx}{(x-1)^3}\right\} = \lim_{\epsilon \to 0+} \left\{\frac{1}{8} - \frac{1}{2\epsilon^2} + \frac{1}{2\epsilon^2} - \frac{1}{32}\right\} = \frac{3}{32}$$

the integral exists in the Cauchy principal value sense. The principal value is 3/32.

12.15. Investigate the convergence of:

(a) $\int_2^3 \frac{dx}{x^2(x^3-8)^{2/3}}$ (c) $\int_1^5 \frac{dx}{\sqrt{(5-x)(x-1)}}$ (e) $\int_0^{\pi/2} \frac{dx}{(\cos x)^{1/n}}, n > 1$

(b) $\int_0^{\pi} \frac{\sin x}{x^3} dx$ (d) $\int_{-1}^{1} \frac{2^{\sin^{-1} x}}{1-x} dx$

(a) $\lim_{x \to 2+} (x-2)^{2/3} \cdot \frac{1}{x^2(x^3-8)^{2/3}} = \lim_{x \to 2+} \frac{1}{x^2}\left(\frac{1}{x^2+2x+4}\right)^{2/3} = \frac{1}{8\sqrt[3]{18}}$. Hence, the integral converges by Theorem 3(i), Page 326.

(b) $\lim_{x \to 0+} x^2 \cdot \frac{\sin x}{x^3} = 1$. Hence, the integral diverges by Theorem 3(ii) on Page 326.

(c) Write the integral as $\int_1^3 \frac{dx}{\sqrt{(5-x)(x-1)}} + \int_3^5 \frac{dx}{\sqrt{(5-x)(x-1)}}$.

Since $\lim_{x \to 1+} (x-1)^{1/2} \cdot \frac{1}{\sqrt{(5-x)(x-1)}} = \frac{1}{2}$, the first integral converges.

Since $\lim_{x \to 5-} (5-x)^{1/2} \cdot \frac{1}{\sqrt{(5-x)(x-1)}} = \frac{1}{2}$, the second integral converges.

Thus, the given integral converges.

(d) $\lim_{x \to 1-} (1-x) \cdot \frac{2^{\sin^{-1} x}}{1-x} = 2^{\pi/2}$. Hence, the integral diverges.

Another method: $\dfrac{2^{\sin^{-1}x}}{1-x} \geqq \dfrac{2^{-\pi/2}}{1-x}$, and $\int_{-1}^{1}\dfrac{dx}{1-x}$ diverges. Hence, the given integral diverges.

(e) $\lim\limits_{x\to 1/2\pi-}(\pi/2-x)^{1/n}\cdot\dfrac{1}{(\cos x)^{1/n}} = \lim\limits_{x\to 1/2\pi-}\left(\dfrac{\pi/2-x}{\cos x}\right)^{1/n} = 1$. Hence, the integral converges.

12.16. If m and n are real numbers, prove that $\int_0^1 x^{m-1}(1-x)^{n-1}\,dx$ (a) converges if $m > 0$ and $n > 0$ simultaneously and (b) diverges otherwise.

(a) For $m \geqq 1$ and $n \geqq 1$ simultaneously, the integral converges, since the integrand is continuous in $0 \leqq x \leqq 1$. Write the integral as

$$\int_0^{1/2} x^{m-1}(1-x)^{n-1}\,dx + \int_{1/2}^{1} x^{m-1}(1-x)^{n-1}\,dx \qquad (1)$$

If $0 < m < 1$ and $0 < n < 1$, the first integral converges, since $\lim\limits_{x\to 0+} x^{1-m}\cdot x^{m-1}(1-x)^{n-1} = 1$, using Theorem 3(i), Page 326, with $p = 1 - m$ and $a = 0$.

Similarly, the second integral converges, since $\lim_{x\to 1-}(1-x)^{1-n}\cdot x)^{n-1}(1-x)^{n-1} = 1$, using 4(i), Page 326, with $p = 1 - n$ and $b = 1$.

Thus, the given integral converges if $m > 0$ and $n > 0$ simultaneously.

(b) If $m \leqq 0$, $\lim\limits_{x\to 0+} x\cdot x^{m-1}(1-x)^{n-1} = \infty$. Hence, the first integral in Equation (1) diverges, regardless of the value of n, by Theorem 3(ii), Page 326, with $p = 1$ and $a = 0$.

Similarly, the second integral diverges if $n \leqq 0$, regardless of the value of m, and the required result follows.

Some interesting properties of the given integral, called the *beta integral* or *beta function*, are considered in Chapter 15.

12.17. Prove that $\int_0^{\pi}\dfrac{1}{x}\sin\dfrac{1}{x}\,dx$ converges conditionally.

Letting $x = 1/y$, the integral becomes $\int_{1/\pi}^{\infty}\dfrac{\sin y}{y}\,dy$ and the required result follows from Problem 12.12.

Improper integrals of the third kind

12.18. If n is a real number, prove that $\int_0^{\infty} x^{n-1}e^{-x}\,dx$ (a) converges if $n > 0$ and (b) diverges if $n \leqq 0$.

Write the integral as

$$\int_0^1 x^{n-1}e^{-x}\,dx + \int_1^{\infty} x^{n-1}e^{-x}\,dx \qquad (1)$$

(a) If $n \geqq 1$, the first integral in Equation (1) converges, since the integrand is continuous in $0 \leqq x \leqq 1$.

If $0 < n < 1$, the first integral in Equation (1) is an improper integral of the second kind at $x = 0$. Since $\lim\limits_{x\to 0+} x^{1-n}\cdot x^{n-1}e^{-x} = 1$, the integral converges by Theorem 3(i), Page 326, with $p = 1 - n$ and $a = 0$.

Thus, the first integral converges for $n > 0$.

If $n > 0$, the second integral in Equation (1) is an improper integral of the first kind. Since $\lim\limits_{x\to\infty} x^2\cdot x^{n-1}e^{-x} = 0$ (by L'Hospital's rule or otherwise), this integral converges by Theorem 1(i), Page 324, with $p = 2$.

Thus, the second integral also converges for $n > 0$, and so the given integral converges for $n > 0$.

(b) If n $\leqq 0$, the first integral of Equation (1) diverges, since $\lim\limits_{x\to 0+} x\cdot x^{n-1}e^{-x} = \infty$ [Theorem 3(ii), Page 326].

If $n \leqq 0$, the second integral of Equation (1) converges, since $\lim\limits_{x\to\infty}\cdot x^{n-1}e^{-x} = 0$ [Theorem 1(i), Page 324].

Since the first integral in Equation (1) diverges while the second integral converges, their sum also diverges; i.e., the given integral diverges if $n \leqq 0$.

Some interesting properties of the given integral, called the *gamma function*, are considered in Chapter 15.

Uniform convergence of improper integrals

12.19. (a) Evaluate $\phi(\alpha) = \int_0^\infty \alpha e^{-\alpha x} dx$ for $\alpha > 0$. (b) Prove that the integral in (a) converges uniformly to 1 for $\alpha \in \alpha 1 > 0$. (c) Explain why the integral does not converge uniformly to 1 for $\alpha > 0$.

(a) $\phi(\alpha) = \lim_{b\to\infty} \int_a^b \alpha e^{-ae} dx = \lim_{b\to\infty} -e^{-\alpha x}\Big|_{x=0}^b = \lim_{b\to\infty} 1 - e^{-ab} = 1$ if $\alpha > 0$

Thus, the integral converges to 1 for all $\alpha > 0$.

(b) **Method 1**, using definition: The integral converges uniformly to 1 in $\alpha \varepsilon \alpha_1 > 0$ if for each $\epsilon > 0$ we can find N, depending on ϵ but not on α, such that $\left|1 - \int_0^u \alpha e^{-\alpha x} dx\right| < \epsilon$ for all $u > N$.

Since $\left|1 - \int_0^u \alpha e^{-\alpha x} dx\right| = |1 - (1 - e^{-\alpha u})| = e^{-au} < e^{-\alpha_1 u} < \epsilon$ for $u > \frac{1}{\alpha_1} \ln \frac{1}{\epsilon} = N$, the result follows.

Method 2, using the Weierstrass M test: Since $\lim_{x\to\infty} x^2 \cdot \alpha e^{-\alpha x} = 0$ for $\alpha\ \alpha_1 > 0$, we can choose $|\alpha e^{-\alpha x}| < \frac{1}{x^2}$ for sufficiently large x—say, $x \geqq x_0$. Taking $M(x) = \frac{1}{x^2}$ and noting that $\int_{x_0}^\infty \frac{dx}{x^2}$ converges, it follows that the given integral is uniformly convergent to 1 for $\alpha \varepsilon \alpha_1 > 0$.

(c) As $\alpha 1 \to 0$, the number N in the first method of (b) increases without limit, so that the integral cannot be uniformly convergent for $\alpha > 0$.

12.20. If $\phi(\alpha) = \int_0^\infty f(x, \alpha)\, dx$ is uniformly convergent for $\alpha_1 \leqq \alpha \leqq \alpha_2$, prove that $\phi(\alpha)$ is continuous this interval.

Let $\phi(\alpha) = \int_u^u f(x, \alpha)\, dx + R(u,\alpha)$, where $R(u,\alpha) = \int_u^\infty f(x, \alpha)\, dx$.

Then $\phi(\alpha + h) = \int_u^u f(x, \alpha + h)\, dx + R(u, \alpha + h)$ and so

$$\phi(\alpha + h) - \phi(\alpha) = \int_u^u \{f(x, \alpha + h) - f(x, \alpha)\}\, dx + R(u, \alpha + h) - R(u, \alpha)$$

Thus,

$$|\phi(\alpha + h) - \phi(\alpha)| \leqq \int_a^u |f(x, \alpha + h) - f(x, \alpha)|\, dx + |R(u, \alpha + h| + |R(u, \alpha)| \tag{1}$$

Since the integral is uniformly convergent in $(\alpha)_1 \leqq (\alpha) \leqq \alpha_2$, we can, for find N independent of (α) such that for $(u) > N$,

$$|R(u, \alpha + h)| < \epsilon/3, \ |R(u, \alpha)| < \epsilon/3 \tag{2}$$

Since $f(x, \alpha)$ is continuous, we can find $\delta > 0$ corresponding to each $\epsilon > 0$ such that

$$\int_a^v |f(x, \alpha + h) - f(x, \alpha)|\, dx < \epsilon/3 \quad \text{for } |h| < \delta \tag{3}$$

Using Equations (2) and (3) in (1), we see that $|\phi(\alpha + h) - \phi(\alpha)| < \epsilon$ for $|(h)| < \delta$, so that $\phi(\alpha)$ is continuous.

Note that in this proof we assume that both α and $\alpha + h$ are in the interval $\alpha_1 \leqq \alpha \leqq \alpha_2$. Thus, if $\alpha = \alpha_1$, for example, $h > 0$ and right-hand continuity is assumed.

Also note the analogy of this proof with that for infinite series.

Other properties of uniformly convergent integrals can be proved similarly.

12.21. (a) Show that $\lim_{\alpha\to 0+} \int_0^\infty \alpha e^{-\alpha x} dx \neq \int_0^\infty \left(\lim_{\alpha\to 0+} \alpha e^{-\alpha x}\right) dx$ and (b) explain the result in (a).

(a) $\lim_{\alpha\to 0+} \int_0^\infty \alpha e^{-\alpha x} dx = \lim_{\alpha\to 0+} = 1$ by Problem 12.19(a).

$\int_0^\infty \left(\lim_{\alpha\to 0+} \alpha e^{-\alpha x}\right) dx = \int_0^\infty 0\, dx = 0$. Thus, the required result follows.

(b) Since $\phi(\alpha) = \int_0^\infty \alpha e^{-\alpha x} dx$ is not uniformly convergent for $\alpha \geqq 0$ (see Problem 12.19), there is no guarantee that $\phi(\alpha)$ will be continuous for $\alpha \geqq 0$. Thus, $\lim_{\alpha\to 0+} \phi(\alpha)$ may not be equal to $\phi(0)$.

12.22. (a) Prove that $\int_0^\infty e^{-\alpha x}\cos rx\,dx = \dfrac{\alpha}{\alpha^2+r^2}$ for $\alpha > 0$ and any real value of r. (b) Prove that the integral in (a) converges uniformly and absolutely for $a \leqq \alpha \leqq b$, where $0 < a < b$ and any r.

(a) From integration formula 34, Page 103, we have

$$\lim_{M\to\infty}\int_0^M e^{-\alpha x}\cos rx\,dx = \lim_{M\to\infty}\left.\frac{e^{-\alpha x}(r\sin rx - \alpha\cos rx)}{\alpha^2+r^2}\right|_0^M = \frac{\alpha}{\alpha^2+r^2}$$

(b) This follows at once from the Weierstrass M test for integrals, by noting that $|e^{-\alpha x}\cos rx| \leqq e^{-\alpha x}$ and $\int_0^\infty e^{-\alpha x}dx$ converges.

Evaluation of definite integrals

12.23. Prove that $\int_0^{\pi/2}\ln\sin x\,dx = \dfrac{\pi}{2}\ln 2$.

The given integral converges [Problem 12.42(f)]. Letting $x = \pi/2 - y$,

$$I = \int_0^{\pi/2}\ln\sin x\,dx = \int_0^{\pi/2}\ln\cos y\,dy = \int_0^{\pi/2}\ln\cos x\,dx$$

Then

$$2I = \int_0^{\pi/2}(\ln\sin x + \ln\cos x)\,dx = \int_0^{\pi/2}\ln\left(\frac{\sin 2x}{2}\right)dx$$
$$= \int_0^{\pi/2}\ln\sin 2x dx - \int_0^{\pi/2}\ln 2\,dx = \int_0^{\pi/2}\ln\sin 2x\,dx - \frac{\pi}{2}\ln 2 \qquad (1)$$

Letting $2x = \upsilon$,

$$\int_{02}^{\pi/2}\ln\sin 2x\,dx = \frac{1}{2}\int_0^{\pi}\ln\sin\upsilon\,d\upsilon = \frac{1}{2}\left\{\int_0^{\pi/2}\ln\sin\upsilon\,d\upsilon\right\}$$
$$= \frac{1}{2}(I+I) = I \quad \text{(letting } \upsilon = \pi - \text{in the last integral)}$$

Hence, Equation (*l*) becomes $2I = I\dfrac{\pi}{2}\ln$ on $I = -\dfrac{\pi}{2}\ln 2$.

12.24. Prove that $\int_0^{\pi} x\ln\sin x\,dx = -\dfrac{\pi^2}{2}\ln 2$.

Let $x = \pi - y$. Then, using the results in the preceding problem,

$$J = \int_0^{\pi} x\ln\sin x\,dx = \int_0^{\pi}(\pi - u)\ln\sin u\,du = \int_0^{\pi}(\pi - x)\ln\sin x\,dx$$
$$= \pi\int_0^{\pi}\ln\sin x\,dx - \int_0^{\pi}x\ln\sin x\,dx$$
$$= -\pi^2\ln 2 - J$$

or $J = -\dfrac{\pi^2}{2}\ln 2$.

12.25. (a) Prove that $\phi(\alpha) = \int_0^\infty \dfrac{dx}{x^{2+\alpha}}$ is uniformly convergent for $\alpha \geqq 1$. (b) Show that $\phi(\alpha) = \dfrac{\pi}{2\sqrt{\alpha}}$.

(c) Evaluate $\int_0^\infty \dfrac{dx}{(x^2+1)^2}$. (d) Prove that $\int_0^\infty \dfrac{dx}{(x^2+1)^{n+1}} = \int_0^{\pi/2}\cos^{2n}\theta\,d\theta = \dfrac{1.3.5\cdots(2n-1)}{2.4.6\cdots(2n)}\dfrac{\pi}{2}$.

(a) The result follows from the Weierstrass test, since $\dfrac{1}{x^2+\alpha} \leqq \dfrac{1}{x^2+1}$ for $a \geqq 1$ and $\int_0^\infty \dfrac{dx}{x^2+1}$ converges.

(b) $\phi(\alpha) = \lim_{b\to\infty}\int_0^b \dfrac{dx}{x^2+\alpha} = \lim_{b\to\infty}\dfrac{1}{\sqrt{\alpha}}\tan^{-1}\left.\dfrac{x}{\sqrt{\alpha}}\right|_0^b = \lim_{b\to\infty}\dfrac{1}{\sqrt{\alpha}}\tan^{-1}\dfrac{b}{\sqrt{\alpha}} = \dfrac{\pi}{2\sqrt{\alpha}}$

(c) From (b), $\int_0^\infty \frac{dx}{x^2+\alpha} = \frac{\pi}{2\sqrt{\alpha}}$. Differentiating both sides with respect to α, we have

$$\int_0^\infty \frac{\partial}{\partial\alpha}\left(\frac{1}{x^2+\alpha}\right) dx = -\int_0^\infty \frac{dx}{(x^2+\alpha)^2} = \frac{\pi}{4}\alpha^{-3/2}$$

the result being justified by Theorem 8, Page 328, since $\int_0^\infty \frac{dx}{(x^2+\alpha)^2}$ is uniformly convergent for $\alpha \geqq 1$

$$\left(\text{because } \frac{1}{(x^2+\alpha)^2} \leqq \frac{1}{(x^2+1)^2} \text{ and } \int_0^\infty \frac{dx}{(x^2+1)^2} \text{ converges}\right).$$

Taking the limit as $\alpha \to 1+$, using Theorem 6, Page 328, we find $\int_0^\infty \frac{dx}{(x^2+1)^2} = \frac{\pi}{4}$.

(d) Differentiating both sides of $\int_0^\infty \frac{dx}{x^2+\alpha} = \frac{\pi}{2}\alpha^{-1/2}$ n times, we find

$$(-1)(-2)\cdots(-n)\int_0^\infty \frac{dx}{(x^2+\alpha)^{n+1}} = \left(-\frac{1}{2}\right)\left(-\frac{3}{2}\right)\left(-\frac{5}{2}\right)\cdots\left(-\frac{2n-1}{2}\right)\frac{\pi}{2}\alpha^{-(2n-1,2)}$$

where justification proceeds as in (c). Letting $\alpha \to 1+$, we find

$$\int_0^\infty \frac{dx}{(x^2+1)^{n+1}} = \frac{1\cdot3\cdot5\cdots(2n-1)}{2^n n!}\frac{\pi}{2} = \frac{1\cdot3\cdot5\cdots(2n-1)}{2\cdot4\cdot6\cdots(2n)}\frac{\pi}{2}$$

Substituting $x = \tan\theta$, the integral becomes $\int_0^{\pi/2} \cos^{2n}\theta\, d\theta$ and the required result is obtained.

12.26. Prove that $\int_0^\infty \frac{e^{-ax} - e^{bx}}{x \sec rx}\, dx = \frac{1}{2}\ln\frac{b^2+r^2}{a^2+e^2}$ where $a, b > 0$.

From Problem 12.22 and Theorem 7, Page 328, we have

$$\int_{x=0}^\infty \left\{\int_{\alpha=a}^b e^{-\alpha x}\cos rx\, d\alpha\right\} dx = \int_{\alpha=a}^b \left\{\int_{x=0}^\infty e^{-\alpha x}\cos rx\, dx\right\} d\alpha$$

or

$$\int_{x=0}^\infty \left.\frac{e^{\alpha x}\cos rx}{-x}\right|_{\alpha=a}^b dx = \int_{\alpha=a}^b \frac{\alpha}{\alpha^2+r^2}\, d\alpha$$

i.e.,

$$\int_0^\infty \frac{e^{-ax} - e^{-bx}}{x\sec rx}\, dx = \frac{1}{2}\ln\frac{b^2+r^2}{a^2+r^2}$$

12.27. Prove that $\int_0^\infty e^{\alpha x}\frac{1-\cos x}{x^2}\, dx = \tan^{-1}\frac{1}{\alpha} - \frac{\alpha}{2}\ln(\alpha^2+1), \alpha > 0$.

By Problem 12.22 and Theorem 7, Page 328, we have

$$\int_0^r \left\{\int_0^\infty e^{-\alpha x}\cos rx\, dx\right\} dr = \int_0^\infty \left\{\int_0^r e^{-\alpha x}\cos rx\, dr\right\} dx$$

or

$$\int_0^\infty e^{-\alpha x}\frac{\sin rx}{x}\, dx = \int_0^r \frac{\alpha}{\alpha^2+r^2} = \tan^{-1}\frac{r}{\alpha}$$

Integrating again with respect to r from 0 to r yields

$$\int_0^\infty e^{-\alpha x}\frac{1-\cos rx}{x^2}\, dx = \int_0^r \tan^{-1}\frac{r}{\alpha}\, dr = r\tan^{-1}\frac{r}{\alpha} - \frac{\alpha}{2}\ln(\alpha^2+r^2)$$

using integration by parts. The required result follows on letting $r = 1$.

12.28. Prove that $\int_0^\infty \frac{1-\cos x}{x^2}\, dx = \frac{\pi}{2}$.

Since $e^{-\alpha x}\dfrac{1-\cos x}{x^2} \leqq \dfrac{1-\cos x}{x^2}$ for $\alpha \geqq 0, x \geqq 0$ and $\int_0^\infty \dfrac{1-\cos x}{x^2}\,dx$ converges [see Problem 12.7(b)], it follows by the Weierstrass test that $\int_0^\infty e^{\alpha x}\dfrac{1-\cos x}{x^2}\,dx$ is uniformly convergent and represents a continuous function of α for $\alpha \geqq 0$ (Theorem 6, Page 328). Then, letting $\alpha \to 0+$, using Problem 12.27, we have

$$\lim_{\alpha\to 0}\int_0^\infty e^{-\alpha x}\frac{1-\cos x}{x^2}\,dx = \int_0^\infty \frac{1-\cos x}{x^2}\,dx = \lim_{\alpha\to 0}\left\{\tan^{-1}\frac{1}{\alpha}-\frac{\alpha}{2}\ln(\alpha^2+1)\right\} = \frac{\pi}{2}$$

12.29. Prove that $\int_0^\infty \dfrac{\sin x}{x} = \int_0^\infty \dfrac{\sin^2 x}{x^2}\,dx = \dfrac{\pi}{2}$.

Integrating by parts, we have

$$\int_\epsilon^M \frac{1-\cos x}{x^2}\,dx = \left(-\frac{1}{x}\right)(1-\cos x)\Bigg|_\epsilon^M + \int_\epsilon^M \frac{\sin x}{x}\,dx = \frac{1-\cos\epsilon}{\epsilon} - \frac{1-\cos M}{M} + \int_\epsilon^M \frac{\sin x}{x}\,dx$$

Taking the limit as $\epsilon \to 0+$ and $M \to \infty$ shows that

$$\int_0^\infty \frac{\sin x}{x}\,dx = \int_0^\infty \frac{1-\cos x}{x}\,dx = \frac{\pi}{2}$$

Since $\int_0^\infty \dfrac{1-\cos x}{x^2}\,dx = 2\int_0^\infty \dfrac{\sin^2(x/2)}{x^2}\,dx = \int_0^\infty \dfrac{\sin^2 u}{u^2}\,du$ on letting $u = x/2$, we also have $\int_0^\infty \dfrac{\sin^2 x}{x^2}\,dx = \dfrac{\pi}{2}$.

12.30. Prove that $\int_0^\infty \dfrac{\sin^3 x}{x}\,dx = \dfrac{\pi}{4}$.

$$\sin^3 x = \left(\frac{e^{ix}-e^{-ix}}{2i}\right)^2 = \frac{(e^{ix})^3 - 3(e^{ix})^2(e^{-ix}) + 3(e^{ix})(e^{ix})^2 - (e^{-ix})^3}{(2i)^3}$$

$$= -\frac{1}{4}\left(\frac{e^{-3ix}-e^{-3ix}}{2i}\right) + \frac{3}{4}\left(\frac{e^{ix}-e^{-ix}}{2i}\right) = -\frac{1}{4}\sin 3x + \frac{3}{4}\sin x$$

Then

$$\int_0^\infty \frac{\sin^3 x}{x}\,dx = \frac{3}{4}\int_0^\infty \frac{\sin x}{x}\,dx - \frac{1}{4}\int_0^\infty \frac{\sin 3x}{x}\,dx = \frac{3}{4}\int_0^\infty \frac{\sin x}{x}\,dx - \frac{1}{4}\int_0^\infty \frac{\sin u}{u}\,du$$

$$= \frac{3}{4}\left(\frac{\pi}{2}\right) - \frac{1}{4}\left(\frac{\pi}{2}\right) = \frac{\pi}{4}$$

Miscellaneous problems

12.31. Prove that $\int_0^\infty e^{-x^2}\,dx = \sqrt{\pi}/2$.

By Problem 12.6, the integral converges. Let $I_M = \int_0^M e^{-x^2}\,dx = \int_0^M e^{-y^2}\,dy$ and let $\lim_{M\to\infty} I_M = I$, the required value of the integral. Then

$$I_M^2 = \left(\int_0^M e^{-x^2}\,dx\right)\left(\int_0^M e^{-x^2}\,dy\right)$$

$$= \int_0^M\int_0^M e^{-(x^2+y^2)}\,dx\,dy$$

$$= \iint_M e^{-(x^2+y^2)}\,dx\,dy$$

where $\Re_M$ is the square $OACE$ of side M (see Figure 12.3). Since the integrand is positive, we have

$$\iint_M e^{-(x^2+y^2)}\,dx\,dy \leqq I_M^2 \leqq \iint_M e^{-(x^2+y^2)}\,dx\,dy \tag{1}$$

where $\Re_1$ and $\Re_2$ are the regions in the first quadrant bounded by the circles having radii M and $M\sqrt{2}$, respectively.

Using polar coordinates, we have, from Equation (1),

$$\int_{\phi=0}^{\pi/2}\int_{\rho=0}^{M} e^{-\rho^2}\rho\,d\rho\,d\phi \leqq I_M^2 \leqq \int_{\phi=0}^{\pi/2}\int_{\pi=0}^{M\sqrt{2}} e^{-\rho^2}\rho\,d\rho\,d\phi \tag{2}$$

or

$$\frac{\pi}{4}(1-e^{M^2}) \leqq I_M^2 \leqq \frac{\pi}{4}(1-e^{-2M^2}) \tag{3}$$

Then, taking the limit as $M \to \infty$ in Equation (3), we find $\lim_{M\to\infty} I^2_M = I^2 = \pi/4$ and $I = \sqrt{\pi}/2$

12.32. Evaluate $\int_0^\infty e^{-x^2} \cos\alpha x\, dx$.

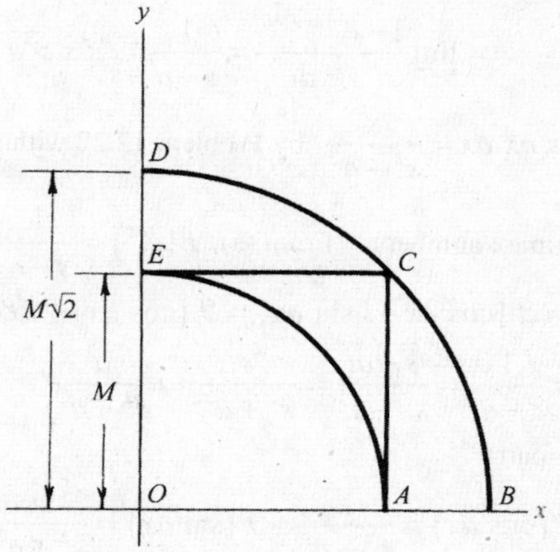

Figure 12.3

Let $I(\alpha) = \int_0^\infty e^{-x^2}\cos\alpha x\, dx$. Then, using integration by parts and appropriate limiting procedures.

$$\frac{dt}{d\alpha} = \int_0^\infty -xe^{-x^2}\sin\alpha x\, dx = \frac{1}{2}e^{-x^2}\sin\alpha x\,|_0^\infty - \frac{1}{2}\alpha\int_0^\infty e^{-x^2}\cos\alpha x\, dx = -\frac{\alpha}{2}I$$

The differentiation under the integral sign is justified by Theorem 8, Page 328, and the fact that $\int_0^\infty xe^{-x^2}$ sin $\alpha x\, dx$ is uniformly convergent for all α (since by the Weierstrass test, $|xe^{-x2}\sin\alpha x| \leqq xe^{-x2}$ and $\int_0^\infty xe^{-x^2}dx$ converges).

From Problem 12.31 and the uniform convergence, and thus continuity, of the given integral (since $|e^{-x2}\cos\alpha x| \leqq e^{-x2}$ and $\int_0^\infty e^{-x2}\,dx$ converges, so that that Weierstrass test applies), we have $I(0) = \lim_{\alpha\to 0} I(\alpha) = \frac{1}{2}\sqrt{\pi}$.

Solving $\frac{dI}{d\alpha} = -\frac{\alpha}{2}I$ subject to $I(0) = \frac{\sqrt{\pi}}{2}$, we find $I(\alpha) = \frac{\sqrt{\pi}}{2}e^{-\alpha^2/4}$

12.33. (*a*) Prove that $I(\alpha) = \int_0^\infty e^{-(x-\alpha/x)^2}dx = \frac{\sqrt{\pi}}{2}$. (*b*) Evaluate $\int_0^\infty e^{-(x^2+x^{-2})}dx$.

(a) We have $I'(\alpha) = 2\int_0^\infty e^{-(x-\alpha/x)^2}(1-\alpha/x^2)\,dx$

The differentiation is proved valid by observing that the integrand remains bounded as $x \to 0+$ and that for sufficiently large x,

$$e^{-(x-\alpha/x)2}(1-\alpha/x^2) = e^{-x2} + 2\alpha - \alpha^{2/x2}(1-\alpha/x^2) \leqq e^{2\alpha}e^{-x^2}$$

so that $I'(\alpha)$ converges uniformly for $\alpha \,\varepsilon\, 0$ by the Weierstrass test, since $\int_0^\infty e^{-x^2}dx$ converges. Now

$$I'(\alpha) = 2\int_0^\infty e^{-(x-\alpha/x)^2}dx - 2\alpha\int_0^\infty \frac{e^{-(x-\alpha/x)^2}}{x^2}dx = 0$$

as seen by letting $\alpha/x = y$ in the second integral. Thus, $I(\alpha) = c$, a constant. To determine c, let $\alpha \to 0+$ in the required integral and use Problem 12.31 to obtain $c = \sqrt{\pi}/2$.

(b) From (a), $\int_0^\infty e^{(x-\alpha/x)^2}dx = \int_0^\infty e^{-(x^2+\alpha^2x^{-2})}dx = e^{2\alpha}\int_0^\infty e^{-(x^2+\alpha^2x^{-2})}dx = \dfrac{\sqrt{\pi}}{2}.$

Then $\int_0^\infty e^{-(x^2+\alpha^2x^{-2})}dx = \dfrac{\sqrt{\pi}}{2}e^{-2\alpha}$. Putting $\alpha = 1$, $\int_0^\infty e^{-(x^2+x^{-2})}dx = \dfrac{\sqrt{\pi}}{2}e^{-2}$.

12.34. Verify the results: (*a*) $\mathcal{L}\{e^{ax}\} = \dfrac{1}{s-\alpha}$, $s > a$ and (*b*) $\mathcal{L}\{\cos az\} = \dfrac{s}{s^2+a^2}$, $s > 0$.

(a)
$$\mathcal{L}\{e^{ax}\} = \int_0^\infty e^{-sx}e^{ax}dx = \lim_{M\to\infty}\int_0^M e^{-(s-a)x}dx$$
$$= \lim_{M\to\infty}\frac{1-e^{-(s-a)M}}{s-a} = \frac{1}{s-a} \quad \text{if } s > a$$

(b) $\mathcal{L}\{\cos ax) = \int_0^\infty e^{-sx}\cos ax\,dx = \dfrac{s}{s^2+a^2}$ by Problem 12.22 with $\alpha = s$, $r = a$.

Another method, using complex numbers. From (a), $\mathcal{L}\{e^{ax}\} = \dfrac{1}{s-a}$. Replace a by ai. Then

$$\mathcal{L}\{e^{aix}\} = \mathcal{L}\{\cos ax + i\sin ax\} = \mathcal{L}\{\cos ax\} + i\mathcal{L}\{\sin ax\}$$
$$= \frac{1}{s-ai} = \frac{s+ai}{s^2+a^2} = \frac{s}{s^2+a^2} + i\frac{a}{s^2+a^2}$$

Equating real and imaginary parts:

$$\mathcal{L}\{\cos ax\} = \frac{s}{s^2+a^2}, \quad \mathcal{L}\{\sin ax\} = \frac{a}{s^2+a^2}$$

This *formal* method can be justified using the methods in Chapter 16.

12.35. Prove that (a) $\mathcal{L}\{Y'(x)\} = (s)\,\mathcal{L}\{Y(x)\} - Y(0)$ and (b) $\mathcal{L}\{Y''(x)\} = s^2\mathcal{L}\{Y(x)\} - sY(0) - Y'(0)$ under suitable conditions on $Y(x)$.

(a) By definition (and with the aid of integration by parts),

$$\mathcal{L}\{Y'(x)\} = \int_0^\infty e^{-sx}Y'(x)dx = \lim_{M\to 0}\int_0^M e^{-sx}Y'(x)dx$$
$$= \lim_{M\to\infty}\left\{e^{-sx}Y(x)\Big|_0^M + s\int_0^M e^{-sx}Y(x)\,dx\right\}$$
$$= s\int_0^\infty e^{-sx}Y(x)dx - Y(0) = s\mathcal{L}\{(x)\} - Y(0)$$

assuming that s is such that $\lim_{M\to\infty} e^{-sM}\,Y(M) = 0$.

(b) Let $U(x) = Y'(x)$. Then by (a), $\mathcal{L}\{U'(x)\} - U(0)$. Thus,

$$\mathcal{L}\{Y'(x)\} = s\,\mathcal{L}\{Y'(x)\} - Y'(0) = s[s\mathcal{L}\{Y(x)\} - Y(0)] - Y'(0)$$
$$= s^2\,\mathcal{L}\{Y(x)\} - sY(0) - Y'(0)$$

12.36. Solve the differential equation $Y''(x) + Y(x) = x$, $Y(0) = 0$, $Y'(0) = 2$.

Take the Laplace transform of both sides of the given differential equation. Then by Problem 12.35.

$$\mathcal{L}\{Y''(x) + Y(x)\} = \mathcal{L}\{x\},\ \mathcal{L}\{Y''(x)\} + \mathcal{L}\{Y(x)\} = 1/s^2$$

and so

$$s^2\,\mathcal{L}\{Y(x)\} - s\mathcal{L}(0) - Y'(0) + \mathcal{L}\{Y(x)\} = 1/s^2$$

Solving for $\mathcal{L}\{Y(x)\}$ using the given conditions, we find

$$\mathcal{L}\{Y(x)\} = \frac{2s^2}{s^2(s^2+1)} = \frac{1}{s^2} + \frac{1}{s^2+1} \tag{1}$$

by methods of partial fractions.

Since $\frac{1}{s^2} = \mathcal{L}\{x\}$ and $\frac{1}{s^2+1} = \mathcal{L}\{\sin x\}$, it follows that $\frac{1}{s^2} + \frac{1}{s^2+1} = \mathcal{L}\{x + \sin x\}$

Hence, from (1), $\mathcal{L}\{Y(x)\} = \mathcal{L}\{x + \sin x\}$, from which we can conclude that $Y(x) = x + \sin x$, which is, in fact, found to be a solution.

Another method: If $\mathcal{L}\{F(x)\} = f(s)$, we call $f(s)$ the *inverse* Laplace transform of $F(x)$ and write $f(s) = \mathcal{L}^{-1}\{F(x)\}$.

By Problem 12.78. $\mathcal{L}^{-1}\{f(s) + g(s)\} = \mathcal{L}^{-1} = \mathcal{L}^{-1}\{f(s) + \mathcal{L}^{-1}\{g(s)\}$. Then, from Equation (1),

$$Y(x) = \mathcal{L}^{-1}\left\{\frac{1}{s^2} + \frac{1}{s^2+1}\right\} = \mathcal{L}^{-1}\left\{\frac{1}{s^2}\right\} + \mathcal{L}^{-1}\left\{\frac{1}{s^2+1}\right\} = x + \sin$$

Inverse Laplace transforms can be read from Table 12.1.

SUPPLEMENTARY PROBLEMS

Improper integrals of the first kind

12.37. Test for convergence:

(a) $\int_0^\infty \frac{x^2+1}{x^4+1}\,dx$ (d) $\int_0^\infty \frac{dx}{x^4+4}$ (g) $\int_0^\infty \frac{x^2\,dx}{(x^2+x+1)^{5/2}}$

(b) $\int_2^\infty \frac{x\,dx}{\sqrt{x^3-1}}$ (e) $\int_{-\infty}^\infty \frac{2+\sin x}{x^2+1}\,dx$ (h) $\int_t^\infty \frac{\ln x\,dx}{x+e^{-x}}$

(c) $\int_t^\infty \frac{dx}{x\sqrt{3x+2}}$ (f) $\int_2^\infty \frac{x\,dx}{(\ln x)^3}$ (i) $\int_0^\infty \frac{\sin^2 x}{x^2}\,dx$

Ans. (a) convergence (b) divergence (c) convergence (d) convergence (e) convergence (*f*) divergence (*g*) convergence (*h*) divergence (*i*) convergence

12.38. Prove that $\int_{-\infty}^\infty \frac{dx}{x^2+2ax+b^2} = \frac{\pi}{\sqrt{b^2-a^2}}$ if $b > |a|$.

12.39. Test for convergence: (*a*) $\int_t^\infty e^{-x}\ln x\,dx$, (*b*) $\int_0^\infty e^{-x}\ln(1+e^x)\,dx$, and (c) $\int_0^\infty e^{-x}\cosh x^2 dx$.

Ans. (a) convergence (b) convergence (c) divergence

12.40. Test for convergence, indicating absolute or conditional convergence where possible: (*a*) $\int_0^\infty \frac{\sin 2x}{x^3+1}\,dx$;

(*b*) $\int_{-\infty}^\infty e^{-ax^2}\cos bx\,dx$, where *a*, *b* are positive constants; (*c*) $\int_0^\infty \frac{\cos x}{\sqrt{x^2+1}}\,dx$; (*d*) $\int_0^\infty \frac{x\sin x}{\sqrt{x^2+a^2}}\,dx$; and

(*e*) $\int_0^\infty \frac{\cos x}{\cosh x}\,dx$.

Ans. (a) absolute convergence (b) absolute convergence (c) conditional convergence (d) divergence (e) absolute convergence

12.41. Prove the quotient tests (b) and (c) on Page 323.

Improper integrals of the second kind

12.42. Test for convergence:

(a) $\int_0^1 \frac{dx}{(x+1)\sqrt{1-x^2}}$ (d) $\int_1^2 \frac{\ln x}{\sqrt[3]{8-x^3}}\,dx$ (g) $\int_0^3 \frac{x^2}{(3-x)^2}\,dx$ (j) $\int_0^3 \frac{dx}{x^x}$

(b) $\int_0^1 \frac{\cos x}{x^2}\,dx$ (e) $\int_0^1 \frac{dx}{\sqrt{\ln(1/x)}}$ (h) $\int_0^{\pi/2} \frac{e^{-x}\cos x}{x}\,dx$

(c) $\int_{-1}^1 \frac{e^{\tan^{-1}x}}{x}\,dx$ (f) $\int_0^{\pi/2} \ln \sin x\,dx$ (i) $\int_0^1 \sqrt{\frac{1-k^2x^2}{1-x^2}}\,dx,\ |k|<1$

Ans. (a) convergence (b) divergence (c) divergence (d) convergence (e) convergence (f) convergence (g) divergence (h) divergence (i) convergence (j) convergence

12.43. (a) Prove that $\int_0^5 \frac{dx}{4-x}$ diverges in the usual sense but converges in the Cauchy principal value senses. (b) Find the Cauchy principal value of the integral in (a) and give a geometric interpretation.

Ans. (b) In 4

12.44. Test for convergence, indicating absolute or conditional convergence where possible: *(a)* $\int_0^1 \cos\left(\frac{1}{x}\right) dx$, *(b)* $\int_0^1 \frac{1}{x}\cos\left(\frac{1}{x}\right) dx$, and *(c)* $\int_0^1 \frac{1}{x^2}\cos\left(\frac{1}{x}\right) dx$.

Ans. (a) absolute convergence (b) conditional convergence (c) divergence

12.45. Prove that $\int_0^{4\pi}\left(3x^2 \sin\frac{1}{x} - x\cos\frac{1}{x}\right) dx = \frac{32\sqrt{2}}{\pi^3}$.

Improper integrals of the third kind

12.46. Test for convergence: *(a)* $\int_0^\infty e^{-x}\ln x\,dx$, *(b)* $\int_0^\infty \frac{e^{-x}dx}{\sqrt{x\ln(x+1)}}$, and *(c)* $\int_0^\infty \frac{e^{-x}dx}{\sqrt[3]{x(3+2\sin x)}}$.

Ans. (a) convergence (b) divergence (c) convergence

12.47. Test for convergence: *(a)* $\int_0^\infty \frac{dx}{\sqrt[3]{x^4+x^2}}$ and *(b)* $\int_0^\infty \frac{e^x\,dx}{\sqrt{\sinh(ax)}}$, $a>0$.

Ans. (a) convergence (b) convergence if a > 2, divergence if $0 < a \leqq 2$.

12.48. Prove that $\int_0^\infty \frac{\sinh(ax)}{\sinh(\pi x)}\,dx$ converges if $0 \leqq |a| < \pi$ and diverges if $|a| \leqq \pi$.

12.49. Test for convergence, indicating absolute or conditional convergence where possible: *(a)* $\int_0^\infty \frac{\sin x}{\sqrt{2}}\,dx$ and *(b)* $\int_0^\infty \frac{\sin\sqrt{x}}{\sinh\sqrt{x}}\,dx$.

Ans. (a) conditional convergence (b) absolute convergence

Uniform convergence of improper integrals

12.50. (a) Prove that $\phi(\alpha) = \int_0^\infty \frac{\cos \alpha x}{1+x^2}\,dx$ is uniformly convergent for all α. (b) Prove that φ(α) is continuous for all α. (c) Find $\lim_{\alpha\to 0}\phi(\alpha)$.

Ans. (c) π/2

12.51. Let $\phi(\alpha) = \int_0^\infty F(x),\alpha)\,dx$, where $F(x,\alpha) = \alpha^2 x e^{-\alpha x^2}$. (a) Show that φ(α) is not continuous at α = 0; i.e., $\lim_{\alpha\to 0}\int_0^\infty F(x,\alpha)\,dx \neq \int_0^\infty \lim_{\alpha\to 0} F(x,\alpha)\,dx$. (b) Explain the result in (a).

12.52. Work Problem 12.51 if $F(x, \alpha) = \alpha^2 xe^{-\alpha x}$.

12.53. If $F(x)$ is bounded and continuous for $-\infty < x < \infty$ and

$$V(x, y) = \frac{1}{\pi}\int_{-\infty}^{\infty} \frac{yF(\lambda)d\lambda}{y^2 + (\lambda - x)^2}$$

Prove that $\lim_{y \to 0} V(x, y) = F(x)$.

12.54. Prove (a) Theorem 7 and (b) Theorem 8 on Page 328.

12.55. Prove the Weierstrass *M* test for uniform convergence of integrals.

12.56. Prove that if $\int_0^{\infty} F(x)\, dx$ converges, then $\int_0^{\infty} e^{-\alpha x} F(x)\, dx$ converges uniformly for $\alpha \geqq 0$.

12.57. Prove that (a) (*a*) $\phi(a) = \int_0^{\infty} e^{-ax}\frac{\sin x}{x}\, dx$ converges uniformly for $a \geqq 0$, (*b*) $\phi(a) = \frac{\pi}{2} - \tan^{-1} a$, and (*c*) $\int_0^{\infty} \frac{\sin x}{x}\, dx = \frac{\pi}{2}$ (compare Problems 12.27 through 12.29).

12.58. State the definition of uniform convergence for improper integrals of the second kind.

12.59. State and prove a theorem corresponding to Theorem 8, Page 328, if α is a differentiable function of α.

Evaluation of definite integrals

Establish each of the following results. Justify all steps in each case.

12.60. $\int_0^{\infty} \frac{e^{-ax} - e^{-bx}}{x}\, dx = \ln(b/a), \qquad a, b > 0$

12.61. $\int_0^{\infty} \frac{e^{-ax} - e^{-bx}}{x \csc rx}\, dx = \tan^{-1}(b/r) - \tan^{-1}(a/r), \qquad a, b, r > 0$

12.62. $\int_0^{\infty} \frac{\sin rx}{x(1+x^2)}\, dx = \frac{\pi}{2}(1 - e^{-r}), \qquad r \geqq 0$

12.63. $\int_0^{\infty} \frac{1 - \cos rx}{x^2}\, dx = \frac{\pi}{2} |r|$

12.64. $\int_0^{\infty} \frac{x \sin rx}{a^2 + x^2}\, dx = \frac{\pi}{2} e^{-ar} \qquad a, r \geqq 0$

12.65. (a) Prove that $\int_0^{\infty} e^{-\alpha x}\left(\frac{\cos ax - \cos bx}{x}\right) dx = \frac{1}{2}\ln\left(\frac{\alpha^2 + b^2}{\alpha^2 + a^2}\right)$, $\alpha \geqq 0$. (b) Use (a) to prove that $\int_0^{\infty} \frac{\cos ax - \cos bx}{x}\, dx = \ln\left(\frac{b}{a}\right)$. [The results of (b) and Problem 12.60 are special cases of *Frullani's integral*, $\int_0^{\infty} \frac{F(ax) - F(bx)}{x}\, dx = F(0)\ln\left(\frac{b}{a}\right)$, where $F(t)$ is continuous for $t < 0$, $t > 0$, $F'(0)$ exists and $\int_1^{\infty} \frac{F(t)}{t}\, dt$ converges.]

12.66. Given $\int_0^{\infty} e^{-\alpha x^2}\, dx = \frac{1}{2}\sqrt{\pi / \alpha}$, $\alpha > 0$, prove that for $p = 1, 2, 3, \ldots$,

$$\int_0^{\infty} x^{2p} e^{-\alpha x^2}\, dx = \frac{1}{2}\cdot\frac{3}{2}\cdot\frac{5}{2}\cdots\frac{(2p-1)}{2}\,\frac{\sqrt{\pi}}{2\alpha^{(2p+1)/2}}.$$

12.67. If $a > 0, b > 0$, prove that $\int_0^\infty (e^{-a/x^2} - e^{-b/x^2})\, dx = \sqrt{\pi b} - \sqrt{\pi a}$.

12.68. Prove that $\int_0^\infty \frac{\tan^{-1}(x/a) - \tan^{-1}(x/b)}{x}\, dx = \frac{\pi}{2}\ln\left(\frac{b}{a}\right)$, where $a > 0, b > 0$.

12.69. Prove that $\int_{-\infty}^\infty \frac{dx}{(x^2+x+1)^3} = \frac{4\pi}{3\sqrt{3}}$. (Hint: Use Problem 12.38.)

Miscellaneous problems

12.70. Prove that $\int_0^\infty \left\{\frac{\ln(1+x)}{x}\right\}^2 dx$ converges.

12.71. Prove that $\int_0^\infty \frac{dx}{1+x^3\sin^2 x}$ converges. $\left[\text{Hint: Consider } \sum_{n=0}^\infty \int_{n\pi}^{(n+1)\pi} \frac{dx}{1+x^3\sin^2 x} \text{ and use the fact that } \int_{n\pi}^{(n+1)\pi} \frac{dx}{1+x^3\sin^2 x} < \int_{n\pi}^{(n+1)\pi} \frac{dx}{1+(n\pi)^3\sin^2 x}\right]$.

12.72. Prove that $\int_0^\infty \frac{x\, dx}{1+x^3\sin^2 x}$ diverges.

12.73. (a) Prove that $\int_0^\infty \frac{\ln(1+\alpha^2 x^2)}{1+x^2}\, dx = \pi\ln(1+\alpha)$, $\alpha \geq 0$. (b) Use (a) to show that $\int_0^{\pi/2} \ln\sin\theta\, d\theta = -\frac{\pi}{2}\ln 2$.

12.74. Prove that $\int_0^\infty \frac{\sin^4 x}{x^4}\, dx = \frac{\pi}{3}$.

12.75. Evaluate (a) $\mathcal{L}(1/\sqrt{x})$, (b) $\mathcal{L}\{\cosh ax\}$, and (c) $\mathcal{L}\{(\sin x)/x\}$.

Ans. (a) $\sqrt{\pi/s}, s > 0$ (b) $\frac{s}{s^2 - a^2}, s > |a|$ (c) $\tan^{-1}\left(\frac{1}{s}\right), s > 0$

12.76. (a) If $\mathcal{L}\{e^{ax}F(x)\}$ prove that $\mathcal{L}\{e^{ax}F(x)\} = f(s-a)$ and (b) evaluate $\mathcal{L}\{e^{ax}\sin bx\}$.

Ans. (b) $\frac{b}{(s-a)^2 + b^2}, s > a$

12.77. (a) If $\mathcal{L}\{F(x)\} = f(s)$, prove that $\mathcal{L}\{x^n F(x)\} = (-1)^n f^{(n)}(s)$, giving suitable restrictions on $F(x)$. (b) Evaluate $\mathcal{L}\{x\cos x]$.

Ans. (b) $\frac{s^2-1}{(s^2+1)^2}, s > 0$

12.78. Prove that $\mathcal{L}^{-1}\{f(s) + g(s)\} = \mathcal{L}^{-1}\{f(s)\} + \mathcal{L}^{-1}[g(s)]$, stating any restrictions.

12.79. Solve using Laplace transforms, the following differential equations subject to the given conditions:

(a) $Y''(x) + 3Y'(x) + 2Y(x) = 0$; $Y(0) = 3, Y'(0) = 0$

(b) $Y''(x) - Y'(x) = x$; $Y(0) = 2, Y'(0) = -3$

(c) $Y''(x) + 2Y'(x) + 2Y(x) = 4$; $Y(0) = 0\ Y'(0) = 0$

Ans. (a) $Y(x) = 6e^{-x} - 3e^{-2x}$ (b) $Y(x) = 4 - 2e^x - \frac{1}{2}x^2 - x$ (c) $Y(x) = 1 - e^{-x}(\sin x + \cos x)$

12.80. Prove that $\mathcal{L}\{F(x)\}$ exists if $F(x)$ is piecewise continuous in every finite interval $[0, b]$ where $b > 0$ and if $F(x)$ is of *exponential order* as $x \to \infty$; i.e., there exists a constant α such that $|e^{-\alpha x}F(x)| < P$ (a constant) for all $x > b$.

12.81. If $f(s) = \mathcal{L}\{F(x)\}$ and $g(s) = \mathcal{L}\{G(x)\}$, prove that $f(s)g(s) = \mathcal{L}\{H(x)\}$ where

$$H(x) = \int_0^x F(u)G(x-u)\, du$$

is called the *convolution* of F and G, written $F*G$. Hint: Write

$$f(s)g(s) = \lim_{M\to\infty}\left\{\int_0^M e^{-su}F(u)\,du\right\}\left\{\int_0^M e^{-se}G(v)\,dv\right\}$$

$$= \lim_{M\to\infty}\int_0^M \int_0^M e^{s(u+v)}\,F(u)\,G(v)\,du\,dv \text{ and then let } u + v = t$$

12.82. (a) Find $\mathcal{L}^{-1}\left\{\dfrac{1}{(s^2+1)^2}\right\}$ (b) Solve $Y''(x) + Y(x) = R(x)$, $Y(0) = Y'(0) = 0$. (c) Solve the integral equation $Y(x) = x + \int_0^x Y(u)\sin(x-u)du$. (Hint: Use Problem 12.81.)

Ans. (a) $\frac{1}{2}(\sin x - x\cos x)$ (b) $Y(x) = \int_0^x R(u)\sin(x-u)du$ (c) $Y(x) = x + x^3/6$

12.83. Let $f(x)$, $g(x)$, and $g'(x)$ be continuous in every finite interval $a \leqq x \leqq b$ and suppose that $g'(x) \leqq 0$. Suppose also that $h(x) = \int_0^x f(x)\,dx$ is bounded for all $x \in a$ and $\lim_{x\to 0} g(x) = 0$.

(a) Prove that $\int_0^\infty f(x)g(x)dx = -\int_a^\infty g'(x)h(x)dx$.

(b) Prove that the integral on the right, and, hence, the integral on the left, are convergent. The result is that under the given conditions on $f(x)$ and $g(x)$, $\int_a^\infty f(x)g(x)dx$ converges and is sometimes called Abel's integral test. [Hint: For (a), consider $\lim_{b\to\infty}\int_a^b f(x)g(x)dx$ after replacing $f(x)$ by $h'(x)$ and integrating by parts. For (b), first prove that if $|h(x)| < H$ (a constant), then $\left|\int_a^b g'(x)h(x)dx\right|$ and then let $b \to \infty$.]

12.84. Use Problem 12.83 to prove that (*a*) $\int_0^\infty \frac{\sin x}{x}\,dx$ and (*b*) $\int_0^\infty \sin x^p\,dx$, $p > 1$ converge.

12.85. (a) Given that $\int_0^\infty \sin x^2 dx = \int_0^\infty \cos x^2 dx = \frac{1}{2}\sqrt{\frac{\pi}{2}}$ [see Problems 15.27 and 15.68(a)], evaluate $\int_0^\infty\int_0^\infty \sin(x^2+y^2)\,dx\,dy$ and (b) Explain why the method of Problem 12.31 cannot be used to evaluate the multiple integral in (a).

Ans. $\pi/4$

CHAPTER 13

Fourier Series

Mathematicians of the eighteenth century, including Daniel Bernoulli and Leonhard Euler, expressed the problem of the vibratory motion of a stretched string through partial differential equations that had no solutions in terms of "elementary functions." Their resolution of this difficulty was to introduce infinite series of sine and cosine functions that satisfied the equations. In the early nineteenth century, Joseph Fourier, while studying the problem of heat flow, developed a cohesive theory of such series. Consequently, they were named after him. Fourier series and Fourier integrals are investigated in this chapter and Chapter 14. As you explore the ideas, notice the similarities and differences with the chapters on infinite series and improper integrals.

Periodic Functions

A function $f(x)$ is said to have a *period* T or to be *periodic* with period T if for all x, $f(x + T) = f(x)$, where T is a positive constant. The least value of $T > 0$ is called the *least period* or simply *the period* of $f(x)$.

EXAMPLE 1. The function $\sin x$ has periods $2\pi, 4\pi, 6\pi, \ldots$, since $\sin(x + 2\pi)$, $\sin(+ 4\pi)$, $\sin(x + 6\pi), \ldots$ all equal $\sin x$. However, 2π is the *least period* or *the period* of $\sin x$.

EXAMPLE 2. The period of $\sin nx$ or $\cos nx$, where n is a positive integer, is $2\pi/n$.

EXAMPLE 3. The period of $\tan x$ is π.

EXAMPLE 4. A constant has any positive number as period.

Other examples of periodic functions are shown in the graphs of Figure 13.1(a), (b), and (c).

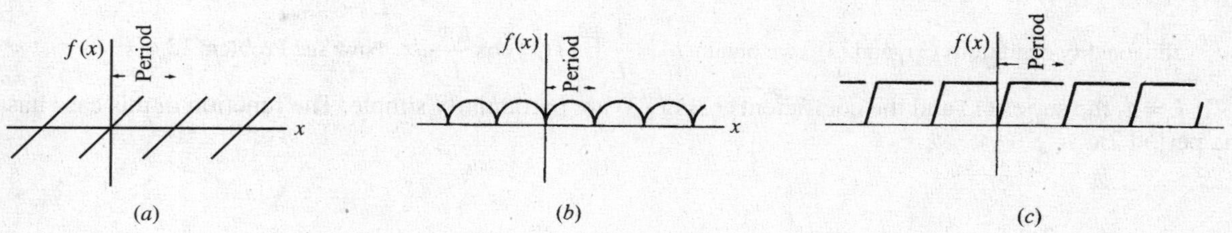

Figure 13.1

Fourier Series

Let $f(x)$ be defined in the interval $(-L, L)$ and outside of this interval by $f(x + 2L) = f(x)$; i.e., $f(x)$ is $2L$-periodic. It is through this avenue that a new function on an infinite set of real numbers is created from the image on $(-L, L)$. The *Fourier series* or *Fourier expansion* corresponding to $f(x)$ is given by

$$\frac{a_0}{2} + \sum_{n=1}^{\infty}\left(a_n \cos\frac{n\pi x}{L} + b_n \sin\frac{n\pi x}{L}\right) \tag{1}$$

where the *Fourier coefficients* a_n and b_n are

$$\begin{cases} a_n = \dfrac{1}{L}\displaystyle\int_{-L}^{L} f(x)\cos\frac{n\pi x}{L}\,dx \\ b_n = \dfrac{1}{L}\displaystyle\int_{-L}^{L} f(x)\sin\frac{n\pi x}{L}\,dx \end{cases} \qquad n = 0, 1, 2, \ldots \tag{2}$$

To correlate the coefficients with the expansion, see the following Examples 1 and 2.

Orthogonality Conditions for the Sine and Cosine Functions

Notice that the Fourier coefficients are integrals. These are obtained by starting with the series (1) and employing the following properties called orthogonality conditions:

(a) $\int_{-L}^{L} \cos\frac{m\pi x}{L}\cos\frac{n\pi x}{L}\,dx = 0$ if $m \neq n$ and L if $m = n$

(b) $\int_{-L}^{L} \sin\frac{m\pi x}{L}\sin\frac{n\pi x}{L}\,dx = 0$ if $m \neq n$ and L if $m = n$ (3)

(c) $\int_{-L}^{L} \sin\frac{m\pi x}{L}\cos\frac{n\pi x}{L}\,dx = 0$. Where m and n assume any positive integer values.

An explanation for calling these orthogonality conditions is given on Page 355. Their application in determining the Fourier coefficients is illustrated in the following pair of examples and then demonstrated in detail in Problem 13.4.

EXAMPLE 1. To determine the Fourier coefficient a_0, integrate both sides of the Fourier series (1) and employ the orthogonality conditions (2).

$$\int_{-L}^{L} f(x)dx = \int_{-L}^{L}\frac{a_0}{2}\,dx + \int_{-L}^{L}\sum_{n=1}^{\infty}\left\{a_n \cos\frac{n\pi x}{L} + b_n \sin\frac{n\pi x}{L}\right\}dx$$

Now

$$\int_{-L}^{L}\frac{a_0}{2}\,dx = a_0 L, \int_{-L}^{L}\sin\frac{n\pi x}{L}\,dx = 0, \int_{-L}^{L}\cos\frac{n\pi x}{L}\,dx = 0$$

therefore

$$a_0 = \frac{1}{L}\int_{-L}^{L} f(x)\,dx$$

EXAMPLE 2. To determine a_1, multiply both sides of series (*1*) by $\cos\frac{\pi x}{L}$ and then integrate. Using the orthogonality conditions $(3)_a$ and $(3)_c$, we obtain $a_1 = \frac{1}{L}\int_{-L}^{L} f(x)\cos\frac{\pi x}{L}\,dx$. Now see Problem 13.4.

If $L = \pi$, the series (1) and the coefficients (2) or (3) are particularly simple. The function in this case has the period 2π.

Dirichlet Conditions

Suppose that

1. $f(x)$ is defined except possibly at a finite number of points in $(-L, L)$.
2. $f(x)$ is periodic outside $(-L, L)$ with period $2L$.
3. $f(x)$ and $f'(x)$ are piecewise continuous in $(-L, L)$.

Then the series (1) with Fourier coefficients converges to

(a) $f(x)$ if x is a point of continuity

(b) $\dfrac{f(x+0)+f(x-0)}{2}$ if x is a point of discontinuity

Here $f(x+0)$ and $f(x-0)$ are the right- and left-hand limits of $f(x)$ at x and represent $\lim_{\epsilon \to 0+} f(x+\epsilon)$ and $\lim_{\epsilon \to 0+} f(x-\epsilon)$, respectively. For a proof, see Problems 13.18 through 13.23.

The conditions 1, 2, and 3 imposed on $f(x)$ are *sufficient* but not necessary, and are generally satisfied in practice. There are at present no known necessary and sufficient conditions for convergence of Fourier series. It is of interest that continuity of $f(x)$ does not *alone* ensure convergence of a Fourier series.

Odd and Even Functions

A function $f(x)$ is called *odd* if $f(-x) = -f(x)$. Thus, x^3, $x^5 - 3x^3 + 2x$, $\sin x$, and $\tan 3x$ are odd functions.

A function $f(x)$ is called *even* if $f(-x) = f(x)$. Thus, x^4, $2x^6 - 4x^2 + 5$, $\cos x$, and $e^x + e^{-x}$ are even functions.

The functions portrayed graphically in Figure 13.1(a) and (b) are odd and even, respectively, but that of Figure 13.1(c) is neither odd nor even.

In the Fourier series corresponding to an odd function, only sine terms can be present. In the Fourier series corresponding to an even function, only cosine terms (and possibly a constant, which we shall consider a cosine term) can be present.

Half Range Fourier Sine or Cosine Series

A half range Fourier sine or cosine series is a series in which only sine terms or only cosine terms are present, respectively. When a half range series corresponding to a given function is desired, the function is generally defined in the interval $(0, L)$ [which is half of the interval $(-L, L)$, thus accounting for the name *half range*] and then the function is specified as odd or even, so that it is clearly defined in the other half of the interval, namely, $(-L, 0)$. In such case, we have

$$\begin{cases} a_n = 0, \quad b_n = \dfrac{2}{L} \displaystyle\int_{-L}^{L} f(x) \sin \frac{n\pi x}{L}\, dx & \text{for } \textit{half range sine series} \\ b_n = 0, \quad a_n = \dfrac{2}{L} \displaystyle\int_{0}^{L} f(x) \cos \frac{n\pi x}{L}\, dx & \text{for } \textit{half range consine series} \end{cases} \tag{4}$$

Parseval's Identity

If a_n and b_n are the Fourier coefficients corresponding to $f(x)$ and if $f(x)$ satisfies the Dirichlet conditions, then

$$\frac{1}{L} \int_{-L}^{L} \{f(x)\}^2\, dx = \frac{a_0^2}{2} + \sum_{n=1}^{\infty} (a_n^2 + b_n^2) \tag{5}$$

(See Problem 13.13.)

Differentiation and Integration of Fourier Series

Differentiation and integration of Fourier series can be justified by using the theorems on Page 285, which hold for series in general. It must be emphasized, however, that those theorems provide sufficient conditions and are not necessary. The following theorem for integration is respecially useful.

Theorem The Fourier series corresponding to $f(x)$ may be integrated term by term from a to and the resulting series will converge uniformly to $\int_a^x f(x)\,dx$ provided that $F(x)$ is piecewise continueus in $-L \leq x \leq L$ and both a and x are in this interval.

Complex Notation for Fourier Series

Using Euler's identities,

$$e^{i\theta} = \cos\theta + i\sin\theta,\ e^{-i\theta} = \cos\theta - i\sin\theta \tag{6}$$

where $i = \sqrt{-1}$ (see Problem 11.48), the Fourier series for $f(x)$ can be written as

$$f(x) = \sum_{n=-\infty}^{\infty} c_n e^{in\pi x/L} \tag{7}$$

where

$$c_n = \frac{1}{2L}\int_{-L}^{L} f(x)e^{-in\pi x/L}\,dx \tag{8}$$

In writing the equality (7), we are supposing that the Dirichlet conditions are satisfied and, further, that $f(x)$ is continuous at x. If $f(x)$ is discontinuous at x, the left side of (7) should be replaced by $\dfrac{(f(x+0)+f(x-0))}{2}$.

Boundary-Value Problems

Boundary value problems seek to determine solutions of partial differential equations satisfying certain prescribed conditions called *boundary conditions*. Some of these problems can be solved by use of Fourier series (see Problem 13.24).

EXAMPLE. The classical problem of a vibrating string may be idealized in the following way. See Figure 13.2.

Suppose a string is tautly stretched between points (0, 0) and $(L, 0)$. Suppose the tension **F** is the same at every point of the string. The string is made to vibrate in the xy plane by pulling it to the parabolic position $g(x) = m(Lx - x^2)$ and releasing it (m is a numerically small positive constant). Its equation will be of the form $y = f(x, t)$. The problem of establishing this equation is idealized by (a) assuming that the constant tension F is so large as compared to the weight wL of the string that the gravitational force can be neglected, (b) the displacement at any point of the string is so small that the length of the string may be taken as L for any of its positions, and (c) the vibrations are purely transverse.

The force acting on a segment PQ is

$$\frac{w}{g}\Delta x \frac{\partial^2 y}{\partial t^2},\quad x < x_1 < x + \Delta x,\ g \approx 32 \text{ ft per sec}^2$$

If α and β are the angles that **F** makes with the horizontal, then the vertical difference in tensions is $F(\sin\alpha - \sin\beta)$. This is the force producing the acceleration that accounts for the vibratory motion.

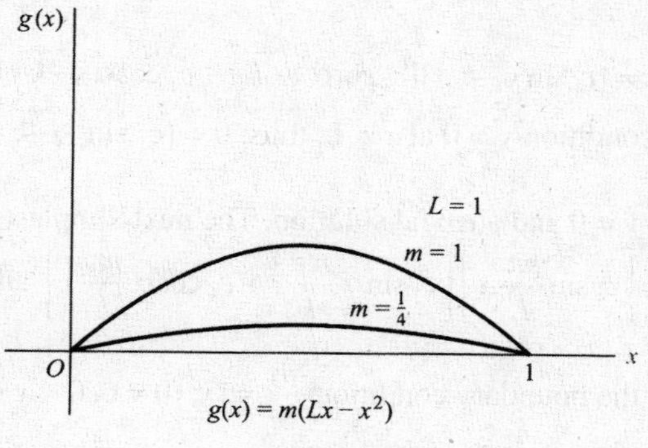

Figure 13.2

Now

$$F\{\sin\alpha - \sin\beta\} = F\left\{\frac{\tan\alpha}{\sqrt{1+\tan^2\alpha}} - \frac{\tan\beta}{\sqrt{1+\tan^2\beta}}\right\} \approx F\{\tan\alpha - \tan\beta\} = F\left\{\frac{\partial y}{\partial x}(x+\Delta x,t) - \frac{\partial y}{\partial x}(x,t)\right\}$$

where the squared terms in the denominator are neglected because the vibrations are small.

Next, equate the two forms of the force, i.e.,

$$F\left\{\frac{\partial y}{\partial x}(x+\Delta x,t) - \frac{\partial y}{\partial x}(x,t)\right\} = \frac{w}{g}\Delta x\frac{\partial^2 y}{\partial t^2}$$

divide by Δx, and then let $\Delta x \to 0$. After letting $\alpha = \frac{\sqrt{Fg}}{w}$, the resulting equation is

$$\frac{\partial^2 y}{\partial t^2} = \alpha^2\frac{\partial^2 y}{\partial x^2}$$

This homogeneous second partial derivative equation is the classical equation for the vibrating string. Associated boundary conditions are

$$y(0, t) = 0,\ y(L, t) = 0,\ t > 0$$

The initial conditions are

$$y(x,0) = m(Lx - x^2), \frac{\partial y}{\partial t}(x,0) = 0, 0 < x < L$$

The method of solution is to separate variables, i.e., assume

$$y(x, t) = G(x)H(t)$$

Then, upon substituting,

$$G(x)\,H''(t) = \alpha^2\,G''(x)H(t)$$

Separating variables yields

$$\frac{G''}{G} = k, \frac{H''}{H} = \alpha^2 k, \text{ where } k \text{ is an arbitrary constant}$$

Since the solution must be periodic, trial solutions are

$$G(x) = c_1\sin\sqrt{-k}\,x + c_2\cos\sqrt{-k}\,x, < 0$$

$$H(t) = c_3\sin\alpha\sqrt{-k}\,t + c_4\cos\alpha\sqrt{-k}\,t$$

Therefore,

$$y = GH = [c_1\sin\sqrt{-k}\,x + c_2\cos\sqrt{-k}\,x][c_3\sin\alpha\sqrt{-k}\,t + c_4\cos\alpha\sqrt{-k}\,t]$$

The initial condition y = 0 at $x = 0$ for all t leads to the evaluation $c_2 = 0$.

Thus,

$$y = [c_1 \sin \sqrt{-k}\, x][c_3 \sin \alpha \sqrt{-k}\, t + c_4 \cos \alpha \sqrt{-k}\, t]$$

Now impose the boundary condition $y = 0$ at $x = L$; thus, $0 = [c_1 \sin \sqrt{-k}\, L]\, [c_3 \sin \alpha \sqrt{-k}\, t + c_4 \cos \alpha \sqrt{-k}\, t]$.

$c_1 \neq 0$, as that would imply $y = 0$ and a trivial solution. The next-simplest solution results from the choice $\sqrt{-k} = \frac{n\pi}{L}$, since $y = \left[c_1 \sin \frac{n\pi}{L} x\right]\left[c_3 \sin \alpha \frac{n\pi}{l} t + c_4 \cos \alpha \frac{n\pi}{L} t\right]$ and the first factor is zero when $x = L$.

With this equation in place, the boundary condition $\frac{\partial y}{\partial t} = (x, 0) = 0, 0 < x < L$ can be considered.

$$\frac{\partial y}{\partial t} = \left[c_1 \sin \frac{n\pi}{L} x\right]\left[c_3 \alpha \frac{n\pi}{L} \cos \alpha \frac{n\pi}{L} t - c_4 \alpha \frac{n\pi}{L} \sin \alpha \frac{n\pi}{L} t\right]$$

At $t = 0$,

$$0 = \left[c_1 \sin \frac{n\pi}{L} x\right] c_3 \alpha \frac{n\pi}{L}$$

Since $c_1 \neq 0$ and $\sin \frac{n\pi}{L} x$ is not identically zero, it follows that $c_3 = 0$ and that

$$y = \left[c_1 \sin \frac{n\pi}{L} x\right]\left[c_4 \alpha \frac{n\pi}{L} \cos \alpha \frac{n\pi}{L} t\right]$$

The remaining initial condition is

$$y(x, 0) = m(Lx - x^2),\ 0 < x < L$$

When it is imposed,

$$m(Lx - x^2) = c_1 c_4 \alpha \frac{n\pi}{L} \sin \frac{n\pi}{L} x$$

However, this relation cannot be satisfied for all x on the interval $(0, L)$. Thus, the preceding extensive analysis of the problem of the vibrating string has led us to an inadequate form:

$$y = c_1 c_4 \alpha \frac{n\pi}{L} \sin \frac{n\pi}{L} x \cos \alpha \frac{n\pi}{L} t$$

and an initial condition that is not satisfied. At this point the power of Fourier series is employed. In particular, a theorem of differential equations states that any finite sum of a particular solution also is a solution. Generalize this to infinite sum and consider

$$y = \sum_{n=1}^{\infty} b_n \sin \frac{n\pi}{L} x \cos \alpha \frac{n\pi}{L} t$$

with the initial condition expressed through a half range sine series, i.e.,

$$\sum_{n=1}^{\infty} b_n \sin \frac{n\pi}{L} x = m(Lx - x^2), \quad t = 0$$

According to the formula on Page 351 for the coefficient of a half range sine series,

$$\frac{L}{2m} b_n = \int_0^L (Lx - x^2) \sin \frac{n\pi x}{L}\, dx$$

That is,

$$\frac{L}{2m} b_n = \int_0^L Lx \sin \frac{n\pi x}{L}\, dx - \int_0^L x^2 \sin \frac{n\pi x}{L}\, dx$$

Application of integration by parts to the second integral yields

$$\frac{L}{2m} b_n = L \int_0^L x \sin \frac{n\pi x}{L}\, dx + \frac{L^3}{n\pi} \cos n\pi + \int_0^L \frac{L}{n\pi} \cos \frac{n\pi x}{L}\, 2x\, dx$$

When integration by parts is applied to the two integrals of this expression and a little algebra is employed, the result is

$$b_n = \frac{4L^2}{(n\pi)^3}(1 - \cos n\pi)$$

Therefore,

$$y = \sum_{n=1}^{\infty} b_n \sin\frac{n\pi}{L}x \cos\alpha\frac{n\pi}{L}t$$

with the coefficients b_n defined previously.

Orthogonal Functions

Two vectors **A** and **B** are called *orthogonal* (perpendicular) if $\mathbf{A} \cdot \mathbf{B} = 0$ or or $A_1B_1 + A_2B_2 + A_3B_3 = 0$, where $\mathbf{A} = A_1\mathbf{i} + A_2\mathbf{j} + A_3\mathbf{k}$ and $\mathbf{B} = B_1\mathbf{i} + B_2\mathbf{j} + B_3\mathbf{k}$. Although not geometrically or physically evident, these ideas can be generalized to include vectors with more than three components. In particular, we can think of a function—say, $A(x)$—as being a vector with an *infinity of components* (i.e., an *infinite dimensional vector*), the value of each component being specified by substituting a particular value of x in some interval (a, b). It is natural in such case to define two functions, $A(x)$ and $B(x)$, as *orthogonal* in (a, b) if

$$\int_a^b A(x)\,B(x)dx = 0 \tag{9}$$

A vector **A** is called a *unit vector* or *normalized vector* if its magnitude is unity, i.e., if $\mathbf{A} \cdot \mathbf{A} = A^2 = 1$. Extending the concept, we say that the function $A(x)$ is *normal* or *normalized* in (a, b) if

$$\int_a^b \{A(x)\}^2\,dx = 1 \tag{10}$$

From this, it is clear that we can consider a set of functions $\{\phi_k(x)\}$, $k = 1, 2, 3, \ldots$, having the properties

$$\int_a^b \phi_m(x)\phi_n(x)dx = 0 \quad m \neq n \tag{11}$$

$$\int_a^b \{\phi_m(x)\}^2\,dx = 1 \quad m = 1, 2, 3, \ldots \tag{12}$$

In such case, each member of the set is orthogonal to every other member of the set and is also normalized. We call such a set of functions an *orthonormal set*.

Equations (11) and (12) can be summarized by writing

$$\int_a^b \phi_m(x)\phi_n(x)\,dx = \delta_{mn} \tag{13}$$

where δ_{mn}, called *Kronecker's symbol*, is defined as 0 if $m \neq n$ and 1 if $m = n$.

Just as any vector **r** in three dimensions can be expanded in a set of mutually orthogonal unit vectors **i**, **j**, **k** in the form $\mathbf{r} = c_1\mathbf{i} + c_2\mathbf{j} + c_3\mathbf{k}$, so we consider the possibility of expanding a function $f(x)$ in a set of orthonormal functions, i.e.,

$$f(x) = \sum_{n=1}^{\infty} c_n\phi_n(x) \quad a \leqq x \leqq b \tag{14}$$

As we have seen, Fourier series are constructed from orthogonal functions. Generalizations of Fourier series are of great interest and utility from both theoretical and applied viewpoints.

SOLVED PROBLEMS

Fourier Series

13.1. Graph each of the following functions.

(a) $f(x) = \begin{cases} 3 & 0 < x < 5 \\ -3 & -5 < x < 0 \end{cases}$ Period = 10

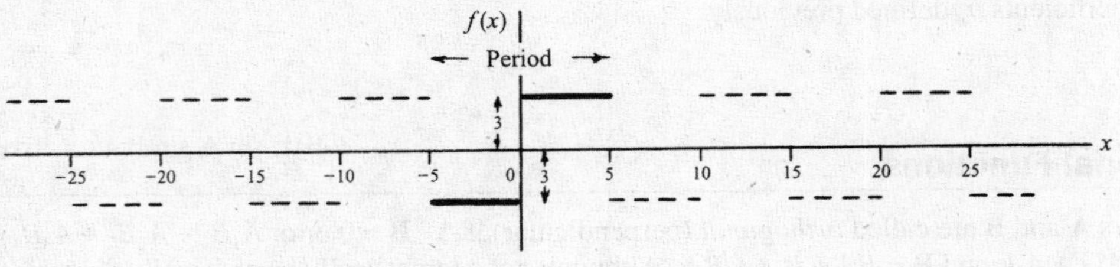

Figure 13.3

Since the period is 10, that portion of the graph in $-5 < x < 5$ (indicated by heavy lines in Figure 13.3) is extended periodically outside this range (indicated by dashed lines). Note that $f(x)$ is not defined at $x = 0, 5, -5, 10, -10, 15, -15$, and so on. These values are the *discontinuities* of $f(x)$.

(b) $f(x) = \begin{cases} \sin x & 0 \leqq x \leqq \pi \\ 0 & \pi < x < 2\pi \end{cases}$ Period = 2π

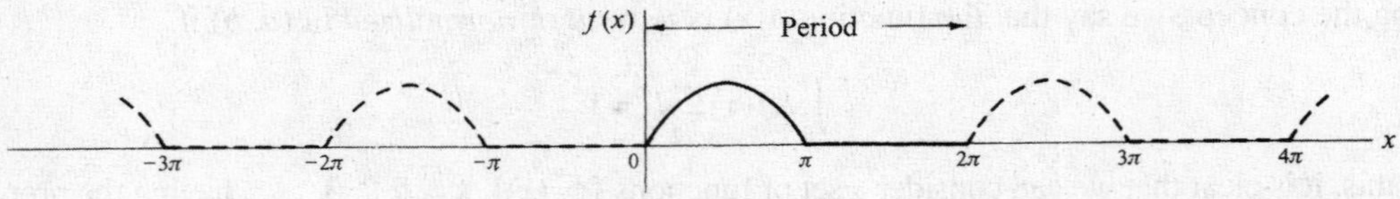

Figure 13.4

Refer to Figure 13.4. Note that $f(x)$ is defined for all x and is continuous everywhere.

(c) $f(x) = \begin{cases} 0 & 0 \leqq x < 2 \\ 1 & 2 \leqq x < 4 \\ 0 & 4 \leqq x < 6 \end{cases}$ Period = 6

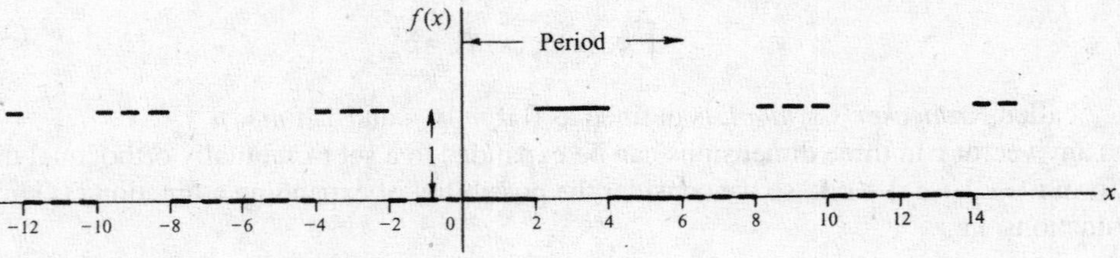

Fig.13.5

Refer to Figure 13.5. Note that $f(x)$ is defined for all x and is discontinuous at $x = \pm 2, \pm 4, \pm 8, \pm 10, \pm 14, \ldots$.

13.2. prove $\int_{-L}^{L} \sin \frac{k\pi x}{L}\, dx = \int_{-L}^{L} \cos \frac{k\pi x}{L}\, dx = 0 \quad if\ k = 1, 2, 3, \ldots$

$$\int_{-L}^{L} \sin\frac{k\pi x}{L}\,dx = -\frac{L}{k\pi}\cos\frac{k\pi x}{L}\Big|_{-L}^{L} = \frac{L}{k\pi}\cos k\pi + \frac{L}{k\pi}\cos(-k\pi) = 0$$

$$\int_{-L}^{L} \cos\frac{k\pi x}{L}\,dx = \frac{L}{k\pi}\sin\frac{k\pi x}{L}\Big|_{-L}^{L} = \frac{L}{k\pi}\sin k\pi - \frac{L}{k\pi}\sin(-k\pi) = 0$$

13.3. Prove (a) $\displaystyle\int_{-L}^{L}\cos\frac{m\pi x}{L}\cos\frac{n\pi x}{L}\,dx = \int_{-L}^{L}\sin\frac{m\pi x}{L}\sin\frac{n\pi x}{L}\,dx = \begin{cases} 0 & m \neq n \\ L & m = n \end{cases}$

(b) $\displaystyle\int_{-L}^{L}\sin\frac{m\pi x}{L}\cos\frac{n\pi x}{L}\,dx = 0$

where m and n can assume any of the values 1, 2, 3, ...

(a) From trigonometry: $\cos A \cos B = \frac{1}{2}\{\cos(A - B) + \cos(A + B)\}$, $\sin A \sin B = \frac{1}{2}\{\cos(A - B) - \cos(A + B)\}$.

Then, if $m \neq n$, by Problem 13.2,

$$\int_{-L}^{L}\cos\frac{m\pi x}{L}\cos\frac{n\pi x}{L}\,dx = \frac{1}{2}\int_{-L}^{L}\left\{\cos\frac{(m-x)\pi x}{L} + \cos\frac{(m+n)\pi x}{L}\right\}dx = 0$$

Similarly, if $m \neq n$,

$$\int_{-L}^{L}\sin\frac{m\pi x}{L}\sin\frac{n\pi x}{L}\,dx = \frac{1}{2}\int_{-L}^{L}\left\{\cos\frac{(m-n)\pi x}{L} - \cos\frac{(m+n)\pi x}{L}\right\}dx = 0$$

If $m = n$, we have

$$\int_{-L}^{L}\cos\frac{m\pi x}{L}\cos\frac{n\pi x}{L}\,dx = \frac{1}{2}\int_{-L}^{L}\left(1 + \cos\frac{2n\pi x}{L}\right)dx = L$$

$$\int_{-L}^{L}\sin\frac{m\pi x}{L}\sin\frac{n\pi x}{L}\,dx = \frac{1}{2}\int_{-L}^{L}\left(1 - \cos\frac{2n\pi x}{L}\right)dx = L$$

Note that if $m = n$ these integrals are equal to $2L$ and 0, respectively.

(b) We have $\sin A \cos B = 1/2\{\sin(A - B) + \sin(A + B)\}$. Then by Problem 13.2, if $m \neq n$,

$$\int_{-L}^{L}\sin\frac{m\pi x}{L}\cos\frac{n\pi x}{L}\,dx = \frac{1}{2}\int_{-L}^{L}\left\{\sin\frac{(m-n)\pi x}{L} + \sin\frac{(m+n)\pi x}{L}\right\}dx = 0$$

If $m = n$,

$$\int_{-L}^{L}\sin\frac{m\pi x}{L}\cos\frac{n\pi x}{L}\,dx = \frac{1}{2}\int_{-L}^{L}\sin\frac{2n\pi x}{L}\,dx = 0$$

The results of (a) and (b) remain valid even when the limits of integration $-L$, L are replaced by c, $c + 2L$, respectively.

13.4. If the series $A + \sum_{n=1}^{\infty}\left(a_n\cos\frac{n\pi x}{L} + b_n\sin\frac{n\pi x}{L}\right)$ converges uniformly to $f(x)$ in $(-L, L)$, show that for $n =$

1, 2, 3, . . . , (a) $a_n = \frac{1}{L}\int_{-L}^{L} f(x)\cos\frac{n\pi x}{L}\,dx$ and (b) $b_n = \frac{1}{L}\int_{-L}^{L} f(x)\sin\frac{n\pi x}{L}\,dx$, (c) $A = \frac{a_0}{2}$.

(a) Multiplying

$$f(x) = A + \sum_{n=1}^{\infty}\left(a_n\cos\frac{n\pi x}{L} + b_n\sin\frac{n\pi x}{L}\right) \tag{1}$$

by $\cos\dfrac{m\pi x}{L}$ and integrating from $-L$ to L, using Problem 13.3, we have

$$\int_{-L}^{L} f(x)\cos\frac{m\pi x}{L}\,dx = A\int_{-L}^{L}\cos\frac{m\pi x}{L}\,dx$$
$$+\sum_{n=1}^{\infty}\left\{a_n\int_{-L}^{L}\cos\frac{m\pi x}{L}\cos\frac{n\pi x}{L}\,dx + b_n\int_{-L}^{L}\cos\frac{m\pi x}{L}\sin\frac{n\pi x}{L}\,dx\right\}$$
$$= a_m L \quad \text{if } m \neq 0$$

Thus,

$$a_m = \frac{1}{L}\int_{-L}^{L} f(x)\cos\frac{m\pi x}{L}\,dx \quad \text{if } m = 1, 2, 3, \ldots$$

(b) Multiplying Equation (1) by $\sin\frac{m\pi x}{L}$ and integrating from –L to L, using Problem 13.3, we have

$$\int_{-L}^{L} f(x)\sin\frac{m\pi x}{L}\,dx = A\int_{-L}^{L}\sin\frac{m\pi x}{L}\,dx$$
$$+\sum_{n=1}^{\infty}\left\{a_n\int_{-L}^{L}\sin\frac{m\pi x}{L}\cos\frac{n\pi x}{L}\,dx + b_n\int_{-L}^{L}\cos\frac{m\pi x}{L}\sin\frac{n\pi x}{L}\,dx\right\}$$

Thus,

$$= b_m L$$

$$b_m = \frac{1}{L}\int_{-L}^{L} f(x)\sin\frac{m\pi x}{L}\,dx \quad \text{if } m = 1, 2, 3, \ldots$$

(c) Integrating Equation (1) from $-L$ to L, using Problem 13.2, gives

$$\int_{-L}^{L} f(x)\,dx = 2AL \quad \text{or} \quad A = \frac{1}{2L}\int_{-L}^{L} f(x)\,dx$$

Putting $m = 0$ in the result of (a), we find $a_0 = \frac{1}{L}\int_{-L}^{L} f(x)\,dx$ and so $A = \frac{a_0}{2}$.

The above results also hold when the integration limits $-L$, L are replaced by c, $c + 2L$.

Note that in (a), (b), and (c), interchange of summation and integration is valid because the series is *assumed* to converge uniformly to $f(x)$ in $(-L, L)$. Even when this assumption is not warranted, the coefficients a_m and b_m as obtained are called *Fourier coefficients* corresponding to $f(x)$, and the corresponding series with these values of a_m and b_m is called the *Fourier series* corresponding to $f(x)$. An important problem in this case is to investigate conditions under which this series actually converges to $f(x)$. Sufficient conditions for this convergence are the *Dirichlet conditions* established in Problems 13.18 through 13.23.

13.5. (a) Find the Fourier coefficients corresponding to the function

$$f(x) = \begin{cases} 0 & -5 < x < 0 \\ 3 & 0 < x < 5 \end{cases} \qquad \text{Period} = 10$$

(b) Write the corresponding Fourier series.

(c) How should $f(x)$ be defined at $x = -5$, $x = 0$, and $x = 5$ in order that the Fourier series will converge to $f(x)$ for $-5 \leqq x \leqq 5$?

The graph of $f(x)$ is shown in Figure 13.6.

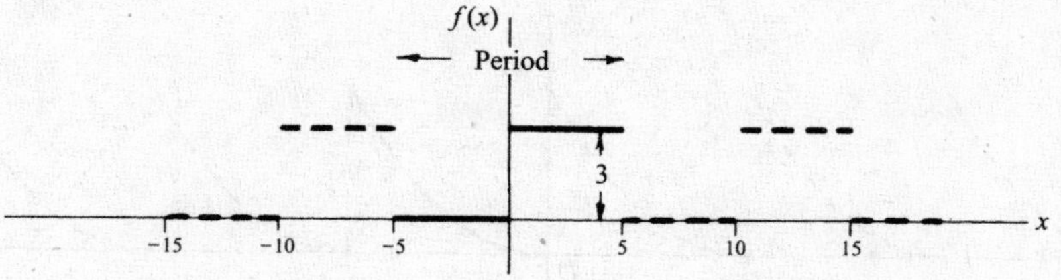

Figure 13.6

(a) Period = $2L = 10$ and $L = 5$. Choose the interval c to $c + 2L$ as −5 to 5, so that $c = -5$. Then

$$a_n = \frac{1}{L}\int_c^{c+2L} f(x)\cos\frac{n\pi x}{L}\,dx = \frac{1}{5}\int_{-5}^{5} f(x)\cos\frac{n\pi x}{5}\,dx$$

$$= \frac{1}{5}\left\{\int_{-5}^{0}(0)\cos\frac{n\pi x}{5}\,dx + \int_0^5 (3)\cos\frac{n\pi x}{5}\,dx\right\} = \frac{3}{5}\int_0^5 \cos\frac{n\pi x}{5}\,dx$$

$$= \frac{3}{5}\left(\frac{5}{n\pi}\sin\frac{n\pi x}{5}\right)\Bigg|_0^5 = 0 \quad \text{if } n \neq 0$$

$$\text{If } n = 0, a_n = a_0 = \frac{3}{5}\int_0^5 \cos\frac{0\pi x}{5}\,dx = \frac{3}{5}\int_0^5 dx = 3.$$

$$b_n = \frac{1}{L}\int_c^{c+2L} f(x)\sin\frac{n\pi x}{L}\,dx = \frac{1}{5}\int_{-5}^{5} f(x)\sin\frac{n\pi x}{5}\,dx$$

$$= \frac{1}{5}\left\{\int_{-5}^{0}(0)\sin\frac{n\pi x}{5}\,dx + \int_0^5 (3)\sin\frac{n\pi x}{5}\,dx\right\} = \frac{3}{5}\int_0^5 \sin\frac{n\pi x}{5}\,dx$$

$$= \frac{3}{5}\left(-\frac{5}{n\pi}\cos\frac{n\pi x}{5}\right)\Bigg|_0^5 = \frac{3(1-\cos n\pi)}{n\pi}$$

(b) The corresponding Fourier series is

$$\frac{a_0}{2} + \sum_{n=1}^{\infty}\left(a_n\cos\frac{n\pi x}{L} + b_n\sin\frac{n\pi x}{L}\right) = \frac{3}{2} + \sum_{n=1}^{\infty}\frac{3(1-\cos n\pi)}{n\pi}\sin\frac{n\pi x}{5}$$

$$= \frac{3}{2} + \frac{6}{\pi}\left(\sin\frac{\pi x}{5} + \frac{1}{3}\sin\frac{3\pi x}{5} + \frac{1}{5}\sin\frac{5\pi x}{5} + \cdots\right)$$

(c) Since $f(x)$ satisfies the Dirichlet conditions, we can say that the series converges to $f(x)$ at all points of continuity and to $\dfrac{f(x+0)+f(x-0)}{2}$ at points of discontinuity. At $x = -5$, 0, and 5, which are points of discontinuity, the series converges to $(3 + 0)/2 = 3/2$, as seen from the graph. If we redefine $f(x)$ as follows,

$$f(x) = \begin{cases} 3/2 & x = -5 \\ 0 & -5 < x < 0 \\ 3/2 & x = 0 \\ 3 & 0 < x < 5 \\ 3/2 & x = 5 \end{cases} \quad \text{Period} = 10$$

then the series will converge to $f(x)$ for $-5 \leqq x \leqq 5$.

13.6. Expand $f(x) = x^2$, $0 < x < 2\pi$ in a Fourier series if (a) the period is 2π and (b) the period is not specified.

(a) The graph of $f(x)$ with period 2π is shown in Figure 13.7.

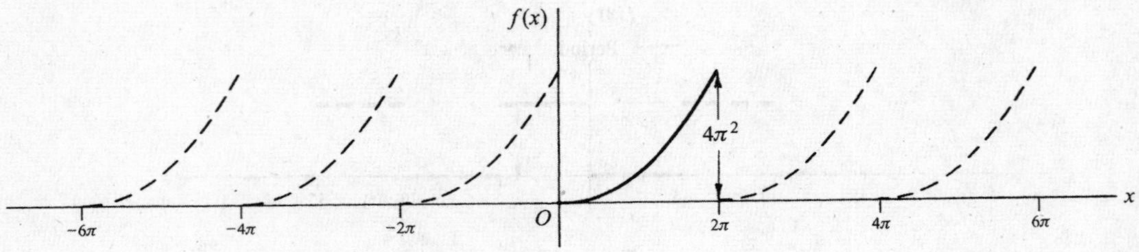

Fig. 13-7

Period = $2L = 2\pi$ and $L = \pi$. Choosing $c = 0$, we have

$$a_n = \frac{1}{L}\int_c^{c+2L} f(x)\cos\frac{n\pi x}{L}\,dx = \frac{1}{\pi}\int_0^{2\pi} x^2 \cos nx\,dx$$

$$= \frac{1}{\pi}\left\{(x^2)\left(\frac{\sin nx}{n}\right) - (2x)\left(\frac{-\cos nx}{n^2}\right) + 2\left(\frac{-\sin nx}{n^3}\right)\right\}\Bigg|_0^{2\pi} = \frac{4}{n^2}, \quad n \neq 0$$

If $n = 0, a_0 = \frac{1}{\pi}\int_0^{2\pi} x^2\,dx = \frac{8\pi^2}{3}$.

$$b_n = \frac{1}{L}\int_c^{c+2L} f(x)\sin\frac{n\pi x}{L}\,dx = \frac{1}{\pi}\int_0^{2\pi} x^2 \sin nx\,dx$$

$$= = \frac{1}{\pi}\left\{(x^2)\left(\frac{\sin nx}{n}\right) - (2x)\left(\frac{-\cos nx}{n^2}\right) + 2\left(\frac{-\sin nx}{n^3}\right)\right\}\Bigg|_0^{2\pi}$$

Then $f(x) = x^2 = \frac{4\pi^2}{3} + \sum_{n=1}^{\infty}\left(\frac{4}{n^2}\cos nx - \frac{4\pi}{n}\sin nx\right)$.

This is valid for $0 < x < 2\pi$. At $x = 0$ and $x = 2\pi$ the series converges to $2\pi^2$.

(b) If the period is not specified, the Fourier series cannot be determined uniquely in general.

13.7. Using the results of Problem 13.6, prove that $\frac{1}{1^2} + \frac{1}{2^2} + \frac{1}{3^2} + \cdots = \frac{\pi^2}{6}$.

At $x = 0$, the Fourier series of Problem 13.6 reduces to $\frac{4\pi^2}{3} + \sum_{n=1}^{\infty}\frac{4}{n^2}$.

By the Dirichlet conditions, the series converges at $x = 0$ to $\frac{1}{2}(0 + 4\pi^2) = 2\pi^2$.

Then $\frac{4\pi^2}{3} + \sum_{n=1}^{\infty}\frac{4}{n^2} = 2\pi^2$, and so $\sum_{n=1}^{\infty}\frac{1}{n^2} = \frac{\pi^2}{6}$.

Odd and even functions, half range Fourier series

13.8. Classify each of the following functions according to whether they are even, odd, or neither even nor odd.

(a) $f(x) = \begin{cases} 2 & 0 < x < 3 \\ -2 & -3 < x < 0 \end{cases}$ Period = 6

From Figure 13.8, it is seen that $f(-x) = -f(x)$, so that the function is odd.

(b) $f(x) = \begin{cases} \cos x & 0 < x < \pi \\ 0 & \pi < x < 2\pi \end{cases}$ Period = 2π

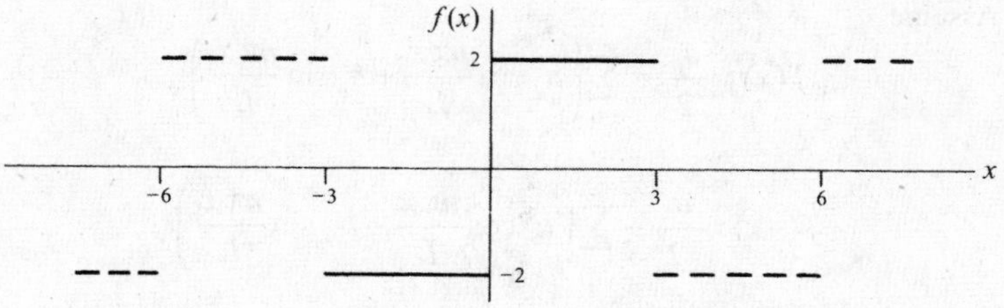

Figure 13.8

From Figure 13.9, it is seen that the function is neither even nor odd.

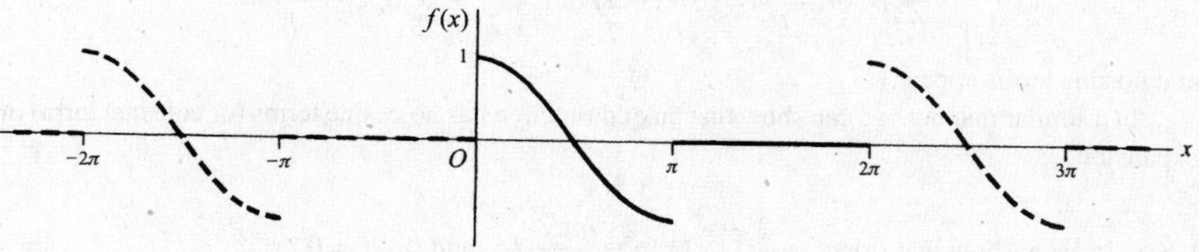

Figure 13.9

(c) $f(x) = x(10 - x),\ 0 < x < 10$ Period $= 10$

From Figure 13.10 below the function is seen to be even.

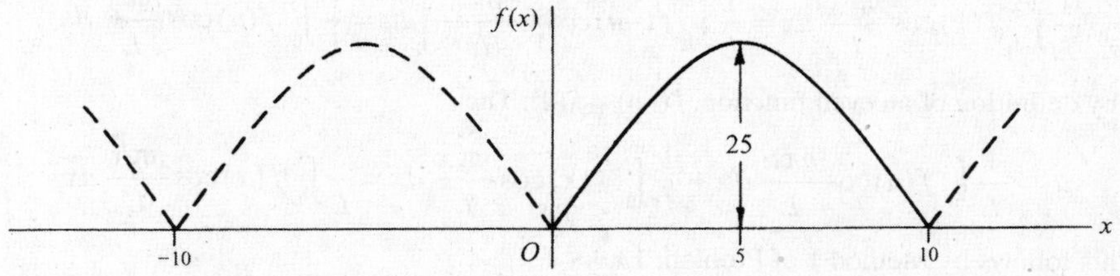

Figure 13.10

13.9. Show that an even function can have no sine terms in its Fourier expansion.

Method 1: No sine terms appear if $b_n = 0$, $n = 1, 2, 3, \ldots$. To show this, let us this, let us write

$$b_n = \frac{1}{L}\int_{-L}^{L} f(x)\sin\frac{n\pi x}{L}\,dx = \frac{1}{L}\int_{-L}^{0} f(x)\sin\frac{n\pi x}{L}\,dx + \frac{1}{L}\int_{0}^{L} f(x)\sin\frac{n\pi x}{L}\,dx \tag{1}$$

If we make the transformation $x = -u$ in the first integral on the right of Equation (1), we obtain

$$\begin{aligned}\frac{1}{L}\int_{-L}^{0} f(x)\sin\frac{n\pi x}{L}\,dx &= \frac{1}{L}\int_{0}^{L} f(-u)\sin\left(-\frac{n\pi u}{L}\right)du = -\frac{1}{L}\int_{0}^{L} f(-u)\sin\frac{n\pi u}{L}\,du \\ &= -\frac{1}{L}\int_{0}^{L} f(u)\sin\frac{n\pi u}{L}\,du = -\frac{1}{L}\int_{0}^{L} f(x)\sin\frac{n\pi u}{L}\,dx\end{aligned} \tag{2}$$

where we have used the fact that for an even function $f(-u) = f(u)$ and in the last last step that the dummy variable of integration u can be replaced by any other symbol, in particular, x. Thus, from Equation (1), using Equation (2), we have

$$b_n = -\frac{1}{L}\int_{0}^{L} f(x)\sin\frac{n\pi x}{L}\,dx + \frac{1}{L}\int_{0}^{L} f(x)\sin\frac{n\pi u}{L}\,dx = 0$$

Method 2: Assume

$$f(x) = \frac{a_0}{2} + \sum_{n=1}^{\infty}\left(a_n \cos\frac{n\pi x}{L} + b_n \sin\frac{n\pi x}{L}\right)$$

Then

$$f(-x) = \frac{a_0}{2} + \sum_{n=1}^{\infty}\left(a_n \cos\frac{n\pi x}{L} - b_N \sin\frac{n\pi x}{L}\right)$$

If $f(x)$ is even, $f(-x) = f(x)$. Hence,

$$\frac{a_0}{2} + \sum_{n=1}^{\infty}\left(a_n \cos\frac{n\pi x}{L} + b_n \sin\frac{n\pi x}{L}\right) = \frac{a_0}{2} + \sum_{n=1}^{\infty}\left(a_n \cos\frac{n\pi x}{L} - b_n \sin\frac{n\pi x}{L}\right)$$

and so

$$\sum_{n=1}^{\infty} b_n \sin\frac{n\pi x}{L} = 0, \quad \text{i.e., } f(x) = \frac{a_0}{2} + \sum_{n=1}^{\infty} a_n \cos\frac{n\pi x}{L}$$

and no sine terms appear.

In a similar manner, we can show that an odd function has no cosine terms (or constant term) in its Fourier expansion.

13.10. If $f(x)$ is even, show that (a) $a_n = \frac{2}{L}\int_0^L f(x)\cos\frac{n\pi x}{L}\,dx$, and (b) $b_n = 0$.

(a) $$a_n = \frac{1}{L}\int_{-L}^{L} f(x)\cos\frac{n\pi x}{L}\,dx = \frac{1}{L}\int_{-L}^{0} f(x)\cos\frac{n\pi x}{L}\,dx + \frac{1}{L}\int_0^L f(x)\cos\frac{n\pi x}{L}\,dx$$

Letting $x = -u$,

$$\frac{1}{L}\int_{-L}^{0} f(x)\cos\frac{n\pi x}{L}\,dx = \frac{1}{L}\int_0^L f(-u)\cos\left(\frac{-n\pi x}{L}\right)du = \frac{1}{L}\int_0^L f(u)\cos\frac{n\pi x}{L}\,du$$

since, by definition of an even function, $f(-u) = f(u)$. Then

$$a_n = \frac{1}{L}\int_0^L f(u)\cos\frac{n\pi u}{L}\,du + \frac{1}{L}\int_0^L f(x)\cos\frac{n\pi x}{L}\,dx = \frac{2}{L}\int_0^L f(x)\cos\frac{n\pi x}{L}\,dx$$

(b) This follows by Method 1 of Problem 13.9.

13.11. Expand $f(x) = \sin x$, $0 < x < \pi$, in a Fourier cosine series.

A Fourier series consisting of cosine terms alone is obtained only for an even function. Hence, we extend the definition of $f(x)$ so that it becomes even (dashed part of Figure 13.11). With this extension, $f(x)$ is then defined in an interval of length 2π. Taking the period as 2π, we have $2L = 2\pi$ so that $L = \pi$.

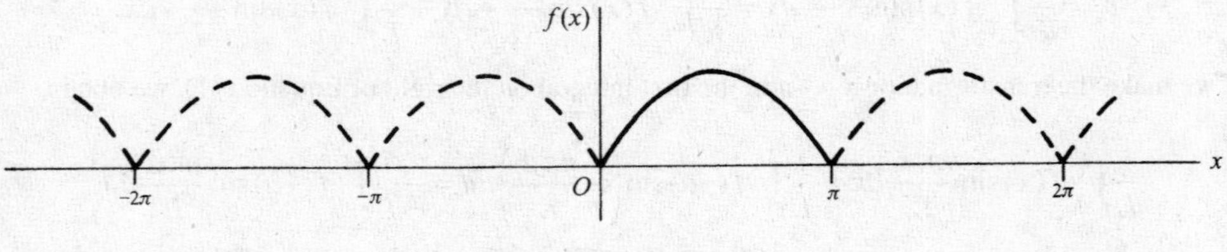

Figure 13.11

By Problem 13.10, $b_n = 0$ and

$$a_n = \frac{2}{L}\int_0^L f(x)\cos\frac{n\pi u}{L}\,dx = \frac{2}{\pi}\int_0^{\pi} \sin x \cos nx\,dx$$

$$= \frac{1}{\pi}\int_0^{\pi}\{\sin(x+nx)+\sin(x-nx)\} = \frac{1}{\pi}\left\{-\frac{\cos(n+1)x}{n+1}+\frac{\cos(n-1)x}{n-1}\right\}\Bigg|_0^{\pi}$$

$$= \frac{1}{\pi}\left\{\frac{1-\cos(n+1)\pi}{n+1}+\frac{\cos(n-1)\pi-1}{n-1}\right\} = \frac{1}{\pi}\left\{\frac{1+\cos n\pi}{n+1}+\frac{1+\cos n\pi}{n-1}\right\}$$

$$= \frac{-2(1+\cos n\pi)}{\pi(n^2-1)} \quad \text{if } n \neq 1.$$

For $n = 1$,

$$n=1, \quad a_1 = \frac{2}{\pi}\int_0^{\pi}\sin x\cos x\,dx = \frac{2}{\pi}\frac{\sin^2 x}{2}\Bigg|_0^{\pi} = 0$$

For $n = 0$,

$$n=0, \quad a_0 = \frac{2}{\pi}\int_0^{\pi}\sin x\,dx = \frac{2}{\pi}(-\cos x)\Bigg|_0^{\pi} = \frac{4}{\pi}$$

Then

$$f(x) = \frac{2}{\pi} - \frac{2}{\pi}\sum_{n=2}^{\infty}\frac{(1+\cos n\pi)}{n^2-1}\cos nx$$

$$= \frac{2}{\pi} - \frac{4}{\pi}\left(\frac{\cos 2x}{2^2-1}+\frac{\cos 4x}{4^2-1}+\frac{\cos 6x}{6^2-1}+\cdots\right)$$

13.12. Expand $f(x) = x$, $0 < x < 2$, in a half range (a) sine series and (b) cosine series.

(a) Extend the definition of the given function to that of the odd function of period 4 shown in Figure 13.12. This is sometimes called the odd extension of $f(x)$. Then $2L = 4$, $L = 2$.

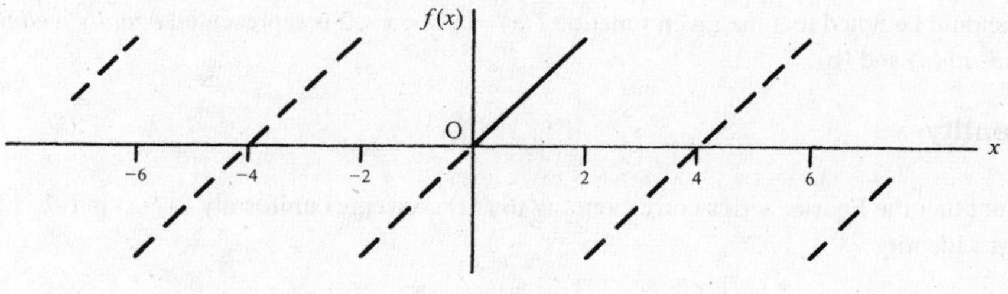

Figure 13.12

Thus, $a_n = 0$ and

$$b_n = \frac{2}{L}\int_0^L f(x)\sin\frac{n\pi x}{L}\,dx = \frac{2}{2}\int_0^2 x\sin\frac{n\pi x}{2}\,dx$$

$$= \left\{(x)\left(\frac{-2}{n\pi}\cos\frac{n\pi x}{2}\right)-(1)\left(\frac{-4}{n^2\pi^2}\sin\frac{n\pi x}{2}\right)\right\}\Bigg|_0^2 = \frac{-4}{n\pi}\cos n\pi$$

Then

$$f(x) = \sum_{n=1}^{\infty}\frac{-4}{n\pi}\cos n\pi\sin\frac{n\pi x}{2}$$

$$= \frac{4}{\pi}\left(\sin\frac{\pi x}{2}-\frac{1}{2}\sin\frac{2\pi x}{2}+\frac{1}{3}\sin\frac{3\pi x}{2}-\cdots\right)$$

(b) Extend the definition of $f(x)$ to that of the even function of period 4 shown in Figure 13.13. This is the even extension of $f(x)$. Then $2L = 4$, $L = 2$.

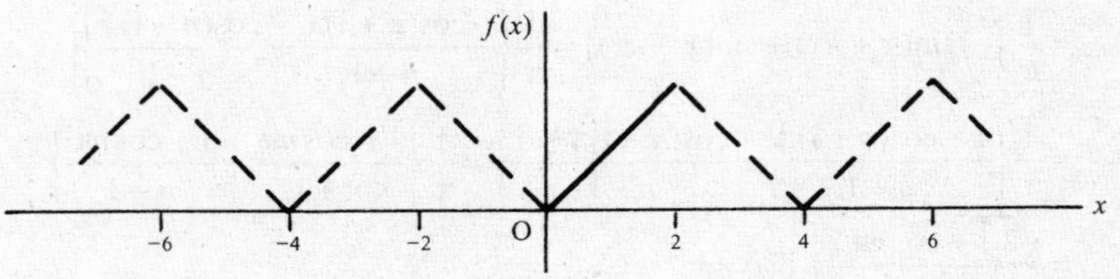

Figure 13.13

Thus, $b_n = 0$,

$$a_n = \frac{2}{L}\int_0^L f(x)\cos\frac{n\pi x}{L}\,dx = \frac{2}{2}\int_0^2 x\cos\frac{n\pi x}{2}\,dx$$
$$= \left\{(x)\left(\frac{2}{\pi}\sin\frac{n\pi x}{2}\right) - (1)\left(\frac{-4}{n^2\pi^2}\cos\frac{n\pi x}{2}\right)\right\}\Bigg|_0^2$$
$$= \frac{4}{n^2\pi^2}(\cos n\pi - 1) \quad \text{If n} \neq 0$$

If $n = 0$, $a_0 = \int_0^2 x\,dx = 2$.

Then

$$f(x) = 1 + \sum_{n=1}^{\infty}\frac{4}{n^2\pi^2}(\cos n\pi - 1)\cos\frac{n\pi x}{2}$$
$$= 1 - \frac{8}{\pi^2}\left(\cos\frac{nx}{2} + \frac{1}{3^2}\cos\frac{3\pi x}{2} = \frac{1}{5^2}\cos\frac{5\pi x}{2} + \cdots\right)$$

It should be noted that the given function $f(x) = x$, $0 < x < 2$ is represented *equally well* by the two *different* series in (a) and (b).

Parseval's identity

13.13. Assuming that the Fourier series corresponding to $f(x)$ converges uniformly to $f(x)$ in $(-L, L)$, prove Parseval's identity

$$\frac{1}{L}\int_{-L}^{L}\{f(x)\}^2\,dx = \frac{a_0^2}{2} + \sum(a_n^2 + b_n^2)$$

where the integral is assumed to exist.

If $f(x) = \frac{a_0}{2} + \sum_{n=1}^{\infty}\left(a_n\cos\frac{n\pi x}{L} + b_n\sin\frac{n\pi x}{L}\right)$, then multiplying by $f(x)$ and integrating term by term from $-L$ to L (which is justified since the series is uniformly convergent), we obtain

$$\int_{-L}^{L}\{f(x)\}^2\,dx = \frac{a_0}{2}\int_{-L}^{L}f(x)dx + \sum_{n=1}^{\infty}\left\{a_n\int_{-L}^{L}f(x)\cos\frac{n\pi x}{L}\,dx + b_n\int_{-L}^{L}f(x)\sin\frac{n\pi x}{L}\,dx\right\}$$
$$= \frac{a_0^2}{2}L + L\sum_{n=1}^{\infty}(a_n^2 + b_n^2) \qquad (1)$$

where we have used the results

$$\int_{-L}^{L}f(x)\cos\frac{n\pi x}{L}\,dx = La_n, \quad \int_{-L}^{L}f(x)\sin\frac{n\pi x}{L}\,dx = Lb_n, \quad \int_{-L}^{L}f(x)dx = La_0 \qquad (2)$$

obtained from the Fourier coefficients.

The required result follows on dividing both sides of Equation (1) by L. Parseval's identity is valid under less restrictive conditions than that imposed here.

13.14. (a) Write Parseval's identity corresponding to the Fourier series of Problem 13.12(b). (b) Determine from (a) the sum S of the series $\frac{1}{1^4}+\frac{1}{2^4}+\frac{1}{3^4}+\cdots+\frac{1}{n^4}+\cdots$

(a) Here $L=2, a_0=2, a_n=\frac{4}{n^2\pi^2}(\cos n\pi-1), n\neq 0, b_n=0.$

Then Parseval's identity becomes

$$\frac{1}{2}\int_{-2}^{2}\{f(x)\}^2\,dx=\frac{1}{2}\int_{-2}^{2}x^2\,dx=\frac{(2)^2}{2}+\sum_{n=1}^{\infty}\frac{16}{n^4\pi^4}(\cos n\pi-1)^2$$

or

$$\frac{8}{3}=2+\frac{64}{\pi^4}\left(\frac{1}{1^4}+\frac{1}{3^4}+\frac{1}{5^4}+\cdots\right).\quad \text{ie., } \frac{1}{1^4}+\frac{1}{3^4}+\frac{1}{5^4}+\cdots=\frac{\pi^4}{96}$$

(b)
$$S=,\frac{1}{1^4}+\frac{1}{2^4}+\frac{1}{3^4}+\cdots=\left(\frac{1}{1^4}+\frac{1}{3^4}+\frac{1}{5^4}+\cdots\right)+\left(\frac{1}{2^4}+\frac{1}{4^4}+\frac{1}{6^4}+\cdots\right)$$
$$=\left(\frac{1}{1^4}+\frac{1}{3^4}+\frac{1}{5^4}+\cdots\right)+\frac{1}{2^4}\left(\frac{1}{1^4}+\frac{1}{2^4}+\frac{1}{3^4}+\cdots\right)$$
$$=\frac{\pi^4}{96}+\frac{S}{16},\quad \text{from which } S=\frac{\pi^4}{90}$$

13.15. Prove that for all positive integers M,

$$\frac{a_0^2}{2}+\sum_{n=1}^{M}(a_n^2+b_n^2)\leqq\frac{1}{L}\int_{-L}^{L}\{f(x)\}^2\,dx$$

where a_n and b_n are the Fourier coefficients corresponding to $f(x)$, and $f(x)$ is assumed piecewise continuous in $(-L, L)$.

Let

$$S_M(x)=\frac{a_0}{2}+\sum_{n=1}^{M}\left(a_n\cos\frac{n\pi x}{L}+b_n\sin\frac{n\pi x}{L}\right)\tag{1}$$

For $M = 1, 2, 3, \ldots$, this is the sequence of partial sums of the Fourier series corresponding to $f(x)$.

We have

$$\int_{-L}^{L}\{f(x)-S_M(x)\}^2\,dx\geqq 0\tag{2}$$

since the integrand is nonnegative. Expanding the integrand, we obtain

$$2\int_{-L}^{L}f(x)S_M(x)dx-\int_{-L}^{L}S_M^2(x)dx\leqq\int_{-L}^{L}\{f(x)\}^2\,dx\tag{3}$$

Multiplying both sides of Equation (1) by $2f(x)$ and integrating from $-L$ to L, using Equations (2) of Problem 13.13, gives

$$2\int_{-L}^{L}f(x)S_M(x)dx=2L\left\{\frac{a_0^2}{2}+\sum_{n=1}^{M}(a_n^2+b_n^2)\right\}\tag{4}$$

Also, squaring Equation (1) and integrating from $-L$ to L, using Problem 13.3, we find

$$\int_{-L}^{L}S_M^2(x)dx=2L\left\{\frac{a_0^2}{2}+\sum_{n=1}^{M}(a_n^2+b_n^2)\right\}\tag{5}$$

Substitution of Equations (4) and (5) into Equation (3) and dividing by L yields the required result.
Taking the limit as $M\to\infty$, we obtain *Bessel's inequality*

$$\frac{a_0^2}{2}+\sum_{n=1}^{\infty}(a_n^2+b_n^2)\leqq\frac{1}{L}\int_{-L}^{L}\{f(x)\}^2\,dx\tag{6}$$

If the equality holds, we have Parseval's identity (Problem 13.13).

We can think of $S_M(x)$ as representing an *approximation* to $f(x)$, while the left-hand side of Equation (2), divided by $2L$, represents the *mean square error* of the approximation. Parseval's identity indicates that as $M \to \infty$, the mean square error approaches zero, while Bessels' inequality indicates the possibility that this mean square error does not approach zero.

The results are connected with the idea of *completeness* of an orthonormal set. If, for example, we were to leave out one or more terms in a Fourier series ($\cos 4\pi x/L$, for example), we could never get the mean square error to approach zero no matter how many terms we took. For an analogy with three-dimensional vectors, see Problem 13.60.

Differentiation and integration of Fourier series

13.16. (a) Find a Fourier series for $f(x) = x^2$, $0 < x < 2$, by integrating the series of Problem 13.12(a). (b) Use (a) to evaluate the series $\sum_{n=1}^{\infty} \frac{(-1)^{n-1}}{n^2}$.

(a) From Problem 13.12(a).

$$x = \frac{4}{\pi}\left(\sin\frac{\pi x}{2} - \frac{1}{2}\sin\frac{2\pi x}{2} + \frac{1}{3}\sin\frac{3\pi x}{2} - \cdots\right) \tag{1}$$

Integrating both sides from 0 to x (applying the theorem of Page 352) and multiplying by 2, we find

$$x^2 = C = \frac{16}{\pi^2}\left(\cos\frac{\pi x}{2} - \frac{1}{2^2}\cos\frac{2\pi x}{2} + \frac{1}{3^2}\cos\frac{3\pi x}{2} - \cdots\right) \tag{2}$$

where $x^2 = C = \frac{16}{\pi^2}\left(1 - \frac{1}{2^2} + \frac{1}{3^2} - \frac{1}{4^2} + \cdots\right)$

(b) To determine C in another way, note that Equation (2) represents the Fourier cosine series for x^2 in $0 < x < 2$. Then, since $L = 2$ in this case,

$$C = \frac{a_0}{2} = \frac{1}{L}\int_0^L f(x) = \frac{1}{2}\int_0^2 x^2\,dx = \frac{4}{3}$$

Then, from the value of C in (a), we have

$$\sum_{n=1}^{\infty} \frac{(-1)^{n-1}}{n^2} = 1 - \frac{1}{2^2} + \frac{1}{3^2} - \frac{1}{4^2} + \cdots = \frac{\pi^2}{16}\cdot\frac{4}{3} = \frac{\pi^2}{12}$$

13.17. Show that term-by-term differentiation of the series in Problem 13.12(a) is not valid.

Term-by-term differentiation yields $2\left(\cos\frac{\pi x}{2} - \cos\frac{2\pi x}{2} + \cos\frac{3\pi x}{2} - \cdots\right)$. Since the nth term of this series does not approach 0, the series does not converge for any value of x.

Convergence of Fourier series

13.18. Prove that

(a) $$\frac{1}{2} + \cos t + \cos 2t + \cdots + \cos Mt = \frac{\sin\left(M + \frac{1}{2}\right)t}{2\sin\frac{1}{2}t}$$

(b) $$\frac{1}{\pi}\int_0^{\pi} \frac{\sin\left(M + \frac{1}{2}\right)t}{2\sin\frac{1}{2}t}\,dt = \frac{1}{2}, \qquad \frac{1}{\pi}\int_{-\pi}^{0} \frac{\sin\left(M + \frac{1}{2}\right)t}{2\sin\frac{1}{2}t}\,dt = \frac{1}{2}$$

(a) We have $\cos nt \sin\frac{1}{2}t = \frac{1}{2}\left\{\sin\left(n+\frac{1}{2}\right)t - \sin\left(n-\frac{1}{2}\right)t\right\}$.

Then, summing from $n = 1$ to M,

$$\sin\frac{1}{2}t\{\cos t + \cos 2t + \cdots + \cos Mt\}. = \left(\sin\frac{3}{2}t - \sin\frac{1}{2}t\right) + \left(\sin\frac{5}{2}t - \sin\frac{3}{2}t\right)$$
$$+ \cdots + (\sin\left(M+\frac{1}{2}\right)t - \sin\left(M - \sin\frac{1}{2}t\right)$$
$$= \frac{1}{2}\left\{\sin\left(M+\frac{1}{2}\right)t - \sin\frac{1}{2}t\right\}$$

On dividing by $\sin\frac{1}{2}t$ and adding $\frac{1}{2}$, the required result follows.

(b) Integrating the result in (a) from $-\pi$ to 0 and 0 to π, respectively, gives the required results, since the integrals of all the cosine terms are zero.

13.19. Prove that $\lim_{n\to\infty}\int_{-\pi}^{\pi} f(x)\sin nx\,dx = \lim_{n\to\infty}\int_{-\pi}^{\pi} f(x)\cos nx\,dx = 0$ if $f(x)$ is piecewise continuous.

This follows at once from Problem 13.15, since if the series $\frac{a_0^2}{2} + \sum_{n=1}^{\infty}(a_n^2 + b_n^2)$ is convergent, $\lim_{n\to\infty} a_n = \lim_{n\to\infty} b_n = 0$.

The result is sometimes called *Riemann's theorem*.

13.20. Prove that $\lim_{m\to\infty}\int_{-\pi}^{\pi} f(x)\sin\left(M+\frac{1}{2}\right)x\,dx = 0$ if $f(x)$ is piecewise continuous.

We have

$$\int_{-\pi}^{\pi} f(x)\sin\left(M+\frac{1}{2}\right)x\,dx = \int_{-\pi}^{\pi}\{f(x)\sin\frac{1}{2}x\}\cos Mx\,dx + \int_{-\pi}^{\pi}\{f(x)\cos\frac{1}{2}x\}\sin Mx\,dx$$

Then the required result follows at once by using the result of Problem 13.19, with $f(x)$ replaced by $f(x)\sin\frac{1}{2}x$ and $f(x)\cos\frac{1}{2}x$, respectively, which are piecewise continuous if $f(x)$ is.

The result can also be proved when the integration limits are a and b instead of $-\pi$ and π.

13.21. Assuming that $L = \pi$, i.e., that the Fourier series corresponding to $f(x)$ has period $2L = 2\pi$, show that

$$S_M(x) = \frac{a_0}{2} + \sum_{n=1}^{M}(a_n\cos nx + b_n\sin nx) = \frac{1}{\pi}\int_{-\pi}^{\pi} f(t+x)\frac{\sin\left(M+\frac{1}{2}\right)t}{2\sin\frac{1}{2}t}\,dt$$

Using the formulas for the Fourier coefficients with $L = \pi$, we have

$$a_n\cos nx + b_n\sin nx = \left(\frac{1}{\pi}\int_{-\pi}^{\pi} f(u)\cos nu\,du\right)\cos nx + \left(\frac{1}{\pi}\int_{-\pi}^{\pi} f(u)\sin nu\,du\right)\sin nx$$
$$= \frac{1}{\pi}\int_{-\pi}^{\pi} f(u)(\cos nu\cos nx + \sin nu\sin nx)du$$
$$= \frac{1}{\pi}\int_{-\pi}^{\pi} f(u)\cos n(u-x)du$$

Also,

$$\frac{a_o}{2} = \frac{1}{2\pi}\int_{-\pi}^{\pi} f(u)du$$

Then

$$
\begin{aligned}
S_M(x) &= \frac{a_0}{2} + \sum_{n=1}^{M}(a_n \cos nx + b_n \sin nx) \\
&= \frac{1}{2\pi}\int_{-\pi}^{\pi} f(u)du + \frac{1}{\pi}\sum_{n=1}^{M}\int_{-\pi}^{\pi} f(u)\cos n(u-x)du \\
&= \frac{1}{\pi}\int_{-\pi}^{\pi} f(u)\left\{\frac{1}{2} + \sum_{n=1}^{M}\cos n(u-x)\right\} du \\
&= \frac{1}{\pi}\int_{-\pi}^{\pi} f(u)\frac{\sin\left(M+\frac{1}{2}\right)(u-x)}{2\sin\frac{1}{2}(u-x)} du
\end{aligned}
$$

using Problem 13.18. Letting $u - x = t$, we have

$$S_M(x) = \frac{1}{\pi}\int_{-\pi-x}^{\pi-x} f(t+x)\frac{\sin\left(M+\frac{1}{2}\right)t}{2\sin\frac{1}{2}t} dt$$

Since the integrand has period 2π, we can replace the interval $-\pi - x, \pi - x$ by any other interval of length 2π, in particular, $-\pi, \pi$. Thus, we obtain the required result.

13.22. Prove that

$$
\begin{aligned}
S_M(x) - \left(\frac{(f(x+0)+f(x-0))}{2}\right) &= \frac{1}{\pi}\int_{-\pi}^{0}\left(\frac{(f(t+x)-f(x-0))}{2\sin\frac{1}{2}t}\right)\sin\left(M+\frac{1}{2}\right)t\,dt \\
&\quad + \frac{1}{\pi}\int_{0}^{\pi}\left(\frac{(f(t+x)-f(x+0))}{2\sin\frac{1}{2}t}\right)\sin\left(M+\frac{1}{2}\right)t\,dt
\end{aligned}
$$

From Problem 13.21,

$$S_M(x) = \frac{1}{\pi}\int_{-\pi}^{0} f(t+x)\frac{\sin\left(M+\frac{1}{2}\right)t}{2\sin\frac{1}{2}t} dt + \frac{1}{\pi}\int_{0}^{\pi} f(t+x)\frac{\sin\left(M+\frac{1}{2}\right)t}{2\sin\frac{1}{2}t} dt \tag{1}$$

Multiplying the integrals of Problem 13.18(b) by $f(x - 0)$ and $f(x + 0)$, respectively,

$$\frac{f(x+0)+f(x-0)}{2} = \frac{1}{\pi}\int_{-\pi}^{0} f(x-0)\frac{\sin\left(M+\frac{1}{2}\right)t}{2\sin\frac{1}{2}t} dt + \frac{1}{\pi}\int_{-\pi}^{0} f(x+0)\frac{\sin\left(M+\frac{1}{2}\right)t}{2\sin\frac{1}{2}t} dt \tag{2}$$

Subtracting Equation (2) from Equation (1) yields the required result.

13.23. If $f(x)$ and $f'(x)$ are piecewise continuous in $(-\pi, \pi)$, prove that

$$\lim_{M\to\infty} S_M(x) = \frac{f(x+0)+f(x-0)}{2}$$

The function $\dfrac{f(t+x)-f(x+0)}{2\sin\frac{1}{2}t}$ is piecewise continuous in $0 < t \leq \pi$ because $f(x)$ is piecewise continuous.

Also, $\lim_{t\to 0+} \frac{f(t+x)-f(x+0)}{2\sin\frac{1}{2}t} = \lim_{t\to 0+} \frac{f(t+x)-f(x+0)}{t} \cdot \frac{t}{2\sin\frac{1}{2}t} = \lim_{t\to 0+} \frac{f(t+x)-f(x+0)}{t}$ exists,

since, by hypothesis, $f'(x)$ is piecewise continuous so that the right-hand derivative of $f(x)$ at each x exists.

Thus, $\frac{f(t+x)-f(x-0)}{2\sin\frac{1}{2}t}$ is piecewise continuous in $0 \leqq t \leqq \pi$.

Similarly, $\frac{f(t+x)-f(x-0)}{2\sin\frac{1}{2}t}$ is piecewise continuous in $-\pi \leqq t \leqq 0$.

Then, from Problems 13.20 and 13.22, we have

$$\lim_{M\to\infty} S_M(x) - \left\{\frac{f(x+0)-f(x-0)}{2}\right\} = 0 \quad \text{or} \quad \lim_{M\to\infty} S_M(x) = \left\{\frac{f(x+0)+f(x-0)}{2}\right\}$$

Boundary value problems

13.24. Find a solution $U(x, t)$ of the boundary value problem

$$\frac{\partial U}{\partial t} = 3\frac{\partial^2 U}{\partial x^2} \qquad t > 0, 0 < x < 2$$

$$U(0,t) = 0, U(2,t) = 0 \qquad t > 0$$

$$U(x,0) = x \qquad 0 < x < 2$$

A method commonly employed in practice is to assume the existence of a solution of the partial differential equation having the particular form $U(x, t) = X(x)\, T(t)$, where $X(x)$ and $T(t)$ are functions of x and t, respectively, which we shall try to determine. For this reason, the method is often called the method of *separation of variables*.

Substitution in the differential equation yields

$$\frac{\partial U}{\partial t}(XT) = 3\frac{\partial^2}{\partial x^2}(XT) \tag{1}$$

or

$$X\frac{dT}{dt} = 3T\frac{d^2 X}{dx^2} \tag{2}$$

where we have written X and T in place of $X(x)$ and $T(t)$.

Equation (2) can be written as

$$\frac{1}{3T}\frac{dT}{dt} = \frac{1}{X}\frac{d^2 X}{dx^2} \tag{3}$$

Since one side depends only on t and the other only on x, and since x and t are independent variables, it is clear that each side must be a constant c.

In Problem 13.47 we see that if $c \geqq 0$, a solution satisfying the given boundary conditions cannot exist.

Let us thus assume that c is a negative constant, which we write as $-\lambda^2$. Then, from Equation (3), we obtain two ordinary differentiation equations

$$\frac{dT}{dt} + 3\lambda^2 T = 0, \qquad \frac{d^2 X}{dx^2} + \lambda^2 X = 0 \tag{4}$$

whose solutions are, respectively,

$$T = C_1 e^{-3\lambda^2 t}, \quad X = A_1 \cos \lambda x + B_1 \sin \lambda x \tag{5}$$

A solution is given by the product of X and T, which can be written

$$U(x,t) = e^{-3\lambda^2 t}\,(A \cos \lambda x + B \sin \lambda x) \tag{6}$$

where A and B are constants.

We now seek to determine A and B so that Equation (6) satisfies the given boundary conditions. To satisfy the condition $U(0, t) = 0$, we must have

$$e^{-s\lambda^2 t}\,(A) = 0 \quad \text{or} \quad A = 0 \tag{7}$$

so that Equation (6) becomes

$$U(x,t) = Be^{-3\lambda^2 t} \sin \lambda x \tag{8}$$

To satisfy the condition $U(2, t) = 0$, we must then have

$$Be^{-s\lambda^2 t} \sin 2\lambda = 0 \tag{9}$$

Since $B = 0$ makes the solution (8) identically zero, we avoid this choice and instead take

$$\sin 2\lambda = 0, \quad \text{i.e.,} \quad 2\lambda = m\pi \quad \text{or} \quad \lambda = \frac{m\pi}{2} \tag{10}$$

where $m = 0, \pm 1, \pm 2, \ldots$.

Substitution in Equation (8) now shows that a solution satisfying the first two boundary conditions is

$$U(x,t) = B_m e^{-3m^2\pi^2 t/4} \sin \frac{m\pi x}{2} \tag{11}$$

where we have replaced B by B_m, indicating that different constants can be used for different values of m.

If we now attempt to satisfy the last boundary condition $U(x, 0) = x$, $0 < x < 2$, we find it to be impossible using Equation (11). However, upon recognizing the fact that *sums* of solutions having the form (11) are also solutions (called the *principle of superposition*), we are led to the possible solution

$$U(x,t) = \sum_{m=1}^{\infty} B_m e^{-3m^2\pi^2 t/4} \sin \frac{m\pi x}{2} \tag{12}$$

From the condition $U(x, 0) = x$, $0 < x < 2$, we see, on placing $t = 0$, that Equation (12) becomes

$$x = \sum_{m=1}^{\infty} B_m \sin \frac{m\pi x}{2} \qquad 0 < x < 2 \tag{13}$$

This, however, is equivalent to the problem of expanding the function $f(x) = x$ for $0 < x < 2$ into a sine series. The solution to this is given in Problem 13.12(a), from which we see that $B_m = \frac{-4}{m\pi} \cos m\pi$ so that Equation (12) becomes

$$U(x,t) = \sum_{m=1}^{\infty} \left(-\frac{4}{m\pi} \cos m\pi \right) e^{-3m^2\pi^2 t/4} \sin \frac{m\pi x}{2} \tag{14}$$

which is a *formal solution*. To check that Equation (14) is actually a solution, we must show that it satisfies the partial differential equation and the boundary conditions. The proof consists in justification of term-by-term differentiation and use of limiting procedures for infinite series and may be accomplished by methods of Chapter 11.

The boundary value problem considered here has an interpretation in the theory of heat conduction. The equation $\frac{\partial U}{\partial t} = k \frac{\partial^2 U}{\partial x^2}$ is the equation for heat conduction in a thin rod or wire located on the x axis between $x = 0$ and $x = L$ if the surface of the wire is insulated so that heat cannot enter or escape. $U(x, t)$ is the temperature at any place x in the rod at time t. The constant $k = K/s\rho$ (where K is the *thermal conductivity*, s is the *specific heat*, and ρ is the *density* of the conducting material) is called the *diffusivity*. The boundary conditions $U(0, t) = 0$ and $U(L, t) = 0$ indicate that the end temperatures of the rod are kept at zero units for all time $t > 0$, while $U(x, 0)$ indicates the initial temperature at any point x of the rod. In this problem the length of the rod is $L = 2$ units, while the diffusivity is $k = 3$ units.

Orthogonal functions

13.25. (a) Show that the set of functions

$$1, \sin\frac{\pi x}{L}, \cos\frac{\pi x}{L}, \sin\frac{2\pi x}{L}, \cos\frac{2\pi x}{L}, \sin\frac{3\pi x}{L}, \cos\frac{3\pi x}{L}, \cdots$$

forms an orthogonal set in the interval $(-L, L)$.

(b) Determine the corresponding normalizing constants for the set in (a) so that the set is orthonormal in $(-L, L)$.

(a) This follows at once from the results of Problems 13.2 and 13.3.

(b) By Problem 13.3,

$$\int_{-L}^{L} \sin^2\frac{m\pi x}{L}\,dx = L, \quad \int_{-L}^{L} \cos^2\frac{m\pi x}{L}\,dx = L$$

Then

$$\int_{-L}^{L}\left(\sqrt{\frac{1}{L}}\sin\frac{m\pi x}{L}\right)^2 dx = 1 \quad \int_{-L}^{L}\left(\sqrt{\frac{1}{L}}\cos\frac{m\pi x}{L}\right)^2 dx = 1$$

Also,

$$\int_{-L}^{L}(1)^2 dx = 2L \quad \text{or} \quad \int_{-L}^{L}\left(\frac{1}{\sqrt{2L}}\right)^2 dx = 1$$

Thus, the required orthonormal set is given by

$$\frac{1}{\sqrt{2L}}, \frac{1}{\sqrt{L}}\sin\frac{\pi x}{L}, \frac{1}{\sqrt{L}}\cos\frac{\pi x}{L}, \frac{1}{\sqrt{L}}\sin\frac{2\pi x}{L}, \frac{1}{\sqrt{L}}\cos\frac{2\pi x}{L}, \ldots$$

Miscellaneous problems

13.26. Find a Fourier series for $f(x) = \cos\alpha x$, $-\pi \leqq x \leqq \pi$, where $\alpha \neq 0, \pm 1, \pm 2, \pm 3, \ldots$.

We shall take the period as 2π so that $2L = 2\pi$, $L = \pi$. Since the function is even, $b_n = 0$ and

$$\begin{aligned} a_n &= \frac{2}{L}\int_0^L f(x)\cos nx\,dx = \frac{2}{\pi}\int_0^\pi \cos\alpha x\cos nx\,dx \\ &= \frac{1}{\pi}\int_0^\pi \{\cos(\alpha - n)x + \cos(\alpha + n)x\}dx \\ &= \frac{1}{\pi}\left\{\frac{\sin(\alpha - n)\pi}{\alpha - n} + \frac{\sin(\alpha + n)\pi}{\alpha + n} + \cdots\right\} = \frac{2\alpha\sin\alpha\pi\cos n\pi}{\pi(\alpha^2 - n^2)} \\ \alpha_0 &= \frac{2\sin\alpha\pi}{\alpha\pi} \end{aligned}$$

Then

$$\begin{aligned} \cos\alpha x &= \frac{\sin\alpha\pi}{\alpha\pi} + \frac{2\alpha\sin\alpha\pi}{\pi}\sum_{n=1}^{\infty}\frac{\cos n\pi}{\alpha^2 - 2^2}\cos nx \\ &= \frac{\sin\alpha\pi}{\pi}\left(\frac{1}{\alpha} - \frac{2\alpha}{\alpha^2 - 1^2}\cos x + \frac{2\alpha}{\alpha^2 - 2^2}\cos 2x - \frac{2\alpha}{\alpha^2 - 3^2}\cos 3x + \cdots\right) \end{aligned}$$

13.27. Prove that $\sin x = x\left(1 - \frac{x^2}{\pi^2}\right)\left(1 - \frac{x^2}{(2\pi)^2}\right)\left(1 - \frac{x^2}{(3\pi)^2}\right)\cdots$

Let $x = \pi$ in the Fourier series obtained in Problem 13.26. Then

$$\cos\alpha = \frac{\sin\alpha\pi}{\pi}\left(\frac{1}{\alpha}+\frac{2\alpha}{\alpha^2-1^2}+\frac{2\alpha}{\alpha^2-2^2}+\frac{2\alpha}{\alpha^2-3^2}+\cdots\right)$$

or

$$\pi\cot\alpha\pi - \frac{1}{\alpha} = \frac{2\alpha}{\alpha^2-1^2}+\frac{2\alpha}{\alpha^2-2^2}+\frac{2\alpha}{\alpha^2-3^2}+\cdots \tag{1}$$

This result is of interest since it represents an expansion of the contangent into partial fractions.

By the Weierstrass M test, the series on the right of Equation (1) converges uniformly for $0 \leq |\alpha| \leq |x| < 1$ and the left-hand side of (1) approaches zero as $\alpha \to 0$, as is seen by using L'Hospital's rule. Thus, we can integrate both sides of (1) from 0 to x to obtain

$$\int_0^x \left(\pi\cot\alpha\pi - \frac{1}{\alpha}\right)d\alpha = \int_0^x \frac{2\alpha}{\alpha^2-1}\,d\alpha + \int_0^x \frac{2\alpha}{\alpha^2-2^2}\,d\alpha + \cdots$$

or

$$\ln\left(\frac{\sin\alpha\pi}{\alpha\pi}\right)\Bigg|_0^x = \ln\left(1-\frac{x^2}{1^2}\right)+\ln\left(1-\frac{x^2}{2^2}\right)+\cdots$$

i.e.,

$$\ln\left(\frac{\sin\pi}{\pi x}\right) = \lim_{n\to\infty}\ln\left(1-\frac{x^2}{1^2}\right)+\ln\left(1-\frac{x^2}{2^2}\right)\Bigg|+\cdots+\ln\left(1-\frac{x^2}{n^2}\right)$$

$$= \lim_{n\to\infty}\ln\left\{\left(1-\frac{x^2}{1^2}\right)\left(1-\frac{x^2}{2^2}\right)\cdots\left(1-\frac{x^2}{n^2}\right)\right\}$$

$$= \ln\left\{\lim_{n\to\infty}\left(1-\frac{x^2}{1^2}\right)\left(1-\frac{x^2}{2^2}\right)\cdots\left(1-\frac{x^2}{n^2}\right)\right\}$$

so that

$$\frac{\sin\pi}{\pi x} = \lim_{n\to\infty}\left(1-\frac{x^2}{1^2}\right)\left(1-\frac{x^2}{2^2}\right)\cdots\left(1-\frac{x^2}{n^2}\right) = \left(1-\frac{x^2}{1^2}\right)\left(1-\frac{x^2}{2^2}\right)\cdots \tag{2}$$

Replacing x by x/π, we obtain

$$\sin x = x\left(1-\frac{x^2}{\pi^2}\right)\left(1-\frac{x^2}{(2\pi)^2}\right)\cdots \tag{3}$$

called the *infinite product for* $\sin x$, which can be shown valid for all x. The result is of interest since it corresponds to a factorization of $\sin x$ in a manner analogous to factorization of a polynomial.

13.28. Prove that $\dfrac{\pi}{2} = \dfrac{2\cdot 2\cdot 4\cdot 4\cdot 6\cdot 6\cdot 8\cdot 8\cdots}{1\cdot 3\cdot 3\cdot 5\cdot 5\cdot 7\cdot 7\cdot 9\cdots}$.

Let $x = 1/2$ in Equation (2) of Problem 13.27. Then,

$$\frac{2}{\pi} = \left(1-\frac{1}{2^2}\right)\left(1-\frac{1}{4^2}\right)\left(1-\frac{1}{6^2}\right)\cdots = \left(\frac{1}{2}\cdot\frac{3}{2}\right)\left(\frac{3}{4}\cdot\frac{5}{4}\right)\left(\frac{5}{6}\cdot\frac{7}{6}\right)\cdots$$

Taking reciprocals of both sides, we obtain the required result, which is often called *Wallis's product*.

SUPPLEMENTARY PROBLEMS

Fourier Series

13.29. Graph each of the following functions and find their corresponding Fourier series using properties of even and odd functions wherever applicable.

(a) $f(x)=\begin{cases} 8 & 0<x<2 \\ -8 & 2<x<4 \end{cases}$ Period 4 (c) $f(x)=4x, 0<x<10$, Period 10

(b) $f(x)=\begin{cases} -x & -4 \leqq x \leqq 0 \\ x & 0 \leqq x \leqq 4 \end{cases}$ Period 8 (d) $f(x)=\begin{Bmatrix} 2x & 0<x<3 \\ 0 & -3<x<0 \end{Bmatrix}$ Period 6

Ans. (a) $\dfrac{16}{\pi}\sum_{n=1}^{\infty}\dfrac{(1-\cos n\pi)}{n}\sin\dfrac{n\pi x}{2}$ (c) $20-\dfrac{40}{\pi}\sum_{n=1}^{\infty}\dfrac{1}{n}\sin\dfrac{n\pi x}{5}$

(b) $2-\dfrac{8}{\pi^2}\sum_{n=1}^{\infty}\dfrac{(1-\cos n\pi)}{n^2}\cos\dfrac{n\pi x}{4}$ (d) $\dfrac{3}{2}\sum_{n=1}^{\infty}+\left\{\dfrac{6(\cos n\pi-1)}{n^2\pi^2}\cos\dfrac{n\pi x}{3}-\dfrac{6\cos n\pi}{n\pi}\sin\dfrac{n\pi x}{3}\right\}$

13.30. In each part of Problem 13.29, tell where the discontinuities of $f(x)$ are located and to what value the series converges at the discontunities.

Ans. (a) $x = 0, \pm 2, \pm 4, \ldots; 0$ (c) $x = 0, \pm 10, \pm 20, \ldots; 20$
(b) no discontinuities (d) $x = \pm 3, \pm 9, \pm 15, \ldots; 3$

13.31. Expand $f(x)=\begin{cases} 2-x & 0<x<4 \\ x-6 & 4<x<8 \end{cases}$ in a Fourier series of period 8.

Ans. $\dfrac{16}{\pi^2}\left\{\cos\dfrac{\pi x}{4}+\dfrac{1}{3^2}\cos\dfrac{3\pi x}{4}+\dfrac{1}{5^2}\cos\dfrac{5\pi x}{4}+\cdots\right\}$

13.32. (a) Expand $f(x) = \cos x$, $0 < x < \pi$, in a Fourier sine series. (b) How should f(x) be defined at x = 0 and x = π so that the series will converge to f(x) for $0 \leqq x \leqq \pi$?

Ans. (a) $\dfrac{8}{\pi}\sum_{n=1}^{\infty}\dfrac{n\sin 2nx}{4n^2-1}$ *(b)* $f(0)=f(\pi)=0$

13.33. (a) Expand in a Fourier series $f(x) = \cos x$, $0 < x < \pi$ if the period is π, and (b) compare with the result of Problem 13.32, explaining the similarities and differences, if any.

Ans. Answer is the same as in Problem 13.32.

13.34. Expand $f(x)=\begin{cases} x & 0<x<4 \\ 8-x & 4<x<8 \end{cases}$ in a series of (a) sines and (b) cosines.

Ans. (a) $\dfrac{32}{\pi^2}\sum_{n=1}^{\infty}\dfrac{1}{n^2}\sin\dfrac{n\pi}{2}\sin\dfrac{n\pi x}{8}$ (b) $\dfrac{16}{\pi^2}\sum_{n=1}^{\infty}\left(\dfrac{2\cos n\pi/2-\cos n\pi-1}{n^2}\right)\cos\dfrac{n\pi x}{8}$

13.35. Prove that for $0 \leq x \leq \pi$,

(a) $x(\pi-x)=\dfrac{\pi^2}{6}-\left(\dfrac{\cos 2x}{1^2}+\dfrac{\cos 4x}{2^2}+\dfrac{\cos 6x}{3^2}+\cdots\right)$

(b) $x(\pi-x)=\dfrac{8}{\pi}\left(\dfrac{\sin x}{1^3}+\dfrac{\sin 3x}{3^3}+\dfrac{\sin 5x}{5^3}+\cdots\right)$

13.36. Use the preceding problem to show that

(a) $\sum_{n=1}^{\infty}\dfrac{1}{n^2}=\dfrac{\pi^2}{6}$ (b) $\sum_{n=1}^{\infty}\dfrac{(-1)^{n-1}}{n^2}=\dfrac{\pi^2}{12}$ (c) $\sum_{n=1}^{\infty}\dfrac{(-1)^{n-1}}{(2n-1)^3}=\dfrac{\pi^3}{32}$

13.37. Show that $\frac{1}{1^3}+\frac{1}{3^3}-\frac{1}{5^3}-\frac{1}{7^3}+\frac{1}{9^3}+\frac{1}{11^3}-\cdots=\frac{3\pi^2\sqrt{2}}{16}$.

Differentiation and integration of Fourier series

13.38. (a) Show that for $-\pi < x < \pi$,

$$x = 2\left(\frac{\sin x}{1}-\frac{\sin 2x}{2}+\frac{\sin 3x}{3}-\cdots\right)$$

(b) By integrating the result of (a), show that for $-\pi \leqq x \leqq \pi$,

$$x^2 = \frac{\pi^2}{3}-4\left(\frac{\cos x}{1^2}-\frac{\cos 2x}{2^2}+\frac{\sin 3x}{3^2}-\cdots\right)$$

(c) By integrating the result of (b), show that for $-\pi \leqq x \leqq \pi$,

$$x(\pi - x)(\pi + x) = 12\left(\frac{\sin x}{1^3}-\frac{\sin 2x}{2^3}+\frac{\sin 3x}{3^3}-\cdots\right)$$

13.39. (a) Show that for $-\pi < x < \pi$,

$$x\cos x = -\frac{1}{2}\sin x + 2\left(\frac{2}{1\cdot 3}\sin 2x-\frac{3}{2\cdot 4}\sin 3x+\frac{4}{3\cdot 5}\sin 4x-\cdots\right)$$

(b) Use (a) to show that for $-\pi \leqq x \leqq \pi$,

$$x\sin x = 1-\frac{1}{2}\cos x - 2\left(\frac{\cos 2x}{1.3}-\frac{\cos 3x}{2.4}+\frac{\cos 4x}{3.5}-\cdots\right)$$

13.40. By differentiating the result of Problem 13.35(b), prove that for $0 \leqq x \leqq \pi$,

$$x = \frac{\pi}{2}-\frac{4}{\pi}\left(\frac{\cos x}{1^2}+\frac{\cos 3x}{3^2}+\frac{\cos 5x}{5^2}+\cdots\right)$$

Parseval's identity

13.41. By using Problem 13.35 and Parseval's identity, show that

$$(a)\ \sum_{n=1}^{\infty}\frac{1}{n^4}=\frac{\pi^4}{90} \qquad (b)\ \sum_{n=1}^{\infty}\frac{1}{n^6}=\frac{\pi^6}{945}$$

13.42. Show that $\frac{1}{1^2\cdot 3^2}+\frac{1}{3^2\cdot 5^2}+\frac{1}{5^2\cdot 7^2}+\cdots=\frac{\pi^2-8}{16}$. (Hint: Use Problem 13.11.)

13.43. Show that

(a) $\sum_{n=1}^{\infty}\frac{1}{(2n-1)^4}=\frac{\pi^4}{96}$

(b) $\sum_{n=1}^{\infty}\frac{1}{(2n-1)^6}=\frac{\pi^6}{960}$

13.44. Show that $\frac{1}{1^2\cdot 2^2\cdot 3^2}+\frac{1}{2^2\cdot 3^2\cdot 4^2}+\frac{1}{3^2\cdot 4^2\cdot 5^2}+\cdots=\frac{4\pi^2-39}{16}$

Boundary value problems

13.45. (a) Solve $\dfrac{\partial U}{\partial t} = 2\dfrac{\partial^2 U}{\partial x^2}$, subject to the conditions $U(0, t) = 0$, $U(4, t) = 0$, $U(x, 0) = 3 \sin \pi x - 2 \sin 5\pi x$, where $0 < x < 4$, $t > 0$. (b) Give a possible physical interpretation of the problem and solution.

Ans. (a) $U(x, t) = 3e^{-2\pi^2 t} \sin \pi x - 2e^{-50\pi^2 t} \sin 5\pi x$

13.46. Solve $\dfrac{\partial U}{\partial t} = \dfrac{\partial^2 U}{\partial x^2}$, subject to the conditions $U(0,t) = 0, U(6,t) = 0, U(x,0) = \begin{cases} 1 & 0 < x < 3 \\ 0 & 3 < x < 6 \end{cases}$, and interpret and interpret physically.

$$Ans.\ U(x,t) = \sum_{m=1}^{\infty} 2\left[\frac{1-\cos(m\pi/3)}{m\pi}\right] e^{-m^2\pi^2 t/36} \sin\frac{m\pi x}{6}$$

13.47. Show that if each side of Equation (3), Page 369, is a constant c, where $c \geqq 0$, then there is no solution satisfying the boundary value problem.

13.48. A flexible string of length π is tightly stretched between points $x = 0$ and $x = = \pi$ on the x axis, its ends are fixed at these points. When set into small transverse vibration, the displacement $Y(x, t)$ from the the x axis of any point x at time t is given by $\dfrac{\partial^2 Y}{\partial t^2} = a^2 \dfrac{\partial^2 Y}{\partial x^2}$, where $a^2 = T/\rho$, T = tension, ρ = mass per unit length.

(a) Find a solution of this equation (sometimes called the wave equation) with $a^2 = 4$ which satisfies the conditions $Y(0, t) = 0$, $Y(\pi, t) = 0$, $Y(x, 0) = 0.1 \sin x + 0.01 \sin 4x$, $Yt(x, 0) = 0$ for $0 < x < \pi$, $t > 0$.

(b) Interpret physically the boundary conditions in (a) and the solution.

Ans. (a) $Y(x, t) = 0.1 \sin x \cos 2t + 0.01 \sin 4x \cos 8t$

13.49. (a) Solve the boundary value problem $\dfrac{\partial^2 Y}{\partial t^2} = 9\dfrac{\partial^2 Y}{\partial x^2}$, subject to the conditions $Y(0, t) = 0$, $Y(2, t) = 0$, $Y(x, 0) = 0.05x(2 - x)$, $Y_t(x, 0) = 0$, where $0 < x < 2$, $t > 0$. (b) Interpret physically.

$$Ans.\ (a)\ Y(x,t) = \frac{1.6}{\pi^3}\sum_{n=1}^{\infty}\frac{1}{(2n-1)^3}\sin\frac{(2n-1)\pi x}{2}\cos\frac{3(2n-1)\pi t}{2}$$

13.50. Solve the boundary value problem $\dfrac{\partial U}{\partial t} = \dfrac{\partial^2 U}{\partial x^2}$, $U(0, t) = 1$, $U(\pi, t) = 3$, $U(x, 0) = 2$. [Hint: Let $U(x, t) = V(x, t) + F(x)$ and choose $F(x)$ so as to simplify the differential equation and boundary conditions for $V(x, t)$.]

$$Ans.\ U(x,t) = 1 + \frac{2x}{\pi} + \sum_{m=1}^{\infty}\frac{4\cos m\pi}{m\pi} e^{-m^2 t} \sin mx$$

13.51. Give a physical interpretation to Problem 13.50.

13.52. Solve Problem 13.49 with the boundary conditions for $Y(x, 0)$ and $Y_t(x, 0)$ interchanged; i.e., $Y(x,) = 0$, $Y_t(x, 0) = 0.05x(2 - x)$, and give a physical interpretation.

$$Ans.\ Y(x,t) = \frac{3.2}{3\pi^4}\sum_{n=1}^{\infty}\frac{1}{(2n-1)^4}\sin\frac{(2n-1)\pi x}{2}\sin\frac{3(2n-1)\pi t}{2}$$

13.53. Verify that the boundary value problem of Problem 13.24 actually has the solution (14), Page 370.

Miscellaneous Problems

13.54. If $-\pi < x < \pi$ and $\alpha \neq 0, \pm 1, \pm 2, \ldots$, prove that

$$\frac{\pi \sin \alpha x}{2 \sin \alpha \pi} = \frac{\sin x}{1^2 - \alpha^2} - \frac{2 \sin 2x}{2^2 - \alpha^2} + \frac{3 \sin 3x}{3^2 - \alpha^2} - \cdots$$

13.55. If $-\pi < x < \pi$, prove that

(a) $$\frac{\pi \sinh \alpha x}{2 \sinh \alpha \pi} = \frac{\sin x}{\alpha^2 + 1^2} - \frac{2 \sin 2x}{\alpha^2 + 2^2} + \frac{3 \sin 3x}{\alpha^2 + 3^2} - \cdots$$

(b) $$\frac{\pi \cosh \alpha x}{2 \sinh \alpha \pi} = \frac{1}{2\alpha} - \frac{\alpha \cos x}{\alpha^2 + 1^2} + \frac{\alpha \cos 2x}{\alpha^2 + 2^2} - \cdots$$

13.56. Prove that $\sinh x = x\left(1 + \frac{x^2}{\pi^2}\right)\left(1 + \frac{x^2}{(2\pi)^2}\right)\left(1 + \frac{x^2}{(3\pi)^2}\right)\cdots$

13.57. Prove that $\cos x = \left(1 - \frac{4x^2}{\pi^2}\right)\left(1 - \frac{4x^2}{(3\pi)^2}\right)\left(1 - \frac{4x^2}{(5\pi)^2}\right)\cdots$. [Hint: $\cos x = (\sin 2x)/(2 \sin x)$.]

13.58. Show that

(a) $$\frac{\sqrt{2}}{\sqrt{2}} = \frac{1 \cdot 3 \cdot 5 \cdot 7 \cdot 9 \cdot 22 \cdot 13 \cdot 15 \ldots}{2 \cdot 2 \cdot 6 \cdot 6 \cdot 10 \cdot 10 \cdot 14 \cdot 14 \ldots}$$

(b) $$\pi\sqrt{2} = 4\left(\frac{4 \cdot 4 \cdot 8 \cdot 8 \cdot 12 \cdot 12 \cdot 16 \cdot 16 \ldots}{3 \cdot 5 \cdot 7 \cdot 9 \cdot 11 \cdot 13 \cdot 15 \cdot 17 \ldots}\right)$$

13.59. Let **r** be any three-dimensional vector. Show that (a) $(\mathbf{r} \cdot \mathbf{i})^2 + (\mathbf{r} \cdot \mathbf{j})^2 \leqq (\mathbf{r})^2$ and (b) $(\mathbf{r} \cdot \mathbf{i})^2 + (\mathbf{r} \cdot \mathbf{j})^2 + (\mathbf{r} \cdot \mathbf{k})^2 = \mathbf{r}^2$ and discuss these with reference to Parseval's identity.

13.60. If $\{\phi_n(x)\}$, $n = 1, 2, 3, \ldots$ is orthonormal in $(a, b$, prove that $\int_a^b \left\{f(x) - \sum_{n=1}^{\infty} c_n \phi_n(x)\right\}^2 dx$ is a minimum when $c_n = \int_a^b f(x)\phi_n(x)dx$. Discuss the relevance of this result to Fourier series.

CHAPTER 14

Fourier Integrals

Fourier integrals are generalizations of Fourier series. The series representation $\frac{a_0}{2}+\sum_{n=1}^{\infty}\left\{a_n \cos\frac{n\pi x}{L}+b_n \sin\frac{n\pi x}{L}\right\}$ of a function is a periodic form on $-\infty < x < \infty$ obtained by generating the coefficients from the function's definition on the least period $[-L, L]$. If a function defined on the set of all real numbers has no period, then an analogy to Fourier integrals can be envisioned as letting $L \to \infty$ and replacing the integer valued index n by a real valued function α. The coefficients a_n and b_n then take the form $A(\alpha)$ and $B(\alpha)$. This mode of thought leads to the following definition. (See Problem 14.8.)

The Fourier Integral

Let us assume the following conditions on $f(x)$:

1. $f(x)$ satisfies the Dirichlet conditions (Page 350) in every finite interval $(-L, L)$.
2. $\int_{-\infty}^{\infty}|f(x)|\,dx$ converges; i.e., $f(x)$ is absolutely integrable in $(-\infty, \infty)$.

Then *Fourier's integral theorem* states that the Fourier integral of a function f is

$$f(x)=\int_0^{\infty}\{A(\alpha)\cos \alpha x+B(\alpha)\sin \alpha x\}\,d\alpha \tag{1}$$

where

$$\left\{\begin{aligned} A(\alpha)&=\frac{1}{\pi}\int_{-\infty}^{\infty} f(x)\cos \alpha x\,dx \\ B(\alpha)&=\frac{1}{\pi}\int_{-\infty}^{\infty} f(x)\sin \alpha x\,dx \end{aligned}\right. \tag{2}$$

$A(\alpha)$ and $B(\alpha)$ with $-\infty < \alpha < \infty$ are generalizations of the Fourier coefficients a_n and b_n. The right-hand side of Equation (1) is also called a *Fourier integral expansion of f*. (Since Fourier integrals are improper integrals, a review of Chapter 12 is a prerequisite to the study of this chapter.) The result (*1*) holds if x is a point of continuity of $f(x)$. If x is a point of discontinuity, we must replace $f(x)$ by $\frac{f(x+0)+f(x-0)}{2}$, as in the case of Fourier series. Note that these conditions are sufficient but not necessary.

In the generalization of Fourier coefficients to Fourier integrals, a_0 may be neglected, since whenever $\int_{-\infty}^{\infty} f(x)\,dx$ exists,

$$|a_0|=\left|\frac{1}{L}\int_{-L}^{L} f(x)\,dx\right|\to 0 \qquad as \qquad L\to\infty$$

Equivalent Forms of Fourier's Integral Theorem

Fourier's integral theorem can also be written in the forms

$$f(x)=\frac{1}{\pi}\int_{\alpha=0}^{\infty}\int_{u=-\infty}^{\infty} f(u)\cos\alpha(x-u)\,du\,d\alpha \tag{3}$$

$$\begin{aligned} f(x)&=\frac{1}{2\pi}\int_{-\infty}^{\infty} e^{i\alpha x}d\alpha\int_{-\infty}^{\infty} f(u)e^{i\alpha u}du \\ &=\frac{1}{2\pi}\int_{-\infty}^{\infty}\int_{-\infty}^{\infty} f(u)e^{i\alpha(u-x)}du\,d\alpha \end{aligned} \tag{4}$$

where it is understood that if $f(x)$ is not continuous at x, the left side must be replaced by $\dfrac{f(x+0)+f(x-0)}{2}$.

These results can be simplified somewhat if $f(x)$ is either an odd or an even function, and we have

$$f(x)=\frac{2}{\pi}\int_0^{\infty}\cos\alpha x\,dx\int_0^{\infty} f(u)\cos\alpha u\,du \qquad \text{if } f(x) \text{ is even} \tag{5}$$

$$f(x)=\frac{2}{\pi}\int_0^{\infty}\sin\alpha x\,dx\int_0^{\infty} f(u)\sin\alpha u\,du \qquad \text{if } f(x) \text{ is odd} \tag{6}$$

An entity of importance in evaluating integrals and solving differential and integral equations is introduced in the next paragraph. It is abstracted from the Fourier integral form of a function, as can be observed by putting Equation (4) in the form

$$f(x)=\frac{1}{\sqrt{2\pi}}\int_{-\infty}^{\infty} e^{-i\alpha x}\left\{\frac{1}{\sqrt{2\pi}}\int_{-\infty}^{\infty} e^{i\alpha u}f(u)du\right\}d\alpha$$

and identifying the parenthetic expression, as $F(\alpha)$. The following Fourier transforms result.

Fourier Transforms

From Equation (4) it follows that

$$F(\alpha)=\frac{1}{\sqrt{2\pi}}\int_{-\infty}^{\infty} f(u)e^{i\alpha u}du \tag{7}$$

then

$$F(x)=\frac{1}{\sqrt{2\pi}}\int_{-\infty}^{\infty} F(\alpha)e^{i\alpha x}d\alpha \tag{8}$$

The function $F(\alpha)$ is called the *Fourier transform* of $f(x)$ and is sometimes written $F(\alpha)=\mathcal{F}\{f(x)\}$. The function $f(x)$ is the *inverse Fourier transform* of $F(\alpha)$ and is written $f(x)=\mathcal{F}^{-1}\{F(\alpha)\}$.

Note: The constants preceding the integral signs in Equations (7) and (8) were here taken as equal to $1/\sqrt{2\pi}$. However, they can be any constants different from zero so long as their product is $1/2\pi$. This is called the *symmetric form*. The literature is not uniform as to whether the negative exponent appears in Equation (7) or in (8).

EXAMPLE. Determine the Fourier transform of f if $f(x)=e^{-x}$ for $x>0$ and e^{2x} when $x<0$.

$$F(\alpha)=\frac{1}{2\sqrt{2\pi}}\int_{-\infty}^{\infty} e^{i\alpha x}f(x)\,dx=\frac{1}{2\pi}\left\{\int_{-\infty}^{0} e^{i\alpha x}e^{2x}dx+\int_0^{\infty} e^{i\alpha x}e^{-x}dx\right\}$$

$$=\frac{1}{\sqrt{2\pi}}\left\{\frac{e^{i\alpha+2}}{i\alpha+2}\Bigg|_{x\to-\infty}^{x\to0-}+\frac{e^{i\alpha-1}}{i\alpha-1}\Bigg|_{x\to0+}^{x\to\infty}\right\}=\frac{1}{\sqrt{2\pi}}\left\{\frac{1}{2+\alpha i}+\frac{1}{1-\alpha i}\right\}$$

If $f(x)$ is an even function, Equation (5) yields

$$\begin{cases} F_c(\alpha)=\sqrt{\dfrac{2}{\pi}}\displaystyle\int_0^{\infty} f(u)\cos\alpha u\,du \\ f(x)=\sqrt{\dfrac{2}{\pi}}\displaystyle\int_0^{\infty} F_c(\alpha)\cos\alpha x\,dx \end{cases} \tag{9}$$

and we call $F_c(\alpha)$ and $f(x)$ *Fourier cosine transforms* of each other.

If $f(x)$ is an odd function, Equation (6) yields

$$\begin{cases} F_s(\alpha) = \sqrt{\dfrac{2}{\pi}} \displaystyle\int_0^\infty f(u) \sin \alpha u \, du \\ f(x) = \sqrt{\dfrac{2}{\pi}} \displaystyle\int_0^\infty F_s(\alpha) \sin \alpha x \, d\alpha \end{cases} \tag{10}$$

and we call $F_s(\alpha)$ and $f(x)$ *Fourier sine transforms* of each other.

Note: The Fourier transforms F_c and F_s are (up to a constant) of the same form as $A(\alpha)$ and $B(\alpha)$. Since f is even for F_c and odd for F_s, the domains can be shown to be $0 < \alpha < \infty$.

When the product of Fourier transforms is considered, a new concept called *convolution* comes into being, and in conjunction with it, a new pair (function and its Fourier transform) arises. In particular, if $F(\alpha)$ and $G(\alpha)$ are the Fourier transforms of f and g, respectively, and the convolution of f and g is defined to be

$$f * g = \frac{1}{\sqrt{\pi}} \int_{-\infty}^{\infty} f(u) g(x-u) \, du \tag{11}$$

then

$$F(\alpha)\, G(\alpha) = \frac{1}{\sqrt{\pi}} \int_{-\infty}^{\infty} e^{i\alpha u} f * g \, du \tag{12}$$

$$f * g = \frac{1}{\sqrt{\pi}} \int_{-\infty}^{\infty} e^{i\alpha x} F(\alpha)\, G(\alpha) \, d\alpha \tag{13}$$

where in both Equations (11) and (13) the convolution $f * g$ is a function of x.

It may be said that multiplication is exchanged with convolution. Also "the Fourier transform of the convolution of two functions, f and g is the product of their Fourier transforms," i.e.,

$$T(f * g) = G(f)\, T(g)$$

[$F(\alpha)\, G(\alpha)$ and $f * g$) are demonstrated to be a Fourier transform pair in Problem 14.29.]

Now equate the representations of $f * g$ expressed in Equations (11) and (13), i.e.,

$$\frac{1}{\pi} \int_{-\infty}^{\infty} f(u)\, g(x-u) \, du = \frac{1}{\sqrt{\pi}} \int_{-\infty}^{\infty} e^{iax} F(\alpha)\, G(\alpha) \, d\alpha \tag{14}$$

and let the parameter x be zero; then

$$\int_{-\infty}^{\infty} f(u)\, g(-u) \, du = \int_{-\infty}^{\infty} F(\alpha)\, G(\alpha) \, d\alpha \tag{15}$$

Now suppose that $g = \bar{f}$ and, thus, $G = \bar{F}$, where the bar symbolizes the complex conjugate function. Then Equation (15) takes the form

$$\int_{-\infty}^{\infty} | f(u) |^2 \, du = \int_{-\infty}^{\infty} | F(\alpha) |^2 \, d\alpha \tag{16}$$

This is Parseval's theorem for Fourier integrals.

Furthermore, if f and g are even functions, it can be shown that Equation (15) reduces to the following Parseval identities:

$$\int_0^\infty f(u)\, g(u) du = \int_0^\infty F_c(\alpha)\, G_c(\alpha) \, d\alpha \tag{17}$$

where F_c and G_c are the Fourier cosine transforms of f and g. If f and g are odd functions, then Equation (15) takes the form

$$\int_0^\infty f(u)\, g(u) du = \int_0^\infty F_s(\alpha)\, G_s(\alpha) \, d\alpha \tag{18}$$

where F_s and G_s are the Fourier sine transforms of f and g. (See Problem 14.3.)

SOLVED PROBLEMS

The Fourier integral and Fourier transforms

14.1. (a) Find the Fourier transform of $f(x)=\begin{cases}1 & |x|<a\\ 0 & |x|>a\end{cases}$. (b) Graph $f(x)$ and its Fourier transform for $a = 3$.

(a) The Fourier transform of $f(x)$ is

$$F(\alpha)=\frac{1}{\sqrt{2\pi}}\int_{-\infty}^{\infty}f(u)e^{i\alpha u}\,du=\frac{1}{\sqrt{2\pi}}\int_{-a}^{a}(1)e^{i\alpha u}\,du=\frac{1}{2\sqrt{\pi}}\frac{e^{i\alpha u}}{i\alpha}\bigg|_{-a}^{a}$$

$$=\frac{1}{\sqrt{2\pi}}\left(\frac{e^{i\alpha a}-e^{-i\alpha a}}{i\alpha}\right)=\sqrt{\frac{2}{\pi}}\frac{\sin\alpha a}{\alpha},\qquad \alpha\neq 0$$

For $\alpha = 0$, we obtain $F(\alpha)=\sqrt{2/\pi}\,a$.

(b) The graphs of $f(x)$ and $F(\alpha)$ for $a = 3$ are shown in Figures 14.1, and 14.2, respectively.

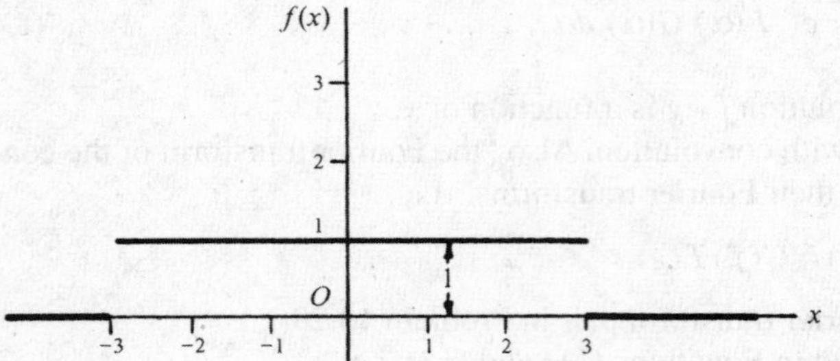

Figure 14.1

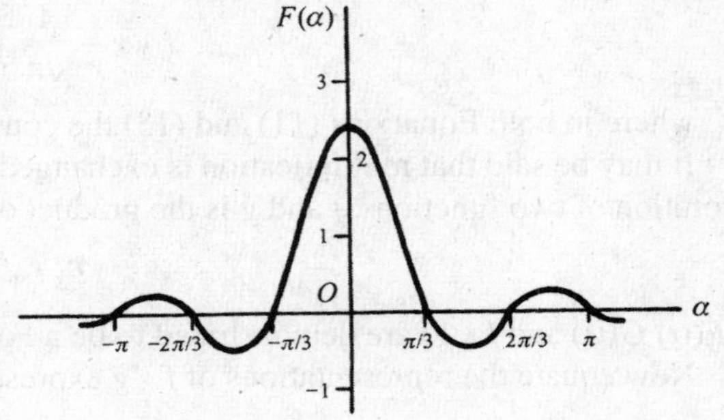

Figure 14.2

14.2. (a) Use the result of Problem 14.1 to evaluate $\int_{-\infty}^{\infty}\frac{\sin\alpha a\cos\alpha x}{\alpha}\,d\alpha$. (b) Deduce the value of $\int_{0}^{\infty}\frac{\sin u}{u}\,du$.

(a) From Fourier's integral theorem, if

$$F(\alpha)=\frac{1}{\sqrt{2\pi}}\int_{-\infty}^{\infty}f(u)e^{i\alpha u}\,du\qquad then\qquad f(x)=\frac{1}{\sqrt{2\pi}}\int_{-\infty}^{\infty}F(\alpha)e^{-i\alpha x}\,d\alpha$$

Then, from Problem 14.1,

$$\frac{1}{\sqrt{2\pi}}\int_{-\infty}^{\infty}\sqrt{\frac{2}{\pi}}\frac{\sin\alpha a}{\alpha}e^{-iax}\,dx=\begin{cases}1 & |x|<a\\ 1/2 & |x|=a\\ 0 & |x|>a\end{cases}\tag{1}$$

The left side of Equation (1) is equal to

$$\frac{1}{\pi}\int_{-\infty}^{\infty}\frac{\sin\alpha a\cos\alpha x}{\alpha}\,d\alpha-\frac{i}{\pi}\int_{-\infty}^{\infty}\frac{\sin\alpha a\sin\alpha x}{\alpha}\,d\alpha\tag{2}$$

The integrand in the second integral of Equation (2) is odd, and so the integral is zero. Then from Equations (1) and (2), we have

$$\int_{-\infty}^{\infty}\frac{\sin\alpha a\cos\alpha x}{\alpha}\,d\alpha=\begin{cases}\pi & |x|<a\\ \pi/2 & |x|=a\\ 0 & |x|>a\end{cases}\tag{3}$$

Alternative solution: Since the function f in Problem 14.1 is an even function, the result follows immediately from the Fourier cosine transform (9).

(b) If x = 0 and a = 1 in the result of (a), we have

$$\int_{-\infty}^{\infty} \frac{\sin \alpha}{\alpha}\, d\alpha = \pi \quad \text{or} \quad \int_0^{\infty} \frac{\sin \alpha}{\alpha}\, d\alpha = \frac{\pi}{2}$$

since the integrand is even.

14.3. If $f(x)$ is an even function, show that (a) $F(\alpha) = \sqrt{\frac{2}{\pi}} \int_0^{\infty} f(u) \cos \alpha u \, du$ and (b) $f(x) = \sqrt{\frac{2}{\pi}} \int_0^{\infty} F(\alpha) \cos \alpha x \, d\alpha$

We have

$$F(\alpha) = \frac{1}{\sqrt{2\pi}} \int_{-\infty}^{\infty} f(u) e^{i\alpha u} du = \frac{1}{\sqrt{2\pi}} \int_{-\infty}^{\infty} f(u) \cos \alpha u \, du + \frac{i}{\sqrt{2\pi}} \int_{-\infty}^{\infty} f(u) \sin \alpha u \, du \qquad (1)$$

(a) If $f(u)$ is even, $f(u) \cos \lambda u$ is even and $f(u) \sin \lambda u$ is odd. Then the second integral on the right of Equation (1) is zero, and the result can be written

$$F(\alpha) = \frac{2}{\sqrt{2\pi}} \int_0^{\infty} f(u) \cos \alpha u \, du = \sqrt{\frac{2}{\pi}} \int_0^{\infty} f(u) \cos \alpha u \, du$$

(b) From (a), $F(-\alpha) = F(\alpha)$ so that $F(\alpha)$ is an even function. Then, by using a proof exactly analogous to that in (a), the required result follows.

A similar result holds for odd functions and can be obtained by replacing the cosine by the sine.

14.4. Solve the integral equation $\int_0^{\infty} f(x) \cos \alpha x \, dx = \begin{cases} 1-\alpha & 0 \leqq \alpha \leqq 1 \\ 0 & \alpha > 1 \end{cases}$.

Let $\sqrt{\frac{2}{\pi}} \int_0^{\infty} f(x) \cos \alpha x \, dx = F(\alpha)$ and choose $F(\alpha) = \begin{cases} \sqrt{2/\pi}(1-\alpha) & 0 < \alpha < 1 \\ 0 & \alpha > 1 \end{cases}$ Then, by Problem 14.3,

$$f(x) = \sqrt{\frac{2}{\pi}} \int_0^{\infty} F(\alpha) \cos \alpha x \, d\alpha = \sqrt{\frac{2}{\pi}} \int_0^1 \sqrt{\frac{2}{\pi}} (1-\alpha) \cos \alpha x \, d\alpha$$
$$= \frac{2}{\pi} \int_0^1 (1-\alpha) \cos \alpha x \, d\alpha = \frac{2(1-\cos x)}{\pi x^2}$$

14.5. Use Problem 14.4 to show that $\int_0^{\infty} \frac{\sin^2 u}{u^2} du = \frac{\pi}{2}$.

As obtained in Problem 14.4,

$$\frac{2}{\pi} \int_0^{\infty} \frac{1-\cos x}{x^2} \cos \alpha x \, dx = \begin{cases} 1-\alpha & 0 \leqq \alpha \leqq 1 \\ 0 & \alpha > 1 \end{cases}$$

Taking the limit as $\alpha \to 0+$, we find

$$\int_0^{\infty} \frac{1-\cos x}{x^2} dx = \frac{\pi}{2}$$

But this integral can be written as $\int_0^{\infty} \frac{2 \sin^2(x/2)}{x^2} dx$, which becomes $\int_0^{\infty} \frac{\sin^2 u}{u^2} du$ on letting $x = 2u$, so that the required result follows.

14.6. Show that $\int_0^{\infty} \frac{\cos \alpha x}{a^2+1} d\alpha = \frac{\pi}{2} e^{-x}$, $x \geqq 0$.

Let $f(x) = e^{-x}$ in the Fourier integral theorem

$$f(x) = \frac{2}{\pi} \int_0^{\infty} \cos \alpha x \, d\alpha \int_0^{\infty} f(u) \cos \lambda u \, du$$

Then

$$\frac{2}{\pi}\int_0^\infty \cos\alpha x\, d\alpha \int_0^\infty e^{-u} \cos \alpha u\, du = e^{-x}$$

But by Problem 12.22, we have $\int_0^\infty e^{-u} \cos \alpha u\, du = \frac{1}{\alpha^2+1}$. Then

$$\frac{2}{\pi}\int_0^\infty \frac{\cos \alpha x}{\alpha^2+1} d\alpha = e^{-x} \quad \text{or} \quad \int_0^\infty \frac{\cos \alpha x}{a^2+1} d\alpha = \frac{\pi}{2} e^{-x}$$

Parseval's identity

14.7. Verify Parseval's identity for Fourier integrals for the Fourier transforms of Problem 14.1.

We must show that

$$\int_{-\infty}^\infty \{f(x)\}^2 dx = \int_{-\infty}^\infty \{F(\alpha)\}^2 d\alpha$$

where

$$f(x) = \begin{cases} 1 & |x| < a \\ 0 & |x| < a \end{cases} \quad \text{and} \quad F(\alpha) = \sqrt{\frac{2}{\pi}} \frac{\sin \alpha a}{\alpha}$$

This is equivalent to

$$\int_{-a}^a (1)^2 dx = \int_{-\infty}^\infty \frac{2}{\pi} \frac{\sin^2 \alpha a}{\alpha^2} d\alpha$$

or

$$\int_{-\infty}^\infty \frac{\sin^2 \alpha a}{\alpha^2} d\alpha = 2\int_0^\infty \frac{\sin^2 \alpha a}{a^2} d\alpha = \pi a$$

i.e.,

$$\int_0^\infty \frac{\sin^2 \alpha a}{\alpha^2} d\alpha = \frac{\pi a}{2}$$

By letting $\alpha a = u$ and using Problem 14.5, it is seen that this is correct. The method can also be used to find $\int_0^\infty \frac{\sin^2 u}{u^2} du$ directly.

Proof of the Fourier integral theorem

14.8. Present a heuristic demonstration of Fourier's integral theorem by use of a limiting form of Fourier series.

Let

$$f(x) = \frac{a_0}{2} + \sum_{n=1}^\infty \left(a_n \cos \frac{n\pi x}{L} + b_n \sin \frac{n\pi x}{L} \right) \tag{1}$$

where

$$a_n = \frac{1}{L}\int_{-L}^L f(u) \cos \frac{n\pi u}{L} du \quad \text{and} \quad b_n = \frac{1}{L}\int_{-L}^L f(u) \sin \frac{n\pi u}{L} du$$

Then, by substitution (see Problem 13.21),

$$f(x) = \frac{1}{2L}\int_{-L}^L f(u)\, du + \frac{1}{L}\sum_{n=1}^\infty f(u) \cos \frac{n\pi}{L}(u-x)\, du \tag{2}$$

If we assume that $\int_{-\infty}^\infty |f(u)|\, du$ converges, the first term on the right of Equation (2) approaches zero as $L \to \infty$, while the remaining part appears to approach

$$\lim_{L\to\infty} \frac{1}{L} \sum_{n=1}^\infty \int_{-\infty}^\infty f(u) \cos \frac{n\pi}{L}(u-x)\, du \tag{3}$$

This last step is not rigorous and makes the demonstration heuristic.

Calling $\Delta\alpha = \pi/L$, Equation (3) can be written

$$f(x) = \lim_{\Delta\alpha\to 0} \sum_{n=1}^{\infty} \Delta\alpha F(n\Delta\alpha) \tag{4}$$

where we have written

$$F(\alpha) = \frac{1}{\pi}\int_0^{\infty} f(u)\cos\alpha(u-x)\,du \tag{5}$$

But the limit (4) is equal to

$$f(x) = \int_0^{\infty} F(\alpha)\,d\alpha = \frac{1}{\pi}\int_0^{\infty} d\alpha \int_{-\infty}^{\infty} f(u)\cos\alpha(u-x)\,du$$

which is Fourier's integral formula.

This demonstration serves only to provide a possible result. To be rigorous, we start with the integral

$$1/\pi \int_0^{\infty} d\alpha \int_{-\infty}^{\infty} f(u)\cos\alpha(u-x)dx$$

and examine the convergence. This method is considered in Problems 14.9 through 14.12.

14.9. Prove that

(a) $\displaystyle\lim_{\alpha\to\infty}\int_0^L \frac{\sin\alpha\upsilon}{\upsilon}\,d\upsilon = \frac{\pi}{2}$,

(b) $\displaystyle\lim_{\alpha\to\infty}\int_{-L}^0 \frac{\sin\alpha\upsilon}{\upsilon}\,d\upsilon = \frac{\pi}{2}$.

(a) Let $\alpha\upsilon = y$. Then $\displaystyle\lim_{\alpha\to\infty}\int_0^L \frac{\sin\alpha\upsilon}{\upsilon}\,d\upsilon = \lim_{\alpha\to\infty}\int_0^{\alpha L} \frac{\sin y}{y}\,dy = \int_0^{\infty}\frac{\sin y}{y}\,dy = \frac{\pi}{2}$. by Problem 12.29.

(b) Let $\alpha\upsilon = -y$. Then $\displaystyle\lim_{\alpha\to\infty}\int_{-L}^0 \frac{\sin\alpha\upsilon}{\upsilon}\,d\upsilon = \lim_{x\to\infty}\int_0^{\alpha L}\frac{\sin y}{y}\,dy = \frac{\pi}{2}$.

14.10. Riemann's theorem states that if $F(x)$ is piecewise continuous in (a, b), then

$$\lim_{\alpha\to\infty}\int_a^b F(x)\sin\alpha x\,dx = 0$$

with a similar result for the cosine (see Problem 14.32). Use this to prove that

(a) $\displaystyle\lim_{\alpha\to\infty}\int_0^L f(x+\upsilon)\frac{\sin\alpha\upsilon}{\upsilon}\,d\upsilon = \frac{\pi}{2}f(x+0)$

(b) $\displaystyle\lim_{\alpha\to\infty}\int_{-L}^0 f(x+\upsilon)\frac{\sin\alpha\upsilon}{\upsilon}\,d\upsilon = \frac{\pi}{2}f(x-0)$

where $f(x)$ and $f'(x)$ are assumed piecewise continuous in $(0, L)$ and $(-L, 0)$, respectively.

(a) Using Problem 14.9(a), it is seen that a proof of the given result amounts to proving that

$$\lim_{\alpha\to\infty}\int_0^L \{f(x+\upsilon) - f(x+0)\}\frac{\sin\alpha\upsilon}{\upsilon}\,d\upsilon = 0$$

This follows at once from Riemann's theorem, because $F(\upsilon) = \dfrac{f(x+\upsilon)-f(x+0)}{\upsilon}$ is piecewise continuous in $(0, L)$, since $\displaystyle\lim_{n\to 0+} F(\upsilon)$ exists and $f(x)$ is piecewise continuous.

(b) A proof of this is analogous to that in (a) if we make use of Problem 14.9(b).

14.11. If $f(x)$ satisfies the additional condition that $\int_{-\infty}^{\infty} |f(x)|\,dx$ converges, prove that

(a) $\lim_{\alpha\to\infty}\int_0^{\infty} f(x+\upsilon)\frac{\sin\alpha\upsilon}{\upsilon}\,d\upsilon = \frac{\pi}{2}f(x+0)$ (b) $\lim_{\alpha\to\infty}\int_{-\infty}^{0} f(x+\upsilon)\frac{\sin\alpha\upsilon}{\upsilon}\,d\upsilon = \frac{\pi}{2}f(x-0)$

We have

$$\int_0^{\infty} f(x+\upsilon)\frac{\sin\alpha\upsilon}{\upsilon}\,d\upsilon = \int_0^{L} f(x+\upsilon)\frac{\sin\alpha\upsilon}{\upsilon}\,d\upsilon + \int_L^{\infty} f(x+\upsilon)\frac{\sin\alpha\upsilon}{\upsilon}\,d\upsilon \quad (1)$$

$$\int_0^{\infty} f(x+0)\frac{\sin\alpha\upsilon}{\upsilon}\,d\upsilon = \int_0^{L} f(x+0)\frac{\sin\alpha\upsilon}{\upsilon}\,d\upsilon + \int_L^{\infty} f(x+0)\frac{\sin\alpha\upsilon}{\upsilon}\,d\upsilon \quad (2)$$

Subtracting,

$$\int_0^{\infty}\{f(x+\upsilon)-f(x+0)\}\frac{\sin\alpha\upsilon}{\upsilon}\,d\upsilon = \int_0^{L}\{f(x+\upsilon)-f(x+0)\}\frac{\sin\alpha\upsilon}{\upsilon}\,d\upsilon + \int_L^{\infty} f(x+\upsilon)\frac{\sin\alpha\upsilon}{\upsilon}\,d\upsilon - \int_L^{\infty} f(x+0)\frac{\sin\alpha\upsilon}{\upsilon}\,d\upsilon \quad (3)$$

Denoting the integrals in Equation (3) by I, I_1, I_2, and I_3, respectively, we have $I = I_1 + I_2 + I_3$ so that

$$|I| \leqq |I_1| + |I_2| + |I_3| \quad (4)$$

Now

$$|I_2| \leqq \int_L^{\infty}\left|f(x+\upsilon)\frac{\sin\alpha\upsilon}{\upsilon}\right|d\upsilon \leqq \frac{1}{L}\int_L^{\infty}|f(x+\upsilon)|\,d\upsilon$$

Also,

$$|I_3| \leqq |f(x+0)|\left|\int_L^{\infty}\frac{\sin\alpha\upsilon}{\upsilon}\,d\upsilon\right|.$$

Since both $\int_0^{\infty}|f(x)\}dx$ and $\int_0^{\infty}\frac{\sin\alpha\upsilon}{\upsilon}\,d\upsilon$ converge, we can choose L so large that $|I_2| \leqq \epsilon/3$, $|I_3| \leqq \epsilon/3$. Also, we can choose α so large that $|I_1| \leqq \epsilon/3$. Then from Equation (4) we have $|I| < \epsilon$ for α and L sufficiently large so that the required result follows.

This result follows by reasoning exactly analogous to that in (a).

14.12. Prove Fourier's integral formula where $f(x)$ satisfies the conditions stated on Page 377.

We must prove that $\lim_{L\to\infty}\frac{1}{\pi}\int_{\alpha=0}^{L}\int_{u=-\infty}^{\infty} f(u)\cos\alpha(x-u)\,du\,d\alpha = \frac{f(x+0)+f(x-0)}{2}$.

Since $\left|\int_{-\infty}^{\infty} f(u)\cos\alpha(x-u)\,du\right| \leqq \int_{-\infty}^{\infty}|f(u)|\,du$, which converges, it follows by the Weierstrass test that $\int_{-\infty}^{\infty} f(u)\cos\alpha(x-u)\,du$ converges absolutely and uniformly for all α. Thus, we can reverse the order of integration to obtain

$$\begin{aligned}\frac{1}{\pi}\int_{\alpha=0}^{L} d\alpha\int_{u=-\infty}^{\infty} f(u)\cos\alpha(x-u)\,du &= \frac{1}{\pi}\int_{u=-\infty}^{\infty} f(u)\,du\int_{\alpha=0}^{L}\cos\alpha(x-u)\,d\alpha \\ &= \frac{1}{\pi}\int_{u=-\infty}^{\infty} f(u)\frac{\sin L(u-x)}{u-x}\,du \\ &= \frac{1}{\pi}\int_{u=-\infty}^{\infty} f(x+\upsilon)\frac{\sin L\upsilon}{\upsilon}\,d\upsilon \\ &= \frac{1}{\pi}\int_{-\infty}^{0} f(x+\upsilon)\frac{\sin L\upsilon}{\upsilon}\,d\upsilon + \frac{1}{\pi}\int_0^{\infty} f(x+\upsilon)\frac{\sin L\upsilon}{\upsilon}\,d\upsilon\end{aligned}$$

where we have let $u = x + \upsilon$.

Letting $L \to \infty$, we see by Problem 14.11 that the given integral converges to $\frac{f(x)+0)+f(x-0)}{2}$ as required.

Miscellaneous problems

14.13. Solve $\dfrac{\partial U}{\partial t} = \dfrac{\partial^2 U}{\partial x^2}$, subject to the conditions $U(0, t) = 0$, $U(x, 0) = \begin{cases} 1 & 0 < x < 1 \\ 0 & x \geqq 1 \end{cases}$, $U(x, 1)$ is bounded where $x > 0, t > 0$.

We proceed as in Problem 13.24. A solution satisfying the partial differential equation and the first boundary condition is given by $Be^{-\lambda^2}t \sin \lambda x$. Unlike Problem 13.24, the boundary conditions do not prescribe the specific values for λ, so we must assume that all values of λ are possible. By analogy with that problem, we sum over all possible values of λ, which corresponds to an integration in this case, and are led to the possible solution

$$U(x, 1) = \int_0^\infty B(\lambda)\, e^{-\lambda^2 t} \sin \lambda x d\lambda \tag{1}$$

where $B(\lambda)$ is undetermined. By the second condition, we have

$$\int_0^\infty B(\lambda) \sin \lambda x d\lambda = \begin{cases} 1 & 0 < x < 1 \\ 0 & x \geqq 1 \end{cases} = f(x) \tag{2}$$

from which we have, by Fourier's integral formula,

$$B(\lambda) = \frac{2}{\pi}\int_0^\infty f(x) \sin \lambda x \, dx = \frac{2}{\pi}\int_0^1 \sin \lambda x \, dx = \frac{2(1-\cos\lambda)}{\pi\lambda} \tag{3}$$

so that, at least formally, the solution is given by

$$U(x, 1) = \frac{2}{\pi}\int_0^\infty \left(\frac{1-\cos\lambda}{\lambda}\right) e^{-\lambda^2 t} \sin \lambda x \, dx \tag{4}$$

See Problem 14.26.

14.14. Show that $e^{-x^2}/2$ is its own Fourier transform.

Since $e^{-x^2}/2$ is even, its Fourier transform is given by $\sqrt{2/\pi} = \int_0^\infty e^{-x^2/2} \cos x\alpha \, dx$.

Letting $x = \sqrt{2u}$ and using Problem 12.32, the integral becomes

$$\frac{2}{\sqrt{\pi}}\int_0^\infty e^{-u^2} \cos(\alpha\sqrt{2}\,u)\, du = \frac{2}{\sqrt{\pi}} \cdot \frac{\sqrt{\pi}}{2} e^{-\alpha^2/2}$$

which proves the required result.

14.15. Solve the integral equation

$$y(x) = g(x) + \int_{-\infty}^{\infty} y(u)\, r(x-u)\, du$$

where $g(x)$ and $r(x)$ are given.

Suppose that the Fourier transforms of $y(x)$, $g(x)$, and $r(x)$ exist, and denote them by $Y(\alpha)$, $G(\alpha)$, and $R(\alpha)$, respectively. Then, taking the Fourier transform of both sides of the given integral equation, we have, by the convolution theorem,

$$Y(\alpha) = G(\alpha) + \sqrt{2\pi}\, Y(\alpha)\, R(\alpha) \qquad or \qquad Y(\alpha) = \frac{G(\alpha)}{1-\sqrt{2\pi}\; R(\alpha)}$$

Then

$$y(x) = \mathscr{F}^{-1}\left\{\frac{G(\alpha)}{1-\sqrt{2\pi}\; R(\alpha)}\right\} = \frac{1}{\sqrt{2\pi}}\int_{-\infty}^{\infty} \frac{G(\alpha)}{1-\sqrt{2\pi}\; R(\alpha)} e^{-i\alpha x} d\alpha$$

assuming this integral exists.

SUPPLEMENTARY PROBLEMS

The Fourier integral and Fourier transforms

14.16. (a) Find the Fourier transform of $f(x)=\begin{cases}1/2\epsilon & |x|\leq\epsilon\\ 0 & |x|>\epsilon\end{cases}$. (b) Determine the limit of this transform as $\epsilon\to 0+$ and discuss the result.

Ans. (a) $\dfrac{1}{\sqrt{2\pi}}\dfrac{\sin\alpha\epsilon}{\alpha\epsilon}$ (b) $\dfrac{1}{\sqrt{2\pi}}$

14.17. (a) Find the Fourier transform of $f(x)=\begin{cases}1-x^2 & |x|<1\\ 0 & |x|>1\end{cases}$. (b) Evaluate $\displaystyle\int_0^\infty\left(\frac{x\cos x-\sin x}{x^3}\right)\cos\frac{x}{2}\,dx$.

Ans. (a) $2\sqrt{\dfrac{2}{\pi}}\left(\dfrac{\alpha\cos\alpha-\sin\alpha}{\alpha^3}\right)$ (b) $\dfrac{3\pi}{16}$

14.18. If $f(x)=\begin{cases}1 & 0<x\leq 1\\ 0 & x\geq 1\end{cases}$, find (a) the Fourier since transform and (b) the Fourier cosine transform of $f(x)$. In each case, obtain the graph of $f(x)$ and its transform.

Ans. (a) $\sqrt{\dfrac{2}{\pi}}\left(\dfrac{1-\cos\alpha}{\alpha}\right)$ (b) $\sqrt{\dfrac{2}{\pi}}\dfrac{\sin\alpha}{\alpha}$

14.19. (a) Find the Fourier sine transform of e^{-x}, $x\geq 0$. (b) Show that $\displaystyle\int_0^\infty\frac{x\sin mx}{x^2+1}\,dx=\frac{\pi}{2}e^{-m}$, $m>0$ by using the result in (a). (c) Explain from the viewpoint of Fourier's integral theorem why the result in (b) does not hold for m = 0.

Ans. (a) $\sqrt{2/\pi}\,[\alpha/(1+\alpha^2)]$

14.20. Solve for $Y(x)$ the integral equation

$$\int_0^\infty Y(x)\sin xt\,dx=\begin{cases}1 & 0\leq t<1\\ 2 & 1\leq t<2\\ 0 & t\geq 0\end{cases}$$

and verify the solution by direction substitution.
Ans. $Y(x) = (2 + 2\cos x - 4\cos 2x)/\pi x$

Parseval's identity

14.21. Evaluate (a) $\displaystyle\int_0^\infty\frac{dx}{(x^2+1)^2}$ and (b) $\displaystyle\int_0^\infty\frac{x^2dx}{(x^2+1)^2}$ by use of Parseval's identity. (Hint: Use the Fourier sine and cosine transforms of e^{-x}, $x>0$.)

Ans. (a) $\pi/4$ (b) $\pi/4$

14.22. Use Problem 14.18 to show that (a) $\displaystyle\int_0^\infty\left(\frac{1-\cos x}{x}\right)^2dx=\frac{\pi}{2}$ and (b) $\displaystyle\int_0^\infty\frac{\sin^4 x}{x^2}\,dx=\frac{\pi}{2}$.

14.23. Show that $\displaystyle\int_0^\infty\frac{(x\cos x-\sin x)^2}{x^6}\,dx=\frac{\pi}{15}$.

Miscellaneous problems

14.24. (a) Solve $\frac{\partial U}{\partial t} = 2\frac{\partial^2 U}{\partial x^2}$, $U(0, t) = 0$, $U(x, 0) = e^{-x}$, $x > 0$, $U(x, t)$ is bounded where $x > 0, t > 0$.
(b) Give a physical interpretation.

Ans. $U(x, t) = \frac{2}{\pi}\int_0^\infty \frac{\lambda e^{-2\lambda^2 t} \sin \lambda x}{\lambda^2 + 1}\, d\lambda$

14.25. Solve $\frac{\partial U}{\partial t} = \frac{\partial^2 U}{\partial x^2}$, $U_x(0, t) = 0$, $U(x, 0) = \begin{cases} x & 0 \leq x \leq 1 \\ 0 & x > 1 \end{cases}$, $U(x, t)$ is bounded where $x > 0, t > 0$.

Ans. $U(x, t) = \frac{2}{\pi}\int_0^\infty \left(\frac{\sin \lambda}{\lambda} + \frac{\cos \lambda - 1}{\lambda^2}\right) e^{-\lambda^2 t} \cos \lambda x \, d\lambda$

14.26. (a) Show that the solution to Problem 14.13 can be written

$$U(x, t) = \frac{2}{\sqrt{\pi}}\int_0^{x/2\sqrt{t}} e^{-v^2}\, dv - \frac{1}{\sqrt{\pi}}\int_{(1-x)/2\sqrt{t}}^{(1+x)/2\sqrt{2}} e^{-v^2}\, dv$$

(b) Prove directly that the function in (a) satisfies $\frac{\partial U}{\partial t} = \frac{\partial^2 U}{\partial x^2}$ and the conditions of Problem 14.13.

14.27. Verify the convolution theorem for the functions $f(x) = g(x) = \begin{cases} 1 & |x| < 1 \\ 0 & |x| > 1 \end{cases}$.

14.28. Establish Equation (4), Page 378, form Equation (3), Page 378.

14.29. Prove the result (12), Page 379. [Hint: If $F(\alpha) = \frac{1}{\sqrt{2\pi}}\int_{-\infty}^{\infty} f(u)e^{i\alpha u} du$ and $G(\alpha) = \frac{1}{\sqrt{2\pi}}\int_{-\infty}^{\infty} g(v)e^{i\alpha v}\, dv$, then $F(\alpha)G(\alpha) = \frac{1}{2\pi}\int_{-\infty}^{\infty}\int_{-\infty}^{\infty} e^{i\alpha(u+v)} f(u)g(v)\, du\, dv$. Now make the transformation $u + v = x$.]

$$F(\alpha)\, G(\alpha) = \frac{1}{\sqrt{\pi}}\int_{-\infty}^{\infty}\int_{-\infty}^{\infty} e^{i\alpha x} f(u)g(x-u)\, du\, dx$$

Define

$$f * g = \frac{1}{\sqrt{\pi}}\int_{-\infty}^{\infty} f(u)g(x-u)\, du \qquad (f * g \text{ is a function of } x)$$

then

$$F(\alpha)\, G(\alpha) = \frac{1}{\sqrt{\pi}}\int_{-\infty}^{\infty}\int_{-\infty}^{\infty} e^{i\alpha x} f * g\, dx$$

Thus, $F(\alpha)\, G(\alpha)$ is the Fourier transform of the convolution $f * g$ and, conversely, as indicated in Equation (13), $f * g$ is the Fourier transform of $F(\alpha)\, G(\alpha)$.

14.30. If $F(\alpha)$ and $G(\alpha)$ are the Fourier transforms of $f(x)$ and $g(x)$, respectively, prove (by repeating the pattern of Problem 14.29) that

$$\int_{-\infty}^{\infty} F(\alpha)\, \overline{G(\alpha)}\, d\alpha = \int_{-\infty}^{\infty} f(x)\, \overline{g(x)}\, dx$$

where the bar signifies the complex conjugate. Observe that if G is expressed as in Problem 14.29, then

$$\bar{G}(\alpha) = \frac{1}{\pi}\int_{-\infty}^{\infty} e^{-i\alpha x} f(u)\, \bar{g}(v)\, dv$$

14.31. Show that the Fourier transform of $g(u-x)$ is $e^{i\alpha x}$; i.e., $e^{i\alpha x}G(\alpha)=\frac{1}{\sqrt{\pi}}\int_{-\infty}^{\infty} e^{i\alpha u} f(u)\, g(u-x)\, du$. (Hint: See Problem 14.29. Let $\upsilon = u - x$.)

14.32. Prove Riemann's theorem (see Problem 14.10).

CHAPTER 15

Gamma and Beta Functions

The Gamma Function

The gamma function may be regarded as a generalization of $n!$ (n-factorial), where n is any positive integer to $x!$, where x is any real number. (With limited exceptions, the discussion that follows will be restricted to positive real numbers.) Such an extension does not seem reasonable, yet, in certain ways, the gamma function defined by the improper integral

$$\Gamma(x) = \int_0^\infty t^{x-1} e^{-t}\, dt \tag{1}$$

meets the challenge. This integral has proved valuable in applications. However, because it cannot be represented through elementary functions, establishment of its properties takes some effort. Some of the important properties are outlined as follows.

The gamma function is convergent for $x > 0$. (See Problem 12.18.)

The fundamental property

$$\Gamma(x + 1) = x\Gamma(x) \tag{2}$$

may be obtained by employing the technique of integration by parts to Equation (1). The process is carried out in Problem 15.1. From the form of Equation (2), the function $\Gamma(x)$ can be evaluated for all $x > 0$ when its values in the interval $1 \leq x < 2$ are known. (Any other interval of unit length will suffice.) Table 15.1 and the graph in Figure 15.1 illustrate this idea.

Tables of Values and Graph of the Gamma Function

TABLE 15.1

N	$\Gamma(N)$
1.00	1.0000
1.10	0.9514
1.20	0.9182
1.30	0.8975
1.40	0.8873
1.50	0.8862
1.60	0.8935
1.70	0.9086
1.80	0.9314
1.90	0.9618
2.00	1.0000

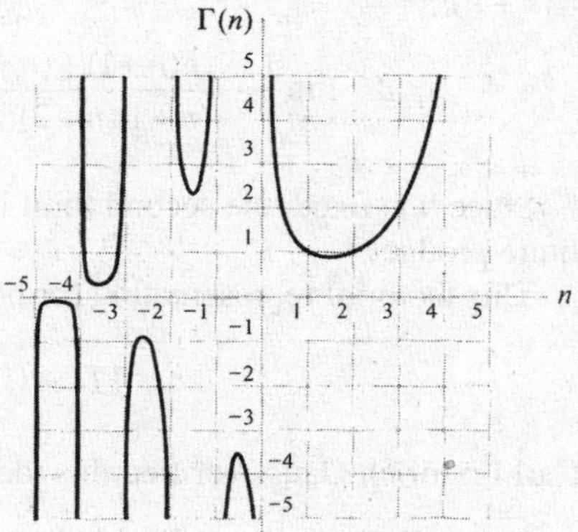

Figure 15.1

Equation (2) is a recurrence relationship that leads to the factorial concept. First observe that if $x = 1$, then Equation (1) can be evaluated and in particular,

$$\Gamma(1) = 1$$

From Equation (2)

$$\Gamma(x+1) = x\Gamma(x) = x(x-1)\Gamma(x-1) = \ldots x(x-1)(x-2)\ldots(x-k)\Gamma(x-k)$$

If $x = n$, where n is a positive integer, then

$$\Gamma(n+1) = n(n-1)(n-2)\ldots 1 = n! \tag{3}$$

If x is a real number, then $x! = \Gamma(x+1)$ is defined by $\Gamma(x+1)$. The value of this identification is in intuitive guidance.

If the recurrence relation (2) is characterized as a differential equation, then the definition of $\Gamma(x)$ can be extended to negative real numbers by a process called *analytic continuation*. The key idea is that even though $\Gamma(x)$ as defined in Equation (1) is not convergent for $x < 0$, the relation $\Gamma(x) = \frac{1}{x}\Gamma(x+1)$ allows the meaning to be extended to the interval $-1 < x < 0$, and from there to $-2 < x < -1$, and so on. A general development of this concept is beyond the scope of this presentation; however, some information is presented in Problem 15.7.

The factorial notion guides us to information about $\Gamma(x+1)$ in more than one way. In the eighteenth century, James Stirling introduced the formula (for positive integer values n)

$$\lim_{n\to\infty} \frac{\sqrt{2\pi}\; n^{n+1}e^{-n}}{n!} = 1 \tag{4}$$

This is called Stirling's formula, and it indicates that $n!$ *asymptotically* approaches $\sqrt{2\pi}\; n^{n+1}e^{-n}$ for large values of n. This information has proved useful, since $n!$ is difficult to calculate for large values of n.

There is another consequence of Stirling's formula. It suggests the possibility that for sufficiently large values of x,

$$x! = \Gamma(x+1) \approx \sqrt{2\pi}\; x^{x+1}e^{-x} \tag{5a}$$

(An argument supporting this is made in Problem 15.20.)

It is known that $\Gamma(x+1)$ satisfies the inequality

$$\sqrt{2\pi}\; x^{x+1}e^{-x} < \Gamma(x+1) < \sqrt{2\pi}\; x^{x+1}e^{-x}\frac{1}{e^{12(x+1)}} \tag{5b}$$

Since the factor $\frac{1}{e^{12(x+1)}} \to 0$ for large values of x, the suggested value (5a) of $\Gamma(x+1)$ is consistent with (5b).

An exact representation of $\Gamma(x+1)$ is suggested by the following manipulation of $n!$. [It depends on $(n+k)! = (k+n)!$.]

$$n! = \lim_{k\to\infty}\frac{12\ldots n(n+1)+(n+2)\ldots(n+k)}{(n+1)(n+2)\ldots(n+k)} = \lim_{k\to\infty}\frac{k!k^n}{(n+1)\ldots(n+k)}\lim_{k\to\infty}\frac{(k+1)(k+2)\ldots(k+n)}{k^n}$$

Since n is fixed, the second limit is one; therefore, $n! = \lim_{k\to\infty}\frac{k!k^n}{(n+1)\ldots(n+k)}$. (This must be read as an infinite product.)

This factorial representation for positive integers suggests the possibility that

$$\Gamma(x+1) = x! = \lim_{k\to\infty}\frac{k!k^x}{(x+1)\ldots(x+k)} \quad x \neq -1, -2, -k \tag{6}$$

Carl Friëdrich Gauss verified this identification back in the nineteenth century.

This infinite product is symbolized by $\Pi(x,k)$; i.e., $\Pi(x,k) = \frac{k!k^x}{(x+1)\cdots(x+k)}$. It is called Gauss's function, and through this symbolism,

$$\Gamma(x+1) = \lim_{k\to\infty}\Pi(x,k) \tag{7}$$

The expression for $\frac{1}{\Gamma(x)}$ [which has some advantage in developing the derivative of $\Gamma(x)$] results as follows. Put Equation (6) in the form

$$\lim_{k\to\infty}\frac{k^x}{(1+x)(1+x/2)\ldots(1+x/k)} \qquad x\neq -\frac{1}{2},\frac{1}{3},\ldots,\frac{1}{k}$$

Next, introduce

$$\gamma_k = 1+\frac{1}{2}+\frac{1}{3}+\cdots\frac{1}{k}-\ln k$$

Then

$$\gamma = \lim_{k\to\infty}\gamma_k$$

is Euler's constant. This constant has been calculated to many places, a few of which are $\gamma \approx 0.57721566\ldots$.

By letting $k^x = e^{x\ln k} = e^{x[-\gamma_k+1+1/2+\cdots+1/k]}$, the representation (6) can be further modified so that

$$\Gamma(x+1) = e^{-\gamma x}\lim_{k\to\infty}\frac{e^x}{1+x}\frac{e^{x/2}}{1+x/2}\cdots\frac{e^{x/k}}{1+x/k} = e^{-\gamma x}\prod_{k=1}^{\infty}e^{\gamma x}e^{x\ln k}\Big/\left(1+\frac{x}{k}\right)$$
$$= \prod_{k=1}^{\infty}k^x k!(k+x) = \lim_{k\to\infty}\frac{1\cdot 2\cdot 3\cdots k}{(x+1)(x+2)\cdots(x+k)}x^x = \lim_{k\to\infty}\Pi(x,k) \tag{8}$$

Since $\Gamma(x+1) = x\Gamma(x)$,

$$\frac{1}{\Gamma(x)} = xe^{\gamma x}\lim_{k\to\infty}\frac{1+x}{e^x}\frac{1+x/2}{e^{x/2}}\cdots\frac{1+x/k}{e^{x/k}} = xe^{\gamma x}\prod_{k=1}^{\infty}(1+x/k)e^{-x/k} \tag{9}$$

Another result of special interest emanates from a comparison of $\Gamma(x)\Gamma(1-x)$ with the well-known formula

$$\frac{\pi x}{\sin\pi x} = \lim_{k\to\infty}\left\{\frac{1}{1-x^2}\frac{1}{1-(x/2)^2}\cdots\frac{1}{(1-x/k^2)}\right\} = \prod_{k=1}^{\infty}\{1-(x/k^2\} \tag{10}$$

[See *Differential and Integral Calculus*, by R. Courant (translated by E. J. McShane), Blackie & Son Limited.]

$\Gamma(1-x)$ is obtained from $\Gamma(y) = \frac{1}{y}\Gamma(y+1)$ by letting $y = -x$; i.e.,

$$\Gamma(-x) = -\frac{1}{x}\Gamma(1-x) \quad \text{or} \quad \Gamma(1-x) = -x\Gamma(-x)$$

Now use Equation (8) to produce

$$\Gamma(x)\Gamma(1-x) = \left(\left\{x^{-1}e^{-\gamma x}\lim_{k\to 1}\prod_{\kappa=\infty}^{\infty}(1+/k)^{-1}e^{x/k}\right\}\right)\left(e^{\gamma x}\lim_{\kappa=1}(1-x/k)^{-1}e^{-x/k}\right) = \frac{1}{x}\lim_{k\to\infty}\prod_{\kappa=1}^{\infty}(1(x/k)^2)$$

Thus,

$$\Gamma(x)\Gamma(1-x) = \frac{\pi}{\sin\pi}, \quad 0<x<1 \tag{11a}$$

Observe that Equation (11a) yields the result

$$\Gamma\left(\frac{1}{2}\right) = \sqrt{\pi} \tag{11b}$$

Another exact representation of $\Gamma(x+1)$ is

$$\Gamma(x+1) = \sqrt{2\pi}\,x^{x+1}e^{-x}\left\{1+\frac{1}{12x}+\frac{1}{288x^2}+\frac{139}{51840x^3}+\cdots\right\} \tag{12}$$

The method of obtaining this result is closely related to Stirling's asymptotic series for the gamma function. (See Problems 15.20 and 15.74.)

The *duplication* formula

$$2^{2x-1}\Gamma(x)\Gamma\left(x+\frac{1}{2}\right)=\sqrt{\pi}\,\Gamma(2x) \tag{13a}$$

is also part of the literature. Its proof is given in Problem 15.24.

The duplication formula is a special case ($m = 2$) of the following product formula:

$$\Gamma(x)\Gamma\left(x+\frac{1}{m}\right)\cdots\Gamma\left(x+\frac{2}{m}\right)\cdots\Gamma\left(X+\frac{m-1}{m}\right)=m^{\frac{1}{2}-mx}(2\pi)^{\frac{m-1}{2}}\Gamma(mx) \tag{13b}$$

It can be shown that the gamma function has continuous derivatives of all orders. They are obtained by differentiating (with respect to the parameter) under the integral sign.

It helps to recall that $\Gamma(x)=\int_0^\infty t^{x-1}e^{-yt}\,dt$ and that if $y = t^{x-1}$, then $\ln y = \ln t^{x-1} = (x-1)\ln t$.

Therefore; $\frac{1}{y}y' = \ln t$.

It follows that

$$\Gamma'(x)=\int_0^\infty t^{x-1}e^{-t}\ln t\,dt. \tag{14a}$$

This result can be obtained (after making assumptions about the interchange of differentiation with limits) by taking the logarithm of both sides of Equation (9) and then differentiating.

In particular,

$$\Gamma'(1) = -\gamma \ (\gamma \text{ is Euler's constant.}) \tag{14b}$$

It also may be shown that

$$\frac{\Gamma'(x)}{\Gamma(x)}=-\gamma+\left(\frac{1}{1}-\frac{1}{x}\right)+\left(\frac{1}{2}-\frac{1}{x+1}\right)+\cdots\left(\frac{1}{n}-\frac{1}{x+n-1}\right) \tag{15}$$

(See Problem 15.73 for further information.)

The Beta Function

The beta function is a two-parameter composition of gamma functions that has been useful enough in application to gain its own name. Its definition is

$$B(x,y)=\int_0^1 t^{x-1}(1-t)^{y-1}\,dt \tag{16}$$

If $x \ge 1$ and $y \ge 1$, this is a proper integral. If $x > 0$, $y > 0$, and either or both $x < 1$ or $y < 1$, the integral is improper but convergent.

It is shown in Problem 15.11 that the beta function can be expressed through gamma functions in the following way

$$B(x,y)=\frac{\Gamma(x)\Gamma(y)}{\Gamma(x+y)} \tag{17}$$

Many integrals can be expressed through beta and gamma functions. Two of special interest are

$$\int_0^{\pi/2}\sin^{2x-1}\theta\cos^{2y-1}\theta\,d\theta=\frac{1}{2}B(x,y)=\frac{1}{2}\frac{\Gamma(x)\Gamma(y)}{\Gamma(x+y)} \tag{18}$$

$$\int_0^\infty\frac{x^{p-1}}{1+x}dx=\Gamma(p)\,\Gamma(p-1)=\frac{\pi}{\sin\pi P}\qquad 0<p<1 \tag{19}$$

See Problem 15.17. Also see Page 391, where a classical reference is given. Finally, see Problem 16.38, where an elegant complex variable resolution of the integral is presented.

Dirichlet Integrals

If V denotes the closed region in the first octant bounded by the surface $\left(\frac{x}{a}\right)^p + \left(\frac{y}{a}\right)^q + \left(\frac{z}{c}\right)^r = 1$ and the coordinate planes, then if all the constants are positive,

$$\iiint_V x^{\alpha-1} y^{\beta-1} z^{\gamma-1} dx\, dy\, dz = \frac{a^\alpha b^\beta c^\gamma}{pqr} \frac{\Gamma\left(\frac{\alpha}{p}\right)\Gamma\left(\frac{\beta}{b}\right)\Gamma\left(\frac{\gamma}{r}\right)}{\Gamma\left(1+\frac{\alpha}{p}+\frac{\beta}{q}+\frac{\gamma}{r}\right)} \tag{20}$$

Integrals of this type are called *Dirichlet integrals* and are often useful in evaluating multiple integrals (see Problem 15.21).

SOLVED PROBLEMS

The gamma function

15.1. Prove (a) $\Gamma(x + 1) = x\Gamma(x)$, $x > 0$ and (b) $\Gamma(n + 1) = n!$, $n = 1, 2, 3, \ldots$

(a)

$$\begin{aligned}\Gamma(v+1) &= \int_0^\infty x^v e^{-x} dx = \lim_{M\to\infty} \int_0^M x^v e^{-x} dx \\ &= \lim_{M\to\infty} \left\{ (x^v)(-e^{-x})\Big|_0^M - \int_0^M (-e^{-x})(vx^{v-1})dx \right\} \\ &= \lim_{M\to\infty} \left\{ -M^v e^{-M} + v\int_0^M x^{v-1} e^{-x} dx \right\} = v\Gamma(v) \quad \text{if } v > 0\end{aligned}$$

(b) $\Gamma(1) = \int_0^\infty e^{-x} dx = \lim_{M\to\infty} \int_0^M e^{-x} dx = \lim_{M\to\infty} (1 - e^{-M}) = 1.$

Put $n = 1, 2, 3, \ldots$ in $\Gamma(n + 1) = n\Gamma(n)$. Then

$$\Gamma(2) = 1\Gamma(1) = 1,\ \Gamma(3) = 2\Gamma(2) = 2 \cdot 1 = 2!\ \Gamma(4) = 3\Gamma(3) = 3 \cdot 2! = 3!$$

In general, $\Gamma(n + 1) = n!$ if n is a positive integer.

15.2. Evaluate each of the following:

(a) $\dfrac{\Gamma(6)}{2\Gamma(3)} = \dfrac{5!}{2 \cdot 2!} = \dfrac{5\cdot4\cdot3\cdot2}{2\cdot2} = 30$

(b) $\dfrac{\Gamma\left(\frac{5}{2}\right)}{\Gamma\left(\frac{1}{2}\right)} = \dfrac{\frac{3}{2}\Gamma\left(\frac{3}{2}\right)}{\Gamma\left(\frac{1}{2}\right)} = \dfrac{\frac{3}{2}\cdot\frac{1}{2}\Gamma\left(\frac{1}{2}\right)}{\Gamma\left(\frac{1}{2}\right)} = \dfrac{3}{4}$

(c) $\dfrac{\Gamma(3)\Gamma(2.5)}{\Gamma(5.5)} = \dfrac{2!(1.5)(0.5)\Gamma(0.5)}{(4.5)(3.5)(2.5)(1.5)(0.5)\Gamma(0.5)} = \dfrac{16}{315}$

(d) $\dfrac{6\Gamma\left(\frac{8}{3}\right)}{5\Gamma\left(\frac{2}{3}\right)} = \dfrac{6\left(\frac{5}{3}\right)\left(\frac{2}{3}\right)\Gamma\left(\frac{2}{3}\right)}{5\Gamma\left(\frac{2}{3}\right)} = \dfrac{4}{3}$

15.3. Evaluate each integral.

(a) $\int_0^\infty x^3 e^{-x} dx = \Gamma(4) = 3! = 6$

(b) $\int_0^\infty x^6 e^{-2x} dx$

Let $2x = 7$. Then the integral becomes

$$\int_0^\infty \left(\frac{y}{2}\right)^6 e^{-y}\frac{dy}{2} = \frac{1}{2^7}\int_0^\infty y^6 e^{-y}dy = \frac{\Gamma(7)}{2^7} = \frac{6!}{2^7} = \frac{45}{8}$$

15.4. Prove that $\Gamma\left(\frac{1}{2}\right) = \sqrt{\pi}$.

$$\Gamma\left(\frac{1}{2}\right) = \int_0^\infty x^{-1/2}e^{-x}dx.$$

Letting $x = u^2$ this integral becomes

$$2\int_0^\infty e^{-u^2}du = 2\left(\frac{\sqrt{\pi}}{2}\right) = \sqrt{\pi}$$

using Problem 12.31. This result also is described in Equation (11a, b) on Page 391.

15.5. Evaluate each integral.

(a) $\int_0^\infty \sqrt{y}\, e^{-y^3}dy$.

Letting $y^3 = x$, the intergral becomes

$$\int_0^\infty \sqrt{x^{1/3}}\ e^{-x}.\frac{1}{3}x^{-2/3}dx = \frac{1}{3}\int_0^\infty x^{-1/2}e^{-x}dx = \frac{1}{3}\Gamma\left(\frac{1}{2}\right) = \frac{\sqrt{\pi}}{3}$$

(b) $\int_0^\infty 3^{-4x^2}dx = \int_0^\infty (e^{\ln 3})^{(-4x^2)}dz = \int_0^\infty (e^{-(4\ln 3)z^2}dz$

Letting $(4 \ln 3)z^2 = x$, the integral becomes

$$\int_0^\infty \mathrm{e}^{-x}d\left(\frac{x^{1/2}}{\sqrt{4\ln 3}}\right) = \frac{1}{2\sqrt{4\ln 3}}\int_0^\infty x^{-1/2}e^{-x}dx = \frac{\Gamma(1/2)}{2\sqrt{4\ln 3}} = \frac{\sqrt{\pi}}{4\sqrt{\ln 3}}$$

(c) $\int_0^1 \frac{dx}{\sqrt{-\ln x}}$.

Let – ln x = u. Then x = e–u. When x = 1, u = 0; when x = 0, u = ∞. The integral becomes

$$\int_0^\infty \frac{e^{-u}}{\sqrt{u}}du = \int_0^\infty u^{-1/2}e^{-u}du = \Gamma(1/2) = \sqrt{\pi}$$

15.6. Evaluate $\int_0^\infty x^m e^{-ax^n}dx$, where m, n, and a are positive constants.

Letting $ax^n = y$, the integral becomes

$$\int_0^\infty \left\{\left(\frac{y}{a}\right)^{1/n}\right\}^m e^{-y}d\left\{\left(\frac{y}{a}\right)^{1/n}\right\} = \frac{1}{na^{(m+1)/n}}\int_0^\infty y^{(m+1)/n-1}e^{-y}dy = \frac{1}{na^{(m+1)/n}}\Gamma\left(\frac{m+1}{n}\right)$$

15.7. Evaluate (a) $(a)\ \Gamma(-1/2)$ (b) $(-5/2)$

We use the generalization to negative values defined by $\Gamma(x) = \dfrac{\Gamma(x+1)}{x}$.

(a) Letting $x = -\frac{1}{2}$, $\Gamma(-1/2) = \dfrac{\Gamma(1/2)}{-1/2} = -2\sqrt{\pi}$.

(b) Letting $x = -3/2$, $\Gamma(-3/2) = \dfrac{\Gamma(-1/2)}{-3/2} = \dfrac{-2\sqrt{\pi}}{-3/2} = \dfrac{4\sqrt{\pi}}{3}$, using (a)

Then $\Gamma(-5/2) = \dfrac{\Gamma(-3/2)}{-5/2} = -\dfrac{8}{15}\sqrt{\pi}$.

15.8. Prove that $\int_0^1 x^m (\ln x)^n\, dx = \dfrac{(-1)^n n!}{(m+1)^{n+1}}$, where n is a positive integer and $m > -1$.

Letting $x = e^{-y}$, the integral becomes $(-1)^n \int_0^\infty y^n e^{-(m+1)y} dy$. If $(m+1)\, y = u$, this last integral becomes

$$(-1)^n \int_0^\infty \frac{u^n}{(m+1)^n} e^{-u} \frac{du}{m+1} = \frac{(-1)^n}{(m+1)^{n+1}} \int_0^\infty u^n e^{-u} du = \frac{(-1)^n}{(m+1)^{n+1}} \Gamma(n+1) = \frac{(-1)^n n!}{(m+1)^{n+1}}$$

Compare with Problem 8.50.

15.9. A particle is attracted toward a fixed point O with a force inversely proportional to its instantaneous distance from O. If the particle is released from rest, find the time for it to reach O

At time $t = 0$, let the particle be located on the x axis at $x = a > 0$ and let O be the origin. Then, by Newton's law,

$$m\frac{d^2x}{dt^2} = -\frac{k}{x} \tag{1}$$

where m is the mass of the particle and $k > 0$ is a constant of proportionality.

Let $\dfrac{dx}{dt} = \upsilon$, the velocity of the particle. Then $\dfrac{d^2x}{dt^2} = \dfrac{d\upsilon}{dt} = \dfrac{d\upsilon}{dx}\cdot\dfrac{dx}{dt} = \upsilon\dfrac{d\upsilon}{dx}$ and Equation (1) becomes

$$m\upsilon\frac{d\upsilon}{dx} = -\frac{k}{x} \quad \text{or} \quad \frac{m\upsilon^2}{2} = -k\ln x + c \tag{2}$$

upon integrating. Since $\upsilon = 0$ at $x = a$, we find $c = k \ln a$. Then

$$\frac{m\upsilon^2}{2} = k\ln\frac{a}{x} \quad \text{or} \quad \upsilon = \frac{dx}{dt} = -\sqrt{\frac{2k}{m}}\sqrt{\ln\frac{a}{x}} \tag{3}$$

where the negative sign is chosen, since x is decreasing as t increases. We thus find that the time T taken for the particle to go from $x = a$ to $x = 0$ is given by

$$T = \sqrt{\frac{m}{2k}}\int_0^a \frac{dx}{\sqrt{\ln a/x}} \tag{4}$$

Letting $\ln a/x = u$ or $x = ae^{-u}$, this becomes

$$T = a\sqrt{\frac{m}{2k}}\int_0^\infty u^{-1/2}e^{-u}du = a\sqrt{\frac{m}{2k}}\Gamma\left(\frac{1}{2}\right) = a\sqrt{\frac{\pi m}{2k}}$$

The Beta Function

15.10. Prove that (a) $B(u, \upsilon) = B(\upsilon, u)$ and (b) $B(u,\upsilon) = 2\int_0^{\pi/2} \sin^{2u-1}\theta \cos^{2\upsilon-1}\theta\, d\theta$.

(a) Using the transformation $x = 1 - y$, we have

$$B(u,\upsilon) = \int_0^1 x^{u-1}(1-x)^{\upsilon-1}dx = \int_0^1 (1-y)^{u-1}y^{\upsilon-1}dy = \int_0^1 y^{\upsilon-1}(1-y)^{u-1}dy = B(\upsilon,u)$$

(b) Using the transformation $x = \sin^2\theta$, we have

$$B(u,\upsilon) = \int_0^1 x^{u-1}(1-x)^{\upsilon-1}dx = \int_0^{\pi/2} (\sin^2\theta)^{u-1}(\cos^2\theta)^{\upsilon-1}\, 2\sin\theta\cos\theta\, d\theta$$
$$= 2\int_0^{\pi/2} \sin^{2u-1}\theta\cos^{2x-1}\theta\, d\theta$$

15.11. Prove that $B(u,\upsilon) = \dfrac{\Gamma(u)\Gamma(\upsilon)}{\Gamma(u+\upsilon)}\quad u,\upsilon > 0.$

Letting $z^2 = x^2$, we have $\Gamma(u) = \int_0^\infty z^{u-1}e^{-z}dx = 2\int_0^\infty x^{2u-1}e^{-x^2}dx.$

Similarly, $\Gamma(\upsilon) = 2\int_0^\infty y^{2\upsilon-1}e^{-y^2}\,dy$. Then

$$\Gamma(u)\Gamma(\upsilon) = 4\left(\int_0^\infty x^{2u-1}e^{-x^2}\,dx\right)\left(\int_0^\infty y^{2\upsilon-1}e^{-y^2}\,dy\right)$$
$$= 4\int_0^\infty\int_0^\infty x^{2u-1}y^{2\upsilon-1}e^{-(x^2+y^2)}\,dx\,dy$$

Transforming to polar coordinates, $x = \rho\cos\phi$, $y = \rho\sin\phi$,

$$\Gamma(u)\Gamma(\upsilon) = 4\int_{\phi=0}^{\pi/2}\int_{\rho=0}^\infty \rho^2(u+v)-1\,e^{-\rho^2}\cos^{2u-1}\phi\sin^{2\upsilon-1}\phi\,d\rho\,d\phi$$
$$= 4\left(\int_{p=0}^\infty p^{2(u+\upsilon)-1}e^{-p^2}\,dp\right)\left(\int_{\phi=0}^{\pi/2}\cos^{2u-1}\phi\sin^{2\upsilon-1}\phi\,d\phi\right)$$
$$= 2\Gamma(u+\upsilon)\int_0^{\pi/2}\cos^{2u-1}\phi\sin^{2\upsilon-1}\phi\,d\phi = \Gamma(u+\upsilon)B(\upsilon,u)$$
$$= \Gamma(u+\upsilon)B(u,\upsilon)$$

using the results of Problem 15.10. Hence, the required result follows.

This argument can be made rigorous by using a limiting procedure as in Problem 12.31.

15.12. Evaluate each of the following integrals.

(a) $\int_0^1 x^4(1-x)^3\,dx = B(5,4) = \dfrac{\Gamma(5)\Gamma(4)}{\Gamma(9)} = \dfrac{4!3!}{8!} = \dfrac{1}{280}$

(b) $\int_0^2 \dfrac{x^2\,dx}{\sqrt{2-x}}$

Letting x = 2υ, the integral becomes

$$4\sqrt{2}\int_0^1 \frac{\upsilon^2}{\sqrt{1-\upsilon}}\,d\upsilon = 4\sqrt{2}\int_0^1 \upsilon^2(1-\upsilon)^{-1/2}\,d\upsilon = 4\sqrt{2}B\left(3,\frac{1}{2}\right) = \frac{4\sqrt{2}\Gamma(3)\Gamma(1/2)}{\Gamma(7/2)} = \frac{64\sqrt{2}}{15}$$

(c) $\int_0^a y^4\sqrt{a^2-y^2}\,dy$

Letting $y^2 = a^2x$ or $y = \sqrt{x}$, the integral becomes

$$a^6\int_0^1 x^{3/2}(1-x)^{1/2}\,dx = a^6B(5/2,3/2) = \frac{a^6\Gamma(5/2)\Gamma(3/2)}{\Gamma(4)} = \frac{\pi a^6}{16}$$

15.13. Show that $\int_0^{\pi/2}\sin^{2u-1}\theta\cos^{2\upsilon-1}\theta\,d\theta = \dfrac{\Gamma(u)\Gamma(\upsilon)}{2\Gamma(u+\upsilon)}$ $u,\upsilon > 0$.

This follows at once from Problems 15.10 and 15.11.

15.14. Evaluate (a) $\int_0^{\pi/2}\sin^6\theta\,d\theta$, (b) $\int_0^{\pi/2}\sin^4\theta\cos^5\theta\,d\theta$, and (c) $\int_0^{\pi/2}\cos^4\theta\,d\theta$.

(a) Let 2u – 1 = 6, 2υ – 1 = 0, i.e., u = 7/2, υ = 1/2, in Problem 15.13. Then the required integral has the value $\dfrac{\Gamma(7/2)\Gamma(1/2)}{2\Gamma(4)} = \dfrac{5\pi}{32}$.

(b) Letting 2u – 1 = 4, 2υ – 1 = 5, the required integral has the value $\dfrac{\Gamma(5/2)\Gamma(3)}{2\Gamma(11/2)} = \dfrac{8}{315}$.

(c) The given integral = $2\int_0^{\pi/2}\cos^4\theta\,d\theta$. Thus, letting $2u - 1 = 0$, $2\upsilon - 1 = 4$ in Problem 15.13, the value is $\dfrac{2\Gamma(1/2)\Gamma(5/2)}{2\Gamma(3)} = \dfrac{3\pi}{8}$.

15.15. Prove $\int_0^{\pi/2} \sin^p\theta\, d\theta = \int_0^{\pi/2} \cos^p\theta\, d\theta = (a) \dfrac{1\cdot 3\cdot 5\cdots(p-1)}{2\cdot 4\cdot 6\cdots p}\dfrac{\pi}{2}$ if p is an even positive integer and (b) $\dfrac{2\cdot 4\cdot 6\cdots(p-1)}{1\cdot 3\cdot 5\cdots p}$ if p is an odd positive integer.

From Problem 15.13, with $2u - 1 = p$, $2\upsilon - 1 = 0$, we have

$$\int_0^{\pi/2} \sin^p\theta\, d\theta = \frac{\Gamma\left[\frac{1}{2}(p+1)\right]\Gamma\left(\frac{1}{2}\right)}{2\Gamma\left[\frac{1}{2}(p+2)\right]}$$

(a) If p = 2r, the integral equals

$$\frac{\Gamma(r+1)\Gamma\left(\frac{1}{2}\right)}{2\Gamma(r+1)} = \frac{\left(r-\frac{1}{2}\right)\left(r-\frac{3}{2}\right)\cdots\frac{1}{2}\Gamma\left(\frac{1}{2}\right)\cdot\Gamma\left(\frac{1}{2}\right)}{2r(r-1)} = \frac{(2r-1)(2r-3)\cdots 1}{2r(2r-2)\cdots 2}\frac{\pi}{2} = \frac{1\cdot 3\cdot 5\cdots(2r-1)}{2\cdot 4\cdot 6\cdots 2r}\frac{\pi}{2}$$

(b) If p = 2r + 1, the integral equals

$$\frac{\Gamma(r+1)\Gamma\left(\frac{1}{2}\right)}{2\Gamma\left(r+\frac{3}{2}\right)} = \frac{r(r-1)\cdots 1\cdot\sqrt{\pi}}{2\left(r+\frac{1}{2}\right)\left(r-\frac{1}{2}\right)\cdots\frac{1}{2}\sqrt{\pi}} = \frac{2\cdot 4\cdot 6\cdots 2r}{1\cdot 3\cdot 5\cdots(2r+1)}$$

In both cases, $\int_0^{\pi/2} \sin^p\theta\, d\theta = \int_0^{\pi/2} \cos^p\theta\, d\theta$, as seen by letting $\theta = \pi/2 - \phi$.

15.16. Evaluate (a) $\int_0^{\pi/2} \cos^6\theta\, d\theta$, (b) $\int_0^{\pi/2} \sin^3\theta \cos^2\theta\, d\theta$, and (c) $\int_0^{2\pi} \sin^8\theta\, d\theta$.

(a) From Problem 15.15, the integral equals $\dfrac{1\cdot 3\cdot 5}{2\cdot 4\cdot 6} = \dfrac{5\pi}{32}$ [compare Problem 15.14(a)].

(b) The integral equals

$$\int_0^{\pi/2} \sin^3\theta(1-\sin^2\theta)\, d\theta = \int_0^{\pi/2} \sin^3\theta\, d\theta - \int_0^{\pi/2} \sin^5\theta\, d\theta = \frac{2}{1\cdot 3} - \frac{2\cdot 4}{1\cdot 3\cdot 5} = \frac{2}{15}$$

The method of Problem 15.14(b) can also be used.

(c) The given integral equals $4\int_0^{\pi/2} \sin^8\theta\, d\theta = 4\left(\dfrac{1\cdot 3\cdot 5\cdot 7}{2\cdot 4\cdot 6\cdot 8}\dfrac{\pi}{2}\right) = \dfrac{35\pi}{64}$.

15.17. Given $\int_0^\infty \dfrac{x^{p-1}}{1+x}\, dx = \dfrac{\pi}{\sin p\pi}$, show that $\Gamma(p)\Gamma(1-p) = \dfrac{\pi}{\sin p\pi}$, where $0 < p < 1$.

Letting $\dfrac{x}{1+x} = y$ or $x = \dfrac{y}{1-y}$, the given integral becomes

$$\int_0^1 y^{p-1}(1-y)^{-p}\, dy = B(p, 1-p) = \Gamma(p)\Gamma(1-p)$$

and the result follows.

15.18. Evaluate $\int_0^\infty \dfrac{dy}{1+y^4}$.

Let $y^4 = x$. Then the integral becomes $\dfrac{1}{4}\int_0^\infty \dfrac{x^{-3/4}}{1+x}\, dx = \dfrac{\pi}{4\sin(\pi/4)} = \dfrac{\pi\sqrt{2}}{4}$ by Problem 15.17, with $p = \dfrac{1}{4}$. The result can also be obtained by letting $y^2 = \tan\theta$.

15.19. Show that $\int_0^2 x\sqrt[3]{8-x^3}\,dx = \dfrac{16\pi}{9\sqrt{3}}$.

Letting $x^3 - 8y$ or $x = 2y^{1/3}$, the integral becomes

$$\int_0^1 2y^{1/3}\sqrt[3]{8(1-y)}.\frac{2}{3}y^{-2/3}dy = \frac{8}{3}\int_0^1 y^{-1/3}(1-y)^{1/3}dy = \frac{8}{3}B\left(\frac{2}{3},\frac{4}{3}\right)$$

$$= \frac{8}{3}\frac{\Gamma\left(\frac{2}{3}\right)\Gamma\left(\frac{4}{3}\right)}{\Gamma(2)} = \frac{8}{9}\Gamma\left(\frac{1}{3}\right)\Gamma\left(\frac{2}{3}\right) = \frac{8}{9}.\frac{\pi}{\sin\pi/3} = \frac{16\pi}{9\sqrt{3}}$$

Stirling's formula

15.20. Show that for large positive integers $n, n! = \sqrt{2\pi n}\; n^n e^{-n}$ approximately.

By definition, $\Gamma(z) = \int_0^\infty t^{z-1}e^{-t}dt$. Let $lfz = x + 1$, then

$$\Gamma(x+1) = \int_0^\infty t^x e^{-t}dt = \int_0^\infty e^{-t+\ln t^x}\,dt = \int_0^\infty e^{-t+x\ln t}dt \tag{1}$$

For a fixed value of x the function x, $\ln t - t$ has a relative maximum for $t = x$ (as is demonstrated by elementary ideas of calculus). The substitution $t = x + y$ yields

$$\Gamma(x+1) = e^{-x}\int_{-x}^\infty e^{x\ln(x+y)-y}dy = x^x e^{-x}\int_{-x}^\infty e^{x\ln\left(1+\frac{y}{x}\right)-y}dy \tag{2}$$

To this point the analysis has been rigorous. The following formal steps can be made rigorous by incorporating appropriate limiting procedures; however, because of the difficulty of the proofs, they have been omitted.

In Equation (2) introduce the logarithmic expansion

$$\ln\left(1+\frac{y}{x}\right) = \frac{y}{x} - \frac{y^2}{2x^2} + \frac{y^3}{3x^3} - + \cdots \tag{3}$$

and also let

$$y = \sqrt{x}\,\upsilon, \quad dy = \sqrt{x}\,d\upsilon$$

Then

$$\Gamma(x+1) = x^x e^{-x}\sqrt{x}\int_{-x}^\infty e^{-\upsilon^2/2+(\upsilon^3/3)\sqrt{x}-\cdots}d\upsilon \tag{4}$$

For large values of x

$$\Gamma(x+1) \approx x^x e^{-x}\sqrt{x}\int_{-x}^\infty e^{-\upsilon^2/2}d\upsilon = x^x e^{-x}\sqrt{2\pi x}$$

When x is replaced by integer values n, then the Stirling relation

$$n! = \Gamma(x+1) \approx \sqrt{2\pi x}\; x^x e^{-x} \tag{5}$$

is obtained.

It is of interest that from Equation (4) we can also obtain the result (12) on Page 391. See Problem 15.72.

Dirichlet integrals

15.21. Evaluate $I = \iiint\limits_V x^{\alpha-1}y^{\beta-1}z^{\gamma-1}dx\,dy\,dz$, where V is the region in the first octant bounded by the sphere $x^2 + y^2 + z^2 = 1$ and the coordinate planes.

Let $x^2 = u$, $y^2 = \upsilon$, $z^2 = w$. Then

$$I = \iiint_{\Re} u^{(\alpha-1)/2}\upsilon^{(\beta-1)/2}w^{(\gamma-1)/2}\frac{du}{2\sqrt{u}}\frac{d\upsilon}{2\sqrt{\upsilon}}\frac{dw}{2\sqrt{w}}$$
$$= \frac{1}{8}\iiint_{\Re} u^{(\alpha/2)-1}\upsilon^{(\beta/2)-1}w^{(\gamma/2)-1}du\,d\upsilon\,dw \qquad (1)$$

where $\Re$ is the region in the $u\upsilon w$ space bounded by the plane $u + \upsilon + w = 1$ and the $u\upsilon$, υw, and uw planes, as in Figure 15.2. Thus,

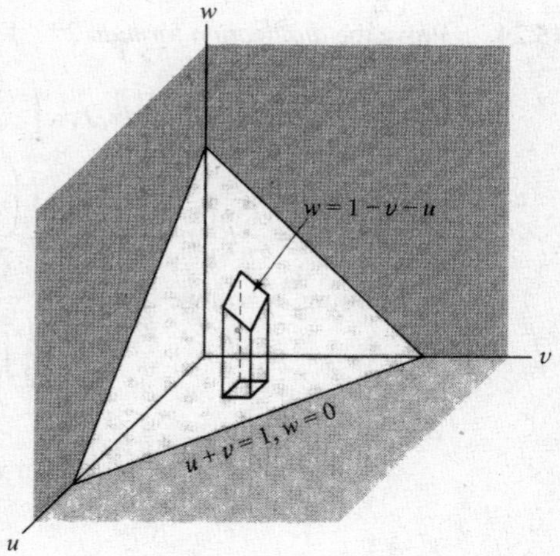

Figure 15.2

$$I = \frac{1}{8}\int_{u=0}^{1}\int_{\upsilon=0}^{1-u}\int_{w=0}^{1-u-\upsilon} u^{(\alpha/2)-1}\upsilon^{(\beta/2)-1}w^{(\gamma-/2)-1}du\,d\upsilon\,dw$$
$$= \frac{1}{4\gamma}\int_{u=0}^{1}\int_{\upsilon=0}^{1-u} u^{(\alpha/2)-1}\upsilon^{(\beta/2)-1}(1-u-\upsilon)^{\gamma/2}\,du\,d\upsilon \qquad (2)$$
$$= \frac{1}{4\gamma}\int_{u=0}^{1} u^{(\alpha/2)-1}\left\{\int_{\upsilon=0}^{1-u}\upsilon^{(\beta/2)-1}(1-u-\upsilon)^{\gamma/2}d\upsilon\right\}du$$

Letting $\upsilon = (1 - u)t$, we have

$$\int_{\upsilon=0}^{1-u}\upsilon^{(\beta/2)-1}(1-u-\upsilon)^{\gamma/2}d\upsilon = (1-u)^{(\beta+\gamma)/2}\int_{t=0}^{1}t^{(\beta/2)-1}(1-t)^{\gamma/2}dt$$
$$= (1-u)^{(\beta+\gamma)/2}\frac{\Gamma(\beta/2)\Gamma(\gamma/2+1)}{\Gamma[(\beta+\gamma)/2+1]}$$

so that Equation (2) becomes

$$I = \frac{1}{4\gamma}\frac{\Gamma(\beta/2)\Gamma(\gamma/2+1)}{\Gamma[(\beta+\gamma)/2+1]}\int_{u=0}^{1}u^{(\alpha/2)^{-1}}(1-u)^{(\beta+\gamma)/2}du$$
$$= \frac{1}{4\gamma}\frac{\Gamma(\beta/2)\Gamma(\gamma/2+1)}{\Gamma[(\beta+\gamma)/2+1]}\cdot\frac{\Gamma(\alpha/2)\Gamma(\beta+\gamma)/2+1)}{\Gamma[(\alpha+\beta+\gamma)/2+1]} = \frac{\Gamma(\alpha/2)\Gamma(\beta/2)\Gamma(\gamma/2)}{8\Gamma[(\alpha+\beta+)/2+1]} \qquad (3)$$

where we have used $(\gamma/2)\,\Gamma(\gamma/2) = \Gamma(\gamma/2 + 1)$.

The integral evaluated here is a special case of the Dirichlet integral Equation (20), Page 393. The general case can be evaluated similarly.

15.22. Find the mass of the region bounded by $x^2 + y^2 + z^2 = a^2$ if the density is $\sigma = x^2y^2z^2$.

The required mass $= 8\iiint_V x^2y^2z^2dx\,dy\,dz$, where V is the region in the first octant bounded by the sphere $x^2 + y^2 + z^2 = a^2$ and the coordinate planes.

In the Dirichlet integral, Equation (20), Page 393, let $b = c = a$, $p = q = r = 2$, and $\alpha = \beta = \gamma = 3$. Then the required result is

$$8.\frac{a^3\cdot a^3\cdot a^3}{2\cdot 2\cdot 2}\frac{\Gamma(3/2)\Gamma(3/2)\Gamma(3/2)}{\Gamma(1+3/2+3/2+3/2)} = \frac{4\pi s^9}{945}$$

Miscellaneous problems

15.23. Show that $\int_0^1\sqrt{1-x^4}\,dx = \frac{\{\Gamma(1/4)\}^2}{6\sqrt{2\pi}}$.

Let $x^4 = y$. Then the integral becomes

$$\frac{1}{4}\int_0^1 y^{-3/4}(1-y)^{1/2}dy = \frac{1}{4}\frac{\Gamma(1/4)\Gamma(3/2)}{\Gamma(7/4)} = \frac{\sqrt{\pi}}{4}\frac{\{\Gamma(1/4)\}^2}{\Gamma(1.4)\Gamma(3/4)}.$$

From Problem 15.17, with $p = 1/4$, $\Gamma(1/4)\Gamma(3/4) = \pi\sqrt{2}$, so that the required result follows.

15.24. Prove the *duplication formula* $2^{2p-1}\,\Gamma(p)\Gamma(p+\frac{1}{2}) = \sqrt{\pi}\;\Gamma(2p)$.

Let $I = \int_0^{\pi/2} \sin^{2p} x\,dx, J = \int_0^{\pi/2} \sin^{2p} 2x\,dx$.

Then $I = \frac{1}{2}B\left(p+\frac{1}{2},\frac{1}{2}\right) = \dfrac{\Gamma\left(p+\frac{1}{2}\right)\sqrt{\pi}}{2\Gamma(p+1)}$.

Letting $2x = u$, we find

$$J = \frac{1}{2}\int_0^{\pi} \sin^{2p} u\,du = \int_0^{\pi/2} \sin^{2p} u\,du = I$$

But

$$J = \int_0^{\pi/2} (2\sin x\cos x)^{2p}\,dx = 2^{2p}\int_0^{\pi/2} \sin^{2p} x\cos^{2p} x\,dx$$

$$= 2^{2p-1}B\left(p+\frac{1}{2},P+\frac{1}{2}\right) = \frac{2^{2p-1}\left\{\Gamma\left(p+\frac{1}{2}\right)\right\}^2}{\Gamma(2p+1)}$$

Then, since $I = J$,

$$\frac{\Gamma\left(p+\frac{1}{2}\right)\sqrt{\pi}}{2p\Gamma(p)} = \frac{2^{2p-1}\left\{\Gamma\left(p+\frac{1}{2}\right)\right\}^2}{2p\Gamma(2p)}$$

and the required result follows. (See Problem 15.74, where the duplication formula is developed for the simpler case of integers.)

15.25. Show that $\displaystyle\int_0^{\pi/2} \frac{d\phi}{\sqrt{1-\frac{1}{2}\sin^2\phi}} = \frac{\{\Gamma(1/4)\}^2}{4\sqrt{\pi}}$.

Consider

$$I = \int_0^{\pi/2} \frac{d\theta}{\sqrt{\cos\theta}} = \int_0^{\pi/2} \cos^{-1/2}\theta\,d\theta = \frac{1}{2}B\left(\frac{1}{4},\frac{1}{2}\right) = \frac{\Gamma\left(\frac{1}{4}\right)\sqrt{\pi}}{2\Gamma\left(\frac{3}{4}\right)} = \frac{\left\{\Gamma\left(\frac{1}{4}\right)\right\}^2}{2\sqrt{2\pi}}$$

as in Problem 15.23.

But $I = \displaystyle\int_0^{\pi/2} \frac{d\theta}{\sqrt{\cos\theta}} = \int_0^{\pi/2} \frac{d\theta}{\sqrt{\cos^2\theta/2-\sin^2\theta/2}} = \int_0^{\pi/2} \frac{d\theta}{\sqrt{1-2\sin^2\theta/2}}$.

Letting $\sqrt{2}\,\sin\theta/2 = \sin\phi$ in this last integral, it becomes $\sqrt{2}\displaystyle\int_0^{\pi/2} \frac{d\phi}{\sqrt{1-\frac{1}{2}\sin^2\phi}}$, from which the result follows.

15.26. Prove that $\displaystyle\int_0^{\infty} \frac{\cos x}{x^p}\,dx = \frac{\pi}{2\Gamma(p)\cos(p\pi/2)}, 0 < p < 1.$

We have $\dfrac{1}{x^p} = \dfrac{1}{\Gamma(p)}\displaystyle\int_0^{\infty} u^{p-1}e^{-xu}\,du$. Then

$$\int_0^{\infty} \frac{\cos x}{x^p}\,dx = \frac{1}{\Gamma(p)}\int_0^{\infty}\int_0^{\infty} u^{p-1}e^{-xu}\cos x\,du\,dx$$

$$= \frac{1}{\Gamma(p)}\int_0^{\infty} \frac{u^p}{1+u^2}\,du \qquad (1)$$

where we have reversed the order of integration and used Problem 12.22.

Letting $u^2 = \upsilon$ in the last integral, we have, by Problem 15.17,

$$\int_0^\infty \frac{u^p}{1+u^2}\,du = \frac{1}{2}\int_0^\infty \frac{\upsilon^{(p-1)/2}}{1+\upsilon}\,d\upsilon = \frac{\pi}{2\sin(p+1)\pi/2} = \frac{\pi}{2\cos p\pi/2} \qquad (2)$$

Substitution of Equation (2) in Equation (1) yields the required result.

15.27. Evaluate $\int_0^\infty \cos x^2\,dx$.

Letting $x^2 = y$, the integral becomes $\frac{1}{2}\int_0^\infty \frac{\cos y}{\sqrt{y}}\,dy = \frac{1}{2}\left(\frac{\pi}{2\Gamma\left(\frac{1}{2}\right)\cos\pi/4}\right) = \frac{1}{2}\sqrt{\pi/2}$ by Problem 15.26.

This integral and the corresponding one for the sine [see Problem 15.68(a)] are called *Fresnel integrals.*

SUPPLEMENTARY PROBLEMS

The gamma function

15.28. Evaluate (a) $\frac{\Gamma(7)}{2\Gamma(4)\Gamma(3)}$, (b) $\frac{\Gamma(3)\Gamma(3/2)}{\Gamma(9/2)}$, and (c) $\Gamma(1/2)\Gamma(3/2)\Gamma(5/2)$.

Ans. (a) 30 (b) 16/105 (c) $\frac{3}{8}\pi^{3/2}$

15.29. Evaluate (a) $\int_0^\infty x^4 e^{-x}\,dx$, (b) $\int_0^\infty x^6 e^{-3x}\,dx$, and (c) $\int_0^\infty x^2 e^{-2x^2}\,dx$.

Ans. (a) 24 (b) $\frac{80}{243}$ (c) $\frac{\sqrt{2\pi}}{16}$

15.30. Find (a) $\int_0^\infty e^{-x^3}\,dx$, (b) $\int_0^\infty \sqrt[4]{x}\,e^{-\sqrt{x}}\,dx$, and (c) $\int_0^\infty y^3 e^{-2y^5}\,dy$.

Ans. (a) $\frac{1}{3}\Gamma\left(\frac{1}{3}\right)$ (b) $\frac{3\sqrt{\pi}}{2}$ (c) $\frac{\Gamma(4/5)}{5\sqrt[5]{16}}$

15.31. Show that $\int_0^\infty \frac{e^{-st}}{\sqrt{t}}\,dt = \sqrt{\frac{\pi}{8}}$, $s > 0$.

15.32. Prove that $\Gamma(\upsilon) = \int_0^1 \left(\ln\frac{1}{x}\right)^{\upsilon-1} dx$, $\upsilon > 0$.

15.33. Evaluate (a) $\int_0^1 (\ln x)^4\,dx$, (b) $\int_0^1 (x\ln x)^3\,dx$, and (c) $\int_0^1 \sqrt[3]{\ln(1/x)}\,dx$.

Ans. (a) 24 (b) –3/128 (c) $\frac{1}{3}\Gamma\left(\frac{1}{3}\right)$

15.34. Evaluate (a) $\Gamma(-7/2)$ and (b) $\Gamma(-1/3)$.

Ans. (a) $(16\sqrt{\pi})/105$ (b) $-3\,\Gamma(2/3)$

15.35. Prove that $\lim_{x\to -m} \Gamma(x) = \infty$, where $m = 0, 1, 2, 3, \ldots$.

15.36. Prove that if m is a positive integer, $\Gamma\left(-m+\frac{1}{2}\right)=\frac{(-1)^m 2^m \sqrt{\pi}}{1\cdot 3\cdot 5\cdots(2m-1)}$.

15.37. Prove that $\Gamma'(1)=\int_0^\infty e^{-x}\ln x\,dx$ is a negative number (it is equal to $-\gamma$, where $\gamma = 0.577215\ldots$ is called *Euler's constant*, as in Problem 11.49).

The beta function

15.38. Evaluate (a) $B(3,5)$, (b) $B(3/2,2)$, and (c) $B(1/3, 2/3)$.

Ans. (a) 1/105 (b) 4/15 (c) $2\pi/\sqrt{3}$

15.39. Find (a) $\int_0^1 x^2(1-x)^3\,dx$, (b) $\int_0^1 \sqrt{(1-x)/x}\,dx$, and (c) $\int_0^1 (4-x^2)^{3/2}\,dx$.

Ans. (a) 1/60 (b) $\pi/2$ (c) 3π

15.40. Evaluate (a) $\int_0^4 u^{3/2}(4-u)^{5/2}\,du$ and (b) $\int_0^3 \frac{dx}{\sqrt{3x-x^2}}$.

Ans. (a) 12π (b) π

15.41. Prove that $\int_0^a \frac{dy}{\sqrt{a^4-y^4}}=\frac{\{\Gamma(1/4)\}^2}{4a\sqrt{2\pi}}$.

15.42. Evaluate (a) $\int_0^{\pi/2}\sin^4\theta\cos^4\theta\,d\theta$ and (b) $\int_0^{2\pi}\cos^6\theta\,d\theta$.

Ans. (a) $3\pi/256$ (b) $5\pi/8$

15.43. Evaluate (a) $\int_0^{\pi}\sin^5\theta\,d\theta$ and (b) $\int_0^{\pi/2}\cos^5\theta\sin^2\theta\,d\theta$.

Ans. (a) 16/15 (b) 8/105

15.44. Prove that $\int_0^{\pi/2}\sqrt{\tan\theta}\,d\theta=\pi/\sqrt{2}$.

15.45. Prove that (a) $\int_0^\infty \frac{x\,dx}{1+x^6}=\frac{\pi}{3\sqrt{3}}$ and (b) $\int_0^\infty \frac{y^2\,dy}{1+y^4}=\frac{\pi}{2\sqrt{2}}$.

15.46. Prove that $\int_{-\infty}^\infty \frac{e^{2x}}{ae^{3x}+b}\,dx=\frac{2\pi}{3\sqrt{3}a^{2/3}b^{1/3}}$. where $a, b > 0$.

15.47. Prove that $\int_{-\infty}^\infty \frac{e^{2x}}{(e^{3x}+1)}\,dx=\frac{2\pi}{9\sqrt{3}}$. [Hint: Differentiate with respect to Problem 15.46(b).]

15.48. Use the method of Problem 12.31 to justify the procedure used in Problem 15.11.

Dirichlet integrals

15.49. Find the mass of the region in the xy plane bounded by $x + y = 1$, $x = 0$, $y = 0$ if the density is $\sigma=\sqrt{xy}$.

Ans. $\pi/24$

15.50. Find the mass of the region bounded by the ellipsoid $\frac{x^2}{a^2}+\frac{y^2}{b^2}+\frac{z^2}{c^2}=1$ if the density varies as the square of the distance from its center.

Ans. $\frac{\pi abck}{30}(a^2+b^2+c^2)$, $k =$ constant of proportionality

15.51. Find the volume of the region bounded by $x^{2/3}+y^{2/3}+z^{2/3}=1$.

Ans. $4\pi/35$

15.52. Find the centroid of the region in the first octant bounded by $x^{2/3}+y^{2/3}+z^{2/3}=1$.

Ans. $\bar{x}=\bar{y}=\bar{z}=21/128$

15.53. Show that the volume of the region bounded by $x^m+y^m+z^m=a^m$, where $m>0$, is given by $\frac{8\{\Gamma(1/m)\}^3}{3m^2\Gamma(3/m)}a^3$.

15.54. Show that the centroid of the region in the first octant bounded by $x^m+y^m+z^m=a^m$, where $m>0$, is given by $\bar{x}=\bar{y}=\bar{z}=\frac{3\Gamma(2/m)\Gamma(3/m)}{4\Gamma(1/m)\Gamma(4/m)}a$.

Miscellaneous problems

15.55. Prove that $\int_a^b (x-a)^p\,(b-x)^q\,dx=(b-a)^{p+q+1}\,\mathrm{B}(p+1,q+1)$, where $p>-1$, $q>-1$, and $b>a$. [Hint: Let $x-a=(b-a)y$.]

15.56. Evaluate (a) $\int_1^3 \frac{dx}{\sqrt{(x-1)(3-x)}}$ and (b) $\int_3^7 \sqrt[4]{(7-x)(x-3)}\,dx$.

Ans. (a) π (b) $\frac{2\{\Gamma(1/4)\}^2}{3\sqrt{\pi}}$

15.57. Show that $\frac{\{\Gamma(1/3)\}^2}{\Gamma(1/6)}=\frac{\sqrt{\pi}\,\sqrt[3]{2}}{3}$.

15.58. Prove that $\mathbf{B}(u,\upsilon)=\frac{1}{2}\int_0^1 \frac{x^{u-1}+x^{\upsilon-1}}{(1+x)^{u+\upsilon}}\,dx$, where $u, \upsilon>0$. [Hint: Let $y=x/(1+x)$.]

15.59. If $0<p<1$, prove that $\int_0^{\pi/2}\tan^p\theta\,d\theta=\frac{\pi}{2}\sec\frac{p\pi}{2}$.

15.60. Prove that $\int_0^1 \frac{x^{u-1}(1-x)^{\upsilon-1}}{(x+r)^{u+\upsilon}}=\frac{\mathrm{B}(u,\upsilon)}{r^u(1+r)^{u+\upsilon}}$, where u, υ and r are positive constants. [Hint: Let $x=(r+1)y/(r+y)$.]

15.61. Prove that $\int_0^{\pi/2}\frac{\sin^{2u-1}\theta\cos^{2\upsilon-1}\theta\,d\theta}{(a\sin^2\theta+b\cos^2\theta)^{u+\upsilon}}=\frac{\mathrm{B}(u,\upsilon)}{2a^{\upsilon}b^{u}}$ where $u, \upsilon>0$. (Hint: Let $x=\sin^2\theta$ in Problem 15.60 and choose r appropriately.)

15.62. Prove that $\int_0^1 \frac{dx}{x^x}=\frac{1}{1^1}+\frac{1}{2^2}+\frac{1}{3^3}+\cdots$

15.63. Prove that for $m = 2, 3, 4, \ldots$, $\sin\frac{\pi}{m}\sin\frac{2\pi}{m}\sin\frac{3\pi}{m}\cdots\sin\frac{(m-1)\pi}{m} = \frac{m}{2^{m-1}}$. [Hint: Use the factored form $x^m - 1 = (x-1)(x-\alpha_1)(x-\alpha_2)\ldots(x-\alpha_{n-1})$, divide both sides by $x - 1$, and consider the limit as $x \to 1$.]

15.64. Prove that $\int_0^{\pi/2} \ln \sin x\,dx = -\pi/2 \ln 2$, using Problem 15.63. (Hint: Take logarithms of the result in Problem 15.63 and write the limit as $m \to \infty$ as a definite integral.)

15.65. Prove that $\Gamma\left(\frac{1}{m}\right)\Gamma\left(\frac{2}{m}\right)\Gamma\left(\frac{3}{m}\right)\cdots\Gamma\frac{(m-1)}{m} = \frac{(2\pi)^{(m-1)/2}}{\sqrt{m}}$. [Hint: Square the left-hand side and use Problem 15.63 and Equation (11a), Page 391.]

15.66. Prove that $\int_0^1 \ln\Gamma(x)\,dx = \frac{1}{2}\ln(2\pi)$. (Hint: Take logarithms of the result in Problem 15.65 and let $m \to \infty$.)

15.67. (a) Prove that $\int_0^\infty \frac{\sin x}{x^p}\,dx = \frac{\pi}{2\Gamma(p)\sin(p\pi/2)}$, $0 < p < 1$. (b) Discuss the cases p = 0 and p = 1.

15.68. Evaluate (a) $\int_0^\infty \sin x^2\,dx$ and (b) $\int_0^\infty x\cos x^3\,dx$

Ans. (a) $\frac{1}{2}\sqrt{\pi/2}$ (b) $\frac{\pi}{3\sqrt{3}\,\Gamma(1/3)}$

15.69. Prove that $\int_0^\infty \frac{x^{p-1}\ln x}{1+x}\,dx = -\pi^2 \csc p\pi \cot p\pi$, $0 < p < 1$.

15.70. Show that $\int_0^\infty \frac{\ln x}{x^4+1}\,dx = \frac{-\pi^2\sqrt{2}}{16}$

15.71. If $a > 0$, $b > 0$, and $4ac > b^2$, prove that $\int_{-\infty}^\infty\int_{-\infty}^\infty e^{-(ax^2+bxy+cy^2)}\,dx\,dy = \frac{2\pi}{\sqrt{4ac-b^2}}$.

15.72. Obtain Equation (12) on Page 391 from the result (4) of Problem 15.20. [Hint: Expand $e^{\upsilon^3/(3\sqrt{n})} + \cdots$ in a power series and replace the lower limit of the integral by $-\infty$.]

15.73. Obtain the result (15) on Page 392. [Hint: Observe that $\Gamma(x) = \frac{1}{x}\Gamma(x+!)$; thus, $\ln\Gamma(x) = \ln\Gamma(x+1) - \ln x$, and $\frac{\Gamma'(x)}{\Gamma(x)} = \frac{\Gamma'(x+1)}{\Gamma(x+1)} - \frac{1}{x}$. Furthermore, according to Equation (6), page 390, $\Gamma(x+!) = \lim_{k\to\infty}\frac{k!k^x}{(x+1)\cdots(x+k)}$. Now take the logarithm of this expression and then differentiate. Also, recall the definition of the Euler constant γ.]

15.74. The duplication formula (13a), Page 392, is proved in Problem 15.24. For further insight, develop it for positive integers; i.e., show that $2^{2n-1}\Gamma\left(n+\frac{1}{2}\right)\Gamma(n) = \Gamma(2n)\sqrt{\pi}$. [Hint: Recall that $\Gamma\left(\frac{1}{2}\right) = \pi$, then show that $\Gamma\left(n+\frac{1}{2}\right)\Gamma\left(\frac{2n+1}{2}\right) = \frac{(2n-1)\cdots 5\cdot 3\cdot 1}{2^n}\sqrt{\pi}$. Observe that $\frac{\Gamma(2n+1)}{2^n\Gamma(n+1)} = \frac{(2n!)}{2^n n!} = (2n-1)\cdots 5\cdot 3\cdot 1$. Now substitute and refine.

CHAPTER 16

Functions of a Complex Variable

Ultimately, it was realized that to accept numbers that provided solutions to equations such as $x^2 + 1 = 0$ was no less meaningful than had been the extension of the real number system to admit a solution for $x + 1 = 0$ or roots for $x^2 - 2 = 0$. The complex number system was in place around 1700, and by the early nineteenth century, mathematicians were comfortable with it. Physical theories took on a completeness not possible without this foundation of complex numbers and the analysis emanating from it. The theorems of the differential and integral calculus of complex functions introduce mathematical surprises as well as analytic refinement. This chapter is a summary of the basic ideas.

Functions

If to each of a set of complex numbers which a variable z may assume there corresponds one or more values of a variable w, then w is called a *function of the complex variable* z, written $w = f(z)$. The fundamental operations with complex numbers have already been considered in Chapter 1.

A function is *single-valued* if for each value of z there corresponds only one value of w; otherwise, it is *multiple-valued* or *many-valued*. In general, we can write $w = f(z) = u(x, y) + i\upsilon(x, y)$, where u and υ are real functions of x and y.

EXAMPLE. $w = z^2 = (x + iy)^2 = x^2 - y^2 + 2ixy = u + i\upsilon$ so that $u(x, y) = x^2 - y^2$, $\upsilon(x, y) = 2xy$. These are called the *real* and *imaginary parts* of $w = z^2$, respectively.

In complex variables, multiple-valued functions often are replaced by a specially constructed single-valued function with *branches*. This idea is discussed in a later paragraph.

EXAMPLE. Since $e^{2\pi ki} = 1$, the general polar form of z is $z = \rho\, e^{i(\theta+2\pi k)}$. This form and the fact that the logarithm and exponential functions are inverse leads to the following definition of ln z:

$$\ln z = \ln \rho + (\theta + 2\pi k)i \quad k = 0, 1, 2, \ldots, n \ldots$$

Each value of k determines a single-valued function from this collection of multiple-valued functions. These are the *branches* from which (in the realm of complex variables) a single-valued function can be constructed.

Limits and Continuity

Definitions of limits and continuity for functions of a complex variable are analogous to those for a real variable. Thus, $f(z)$ is said to have the *limit* l as z approaches z_0 if, given any $\epsilon > 0$, there exists a $\delta > 0$ such that $|f(z) - l| < \epsilon$ whenever $0 < |z - z_0| < \delta$.

Similarly, $f(z)$ is said to be *continuous* at z_0 if, given any $\epsilon > 0$, there exists a $\delta > 0$ such that $|f(z) - f(z_0)| < \epsilon$ whenever $|z - z_0| < \delta$. Alternatively, $f(z)$ is continuous at z_0 if $\lim_{z \to z_0} f(z) = f(z_0)$.

Note: While these definitions have the same appearance as in thereal variable setting, remember that $|z - z_0| < \delta$ means

$$|(x - x_0| + i(y - y_0)| = \sqrt{(x - x_0)^2 \ (y - y_0)^2} < \delta$$

Thus, there are two degrees of freedom as $(x, y) \to (x_0, y_0)$.

Derivatives

If $f(z)$ is single-valued in some region of the z plane, the *derivative* of $f(z)$, denoted by $f'(z)$, is defined as

$$\lim_{\Delta z \to 0} \frac{(f(z + \Delta z) - f(z)}{\Delta z} \tag{1}$$

provided the limit exists independent of the manner in which $\Delta z \to 0$. If the limit (*1*) exists for $z = z_0$, then $f(z)$ is called *analytic* at z_0. If the limit exists for all z in a region $\Re$, then $f(z)$ is called *analytic in* $\Re$. In order to be analytic, $f(z)$ must be single-valued and continuous. The converse, however, is not necessarily true.

We define elementary functions of a complex variable by a natural extension of the corresponding functions of a real variable. Where series expansions for real functions $f(x)$ exist, we can use as definition the series with x replaced by z. The convergence of such complex series has already been considered in Chapter 11.

EXAMPLE 1. We define $e^x = 1 + z + \frac{z^2}{2!} + \frac{z^3}{3!} + \cdots$, $\sin z = z - \frac{z^3}{3!} + \frac{z^5}{5!} - \frac{z^7}{7!} + \cdots$, and $\cos z = 1 - \frac{z^2}{2!} + \frac{z^4}{4!} - \frac{z^6}{6!} + \cdots$. From these we can show that $e^x = e^{x+iy} = e^x(\cos y + \sin y)$, as well as numerous other relations.

Rules for differentiating functions of a complex variable are much the same as for those of real variables. Thus, $\frac{d}{dz}(z^n) = nz^{n-1}$, $\frac{d}{dz}(\sin z) = \cos z$ and so on.

Cauchy-Riemann Equations

A necessary condition that $w = f(z) = u(x, y) + i\upsilon(x, y)$ be analytic in a region $\Re$ is that u and υ satisfy the *Cauchy-Riemann equations*

$$\frac{\partial u}{\partial x} = \frac{\partial \upsilon}{\partial y}, \qquad \frac{\partial u}{\partial y} = -\frac{\partial \upsilon}{\partial x} \tag{2}$$

(see Problem 16.7). If the partial derivatives in Equations (2) are continuous in $\Re$, the equations are sufficient conditions that $f(z)$ be analytic in $\Re$.

If the second derivatives of u and υ with respect to x and y exist and are continuous, we find by differentiating Equations (2) that

$$\frac{\partial^2 u}{\partial x^2} = \frac{\partial^2 u}{\partial y^2} = 0, \qquad \frac{\partial^2 \upsilon}{\partial x^2} + \frac{\partial^2 \upsilon}{\partial y^2} = 0 \tag{3}$$

Thus, the real and imaginary parts satisfy Laplace's equation in two dimensions. Functions satisfying Laplace's equation are called *harmonic functions*.

Integrals

Let $f(z)$ be defined, single-valued, and continuous in a region $\Re$. We define the *integral* of of $f(z)$ along some path C in $\Re$ from point z_1 to point z_2, where $z_1 = x_1 + iy_1$, $z_2 = x_2 + iy_2$, as

$$\int_c f(z)\,dz = \int_{(x_1, y_1)}^{(x_2, y_2)} (u + i\upsilon)(dx + i\,dy) = \int_{(x_1, y_1)}^{(x_2, y_2)} u\,dx - \upsilon\,dy + i\int_{(x_1, y_1)}^{(x_2, y_2)} \upsilon\,dx + u\,dy$$

With this definition, the integral of a function of a complex variable can be made to depend on line integrals for real functions already considered in Chapter 10. An alternative definition based on the limit of a sum, as for functions of a real variable, can also be formulated and turns out to be equivalent to the one aforementioned.

The rules for complex integration are similar to those for real integrals. An important result is

$$\left|\int_C f(z)\,dz\right| \leqq \int_C |f(z)||dz| \leqq M \int_C ds = ML \tag{4}$$

where M is an upper bound of $|f(z)|$ on C; i.e., $|f(z)| \leqq M$, and L is the length of the path C.

Complex function integral theory is one of the most esthetically pleasing constructions in all of mathematics. Major results are outlined as follows.

Cauchy's Theorem

Let C be a simple closed curve. If $f(z)$ is analytic within the region bounded by C as well as on C, then we have *Cauchy's theorem* that

$$\int_C f(z)\,dz \equiv \oint_C f(z)\,dz = 0 \tag{5}$$

where the second integral emphasizes the fact that C is a simple closed curve.

Expressed in another way, Equation (5) is equivalent to the statement that $\int_{z_1}^{z_2} f(z)\,dz$ has a value *independent of the path* joining z_1 and z_2. Such integrals can be evaluated as $F(z_2) - F(z_1)$, where $F'(z) = f(z)$. These results are similar to corresponding results for line integrals developed in Chapter 10.

EXAMPLE. Since $f(z) = 2z$ is analytic everywhere, we have for any simple closed curve C

$$\oint_C 2z\,dz = 0$$

Also,

$$\int_{2i}^{1+i} 2z\,dz = z^2\Big|_{2i}^{1+i} = (1+i)^2\ (2i)^2 = 2i + 4$$

Cauchy's Integral Formulas

If $f(z)$ is analytic within and on a simple closed curve C and a is any point interior to C, then

$$f(a) = \frac{1}{2\pi i} \oint_C \frac{f(z)}{z-a}\,dz \tag{6}$$

where C is traversed in the positive (counterclockwise) sense.

Also, the nth derivative of $f(z)$ at $z = a$ is given by

$$f^{(n)}(a) = \frac{n!}{2\pi i} \oint_C \frac{f(z)}{(z-a)^{n+1}}\,dz \tag{7}$$

These are called *Cauchy's integral formulas*. They are quite remarkable because they show that if the function $f(z)$ is known *on* the closed curve C then it is also known *within* C, and the various derivatives at points within C can be calculated. Thus, if a function of a complex variable has a first derivative, it has all higher derivatives as well. This, of course, is not necessarily true for functions of real variables.

Taylor's Series

Let $f(z)$ be analytic inside and on a circle having its center at $z = a$. Then for all points z in the circle we have the *Taylor series* representation of $f(z)$ given by

$$f(z) = f(a) + f'(a)(z-a) + \frac{f''(a)}{2!}(z-a)^2 + \frac{f'''(a)}{3!}(z-a)^3 + \cdots \tag{8}$$

See Problem 16.21.

Singular Points

A singular point of a function $f(z)$ is a value of z at which $f(z)$ fails to be analytic. If $f(z)$ is analytic everywhere in some region except at an interior point $z = a$, we call $z = a$ an *isolated singularity* of $f(z)$.

EXAMPLE. $f(z) = \dfrac{1}{(z-3)^2}$, then $z = 3$ is an isolated singularity of $f(z)$.

EXAMPLE. The function $f(z) = \dfrac{\sin z}{z}$ has a singularity at $z = 0$. Because $\lim\limits_{z\to 0}$ is finite, this singularity is called a removable singularity.

Poles

If $f(z) = \dfrac{\phi(z)}{(z-a)^n}$, $\phi(a) \neq 0$, where $\phi(z)$ is analytic everywhere in a region including $z = a$, and if n is a positive integer, then $f(z)$ has an isolated singularity at $z = a$, which is called a *pole of order n*. If $n = 1$, the pole is often called a *simple pole*; if $n = 2$, it is called a *double pole*, and so on.

Laurent's Series

If $f(z)$ has a pole of order n at $z = a$ but is analytic at every other point inside and on a circle C with center at a, then $(z-a)^n f(z)$ is analytic at all points inside and on C and has a Taylor series about $z = a$ so that

$$f(z) = \frac{a_{-n}}{(z-a)^n} + \frac{a_{-n+1}}{(z-a)^{n-1}} + \cdots + \frac{a_{-1}}{z-a} + a_0 + a_1(z-a) + a_2(z-a)^2 + \cdots \tag{9}$$

This is called a *Laurent series* for $f(z)$. The part $a_0 + a_1(z-a) + a_2(z-a)^2 + \ldots$ is called the *analytic part*. while the remainder consisting of inverse powers of $z - a$ is called the *principal part*. More generally, we refer to the series $\sum\limits_{k=-\infty}^{\infty} a_k(z-a)^k$ as a Laurent series, where the terms with $k < 0$ constitute the principal part.

A function which is analytic in a region bounded by two concentric circles having center at $z = a$ can always be expanded into such a Laurent series (see Problem 16.92).

It is possible to define various types of singularities of a function $f(z)$ from its Laurent series. For example, when the principal part of a Laurent series has a finite number of terms and $a_{-n} \neq 0$ while $a_{-n-1}, a_{-n-2}, \ldots$ are all zero, then $z = a$ is a pole of order n. If the principal part has infinitely many terms, $z = a$ is called an *essential singularity* or sometimes a *pole of infinite order*.

EXAMPLE. The function $e^{1/z} = 1 + \dfrac{1}{z} + \dfrac{1}{2!\,z^2} + \cdots$ has an essential singularity at $z = 0$.

Branches and Branch Points

Another type of singularity is a branch point. These points play a vital role in the construction of single-valued functions from ones that are multiple-valued, and they have an important place in the computation of integrals.

In the study of functions of a real variable, domains were chosen so that functions were single-valued. This guaranteed inverses and removed any ambiguities from differentiation and integration. The applications of complex variables are best served by the approach illustrated here. It is in the realm of real variables and yet illustrates a pattern appropriate to complex variables.

Let $y^2 = x$, $x < 0$, then $y = \pm\sqrt{x}$. In real variables, two functions f_1 and f_2 are described by $y = +\sqrt{x}$ on $x > 0$ and $y = -\sqrt{x}$ on $x > 0$, respectively. Each of them is single-valued.

An approach that can be extended to complex variables results by defining the positive x axis (not including zero) as a *cut* in the plane. This creates two *branches* f_1 and f_2 of a new function on a domain called the *Riemann axis*. The only passage joining the spaces in which the branches f_1 and f_2, respectively, are defined is through 0. This connecting point, zero, is given the special name *branch point*. Observe that two points x^* in the space of f_1 and x^{**} in that of f_2 can appear to be near each other in the ordinary view but, from the Riemannian perspective, are not. (See Figure 16.1.)

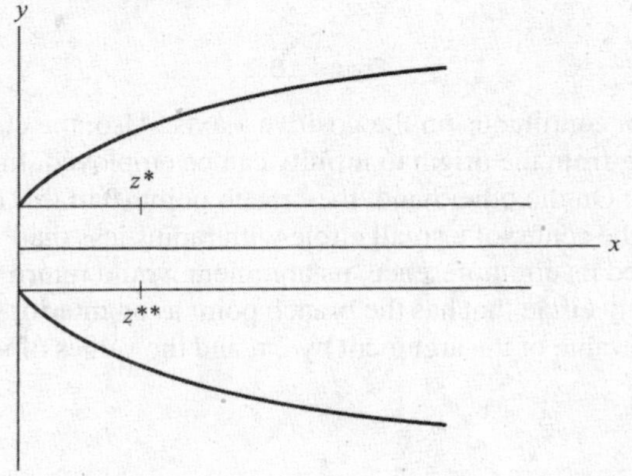

Figure 16.1

The preceding real variables construction suggests one for complex variables illustrated by $w = z^{1/2}$.

In polar coordinates, $e^{2\pi i} = 1$; therefore, the general representation of $w = z^{1/2}$ in that system is $w = \rho^{1/2}\, e^{i(\theta + 2\pi k)/2}$, $k = 0, 1$.

Thus, this function is double-valued.

If $k = 0$, then $w_1 = \rho^{1/2} \cdot e^{i\theta/2}$, $0 < \theta \le 2\pi$, $\rho > 0$

If $k = 1$, then $w_2 = \rho^{1/2} \cdot e^{i(\theta + 2\pi)/2} = \rho^{1/2} \cdot e^{i\theta/2} e^{i\pi} = -\rho^{1/2}$, $2\pi < \theta \le 4\pi$, $\rho > 0$.

Thus, the two branches of w are w_1 and w_2, where $w_1 = -w_2$. (The double-valued characteristic of w is illustrated by noticing that as z traverses a circle, C: $|z| = \rho$ through the values ϵ to 2π; the functional values run from $\rho^{1/2}\, e^{i\epsilon/2}$ to $\rho^{1/2} e^{\pi i}$. In other words, as z navigates the entire circle, the range variable only moves halfway around the corresponding range circle. In order for that variable to complete the circuit, z would have to make a second revolution. Thus, we would have coincident positions of z giving rise to distinct values of w. For example, $z_1 = e^{(\pi/2)i}$ are and $z_2 = e^{(\pi/2 + 2\pi)i}$ are coincident points on on the unit circle. The distinct functional values are $z_1^{1/2} = \frac{\sqrt{2}}{2}(1 + i)$ and $z_2^{1/2} = -\frac{\sqrt{2}}{2}(1 + i)$.

The following abstract construction replaces the multiple-valued function with a new single-valued one.

Make a cut in the complex plane that includes all of the positive x axis except the origin. Think of two planes P_1 and P_2, the first one of infinitesimal distance above the complex plane and the other infinitesimally below it. The point 0 which connects these spaces is called a *branch point*. The planes and the connecting point constitute a *Riemann surface*, and w_1 and w_2 are the branches of the function each defined in one of the planes. (Since the space of complex variables is the complex plane, this Riemann surface may be thought of as a flight of fancy that supports a rigorous analytic construction.)

To visualize this Riemann surface and perceive the single-valued character of the new function in it, first think of duplicates C_1 and C_2 of the domain circle C: $|z| = \rho$ in the planes P_1 and P_2, respectively. Start at $\theta = \epsilon$ on C_1, and proceed counterclockwise to the edge U_2 of the cut of P_1. (This edge corresponds to $\theta = 2\pi$.) Paste U_2 to L_1, the initial edge of the cut on P_2. Transfer to P_2 through this join and continue on C_2. Now, after a complete counterclockwise circuit of C_2, we reach the edge L_2 of the cut. Pasting L_2 to U_1 provides passage back to P_1 and makes it possible to close the curve in the Riemann plane. See Figure 16.2.

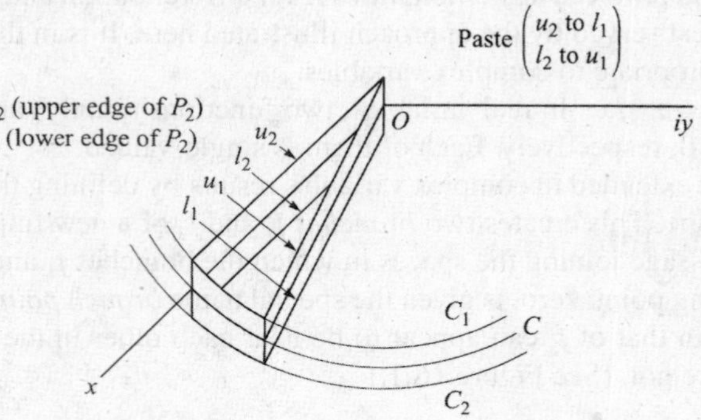

Figure 16.2

Note that the function is not continuous on the positive x axis. Also, the cut is somewhat arbitrary. Other rays and even curves extending from the origin to infinity can be employed. In many integration applications the cut $\theta = \pi i$ proves valuable. On the other hand, the branch point (0 in this example) is special. If another point, $z_0 \neq 0$, were chosen as the center of a small circle with radius less than $|z_0|$, then the origin would lie outside it. As a point z traversed its circumference, its argument would return to the original value, as would the value of w. However, for any circle that has the branch point as an interior point, a similar traversal of the circumference will change the value of the argument by 2π, and the values of w_1 and w_2 will be interchanged. (See Problem 16.37.)

Residues

The coefficients in Equation (9) can be obtained in the customary manner by writing the coefficients for the Taylor series corresponding to $(z - a)^n f(z)$. In further developments, the coefficient a_{-1}, called the formula *residue* of $f(z)$ at the pole $z = a$, is of considerable importance. It can be found from the formula

$$a_{-1} = \lim_{z \to a} \frac{1}{(n-1)!} \frac{d^{n-1}}{dz^{n-1}} \{(z-a)^n f(z)\} \tag{10}$$

where n is the order of the pole. For simple poles, the calculation of the residue is of particular simplicity since it reduces to

$$a_{-1} = \lim_{z \to a} (z - a) f(z) \tag{11}$$

Residue Theorem

If $f(z)$ is analytic in a region $\Re$ except for a pole of order n at $z = a$ and if C is any simple closed curve in $\Re$ containing $z = a$, then $f(z)$ has the form of Equation (9). Integrating Equation (9), using the fact that

$$\oint_C \frac{dz}{(z-a)^n} = \begin{cases} 0 & \text{if } n \neq 1 \\ 2\pi i & \text{if } n \neq 1 \end{cases} \tag{12}$$

(see Problem 16.13), it follows that

$$\oint_C f(z)\, dz = 2\pi i a_{-1} \tag{13}$$

i.e., the integral of $f(z)$ around a closed path enclosing a single pole of $f(z)$ is $2\pi i$ times the residue at the pole.

More generally, we have the following important theorem.

Theorem. If $f(z)$ is analytic within and on the boundary C of a region $\Re$ except at a finite number of poles $a, b, c, \ldots$ within $\Re$, having residues $a_{-1}, b_{-1}, c_{-1}, \ldots$, respectively, then

$$\oint_C f(z)\,dz = 2\pi i\,(a_{-1} + b_{-1} + c_{-1} + \cdots) \tag{14}$$

i.e., the integral of $f(z)$ is $2\pi i$ times the sum of the residues of $f(z)$ at the poles enclosed by C. Cauchy's theorem and integral formulas are special cases of this result, which we call the *residue theorem*.

Evaluation of Definite Integrals

The evaluation of various definite integrals can often be achieved by using the residue theorem together with a suitable function $f(z)$ and a suitable path or *contour* C, the choice of which may require great ingenuity. The following types are most common in practice.

1. $\int_0^\infty F(x)\,dx$, $F(x)$ is an even function.

   Consider $\oint_C F(z)\,dz$ along a contour C consisting of the line along the x axis from $-R$ to $+R$ and the semicircle above the x axis having this line as diameter. Then let $R \to \infty$. See Problems 16.29 and 16.30.

2. $\int_0^{2\pi} G(\sin\theta, \cos\theta)\,d\theta$, G is a rational function of $\sin\theta$ and $\cos\theta$.

   Let $z = e^{i\theta}$. Then $\sin\theta = \dfrac{z - z^{-1}}{2i}$, $\cos\theta = \dfrac{z + z^{-1}}{2}$ and $dz = ie^{\theta}\,d\theta$ or $d\theta = dz/iz$. The given integral is equivalent to $\oint_C F(z)\,dz$, where C is the unit circle with center at the origin. See Problems 16.31 and 16.32.

3. $\int_{-\infty}^{\infty} F(x)\begin{Bmatrix}\cos mx\\ \sin mx\end{Bmatrix} dx$, $F(x)$ is a rational function.

   Here we consider $\oint_C F(z)e^{imz}\,dz$, where C is the same contour as that in Type 1. See Problem 16.34.

4. Miscellaneous integrals involving particular contours. See Problems 16.35 and 16.38. In particular, Problem 16.38 illustrates a choice of path for an integration about a branch point.

SOLVED PROBLEMS

Functions, limits, continuity

16.1. Determine the locus represented by (a) $|z-2| = 3$, (b) $|z-2| = |z+4|$, and (c) $|z-3| + |z+3| = 10$.

(a) **Method 1:** $|z-2| = |x+iy-2| = |x-2+iy| = \sqrt{(x-2)^2+y^2} = 3$ or $(x-2)^2 + y^2 = 9$, a circle with center at (2, 0) and radius 3.

Method 2: $|z-2|$ is the distance between the complex numbers $z = x+iy$ and $2+0i$. If this distance is always 3, the locus is a circle of radius 3 with center at $2+0i$ or (2, 0).

(b) **Method 1:** $|x+iy-2| = |x+iy+4| =$ or $\sqrt{(x-2)^2+y^2} = \sqrt{(x+4)^2+y^2}$, Squaring, we find $x = -1$, a straight line.

Method 2: The locus is such that the distances from any point on it to (2, 0) and (–4, 0) are equal. Thus, the locus is the perpendicular bisector of the line joining (2, 0) and (–4, 0), or $x = -1$.

(c) **Method 2:** The locus is given by $\sqrt{(x-3)^2+y^2} + \sqrt{(x+3)^2+y^2} = 10$ or $\sqrt{(x-3)^2+y^2}$ $= 10 - \sqrt{(x+3)^2+y^2}$. Squaring and simplifying, $25 + 3x = 5\sqrt{(x+3)^2+y^2}$. Squaring and simplifying again yields $\dfrac{x^2}{25} + \dfrac{y^2}{16} = 1$, an ellipse with semimajor and semiminor axes of lengths 5 and 4, respectively.

Method 2: The locus is such that the sum of the distances from any point on it to (3, 0) and (–3, 0) is 10. Thus, the locus is an ellipse whose foci are at (–3, 0) and (3, 0) and whose major axis has length 10.

16.2. Determine the region in the z plane represented by each of the following.

(a) $|z| < 1$.

Interior of a circle of radius 1. See Figure 16.3(*a*).

(b) $1 < |z + 2i| \leqq 2$.

$|z + 2i|$ is the distance from z to $-2i$, so that $|z + 2i| = 1$ is a circle of radius 1 with center at $-2i$; Then $1 < |z + 2i| \leq 2$ represents the region *exterior* to $|z + 2i| = 1$ but *interior* to or *on* $|z + 2i| = 2$. See Figure 16.3(*b*).

(c) $\pi/3 \leqq \arg z \leqq \pi/2$.

Note that $\arg z = \phi$, where $z = \rho e^{i\phi}$. The required region is the infinite region bounded by the lines $\phi = \pi/3$ and $\phi = \pi/2$, including these lines. See Figure 16.3(*c*).

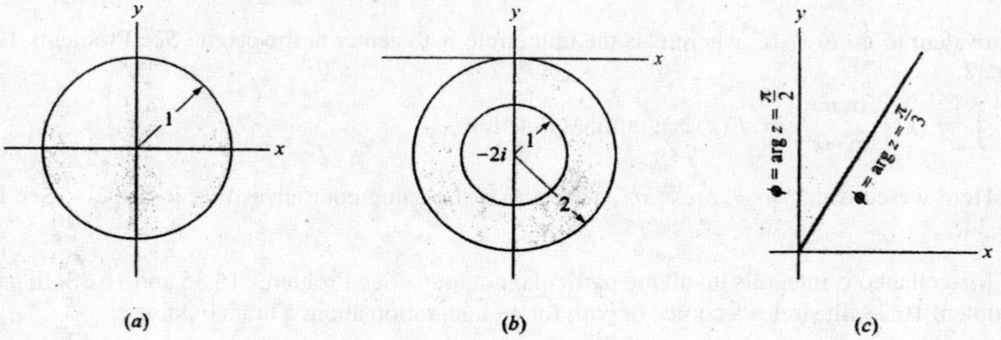

Figure 16.3

16.3. Express each function in the form $u(x, y) + i\upsilon(x, y)$, where u and υ are real: (a) z^3, (b) $1/(1-z)$, (c) e^{3z}, and (d) $\ln z$.

(a) $w = z^3 = (x+iy)^3 = x^3 + 3x^2(iy) + 3x(iy)^2 + (iy^3 = x^3 + 3ix^2y - 3xy^2 - iy^2$
$= x^3 - 3xy^2 + i(3x^2y - y^3)$

Then $u(x, y) = x^3 - 3xy^2$, $\upsilon(x, y) = 3x\ 3x^2y - y^3$.

(b) $$w = \frac{1}{1-z} = \frac{1}{1-(x+iy)} = \frac{1}{1-x-iy} \cdot \frac{1-x+iy}{1-x+iy} = \frac{1-x+iy}{(1-x)^2+y^2}$$

Then $u(x, y) = \dfrac{1-x}{(1-x)^2+y^2}$, $\upsilon(x, y) = \dfrac{y}{(1-x)^2+y^2}$.

(c) $e^3z = e^3(x + iy) = e^3x = e^3x\, e^3iy = e^3x\,(\cos 3y + i \sin 3y)$ and $u = e^3x \cos 3y$, $\upsilon = e^3x \sin 3y$

(d) $\ln z = \ln(\rho e^{i\phi}) = \ln \rho + i\phi = \ln \sqrt{x^2 + y^2} + i \tan^{-1} y/x$ and

$$u = \frac{1}{2} \ln (x^2 + y^2), \qquad \upsilon = \tan^{-1} y/x$$

Note that In z is a multiple-valued function (in this case it is *infinitely* many-valued), since ϕ can be increased by any multiple of 2π. The *principal value* of the logarithm is defined as that value for which $0 \leqq \phi < 2\pi$ and is called the *principal branch* of In z.

16.4. Prove (a) $\sin(x + iy) = \sin x \cosh y + i \cos x \sinh y$ and (b) $\cos(x + iy) = \cos x \cosh y - i \sin x \sinh y$.

We use the relations $e^{ix} = \cos z + i \sin z$, $e^{-ix} = \cos z - i \sin z$, from which

$$\sin z = \frac{e^{iz} - e^{-iz}}{2i}, \qquad \cos z = \frac{e^{iz} + e^{-iz}}{2}$$

Then

$$\sin z = \sin (x + iy) = \frac{e^{i(x+iy)} - e^{-i(x+iy)}}{2i} = \frac{e^{ix-y} - e^{-ix+y}}{2i}$$

$$= \frac{1}{2i}\{e^{-y} (\cos x + i \sin x) - e^{y} (\cos x - i \sin x)\}$$

$$= (\sin x)\left(\frac{e^{y} + e^{-y}}{2}\right) + i (\cos x)\left(\frac{e^{y} - e^{-y}}{2}\right)$$

$$= \sin x \cosh y + i \cos x \sinh y$$

Similarly,

$$\cos z = \cos(x + iy) = \frac{e^{i(x+iy)} + e^{-i(x+iy)}}{2}$$

$$= \frac{1}{2}\{e^{ix-y} + e^{-ix+y}\} = \frac{1}{2}\{e^{-y} (\cos x + i \sin x) + e^{y} (\cos x - i \sin x)\}$$

$$= (\cos x)\left(\frac{e^{y} + e^{-y}}{2}\right) - i (\sin x)\left(\frac{e^{y} - e^{-y}}{2}\right) = \cos x \cosh y - i \sin x \sinh y$$

Derivatives, cauchy-riemann equations

16.5. Prove that $\dfrac{d}{dz}\,\bar{z}$, where $\bar{z}$ is the conjugate of z, does not exist anywhere.

By definition, $\dfrac{d}{dz} f(z) = \lim\limits_{\Delta z \to 0} \dfrac{f(z + \Delta z) - f(z)}{\Delta z}$ if this limit exists independent of the manner in which $\Delta z = \Delta x + i\,\Delta y$ approaches zero. Then

$$\frac{d}{dz}\bar{z} = \lim_{\Delta z \to 0} \frac{\overline{z + \Delta z} - \bar{z}}{\Delta z} = \lim_{\substack{\Delta x \to 0 \\ \Delta y \to 0}} \frac{\overline{x + iy + \Delta x + i\,\Delta y} - \overline{x + iy}}{\Delta x + i\,\Delta y}$$

$$= \lim_{\substack{\Delta x \to 0 \\ \Delta y \to 0}} \frac{x - iy + \Delta x + i\,\Delta y - (x - iy)}{\Delta x + i\,\Delta y} = \lim_{\substack{\Delta x \to 0 \\ \Delta y \to 0}} \frac{\Delta x - i\,\Delta y}{\Delta x + i\,\Delta y}$$

If $\Delta y = 0$, the required limit is $\lim\limits_{\Delta x \to 0} \dfrac{\Delta x}{\Delta x} = 1$.

If $\Delta x = 0$, the required limit is $\lim\limits_{\Delta y \to 0} \dfrac{-i\,\Delta y}{i\,\Delta y} = -1$.

These two possible approaches show that the limit depends on the manner in which $\Delta z \to 0$, so that the derivative does not exist; i.e., $\bar{z}$ is *nonanalytic* anywhere.

16.6. (a) If $w = f(z) = \dfrac{1+z}{1-z}$, find $\dfrac{dw}{dz}$. (b) Determine where w is nonanalytic.

(a) **Method 1:**
$$\frac{dw}{dz} = \lim_{\Delta z \to \infty} \frac{\dfrac{1+(z+\Delta z)}{1-(z+\Delta z)} - \dfrac{1+z}{1-z}}{\Delta z} = \lim_{\Delta z \to 0} \frac{2}{(1-z-\Delta z)(1-z)}$$
$$= \frac{2}{(1-z)^2}$$

provided $z \neq 1$, independent of the manner in which $\Delta z \to 0$.

Method 2: The usual rules of differentiation apply provided $z \neq 1$. Thus, by the quotient rule for differentiation,

$$\frac{d}{dz}\left(\frac{1+z}{1-z}\right) = \frac{(1-z)\dfrac{d}{dz}(1+z) - (1+z)\dfrac{d}{dz}(1-z)}{(1-z)^2} = \frac{(1-z)(1)-(1+z)(-1)}{(1-z)^2} = \frac{2}{(1-z)^2}$$

(b) The function is analytic everywhere except at $z = 1$, where the derivative does not exist; i.e., the function is nonanalytic at $z = 1$.

16.7. Prove that a necessary condition for $w = f(z) = u(x, y) + i\,\upsilon(x, y)$ to be analytic in a region is that the Cauchy-Riemann equations $\dfrac{\partial u}{\partial x} = \dfrac{\partial \upsilon}{\partial y}, \dfrac{\partial u}{\partial x} = -\dfrac{\partial \upsilon}{\partial x}$ be satisfied in the region.

Since $f(z) = f(x + iy) = u(x, y) + i\,\upsilon(x, y)$, we have

$$f(z + \Delta z) = f[x + \Delta x + i(y + \Delta y)] = u(x + \Delta x, y + \Delta y) + i\,\upsilon(x + \Delta x, y + \Delta y)$$

Then

$$\lim_{\Delta z \to 0} \frac{f(z+\Delta z) - f(z)}{\Delta z}$$
$$= \lim_{\substack{\Delta x \to 0 \\ \Delta y \to 0}} \frac{u(x+\Delta x, y+\Delta y) - u(x, y) + i\{\upsilon(x+\Delta x, y+\Delta y) - \upsilon(x, y)\}}{\Delta x + i\,\Delta y}$$

If $\Delta y = 0$, the required limit is

$$\lim_{\Delta x \to 0} \frac{u(x+\Delta x, y) - u(x, y)}{\Delta x} + i\left\{\frac{\upsilon(x+\Delta x, y) - \upsilon(x, y)}{\Delta x}\right\} = \frac{\partial u}{\partial x} + i\frac{\partial \upsilon}{\partial x}$$

If $\Delta x = 0$, the required limit is

$$\lim_{\Delta y \to 0} \frac{u(x, y+\Delta y) - u(x, y)}{i\,\Delta y} + \left\{\frac{\upsilon(x, y+\Delta y) - \upsilon(x, y)}{\Delta y}\right\} = \frac{1}{i}\frac{\partial u}{\partial y} + \frac{\partial \upsilon}{\partial y}$$

If the derivative is to exist, these two special limits must be equal, i.e.,

$$\frac{\partial u}{\partial x} + i\frac{\partial \upsilon}{\partial x} = \frac{1}{i}\frac{\partial u}{\partial y} + \frac{\partial \upsilon}{\partial y} = -1\frac{\partial u}{\partial y} + \frac{\partial \upsilon}{\partial y}$$

so that we must have $\dfrac{\partial u}{\partial x} = \dfrac{\partial \upsilon}{\partial x}$ and $\dfrac{\partial \upsilon}{\partial x} = -\dfrac{\partial u}{\partial y}$.

Conversely, we can prove that if the first partial derivatives of u and υ with respect to x and y are continuous in a region, then the Cauchy-Riemann equations provide sufficient conditions for $f(z)$ to be analytic.

16.8. (a) If $f(z) = u(x, y) + i\,\upsilon(x, y)$ is analytic in a region $\Re$, prove that the one-parameter families of curves $u(x, y) = C_1$ and $\upsilon(x, y) = C_2$ are orthogonal families. (b) Illustrate by using $f(z) = z^2$.

(a) Consider any two particular members of these families u(x, y) = u0, υ(x, y) = υ0 which intersect at the point (x0, y0).

Since $du = u_x\,dx + u_y\,dy = 0$, we have $\dfrac{dy}{dx} = \dfrac{u_x}{u_y}$.

Also, since $d\upsilon = \upsilon_x\,dx + \upsilon_y\,dy = 0$, $\dfrac{dy}{dx} = \dfrac{\upsilon_x}{\upsilon_y}$.

When evaluated at (x_0, y_0), these represent, respectively, the slopes of the two curves at this point of intersection.

By the Cauchy-Riemann equations, $u_x = \upsilon_y$, $u_y = -\upsilon_x$, we have the product of the slopes at the point (x_0, y_0) equal to

$$\left(-\frac{u_x}{u_y}\right)\left(-\frac{\upsilon_x}{\upsilon_y}\right) = -1$$

so that any two members of the respective families are orthogonal, and thus the two families are orthogonal.

(b) If $f(z) = z2$, then $u = x2 - y2$, $\upsilon = 2xy$. The graphs of several members of $x2 - y2 = C1$, $2xy = C2$ are shown in Figure 16.4.

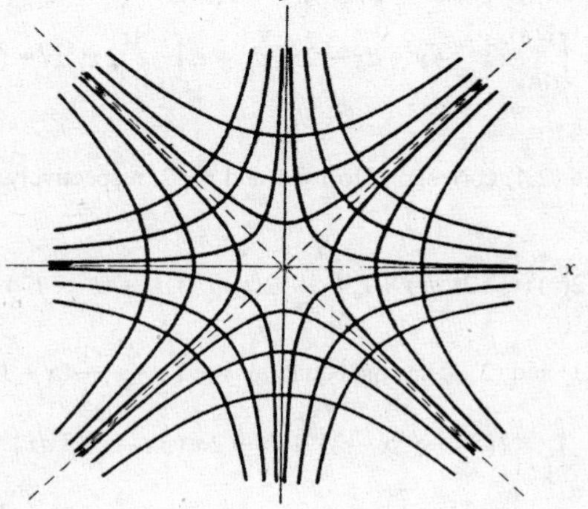

Figure 16.4

16.9. In aerodynamics and fluid mechanics, the functions ϕ and ψ in $f(z) = \phi + i\psi$, where $f(z)$ is analytic, are called the *velocity potential* and *stream function*, respectively. If $\phi = x^2 + 4x - y^2 + 2y$, (a) find ψ and (b) find $f(z)$.

(a) By the Cauchy-Riemann equations, $\dfrac{\partial\phi}{\partial x} = \dfrac{\partial\psi}{\partial y}$, $\dfrac{\partial\psi}{\partial x} = -\dfrac{\partial\phi}{\partial y}$. Then

$$\frac{\partial\psi}{\partial y} = 2x + 4 \qquad (1)$$

$$\frac{\partial\psi}{\partial x} = 2x - 2 \qquad (2)$$

Method 1: Integrating Equation (1), $\psi = 2xy + 4y + F(x)$.
Integrating Equation (2), $\psi = 2xy - 2x + G(y)$.

These are identical if $F(x) = -2x + c$, $G(y) = 4y + c$, where c is a real constant. Thus, $\psi = 2xy + 4y - 2x + c$.

Method 2: Integrating Equation (1), $\psi = 2xy + 4y + F(x)$. Then substituting in Equation (2), $2y + F'(x) = 2y - 2$ or $F'(x) = -2$ and $F(x) = -2x + c$. Hence, $\psi = 2xy + 4y - 2x + c$.

(b) From (a),

$$f(z) = \phi + i\psi = x2 + 4x - y2 + 2y + i(2xy + 4y - 2x + c)$$
$$= (x^2 - y^2 + 2ixy) + 4(x + iy) - 2i(x + iy) + ic = z^2 + 4z - 2iz + c_1$$

where c_1 is a pure imaginary constant.

This can also be accomplished by noting that $z = x + iy$, $\overline{z} = x - iy$ so that $x = \dfrac{z+\overline{z}}{2}$, $y = \dfrac{z-\overline{z}}{2i}$. The result is then obtained by substitution; the terms involving $\overline{z}$ drop out.

Integrals, Cauchy's theorem, Cauchy's integral formulas

16.10. Evaluate $\int_{1+i}^{2+4i} z^2\,dz$

(a) along the parabola $x = t$, $y = t^2$, where $1 \leqq t \leqq 2$

(b) along the straight line joining $1 + i$ and $2 + 4i$

(c) along straight lines from $1 + i$ to $2 + i$ and then to $2 + 4i$

We have

$$\int_{1+i}^{2+4i} z^2\,dz = \int_{(1.1)}^{(2.4)} (x+iy)^2\,(dx + idy) = \int_{(1.1)}^{(2.4)} (x^2 - y^2 + 2\,ixy)\,(dx + i\,dy)$$

$$= \int_{(1.1)}^{(2.4)} (x^2 - y^2)\,dx - 2xy\,dy + i\int_{(1.1)}^{(2.4)} 2xy\,dx + (x^2 - y^2)\,dy$$

Method 1:

(a) The points (1,1) and (2,4) correspond to $t = 1$ and $t = 2$, respectively. Then the preceding line integrals become

$$\int_{t=1}^{2} \{(t^2 - t^4)\,dt - 2(t)\,(t)^2\,2t\,dt\} + i\int_{t=1}^{2} \{2(t)\,(t^2)\,dt + (t^2 - t^4)\,(2t)\,dt\} = -\frac{86}{3} - 6i$$

(b) The line joining (1,1) and (2, 4) has the equation $y - 1 = \dfrac{4-1}{2-1}(x-1)$ or $y = 3x - 2$. Then we find

$$\int_{x=1}^{2} \{[x^2 - (3x-2)^2]\,dx - 2x\,(3x-2)\,3\,dx\}$$

$$+\,i\int_{x=1}^{2} \{2x\,(3x-2)\,dx + [x^2 - (3x-2)^2]\,3\,dx\} = -\frac{86}{3} - 6i$$

(c) From $1 + i$ to $2 + i$ [or (1, 1) to (2, 1)], $y = 1$, $dy = 0$ and we have

$$\int_{x=1}^{2} (x^2 - 1)\,dx + i\int_{x=1}^{2} 2x\,dx = \frac{4}{3} + 3i$$

From $2 + i$ to $2 + 4i$ [or (2, 1) to (2, 4)], $x = 2$, $dx = 0$ and we have

$$\int_{y=1}^{4} -4y\,dy + i\int_{y=1}^{4} (4 - y^2)\,dy = -\,30 - 9i$$

Adding, $(\frac{4}{3} + 3i) + (-30 - 91) = -\frac{86}{3} - 6i$.

Method 2: By the methods of Chapter 10 it is seen that the line integrals are independent of the path, thus accounting for the same values obtained in (a), (b), and (c). In such case the integral can be evaluated directly, as for real variables, as follows:

$$\int_{1+i}^{2+4i} z^2\,dz = \frac{z^3}{3}\Big|_{1i}^{2+4i} = \frac{(2+4i)^3}{3} - \frac{(1+i)^3}{3} = -\frac{86}{3} - 6i.$$

16.11. (a) Prove Cauchy's theorem: If $f(z)$ is analytic inside and on a simple closed curve C, then $\oint_C f(z)\,dz = 0$.

(b) Under these conditions prove that $\int_{P_1}^{P_2} f(z)\,dz$ is independent of the path joining P_1 and P_2.

(a) $\oint_C f(z)\,dz = \oint_C (u + i\upsilon)\,(dx + i\,dy) = \oint_C u\,dx - \upsilon\,dy + i\oint_C \upsilon\,dx + u\,dy$

By Green's theorem (Chapter 10),

$$\oint_C u\,dx - \upsilon\,dy = \iint_{\Re}\left(-\frac{\partial u}{\partial x} - \frac{\partial \upsilon}{\partial y}\right)dx\,dy, \qquad \oint_C \upsilon\,dx + u\,dy = \iint_{\Re}\left(\frac{\partial u}{\partial x} - \frac{\partial \upsilon}{\partial y}\right)dx\,dy$$

where $\Re$ is the region (simply-connected) bounded by C.

Since $f(z)$ is analytic, $\dfrac{\partial u}{\partial x} = \dfrac{\partial \upsilon}{\partial y}$, $\dfrac{\partial \upsilon}{\partial x} = -\dfrac{\partial u}{\partial y}$ (Problem 16.7), and so these integrals are zero. Then $\oint_C f(z)\,dz = 0$, assuming $f'(z)$ (and, thus, the partial derivatives) to be continuous.

(b) Consider any two paths joining points P_1 and P_2 (see Figure 16.5).
By Cauchy's theorem,

$$\int_{P_1AP_2BP_1} f(z)\,dz = 0$$

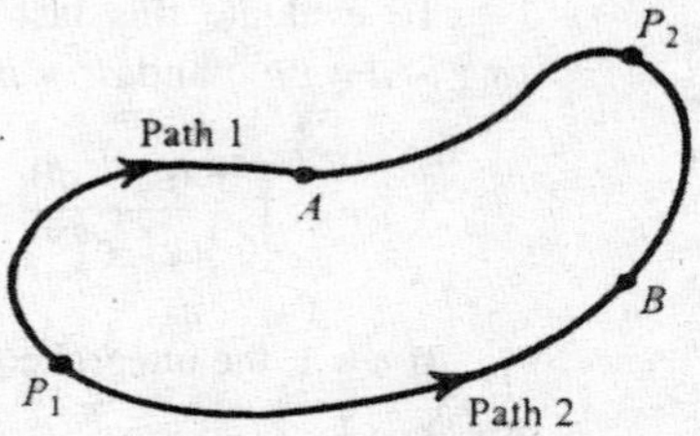

Figure 16.5

Then

$$\int_{P_1AP_2} f(z)\,dz + \int_{P_2BP_1} f(z)\,dz = 0$$

or

$$\int_{P_1AP_2} f(z)\,dz = -\int_{P_2BP_1} f(z)\,dz = \int_{P_1BP_2} f(z)\,dz$$

i.e., the integral along P_1AP_2 (path 1) = integral along P_1BP_2 (path 2), and so the integral is independent of the path joining P_1 and P_2.

This explains the results of Problem 16.10, since $f(z) = z^2$ is analytic.

16.12. If $f(z)$ is analytic within and on the boundary of a region bounded by two closed curves C_1 and C_2 (see Figure 16.6), prove that

$$\oint_{C_1} f(z)\,dz = \oint_{C_2} f(z)\,dz$$

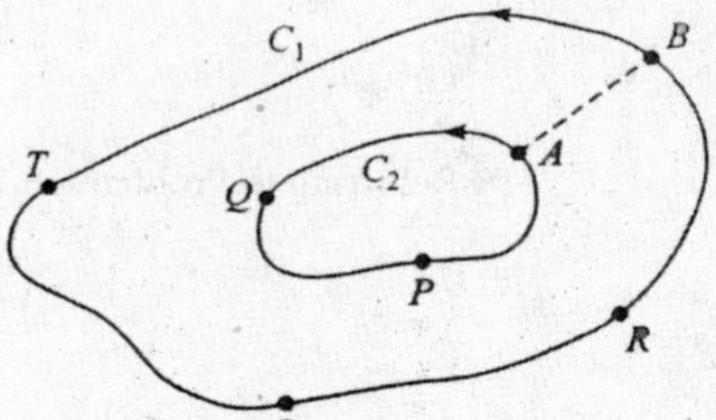

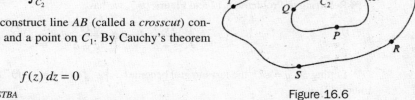

Figure 16.6

As in Figure 16.6, construct line AB (called a *crosscut*) connecting any point on C_2 and a point on C_1. By Cauchy's theorem (Problem 16.11),

$$\int_{AQPABRSTBA} f(z)\,dz = 0$$

since $f(z)$ is analytic within the region shaded and also on the boundary. Then

$$\int_{AQPA} f(z)\,dz + \int_{AB} f(z)\,dz + \int_{BRSTB} f(z)\,dz + \int_{BA} f(z)\,dz = 0 \qquad (1)$$

But $\int_{AB} f(z)\,dz = -\int_{BA} f(z)\,dz$. Hence, (1) gives

$$\int_{AQPA} f(z)\,dz = -\int_{BRSTB} f(z)\,dz = \int_{BTSRB} f(z)\,dz$$

i.e.,

$$\oint_{C_1} f(z)\,dz = \oint_{C_2} f(z)\,dz$$

Note that $f(z)$ need not be analytic *within* curve C_2.

16.13. (a) Prove that $\oint_C \frac{dz}{(z-a)^n} = \begin{cases} 2\pi i & \text{if } n = 1 \\ 0 & \text{if } n = 2, 3, 4, \ldots \end{cases}$, where C is a simple closed curve bounding a region having $z = a$ as interior point. (b) What is the value of the integral if $n = 0, -1, -2, -3, \ldots$?

(a) Let C_1 be a circle of radius ϵ having center at $z = a$ (see Figure 16.7). Since $(z-a)^{-n}$ is analytic within and on the boundary of the region bounded by C and C_1, we have, by Problem 16.12,

$$\oint_C \frac{dz}{(z-a)^n} = \oint_{C_1} \frac{dz}{(z-a)^n}$$

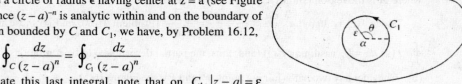

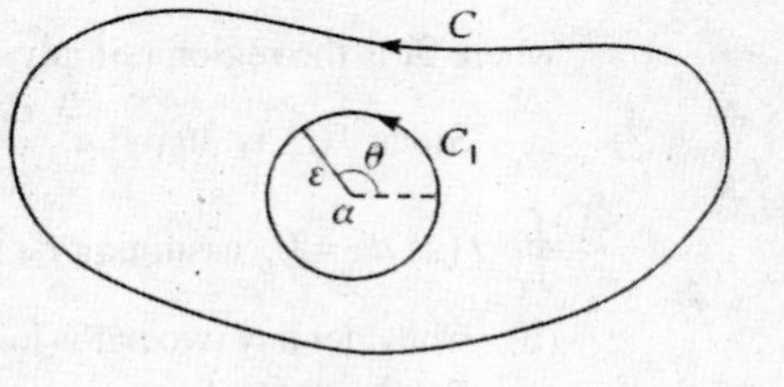

Figure 16.7

To evaluate this last integral, note that on C_1, $|z-a| = \varepsilon$ or $z - a = \varepsilon e^{i\theta}$ and $dz = i\varepsilon e^{i\theta}\, d\theta$. The integral equals

$$\int_0^{2\pi} \frac{i\varepsilon e^{i\theta}\, d\theta}{\varepsilon^n\, e^{in\theta}} = \frac{i}{\varepsilon^{n-1}} \int_0^{2\pi} e^{(1-n)i\theta}\, d\theta = \frac{i}{\varepsilon^{n-1}} \left. \frac{e^{(1-n)i\theta}}{(1-n)i} \right|_0^{2\pi} = 0 \quad \text{if } n \neq 1$$

If $n = 1$, the integral equals $i \int_0^{2\pi} d\theta = 2\pi i$.

(b) For $n = 0, -1, -2, \ldots$, the integrand is $1, (z-a), (z-a)^2, \ldots$ and is analytic everywhere inside C_1, including $z = a$. Hence, by Cauchy's theorem, the integral is zero.

16.14. Evaluate $\oint_C \frac{dz}{z-3}$, where C is (a) the circle $|z| = 1$ and (b) the circle $|z+i| = 4$.

(a) Since $z = 3$ is not interior to $|z| = 1$, the integral equals zero (Problem 16.11).

(b) Since $z = 3$ is interior to $|z+i| = 4$, the integral equals $2\pi i$ (Problem 16.13).

16.15. If $f(z)$ is analytic inside and on a simple closed curve C, and a is any point within C, prove that

$$f(a) = \frac{1}{2\pi i} \oint_C \frac{f(z)}{z-a}\, dz$$

Referring to Problem 16.12 and Figure 16.7, we have

$$\oint_C \frac{f(z)}{z-a}\, dz = \oint_{C_1} \frac{f(z)}{z-a}\, dz$$

Letting $z - a = \varepsilon e^{i\theta}$, the last integral becomes $i \int_0^{2\pi} f(a + \varepsilon e^{i\theta})\, d\theta$. But since $f(z)$ is analytic, it is continuous. Hence,

$$\lim_{\varepsilon \to 0} \int_0^{2\pi} f(a + \varepsilon e^{i\theta})\, d\theta = i \int_0^{2\pi} \lim_{\varepsilon \to 0} f(a + \varepsilon e^{i\theta})\, d\theta = i \int_0^{2\pi} f(a)\, d\theta = 2\pi i\, f(a)$$

and the required result follows.

16.16. Evaluate (a) $\oint_C \frac{\cos z}{z - \pi}\, dz$ and (b) $\oint_C \frac{e^x}{z(z+1)}\, dz$, where C is the circle $|z-1| = 3$.

(a) Since $z = \pi$ lies within C, $\frac{1}{2\pi i} \oint_C \frac{\cos z}{z-\pi}\, dz = \cos \pi = -1$ by Problem 16.15 with $f(z) = \cos z$, $a = \pi$.

Then $\oint_C \frac{\cos z}{z - \pi}\, dz = -2\pi i$.

(b)
$$\oint_C \frac{e^z}{z(z+1)}\, dz = \oint_C e^z \left(\frac{1}{z} - \frac{1}{z+1} \right) dz = \oint_C \frac{e^z}{z}\, dz - \oint_C \frac{e^z}{z+1}\, dz$$
$$= 2\pi i e^0 - 2\pi i e^{-1} = 2\pi i (1 - e^{-1})$$

by Problem 16.15, since $z = 0$ and $z = -1$ are both interior to C.

16.17. Evaluate $\oint_C \frac{5z^2 - 3z + 2}{(z-1)^3}\, dz$, where C is any simple closed curve enclosing $z = 1$.

Method 1: By Cauchy's integral formula, $f^{(n)}(a) = \frac{n!}{2\pi i} \oint_C \frac{f(z)}{(z-a)^{n+1}}\, dz$.

If $n = 2$ and $f(z) = 5z^2 - 3z + 2$, then $f''(1) = 10$. Hence,

$$10 = \frac{2!}{2\pi i} \oint_C \frac{5z^2 - 3z + 2}{(z-1)^3}\, dz \quad \text{or} \quad \oint_C \frac{5z^2 - 3z + 2}{(z-1)^3}\, dz = 10\pi i$$

Method 2: $5z^2 - 3z + 2 = 5(z-1)^2 + 7(z-1) + 4$. Then

$$\oint_C \frac{5z^2 - 3z + 2}{(z-1)^3}\, dz = \oint_C \frac{5(z-1)^2 + 7(z-1) + 4}{(z-1)^3}\, dz$$

$$= 5\oint_C \frac{d}{z-1} + 7\oint_C \frac{dz}{(z-1)^2} + 4\oint_C \frac{dz}{(z-1)^3} = 5(2\pi i) + 7(0) + 4(0)$$

$$= 10\pi i$$

by Problem 16.13.

Series and singularities

16.18. For what values of z does each series converge?

(a) $\sum_{n=1}^{\infty} \frac{z^n}{n^2\, 2^n}$. The nth term $= u_n = \frac{z^n}{n^2\, 2^n}$. Then

$$\lim_{n\to\infty} \left| \frac{u_{n+1}}{u_n} \right| = \lim_{n\to\infty} \left| \frac{z^{n+1}}{(n+1)^2\, 2^{n+1}} \cdot \frac{n^2\, 2^n}{z^n} \right| = \frac{|z|}{2}$$

By the ratio test, the series converges if $|z| < 2$ and diverges if $|z| > 2$. If $|z| = 2$, the ratio test fails.

However, the series of absolute values $\sum_{n=1}^{\infty} \left| \frac{z^n}{n^2\, 2^n} \right| = \sum_{n=1}^{\infty} \frac{|z|^n}{n^2\, 2^n}$ converges if $|z| = 2$, since $\sum_{n=1}^{\infty} \frac{1}{n^2}$ converges.

Thus, the series converges (absolutely) for $|z| \leqq 2$, i.e., at all points inside and on the circle $|z| = 2$

(b) $\sum_{n=1}^{\infty} \frac{(-1)^{n-1} z^{2n-1}}{(2n-1)!} = z - \frac{z^3}{3!} + \frac{z^5}{5!} - \cdots$. We have

$$\lim_{n\to\infty} \left| \frac{u_{n+1}}{u_n} \right| = \lim_{n\to\infty} \left| \frac{(-1)^n z^{2n+1}}{(2n+1)!} \cdot \frac{(2n-1)!}{(-1)^{n-1} z^{2n-1}} \right| = \lim_{n\to\infty} \left| \frac{-z^2}{2n(2n+1)} \right| = 0$$

Then the series, which represents $\sin z$, converges for all values of z.

(c) $\sum_{n=1}^{\infty} \frac{(z-i)^n}{3^n}$. We have $\lim_{n\to\infty} \left| \frac{u_{n+1}}{u_n} \right| = \lim_{n\to\infty} \left| \frac{(z-i)^{n+1}}{3^{n+1}} \cdot \frac{3n}{(z-i)^n} \right| = \frac{|z-i|}{3}$.

The series converges if $|z - i| < 3$, and diverges if $|z - i| > 3$.

If $|z - i| = 3$, then $z - i = 3e^{i\theta}$, and the series becomes $\sum_{n=1}^{\infty} e^{in\theta}$. This series diverges, since the nth term does not approach zero as $n \to \infty$.

Thus, the series converges within the circle $|z - i| = 3$ but not on the boundary.

16.19. If $\sum_{n=0}^{\infty} a_n z^n$ is absolutely convergent for $|z| \leqq R$, show that it is uniformly convergent for these values of z.

The definitions, theorems, and proofs for series of complex numbers are analogous to those for real series.

In this case we have $|a_n z^n| \le |a_n| R^n = M_n$. Since, by hypothesis, $\sum_{n=1}^{\infty} M_n$ converges, it follows by the Weierstrass M test that $\sum_{n=0}^{\infty} a_n z^n$ converges uniformly for $|z| \le R$.

16.20. Locate in the finite z plane all the singularities, if any, of each function and name them.

(a) $\dfrac{z^2}{(z+1)^3}$. $z = -1$ is a pole of order 3.

(b) $\dfrac{2z^3 - z + 1}{(z-4)^2\,(z-i)\,(z-1+2i)}$. $z = 4$ is a pole of order 2 (double pole); $z = i$ and $z = 1 - 2i$ are poles of order 1 (simple poles).

(c) $\dfrac{\sin mz}{z^2 + 2z + 2}$, $m \ne 0$. Since $z^2 + 2z + 2 = 0$ when $z = \dfrac{-2 \pm \sqrt{4-8}}{2} = \dfrac{-2 \pm 2i}{2} = 1 \pm i$, we can write

$$z^2 + 2z + 2 = \{z - (-1+i)\}\{z - (-1-i)\} = (z+1-i)(z+1+i).$$

The function has the two simple poles: $z = -1 + i$ and $z = -1 - i$.

(d) $\dfrac{1-\cos z}{z}$. $z = 0$ appears to be a singularity. However, since $\lim\limits_{x \to 0} \dfrac{1-\cos z}{z} = 0$, singularity.

Another method: Since $\dfrac{1-\cos z}{z} = \dfrac{1}{z}\left\{1 - \left(\dfrac{z^2}{2!} + \dfrac{z^4}{4!} - \dfrac{z^6}{6!} + \cdots\right)\right\} = \dfrac{z}{2!} - \dfrac{z^3}{4!} + \cdots$, we see that $z = 0$ is a removable singularity.

(e) $e^{-1/(x-1)2} = 1 - \dfrac{1}{(z-1)^2} + \dfrac{1}{2!(z-1)^4} - \cdots$. This is a Laurent series where the principal part has an infinite number of nonzero terms. Then $z = 1$ is an *essential singularity.*

(f) e^z. This function has no finite singularity. However, letting $z = 1/u$, we obtain $e^{1/u}$, which has an essential singularity at $u = 0$. We conclude that $z = \infty$ is an essential singularity of e^z.

In general, to determine the nature of a possible singularity of $f(z)$ at $z = \infty$, we let $z = \infty$, we let $z = 1/u$ and then examine the behavior of the new function at $u = 0$.

16.21. If $f(z)$ is analytic at all points inside and on a circle of radius R with center at a, and if $a + h$ is any point inside C, prove *Taylor's theorem* that

$$f(a+h) = f(a) + hf'(a) + \frac{h^2}{2!} f''(a) + \frac{h^3}{3!} f'''(\alpha) + \cdots$$

By Cauchy's integral formula (Problem 16.15), we have

$$f(a+h) = \frac{1}{2\pi i} \oint_C \frac{f(z)\,dz}{z-a-h} \tag{1}$$

By division,

$$\begin{aligned}\frac{1}{z-a-h} &= \frac{1}{(z-a)[1-h/(z-a)]} \\ &= \frac{1}{(z-a)}\left\{1 + \frac{h}{(z-a)} + \frac{h^2}{(z-a)^2} + \cdots + \frac{h^n}{(z-a)^n} + \frac{h^{n+1}}{(z-a)^n\,(z-a-h)}\right\}\end{aligned} \tag{2}$$

Substituting Equation (2) in Equation (1) and using Cauchy's integral formulas, we have

$$\begin{aligned}f(a+h) &= \frac{1}{2\pi i}\oint_C \frac{f(z)\,dz}{z-a} + \frac{h}{2\pi i}\oint_C \frac{f(z)\,dz}{(z-a)^2} + \cdots + \frac{h^n}{2\pi i}\oint_C \frac{f(z)\,dz}{(z-a)^{n+1}} + R_n \\ &= f(a) + hf'(a) + \frac{h^2}{2!} f''(a) + \cdots + \frac{h^n}{n!} f^{(n)}(a) + R_n\end{aligned}$$

where

$$R_n = \frac{h^{n+1}}{2\pi i}\oint_C \frac{f(z)\,dz}{(z-a)^{n+1}\,(z-a-h)}$$

Now when z is on C, $\left|\dfrac{f(z)}{z-a-h}\right| \leq M$ and $|z-a| = R$, so that by Equation (4), Page 407, we have, since $2\pi R$ is the length of C,

$$|R_n| \leq \frac{|h|^{n+1}M}{2\pi\;R^{n+1}}\cdot 2\pi R$$

As $n \to \infty$, $|R_n| \to 0$. Then $R_n \to 0$, and the required result follows.

If $f(z)$ is analytic in an annular region $r_1 \leqq |z-a| \leqq r_2$, we can generalize the Taylor series to a Laurent series (see Problem 16.92). In some cases, as shown in Problem 16.22, the Laurent series can be obtained by use of known Taylor series.

16.22. Find Laurent series about the indicated singularity for each of the following functions. Name the singularity in each case and give the region of convergence of each series.

(a) $\dfrac{e^z}{(z-1)^2}$; $z = 1$. Let $z - 1 = u$. Then $z = 1 + u$ and

$$\frac{e^z}{(z-1)^2} = \frac{e^{1+u}}{u^2} = e\cdot\frac{e^u}{u^2} = \frac{e}{u^2}\left\{1 + u + \frac{u^2}{2!} + \frac{u^3}{3!} + \frac{u^4}{4!} + \cdots\right\}$$

$$= \frac{e}{(z-1)^2} + \frac{e}{z-1} + \frac{e}{2!} + \frac{e(z-1)}{3!} + \frac{e(z-1)^2}{4!} + \cdots$$

$z = 1$ is a *pole of order 2*, or *double pole*.
The series converges for all values of $z \neq 1$.

(b) $z\cos\dfrac{1}{z}$; $z = 0$.

$$z\cos\frac{1}{z} = z\left(1 - \frac{1}{2!\,z^2} + \frac{1}{4!\,z^4} - \frac{1}{6!\,z^6} + \cdots\right) = z - \frac{1}{2!\,z} + \frac{1}{4!\,z^3} - \frac{1}{6!\,z^5} + \cdots$$

$z = 0$ is an *essential singularity*.
The series converges for all values of $z \neq 0$.

(c) $\dfrac{\sin z}{z-\pi}$; $z = \pi$. Let $z - \pi = u$. Then $z = \pi + u$ and

$$\frac{\sin z}{z-\pi} = \frac{\sin(u+\pi)}{u} = -\frac{\sin u}{u} = -\frac{1}{u}\left(u - \frac{u^3}{3!} + \frac{u^5}{5!} - \cdots\right)$$

$$= -1 + \frac{u^2}{3!} + \frac{u^4}{5!} + \cdots = -1 + \frac{(z-\pi)^2}{3!} - \frac{(z-\pi)^4}{5!} + \cdots$$

$z = \pi$ is a *removable singularity*.
The series converges for all values of z.

(d) $\dfrac{z}{(z+1)(z+2)}$; $z = -1$. Let $z + 1 = u$. Then

$$\frac{z}{(z+1)(z+2)} = \frac{u-1}{u(u-1)} = \frac{u-1}{u}(1 - u + u^2 - u^3 + u^4 - \cdots)$$

$$= -\frac{1}{u} + 2 - 2u + 2u^2 - 2u^3 + \cdots$$

$$= -\frac{1}{z+1} + 2 - 2(z+1) + 2(z+1)^2 - \cdots$$

$z = -1$ is a *pole of order 1*, or *simple pole*.
The series converges for values of z such that $0 < |z+1| < 1$.

(e) $\dfrac{1}{z(z+2)^3}$; $z = 0, -2$.

Case 1, $z = 0$. Using the binomial theorem,

$$\frac{1}{z(z+2)^3} = \frac{1}{8z\,(1+z/2)^3} = \frac{1}{8z}\left\{1+(-3)\left(\frac{z}{2}\right)+\frac{(-3)(-4)}{2!}\left(\frac{z}{2}\right)^2+\frac{(-3)(-4)(-5)}{3!}\left(\frac{z}{2}\right)^3+\cdots\right\}$$

$$= \frac{1}{8z} - \frac{3}{16} + \frac{3}{16}z - \frac{5}{32}z^2 + \cdots$$

$z = 0$ is a *pole of order 1*, or *simple pole*.
The series converges for $0 < |z| < 2$.
Case 2, $z = -2$. Let $z + 2 = u$. Then

$$\frac{1}{z(z+2)^3} = \frac{1}{(u-2)\,u^3} = \frac{1}{-2u^3\,(1-u/2)} = -\frac{1}{2u^3}\left\{1+\frac{u}{2}+\left(\frac{u}{2}\right)^2+\left(\frac{u}{2}\right)^3+\left(\frac{u}{2}\right)^4+\cdots\right\}$$

$$= -\frac{1}{2u^3} - \frac{1}{4u^2} - \frac{1}{8u} - \frac{1}{16} - \frac{1}{32}u - \cdots$$

$$= -\frac{1}{2(z+2)^3} - \frac{1}{4(z+2)^2} - \frac{1}{8(z+2)} - \frac{1}{16} - \frac{1}{32}(z+2) - \cdots$$

$z = -2$ is a *pole of order 3*.
The series converges for $0 < |z+2| < 2$.

Residues and the residue theorem

16.23. Suppose $f(z)$ is analytic everywhere inside and on a simple closed curve C except at $z = a$, which is a pole of order n. Then

$$f(z) = \frac{a_{-n}}{(z-a)^n} + \frac{a_{-n+1}}{(z-a)^{n-1}} + \cdots + a_0 + a_1\,(z-a) + a_2\,(z-a)^2 + \cdots$$

where $a_{-n} \neq 0$. Prove that

(a) $\oint_C f(z)\,dz = 2\pi i a_{-1}$

(b) $a_{-1} = \lim_{z\to a} \dfrac{1}{(n-1)!}\dfrac{d^{n-1}}{dz^{n-1}}\{(z-a)^n\,f(z)\}$

(a) By integration, we have, on using Problem 16.13,

$$\oint_C f(z)\,dz = \oint_C \frac{a_{-n}}{(z-a)^n}\,dz + \cdots + \oint_C \frac{a_{-1}}{z-a}\,dz + \oint_C \{a_0 + a_1\,(z-a) + a_2\,(z-a)^2 + \cdots\}\,dz$$
$$= 2\pi i a_{-1}$$

Since only the term involving a_{-1} remains, we call a_{-1} the *residue* of $f(z)$ at the pole $z = a$.

(b) Multiplication by $(z-a)^n$ gives the Taylor series

$$(z-a)^n\,f(z) = a_{-n} + a_{-n+1}\,(z-a) + \cdots + a_{-1}\,(z-a)^{n-1} + \cdots$$

Taking the $(n-1)$st derivative of both sides and letting $z \to a$, we find

$$(n-1)!\,a_{-1} = \lim_{z\to a}\frac{d^{n-1}}{dz^{n-1}}\{(z-a)^n\,f(z)\}$$

from which the required result follows.

16.24. Determine the residues of each function at the indicated poles.

(a) $\dfrac{z^2}{(z-2)(z^2+1)}$; $2, i, -i$. These are simple poles. Then:

$$\text{Residue at } z = 2 \text{ is} \quad \lim_{z\to 2}(z-2)\left\{\frac{z^2}{(z-2)(z^2+1)}\right\} = \frac{4}{5}$$

$$\text{Residue at } z = i \text{ is} \quad \lim_{z\to i}(z-i)\left\{\frac{z^2}{(z-2)(z-i)(z+i)}\right\} = \frac{i^2}{(i-2)(2i)} = \frac{1-2i}{10}$$

$$\text{Residue at } z = -i \text{ is} \quad \lim_{z\to -i}(z+i)\left\{\frac{z^2}{(z-2)(z-i)(z+i)}\right\} = \frac{i^2}{(-i-2)(-2i)} = \frac{1+2i}{10}$$

(b) $\dfrac{1}{z(z+2)^3}$; $z = 0, -2$. $z = 0$ is a simple pole, $z = -2$ is a pole of order 3. Then:

$$\text{Residue at } z = 0 \text{ is} \quad \lim_{z\to 0} z\cdot\frac{1}{z(z+2)^3} = \frac{1}{8}$$

$$\text{Residue at } z = -2 \text{ is} \quad \lim_{z\to -2}\frac{1}{2!}\frac{d^2}{dz^2}\left\{(z+2)^3\cdot\frac{1}{z(z+2)^3}\right\}$$

$$= \lim_{z\to -2}\frac{1}{2}\frac{d^2}{dz^2}\left(\frac{1}{z}\right) = \lim_{z\to -2}\frac{1}{2}\left(\frac{2}{z^3}\right) = -\frac{1}{8}$$

Note that these residues can also be obtained from the coefficients of $1/z$ and $1/(z+2)$ in the respective Laurent series [see Problem 16.22(e)].

(c) $\dfrac{ze^{zt}}{(z-3)^2}$; $z = 3$, a pole of order 2 or double pole. Then:

$$\text{Residue is} \quad \lim_{z\to 3}\frac{d}{dz}\left\{(z-3)^2\cdot\frac{ze^{zt}}{(z-3)^2}\right\} = \lim_{z\to 3}\frac{d}{dz}(ze^{zt}) = \lim_{z\to 3}(e^{zt} + zte^{zt})$$

$$= e^{3t} + 3te^{3t}$$

(d) $\cot z$; $z = 5\pi$, a pole of order 1. Then:

$$\text{Residue is} \quad \lim_{z\to 5\pi}(z-5\pi)\cdot\frac{\cos z}{\sin z} = \left(\lim_{z\to 5\pi}\frac{z-5\pi}{\sin z}\right)\left(\lim_{z\to 5\pi}\cos z\right) = \left(\lim_{z\to 5\pi}\frac{1}{\cos z}\right)(-1)$$

where we have used L'Hospital's rule, which can be shown to be applicable for functions of a complex variable.

16.25. If $f(z)$ is analytic within and on a simple closed curve C except at a number of poles $a, b, c, \ldots$ interior to C, prove that

$$\oint_C f(z)\,dz = 2\pi i\ \{\text{sum of residues of } f(z) \text{ at poles } a, b, c, \text{ etc.}\}$$

Refer to Figure 16.8.

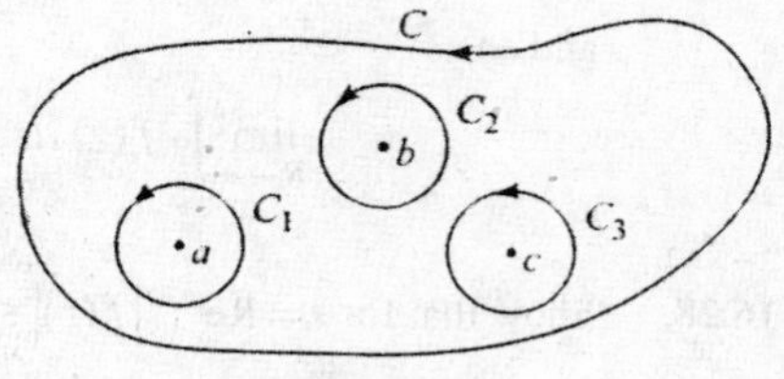

Figure 16.8

By reasoning similar to that of Problem 16.12 (i.e., by constructing crosscuts from C to C_1, C_2, C_3, etc.), we have

$$\oint_C f(z)\,dz = \oint_{C_1} f(z)\,dz + \oint_{C_2} f(z)\,dz + \cdots$$

For pole a,

$$f(z)\frac{a_{-m}}{(z-a)^m} + \cdots + \frac{a_{-1}}{(z-a)} + a_0 + a_1(z-a) + \cdots$$

hence, as in Problem 16.23,

$$\oint_{C_1} f(z)\,dz = 2\pi i a_{-1}$$

Similarly for pole b,

$$f(z) = \frac{b_{-n}}{(z-b)^n} + \cdots + \frac{b_{-1}}{(z-b)} + b_0 + b_1(z-b) + \cdots$$

so that

$$\oint_{C_2} f(z)\,dz = 2\pi i b_{-1}$$

Continuing in this manner, we see that

$$\oint_{C_2} f(z)\,dz = 2\pi i\,(a_{-1} + b_{-1} + \cdots) = 2\pi i \text{ (sum of residues)}$$

16.26. Evaluate $\oint_C \frac{e^z\,dz}{(z-1)(z+3)^2}$, where C is given by (a) $|z| = 3/2$ and (b) $|z| = 10$.

Residue at simple pole $z = 1$ is $\lim_{z\to 1}\left\{(z-1)\frac{e^2}{(z-1)(z+3)^2}\right\} = \frac{e}{16}$

Residue at double pole $z = -3$ is $\lim_{z\to -3}\frac{d}{dz}\left\{(z+3)^2\frac{e^z}{(z-1)(z+3)^2}\right\} = \lim_{z\to -3}\frac{(z-1)e^z - e^z}{(z-1)^2} = \frac{-5e^{-3}}{16}$

(a) Since $|z| = 3/2$ encloses only the pole $z = 1$,

$$\text{the required integral } = 2\pi i\left(\frac{e}{16}\right) = \frac{\pi i e}{8}$$

(b) Since $|z| = 10$ encloses both poles $z = 1$ and $z = -3$,

$$\text{the required integral} = 2\pi i\left(\frac{e}{16} - \frac{5e^{-3}}{16}\right) = \frac{\pi i(e - 5e^{-3})}{8}$$

Evaluation of definite integrals

16.27. If $|f(z)| \leqq \frac{M}{R^k}$ for $z = Re^{i\theta}$, where $k > 1$ and M are constants, prove that $\lim_{R\to\infty}\int_\Gamma f(z)\,dz = 0$, where Γ is the semicircular arc of radius R shown in Figure 16.9.

By the result (4), Page 407, we have

$$\left|\int_\Gamma f(z)\,dz\right| \leqq \int_\Gamma |f(z)|\,|dz| \leqq \frac{M}{R^k}\cdot \pi R + \frac{\pi M}{R^{k-1}}$$

since the length of arc $L = \pi R$. Then

$$\lim_{R\to\infty}\left|\int^\Gamma f(z)\,dz\right| = 0$$

and so

$$\lim_{R\to\infty}\int_\Gamma f(z)\,dz = 0$$

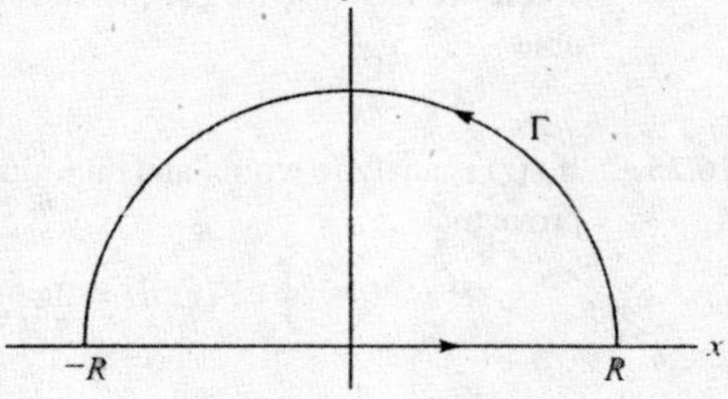

Figure 16.9

16.28. Show that for z = $Re^{i\theta}$, $|f(z)| \leqq \frac{M}{R^k}$, $k > 1$ if $f(z) = \frac{1}{1+z^4}$.

If $z = Re^{i\theta}, |f(z)| = \left|\frac{1}{1+R^4\,e^{4i\theta}}\right| \leqq \frac{1}{|R^4\,e^{4i\theta}| - 1} = \frac{1}{R^4 - 1} \leqq \frac{2}{R^4}$ if R is large enough ($R > 2$, for example) so that $M = 2$, $k = 4$.

Note that we have made use of the inequality $|z_1 + z_2| \geq |z_1| - |z_2|$ with $z_1 = R^4 e^{4i\theta}$ and $z_2 = 1$.

16.29. Evaluate $\displaystyle\int_0^\infty \frac{dx}{x^4+1}$.

Consider $\displaystyle\oint_C \frac{dz}{z^4+1}$, where C is the closed contour of Problem 16.27 consisting of the line from $-R$ to R and the semicircle Γ, traversed in the positive (counterclockwise) sense.

Since $z^4 + 1 = 0$ when $z = e^{\pi i/4}, e^{3\pi i/4}, e^{5\pi i/4}, e^{7\pi i/4}$, these are simple poles of $1/(z^4 + 1)$. Only the poles $e^{\pi i/4}$ and $e^{3\pi i/4}$ lie within C. Then, using L'Hospital's rule,

$$\begin{aligned}\text{Residue at } e^{\pi i/4} &= \lim_{z\to e^{\pi i/4}} \left\{(z - e^{\pi i/4})\frac{1}{z^4+1}\right\} \\ &= \lim_{z\to e^{\pi i/4}} \frac{1}{4z^3} = \frac{1}{4}e^{-3\pi i/4} \\ \text{Residue at } e^{3\pi i/4} &= \lim_{z\to e^{3\pi i/4}} \left\{(z - e^{3\pi i/4})\frac{1}{z^4+1}\right\} \\ &= \lim_{z\to e^{3\pi i/4}} \frac{1}{4z^3} = \frac{1}{4}e^{-9\pi i/4}\end{aligned}$$

Thus,

$$\oint_C \frac{dz}{z^4+1} = 2\pi i\left\{\frac{1}{4}e^{-3\pi i/4} + \frac{1}{4}e^{-9\pi i/4}\right\} = \frac{\pi\sqrt{2}}{2} \tag{1}$$

i.e.,

$$\int_{-R}^{R} \frac{dx}{x^4+1} + \int_\Gamma \frac{dz}{z^4+1} = \frac{\pi\sqrt{2}}{2} \tag{2}$$

Taking the limit of both sides of Equation (2) as $R \to \infty$ and using the results of Problem 16.28, we have

$$\lim_{R\to\infty} \int_{-R}^{R} \frac{dx}{x^4+1} = \int_{-\infty}^{\infty} \frac{dx}{x^4+1} = \frac{\pi\sqrt{2}}{2}$$

Since $\displaystyle\int_{-\infty}^{\infty} \frac{dx}{x^4+1} = \int_0^\infty \frac{dx}{x^4+1}$, the required integral has the value $\dfrac{\pi\sqrt{2}}{4}$.

16.30. Show that $\displaystyle\int_{-\infty}^{\infty} \frac{x^2\,dx}{(x^2+1)^2(x^2+2x+2)} = \frac{7\pi}{50}$.

The poles of $\dfrac{z^2}{(z^2+1)^2(z^2+2z+2)}$, enclosed by the contour C in Problem 16.27, are $z = i$ of order 2 and $z = -1 + i$ of order 1.

$$\text{Residue at } z = i \text{ is } \lim_{z\to i} \frac{d}{dz}\left\{(z-i)^2 \frac{z^2}{(z+i)^2(z-i)^2(z^2+2z+2)}\right\} = \frac{9i-12}{100}$$

$$\text{Residue at } z = -1+i \text{ is } \lim_{z\to -1+i} (z+1-i)\frac{z^2}{(z^2+1)^2(z+1-i)(z+1+i)} = \frac{3-4i}{25}$$

Then

$$\oint_C \frac{z^2 \cdot dz}{(z^2+1)^2(z^2+2z+2)} = 2\pi i\left\{\frac{9i-12}{100} + \frac{3-4i}{25}\right\} = \frac{7\pi}{50}$$

or

$$\int_{-R}^{R} \frac{x^2\,dx}{(x^2+1)^2(x^2+2x+2)} + \int_\Gamma \frac{z^2\,dz}{(z^2+1)^2(z^2+2z+2)} = \frac{7\pi}{50}$$

Taking the limit as $R \to \infty$ and noting that the second integral approaches zero, by Problem 16.27, we obtain the required result.

16.31. Evaluate $\displaystyle\int_0^{2\pi} \frac{d\theta}{5 + 3\sin\theta}$.

Let $z = e^{i\theta}$. Then $\sin\theta = \dfrac{e^{i\theta} - e^{-i\theta}}{2i} = \dfrac{z - z^{-1}}{2i}$, $dz = ie^{i\theta}\, d\theta = iz\, d\theta$ so that

$$\int_0^{2\pi} \frac{d\theta}{5 + 3\sin\theta} = \oint_C \frac{dz/iz}{5 + 3\left(\dfrac{z - z^{-1}}{2i}\right)} = \oint_C \frac{2\, dz}{3z^2 + 10iz - 3}$$

where C is the circle of unit radius with center at the origin, as shown in Figure 16.10.

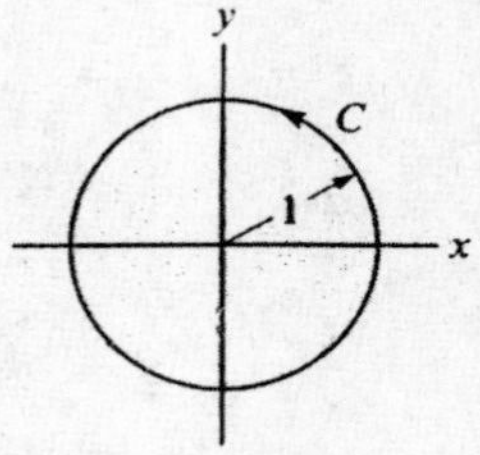

Figure 16.10

The poles of $\dfrac{2}{3z^2 + 10iz - 3}$ are the simple poles

$$z = \frac{-10i \pm \sqrt{-100 + 36}}{6} = \frac{-10i \pm 8i}{6} = -3i, -i/3$$

Only $-i/3$ lies inside C.

$$\text{Residue at } -i/3 = \lim_{z \to i/2} \left(z + \frac{i}{3}\right)\left(\frac{2}{3z^2 + 10iz - 3}\right) = \lim_{z \to i/2} \frac{2}{6z + 10i} = \frac{1}{4i}$$

Then $\displaystyle\oint_C \frac{2\, dz}{3z^2 + 10iz - 3} = 2\pi i\left(\frac{1}{4i}\right) = \frac{\pi}{2}$, the required value.

16.32. Show that $\displaystyle\int_0^{2\pi} \frac{\cos 3\theta}{5 - 4\cos\theta}\, d\theta = \frac{\pi}{12}$.

If $z = e^{i\theta}$, $\cos\theta = \dfrac{z + z^{-1}}{2}$, $\cos 3\theta = \dfrac{e^{3i\theta} + e^{-3i\theta}}{2} = \dfrac{z^3 + z^{-3}}{2}$, $dz = iz\, d\theta$.

Then

$$\int_0^{2\pi} \frac{\cos 3\theta}{5 - 4\cos\theta}\, d\theta = \oint_C \frac{(z^3 + z^{-3})/2}{5 - 4\left(\dfrac{z + z^{-1}}{2}\right)} \frac{dz}{iz} = -\frac{1}{2i} \oint_C \frac{z^6 + 1}{z^3(2z - 1)(z - 2)}\, dz$$

where C is the contour of Problem 16.31.

The integrand has a pole of order 3 at $z = 0$ and a simple pole $z = \dfrac{1}{2}$ within C.

$$\text{Residue at } z = 0 \text{ is } \lim_{z \to 0} \frac{1}{2!} \frac{d^2}{dz^2}\left\{z^3 \cdot \frac{z^6 + 1}{z^3(2z - 1)(z - 2)}\right\} = \frac{21}{8}.$$

$$\text{Residue at } z = \frac{1}{2} \text{ is } \lim_{z \to 1/2} \left\{\left(z - \frac{1}{2}\right) \cdot \frac{z^6 + 1}{z^3(2z - 1)(z - 2)}\right\} = \frac{65}{24}.$$

Then $\displaystyle -\frac{1}{2i} \oint_C \frac{z^6 + 1}{z^3(2z - 1)(z - 2)}\, dz = -\frac{1}{2i}(2\pi i)\left\{\frac{21}{8} - \frac{65}{24}\right\} = \frac{\pi}{12}$ as required.

16.33. If $|f(z)| \le \dfrac{M}{R^k}$ for $z = Re^{i\theta}$, where $k > 0$ and M are constants, prove that

$$\lim_{R\to\infty} \int_\Gamma e^{imz} f(z)\, dz = 0$$

where Γ is the semicircular arc of the contour in Problem 16.27 and m is a positive constant.

If $z = Re^{i\theta}$, $\displaystyle\int_\Gamma e^{imz} f(z)\, dz = \int_0^\pi e^{imRe^{i\theta}} f(Re^{i\theta})\, iRe^{i\theta}\, d\theta$.

Then

$$\left|\int_0^\pi e^{imRe^{i\theta}} f(Re^{i\theta})\, iRe^{i\theta}\, d\theta\right| \le \int_0^\pi \left|e^{imRe^{i\theta}} f(Re^{i\theta})\, iRe^{i\theta}\right| d\theta$$

$$= \int_0^\pi \left|e^{imR\cos\theta - mR\sin\theta} f(Re^{i\theta})\, iRe^{i\theta}\right| d\theta$$

$$= \int_0^\pi e^{-mR\sin\theta} \left|f(Re^{i\theta})\right| R\, d\theta$$

$$\le \frac{M}{R^{k-1}} \int_0^\pi e^{-mR\sin\theta}\, d\theta = \frac{2M}{R^{k-1}} \int_0^{\pi/2} e^{-mr\sin\theta}\, d\theta$$

Now $\sin\theta\ \varepsilon\ 2\theta/\pi$ for $0 \le \theta \le \pi/2$ (see Problem 4.73). Then the last integral is less than or equal to

$$\frac{2M}{R^{k-1}} \int_0^{\pi/2} e^{-2mR\theta/\pi}\, d\theta = \frac{\pi M}{mR^k}(1 - e^{-mR})$$

As $R \to \infty$, this approaches zero, since m and k are positive, and the required result is proved.

16.34. Show that $\displaystyle\int_0^\infty \frac{\cos mx}{x^2+1}\, dx = \frac{\pi}{2} e^{-m}$, $m > 0$.

Consider $\displaystyle\oint_C \frac{e^{imz}}{z^2+1}\, dz$, where C is the contour of Problem 16.27.

The integrand has simple poles at $z = \pm i$, but only $z = i$ lies within C.

Residue at $z = i$ is $\displaystyle\lim_{z\to i}\left\{(z-i)\frac{e^{imz}}{(z-i)(z+i)}\right\} = \frac{e^{-m}}{2i}$.

Then

$$\oint_C \frac{e^{imz}}{z^2+1}\, dz = 2\pi i\left(\frac{e^{-m}}{2i}\right) = \pi e^{-m}$$

or

$$\int_{-R}^R \frac{e^{imx}}{x^2+1}\, dx + \int_\Gamma \frac{e^{imz}}{z^2+1}\, dz = \pi e^{-m}$$

i.e.,

$$\int_{-R}^R \frac{\cos mx}{x^2+1}\, dx + i\int_{-R}^R \frac{\sin mx}{x^2+1}\, dx + \int_\Gamma \frac{e^{imz}}{z^2+1}\, dz = \pi e^{-m}$$

and so

$$2\int_0^R \frac{\cos mx}{x^2+1}\, dx + \int_\Gamma \frac{e^{imz}}{z^2+1}\, dz = \pi e^{-m}$$

Taking the limit as $R \to \infty$ and using Problem 16.33 to show that the integral around Γ approaches zero, we obtain the required result.

16.35. Show that $\displaystyle\int_0^\infty \frac{\sin x}{x}\, dx = \frac{\pi}{2}$.

The method of Problem 16.34 leads us to consider the integral of e^{iz}/z around the contour of Problem 16.27. However, since $z = 0$ lies on this path of integration and since we cannot integrate through a singularity, we modify that contour by indenting the path at $z = 0$, as shown in Figure 16.11, which we call contour C' or $ABDEFGHJA$.

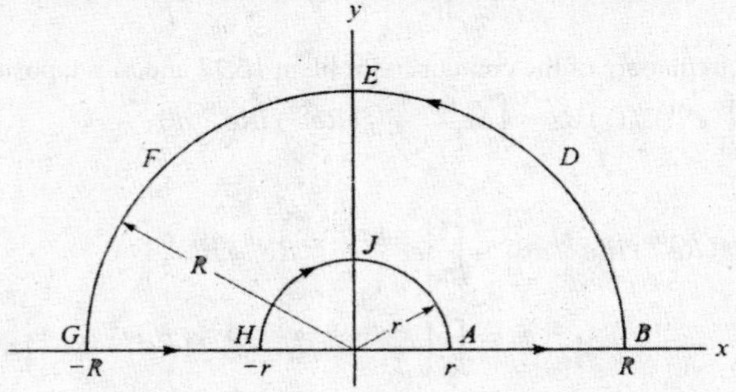

Figure 16.11

Since $z = 0$ is outside C', we have

$$\int_{C'} \frac{e^{iz}}{z}\,dz = 0$$

or

$$\int_{-R}^{-r} \frac{e^{ix}}{x}\,dx + \int_{HJA} \frac{e^{iz}}{z}\,dz + \int_{r}^{R} \frac{e^{ix}}{x}\,dx + \int_{BDEFG} \frac{e^{iz}}{z}\,dz = 0$$

Replacing x by $-x$ in the first integral and combining with the third integral, we find

$$\int_{r}^{R} \frac{e^{ix} - e^{-ix}}{x}\,dx + \int_{HJA} \frac{e^{iz}}{z}\,dz + \int_{BDEFG} \frac{e^{iz}}{z}\,dz = 0$$

or

$$2i\int_{r}^{R} \frac{\sin x}{x}\,dx = -\int_{HJA} \frac{e^{iz}}{z}\,dz - \int_{BDEFG} \frac{e^{ix}}{z}\,dz$$

Let $r \to 0$ and $R \to \infty$. By Problem 16.33, the second integral on the right approaches zero. The first integral on the right approaches

$$-\lim_{r\to 0}\int_{\pi}^{0} \frac{e^{ire^{i\theta}}}{re^{i\theta}}\,ire^{i\theta}\,d\theta = -\lim_{r\to 0}\int_{\pi}^{0} ie^{ire^{i\theta}}\,d\theta = \pi i$$

since the limit can be taken under the integral sign.

Then we have

$$\lim_{\substack{R\to\infty \\ r\to 0}} 2i\int_{r}^{R} \frac{\sin x}{x}\,dx = \pi i \quad \text{or} \quad \int_{0}^{\infty} \frac{\sin x}{x}\,dx = \frac{\pi}{2}$$

Miscellaneous problems

16.36. Let $w = z^2$ define a transformation from the z plane (xy plane) to the w plane ($u\upsilon$ plane). Consider a triangle in the z plane with vertices at $A(2, 1)$, $B(4,1)$, $C(4, 3)$. (a) Show that the *image* or *mapping* of this triangle is a curvilinear triangle in the $u\upsilon$ plane. (b) Find the angles of this curvilinear triangle and compare with those of the original triangle.

(a) Since $w = z^2$, we have $u = x^2, -y^2$, $\upsilon = 2xy$ as the transformation equations. Then point $A(2, 1)$ in the xy plane maps into point A' (3, 4) of the $u\upsilon$ plane (see Figure 16.12). Similarly, points B and C map into points B' and C', respectively. The line segments AC, BC, AB of triangle ABC map, respectively, into parabolic segments $A'C'$, $B'C'$, $A'B'$ of curvilinear triangle $A'B'C'$ with equations as shown in Figure 16.12 (*a*) and (*b*).

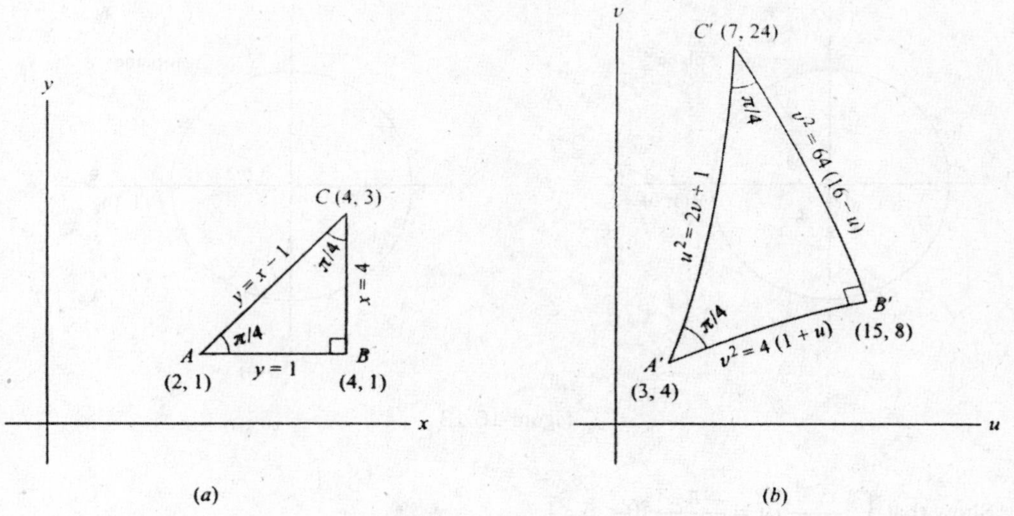

Figure 16.12

(b) The slope of the tangent to the curve $\upsilon^2 = 4(1+u)$ at (3, 4) is $m_1 = \left.\dfrac{d\upsilon}{du}\right|_{(3,4)} = \left.\dfrac{2}{\upsilon}\right|_{(3,4)} = \dfrac{1}{2}$.

The slope of the tangent to the curve $u^2 = 2\upsilon + 1$ at (3, 4) is $m_2 = \left.\dfrac{d\upsilon}{du}\right|_{(3,4)} = u = 3$.

Then the angle θ between the two curves at A' is given by

$$\tan\theta = \frac{m_2 - m_1}{1 + m_1 m_2} = \frac{3 - \frac{1}{2}}{1 + (3)\left(\frac{1}{2}\right)} = 1, \text{ and } \theta = \pi/4$$

Similarly, we can show that the angle between $A'C'$ and $B'C'$ is $\pi/4$, while the angle between $A'B'$ and $B'C'$ is $\pi/2$. Therefore, the angles of the curvilinear triangle are equal to the corresponding ones of the given triangle. In general, if $w = f(z)$ is a transformation where $f(z)$ is analytic, the angle between two curves in the z plane intersecting at $z = z_0$ has the same magnitude and sense (orientation) as the angle between the images of the two curves, so long as $f'(z_0) \neq 0$. This property is called the *conformal property* of *analytic functions*, and, for this reason, the transformation $w = f(z)$ is often called a *conformal transformation* or *conformal mapping function*.

16.37. Let $w = \sqrt{z}$ define a transformation from the z plane to the w plane. A point moves counterclockwise along the circle $|z| = 1$. Show that when it has returned to its starting position for the first time, its image point has not yet returned, but that when it has returned for the second time, its image point returns for the first time.

Let $z = e^{i\theta}$. Then $w = \sqrt{z} = e^{i\theta/2}$. Let $\theta = 0$ correspond to the starting position. Then $z = 1$ and $w = 1$ [corresponding to A and P in Figure 16.13(*a*) and (*b*)].

When one complete revolution in the z plane has been made, $\theta = 2\pi$, $z = 1$, but $w = e^{i\theta/2} = e^{i\pi} = -1$, so the image point has not yet returned to its starting position.

However, after two complete revolutions in the z plane have been made, $\theta = 4\pi$, $z = 1$ and $w = e^{i\theta/2} = e^{2\pi i} = 1$, so the image point has returned for the first time.

It follows from this that w is not a single-valued function of z but is a *double-valued function* of valued function, we must restrict θ. We can, for example, choose $0 \leq \theta < 2\pi$, although other possibilities exist. This represents one branch of the double-valued function $w = \sqrt{z}$. In continuing beyond this interval we are on the second branch, e.g., $2\pi \leq \theta < 4\pi$. The point $z = 0$ about which the rotation is taking place is called a *branch point*. Equivalently, we can ensure that $f(z) = \sqrt{z}$ will be single-valued by agreeing not to cross the line Ox, called a *branch line*.

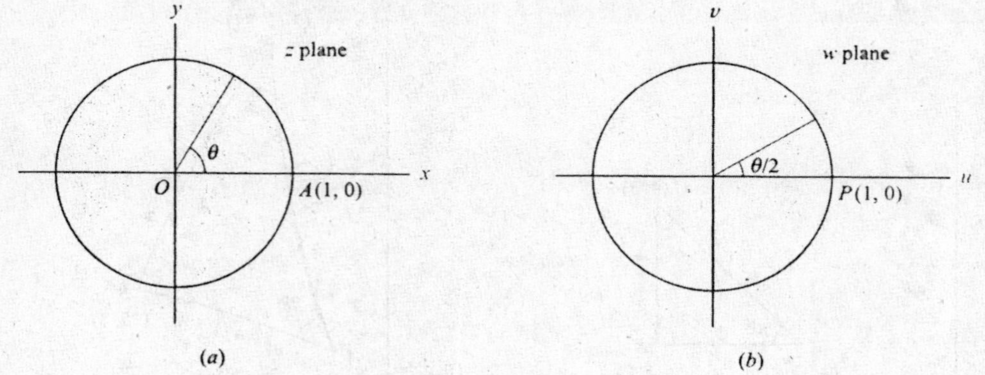

Figure 16.13

16.38. Show that $\int_0^\infty \frac{x^{p-1}}{1+x}\,dx = \frac{\pi}{\sin p\pi}, 0<p<1.$

Consider $\oint_C \frac{z^{p-1}}{1+z}\,dz$. Since $z = 0$ is a branch point, choose C as the contour of Figure 16.14, where AB and GH are actually coincident with the x axis but are shown separated for visual purposes.

The integrand has the pole $z = -1$ lying within C.

Residue at $z = -1 = e^{\pi i}$ is

$$\lim_{z\to -1} (z+1)\frac{z^{p-1}}{1+z} = (e^{\pi i})^{p-1} = e^{(p-1)\pi i}$$

Then

$$\oint_C \frac{z^{p-1}}{1+z}\,dz = 2\pi i e^{(p-1)\pi i}$$

or, omitting the integrand,

$$\int_{AB} + \int_{BDEFG} + \int_{GH} + \int_{HJA} = 2\pi i e^{(p-1)\pi i}$$

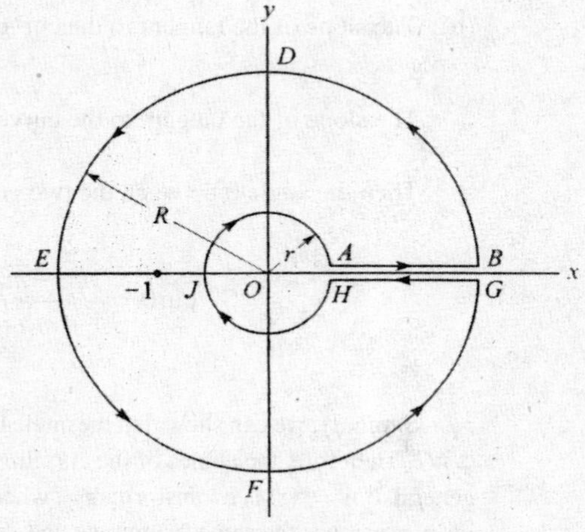

Figure 16.14

We thus have

$$\int_r^R \frac{x^{p-1}}{1+x}\,dx + \int_0^{2\pi} \frac{(Re^{i\theta})^{p-1}\, iRe^{i\theta}\,d\theta}{1+Re^{i\theta}} + \int_R^r \frac{(xe^{2\pi i})^{p-1}}{1+xe^{2\pi i}}\,dx + \int_{2\pi}^0 \frac{(re^{i\theta})^{p-1}\, ire^{i\theta}\,d\theta}{1+re^{i\theta}} = 2\pi i e^{(p-1)\pi i}$$

where we have to use $z = xe^{2\pi i}$ for the integral along GH, since the argument of z is increased by 2π in going round the circle $BDEFG$.

Taking the limit as $r \to 0$ and $R \to \infty$ and noting that the second and fourth integrals approach zero, we find

$$\int_0^\infty \frac{x^{p-1}}{1+x}\,dx + \int_\infty^0 \frac{e^{2\pi i(p-1)}\,x^{p-1}}{1+x}\,dx = 2\pi\, e^{(p-1)\pi i}$$

or

$$(1-e^{2\pi i(p-1)})\int_0^\infty \frac{x^{p-1}}{1+x}\,dx = 2\pi i\, e^{(p-1)\pi i}$$

so that

$$\int_0^\infty \frac{x^{p-1}}{1+x}\,dx = \frac{2\pi i\, e^{(p-1)\pi i}}{1-e^{2\pi i(p-1)}} = \frac{2\pi i}{e^{p\pi i} - e^{-p\pi i}} = \frac{\pi}{\sin p\pi}$$

SUPPLEMENTARY PROBLEMS

Functions, limits, continuity

16.39. Describe the locus represented by (a) $|z+2-3i|=5$, (b) $|z+2|=2|z-1|$, and (c) $|z+5|-|z-5|=6$. Construct a figure in each case.

Ans. (a) Circle $(x+2)^2+(y-3)^2=25$, center $(-2, 3)$, radius 5
(b) Circle $(x-2)^2+y^2=4$, center $(2,0)$, radius 2
(c) Branch of hyperbola $x^2/9-y^2/16=1$, where $x \geqq 3$

16.40. Determine the region in the z plane represented by each of the following: (a) $|z-2+i| \geqq 4$, (b) $|z| \leqq 3, 0 \leqq \arg z \leqq \dfrac{\pi}{4}$, and (c) $|z-3|+|z+3|<10$.Construct a figure in each case.

Ans. (a) Boundary and exterior of circle $(x-2)^2+(y+1)^2=16$
(b) Region in the first quadrant bounded by $x^2+y^2=9$, the x axis, and the line $y=x$
(c) Interior of ellipse $x^2/25+y^2/16=1$

16.41. Express each function in the form $u(x, y)+i\upsilon(x, y)$, where u and υ are real: (a) z2 + 2iz, (b) z/(3 + z), (c) e2, and (d) ln(1 + z).

Ans. (a) $u=x^3-3xy^2-2y, \upsilon=3x^2y-y^2+2x$

(b) $u=\dfrac{x^2+3x+y^2}{x^2+6x+y^2+9}, \upsilon=\dfrac{3y}{x^2+6x+y^2+9}$

(c) $u=e^{x^2-y^2}\cos 2xy$, $\upsilon=e^{x^2}-y^2\sin 2xy$

(d) $u=\dfrac{1}{2}\ln\{(1+x)^2+y^2\}, \upsilon=\tan^{-1}\dfrac{y}{1+x}+2k\pi, k=0, \pm 1, \pm 2, \ldots$

16.42. Prove that (a) $\lim\limits_{z\to x_0} z^2=z_0^2$ (b) $f(z)=z^2$ is continuous at $z=z_0$ directly

16.43. (a) If $z=\omega$ is any root of $z^5=1$ different from 1, prove that all the roots are $1, \omega, \omega^2, \omega^3, \omega^4$. (b) Show that $1+\omega+\omega^2+\omega^3+\omega^4=0$. (c) Generalize the results in (a) and (b) to the equation $z^n=1$.

Derivatives, Cauchy-Riemann equations

16.44. (a) If If $w=f(z)=z+\dfrac{1}{z}$, find $\dfrac{dw}{dz}$ directly from the definition. (b) For what finite values of z is $f(z)$ nonanalytic?

Ans. (a) $1-1/z^2$ (b) $z=0$

16.45. Given the function $w=z^4$, (a) find real functions u and υ such that $w=u+i\upsilon$, (b) Show that the Cauchy-Riemann equations hold at all points in the finite z plane, (c) prove that u and υ are harmonic functions, and (d) Determine dw/dz.

Ans. (a) $u=x^4-6x^2y^2+y^4, \upsilon=4x^3y-4xy^2$ (d) $4z^3$

16.46. Prove that $f(z)=z|z|$ is not analytic anywhere.

16.47. Prove that $f(z) = \dfrac{1}{z-2}$ is analytic in any region not including $z = 2$.

16.48. If the imaginary part of an analytic function is $2x(1 - y)$, determine (a) the real part and (b) the function.

Ans. (a) $y^2 - x^2 - 2y + c$ (b) $2iz - z^2 + c$, where c is real

16.49. Construct an analytic function $f(z)$ whose real part is $e^{-x}(x \cos y + \sin y)$ and for which $f(0) = 1$.

Ans. $ze^{-z} + 1$

16.50. Prove that there is no analytic function whose imaginary part is $x^2 - 2y$.

16.51. Find $f(z)$ such that $f'(z) = 4z - 3$ and $f(1 + i) = -3i$.

Ans. $f(z) = 2z^2 - 3z + 3 - 4i$

Integrals, Cauchy's theorem, Cauchy's integral formulas

16.52. Evaluate $\int_{1-2i}^{3+i} (2z + 3)\, dz$: (a) along the path $x = 2t + 1$, $y = 4t2 - t - 2$ $0 \leqq t \leqq 1$, (b) along the straight line joining $1 - 2i$ and $3 + i$, and (c) along straight lines from $1 - 2i$ to $1 + i$ and then to $3 + i$.

Ans. $17 + 19i$ in all cases

16.53. Evaluate $\int_C (z^2 - z + 2)\, dz$, where C is the upper half of the circle $|z| = 1$ tranversed in the positive sense.

Ans. $-14/3$

16.54. Evaluate $\oint_C \dfrac{z,}{2z + 5}$, where C is the circle (a) $|z| = 2$ and (b) $|z| = 2$, (b) $|z - 3| = 2$.

Ans. (a) 0 (b) $5\pi i/2$

16.55. Evaluate $\oint_c \dfrac{z^2}{(z+2)(z-1)}\, dz$ where C is (a) a square with vertices at $-1, -i, -1 + i, -3 + i, -3 - i$; (b) the circle $|z + i| = 3$; and (c) the circle $|z| = \sqrt{2}$.

Ans. (a) $-8\pi i/3$ (b) $-2\pi\, i$ (c) $-2\pi i/3$

16.56. Evaluate (a) $\oint_C \dfrac{\cos \pi z}{z-1}\, dz$ and (b) $\oint_C \dfrac{e^2 + z}{(z-1)^4}\, dz$, where C is any simple closed curve enclosing $z = 1$.

Ans. (a) $-2\pi i$ (b) $\pi i e/3$

16.57. Prove Cauchy's integral formulas. (Hint: Use the definition of derivative and then apply mathematical induction.)

Series and singularities

16.58. For what values of z does each series converge?

(a) $\displaystyle\sum_{n=1}^{\infty} \frac{(z+2)^n}{n!}$ (b) $\displaystyle\sum_{n=1}^{\infty} \frac{n(z-1)^n}{n+1}$ (c) $\displaystyle\sum_{n=1}^{\infty} (-1)^n (z^2 + 2z + 2)^{2n}$

Ans. (a) all z (b) $|z - i| < 1$ (c) $z = -1 \pm i$

16.59. Prove that the series $\sum_{n=1}^{\infty} \frac{z^n}{n(n+1)}$ is (a) absolutely convergent and (b) uniformly convergent for $|z| \leq 1$.

16.60. Prove that the series $\sum_{n=0}^{\infty} \frac{(z+i)^n}{2^n}$ converges uniformly within any circle of radius R such that $|z+i| < R < 2$.

16.61. Locate in the finite z plane all the singularities, if any, of each function and name them:

(a) $\dfrac{z-2}{(2z+1)^4}$ (d) $\cos \dfrac{1}{z}$

(b) $\dfrac{z}{(z-1)(z+2)^2}$ (e) $\dfrac{\sin(z-\pi/3)}{3z-\pi}$

(c) $\dfrac{z^2+1}{z^2+2z+2}$ (f) $\dfrac{\cos z}{(z^2+4)^2}$

Ans. (a) $z = -1/2$, pole of order 4 (d) $z = 0$, essential singularity
(b) $z = 1$, simple pole; $z = -2$, double pole (e) $z = \pi/3$, removable singularity
(c) simple poles $z = -1 \pm i$ (f) $z = \pm 2i$, double poles

16.62. Find Laurent series about the indicated singularity for each of the following functions, naming the singularity in each case. Indicate the region of convergence of each series.

(a) $\dfrac{\cos z}{z-\pi}$; $z = \pi$ (b) $z^2 e^{-1/z}$; $z = 0$ (c) $\dfrac{z^2}{(z-1)^2(z+3)}$; $z = 1$

Ans. (a) $-\dfrac{1}{z-\pi} + \dfrac{z-\pi}{2!} - \dfrac{(z-\pi)^3}{4!} + \dfrac{(z-\pi)^5}{6!} - \cdots$, simple pole, all $z \neq \pi$

(b) $z^2 - z + \dfrac{1}{2!} - \dfrac{1}{3!z} + \dfrac{1}{4!z^2} - \dfrac{1}{5!z^3} + \cdots$, essential singularity, all $z \neq 0$

(c) $\dfrac{1}{4(z-1)^2} + \dfrac{7}{16(z-1)} + \dfrac{9}{64} - \dfrac{9(z-1)}{256} + \cdots$, double pole, $0 < |z-1| < 4$

Residues and the residue theorem

16.63. Determine the residues of each function at its poles: (a) $\dfrac{2z+3}{z^2-4}$, (b) $\dfrac{z-3}{z^3+5z^2}$, (c) $\dfrac{e^{zt}}{(z-2)^3}$, and (d) $\dfrac{z}{(z^2+1)^2}$.

Ans. (a) $z = 2$; 7/4, $z = -2$; 1/4 (c) $z = 2$; $\dfrac{1}{2} t^2 e^{2t}$

(b) $z = 0$; 8/25, $z = -5$; $-8/25$ (d) $z = i$; 0, $z = -i$; 0

16.64. Find the residue of $e^{zt} \tan z$ at the simple pole $z = 3\pi/2$.

Ans. $-e^{3\pi t/2}$

16.65. Evaluate $\oint_C \dfrac{z^2\, dz}{(z+1)(z+3)}$, where C is a simple closed curve enclosing all the poles.

Ans. $-8\pi i$

16.66. If C is a simple closed curve enclosing $z = \pm i$, show that

$$\oint_C \frac{ze^{zt}}{(z^2+1)^2}\, dz = \frac{1}{2} t \sin t$$

16.67. If $f(z) = P(z)/Q(z)$, where $P(z)$ and $Q(z)$ are polynomials such that the degree of $P(z)$ is at least two less than the degree of $Q(z)$, prove that $\oint_C f(z)\,dz = 0$, where C encloses all the poles of $f(z)$.

Evaluation of definite integrals

Use contour integration to verify each of the following

16.68. $\displaystyle\int_0^\infty \frac{x^2\,dx}{x^4+1} = \frac{\pi}{2\sqrt{2}}$

16.69. $\displaystyle\int_{-\infty}^\infty \frac{dx}{x^6+a^6} = \frac{2\pi}{3a^5}, \quad a>0$

16.70. $\displaystyle\int_0^\infty \frac{dx}{(x^2+4^2)} = \frac{\pi}{32}$

16.71. $\displaystyle\int_0^\infty \frac{\sqrt{x}}{x^3+1}\,dx = \frac{\pi}{3}$

16.72. $\displaystyle\int_0^\infty \frac{dx}{(x^4+a^4)^2} = \frac{3\pi}{8\sqrt{2}}a^{-7}, \quad a>0$

16.73. $\displaystyle\int_{-\infty}^\infty \frac{dx}{(x^2+1)^2\,(x^2+4)} = \frac{\pi}{9}$

16.74. $\displaystyle\int_0^{2\pi} \frac{d\theta}{2-\cos\theta} = \frac{2\pi}{\sqrt{3}}$

16.75. $\displaystyle\int_0^{2\pi} \frac{d\theta}{(2+\cos\theta)^2} = \frac{4\pi\sqrt{3}}{9}$

16.76. $\displaystyle\int_0^{\pi} \frac{\sin^2\theta}{5-4\cos\theta}\,d\theta = \frac{\pi}{8}$

16.77. $\displaystyle\int_0^{2\pi} \frac{d\theta}{(1+\sin^2\theta)^2} = \frac{3\pi}{2\sqrt{2}}$

16.78. $\displaystyle\int_0^{2\pi} \frac{\cos n\theta\,d\theta}{1-2a\cos\theta+a^2} = \frac{2\pi a^n}{1-a^2}, \quad n = 0, 1, 2, 3, \ldots, \quad 0<a<1$

16.79. $\displaystyle\int_0^{2\pi} \frac{d\theta}{(a+b\cos\theta)^3} = \frac{(2a^2+b^2)\pi}{(a^2-b^2)^{5/2}}, \quad a>|b|$

16.80. $\displaystyle\int_0^\infty \frac{x\sin 2x}{x^2+4}\,dx = \frac{\pi e^{-4}}{4}$

16.81. $\displaystyle\int_0^\infty \frac{\cos 2\pi x}{x^4+4}\,dx = \frac{\pi c^{-\pi}}{8}$

16.82. $\displaystyle\int_0^\infty \frac{x\sin\pi x}{(x^2+1)^2}\,dx = \frac{\pi^2 e^{-\pi}}{4}$

16.83. $\displaystyle\int_0^\infty \frac{\sin x}{x(x^2+1)^2}\,dx = \frac{\pi\,(2e-3)}{4e}$

16.84. $\displaystyle\int_0^\infty \frac{\sin^2 x}{x^2}\,dx = \frac{\pi}{2}$

16.85. $\displaystyle\int_0^\infty \frac{\sin^3 x}{x^3}\,dx = \frac{3\pi}{8}$

16.86. $\displaystyle\int_0^\infty \frac{\cos x}{\cosh x}\,dx = \frac{\pi}{2\cosh(\pi/2)}$

[Hint: Consider $\displaystyle\oint_C \frac{e^{iz}}{\cosh z}\,dz$, where C is a rectangle with vertices at $(-R, 0)$, $(R, 0)$, (R, π), $(-R, \pi)$. Then let $R \to \infty$.]

Miscellaneous problems

16.87. If $z = \rho e^{i\phi}$ and $f(z) = u(\rho, \phi) + i\upsilon(\rho, \phi)$, where ρ and ϕ are polar coordinates, show that the Cauchy-Riemann equations are

$$\frac{\partial u}{\partial \rho} = \frac{1}{\rho}\frac{\partial \upsilon}{\partial \phi}, \qquad \frac{\partial \upsilon}{\partial \rho} = -\frac{1}{\rho}\frac{\partial u}{\partial \phi}$$

16.88. If $w = f(z)$, where $f(z)$ is analytic, defines a transformation from the z plane to the w plane where $z = x + iy$ and $w = u + i\upsilon$, prove that the Jacobian of the transformation is given by

$$\frac{\partial(u, \upsilon)}{\partial(x, y)} = |f'(z)|^2$$

16.89. Let $F(x, y)$ be transformed to $G(u, \upsilon)$ by the transformation $w = f(z)$. Show that if $\dfrac{\partial^2 F}{\partial x^2} + \dfrac{\partial^2 F}{\partial y^2} = 0$, then at all points where $f'(z) \neq 0$, $\dfrac{\partial^2 G}{\partial u^2} + \dfrac{\partial^2 G}{\partial \upsilon^2} = 0$.

16.90. Show that by the *bilinear transformation* $w = \dfrac{az + b}{cz + d}$, where $ad - bc \neq 0$, circles in the z plane are transformed into circles of the w plane.

16.91. If $f(z)$ is analytic inside and on the circle $|z - a| = R$, prove *Cauchy's inequality*, namely,

$$|f^{(n)}(a)| \leqq \frac{n!M}{R^n}$$

where $|f(z)| \leqq M$ on the circle. (Hint: Use Cauchy's integral formulas.)

16.92. Let C_1 and C_2 be concentric circles having center a and radii r_1 and r_2, respectively, where $r_1 < r_2$. If $a + h$ is any point in the annular region bounded by C_1 and C_2, and $f(z)$ is analytic in this region, prove *Laurent's theorem* that

$$f(a + h) = \sum_{-\infty}^{\infty} a_n h^n$$

where

$$a_n = \frac{1}{2\pi i} \oint_C \frac{f(z)\, dz}{(z - a)^{n+1}}$$

C being any closed curve in the angular region surrounding C_1. [Hint: Write $f(a + h) = \dfrac{1}{2\pi i} \oint_C \dfrac{f(z)\, dz}{z - (a + h)} - \dfrac{1}{2\pi i} \oint_{C_1} \dfrac{f(z)\, dz}{z - (a + h)}$ and expand $\dfrac{1}{z - a - h}$ in two different ways.]

16.93. Find a Laurent series expansion for the function $f(z) = \dfrac{z}{(z + 1)(z + 2)}$ which converges for $1 < |z| < 2$ and diverges elsewhere. $\left[\text{Hint: Write } \dfrac{z}{(z + 1)(z + 2)} = \dfrac{-1}{z + 1} + \dfrac{2}{z + 2} = \dfrac{-1}{z(1 + 1/z)} + \dfrac{1}{1 + z/2)}.\right]$

Ans. $\cdots - \dfrac{1}{z^5} + \dfrac{1}{z^4} - \dfrac{1}{z^3} + \dfrac{1}{z^2} - \dfrac{1}{z} + 1 - \dfrac{z}{2} + \dfrac{z^2}{4} - \dfrac{z^3}{8} + \cdots$

16.94. Let $\int_0^\infty e^{-st} F(t)\, dt = f(s)$, where $f(s)$ is a given rational function with numerator of degree less than that of the denominator. If C is a simple closed curve enclosing all the poles of $f(s)$, we can show that $F(t) = \dfrac{1}{2\pi i} \oint_C e^{zt} f(z)\, dz =$ sum of residues of $e^{zt} f(z)$ at its poles. Use this result to find $F(t)$ if $f(s)$ is (a) $\dfrac{s}{s^2 + 1}$, (b) $\dfrac{1}{s^2 + 2s + 5}$, (c) $\dfrac{s^2 + 1}{s(s - 1)^2}$, and (d) $\dfrac{1}{(s^2 + 1)^2}$, and check results in each case. [Note that $f(s)$ is the *Laplace transform* of $F(t)$, and $F(t)$ is the *inverse Laplace transform* of $f(s)$ (see Chapter 12). Extensions to other functions $f(x)$ are possible.]

Ans. (a) cos t, (b) $\frac{1}{2} e^{-t} \sin 2t$, (c) $\frac{1}{4} + \frac{5}{2} te^{2t} + \frac{3}{4} e^{2t}$, (d) $\frac{1}{2}(\sin t - t \cos t)$

INDEX